T0135846

THE COLONIAL MACHINE:

FRENCH SCIENCE AND OVERSEAS EXPANSION IN THE OLD REGIME

DE DIVERSIS ARTIBUS

COLLECTION DE TRAVAUX
DE L'ACADÉMIE INTERNATIONALE
D'HISTOIRE DES SCIENCES

COLLECTION OF STUDIES
FROM THE INTERNATIONAL ACADEMY
OF THE HISTORY OF SCIENCE

DIRECTION
EDITORS

EMMANUEL
POULLE (†)

ROBERT
HALLEUX

TOME 87 (N.S. 50)

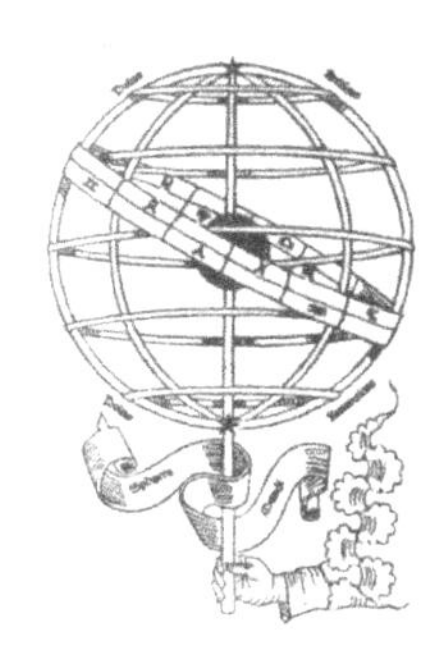

BREPOLS

The Colonial Machine:

French Science and Overseas Expansion in the Old Regime

James E. McClellan III
François Regourd

Brepols

COMITE SLUSE

Publié avec le soutien de la Région Wallonne,
le concours de l'ESNA
(Empire, Sociétés Nations Amériques-UMR 8168 MASCIPO)
et de l'Université Paris Ouest-Nanterre La Défense.

D/2011/0095/220
ISBN 978-2-503-53260-8

Printed on acid-free paper

For
Charles C. Gillispie
and
Daniel Roche

If the principles and genius of the French government are at all conspicuous in the preceding example, which has given of their civil and political ordinances respecting their Negroe slaves, and slaveowners; they are still more so, in the other department of their colony-system. These manifest a degree of forecast, prudence, and vigor, that are not so observable in any movement of our own torpid machine. There is a spirit in the French monarchy, which pervades every part of their empire; it has select objects perpetually in view, which are steadily and consistently pursued; in their system the state is at once the sentient and the executive principle. It is in short, *all soul*; motion corresponds with will; action treads on the heels of contrivance; and sovereign power usefully handled and directed, hurries on, in full career to attain its end. With us, the liberty to which every corporate society, and every individual member of those societies, lays claim, of independent thinking and acting, excludes almost a possibility of concurrent exertion, to any one finite and determinate point.

Edward Long, *The History of Jamaica*
vol. III, p. 941 (1774)

L'historien, épris de sources originales, entretient avec l'archive un rapport de fascination tel qu'il ne cesse de s'en justifier et de contrôler chez lui et chez les autres tout ce qui à partir d'elle, pourrait faire croire à une dépossession de son système de rationalité. L'impact que l'archive a sur lui – et qu'il ne reconnaît presque jamais explicitement – a parfois pour conséquence un déni de sa valeur: belle mais piégée, elle aurait pour corollaire de sa beauté toute une mise en scène de l'illusion. Elle attire, mais trompe, et l'historien la prenant comme compagne ne se méfierait pas assez de l'improbable tracé des images qu'elle sécrète.

Arlette Farge, *La vie fragile* (1986)

Table of Contents

Preface and Acknowledgments

This study examines a set of institutions and scientific and technical experts devoted to French colonization and overseas expansion in the Old Regime. This unique and historically significant conglomeration of men and institutions we are calling the Colonial Machine. The introduction and prologue that follow explain more exactly what we mean by the Colonial Machine, and the succeeding pages lead us through successive layers and dimensions of its fascinating, untold story.

This project has unfolded over an unusually long period and has accumulated more than its share of scholarly debts. It originated over lunch at the Bistrot Mazarin in Paris in January of 1999. The authors had corresponded a time or two previously, but this was their first meeting. Even though it was January, we ate outside in the warmth of a bright winter sun. We compared historiographical notes on the literature labeled "Science and Empire," and we discussed each other's work to date, notably McClellan's 1992 book, *Colonialism and Science: Saint Domingue in the Old Regime* and Regourd's dissertation then in progress, *Sciences et colonisation sous l'Ancien Régime. Le cas de la Guyane et des Antilles françaises* XVIIe-XVIIIe *siècles.* At one point in the conversation, we surprised one another by simultaneously using the expression, *une machine coloniale* (a colonial machine), to sum up how we understood science and the colonial and overseas endeavors of the Bourbon monarchy and the French state in the seventeenth and eighteenth centuries. Beyond what each of us had done to date, we both immediately recognized the need and utility of examining the workings of this colonial "machine" in its totality and from an overarching point of view. At what might have been an awkward moment involving scholarly turf, speaking in French, we switched to the familiar form, and so began this collaboration.

One of us (McClellan) enjoyed a sabbatical leave from his home institution, Stevens Institute of Technology, during the 1999-2000 academic year; and, thanks to a complementary sabbatical fellowship from the American Philosophical Society in Philadelphia, he was able to devote full-time to archival research in support of this project alongside his collaborator. The other of us (Regourd) also used this period to complete the dissertation he defended at Université Bordeaux III in December of 2000. The extent to which Regourd's dissertation and McClellan's book on Saint Domingue figure in what follows will be obvious. By sharing our research, we extended the range and depth of our knowledge, particularly through more extensive coverage of archival sources. On a regular basis in the 1999-2000 academic year, we labored side by side, usually daily, in various archives. Most agreeably at the cafés *Le Près aux Clercs* and *Le Rostand* we explored what we knew and what we did not know about our subject.

From there, the opportunity arose to participate in the 2000 thematic issue of *Osiris* devoted to updating the literature on science and colonialism-imperialism (MacLeod, ed., 2000). In helping shape our interpretation at that early phase we wish to thank Roy McLeod, Michael Osborne, Kapil Raj, and Christophe Bonneuil who gathered at Professor McLeod's quarters on the rue Suger to hammer out and mutually evaluate the papers that went into that special number of *Osiris*. For better and worse, the result was an initial article that sketched the territory as we saw it at the time.

Over the several years that followed we have labored to fill the gaps in our research and to cast our full story into words. Together and separately we have published articles on aspects of this history, parts of which we incorporate in our text here. Over this period we spent many more days individually and collectively in the archives, working together more recently at the *Café Lili et Marcel* near the Bibliothèque Nationale François Mitterand and at our homes in Paris, Palaiseau, and Hoboken. A presentation to the valedictory seminar of Daniel Roche at the Collège de France in March, 2005 provided valuable input from a seasoned critical audience. Similarly, in May of 2006 and, thanks to another leave from Stevens for McClellan, we benefited from the opportunity to present this material to several sessions of Kapil Raj's seminar at the École des Hautes Études en Sciences Sociales in Paris, and we thank Kapil Raj and the auditors on that occasion for their valuable input, notably Éric Brian and Bertrand Daugeron. In 2005 a workshop on "European Capitals and the Globalization of Knowledge in the 17th and 18th Centuries" at the Maison Française d'Oxford was for Regourd an occasion of constructive debates with an impressive group of specialists in the field, as was his participation in the effort that gave rise to the special issue in 2008 of the *Revue d'Histoire moderne et contemporaine* on *Sciences et Villes-Mondes XVIe-XVIIIe Siècles*. In April 2007, thanks to the "French Atlantic History Group" based at McGill University, our topic came up in papers we presented to another workshop devoted to "Knowledge and Science in the French Atlantic, 1500-1800." We thank the respondents at the meeting for a lively debate that led us to hone even further our presentation of the Colonial Machine. In particular, we wish to acknowledge Catherine Desbarats, Nicholas Dew, Daniela Bleichmar, Marie-Noëlle Bourguet, James Delbourgo, Jordan Kellman, Karol Weaver, Neil Safier, and Thomas Wien. For Regourd especially, but not only, Marie-Noëlle Bourguet, Danielle Bégot, Patrice Bret, Paul Butel, Marcel Dorigny, Serge Gruzinsky, Silvia Marzagalli, Daniel Nordman, Antonella Romano, and

Stéphane Van Damme played important roles in the making of this book, offering their friendship and wise advice from the outset of this research work. We thank Boris Lesueur for his additional commentaries at late stages, and Charlie Williams, Eliane Norman, and Walter Taylor for sharing their knowledge of André Michaux. Harold Dorn once again contributed time and his red pen to our project. We owe a special debt to our friend and colleague, April Shelford, not least for providing us the epigram from Edward Long that heads this work and that so captures what we mean by the Colonial Machine.

We gratefully acknowledge the many people who aided us in our archival and library research. We spent many hours in the archives of the Académie des Sciences of the Institut de France, where Claudine Pouret and the head of that service, Florence Greffe, were of invaluable assistance, as was the previous archivist, Christiane Demeulenaere-Douyère. Similarly we have to thank Josette Alexandre at the Bibliothèque de l'Observatoire de Paris; René d'Estienne of the Services Historiques de la Marine at Lorient; Roy Goodman and friends at the American Philosophical Society; Angela Todd at the Hunt Botanical Institute in Pittsburgh; Inès Villela-Petit and Thierry Sarmant at the Cabinet de Monnaies of the Bibliothèque Nationale de France; Gabriela Lamy at the Petit Trianon; Jacques Dion at the Centre des Archives d'Outre-Mer in Aix-en-Provence; Jean-Louis Sarrazin and Monique Pelletier at the Dépôt des Cartes et Plans. Similarly, we owe debts of gratitude to all the helpful, but anonymous people in the archives, libraries, and research centers where we toiled.

Finally in this connection, we owe a special debt to Professor Robert Halleux and Brepols Publishers for accepting this work for publication in the series, *De diversis artibus*. Two anonymous referees made many helpful suggestions, which we hope to have incorporated into this final version. We thank Eleanor Wedge for her careful copyediting and Andrew Rubenfeld for his advice on preparing the indices. We also warmly thank Thierry Mozdziej of the publications department of the Centre d'Histoire des Sciences et des Techniques at the Université de Liège for his expertise in turning our manuscript into a real book and for putting up with us as we made corrections.

In the many details of its conception and the research that went into the making of this book, the result is inextricably the work of the two of us functioning throughout as one, and bringing our best, most critical abilities to the effort. Nevertheless, Regourd has said that, if this were a piece for piano four hands, in the composition and actual writing of this book it fell to McClellan to play three of the four hands. Even so, the result embodies a seamless intellectual collaboration and fulfills the unifying idea we had in starting out. This book is a mature piece of scholarship that presents a comprehensive history of French science and overseas expansion in the Old Regime – or at least represents our best efforts to do so, and it culminates a most amicable and fruitful collaboration. At the least, we hope that this study makes a contribution to the literature and is of use and interest to those who run across it. As the stock phrase has it, we alone are responsible for its errors and shortcomings.

Each of us wish to thank our wives and families for their patience and support as we toiled on this project. Marie-Claire Regourd, Lucile, and Lisa, François' wife and daughters, have been wonderful in putting up with short weekends sometimes spoiled by the crush of work. Annie Regourd regularly helped by providing on the fly advice and smart suggestions for some of the translations. Jackie McClellan again proved a stalwart companion; we owe her deep thanks for the untold hours she devoted to reviewing our manuscript and checking the page proofs; in addition, we cannot adequately express our gratitude and our indebtedness to her for taking on the preparation of the name index.

Charles C. Gillispie, Professor Emeritus of History of Science at Princeton University, and Daniel Roche, Professor Emeritus of the History of France in the Enlightenment (*la Chaire des Lumières*) at the Collège de France, are respectively the preeminent scholars of the history of French science and French social history in the eighteenth century. No two guiding lights have been more influential for the authors or this work, and this book is warmly dedicated to these two world-class *Doktorfaters* from two grateful sons.

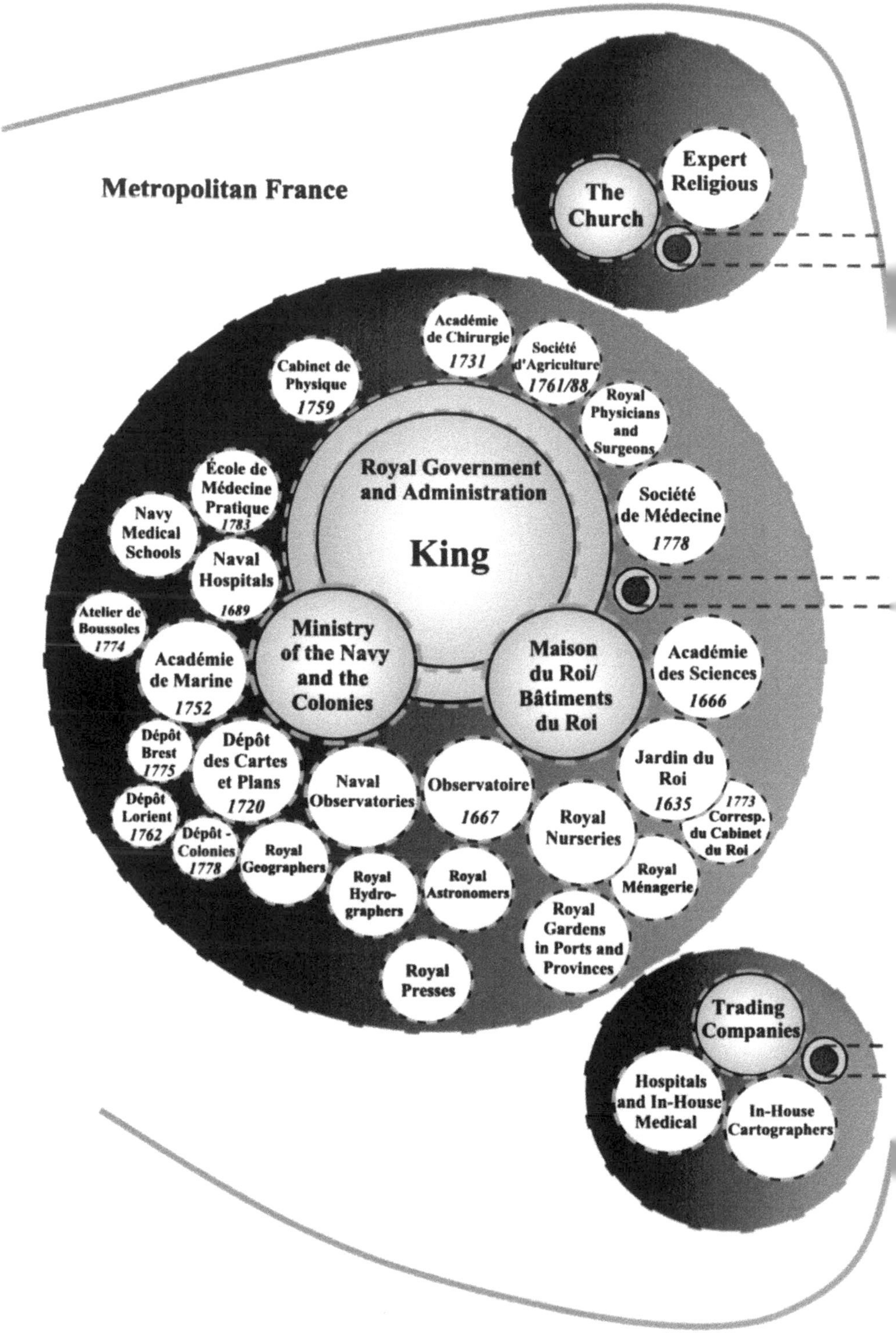

Metropolitan France

The Church
Expert Religious

Royal Government and Administration
King

Ministry of the Navy and the Colonies
Maison du Roi/ Bâtiments du Roi

Académie de Chirurgie 1731
Société d'Agriculture 1761/88
Royal Physicians and Surgeons
Cabinet de Physique 1759
Société de Médecine 1778
École de Médecine Pratique 1783
Navy Medical Schools
Naval Hospitals 1689
Atelier de Boussoles 1774
Académie de Marine 1752
Académie des Sciences 1666
Dépôt Brest 1775
Dépôt des Cartes et Plans 1720
Dépôt Lorient 1762
Dépôt - Colonies 1778
Naval Observatories
Observatoire 1667
Jardin du Roi 1635
1773 Corresp. du Cabinet du Roi
Royal Nurseries
Royal Geographers
Royal Hydrographers
Royal Astronomers
Royal Ménagerie
Royal Gardens in Ports and Provinces
Royal Presses

Trading Companies
Hospitals and In-House Medical
In-House Cartographers

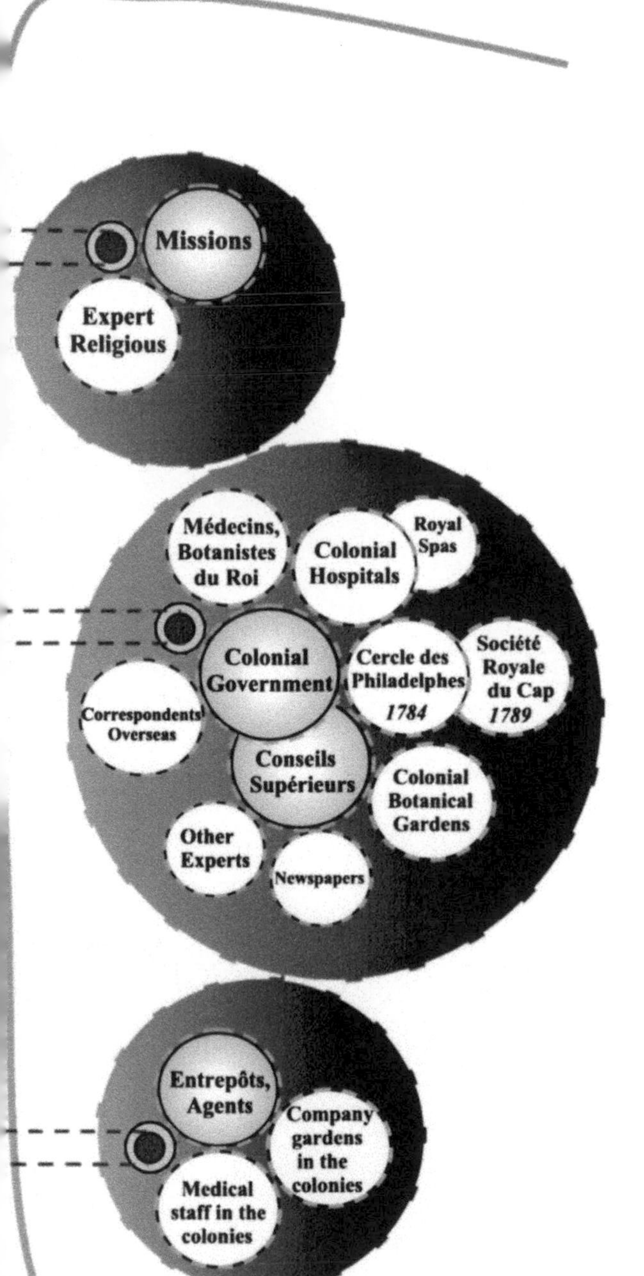
Missions
Expert Religious
Médecins, Botanistes du Roi
Colonial Hospitals
Royal Spas
Colonial Government
Cercle des Philadelphes 1784
Société Royale du Cap 1789
Correspondents Overseas
Conseils Supérieurs
Colonial Botanical Gardens
Other Experts
Newspapers
Entrepôts, Agents
Company gardens in the colonies
Medical staff in the colonies
Overseas/Colonies

INTRODUCTION

This is a big book about a big subject – French science and overseas expansion in the Old Regime. Our analysis centers on a key set of individuals and institutions that we are labeling the Colonial Machine. The prologue that follows lays out its component parts. In this Introduction we set our subject in larger historical and historiographical contexts, and we hope to explain why the Colonial Machine merits attention and why one would be foolhardy enough to take on such a big topic in the first place.

The Big-Picture Background to this Study

Modern world history has been significantly affected by two great historical developments: European colonial expansion since the fifteenth century and the advent of modern science since the scientific revolution of the sixteenth and seventeenth centuries. The world-historical significance of European colonialism and imperialism on the one hand and of modern science on the other pose the question of their historical interconnections. How did European science affect the course of European colonialism? What was the impact of the colonial experience on the development of the natural sciences in the period after Columbus? What explains the mutual interactions and reciprocal influences of science and overseas expansion in political, sociological, and intellectual perspective? These questions and the promise of what historical research might uncover provide a compelling rationale for what is now a mature field of scholarly work and research, and from the most general point of view these questions drive this inquiry.

Juxtaposing science and colonialism in this way immediately situates our subject around the theme of power. On the one hand, we are dealing with political power and the violence of state authority, especially as projected overseas. In this sense, ours is the story of the relationship of science and power. But in addition we have to confront the intellectual and moral authority of science and natural philosophy as these developed over time, both as sources of reliable knowledge about the world and as practical instruments for human mastery over nature. The story of the Colonial Machine is thus more than a history of the marriage of science and the state. It involves, rather, a dialectic of powers, political and scientific.

Two simple premises argue for the importance of looking at the history of the French colonial experience and French science in the Old Regime. The first premise places Bourbon France among the leading colonial powers of Europe. The Spanish and Portuguese ran huge empires of long standing, of course, but, while these powers were still to be contended with in the eighteenth century, they were in decline, having flourished in the sixteenth and seventeenth centuries. Beginning in the seventeenth century Holland, England, and France were potent new entrants into the global competition for colonies and wealth. The Dutch colonial empire centered on the spice trade and Batavia in the East Indies. The Dutch were dominant through the seventeenth and early eighteenth centuries and, while a seafaring force to be reckoned with and full participants in the contemporary global colonial struggle, as the eighteenth century unfolded they did not grow as did their English and French rivals.

The extraordinary productivity of its colonial possessions made France a major, perhaps the major colonial power of the day. That productivity was based primarily on its West Indian possessions, on the highly profitable commodity production of sugar and coffee, on the productive labor of legions of slaves who toiled on French plantations, and on the triangular trade that drove Atlantic colonization, slavery, and merchant trading activity. For a while, circumstances put the French in the vanguard of contemporary European colonial development.

In the Anglo-American world especially, and from a retrospective point of view, the success and superiority of the English in eighteenth-century matters naval and colonial is too easily taken for granted. In fact, the French closely challenged the English at sea and in developing colonies, and at various points in the century the outcome of their struggle was not clear. To be sure, in the Seven Years War the French suffered devastating losses in Canada, Louisiana, and India, which were ratified in the Treaty of Paris of 1763. Those losses tend to obscure the greater parity of the two nations up until then. By the same token, the French rebounded, rebuilt, and upgraded their navy. And then, after 1763, the English had trouble with and ultimately lost their great enclave of colonies along the Atlantic coast of North America, thanks in large measure to the French navy and the French purse. In the 1770s and 1780s France was a seafaring and military power at least the equal of the English, and she possessed not only a comparable colonial empire, but the single most productive of all of Europe's overseas colonies, the Caribbean crown jewel of Saint Domingue, modern-day Haiti. All this is only to say that the case of France in the Old Regime is key to thinking about contemporary European colonialism and the ongoing course of Atlantic and world history.

The second, complementary premise on which this study is built concerns the high importance of contemporary French science. Here there was no real rivalry, as no other European nation could match organized science in France.[1] It was not the quality of natural science in contemporary Britain and elsewhere, but rather the heavily institutionalized and state-subsidized character of science on the French side of the Channel that gave the categorical advantage to the French. In this connection one thinks first and foremost of the state-funded French Académie Royale des Sciences, with its preeminent membership of scientific luminary academicians, its famous scientific publications, its expeditions, and related attendant glory. The Académie Royale des Sciences naturally figures a great deal in what follows, but other contemporary French scientific institutions such as the Observatoire Royal, the Jardin du Roi, and the Société Royale de Médecine also contributed to making France the top scientific nation of the day. These and other related institutions appear prominently in our account; and, indeed, to state the main point of this study, the coordinated whole of organized and institutionalized French science and medicine, when turned toward colonial ends, constituted the Colonial Machine in action.

The Colonial Machine of the Old Regime did have organizational and thematic antecedents, particularly in earlier Spanish and Portuguese efforts to tap science and expert knowledge in order to advance their colonial and overseas objectives. Here, one has in mind a range of Iberian institutions that arose in the sixteenth century: the Portuguese Casa da Índia (1501); the Spanish Casa de la Contratación (1503) with its positions of Pilot-Major (1508), Cosmographer Royal (1523), and a royal chair in navigation (1552); the Spanish Council of the Indies (*Consejo De Indias*, 1524) with its corps of royal cosmographers and others charged with various scientific and practical duties in the expansion of the Spanish empire; and the Academy of Mathematics founded in Madrid by Philip II in 1582.[2]

The example of the union of science and the colonizing state was thus already well established by the time the French came on the scene. European powers were all deadly serious players in a fierce international struggle, and not surprisingly France's contemporary rivals, the Dutch and the English, adopted similar institutions and approaches toward tapping science and expert knowledge in the service of global empire. For the British, one would begin with the Royal Society of London (analogous in role to the Parisian Academy of Science) and the analogy continues through the Royal Gardens at Kew, the Royal Observatory at

1. The point is not in dispute. Charles Gillispie (1980), p. 74, is explicit about the matter: "[From 1774 through 1830] the French community of science predominated in the world to a degree that no other national complex has since done or had ever done. In its eminence, French cultural leadership in Europe reached a climax." David Knight echoed this point in 2004, writing, pp. 483, 485: "...France had for these sixty years [prior to 1833] been in a different scientific league from everybody else. Paris was the world's centre of excellence in science, from pure mathematics through to botany, and Britain was provincial...France was the top nation scientifically."

2. Gerardus Mercator's new cartographical projection of 1569 can be seen in this light. See Cañizares-Esguerra (2005) and sources cited in his valuable review article; for further on colonialism and science in the case of Spain and Latin America, see also Cañizares-Esguerra (2001) and (2006), esp. chapt. 3; Pimentel; Saldaña, ed. (2006); McCook; Barrera-Osorio (2006); Bleichmar et al. (2009).

Greenwich, the Board of Longitude, the Admiralty and the Royal Navy, botanical gardens in Saint Vincent and Jamaica in the West Indies, Calcutta, and Sydney; the Asiatic Society of Bengal, and the East India and Hudson Bay companies.[3] For the Dutch one could point to the numerous scientific societies in the United Provinces themselves, the Amsterdam Botanic Garden and colonial gardens in Capetown, Sri Lanka, and Java; the Dutch East India Company (*Verenigde Oost-Indische Compagnie* - the VOC), and the Dutch colonial science academy, the *Bataviaasch Genootschap van Kunsten en Wetenschappen* (Batavia, Java, the East Indies), that antedated the French scientific society, the Cercle des Philadelphes, which arose in the Caribbean colony of Saint Domingue in 1784.[4]

Still, as Edward Long makes plain in the epigram of 1774 that heads this work, no country could match the French Colonial Machine for its complexity or the extent to which officialdom institutionalized expert knowledge and directed it toward colonial ends. The lack of a comparable colonial-science bureaucracy in contemporary Britain or Holland until the 1780s further underscores the rationale for looking at France.[5] A revitalized set of Spanish institutions and the creation of a comparable Spanish "colonial machine" in the second half of the eighteenth century makes the same point.[6]

Balanced as it was, then, as a scientific and as a colonial power, France represents a potent case for investigating the historical interactions between the two great world-historical forces of modern science and European colonial expansion in the seventeenth and eighteenth centuries. It is not too early to suggest that, not only were contemporary French science and French colonial and expansionist efforts overseas deeply intertwined, but that, to an extent at least, they depended on one another for their mutual success.

3. Citing a larger literature, Stuchtey, p. 35, refers to the "informality of institutional linkages" that held together the scientific network of the British Empire. Be it noted that at that time, the Sydney garden was not much more than a kind of kitchen garden.

4. The Dutch case regarding medicine and natural history is now remarkably treated for an earlier period by Harold J. Cook; on the VOC, see especially pp. 62-68, 177-91; on botanical gardens, see esp. pp. 305-29.

5. John Gascoigne documents the gradual appearance of analogous structures in Britain, but only in the late eighteenth century and only after Joseph Banks became President of the Royal Society in 1778. The French Colonial Machine had been maturing for over a century at that point. Mark Harrison makes this same point, writing, p. 56: "...the British Empire stands in marked contrast to the French, where the state was involved actively in science from the eighteenth century." Stuchtey, p. 2, touches on this theme also.

6. Cañizares-Esquerra (2006), pp. 56-60, and McCook, pp. 11-13, point to the revitalized scientific interest of the *Spanish* crown in its American colonies from the 1780s; Bleichmar (2008), p. 234, makes the same point, noting the foundation of the Spanish Royal Natural History Cabinet in 1776 and the organization of an equivalent Spanish "colonial machine" at this point in time; see also Bleichmar (2009), p. 299, where one reads "The scientific 'colonial machine' at work in the Spanish world in the second half of the eighteenth century aimed to produce useful and profitable information and commodities." Saldaña (2006b), pp. 53-54, 85, and 87, similarly points to "a new social framework" and the implantation of new scientific institutions in the Spanish colonies, largely in the 1780s, e.g., the Jardín Botánico in New Spain in 1788 and the Jardín Botánico in Guatemala in 1796.

Historiographical Background

A detailed review of the historiography out of which the present work has emerged is not necessary, but a general word is in order. As the accompanying notes and bibliography make plain, the role of science in European colonial and imperial expansion has been at the heart of an expanding body of work by historians of science since the end of the 1960s.[7] The article by George Basalla appearing in *Science* in 1967, "The Spread of Western Science," is usually acknowledged as the Ur-source of this literature.[8] Basalla's theoretical model triggered a substantial critical response in the years that followed, as scholars sought to add cultural nuance and historical complexity to the topic. Their work and reflections produced a variety of analytical novelties: concepts of "metropolis/colony," "center/periphery," and other models; calls for case studies and comparative national and transnational studies; and approaches differentiating scientific disciplines, chronological disjunctures, and different national styles in colonial science.[9] A series of international conferences in Australia in 1981, Paris in 1990, and Madrid in 1991 gave rise to landmark volumes critically exploring and developing these themes.[10] And since then there has been an explosion of works of ever greater sophistication. These have expanded the vistas of "colonial science" beyond the literature's original Anglo-American and Anglo-Australian orientation, to include India and Asia, the French colonial experience, and, even more notably, Latin America.[11] They have also broadened the chronological scope of investigations beyond the heyday of European colonialism and imperialism in the nineteenth and twentieth

7. This is not to overlook an earlier, somewhat antiquarian literature. The otherwise useful and informative Alfred Lacroix (1932-1938) is the paradigmatic example; Fournier also fits here. Petitjean (2005), p. 124, situates Lacroix in the institutional construction of pro-colonial propaganda in the early decades of the twentieth century.

8. See Basalla (1967) and his reflections of 1993; see also Stearns (1970) and Brockway (1979) for other early landmark studies; McClellan (1992), p. 6, references some other early literature.

9. See especially the works of Lewis Pyenson (1985, 1989a, 1993) on national traditions and the exact sciences. The volume edited by Marie-Noëlle Bourguet and Christophe Bonneuil in 1999 focused on a particular discipline, botany. Saldaña, ed., published originally in 1999, offered an overview of the Latin American scientific theater. The 2008 volume edited by Delbourgo and Dew on science and the Atlantic world may be grouped with this regional literature. In 2005 Stuchtey observed, p. 21, "... it is now almost a commonplace of modern imperial historiography [of science] to call for disciplinary and national boundaries to be crossed...."

10. See Reingold and Rothenberg, eds., with the classic article by MacLeod (1987) on the "moving metropolis"; Petitjean et al., eds.; and Lafuente et al., eds. The 1991 volume edited by Teresa Meade and Mark Walker is another early work that deserves mention here.

11. On the Indian case, see Sangwan; Kumar; Kumar, ed.; Raj (1997) and (2000), and further references in Stuchtey, p. 19. For the French, see McClellan (1992); Osborne (1995) and (2005); Regourd (2000); Daugeron (2007) and (2009); Lesueur (2007); Boucher for an earlier period; Marchand adds a popular account. For the Spanish, see Polanco; Struik; Bleichmar (2008); Bleichmar et al. (2009); and above note 2. Benedikt Stuchtey, ed., (2005) adds valuable comparative perspectives for a later period. MacLeod and Kumar, eds. (1995) and the 2005 volume edited by Sylviane Llinares and Philippe Hroděj develop the theme of technology and colonization and complements the classic works by Daniel Headrick (1981) and (1988).

centuries. In the last twenty years individual, collective, and comparative works have produced an ever more sophisticated and nuanced understanding of the heterogeneous realities of European science vis-à-vis local conditions on the ground in the colonies and the variegated relations of the colonies with the metropolis. Colonial settings are no longer seen as passive environments, but sites for complex interactions of local cultures with exogenous forces that produced a blend of responses. In particular, recent work has exposed that it is no longer possible to envision the topic from a wholly Eurocentric point of view; rather, recognizing the importance of non-European actors and the knowledge systems they brought to the encounter with Europeans has engendered fruitful insights regarding the dynamics of cultural exchange and the development of local knowledge communities.[12] Europe remains a reference point for the organization and diffusion of scientific knowledge, but historiographically Europe is now integrated into a global and multipolar approach to the interpretative problems taken up to date.

These analytical themes and approaches are articulated in a literature that is unusually self-reflective, and, indeed, the history of the historiography of "colonial science" can now be thought of as a subject of its own. Patrick Petitjean's 1992 *Science et Empires: Un thème prometteur, des enjeux cruciaux* (Science and Empires: A Promising Theme, Crucial Issues) led the way among more general historiographical surveys.[13] Marie-Noëlle Bourguet and Christophe Bonneuil's 1999 presentation provided another landmark early review of the literature. The same year saw surveys by Richard Drayton on the British overseas experience and by Juan José Saldaña on the historiography of science in Latin America.[14] The special number of *Osiris* in 2000, devoted to "Nature and Empire: Science and the Colonial Enterprise," included an in-depth overview of the field by Roy MacLeod; in the same year James McClellan likewise surveyed the literature through 1999 in *Reader's Guide to the History of Science*.[15] Similarly, the 2002 conference at held at the University of Paris-Ouest, Nanterre, and the resulting 2005 volume, *Connaissances et pouvoirs. Les espaces impériaux (XVIe-XVIII^e^s): Espagne, France, Portugal*, can be highlighted in the context of the literature review by the editors.[16] The special "Forum on Colonial Science," edited by Londa Schiebinger and appearing in *ISIS* in 2005, presented historiographical synopses of colonial science and the British, French, and Spanish cases.[17] Also in 2005, Benedikt Stuchtey added another substantial review heading yet another volume of special studies;

12. On this theme, see, among others, the more recent Raj (2006). We take up the European encounter with the "other" further below in this introduction.

13. Petitjean (1992), pp. 3-12.

14. Bourguet and Bonneuil; Drayton; Saldaña (2006a), originally published in Spanish in 1999.

15. See MacLeod, ed. (2000) and MacLeod (2000); McClellan (2000b).

16. See de Castelnau-L'Estoile and Regourd, eds., and especially their introduction, pp. 11-22.

17. See Schiebinger, ed. (2005) with the editor's introduction and the included articles by Harrison, Osborne (2005), and Cañizares-Esguerra (2005).

Stuchtey's survey is noteworthy for treating a neglected swath of literature in German and for incorporating thinking about science into postcolonial studies and vice versa.[18] In their introduction to the 2008 volume *Science and Empire in the Atlantic World*, James Delbourgo and Nicholas Dew updated these historiographical reflections, further evidence of the continuing scholarly interest in the topic and the development of a mature scholarly discipline, variously incorporated today under the rubric of Science and Empire Studies.[19] The 2009 book *The Brokered World. Go-Betweens and Global Intelligence, 1770-1820*, published under the direction of Simon Schaffer, Lissa Roberts, Kapil Raj, and James Delbourgo, confirms the creative capacity of the field and displays a variety of complementary approaches.[20] More works like these will undoubtedly follow.[21]

Metaphors and Reality

Our term, "Colonial Machine," is a metaphor, of course, and we have been asked why we picked so mechanical a metaphor. Why not something biological, say, like Behemoth or Leviathan? What is it about the agglomeration we have under the historical microscope that makes it seem particularly mechanical in its operations? The short answer is that the metaphor seems to fit. The crazy whole of our Colonial Machine functioned something like a machine, albeit a creaky one, particularly as it was fundamentally a bureaucracy burdened by routine and dusty files.

The Colonial Machine is a metaphor to be sure, but it would not surprise contemporaries because it fits contemporary usage. As Jean-Louis Martine notes in his thesis of 2003 on the very topic of the historical uses of the term "machine": "Mechanical analogies [are] so numerous and important in the seventeenth and eighteenth centuries that they ended up carving out a space that went well beyond the technical – the legitimate domain of the machine, it would seem – to permeate the whole of human discourse and knowledge."[22] After

18. Stuchtey, passim and p. 22, where he notes one thousand titles on science, technology, and medicine listed in The Royal Historical Society's 2002 *Bibliography of Imperial, Colonial, and Commonwealth History.* The article by Roy MacLeod in the volume edited by Stuchtey (MacLeod, 2005) likewise sets the historiographical stage. The papers in Stuchtey's volume were themselves the result of yet another conference held on the topic of science and imperialism.

19. Delbourgo and Dew (2008); the article by Daniela Bleichmar in this same volume, esp. pp. 237-39, further pushes the historiographical envelope.

20. Schaffer et al. (2009).

21. The 2008 volume edited by Antonella Romano and Stéphane Van Damme on science and world cities should likewise be considered in this context; see the editors' introduction, esp. pp. 14-16. Vlahakis et al. offer a textbook treatment of the subject.

22. Martine (2003), p. 22; on "machine" in the seventeenth and eighteenth centuries, see also synthesis by Halleux (1998). Revel (1997) offers historiographical insights on this point. The 2009 number of the *Revue de Synthèse* (Le Roux et al.) devoted to *Les machines: Objets de connaissance* is a recent, signal contribution to these kinds of discussions.

Descartes, the notion that the human body is a grand machine was a common trope among eighteenth-century authors, one taken to an extreme in Julien Offray de La Mettrie's *L'homme machine* (1748). Already in 1666 the poet Nicolas Boileau could write, "For to think, then, that God turns the world and suspends the springs of the spherical machine."[23] For Voltaire and eighteenth-century Deists, God was the Great Clockmaker; others, such as Charles Bonnet and later Bernardin de Saint-Pierre, also extended the metaphor to the wider world, the former writing in 1754 of "the great machine of the world" and the latter in 1815 of the "vast machine of the universe."[24] The mechanical image rapidly extended to politics. In his *L'Ami des hommes* of 1756, Mirabeau brought the notion down to earth in referring to "the great springs of the political machine," and Rousseau similarly wrote in 1762 and 1764 of "the political machine" and "the machine of government."[25] The Abbé Raynal, author of the controversial *Histoire philosophique et politique des établissements et du commerce des Européens dans les deux Indes* (1770), likewise used the latter expression, as did Helvétius in his *De l'esprit* of 1758.[26] As the nineteenth century unfolded, Tocqueville made the concept even more modern in discuss-

23. "...Car de penser alors que Dieu tourne le monde, Et règle les ressorts de la machine ronde," *Satire* I (1666).

24. See Voltaire, *Les cabales* (1772), lines 111-112, where he writes, "L'univers m'embarrasse, et je ne puis songer / Que cette horloge existe et n'ait point d'horloger." Bonnet: "C'est ainsi que le mouvement imprimé dès le commencement à la grande machine du monde continue suivant les loix établies par le premier moteur," *Essai de psychologie* (1754), p. 4. Bernadin de Saint-Pierre: "Il est [le soleil], dans cette vaste machine de l'univers, comme une grande roue qui communique le mouvement à une infinité de petites bobèches, non à toutes à la fois, mais successivement...," *Harmonies de la nature* (1815), p. 320. These references are taken from the ARTFL electronic database, University of Chicago. Martine, pp. 168-80, offers a complete chapter on this subject.

25. "...je ne m'écarterai pas dans la partie que je traite actuellement, de mon principe général, qui est que le gouvernement ne doit se réserver que les grands ressorts de la machine politique, persuadé que quand ceux-là seront en régle dans ses mains, les détails iront d'eux-mêmes," Mirabeau, p. 138; "Afin donc que le pacte social ne soit pas un vain formulaire, il renferme tacitement cet engagement qui seul peut donner de la force aux autres, que quiconque refusera d'obéir à la volonté générale y sera contraint par tout le corps: ce qui ne signifie autre chose sinon qu'on le forcera à être libre; car telle est la condition qui donnant chaque citoyen à la patrie le garantit de toute dépendance personnelle; condition qui fait l'artifice et le jeu de la machine politique, et qui seule rend légitimes les engagements civils, lesquels sans cela seroient absurdes, tyranniques, et sujets aux plus énormes abus," Rousseau, *Du Contrat social* (1762), Book 1, chapter 8, p. 197. And, speaking of supporters of a state religion, "Ils n'ont pas moins blessé les saines maximes de la politique, puisqu'au lieu de simplifier la machine du gouvernement, ils l'ont composée, ils lui ont donné des ressorts étrangers superflus, et l'assujetissant à deux mobiles différens, souvent contraires, ils ont causé les tiraillemens qu'on sent dans tous les états chrétiens où l'on a fait entrer la religion dans le systême politique," *Lettres écrites de la montagne* (1764), p. 704.

26. "La machine du gouvernement étoit compliquée. Il falloit, pour la conduire, manier une multitude de ressorts délicats," Raynal, vol. 6, p. 4. "S'il est très-dangereux de toucher trop souvent à la machine du gouvernement, je sais aussi qu'il est des temps où la machine s'arrête, si l'on n'y remet de nouveaux ressorts. L'ouvrier ignorant n'ose l'entreprendre; et la machine se détruit d'elle-même. Il n'en est pas ainsi de l'ouvrier habile; il sait, d'une main hardie, la conserver en la réparant," Helvétius, *De l'Esprit* (1758), p. 628.

ing "the social machine" and "an administrative machine."[27] It is not anachronistic to suggest that Raynal and other contemporary actors would resonate to our metaphor – as Edward Long did.

In the end we use the term Colonial Machine as a not inapt convenience for encapsulating a complex and multifaceted historical subject related to expert knowledge and the nation-state in French overseas expansion and colonization in the seventeenth and eighteenth centuries. The term is a helpful abstraction and a model concocted by us as historians. We recognize that the danger exists of reifying this Colonial Machine and making it into an abstract entity it was not. Therefore, we start from the premise that our Colonial Machine existed only in action and needs to be studied from a dynamical point of view.

By the same token, we are self-consciously capitalizing Colonial Machine. Beyond being simply a metaphor, the term captures a dimension of historical reality because the actors and institutions documented in this account and grouped under this label existed and functioned sociologically and bureaucratically as a distinct entity. Our "machine" had distinct and identifiable parts that functioned together toward particular ends. Parts interacted for common purposes, and they formed a collective unity that was real with a history of its own spanning the seventeenth and eighteenth centuries. Not a figment of our imaginations, the Colonial Machine was a dynamic agglomeration of men and institutions that was historically significant both for the history of science and for the history of European expansion. The notion of the Colonial Machine of Old Regime France should come as no surprise, even if its history, told here, is full of surprises.[28]

Utility as the Driving Force

When we seek to understand the dynamics behind the Colonial Machine and attempt to explain why it came into existence and why it acted in the ways it did, we encounter several related explanatory themes: the inherent demands

27. "En AMÉRIQUE, on voit des lois écrites; on en aperçoit l'exécution journalière; tout se meut autour de vous, et on ne découvre nulle part le moteur. La main qui dirige la machine sociale échappe à chaque instant," Tocqueville, *De la démocratie en Amérique* (1835), p. 57; "Richelieu détruisit, dit-on, cent mille offices. Ceux-ci renaissaient aussitôt sous d'autres noms. Pour un peu d'argent on s'ôta le droit de diriger, de contrôler et de contraindre ses propres agents. Il se bâtit de cette manière peu à peu une machine administrative si vaste, si compliquée, si embarrassée et si improductive, qu'il fallut la laisser en quelque façon marcher à vide, et construire en dehors d'elle un instrument de gouvernement qui fût simple et mieux à la main, au moyen duquel on fit en réalité ce que tous ces fonctionnaires avaient l'air de faire," *L'Ancien Régime et la Révolution* (1856), p. 187.

28. Based on our earlier published work, the concept of the "Colonial Machine" seems to have been taken up in some quarters; see Romano and Van Damme, p. 15; Daugeron (2007), pp. 362-65; Daugeron (2009), pp. 378-81; Touchet, Part 1; Bleichmar (2008), p. 234, quoted in note 6 above; regarding Spanish expeditions at the end of the eighteenth century, Bleichmar (2009), p. 299, also writes, "These expeditions functioned within a complex system of institutions and personnel, a 'colonial machine' – to use James McClellan and François Regourd's evocative phrase – comprising botanical gardens, natural history cabinets, pharmacies, observatories, naval schools, and a widespread network of contributors adhering to shared rules of correspondence and exchange that determined the types of materials to be sent and the way to do so."

of the state, a dialectic of power and knowledge, administrative history and the inertia of institutions, the sociology of patronage and patronage systems, and notions of expertise and utility. All of these themes and explanations weave their way through this account, but that of utility takes priority in explaining why the French state would go to such unprecedented lengths to institutionalize and subsidize experts and specialized knowledge in support of its overseas initiatives and activities.

The power and policies of national governments and the existence of the bitter colonial rivalries mentioned above offer a first take on why resources went to support science in the service of colonial development.[29] Governments – the French government in our case – deployed experts to work on colonial problems where science and expert knowledge offered (or seemed to offer) the possibility of solutions. The examples of geodesy or what we would call tropical medicine make the point. For clients as well as for patrons, expert knowledge was useful knowledge worthy of patronage.

The union of science and the French state in the Old Regime has long been recognized as a key development in the social history of early modern science. Charles Gillispie's monumental work, *Science and Polity in France*, stands as the historiographical touchstone in this regard.[30] Writing long ago, in 1970, about the great scientific bureaucrat of the Old Regime, Duhamel du Monceau, and his connections to the French Ministry of the Navy and the Colonies, Michel Allard hit on this understanding as well. Allard pointed to "a context integrating French political and scientific institutions," and he elaborated:

> Let us in addition underscore the interrelation that developed between science and the State. The two mutually relied on one another. If the progress of science could contribute to augment the power of the State, science itself depended on the State...The administrative machine of the State was put in the service of an organization capable of bringing together a large mass of information and observations and to channel the efforts of any and all as a function of attaining a common goal.[31]

29. Regarding the Spanish case, Bleichmar (2009), notes, p. 299, "Government officials, colonial administrators, and naturalists alike tirelessly invoked utility, repeating the word as a mantra. It appears again and again in official correspondence...."

30. See Gillispie (1980), esp. chapt. 2, and Gillispie (2004); see also the early article on this subject by Rhoda Rappaport (1969). The literature on the making of the modern state is substantial and beyond our ken; Strayer is a classic, as is Tilly; the seven-volume series under the general editorship of Blockmans and Genet is the starting point for any inquiry; see also Bonney and more recently Nelson; Sarmant and Stoll; Collins. On the "military revolution" of the early modern period, see Parker; McNeill; Hall; Crosby (2002); Bret (2002). A broad overview of historiography of the links between and among the state, power, society, and production of knowledge is given by Jasanoff (2004a) and Jasanoff 2004b, esp. pp. 39-43.

31. Allard (1969), pp. 94 and 35-36. Allard returned to this theme at the end of his study, p. 93, writing "...des relations très étroites sont ainsi établies entre l'Académie des Sciences et le Ministère de la Marine...." Writing from the perspective of the nineteenth century, Patrick Petitjean (2005) noted, p. 119, "The strong state was not a revolutionary creation. And a marriage of interests between the state, the [military], and the pursuit of science existed long before the French Revolution."

Allard alludes to the point, but plainly we are not the first to note the melding of French science and the contemporary state in French colonial and overseas expansion in the period. Edward Long saw as much in 1774.[32] Historians of contemporary geography and cartography have regularly pointed to the power of state-supported science as an instrumentality for France's expansion.[33] And in 2003, E. C. Spary remarked explicitly that in the first half of the eighteenth century, "...relations between the sciences and colonialism would begin a process of transformation that would render scientific activities inseparable from the colonial enterprise and would make the colonial enterprise itself a measure of national strength."[34]

Key for us, the French state *institutionalized* expert knowledge into a range of formal organizations. On that general basis organized French science achieved its remarkably broad and institutionalized character that made it preeminent among nations of the day. Science and expert knowledge thus became formal instruments of French colonization. The scientifico-medico-technical Colonial Machine examined here was bureaucratically based and highly centralized. Its actions were driven by and supported the interests of the central government, the Bourbon monarchy, and their mercantilist economic policies.

The perceived utility of science and expert knowledge thus provided the primary motive force that drove the Colonial Machine. This force manifested itself in the unrelenting efforts of all concerned, scientific experts as well as governing authorities, to apply or extract anything of any possible use from science in the service of the colonial effort. As far as science and the French colonial experience are concerned, the French scientific enterprise faithfully served French colonial development in the Old Regime. In this sense, then, science and organized knowledge were not separate from the rest of the colonizing process but formed an integral part of French colonialism from the beginning.

32. While not directly dealing with science, Banks adds a useful perspective on the French state and its colonies overseas in the first half of the eighteenth century. James Pritchard, however, controversially challenges accepted notions that the French state or government policies were effective in shaping the course of French colonization before about 1730; see Pritchard (2004), passim and pp. xx, 72-73, 187-88, 190-93, 230-31, 260-63, 402-403, 420-22. About these "scathing attacks," Boucher, p. 256, retorts that Pritchard "swings the pendulum too far in the anti-statist direction." Pritchard's revisionist critique does not seem germane to our treatment of the Colonial Machine.

33. Pedley, Godlewska, and Petto each underscore the connection between institutionalized expertise and state power in the case of contemporary geography and cartography; see Godlewska, chapt. 5; Petto, p. 59; Pedley (2005), p. 80 writes: "Much research on eighteenth-century cartography has focused, naturally, on the state, analyzing the extent of government support and degree of interest in mapmaking endeavors. Certainly one finds centralized state support for reconnaissance surveys and large-scale manuscript mapmaking of small areas, largely performed by the military."

34. Spary (2003), p. 14.

It is not enough, however, to point to utility as the driving force behind the Colonial Machine and leave it at that. We need to pinpoint utility for whom and for what exact ends and purposes. As for whom, the interventionist French state invested in science and experts, and the state was the principal beneficiary. As the present account shows, the French government used expert knowledge virtually as a "productive force" in building its colonies and recruiting science and medicine in the cause of colonial development and the French reach overseas. In the final analysis, the Colonial Machine bolstered the state's power and its solidity internally in France and externally in the geopolitical global combat in which it was engaged.

The French state in the eighteenth century was a large and complicated affair. The monarchy per se or the royal court were only small parts of the contemporary French government and the state largely conceived. In fact, king and court are mostly peripheral to the main story of the Colonial Machine. By the same token, the Colonial Machine offered the monarchy particular utilities that cannot be overlooked. One was glory, the *gloire* that reflected on the power and majesty of France's Most Christian King. Another was the king's pleasure and serving royal and noble identities and sensibilities. We will see several examples of what is meant here, but one concerns pineapples! Already in the seventeenth century pineapples were exotic fruits for the king's table, and all through the eighteenth century pineapples symbolized the exotic and were rare gifts and precious cultivars in royal and aristocratic hothouses. As late as 1787 the French royal gardener on the remote island of Île de France in the Indian Ocean sent pineapples to a royal administrator in Versailles, along with a note saying, "My hope is that this shipment might please you in that it could provide you the occasion one day to present to the King, the Queen, and the Royal Family pineapples different from the ones they know, but not inferior in taste or quality from those in France."[35] This small example is a reminder that Old Regime France was a preindustrial and socially ordered world far different from ours today and that we need to be mindful of not projecting our modernity onto the period under consideration.

More significantly, the Colonial Machine also offered economic benefits for the state in support of mercantilist policies and for the economy in general. The principles and practice of mercantilism rather than capitalism or industrial expansion focused the government's colonial and overseas vision virtually down to 1789. The mercantilist French state was the primary agent in control of colonial development, and the French state alone possessed the resources to enlist science effectively in support of its larger mercantilist policies.

35. "J'ai supposé que cet envoi pourroit vous faire plaisir, en ce qu'il vous procureroit le moyen de présenter un jour au Roi, à la Reine et à la Famille Royale des Ananas différens de ceux qu'il connoissent, et qui ne sont pas inférieurs en goût et en qualité à ceux de France." ALS "Cossigny" dated "Au Port Louis Ile de France le 9 Mars 1787" to "M. le C.te d'Angivillers...," AN-O[1] 2113A *Pépinières de 1788*.

The Colonial Machine rode the waves aboard ships of the Royal Navy and those of the Compagnies des Indes. The French Royal Navy – La Marine Royale – was itself an integral element of the Colonial Machine and an economic engine for Old Regime France. For the various Compagnies des Indes and their investors, the economic utility of the Colonial Machine and expanding colonial products was also paramount. Notably, however, the French army is almost entirely absent from the story of the Colonial Machine. The army had comparatively little presence overseas.[36] The Colonial Machine was also hardly engaged with merchants, with the average planter or colonist overseas, with nobles or the nobility, and much less with the peasants of France. Indeed, the Colonial Machine flourished not so much from private or strictly colonial interests or initiatives as from metropolitan imperatives. Although government colonial policy worked to promote productivity and profit in the plantation system, how much colonists, plantation owners and operators, or merchants concerned themselves directly with the application of science and learning is another question. Large state-chartered trading companies aside, there is little to suggest that merchant or trading interests not connected to the Colonial Machine looked to science or medicine to improve their endeavors.

As useful as is the concept of utility and as key as it is to identifying the interested parties involved in the institutions and infrastructures incorporated into the Colonial Machine, we need also to identify the arenas and domains for which the Colonial Machine was thought to be useful. Utility for what ends and purposes? As it turns out, the scientific and technical problems encountered in overseas expansion fall naturally into three broad categories. The first concerns navigation, geography, and cartography. The second centers on medical issues, the health of colonists and slaves, and medical problems associated with long sea voyages and establishing colonies in exotic locales. The third area involves botany and related natural sciences, and improvements to agricultural production and the colonial system of production. For each of these areas the *savoir* and *savoir-faire* of experts were perceived to be useful in solving particular technical problems for more general strategic ends. That explains why civil and lay authorities underwrote experts and expert institutions.

The state and the monarchy benefited from supporting the Colonial Machine, but so too did scientific specialists, physicians, and other experts recruited into its service. These men had a vested self-interest in the Colonial Machine and its expansion. This interconnected and interacting set of specialist and expert institutions arose for and was put to the service of colonization and French national interests; in turn, the horizons of men of science and their reach expanded, and the content of the sciences became enlarged. Astronomy, botany, natural history, cartography, geology, medicine, and a variety of other

36. On the army in the colonies, see especially the recent thesis by Boris Lesueur; also McClellan (1992), p. 41.

scientific specialties advanced over the eighteenth century by dint of experience with the colonial and overseas worlds. By operating on an essentially global scale, learned experts and organized science became even more expert and better organized. And careers and patronage explain why savants were eager to participate in the affairs of the Colonial Machine – that is, because technical and scientific work could pay the bills and simultaneously advance fields of science and related expertise.

Having said all that, the theoretical and dynamical model of the Colonial Machine should not be naively thought of along the single axis of utility. Once employed, scientific experts enjoyed a measure of freedom to develop according to their own lights; the internal logic of science and research does not correspond totally with the logic of power. Within its own sphere and out of the control of the sponsoring authorities the world of science maintained its own dialogues, conversations, and inquiries, showing once again that if science can serve power, it is not without its own leverage. And, thus, a negotiation between science and power is at the heart of our story.

Expertise, Careers, and Wanderlust

This book is about the state tapping expert knowledge. Along with utility, the concept of expertise – scientific, medical or technical – needs to be highlighted as an especially useful analytical tool. The Colonial Machine consisted of two parts: one, a lay bureaucratic core (government, church, trading companies), and secondly a set of individuals and institutions possessed of specialized knowledge. The latter (including astronomers, cartographers, physicians, engineers) were masters of specialized bodies of knowledge and expertise; the former felt in need of same and had the resources to sponsor them in a grand way. This dichotomy between experts and lay authority, between clients and patrons, and between institutionalized and institutionalizing power define the Colonial Machine's specificity; expertise is what defines the scientific, medical, and technical components of our Colonial Machine.[37]

The crown and its agents sought to apply science and expertise everywhere they could, not just in the natural sciences. We will see many examples throughout, but a small one here makes the point. French bureaucrats in the 1720s and 1730s sought to create Franco-Indian vocabularies and to train translators to communicate with Indian tribes in Louisiana. To this end, they proposed a plan of having French children raised in Indian communities to develop this competence![38]

37. The 2007 volume edited by Christelle Rabier and especially her introduction provide an entrée into the topic of expertise in historical context.

38. AN-MAR G[62] fol. 141v: Louisiane. 1732. "Dessein dc former des Interpretes des langues sauvages." See also correspondence related to this matter AN-COL C[13A] 13 (1731) fols. 43, 124; C[13A] 16 (1733) fol. 31; C[13C] 4 fol. 59.

One of the themes we sound in this volume is that the Colonial Machine provided employment and recognized careers in science, medicine, and colonial administration. Plainly, a defining feature of the Colonial Machine was that it offered jobs and livelihoods. It did so for dozens, if not hundreds of men and their families for more than a century. Patronage networks and clientelism operated sub rosa, to be sure, but it was the *institutions* of the Colonial Machine that provided salaried positions and opportunities for advancement.[39] We will see many examples of careers great and small within the Colonial Machine. From the great scientific academicians and administrators of the day such as Duhamel du Monceau, the Count de Buffon, and Pierre-Isaac Poissonnier, whose names will recur again and again, down to a royal doctor or botanist slogging away in some god-forsaken colonial outpost, the Colonial Machine was a large bureaucracy that provided careers and career pathways.

The minister of the navy in 1783, De Castries, recognized the extent to which colonial administration itself had become a civil service career. In a remarkable communication to the rector of the University of Paris he sought recruits for colonial administration:

> The new structure that His Majesty just created for the administration of colonies in America and in India was an opportunity for Him to apply these just and beneficent dispositions. This administration is now a career in which subjects who display an early aptitude for higher employment will be successively promoted and can then attain the most honorable positions in the political order.
>
> This career thus offering a satisfying prospect to merit and talent, His Majesty in His Wisdom thought that it would serve His Colonies to post there, preferentially, selected subjects who could uphold the consideration it is right and useful to extend.
>
> It is for these reasons and from this point of view that His Majesty has charged me, Sir, to inform you that He wishes to take under His special Protection those graduates crowned each year by the university but who have no fixed plans and who wish to attach themselves to the Administration of the Colonies. Immediately on completion of their studies of Rhetoric and Philosophy, they will pass at His Majesty's expense to various French establishments in America and in India. Honored before their departure by direct testimony of His royal bounty [i.e., they will be paid in advance], these subjects will be given letters with strong recommendations to the Administrators of these establishments that they

39. See especially McClellan (2003a) on this point; see also McClellan (2003b), and Banks, pp. 187-88, 194-97.

> be placed on their arrival in the first open position, and these Administrators will be charged to write an annual report about these subjects.[40]

Scientific posts and specialist career paths, no less than strictly administrative positions likewise opened up in the Colonial Machine to experts in the sciences and medicine. One example can illustrate the specialist career opportunities available within the Colonial Machine. The case concerns a certain Jean-Guillaume Bruguière (1749-1798).[41] With an M.D. from Montpellier in 1770, Bruguière studied botany in Paris before sailing as a naturalist with Yves-Joseph de Kerguelen on the latter's second voyage to the Indian Ocean in 1773. Back in France, Bruguière was having trouble getting himself further established. He complained that, lacking fortune or connections, private medical practice seemed out of the question, as did the university (which, in any event "is a real dead end and low paying and has no prestige to boot"). So he decided "to go to America as soon as possible."[42] He proposed a simple, but for the day not unreasonable research program: general natural history research, the introduction of vanilla cultivation in the islands, searching for cinchona (the plant containing quinine), which he thought might grow on Martinique or even on Saint Domingue, and raising the cochineal insect as a dyestuff. As it turns out, a position was open to succeed Joubert de la Mothe as official royal botanist (*botaniste du roi*) and head of the royal gardens, the Jar-

40. The Marquis de Castries, "Projet de lettre du Ministre au Recteur de L'Université," AN-MAR G 165, #39: "La nouvelle organisation que Sa Majesté vient d'Etablir pour l'administration de ces Colonies de L'Amérique & L'Inde a été pour Elle l'occasion d'appliquer ces Dispositions aussi justes que bienfaisantes; cette Administration est maintenant une Carriere dans la quelle les Sujets qui dès les premiers pas annonceront une aptitude à des Emplois superieurs y seront succéssivement promu et pouroient parvenir ensuite aux places les plus honorables dans l'ordre politique.

"Cette carriere offrant ainsy au merite et aux talens une Perspective Satisfaisante, Sa Majesté a pensé qu'il étoit de Sa Sagesse et du bien de ses Colonies, d'y introduire de préférence des sujets de choix propres à soutenir la Considération qu'il est juste et utile d'y attacher.

"C'est par ces motifs et dans ce point de vüe qu'Elle m'a chargé, Monsieur, de vous annoncer qu'Elle veut bien prendre sous sa Protection spéciale ceux des Rhetoriciens couronnés chaque année par L'université qui n'auront point de Destination fixe et qui voudront s'attacher à L'Administration des Colonies, qu'en conséquence immediatement après qu'ils auront terminés leurs Etudes par celle & La Philosophie, ils passeront aux fraix de Sa M.é dans les divers Etablissemens françois de L'amérique et de l'Inde, et qu'honorés avant leur départ des premiers temoignages de ses bontés ils seront accompagnés des Recommandations les plus spéciales pour les Administrateurs de ces divers Etablissemens qu'ils y seront placés à leur arrivée aux 1eres places vacantes, et que ces Administateurs seront chargés de rendre chaque année un compte particulier de ces sujets...."

41. On Bruguière, see Théodoridès (1962); Allorge, pp. 483-86; Lacroix (1938), vol. 4, pp. 56-57; see also PA-DB (Bruguière) and his personnel file, AN-COL E 54 (Bruguière).

42. "...sans fortune mes tentatives seroient vaines tant aux places de luniversité,...mais qui est un vray cul de sac, au surplus pas payée et sans honneur. la pratique est l'affaire du hazard dans notre pays, ceux qui y compte ont assés de commencer a la bavette....le conseil est donc pris, j'irai en Amérique et le plus tôt possible," ALS Bruguière to Broussonnet of the Société Royale d'Agriculture, de Monpellier le 8e mars 1784, PA-DB (Bruguière).

din du Roi, in Port-au-Prince in Saint Domingue. Bruguière's background and program made him a strong candidate. He received support from the state minister without portfolio, Lamoignon de Malesherbes, but notably only after Malesherbes had consulted the botanists and chief gardener at the Jardin du Roi in Paris (Thouin, Daubenton, Jussieu, and Desfontaines) and received a positive response; that is, Bruguière was vetted by the relevant scientific parties in the Colonial Machine. Unfortunately for him but revealing for us, the minister directly responsible for the appointment, the Count de La Luzerne at the Ministry of the Navy and the Colonies explicitly wanted a botanical specialist, not a physician who might wander off into private practice. Malesherbes wrote to La Luzerne: "It's too bad for poor Bruguière who had announced his voyage and had made all his arrangements for leaving, but particular interest must give way to the choice made by the public man."[43] All was not lost for Bruguière, however; he found supporters in Auguste Broussonnet of the Société Royale d'Agriculture and Daubenton at the Jardin du Roi, and from 1792 to 1798 Bruguière undertook a state-sponsored botanical mission to the Ottoman empire, Persia, and Egypt. He died on the return trip. He never made it to America, but the ups and downs of his career illustrate the range of opportunities within the Colonial Machine and the paths one could follow.

We should not overlook something else that pops up in the archives: Wanderlust and the lure of the "overseas." The foreign and the exotic motivated many of the individuals we encounter in the following pages.[44] In this connection we wanted to present the remarkable example of Pierre-Marie-François, the Viscount de Pagès (1740-1792).[45] Young de Pagès became a naval officer and saw service in the Seven Years War. Supposedly bitten by the bug to travel the world and disappointed at not sailing in eastern oceans, in 1767, while in Saint Domingue, de Pagès deserted his ship and the navy. He slipped off to New Orleans, traveled up the Mississippi and then overland to Acapulco, whence it was on to Manila, Batavia, the Indian Ocean, and finally overland through India, returning to Marseilles in 1771. Upon his desertion, his name was immediately stripped from the navy's rolls. But after his return, in 1772 the minister approved de Pagès's rehabilitation and reinstatement as *Enseigne de Vaisseau:* "Against every expectation and despite all the obstacles M. de Pagès completed his undertaking... The daring and the success of the undertaking seem to expunge the mistake M. de Pagès made in abandoning his ves-

43. "Cela est facheux pour le pauvre Bruguere qui avait annoncé son voyage et fait tous ses arrangemens pour partir, mais l'intéret du particulier doit céder au choix fait par l'homme public...," ALS from Malesherbes to La Luzerne, dated "à Versailles le 3 juillet 1787," AN-COL E 350 Dossier Richard (Louis). We take up this appointment again below, Part IIC.

44. A related theme not manifested in the present account concerns escaping to the colonies to avoid legal or other personal problems in France; see McClellan (1992), p. 57.

45. On de Pagès see Broc, "Introduction," pp. 15-25 in Broc (1991); Taillemite (2002), p. 402; Lacroix (1938), vol. 3, p. 131; vol. 4, pp. 53-56, and materials in PA-DB (Pagès).

sel and his duty, and this adventure announces in this officer a strength of character and self-reliance that are not indifferent in a military man."[46]

As it turns out, de Pagès's wanderlust may have been a cover story for a spying mission sponsored by the navy, but, if true, that this cover story was successful testifies to its currency and plausibility.[47] What is certain is that de Pagès's career made him someone of note within the Colonial Machine. He continued as a naval officer, sailing with Kerguelen to Antarctic waters in 1773, and another major voyage followed in 1776 to high in the North Atlantic. He published his two-volume *Voyages autour du monde et vers les deux pôles par terre et par mer pendant les années 1767-1776* in 1782.[48] A high-ranking and highly decorated noble officer, de Pagès was a *membre titulaire* of the Académie Royale de Marine at Brest.[49] He was elected a correspondent of the Académie Royale des Sciences in Paris in 1781, and he later reported to the academy from the New World.[50] He retired to Saint Domingue in 1782 with a 1200-livre pension and an annual gratuity of 4000 livres.[51] It says something about our story that de Pagès was murdered in 1792 by slaves in revolt in Saint Domingue.

De Pagès's wanderlust may not have been entirely genuine, but for others, especially the botanists, the world beyond France and Europe offered a scintillating allure. The academician and botanist at the Jardin du Roi, Sébastien Vaillant (1669-1722), had this point in mind when he commented early in the eighteenth century:

46. See copy of ministerial document dated 1 March 1772, PA-DB (Pagès): "Cependant le sieur de Pagès contre toute espérance, a rempli son entreprise avec la plus grande étendue malgré les obstacles.... La hardiesse et le succès de l'entreprise semblent bien effacer la faute que le sieur de Pagès a commise en abandonnant son vaisseau et son service et cette avanture annonce dans cet officier une fermeté d'âme et des resources qui ne sont pas indiférentes dans un militaire. On propose de *le rétablir* en sa qualité d'enseigne de vaisseau au rang qu'il occupoit...," marked "approuvé."

47. Bertrand Daugeron pointed out this possibility to us. This action would not be unprecedented; in 1729, a *lettre de cachet* was issued against an army officer in Louisiana, one Livaudais, as a cover for a secret commercial mission to Mexico; see ALS Le Peletier to "M Le Comte de Maurepas," dated "A Versailles le 8 Juin 1729" in AN-MAR B^3 334 fols. 109-110.

48. This work was approved for publication by the Académie des Sciences; PA-PV-110: Pagès/ 2 mai 1781 Manuscrit contenant la relation de trois voyages; le 1.er autour du monde fait pour l'amérique Septm.le &c. le 2.e aux terres australes et le 3.e au nord dans la mer glaciale &c./App.é et imp.é sous privilège/Lalande, Borda, Jussieu. See also the critical edition of this work by Broc (1991).

49. His 1772 paper to the Academy at Brest dealt with native canoes used in the Philippines; DDP III: 54.

50. See Appendix 1 and PA-PV-110, Pagès/Comte de/31 Janvier 1789/Extrait d'un Voyage dans l'Amérique Méridionale/Rapport favorable/Tillet, Thouin, Jussieu.

51. The French pound or *livre tournois* was the standard unit of currency in contemporary France. The range of five to ten U.S. dollars provides a helpful rule of thumb in interpreting the figures. In making conversions, one recalls that the salary of a skilled worker at the end of the Old Regime was on the order of 500 livres annually; on contemporary weights, measures, and currency, see McClellan (1992), pp. xvii-xviii; Pedley (2005), p. xvii; Pritchard (2004), pp. xxiii-xxvi.

> ...who can doubt that Botany will not attain a state of perfection from those who, by dint of recently established rules, will henceforth sojourn in foreign lands where in ten months a Botanist can boast of making more progress than in an equal number of years working in the best gardens of Europe....On the basis of so many beautiful observations and marvelous discoveries, who would not be tempted to voyage? We are told, and it should be believed, that it is to these discoveries and an infinity of similar ones made in various voyages to foreign lands that we owe the greatest part of the perfection of the canvas of a new Method in Botany.[52]

The great, if idiosyncratic botanist Michel Adanson spent the years 1748-1754 in Senegal, and he commented to similar effect in 1759:

> Botany cannot be learned in *cabinets* or in a single country, and because the perfection of this science depends on observation and the knowledge of the greatest number of objects, it is also only through long-distance voyages that a great botanist can hope to be formed. This point is too much neglected in the natural sciences today, and we can say that this is the surest means they have to perfect themselves.[53]

The call of the wild was still present in the 1780s, as it would be for Darwin a half a century later. Writing in 1804, the botanist and explorer Du Petit-Thouars (he traveled to Île de France and Madagascar in 1792-1794) makes clear the passion and excitement surrounding botany and botanical research overseas in the 1780s:

> Effectively, as soon as a young man has tasted the charms of this science, he experiences a kind of whirlwind that gradually propels him to great travel. His ambition develops, and finally the taste, or rather the

52. Ms., "Characteres de 14 genres de plantes," PA-DB (Vaillant). "Voila le point de progrés où la Botanique est parvenue par les observations de ces grands maîtres. Qui peut douter, aprés celà, qu'elle ne doive attendre son état de perfection de celles qui, par le moïen des régles établies depuis peu, se feront dorénavant dans les pays étrangers, où en dix mois de temps, un Botaniste peut se vanter hardiment de faire plus de chemin qu'en un pareil nombre d'années, il n'en feroit dans les Jardins les mieux peuplés...Au recit de tant de belles observations et de si merveilleuses découvertes, qui ne seroit pas tenté de voyager; car on nous asseûre, et on le doit croire, que c'est à elles et a une infinité d'autres semblables faites en divers voyages aux pays étrangers, que nous devons la plus grande partie de l'état de perfection du canevas d'une nouvelle Methode...."

53. "Plan d'un ouvrage général sur la botanique," HI, AD 263: "La Botanique ne peut s'aprendre ny dans les cabinets n'y dans un seul paÿ et comme la perfection de cette science dépend de l'observation et de la conaissance du plus grand nombre des objets ce n'est aussy que par les voïages de long cours qu'on peut esperer de former un grand botaniste. C'est un point que l'on néglige trop aujourd'huy dans les sciences naturelles et je puis dire que c'est le plus sur moyen qu'elles aïent pour se perfectioner...."

> passion for exotic travel awakens in him. He believes he can transport himself to anchorages one hardly knows the name of; a discovery will mark each of his steps. That is the series of sensations I have felt since 1780 when I gave myself over to the study of Botany.[54]

Conquest, Overseas, and the Other

Ours is a focused study of the unparalleled scientifico-administrative structures that facilitated French colonization and overseas expansion in the Old Regime. To conclude this introduction, we need to touch on a few further parameters that allow us to better situate our subject historiographically and analytically.

The Machine within the Machine. As we have already suggested, the army, the slave trade, the merchant marine, and private travelers and colonists formed part of French expansion and colonization in the seventeenth and eighteenth centuries. Our specialist scientific-technico-medical Colonial Machine has to be subsumed into a larger, more grand contemporary French colonial and overseas outreach, both formal and informal. Scientific specialists and other experts formed only part of the totality of that larger enterprise. Other elements included most of the governing colonial bureaucracy in the metropolis and in the colonies, most aspects of the army, the navy, trading companies, and missionary institutions, and, in short, a good bit of what went on in establishing and maintaining overseas colonies and trading posts. In other words, our focus is on the scientific gears and wheels of an even larger national enterprise; so technically it might be more correct to label the object of our study the scientific-technico-medical Colonial Machine of the Old Regime. Although not the whole story of contemporary French colonialism, the subset of expert institutions and individuals examined here represents the key intersection of government and political authority on the one hand (shall we say the state, in short?) and scientific, medical, and technical experts and expert knowledge systems on the other. This subset was an essential element of the total machinery behind French colonization and overseas expansion, and it makes plain the important ways in which officialdom tapped expert scientific, medical, and technical knowledge in the service of colonial development.

54. Du Petit-Thouars (1804), "Discours préliminaire," copy HI AD 53: "Effectivement, dès qu'un jeune homme a goûté les charmes de cette science, il éprouve une espèce d'excentricité qui l'entraîne peu-à-peu dans les courses les plus lontaines ....son ambition se développe.... Enfin, le goût, ou plutôt la passion des voyages lointains, s'éveille en lui; il croit que s'il pouvoit se transporter dans ces parages que l'on connoît à peine de nom, chacun de ses pas seroit marqué par une découverte. Telle est la série de sensations que j'ai éprouvées depuis 1780, que je me livrai à l'étude de la Botanique...."

Expeditions, Colonies, and the World Overseas. We should also clarify the "overseas" of our title. We strive to be inclusive and complete in treating the Colonial Machine. Thus, our story primarily concerns the establishment and maintenance of formal colonies and outposts overseas, notably those in Canada, Louisiana, Martinique, Guadeloupe, Saint Domingue, and Guiana in the New World; Senegal, Madagascar, Île de France, and Île Bourbon in Africa and the Indian Ocean; and commercial and diplomatic extensions into India, notably at Pondicherry, and on to China through the Jesuits stationed at Beijing and connections to the Compagnie des Indes in Canton. The practical and technical problems facing French agents in creating and maintaining these far-flung entrepôts and settlements occupied the Colonial Machine to a high degree.

By the same token, we do not want to lose sight of exploration, voyages of discovery, and individual travelers, all of which also played a central role in the history of the Colonial Machine and French overseas expansion in the Old Regime.

French voyages to the New World, such as those of Jacques Cartier between 1534 and 1541, preceded the foundation of the first French overseas colony (Quebec, 1608) by many decades. Voyages represent the leading edge of expansion and colonization. Voyages were the mechanism for the initial charting of physical and intellectual spaces. Voyages and colonization efforts were intimately connected, and expeditions of all sorts continued to be mounted throughout the period of concern here. The sea was awash with ships, and ideas and artifacts sailed back and forth with them. Indeed, the colonies and the "overseas" in general were accessible only via the watery main. Be it noted, it took weeks and months at sea to reach the colonies of France or to return, and these voyages involved a nontrivial mortality of their own.[55] We will want to distinguish between publicly and privately sponsored voyages. And in this connection, the great, more purely scientific voyages of discovery – those to Lapland and Peru led by Maupertuis and La Condamine in the 1730s, the voyages to observe the transits of Venus in 1761 and 1769, and the circumnavigations by Bougainville (1766-1769) and by La Pérouse (1785-1788) – have bedazzled historians. But these great endeavors were merely the most refined and glorious manifestations of the French overseas outreach in the two centuries leading up to the French Revolution. These expeditions need to be seen as outgrowths of the Colonial Machine, and they, too, evidence the Colonial Machine in action. And then, individual French voyagers – some private, many sponsored by the Colonial Machine – continued to travel worldwide throughout the later seventeenth and eighteenth centuries. All of these elements – voyages great and small, individual travel, and colonization per se

55. McClellan (1992), pp. 23-24; Havard and Vidal, pp. 205-207; Banks, p. 86, reports that sailing to Louisiana from France took an average of seventeen weeks.

intertwined in the French overseas experience in the Old Regime. The Colonial Machine is visible in them all.[56]

The Colonial Machine, Race, and Slavery. Slavery permeates and sullies the story of the Colonial Machine. The French slave trade was the largest in the world in the 1780s, and slavery and the triangular Atlantic slave trade propelled French colonization in the Caribbean and South America. Issues touching on slavery run throughout our narrative, and we are far from writing slavery out of the story. Indeed, official eighteenth-century French science and medicine bolstered slavery and the slave system.

Many telling examples support this unfortunate, but unsurprising fact. The best known perhaps is the 1772 prize offered by the Bordeaux Academy for the best means of preserving the health of slaves on the long Middle Passage from Africa to the American colonies.[57] Bordeaux (along with Nantes) was the main center of the French slave trade, so it is no wonder the Bordeaux Academy posed this question and others like it.[58] Scions of the great commercial house in Nantes, the Montaudouin brothers Daniel-René (1715-1754) and Jean-Gabriel (1722-1780), were themselves slave merchants who were also elected and active *correspondants* of the Paris Academy of Sciences; Jean-Gabriel was also a property holder in Saint Domingue. Books and dissertations devoted to the health of slaves crop up in the context of the Colonial Machine, such as the *Observations sur les Maladies des Nègres* published in 1776 by the royal physician, inspector of colonial hospitals, and *correspondant* of the Société Royale de Médecine, Jean-Barthélemy Dazille.[59] In Saint Domingue, the Cercle des Philadelphes categorized Africans according to kinds of work that best suited them. French botanists imported the "precious" breadfruit from the Indian Ocean to feed slaves. Inoculation against smallpox, introduced in Saint Domingue in the 1760s and 1770s, proved a boon to the slave system! In 1778 a colonist in Saint Domingue, one Lefebvre-Deshayes, wrote to the naturalist Michel Adanson at the Academy of Sciences in Paris about an ambitious natural history collection he and friends were assembling, and he was explicit about a connection to slavery: "When I undertook this collection my first

56. In our 2000 *OSIRIS* article, our focus was more on colonization per se than on voyages of discovery or individual travelers. This was probably a useful corrective at the time to counterbalance the overpowering examples of scientific voyages such as those for the transits of Venus or of Bougainville or La Pérouse. A number of colleagues, notably Bertrand Daugeron, have pointed out the restrictive nature of our initial approach and the necessity of including such voyages in our account. The French overseas adventure in the seventeenth and eighteenth centuries was multifaceted, and so was the involvement of the Colonial Machine, and our present account strives to be more inclusive than our initial sketch.

57. BM., Bordeaux, ms. 1696 (XXXIV) and ms. 828 LXXXIX. See also Barrière, p. 302-303. The Academy received only a few submissions and only a few of these it found acceptable. In 1777 the Bordeaux Academy abandoned the prize in favor of a new one on nursing abandoned babies.

58. See materials, BM Bordeaux, ms. 828, XLV and XLVI and prize question: *Quelle est la cause phisique de la couleur des Nègres, de la qualité de leurs cheveux, et de la dégénération de l'une et de l'autre.*

59. A second, expanded edition of this work appeared in 1792; see Bibliography (Contemporary Sources). For other examples of this literature, see Guerra (1994).

thoughts were for my own personal benefit and the conservation of my slaves."[60] In 1779, over on Île de France in the Indian Ocean government authorities rewarded the notable colonist and royal engineer, Cossigny de Palma, for his innovations in indigo production by allowing him to import two hundred slaves from Madagascar, eighty for his own profit.[61]

Directly and indirectly, then, the Colonial Machine lent its support to slavery and the slave system. Slavery was not a prime preoccupation of the Colonial Machine, but the Colonial Machine cannot be separated from the institutions of slavery in the eighteenth century, and the theme of science and slavery provides a sober reminder that in some instances at least science has served "unfreedom" and human oppression. When one considers the effect of eighteenth-century science and medicine in underpinning the ultimately retrogressive policies of both mercantilism and slavery, one has at least to temper the Enlightenment's and our assumption that science and medicine inevitably have been agents of progress in modern history.

But *slavery* is not entirely congruent with *race*, and we need to recognize that the story of the Colonial Machine unfolds in a pluralistic and multiracial world in which whites, by definition legally free, constituted only one element of the human landscape. Slaves were another component, of course, but they could be black, red, or mulatto, and with many racial shades in-between.[62] *Free* blacks and mulattoes populated this world, too, as did Asians from various regions. Then, of course, mixed-race individuals, some free and some slave, also populated the contemporary colonial world. In this context, the case of Jean-Baptiste Lislet (1755-1836) is an exceptional indicator of race in the Colonial Machine. Lislet was the illegitimate, mulatto son of a white planter in the Indian Ocean colony of Île Bourbon and an enslaved African princess from Guinea who had been sold to the Compagnie des Indes and later freed by her owner-consort. The young Lislet was tutored by his white father, eventually becoming an expert engineer, surveyor, and government cartographer of the Indian Ocean region. In a self-conscious nod to racial liberalism, in 1786 the Académie Royale des Sciences elected Lislet its *correspondant*, attaching him to academy Honorary and very noble and very white Peer of France, Louis-Alexandre, Duc de La Rochefoucauld d'Enville (1743-1792).[63] This example

60. "...mon utilité personnelle, et la conservation de mes esclaves, ont été les premiers objets que j'ai eu encore lorsque j'ai entrepris cette collection." ALS, Lefebvre Deshayes to Adanson dated "a Tivoli, quartier de la Nouvelle Plymouth, par la ville des Cayes du fond-de-lisle-de vache, partie du Sud de St. Domingue le 16 Aoust 1778," HI, AD 208.

61. ALS, Cossigny to Le Monnier dated "Palma le 1 9bre 1779," BCMNHN, Ms. 1995.

62. McClellan (1992), pp. 61-62, points to enslaved Native Americans in Saint Domingue. In terms of coerced labor one should not forget indentured servants (the *engagés*), more important in the seventeenth century, but who still populated the colonies in the eighteenth.

63. Lislet held these positions first under the French and then as Hydrographical Engineer under the English after the secession of Île de France to the English in 1814 and its renaming as Mauritius. Lislet was legitimated by his father in the Revolution, becoming Lislet-Geoffroy; he was subsequently elected to the Académie des Sciences of the Institut de France in the section of geography and navigation; see Schiebinger (1995), p. 195 and notes; Jameson (1837).

and one or two others like it do not take away from the fact that the Colonial Machine was dominated by white French men.

The Colonial Machine as Male. Another striking point we observed in mining the archives is the general absence of women in the record of the Colonial Machine. To say that the public sphere in Old Regime France was dominated by men is to utter an obvious truism, but the colonial world of the day – the world overseas, the world of shipboard life – was especially a world of men.[64] The bureaucrats, administrators, officeholders, and agents who staffed and ran the Colonial Machine at home and abroad were likewise all male. It was a man's world, and the Colonial Machine was a manly enterprise. And thus, as historians we are saddled with expressions like "the men and institutions" of the Colonial Machine.

Such statements, as true as they are, pose the question of the role of women in the story of the Colonial Machine, and, as soon as we look, we find women everywhere. Women are not central, and they do not play a significant direct part in the overall story, but they are there, and they need to be recognized, if only to emphasize what a man's world it was.

Wives, lovers, companions, mothers, daughters, nuns, female slaves, and other female relatives, friends, and acquaintances were inevitably part of the lives of the men of the Colonial Machine.[65] The categories of gender, race, and slavery became blurred in the case of the French government botanist Fusée-Aublet, who freed his consort and the mother of his three children, a Senegalese slave named Armelle.[66] On occasion, the presence of women was more than simply ancillary to the science of learned men. Disguised as a man, Jeanne Baret (1740-1803) famously accompanied and assisted the naturalist Philibert Commerson on Bougainville's circumnavigation in the years 1766-69. The wife of the royal botanist in Cayenne superintended the royal spice nurseries alongside her husband, Joseph Martin.[67] Similarly, J.-N. Céré's wife seems to have aided him at the Jardin du Roi on Île de France in the Indian Ocean.[68] Aristocratic and well-connected women show up here and there, as with Mme Des Tauches who forwarded botanical samples from Africa to the Jardin du Roi in Paris or Mme Le Couteux du Moley who passed on specimens from North America.[69] A certain Mme de Bonneuil, a Creole (that is, an island-born, presumably white woman) from Île de France, was a correspondent of Thouin and the Jardin du Roi.[70] The estate of a certain Mlle Guérin funded a prize on

64. On gender divisions in the colonies, see McClellan (1992), pp. 49, 56-57.

65. The 2005 volume edited by Ballantyne and Burton richly pursues the themes of women, gender, and sexuality in the global colonial encounter.

66. Schiebinger (2004), p. 55.

67. BCMNHN, Ms. 2310, "Liste des Correspondans du Museum."

68. Laissus (1973), p. 39n.

69. BCMNHN, Ms. 691, "Catalogues des graines semées au Muséum pendant les années *1790-1801*," fols. 119-20, 129.

70. BCMNHN, Ms. 2310, "Liste des Correspondans du Museum."

antiscorbutics at the Royal Society of Medicine in Paris in 1783.[71] And in this connection we cannot omit mentioning Mlle Marie Le Masson Le Golft, the provincial amateur from Normandy whose efforts at botany and natural history drew the attention of and membership in the Academy in Rouen and the Cercle des Philadelphes in Saint Domingue.[72] She even had some positive contact with the great Academy of Sciences in Paris.[73] Even more remarkable was Charlotte Dugée, a mulatto woman from Saint Domingue who became a licensed botanical artist working alongside the botanist J.-B. Patris in the forests of Guiana in 1764-67; she produced more than six hundred sketches of plants before apparently going insane and disappearing into the Guyanese jungle.[74]

We do not wish to write these women out of our history, yet these and other women, notable as they are, remain marginal to the Colonial Machine, and in this connection the maleness of the Colonial Machine contrasts sharply with other aspects of intellectual life in France at the time, notably salon culture, which was dominated by sophisticated women.[75]

Global Perspectives and the "Other." Our focus is on the nuts and bolts of the Colonial Machine, but we need to avoid the pitfalls of Eurocentrism and any sense of triumphalism about the French or their European competitors. Historiographically, it is not possible or desirable to cast this story among those that unselfconsciously recount the heroics of European explorers and colonists in planting the flag or founding hardy colonies in faraway lands. The history of European expansion and globalization since Columbus is a more complex and multidimensional subject than once thought. We have to look closely at Europe, of course, and understand Europe's outward push after 1492 because that expansion was of such great historical importance, but in so doing our perspective still has to be global. The European center remains a reference point and a locus of activity for the organization, production, and diffusion of scientific knowledge in the West, but that center must be integrated into a planet-wide and multipolar vision of contemporary European colonialism and the scientific and technical problems of the day.[76] Any alternative approach is parochial and poor history.

71. BAM, SRdM, Ms. #9, p. 549: "Séance publique 26 aoust 1783."

72. Poirier, pp. 350-355; McClellan (1992), p. 237.

73. PA-PV-110: "Masson /M.elle/1.er Juin 1791/Envoi d'un manuscrit intitulé: Portfeuille de feu L'abbé Dicquemare relatif à la Connoissance des êtres et a quelques autres parties de la philosophie. /Rap: favorable/ Vicq-d'azir, Daubenton, Broussonnet."

74. Chaïa (1975), p. 193; Patris, p. 144; Touchet, pp. 267-68. Her drawings are now lost. The draft of her "Brevet de dessinatrice du roi pour Cayenne" is dated 1 January 1764 and is kept in the personnel file of Jean-Étienne Montucla, COAM, COL E 315. Her disappearance in the depths of the forest with no witnesses other than the man she worked with may seem rather odd.

75. On this point see especially Lilti, pp. 110-120, and Sutton; see also Terrall (1995) and (1996), and Andrew, chapt. 5.

76. On this point, see Gruzinski (2004); Polanco; Raj (2006); the thematic issue of the *Revue française d'Histoire d'Outre-mer* edited by Marie-Noëlle Bourguet and Christophe Bonneuil; and Stuchtey, pp. 29-34; the volumes edited by Delbourgo and Dew (2008) and by Antonella Romano and Stéphane Van Damme (2008) likewise bring this perspective to bear.

The French were just one set of actors in a developing global dynamic, then, and the Colonial Machine is just one part of a much larger story. Along these lines, we cannot lose sight of the "other" – the peoples, countries, cultures, and diverse traditions – encountered by the French as their colonies and overseas efforts grew. The English, Dutch, Spanish, and Portuguese might figure in this category, as suggested above, but here one has more in mind the broad array of aboriginal peoples in North and South America, the diversity of Africans (in East and West Africa and elsewhere as slaves and free people of color), and the great range of societies and cultures in South and East Asia that the French encountered. In India, the French had to negotiate with sophisticated local rulers and literate high cultures that rivaled their own. In China, French traders in Canton or French Jesuits in Beijing faced similar cultural conditions. Furthermore, this "other" was not passive. Science and learning in the context of European expansion grew through a dynamic interaction with indigenous peoples and native knowledge. The history is an entangled one, with local informants, usually written out of the story, playing decisive roles.[77] It cannot be said that there was a single frontier or contact point, however. Rather, networks of exchange that only partly included the French extended across wide areas in Asia, Africa, and the New World. Knowledge and expertise circulated freely across and through these zones a circulation the French were anxious to tap.[78]

Finally in this connection, the French were racist and sexist, to be sure, but they did not display the kind of imperialist mentality common to nineteenth- and twentieth-century European colonialism and imperialism. Old Regime France did not have a "civilizing mission."[79] While thinking themselves superior, the men of the Colonial Machine did not have complete confidence in their superiority, most likely because of the fragility of the overseas endeavor in general and the evident shortcomings in what they then knew of the world. As a result, the French in the Old Regime were more open than they would later become to the "other" and what the "other" might offer.

* * *

77. The notion of the "strategic effacement" of non-Europeans in the making of contemporary European science is a theme productively explored by Neil Safier; see Safier (2008b), passim; and Petitjean (2005), p. 111.

78. On this point see Pratt, pp. 6-7, who articulated the notion of a "contact zone," and critical reflections by Raj (2006), esp. his introduction and conclusion. On this enlarged perspective into which our story fits, see also Bravo (1996) and (1999); Kumar (1994); Kumar, ed. (1991); Agarwal; Schiebinger (2004); Schiebinger (2005); Schiebinger and Swan, eds. (2004); and Subrahmanyam (2007).

79. See Pyenson (1993) in particular; Petitjean (2005) details the novelties in the "civilizing mission" of the French in the nineteenth and twentieth centuries; Stuchtey, pp. 36-37, outlines the point.

Given the complexity of our subject and the various things we have to do to dismantle the Colonial Machine and to present all that it was and did, the organization of our presentation is not cumbersome, we hope, but it is somewhat complex. We divide our account into several parts of unequal length and serving different purposes. To orient readers, a brief prologue offers a snapshot of the Colonial Machine and how it came to be on the eve of the French Revolution. Part I (Metropolitan Wheels & Gears of a Colonial Scientific Bureaucracy) unpacks the main constituent elements of the Colonial Machine in the metropolis and unveils the interconnections of these elements and their overseas orientations and colonial concerns. Part II (The Colonial Machine in Action) proceeds to analyze how in fact the Colonial Machine facilitated contemporary French overseas expansion and how that experience impacted contemporary science. This second part is made up of three substantial subparts or chapters that probe the Colonial Machine's accomplishments across three broad subject areas. Part IIA (The Conquest of Space) examines work done in astronomy and cartography. Part IIB (Living and Dying in the Colonies) delves into expert medical knowledge, its transformations, and its role in the maintenance and administration of the colonies. And Part IIC (Cultivating an Empire) similarly explores the place and role of botany and agronomy for the evolution of French science and the colonial economy. All this, with a constant focus on the machine in action. We need to avoid making more of our Colonial Machine than is merited, so a final Part III (Limits and Alternatives) assesses the limits and failures of the Colonial Machine, obstacles to its success, and scientific and political alternatives that help define the Colonial Machine by what it was not.

The colonial side of eighteenth-century French science is a story yet to be fully told, and our account should modify received views of the history of science of the period, particularly in arguing that the colonies and the world overseas formed more a part of contemporary French (and European) science than has been generally recognized. Conversely, increased awareness of the scientific side of the French colonial experience extends our grasp of contemporary French colonial and overseas history, and we hope it will draw more attention to the importance of colonies and the overseas in the general history of prerevolutionary France.

Prologue: The Colonial Machine – 1789

Three cheers for 1789! The great French Revolution that began in July of that year is a godsend for the historian because the event cuts so deeply across modern European history and permits historians to sharply differentiate periods before and after the fall of the Bastille. The Revolution in France unfolded over a number of years, of course, and the Old Regime and the colonial world of Old Regime France did not come to a crashing halt on 14 July 1789. But, simply put, on one side of 1789 is France of the Old Regime, and on the other side come the French Republic, Napoleon, and the nineteenth-century. The date marks the high point of the Old Regime in France. Political, colonial, and scientific continuities notwithstanding, 1789 serves as an unusually clear and convenient marker of a major break in the landscape of history.

So it is with our story. The turn of the French Revolution represents the high-water mark of an elaborate state-level scientifico-medical-technical bureaucracy set in place to support and advance the French presence overseas, the Colonial Machine. Before entering into the details, it will repay us to offer an overview of the Colonial Machine at its prime on the eve of the revolution in France. The accompanying figure presents a visual representation, albeit an idealized one, of the Colonial Machine as it existed at the end of the Old Regime.[1]

First and foremost, the illustration depicts a great bureaucratic structure, and our subject is as much the administrative history of the Old Regime as anything else. Royal authority and government are largely at the center, and this story primarily concerns the central, monarchical state. Not everything, but much of importance, revolved around Versailles, the king (Louis XVI in 1789), and the vast offices of royal government and administration there, in Paris, elsewhere in France, and in the colonies. The Colonial Machine consisted of two distinct components: a set of nonexpert government, commercial, and religious powers on the one hand that provided the resources and sponsorship, and an accompanying set of largely-royal scientific, medical, and technical institutions and experts on the other hand that served their patrons, their science, and themselves. Of particular note on the lay side of the royal coin are the Ministry

1. An earlier schematic appears in McClellan and Regourd (2000).

of the Navy and the Colonies (*Ministère de la Marine et des Colonies*, one of a shifting handful of state ministries) and another large administrative department, the King's Household (the *Maison du Roi*) along with its subdivision, the King's Buildings (the *Bâtiments du Roi*). But, generally speaking, royal government and administration did not per se embody scientific or technical expertise of the sort the state needed or that is of concern here. And so, connected to and out of these sources of royal authority and patronage spun a host of specialized scientific and technical institutions and offices that provided specialized expertise in the service of French colonial and overseas expansion. (These expert organizations are shown as white circles to underscore the difference between them and the 'secular' supporting authorities in gray.) Notable within the enveloping circle of specialist royal institutions on the left of the diagram are the Académie Royale des Sciences (1666), the Observatoire Royal (1667), the Jardin du Roi (1626/1635) and related nurseries and gardens, the Société Royale de Médecine (1776/1778), and the Société Royale d'Agriculture (1761/1788). Notable around the Ministry of the Navy and the Colonies are the Académie Royale de Marine (1752) and the navy's map depository, the Dépôt des Cartes et Plans (1720). Among other royal specialist institutions in the colonies in the 1780s one finds (on the right) the Société Royale des Sciences et des Arts du Cap-François in the preeminent French Caribbean colony of Saint Domingue (modern Haiti). This royal society was originally founded in 1784 as the Cercle des Philadelphes, and the signing of its formal letters patent by Louis XVI in May of 1789 is emblematic of the Colonial Machine at its height.

These royal, government institutions constituted the core of the Colonial Machine. Two additional sources of authority and institutional support were French Catholic religious orders and various state-chartered trading companies. Not government or royal per se, but strongly tied to officialdom, French trading companies along with Catholic missions and missionaries overseas are to be included as distinct parts of the Colonial Machine. These agencies deployed administrative structures in the motherland and overseas analogous to those of government, and they represent powers corresponding to those of government. To a lesser, but still notable degree also, both trading companies and church authorities created a corresponding set of specialized institutions and subsidized specialists, notably missionary experts for the church, and medical, cartographical, and botanical infrastructures of the various Compagnies des Indes. All of these elements – royal-governmental, the semi-autonomous entities of religion and state-controlled commerce, and the expert institutions they created – combined. Through their interactions they completed and animated the Colonial Machine of Old Regime France.

We would underscore the deep geographical divide incorporated into the diagram. On the left is metropolitan France and on the right is "outre-mer," the colonial world overseas. The oceans, the necessity of crossing them, and the technology of sailing constituted defining elements of the Colonial Machine and what it faced in action.[2] The gearing in the diagram – particularly the symbolic conveyor belts indicating the links between the overseas world and the metropolis – seeks to depict exchanges across the oceans and, more generally, the dynamic interactions of the various parts of this larger assemblage. In, around, and through the Colonial Machine people, things, and ideas ceaselessly circulated.

* * *

The Colonial Machine that was in place in the 1780s arose in several stages over what is known as the "long eighteenth century," that is, from the era of Louis XIV and Colbert in the 1660s through the fall of Louis XVI in the course of the French Revolution.

An antecedent stage in the first half of the seventeenth century is noteworthy for the beginnings of formal French colonization (Quebec, 1608; Guadeloupe and Martinique, 1635; Senegal, 1638/1659; Madagascar, 1643; Île Bourbon, 1649/1652). But through the 1660s French colonization remained modest. In addition to the first of many thousands of black slaves who were key to the production of sugar as it began in the colonies, we can estimate that in the 1660s white colonists numbered only 15,000 to 20,000 souls overseas. The Colonial Machine itself cannot be said to have existed prior to the 1660s, however, particularly as the French state played second fiddle to trading companies and the church in the prior period.[3]

The first real chapter in the history of the Colonial Machine runs from the 1660s through Louis XIV's death in 1715. With Louis XIV in full charge in 1661 and with Jean-Baptiste Colbert (1619-1683) as pluripotent minister, France's colonial and scientific endeavors became transformed, as did so much of France under the Sun King. All but one of France's Old Regime colonies were settled by the end of the seventeenth century under more direct royal administration (Cayenne and Guiana, 1664; Saint Domingue, 1665; Pon-

2. The comment made in 1797 by Moreau de Saint-Méry, the colonial lawyer and historian, that the great tall ships are "the most astonishing machines created by the genius of man" never fails to be apt in this regard; DPF, p. 465.

3. Discounting a disconcerting disagreement in the secondary literature, the dates given for the foundation of colonies need to be treated with caution, as there are disjunctures between and among dates for initial territorial claims, the issuance of legal charters to found colonies, sailing versus arriving, failed initiatives, and the spordic beginnings of several colonies, indicated by the dual dates. Boucher provides the most recent entrée into the French in the Americas in the seventeenth century; Pritchard (2004) provides a comprehensive history of the French in the Americas from 1670 through 1730; on French colonization in the Old Regime generally, see Abénon and Dickinson; Havard and Vidal; Pluchon (1991); Devaux; Huetz de Lemps; and Meyer et al. Formal French colonization of Madagascar lasted from 1643 until 1674.

dicherry, 1674; Louisiana, 1682).[4] More to the point, the core bureaucratic and scientific institutions of the Colonial Machine took shape in the last decades of the seventeenth century. Colbert solidified court and government positions (including the Bâtiments du Roi and the posts of royal physician). He reestablished the moribund *Marine Royale* – the Royal Navy, to which several medical and technical bureaux were added later in the century. Colbert created the great French trading company, the Compagnie des Indes in 1664. He took administrative control of the Jardin du Roi in 1671, and he was responsible for the foundation of the key institutions of the Académie Royale des Sciences in 1666 and the Observatoire Royal in 1667 with their paid staffs of scientific experts. In this way, Colbert, the "Great Clockmaker" of the Colonial Machine, laid the technical and administrative foundations of an aggressive and global overseas policy geared to the potentialities of the centralizing machinery of the absolutist state.[5]

The first half of the eighteenth century – to the beginning of the Seven Years War in 1756 – marks another distinct phase in the assembling of the Colonial Machine. As depicted on the accompanying map, French colonization reached its geographical limits in this period with substantial economic growth of the Caribbean colonies and the Compagnie des Indes' holdings on Île Bourbon and Île de France. The French empire at this time counted roughly 100,000 white colonists, a few thousand free people of color, and more than 250,000 black slaves, the greater part being in the West Indies.

On the scientific and specialist side of the coin in this period, the formal letters patent of 1699 restructured the Académie Royale des Sciences, making it an ever more formidable institution concerned with science and with the world overseas. The Observatoire and the Jardin du Roi continued to flourish as active centers of science. But the first half of the eighteenth century evidences a trend toward the creation of more specialized units within the Colonial Machine, complementary to the more general expert institutions that saw the light of day in the seventeenth century. Royal botanical gardens appeared strategically in Brest (1694/1740), Nantes (1709), and Rochefort (1738) along with nascent botanical gardens in the colonies in Quebec (circa 1700), New Orleans (circa 1725), Guadeloupe (1707), and Île de France (1735 and 1748).[6] Further spe-

4. The exception was Île de France in the Indian Ocean, nominally founded in 1715, with colonization seemingly to have begun in 1721, and with the colony's take-off in 1734 with the arrival of a new governor, Mahé de La Bourdonnais. After claiming Louisiana in 1682, La Salle's attempt to establish a colony there in 1684 was short-lived; the French colony in Louisiana may more properly be said to have begun in 1699 with settlements established in Biloxi and then Mobile by Jean-Baptiste Le Moyne d'Iberville; the city of New Orleans dates from 1718. Pritchard (2004), p. xvii, notes fourteen French colonies in the Americas alone.

5. On Colbert's reforms, see below Part I at note 10 and passim. The expression, *grand horloger*, alludes to Voltaire's deist reference to God, as noted above, Introduction at note 24.

6. The general hospital in Quebec dates from 1698 with a garden soon attached; there were two botanical gardens in New Orleans, one dating from the mid-1720s, the other from 1738; see below, Part IIC.

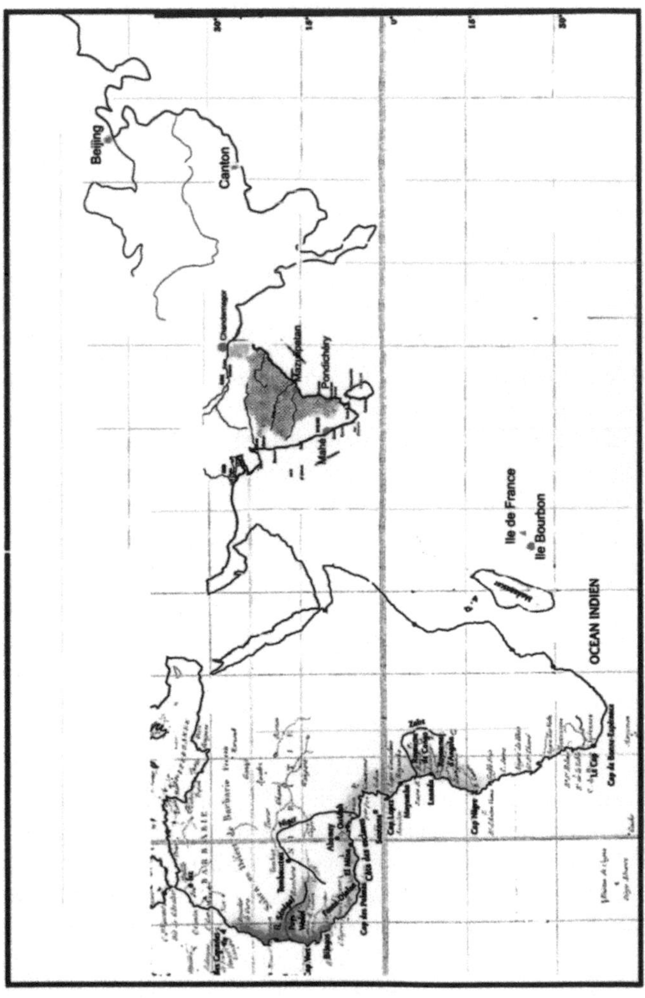

[Map: The French Colonial World – 1756]

cialized additions to the Colonial Machine arose within the Royal Navy in the first half of the eighteenth century, notably the navy's map depository, the Dépôt des Cartes et Plans (1720), and naval observatories in Marseilles (1702) and Toulon (1720), the former elevated to the rank of Observatoire Royal de la Marine in 1749. A related specialist institution, the Académie de Chirurgie, emerged in 1731, ultimately developing its own colonial dimension. And, given a royal charter in 1752, the Académie de Marine, or Academy of Naval Sciences, at Brest was another notable new cog in the Colonial Machine that appeared at the end of this period.

The French defeat in the Seven Years War (1756-1763) was a major reversal for contemporary French colonization, but not so much of one for the Colonial Machine. The Treaty of Paris of 1763 ratified the losses, notably of Canada, Louisiana, and India (except for five small *comptoirs*). Yet this setback was not the long-term disaster it is often portrayed to be, as the remaining colonies – particularly Saint Domingue – flourished spectacularly through the 1780s. Two components of the Colonial Machine did prove decidedly less important in the second half of the eighteenth century: the Roman Catholic church and the Compagnie des Indes; but rather than contracting, the Colonial Machine as a whole expanded dramatically after 1763, reaching its peak in the 1780s by dint of a crescendo of secondary and tertiary additions to its structure.

The second half of the century represents the apogee of the Académie Royale des Sciences in the Old Regime. The Observatoire Royal likewise continued to flourish, not least because of a reform in 1785 that installed new instruments and a new class of astronomer-trainees. The botanical establishment, under the aegis of Buffon, became greatly strengthened in this period. To wit, several changes upgraded the Jardin du Roi in Paris; and with horticultural stations at Trianon/Versailles and then at Rambouillet (1784), the system of royal nurseries expanded considerably, as did the number of provincial gardens, notably the one at Lorient (1768). Even more tellingly, a functioning system of royal botanical gardens in the colonies coalesced in this period with new or upgraded outposts in Île Bourbon (1769), Île de France (previous gardens revived in 1772), Guadeloupe (1775), Port-au-Prince (1777), Cayenne (1778), and even New Jersey (1786) and South Carolina (1787) in the United States!

Within the navy, the Académie de Marine came into its own after the Seven Years War, becoming a royal academy in 1769. The Dépôt des Cartes et Plans continued to grow, ultimately sprouting subdivisions, including a special colonial division, the Dépôt des Cartes et Plans des Colonies (1778). The navy's already elaborate medical infrastructure expanded in 1763 with the new posts of inspector (*Inspecteur*) and medical director (*Directeur général de la médecine de la marine et des colonies*). The creation of a specialized school for colonial medicine in Brest in 1783 (*École de médecine pratique*) is a telling, if ephemeral, addition to the overall structure. A milestone in the development of the Colonial Machine, the Société Royale de Médecine received royal letters-patent in 1778, immediately assuming a central role for all things medical within the Colonial Machine. The same can be said of the Société Royale d'Agriculture in its domain; this organization grew out of an antecedent Parisian agricultural society of 1761, but took off in the 1780s, receiving its own royal letters patent as a national organization in 1788. To repeat, the emergence of the Cercle des Philadelphes in Saint Domingue in 1784 and the elevation of this colonial scientific society to the status of Société Royale des Sciences et Arts du Cap François in 1789 crowns the overall structure of the Colonial Machine before the French Revolution saw the dramatic end of the Colonial Machine and the Old Regime itself.

* * *

On the eve of the French Revolution, then, the French state possessed a huge bureaucracy devoted to overseas expansion and colonial development. A noteworthy set of scientific and technical institutions functioned within a larger administrative structure, and these specialized institutions collectively constituted the largest and most complex "scientifico-medico-technical" bureaucracy in the history of the world to that point. Not ignoring precedents set by the Spanish and Portuguese in the sixteenth and seventeenth centuries to utilize science and expert knowledge for the advancement of their overseas empires, prior to 1789 no nation had so institutionalized scientific and technical expertise nor so tapped the power of specialized knowledge of nature for the establishment and maintenance of extraterritorial colonies as France under the Bourbon monarchy. The scientific and technical wheels and gears of the colonial bureaucracy turned deep in the engine room of the contemporary French colonial enterprise. This dynamic union of men and institutions we are calling the Colonial Machine, and this is its story.

Part I

The Metropolitan Wheels & Gears of a Colonial Scientific Bureaucracy

Si on voulait donner à quelqu'un l'idée d'une machine un peu compliquée, on commencerait par démonter cette machine, par en faire voir séparément & distinctement toutes les pièces, & ensuite on expliquerait le rapport de chacune de ces pièces à ses voisines; & en procédant ainsi, on feroit entendre clairement le jeu de toute la machine.

– d'Alembert, *Encyclopédie*, "Dictionnaire"[1]

The Colonial Machine was a complex and fascinating concatenation. It embodied a lay bureaucratic core and a constellation of dependent specialized institutions. The sponsoring authorities consisted of elements of the central royal government, complemented by the church and trading companies. These powers chartered, funded, and superintended an extensive set of expert/specialist institutions and bureaux, and they likewise recruited and underwrote cadres of scientific, medical, and technical experts to run these specialized units. In this opening Part we begin by reviewing the political powers that undergirded the expert institutions of the Colonial Machine, and we move on to survey the constitutions and colonial character of its constituent institutions.

The Sponsoring Authorities

Royal Government. As it came to be, the Colonial Machine was first and foremost a creation of the centralized French state under the Bourbon monarchs of the seventeenth and eighteenth centuries – the Old Regime. The political and bureaucratic organization of the Colonial Machine centered on the king, on royal government, and in particular on two great bureaucratic agencies: the

1. Cited in Martine (2003), p. 16.

Maison du Roi and the Ministry of the Navy and the Colonies (the Ministère de la Marine et des Colonies).[2]

Although organizational variations arose over the course of the seventeenth and eighteenth centuries, the state secretary for the Ministry of the Navy and the Colonies was one of only a handful (usually four) of chief ministers who customarily made up the king's innermost cabinet (the *Conseil d'en haut* or later the *Conseil d'État*), the others being the controller general of finances and the ministers of foreign affairs and the army.[3] It is telling that colonial affairs ranked at the top of government priorities and administration. The impulse driving the Colonial Machine originated in absolutist royal power and priorities and radiated outward from Versailles.

The Ministry of the Navy and the Colonies formed a complex bureaucratic world unto itself. There, at Versailles, its chief minister reviewed correspondence and made decisions or transmitted the most weighty matters to and from the King's Council.[4] In considering the colonial and naval bureaucracies, the mind's eye conjures up offices and antechambers and the scurryings of secretaries, assistants, minor functionaries, copyists, porters, grooms, servants, and guards, not to mention carriages and postal traffic back and forth from Versailles to Paris and the various ports of France and thence back and forth across the oceans to the colonies and outposts of France. An unintended consequence of this elaborate and finely honed bureaucracy was the creation of a rich set of archives that still exists and that provides an unsurpassed resource for historians generally and for the present study in particular.

The actual governance of the colonies and much of the administration of the Colonial Machine went through the Ministry of the Navy and the Colonies. With the same style of "dual government" imposed on the French provinces, the colonies were governed locally by royally appointed governors and intendants who reported to the Navy ministry and received their instructions from Versailles.[5] Governors and Intendants, not to mention subsidiary agents, had great power as appointees of the king and the ministry. We will encounter more than one activist administrator, such as Roland-Michel Barrin, the Marquis de

2. See Marion; Richet; Mousnier; Bély; and Barbiche for entrées into the bureaucratic worlds of the Old Regime. Desbarats provides an overview of the historiography of the state in a colonial context, as well as Sarmant and Stoll, pp. 603-606.

3. The numbers of State Secretaries fluctuated, especially under the regency of Philippe-d'Orléans and the committee structures that then governed (*polysynodie*); see esp. Barbiche, pp. 173-94.

4. Dessert (2003); for a list of "Secrétaires d'État et Ministres de la Marine et des Colonies," see Henrat et al. pp. 12-13 and Chapuis, pp. 758-61; Banks, Appendices; see also Barbiche, pp. 209-228; Pritchard (1987a), chapt. 2; in his chapt. 1, Pritchard details the activities of seven state secretaries of the navy and colonies in the period 1748-1762. Banks, pp. 9-10, makes the case for a strong colonial bureaucracy, whereas Pritchard (2004), pp. 233-34, emphasizes the relative weakness of the navy and colonial administration in the early years.

5. Pritchard (2004), pp. 241-54 presents the organization of colonial government and justice; Banks, pp. 188-94, also outlines colonial administration and "a well-articulated, integrated, and hierarchical organization in each colony to ensure the smooth collection and distribution of information and knowledge." Further, pp. 26-27, Banks details the career of Michel Bégon, at one point Intendant at Martinique and generally one of the great administrators in the workings of the Colonial Machine.

La Galissonnière (1693-1756), governor-general of Canada (1746-1749) and César-Henri, the count de La Luzerne (1737-1799), former ambassador to the United States and governor-general of Saint Domingue (1786-1787) who went on to become minister of the navy and the colonies from 1788 through 1790.[6]

Colonial affairs became more elaborate as time went on, and the organization of the navy ministry adapted as a result. A formal colonial department (the *Bureau des Colonies*) appeared in 1710 with its own administrative commissioner, the *Chef du Bureau des Colonies*. In 1770, following the royal takeover of the French East India Company (the *Compagnie des Indes Orientales*), the colonial department split into two divisions, one charged to oversee French colonies in the Americas and West Africa, the other to superintend the Indian Ocean colonies and Asian entrepôts. Finally, capping the development of this essentially separate colonial ministry within the larger Ministry of the Navy, a new office, the post of intendancy-general for the colonies, came into being in 1783.[7]

The Ministry of the Navy and the Colonies did not itself harbor scientific or technical experts. Rather, as an arm of central government, it paid for this expertise, and it served as a coordinating institution evaluating needs and enlisting specialists and bureaux elsewhere in the establishment to serve its ends. One notable function of the ministry was to provide postal cover.[8] In particular for present purposes, the ministry also ran the Royal Navy.

[Medal: RES NAVALIS INSTAURATA – 1670][9]

6. Both of these individuals will reappear often in this account; on de La Galissonnière in particular see Taillemite (2008), pp. 117-24, (2002), pp. 290-91; Lamontagne (1960), (1961), and (1962a); Bonnault; Rousseau (1957), pp. 156-57; Lortie, pp. 12-15; on La Luzerne, see Taillemite (2002), p. 295.

7. On these points see, Archives Nationales (1984), p. 625; Krakovitch, p. 7; Taillemite in DPF, p. viii; Banks, p. 28.

8. For example, see reply to letter dated "Nantes 20 mars 79," where Michel Adanson asks his correspondent, "Lorsque vous aures quelque Lettres ou paquets à me faire passer Veuillez bien les mettre sous l'enveloppe du Ministre de la Marine, en ecrivant en gros caracters sur l'adresse de mon paquet le mot *Botanique* pour indique que les objets i continues ne concerne que les sciences relativ. à mes travaux académiques," HI, AD 220.

9. This medal was struck to memorialize Colbert's reorganization of the navy. Harking back to practices in antiquity, the French struck such medals to commemorate major events in the reigns of the Bourbon monarchs. In addition, many organizations of the Old Regime – particularly agencies of the royal bureaucracy – regularly produced and distributed ceremonial tokens or *jetons*. We reproduce several jetons and medals here to underscore the bureaucratized and institutionalized nature of the Colonial Machine.

The Royal Navy (*La Marine Royale*) was a vast organization that generated phenomenal commercial, industrial, and human activity in pre-Revolutionary France.[10] From its modest roots early in the seventeenth century, the French Navy became thoroughly reorganized and established on a permanent basis by Colbert only in 1669. It came to number 35,000 officers and sailors, seventy ships of the line, and as many frigates at its height in 1779 during the American War.[11] Naval operations centered around the three major home ports of Brest in Brittany, Rochefort on the mid-Atlantic coast, and Toulon in the Mediterranean; and the staggering material needs of the Navy stimulated enormous economic activity throughout France, making it "by far" the largest enterprise in the kingdom.[12] French colonial administration and the Colonial Machine primarily and literally rode on the bottoms of the French Navy.

In addition to the navy and the government per se, we need to take account of the separate court offices of the *Bâtiments du Roi* (the department of the king's buildings) and its superintending *Maison du Roi* (the King's Household).[13] These were not minor court offices, but elaborate bureaucracies in and of themselves. The Bâtiments du Roi was a subdivision of the more complex bureaucratic entity, the Maison du Roi, that Byzantine array of offices and functions that arose around the court. The Maison du Roi included major chaplaincies, chamberlains, equerries, down to the king's musicians and those in charge of his silverware. The subdivision of the Bâtiments du Roi was itself an umbrella organization with over a dozen departments responsible for staffing and maintaining the king's buildings and gardens. Over time, legislation shifted the lines of administrative authority between the Maison du Roi and the Bâtiments du Roi, but in one guise or another these bureaux superintended most of the royal academies, the *Observatoire*, the *Jardin du Roi* and other royal parks and gardens, the royal library, royal presses, and royal manufactures, such as the Gobelins dye works. The Maison du Roi and the Bâtiments du Roi functioned something like a Ministry of Culture or Ministry of the Arts, and they constituted a major locus of state expenditures. Under Louis XIV, for

10. On La Marine Royale, see especially works by Michel Vergé-Franceschi and Dessert (1996); see also Pritchard (1987a) for a detailed presentation of the organization of the navy and naval administration at mid-century; Haudrère (1997), Taillemite (2003), and Polak.

11. Taillemite (1991), p. 123; Devaux, p. 288, who puts the total number of sailors in the French navy and merchant marine at 70,000. Pritchard (1987a), Table 4 and p. 80, gives an even higher figure of 100,000 sailors in the navy, but with a fifth of these diverted into service of the Compagnie des Indes. See also detailed description of naval manpower, SRdM, Ms. 7, pp. 352-3 ("Séance du mardi 9 9bre 1779"). On Colbert's reforms, see Taillemite (2008), pp. 52-61; Petto, p. 33; Banks, pp. 22-27; Dessert (1996), pp. 180-84; Pritchard (1987a), p. 19; Pritchard (2004), pp. 234-41, discusses, but plays down Colbert's effectiveness.

12. Taillemite (1991), p. 125; Pritchard (1987a), pp. 89-92.

13. On the Maison du Roi and the Bâtiments du Roi, see Guillemet, passim and pp. 177-78, 221; Marion, p. 41; Mousnier, vol. 2, pp. 115-31, esp. p. 121; Sarmant and Stoll, pp. 220-22, 236-39; Barbiche and Rostaign, p. 141; Barbiche, pp. 38-41 and 239-52, esp. pp. 242-44, where he signals the subordination of the Bâtiments du Roi to the Maison du Roi after 1691.

example, four million livres a year were channeled through the Bâtiments du Roi, mostly to build Versailles. The Bâtiments du Roi also possessed a large and expensive technical staff of its own, including royal architects. Not counting its component parts, between two and three hundred technical positions were to be found within the Bâtiments du Roi itself (145 in 1706). The position of director-general (sometimes superintendent) of the Bâtiments du Roi was worth 150,000 livres a year all in. Like the Ministry of the Navy and the Colonies, the Maison du Roi/Bâtiments du Roi is a major player in our story.

[Jeton: Bâtiments du Roi, 1712]

We should note en passant that other departments of government played their part in the functioning of the Colonial Machine. The Ministries of Foreign Affairs and State Finances (the *Ministère des Affaires Étrangères* and the *Contrôle Général*) make cameo appearances in this account. The *Société Royale de Médecine* and the *Société Royale d'Agriculture*, for example, technically fell under the Contrôle Général. To pick two small examples: in 1774 the minister of the navy contacted the French ambassador to Sweden to secure a copy of a Swedish book on naval construction.[14] In 1779 the head of French nurseries tapped the ministry of foreign affairs to distribute a pamphlet of botanical desiderata to French consuls in the Levant.[15] Other diplomatic consuls of France will make appearances in our story.

Finally, we also should not lose sight of the court itself and the person of the king. The entourage around the king constituted an elaborate, if private, little world, one highly developed and ritualized that came to have its own specialists, notably royal physicians and gardeners. The royal press is to be noted in this regard; we will point out many scientific and technical books and pamphlets that were produced on royal order and at royal expense. All the basic institutions of government and administration were in place under Louis XIV

14. SHMV, Ms. 88, Boynes letter dated "Versailles le 31 Janvier 1774."

15. Draft letter in Angiviller's hand to "M. de sartine," dated "Vers.les le 24 aoust 1779," AN O[1] 2111: Pépinières, 1778-1782.

and Colbert in the latter decades of the seventeenth century; they solidified and expanded thereafter.

Trading Companies. The next sponsoring institution to come into sight, the Compagnie des Indes, was a lesser, but still important, component of the Colonial Machine.[16] To speak of *the* or *a* Compagnie des Indes is, of course, a misnomer, not least because over the course of the seventeenth and eighteenth centuries the crown chartered dozens of ventures to establish or expand overseas colonies and develop colonial trade.[17] Usually, however, one has in mind the three great royal companies that succeeded one another across the seventeenth and eighteenth centuries. Colbert founded the first of these, the Compagnie des Indes Orientales, in 1664, and in 1666 he established a new town and port at Lorient in Brittany to serve as its commercial hub. In 1719 Colbert's original organization gave way to a renewed Compagnie, organized by John Law. The Compagnie of John Law outlasted its founder and held monopolies on trade in the east and west until it was "suspended" in 1769 and then disbanded in favor of free trade after the Seven Years War. The third great Compagnie des Indes began in 1785 to revive trade with India and China, and lasted until 1791.

These Compagnies were hardly independent institutions, but were "tied tightly to the State."[18] The merchant marine connected closely to the Royal Navy, despite tensions between military and commercial interests that sometimes arose, especially in times of war.[19] At other times, the Compagnie exhorted its agents to "ally the interests of the King with those of the Compagnie."[20] All three of the great French Compagnies des Indes were run by government insiders and elite investors, led by the king. These Compagnies des Indes collectively constituted a strong French equivalent of the rival East India Company based in London and the Dutch *Vereenigde Oost Indische Compagnie*. The crown controlled the Compagnie through the shares it held, through its ability to appoint directors, and through a royal commissioner serving de jure on the governing board. There is no question that the Compagnie des Indes was a state-chartered institution closely monitored and controlled by the crown. Nonetheless, the Compagnie des Indes was a mighty institution of its own, and represented a huge state within the state. From its administrative

16. The standard work on the Compagnie des Indes remains Haudrère (1989), 2nd ed., 2005; see also Wellington; Gunny (1981); Broc (1975), pp. 22-23; Chaligne, chapt. 1; Boucher, pp. 172-76. For a clear overview of the other French trading companies and an analysis of the role of the state in these organizations, see Pétré-Grenouilleau, esp. pp. 35-54.

17. On these companies, see Chailley-Bert; and partial list in Boucher, p. xiii.

18. "...étroitement liées à l'État," Lespagnol, p. 285. Haudrère (2005) makes the same point, pp. 147ff, writing, p. 153: "Ainsi, la Compagnie est véritablement la chose du roi, qui la gouverne par le biais du contrôleur général et du commissaire."

19. Llinares (1996), p. 47, Taillemite (1996), p. 320.

20. "allier les interets du Roy avec ceux de la Compagnie," SHML, IP 284, Liasse 108, #29: Syndics to Buisson, "A Paris Le 4 mars 1767."

headquarters in the Hôtel Tubœuf in Paris, the Compagnie extended its reach to North America, Africa, the Indian Ocean, India, and China.[21] The Compagnie began as a strictly commercial enterprise, but it quickly and necessarily became a territorial and political power.[22] As essential way-stations for trade with India and China, the Mascarene Islands in the Indian Ocean were notable fiefs of the Compagnie.[23] French claims to the Indian Ocean islands of Île Bourbon (modern Réunion) and Île de France (today's Mauritius) date from 1649 and 1715, respectively. As governor appointed by the crown and the Compagnie, Mahé de la Bourdonnais led the real settlement of these islands beginning in 1734, and until it relinquished control to the government in 1769, the Compagnie des Indes was a direct player in French colonization along with the crown.[24] In addition to these true colonies, at various points in time the Compagnie ran major *comptoirs* in Louisiana; in Senegal and Juda in West Africa; in Pondicherry, Mazulipatam, Yanaon, Chandernagor, and Mahé in India; and in Canton in China. The Compagnie administered sovereign justice in the areas of its control, it maintained its own army of 1200-1500 soldiers, and in India at least it minted its own money.

[Jeton: Compagnie des Indes, 1785]

The scale and complexity of the Compagnie des Indes deserves emphasis. The size of its navy rivaled that of the Marine Royale itself.[25] It outfitted approximately forty ships a year at 1 million livres per expedition, and it kept

21. The internal organization of the Compagnie des Indes was complex and included separate offices, among others, for accounting, purchasing, spending, litigation, buildings, armaments, military affairs, the beaver trade, and colonial administration. See SHML, 1P 305 Liasse 70** Extraits du registre, 1723-1785; Taillemite (1996), p. 320; Haudrère (2005), pp. 541-80.

22. Taillemite (1996), p. 320, makes the point about the Compagnie des Indes shifting from a commercial enterprise to a political one.

23. Delaporte, p. 224, and Haudrère (2005), pp. 200-206 make this point.

24. See Haudrère (1992), chapt. 4; Taillemite (2002), p. 282; Gunny.

25. Estienne, p. 131.

20-25 thousand seamen on its rolls at any one time.[26] In the period 1780-1789, 116 expeditions wended their way to Île de France from Bordeaux alone.[27] In 1752 the cost of Compagnie operations in Lorient itself ran to 4,663,000 livres.[28] The volume of its trade staggers the imagination: The Compagnie imported up to 2-½ million pounds of coffee annually, 2 million pounds of tea, 500 thousand pounds of pepper, 700 thousand pounds of wood for dyeing, 300 thousand pieces of porcelain, and 300 thousand bolts of cotton cloth.[29] One shipment arriving in Lorient from Canada in 1747 brought 80 thousand beaver pelts; another in 1749 unloaded 625 bales of beaver pelts weighing over 7 thousand pounds and worth over a quarter of a million livres.[30] In 1767 the Compagnie received over 400 thousand linear meters of sailcloth.[31] The annual sales that took place, first at Nantes and then at Lorient, constituted major economic events in the life of the nation. The value of goods inventoried in the stores of the Compagnie in 1769, for example, amounted to nearly 7.5 million livres.

As it developed, the town of Lorient became a major and unheralded French industrial center, not least for ship construction, with the Compagnie employing 3000 men in its dry docks there. In and around Lorient sprang up forges, forgers and foundries, glassworks, paper mills, hardware shops, weavers and cloth mills, painters, joiners, locksmiths, coopers, and others who lived off commercial orders from the Compagnie.[32] Over 150 different commercial and industrial enterprises contracted with the Compagnie.[33] The demands of the Compagnie for timber, rope, sails, nails, and the myriad other items needed for outfitting its vessels, not to mention items for export, stimulated artisanal and industrial production across the whole of France.[34]

French Religious Orders. There was not a single institution known as the French Catholic Church, but collectively various Catholic religious orders complemented the royal government institutions and trading companies mentioned thus far and rounded out the supporting authorities of the Colonial

26. See previous sources and Haudrère and Le Bouëdec, p 111; Lespagnol, p. 290; see also "Tableau général de la navigation des vaisseaux de la Compagnie des Indes, 1685-1767," SHMV, Manuscrits de la Bibliothèque du Dépôt des cartes et plans de la marine, SH 45 230.

27. Butel (1996), p. 328.

28. Legrand and Marec, p. v.

29. Haudrère and Le Bouëdec, p. 78-80; SHML, 1P 278, Liasse 14, #5: Bureau de Castor; Syndics to Godeheu, "A Paris le 4 mars 1750." On trade in general in the period, see Haudrère (1997), esp. pp. 81-82, 90-100.

30. SHML, 1P 278 Liasse 13 "Lettres du Bureau du Castor," #2, letter from Syndics to Godeheu, "A Paris le 16 X. 1747."

31. Legrand and Marec, p. xxii.

32. Haudrère and Le Bouëduc, p. 108; Legrand and Marec, p. v; Pritchard (1987a), chapt. 7 and Table 5, makes the same point regarding crafts and artisanal trades in the Marine Royale itself.

33. Legrand and Marec, p. xxiv.

34. Legrand and Marec, pp. v, xi, xviii, xxii, xxiv.

Machine.[35] These included the Jesuits as well as Dominicans, Franciscans, Minims, Brothers and Sisters of Charity, and other groups. Each of these orders staffed missions overseas, and the creation of the *Missions Étrangères de Paris* in 1663 added to the outreach efforts of the French Catholicism overseas and in the colonies.[36] The creation of the Jesuit mission in Beijing at the end of the seventeenth century represents the greatest geographical extent of French overseas influence and a notable outpost of the Colonial Machine. Religious orders and missionaries were not monolithic in character or role, and neither were they wholly independent of the state or its goals for French colonization. From the outset, religious orders and their members were thoroughly enmeshed with rest of Colonial Machine, both institutionally and as individuals.

* * *

Such was the lay bureaucratic core of the Colonial Machine – the supporting authorities of church, state, and formalized trade. The government, trading companies, and the church were not per se possessed of scientific or technical expertise (discounting the nontrivial expertise of bureaucrats, sailors, and perhaps priests), and so these powers supported their own specialist extensions in order to fulfill their respective political, commercial, and spiritual missions. This support involved more than just patronage. The key here is the institutionalization of expertise, and in the rest of this Part I we put on view the scientific, medical, and technical components institutionalized in the Colonial Machine, beginning with the Royal Navy.

The Royal Navy – La Marine Royale

[Jeton: La Marine Royale, 1712]

35. Pritchard (2004), p. 262; Pluchon (1991), pp. 650-660; Banks, p. 193, points out that "in all colonies the state paid the clergy's bills." Boucher, pp. 141-42, discusses early missionaries.

36. On the Missions Étrangères de Paris, see Guennou.

A ship's captain steering his vessel across the open ocean was obviously someone in possession of considerable technical and practical expertise, and to this extent the Marine Royale needs to be considered a specialist and expert institution underwritten by the state. What needs emphasis here is that, in addition, the navy incorporated cadres of experts charged with developing and transmitting specialized scientific and technical knowledge throughout an institution oriented toward the sea and overseas.

Training Schools. At the lower end of the navy's scientific and technical infrastructure stood the three officer training schools, the *Écoles des Gardes du Pavillon et de la Marine*, formally established in 1689 in Brest, Rochefort, and Toulon.[37] Through to 1762, Jesuit professors of mathematics and hydrography were charged with the scientific instruction of cadets, a revealing indication of the ways in which religious orders meshed with the state in the Colonial Machine. Father Du Chatelard, S. J., for example, instructed cadets for thirty years at Toulon in "arithmetic, navigation, naval maneuvers, geometry, fortification, military maneuvers, calculus, astronomy, etc."[38] Chatelard likewise kept meteorological and astronomical diaries, which he communicated to the *Académie Royale des Sciences* and to the *Journal de Trévoux*. His textbook on mathematics for use by students was published at state expense.[39] Over at Rochefort, Jean Dizard de Kerguette was the longtime professor of mathematics and hydrography, while La Roche held sway at Brest and one Advizard at Arles.[40] The Jesuits were replaced by no less expert secular professors after that order was closed in 1763.

Subjects taught in these schools included geometry, mechanics, astronomy, optics, hydrography, experimental physics, architecture, and design, as well as practical shipboard training. English language courses were occasionally included.[41] A significant normalizing of instruction took place in 1764. In that year the academician Étienne Bézout (1739-1783) received appointment as examiner of the Gardes de la Marine (ultimately with a salary of 7200 livres). In 1764 the first of many editions of his textbook, *Cours de mathématiques à*

37. Antecedents to these schools hark back to the time of Colbert. We pass over the several organizational reforms that modified these institutions over the course of the eighteenth century. On these schools, see Vergé-Franceschi (1991) and (1999); Dhombres (2003b); and Hahn (1964a); Neuville (1882), passim; for ms. materials related to the schools discussed here, see AN-MAR G86 (dossiers 1-2) and AN-MAR G87-90.

38. "d'arithmetique, de Navigation, d'Evolutions navales, de Geometrie, de Mechanique, de Fortifications, d'Evolutions militaires, d'analyse, d'astronomie &c...." See AN-MAR C^7 92 (Chatelard).

39. See previous note. In one instance at least the town of La Rochelle covered the 500 livre salary of the Jesuit professor there; see AD/La Rochelle, B225, Registres de l'Amirauté, 29 Octobre 1732.

40. AN-MAR C^7 87 (Dizard de Kerguette); see also Neuville (1882), pp. 41-43, who has him as Digard de Kerguette, AN-MAR C^7 2 (Advizard).

41. See Goldlewska, pp. 31-32, and materials, AN-MAR, G86; on instruction in cadet training schools and in hydrography schools, see also Neuville (1882), pp. 14-29.

l'usage des Gardes du Pavillon et de la Marine (all sanctioned by the Académie des Sciences), appeared. It was mandatory.[42] (Elected in 1758, Bézout himself rose through the ranks in the Académie des Sciences, becoming a *pensionnaire* in 1768.) It cannot be said that the science taught in any of these naval training or related hydrography or engineering schools was of an especially high level, or that students themselves went on to do notable scientific work; but these institutions and the teaching and examining positions they supported added to the overall weight of the scientific establishment of the Navy and the Colonial Machine.[43]

Schools of hydrography arose in the sixteenth century in France. They were codified for the first time in 1629 with new legislation issued by Colbert in 1681. By 1785, hydrography schools staffed by royal professors of hydrography cropped up in twenty-four port towns, including major ones such as Le Havre and obscure ones such as Saint-Valéry-sur-Somme.[44] One professor, Fouray, held the position of royal hydrographer at Dieppe for twenty years and was in contact with the Académie des Sciences in Paris.[45] From 1685, in order to strengthen the navy, the king supported two royal hydrographers – both Jesuits – at Marseilles.[46] A nontrivial figure in the Colonial Machine, Father Esprit Pézenas, was royal hydrographer at the *École Royale d'Hydrographie* at Marseilles from 1728 until the school was closed in 1749; Pézenas was also a correspondent of the Academy of Sciences in Paris and the Marine Academy at Brest, and he published volumes on piloting, gauging, and nautical astronomy.[47] Until their suppression, Jesuits not only monopolized this teaching in port towns in France, the professorship of hydrography at Quebec was likewise reserved for them; and the Jesuit college at Quebec, in existence from 1635 through 1759, thus became a colonial center for royal hydrographers and hydrographical instruction.[48] Various Jesuits and lay personnel held these posts, teaching five to ten students a year, training pilots, accrediting survey-

42. On Bézout, see Grabiner; Taillemite (2002), p. 47; and dossier AN-MAR C^7 29 (Bézout). On approbation of the Paris Academy of Sciences, see PA-PV 84 (1765), fol. 137v ("Mercredi 20. Mars 1765"). Bézout's *Cours de Mathématiques* went through many editions and remained the official training manual for naval cadets until 1864!

43. See Boyer et al. The personnel files of the Marine Royale – AN-MAR Series C^7 – list hundreds of specialist positions. One might note in passing that a 1779 proposal to establish four free schools of navigation and commerce went nowhere; see AR-MAR G 146 "Projets d'etablissement de 4 Ecoles de Marine." There was also, apparently, a royal merchant marine academy at least under discussion in 1771; see AR-MAR G89 (dossier 2).

44. Neuville (1882), pp. 1-13; Russo; see also Anthiaume, and SHMV - Manuscrits de la Bibliothèque historique de la Marine. Catalogue général, 0189 XX. These positions were sometimes underwritten by local towns, chambers of commerce, and/or Intendants.

45. AN-MAR C^7 109 (Fouray).

46. Bigourdan (1923), p. 2. Bigourdan reports a special school in Marseilles to train pilots for the galleys!

47. J. Taton; Taillemite (2002), p. 419; AN-MAR C^7 245 (Pezenas).

48. Chartrand et al., pp. 22-34; Audet, pp. 20-37; Banks, p. 72; Havard and Vidal, p. 181; Lortie provides a valuable overview of organized science in Canada.

ors, making maps, and contributing to the defense of the colony. The course in *physique* lasted two years. Among the more notable individuals to hold the position were Louis Jolliet (1645-1700), the explorer of the Mississippi, and Father Joseph-Pierre Bonnecamps (1707-1790), professor of hydrography and mathematics at Quebec from 1741 through 1759.[49] The royal hydrographer in Canada in 1687, Jean-Baptiste Franquelin, received a mere 400 livres for his services; Father Charlevoix did better, receiving 800 livres.[50] For a while at least, the Jesuits also ran a hydrography school in Montreal.[51]

The trajectory of the astronomer and academician Pierre Bouguer (1698-1758) illustrates how the position of royal hydrographer could become a springboard to a career in science.[52] Bouguer succeeded his father as royal hydrographer in out-of-the way Croisic in Brittany. After a few years he transferred and became royal hydrographer at Le Havre. He went on to study in Paris and devote himself to science; he won prize contests of the Académie des Sciences in 1727, 1729, and 1731 and was elected to the institution, becoming *pensionnaire astronome* in 1735. He voyaged to Peru with La Condamine in 1735-1744, and he ended up writing books on ship construction, navigation, and pilotage, works reviewed and approved by the Paris Académie des Sciences.

The Dépôt des Cartes, Plans et Journaux de la Marine. Established in Paris in 1720, the navy's repository of maps, charts, and ships' logs – the *Dépôt des Cartes, Plans et Journaux de la Marine* – was the first official hydrographical bureau of its kind and a major technical division within the larger organization of the Marine Royale. It was a key, if secondary, institution in the whole of the Colonial Machine.[53] The Dépôt served as the central clearinghouse for cartographical, navigational and hydrographic information. Naval captains were

49. See previous note and AN-COL B[19] (1697), fol. 275 ("Brevet de maître d'hydrographie à Québec pour le sr Jolliet") and B[22] (1701), fol. 215v ("Brevet de maître d'hydrographie à Québec pour le sr Franquelin à la place du feu sr Joliet"); Neuville (1895-1913) provides an entrée to these archival sources.

50. See Audet, p. 20 and correspondence related to Franquelin, AN-COL B[15] (1689), fol. 70 and B[16] (1693), fol. 286 v. On Franquelin, see also Palomino; and Rousseau (1969), p. 625. Regarding Charlevoix, see AN-COL B[75] (1742), fol. 44.

51. See above, note 48. For other hydrographers and related technical personnel in Canada in this period, see AN-COL B[7] (1678), fol. 186 ("Brevet d'ingénieur au Canada pour le Sr Martin Boutet qui enseignera l'hydrographie, le pilotage et les mathématiques") and AN-COL B[39] (1717), fol. 252 ("Conseil de Marine à MM. de Vaudreuil et Bégon.... Le Père Leblanc, qui enseigne l'hydrographie à Québec, aura le droit de donner des certificats à ses élèves les qualifiant comme pilots...").

52. Middleton; Taillemite (2002), p. 64; AN-MAR C[7] 40 (Bouguer). Boistel (2005), pp. 113-15, also discusses Bouguer and his career.

53. On the Dépôt des Cartes et Plans, see Chapuis, pp. 159-68; Pritchard (1987a), pp. 27-28; Pedley (2005), pp. 83-84, 194-95; Filliozat-Restif (2005), p. 141; Le Guisquet; appropriate sections in Taillemite (1991); and La Roncière. A full modern study of this institution is a desideratum. Manuscript sources are to be found AN-MAR G 234 (Divers: Dépôt des Cartes et Plans) and BnF, N.A.F. 9498, fol. 154, Ms. "Notice chronologique et historique sur le Dépôt des cartes et plans de la Marine."

required to submit copies of their journals to the Dépôt.[54] By 1759 its collections had grown to 13,000 maps and 5000 other assorted journals and reports.[55] By the 1770s, if not before, the Dépôt was reputed to house the most significant repository of cartographical and up-to-date voyaging information in the world.[56] Expenses of the Dépôt amounted to over 10,000 livres a year.[57] After trying to market its maps through booksellers, the Dépôt took to selling its maps directly to the public.[58] The colonial dimension to the Dépôt grew and in 1778, authorities created a special colonial division within the larger organization, the *Dépôt des Cartes et Plans des Colonies*, renamed the *Dépôt des Cartes et Plans des Colonies Orientales et Occidentales* in 1782.[59]

The Dépôt des Cartes et Plans incorporated a sizeable staff of technical administrators, geographers, cartographers, hydrographers, and astronomers, including Jacques-Nicolas Bellin (1703-1772).[60] In 1741, through connections to La Galissonnière, Bellin secured a post as a navy hydrographer and engineer (*Ingénieur-Hydrographe de la Marine*) at the Dépôt with a salary of 2400 livres.[61] In 1762 he became a naval commissioner (*Commissaire de la Marine*) with emoluments rising to 3000 livres, and he remained at the Dépôt until his death in 1772. Displaying an early interest in navigation and mapping of the American possessions, Bellin produced sixty-three such maps for the Dépôt des Cartes et Plans from 1737 through 1759.[62] His most important work, *Neptune françois*, came out in 1753. (It hardly needs remarking that *publishing* maps and collections offered important and obvious advantages over manuscript maps that seem to have been more the norm in the seventeenth century.) Bellin was also a royal censor and a member of the Marine Academy at Brest and the Royal Society of London. He was a typical scholarly compiler, a *géographe de cabinet*, assembling information coming in from others. As much as they were positive accomplishments, Bellin's maps came in for heavy criticism in the second half of the eighteenth century, and their shortcomings in many ways

54. On the 1757 effort to enforce these regulations through the Académie Royale de Marine, see SHMB, Série A: Sous-série 1A, 1757: fol. 24, letter dated, "A Versailles le 17 J.er 1757."

55. AN MAR G 234, #15; see also draft Delisle letter "au P. Cœurdoux à Pondichery," dated "de Paris le 24 Janvier 1756," AN-MAR 2JJ 68.

56. Chapuis (1999), p. 164; "Mémoire Concernant le Dépôt des cartes De la Marine" by the Comte de Narbonne dated "a Paris Le 12 Mars 1771," AN-MAR G 234, #10.

57. Undated "Etat de ce que coute le Dépôt des plans des Colonies, par année," AN-MAR G 234, #21. Pedley (2005), pp. 210-11 lists other expenses for the Dépôt, including sizeable sums for copper plates for engraving.

58. Pedley (2005), pp. 78-79, 234.

59. Rinckenbach, ed., pp. 7-8; also Barbiche, p. 223.

60. Filliozat-Restif (2005), p. 141, puts the total technical staff at sixteen in 1776, a number she calls small.

61. On Bellin, see Chapuis, pp. 165-68 and passim; Taillemite (2002), pp. 39-40; AN-MAR C^7 24 (Bellin); see also, Polak #565-#602. Bellin also figures largely in Petto's study; see Petto, pp. 69ff.

62. AN-MAR G 234 #15.

propelled further French cartographical work in the Caribbean and South America in the decades prior to the Revolution.[63]

Other specialists were there, too. Giovanni Antonio Rizzi Zanonni (1736-1814) succeeded Bellin as *Ingénieur Géographe et Hydrographe de la Marine* from 1772 through 1775.[64] Pierre-François-André Méchain (1744-1804) followed as *Astronome Hydrographe* in the Dépôt from 1775 through at least 1789; Méchain was also a junior astronomer in the Academy of Sciences.[65] The royal geographers Philippe Buache (1700-1773) and Jean-Nicolas Buache de la Neuville (1741-1825) were likewise associated with this Dépôt, as were other, lesser technical specialists.[66] For a brief period after his stint as Governor-General in Canada, the Marquis de La Galissonnière parked there as director of the Dépôt.[67] Charles-Pierre Claret, the count de Fleurieu (1738-1810), a scientific naval officer of note and ultimately Minister of the Navy, served as adjunct director in 1776.[68]

The official position of naval astronomer (*Astronome de la Marine*) was generally associated with the Dépôt des Cartes et Plans, and through the tradition of the *cumul*, or overlapping appointments, naval astronomers effected ties with the rest of the scientific establishment, notably the Academy of Sciences. For example, the academician Pierre-Louis Maupertuis (1698-1759) was appointed as naval astronomer "to perfect navigation and instruments pertaining thereto" with a stipend of 3000 livres, and he held the post for the period 1740-1744 after his triumphant return from his and the Academy's expedition to Lapland.[69] Coming back to France after two decades in Russia, the astronomer Joseph-Nicolas Delisle (1686-1768) sold his books and papers to the crown in 1754 in return for a lifetime annual annuity of 3000 livres and an

63. DDP VI:48; McClellan (1992), p. 125. Pedley (2005), pp. 25-26 elucidates what is meant by *géographe de cabinet*. About Bellin in particular, Goimpy of the Marine Academy complained about Bellin's accuracy in 1771, and in 1786 complaints were serious enough for the Marine Academy to alert the minister; see Chapuis (1999), p. 168; SMHV, Ms. 89 ("Extrait d'une lettre a M.r de Briqueville par M. de Goimpy dattée de Paris le 1.er 8.bre 1771") and SHMV, Ms. 74 ("Mémoire sur la carte de Marie-Galante et de la Désirade").

64. Chapuis, pp. 167-68; AN-MAR C^7 277 (Rizzi-Zanonny).

65. Chapuis, passim; AN-MAR C^7 207 (Méchain). See reference to Méchain as "astronôme hydrographe du dépôt général de la Marine," PA-PV 99 (1780), fol. 98 ("Du Samedi 15. Avril 1780").

66. Taillemite (2002), p. 78. Other specialists included Pierre-Nicolas Leroy, Ingénieur au Dépôt des Cartes et Plans from 1772; AN-MAR C^7 182 (Leroy); AN-MAR C^7 132 (Grognard du Justin, Benoit. Ingénieur au Dépôt des Cartes et Plans). Pedley (2005), p. 207, reports that a draftsman at the Dépôt earned 600-700 livres a year; she gives figures, p. 239, for paper, pencils, wood, matches, candles, and nine livres a year to dust the books.

67. Lamontagne (1961); Rousseau (1966); AN-MAR C^7 159 (La Galissonnière). For Buache presenting maps based on holdings in the Dépôt, see PA-PV 49 (1730), fol. 122 ("Le Samedi 6.e May"). Petto, pp. 107-108 and note, gives a more positive account of Galissonnière's tenure as head of the Dépôt des Cartes et Plans. Pedley (2005), pp. 146-53, probes the example of the Dépôt map of Narraganset Bay made in 1780 as part of a projected *Neptune Americo-septentrional*.

68. Taillemite (2002), pp. 185-86.

69. AN-MAR C^7 219 (Maupertuis).

appointment as naval astronomer (*Astronome-Géographe de la Marine*) within the Dépôt des Cartes et Plans.[70] The deal included another 600 livres for a secretary and 500 livres for his student, Messier. Furthermore, with instruments from the Academy of Sciences, Delisle established a formal observatory for the Marine at the Hôtel de Clugny on the rue des Mathurins in Paris, an observatory made available to naval officers interested in astronomy.[71] Regarding his assimilation into the Dépôt des Cartes et Plans, Delisle seemed to recognize its institutional significance for the Colonial Machine:

> All my manuscripts and the books from my astronomical and geographical library have been transferred to the Navy's depository of journals, maps, and memoirs. This is a depository attached to the Ministry of the Navy and is under the direction of a Commodore or a Navy Lieutenant General. There will always be an astronomer and a geographer attached to this naval repository, and I am such presently. After my death this astronomer can continue the correspondence with the Jesuit missionaries who will undertake astronomical observations.[72]

Delisle was followed in the post as official naval astronomer by Joseph-Jérôme le François de Lalande and Charles Messier, all with overlapping connections to other parts of the Colonial Machine.[73] At one point, the academician and astronomer Pierre Le Monnier was seemingly also attached to the Marine Royale.[74] Jesuit astronomers and their successors manned naval observatories in Marseilles and Toulon, and we will say more about them further on.

70. E. Doublet, (1910), p. 7; Jaquel; Chapin (1971).

71. AN, MAR C[7] 186 (Lisle); esp. 6pp. ms memoir to king: "Acquisition du cabinet de Mrs. de L'isle pour le depôt des Cartes, plans et Journeaux de la Marine." "Le S.r de L'Isle depuis son retour de Russie a établi à l'hotel de Clugny ruë des Mathurins un petit observatoire muni de tous les instruments nécessaires, les quels instruments appartiennent à l'Académie des sciences. Les officiers de la marine qui se sont attaché à l'Astronomie joüissent de cet observatoire..." This hôtel is known today as the Hôtel de Cluny.

72. "tous mes autres manuscrits et les livres de ma bibliotheque astronomique et Geographique [sont mis] au depost des journaux plans et memoires de la marine qui est un depost sous la dependance du ministre de la marine et sous la direction d'un chef d'escadre ou lieutenant General de la marine; il y aura toujours un astronome et Geographe attaché à ce depost de la marine, qui je le suis à present, et cet astronome pourra continuer aprez ma mort la correspondance avec les missionaires de la Compagnie de Jesus qui s'appliqueront aux observations astronomiques," Delisle letter "au P. Coeurdoux à Pondichery," dated "de Paris le 24 Janvier 1756," AN-MAR 2JJ 68.

73. See AN-MAR C[7] 161 (Lalande); C[7] 207 (Messier). A student of Lalande, Pierre-Antoine Véron was appointed as "astronome de la Marine" to accompany Bougainville, only to die in 1770 on the Ile de France in the Indian Ocean; AN-Marine C7 344 (Véron). Further on the Observatoire de la Marine à Paris, see AN-MAR G92 (dossier 2).

74. Llinares (1996), p. 45.

The Compagnie des Indes felt the same necessity as the Navy for a Dépôt des Cartes et Plans of its own, and in 1762 it created one in Lorient built around the collections of the great cartographer and ship's captain for the Compagnie des Indes, Jean-Baptiste-Nicolas-Denis d'Après de Mannevillette (1707-1780). D'Après de Mannevillette was appointed as *Garde du Dépôt des Cartes et Journaux* of the Compagnie's hydrological office at 2400 livres.[75] With the suppression of the Compagnie des Indes in 1769 the Compagnie's Dépôt in Lorient became an official subsidiary of its Navy homologue in Paris, and d'Après de Mannevillette stayed on to head the *Dépôt des Cartes, Plans et Journaux de la Marine à Lorient*. D'Après maintained active contact with the Dépôt in Paris and with other institutions of the Colonial Machine. After his death, the Dépôt in Lorient was folded in with the Dépôt des Cartes et Plans de la Marine in Paris. On a lesser scale, but likewise indicative of the felt importance of institutionalizing cartographical and navigational information, it seems that French officials maintained a chart house on Île de France.[76]

A small example illustrates the ordinary operations of the Dépôt des Cartes et Plans de la Marine within the Colonial Machine. In 1735 the president of what was at the time the ministerial council superintending the royal navy wrote to the count de Vaudreuil, who had just returned from Louisiana with a map of his travels prepared for him by a local pilot: The council reported itself satisfied with the map sent to it of the Gulf of Mexico and the coast of Louisiana. It forwarded Vaudreuil's map to the Dépôt des Cartes of the Navy and sent an annotated copy to be deposited at navy headquarters at Rochefort, where officers needing it could copy it.[77]

The Navy's Inspector General: Duhamel du Monceau. The post of inspector general of the navy topped the navy's scientific and technical departments, at least as Henri-Louis Duhamel du Monceau came to define the position.[78] The polymath Duhamel du Monceau (1700-1782) was named inspector general of the Navy in 1739, in part as recompense for losing out to Buffon as intendant of the Jardin du Roi, and for over four decades until his death in 1782 Duhamel labored tirelessly not just as inspector but as the navy's chief scientist and technical consultant, for which he came to earn the handsome salary of 6200 livres annually.

75. Filliozat (1993), pp. 92-102.

76. AN-MAR C^7 5 (André, Michel: garde des plans de la Marine à l'Ile de France).

77. See letter, AN-COL B^{62} (1735), fol. 216v: Le président du conseil de marine à M. le comte de Vaudreuil, 27 September, 1735.

78. On Duhamel du Monceau and his monumental opus, see Dupont de Dinechin, esp. chapts. 3-5 and Appendix I; Allard (1970), esp. chapt. 2; Demeulenaere-Douyère (2000b); see also Boudriot (2000); Eklund; Viel; Sautereau; Raunet; Plantefol (1969); Jaoul and Pinault; Chaïa (1977); Allorge, pp. 461-65; and further articles in the volume edited by Corvol. See also his naval personnel dossier AN-MAR, C^7 93 (Duhamel du Monceau); his éloge, SRdM-H&M, vol. 4 1780-1781), Histoire, pp. 101-135, esp. 126-32; and other materials PA-DB (Duhamel du Monceau).

Duhamel visited French ports almost annually through the later 1760s, and he took trips to England and the United Provinces to observe British and Dutch naval practices. He collected elaborate models of ships, not as a hobby but as an aid to his work and for the instruction of naval engineering cadets. He gifted his models to the kingdom; they were displayed in the Louvre in the hallway outside the rooms of the Académie des Sciences, and became the basis of today's *Musée de la Marine.*[79] In 1740 Duhamel was instrumental in organizing the *École de Chirurgie* at Brest; in 1741 he started a school in Paris to train naval architects and engineers; and in 1765 he secured official foundation of this institution as the *École des Ingénieurs Constructeurs de la Marine* (also known as the *École du Génie Maritime*).[80] As in the navy's cadet training schools, instruction for these future *Ingénieurs Constructeurs de la Marine* likewise centered on mathematics and *physique*.

Duhamel's published scientific output was considerable, and much if not most of it pertained to matters naval and colonial. In this connection we can signal his *Traité de la fabrique des manoeuvres pour les vaisseaux, ou l'art de la corderie perfectionné* (1747); *Élémens de l'architecture navale, ou traité pratique de la construction des vaisseaux* (1752); *Avis pour le transport par mer des arbres, des plantes vivaces, des semences, et de diverses autres curiosités d'histoire naturelle* (1752); *Moyens de conserver la santé aux équipages des vaisseaux* (1759); *Art du sucrier* (1764); and his *Du transport, de la conservation et de la force des bois ... faisant la conclusion du traité complet des arbres et des forêts* (1767).[81]

Duhamel du Monceau became a central figure in the Colonial Machine in part because he accumulated additional posts, and he is the first of a handful of powerful individuals we will encounter who informally coordinated colonial-scientific affairs by dint of overlapping, behind-the-scenes roles. In addition to being inspector-general of the navy, Duhamel was an academician to be reckoned with in the Académie Royale des Sciences. He joined the Academy in 1728 and became a *pensionnaire* botanist a decade later. He was editor of the Academy's famous *Description des arts et métiers*, himself writing 20 of the 113 technical descriptions comprising this important collection. He served

79. AN-MAR C^7 194 (Manoury: création d'un cabinet de marine, au Louvre, 1778); Allard (1970), p. 62; McClellan (2007), p. 723.

80. Demarcq (2005), p. 103; Gille; Gay (1999); Allard (1970), pp. 55-56; Boudriot (2000), pp. 181-83; Vérin (2000), p. 157. For archival sources related to this school and related proposals, see AN-MAR, G87 and G89 (dossier 3); SHMV, Manuscrits de la Bibliothèque historique de la Marine, Catalogue général, 0241 1759, "...les avantages d'une école de Marine à Paris," by de Kerguette, professor of hydrography at Croisic. For something of the high standards demanded of students at this school, see 1771 letter from Duhamel and Bézout, AN-MAR C^7 227 (Rivoal).

81. See Bibliography, Contemporary Sources; Dupont de Dinechin, Appendix I, and Allard (1970), Appendix II, provide bibliographies of Duhamel's major published works. Demeulenaere-Douyère, Appendix, gives the details of his papers that appeared in the *Mémoires* of the Paris Academy and the scientific commissions on which he served in the Paris Academy of Sciences.

a record thirty-nine years on the Academy's publications committee, the *Comité de Librairie*, and between 1728 and 1781 he published ninety-seven papers in the Academy's *Histoire et Mémoires* series; most of these were undramatic meteorological reports, but they did include *Essais sur la conservation des grains* ("Essays on the Conservation of Grain," 1745) and *Differens moyens pour perfectionner la boussole* ("Different Ways to Perfect the Compass," 1750). Duhamel was thrice elected as vice director and then director of the institution, and he seemed to collect colonial correspondents, notably (from 1745) Jean-François Gaultier (1708-1756), *médecin du roi* in Quebec – from whom he received Canadian meteorological observations and botanical shipments; a certain M. Fage in Saint Domingue (from 1760 to 1767); Jacques-Alexandre Barbotteau in Guadeloupe (from 1776); and the previously mentioned Viscount de Pagès (from 1781) in Saint Domingue. Bigot de Morogues of the Académie Royale de Marine at Brest was also an official correspondent, and from 1739 to 1749 Duhamel carried on a scientific exchange with Artur, the médecin du roi in Cayenne.

In addition, Duhamel was instrumental in founding the Académie de Marine in Brest, where he became honorary academician and a key intermediary with the Academy of Sciences in Paris.[82] Duhamel was elected a Fellow of the Royal Society of London in 1735, and he was an original member of the *Société d'Agriculture* in Paris in 1761. He had a hand in coordinating astronomical observations of the passage of Mercury in 1752 and French chronometer tests in the 1770s, and he wrote instructions for colonial botanists on how to collect plants. He received botanical specimens from all over the world, specimens he then passed on to the Jardin du Roi and the royal Trianon gardens.[83] The record is replete with instances of Duhamel being called upon by the Minister and/or the Academy of Sciences to serve as an expert consultant on matters naval and colonial. In 1764, for example, he reported to the navy minister the judgment of an Academy committee appointed to investigate a method for desalinizing seawater.[84] In 1748, on instructions from the ministry, Duhamel sent information to the Marquis de la Galissonnière, then governor-general in Canada, on how to extract salt from salt marshes.[85] In his connection with de la Galissonnière, the two friends often worked together in one capacity or another after the latter's return from Canada in 1749. For example,

82. See correspondence to this effect, SHMV, Ms. de Brest 87: "Registre de la correspondance de l'Académie de Marine 1752-1765."

83. See APS Duhamel papers B, D87 Group 24 and APS Collection of Botanical Manuscripts 580 D881. See also Duhamel entries, BCMNHN, Ms. 1327: "Journal des Envois de Plantes, Arbres et Graines faits au Jardin du Roy par ses Correspondants."

84. AN-MAR D^3 40, "Divers procédés pour dessaler l'eau de mer: 1670-1800" ALS, Duhamel du Monceau to "Monseigneur" dated "a Paris ce 20 Juin 1764."

85. AN-COL B^{87} (1748), fol. 48. ALS dated 6 mars 1748. La Galissonnière and Duhamel du Monceau were closely connected; Allard (1970), pp. 74-76, and above at note 6.

in 1751 the minister named them to supervise the work of the royal engineer, Magin, in making a map of the port of Bordeaux.[86] In a word, Duhamel du Monceau was a great human pivot around which the Colonial Machine and much of contemporary French science turned, and his name will recur regularly throughout this account.

Naval Medical Infrastructures: The Marine Royale had many sick and injured to deal with both in France and in the colonies, and institutionalized medicine and surgery in the navy constituted a major component of the entire Colonial Machine.[87] Already in 1686 the navy had built hospitals in its home ports of Brest, Rochefort, and Toulon, and these hospitals were the initial institutional manifestations of what became a substantial colonial medical bureaucracy. In addition to hospitals, the navy needed trained physicians and surgeons for service at sea and on land; and in the early decades of the eighteenth century, to meet that need, the Ministry of the Navy established three complementary medical schools alongside the naval hospitals in each of these same port cities: Rochefort (1722), Toulon (1725), Brest (1731).[88] The last officially became the *Collège Royal de Chirurgie de la Marine* in 1775. To superintend medical care in its hospitals, to teach in its medical schools, and to serve at sea and in the colonies, the navy likewise created, funded, and staffed the posts of Royal Navy physician (*Médecin de la Marine*) and Royal Naval surgeon (*Chirurgien de la Marine*). There were four official naval physicians at Brest in 1783.[89] These were high-ranking, salaried positions, and the medical cadres that filled them at Brest and elsewhere in France and overseas added to the weight of the contemporary medical establishment of the navy.

Étienne Chardon de Courcelles (1705-1775), for example, served as chief naval physician at Brest (*Premier Médecin de la Marine*) for two decades.[90] He also directed the naval medical school there. His total compensation rose to 3600 livres, including an extraordinary bonus of 1000 livres he received in 1752 for "the book he wrote for instructing surgery students."[91] Jean Cochon Dupuys (also du Puy and Dupuis) was the longtime holder of the post of chief

86. AN-MAR D^2 37, ALS, Duhamel du Monceau to "Monseigneur," dated "a Paris ce 28 Mars, 1772." See also AN-MAR C^7 159 (La Galissonnière).

87. Pluchon, ed. (1985) remains the Ur-source for French naval and colonial medicine; see also McClellan (1992), chapt. 8; Hannaway; Roussel et al., pp. 21-34.

88. See Pluchon, ed., pp. 73-75; Lefèvre; Suberchiot (1998); and AN-MAR G90 "Écoles de chirurgie de la marine." The naval medical schools had periods of greater or lesser vitality and were subject to occasional reform; their full story remains to be told. On the École de Chirurgie de la Marine at Brest, see Hamet; on Toulon, see Querangal des Essarts.

89. See SRdM, Archives, 132 dr. 59 ("Médecins de la marine à Brest"), #2.

90. Taillemite (2002), p. 113; AN-MAR C^7 60 (Chardon de Courcelles).

91. "Gratification extraordinaire de 1000 livres à Mr de Courcelles médecin de la marine en considération du livre composé pour l'instruction des élèves chirurgiens," SHMB, Série A: Sous-série 1A, fol. 193. This bonus was presumably for his *Abrégé d'anatomie pour l'instruction des élèves-chirurgiens de la marine de l'École de Brest.* It might also have been for his *Traité abrégé sur l'usage des différentes saignées* first published in 1747 and again in 1751.

naval physician at Rochefort until his death in 1757.[92] There, he established an *École d'Anatomie* that became the model for other schools at Brest and Toulon.[93] His son, Gaspard Cochon Dupuys, succeeded his father in the position at Rochefort, receiving 5000 livres that included bonuses and a housing allowance. In 1783 Antoine-Chaumont Sabatier (1740-1798) was second naval physician at Brest, member of the Académie Royale de Marine, and correspondent of the *Société Royale de Médecine* in Paris. His commission as professor of medical practice (médecine pratique), says something about what these doctors were expected to do:

> ...the said professor is employed to teach physicians destined to serve in naval military hospitals both in Europe and in the colonies, to give them lessons on the best treatments for the illnesses of seamen, to have them recognize the symptoms of these illnesses, to lead students through naval hospitals on the rounds he will make to patients being treated there, to see that these students make their own hospital visits as often as possible to observe the course of diseases, to teach the different parts of medicine, *and to report every quarter to the State Secretary for the Navy.*[94]

The bureaucratization of naval medicine culminated in 1764 with the creation of the powerful posts of *Inspecteur* and *Directeur général de la médecine de la marine et des colonies*, posts held by Pierre-Issac Poissonnier (1720-1798) until 1791.[95] Like Duhamel du Monceau, Poissonnier was a major player, who juggled several important positions in the scientific and colonial bureaucracies and whose name will crop up again and again. He was connected to the Paris medical faculty as *docteur-régent*. He was a counselor of state (*Conseiller d'État*) ennobled by Louis XV, and was called on as royal medical consultant and as a royal censor. He was an active, original member

92. Taillemite (2002), pp. 104-105; AN-MAR C^7 70 (Cochon Dupuy).

93. At Brest at least, executed military deserters were discretely dissected at the anatomy school; see SHMB, Série A: 1A-1, letter dated 6 avril 1738.

94. "...le dit professeur employé à l'instruction des Médecins destinés à servir dans les hopitaux militaires de la Marine tant en Europe que dans les Colonies...leur donner des leçons sur le traitemens le plus convenable aux maladies des Gens de mer, leur en faire connoitre les symptomes, conduire les dits Eleves dans les hopitaux de la Marine pour assister aux visites qu'il fera des malades qui y seront traités; Veiller à ce que les dits Eleves fassent euxmêmes et le plus fréquemment qu'il sera possible des visites dans les hopitaux pour y observer les différents périodes des maladies; faire des Cours sur les différentes parties de la medecine; Rendre compte tous les Trois mois au Secretaire d'Etat ayant le Départment de la Marine...," our emphasis; AN-MAR C^7 290 (Sabatier, Antoine-Chaumont), "Brevet de Professeur de Medecine pratique Pour Le S.r Sabatier second Medecin de la Marine à Brest," dated 7 Mars 1783.

95. On Poissonnier see Valléry-Radot; Gillispie (1980), pp. 134-35ff; Taillemite (2002), p. 427; Simonetta, p. 183; AN-MAR C^7 253 (Poissonnier).

and officer of the *Société Royale de Médecine*. He held a professorship of chemistry and served as the dean of the prestigious Collège Royal. He was an *académicien libre* at the Académie des Sciences from 1765. He was an active member of both the *Société Royale d'Agriculture* and the *Comité d'Agriculture*. He was named an honorary academician in the Marine Academy. Poissonnier was elected a Fellow of the Royal Society of London (F.R.S.) and wrote a two-volume *Abrégé d'anatomie, à l'usage des élèves en chirurgie dans les écoles royales de marine* (1783).[96] His powerful posts in the naval medical establishment ultimately earned him the fabulous sum of 20,000 livres per annum.[97]

A few small examples illustrate the elder Poissonnier's multifaceted role in the Colonial Machine. In 1779 the Société Royale de Médecine received a report from a doctor in Bordeaux about a new method for preserving meat. Poissonnier volunteered to see that trials were made by sending samples to the colonies; in 1781 he passed on to the Société for testing some seeds that supposedly prevented scurvy.[98] Poissonnier was also in regular contact with André Thouin, the chief gardener at the Jardin du Roi, and he served as intermediary in transmitting botanical samples back and forth between the Jardin in Paris and the royal naval botanical garden in Brest.[99] At another point he maneuvered behind the scenes to get the botanist Commerson recalled from Île de France in order to save the latter's 3000 livre salary.[100] Poissonnier was clearly someone to be reckoned with, not least as chief of the naval medical establishment over which he seemingly held full powers of appointment.[101]

In 1775, in a not insignificant move augmenting naval medical administration, Poissonnier appointed his brother, Antoine Poissonnier-Desperrières, as inspector of naval and colonial hospitals (*Inspecteur des hôpitaux de la marine et des colonies*) with a salary of 6000 livres.[102] The younger Poissonnier-Desperrières already had twenty-five years of experience with naval and colonial medicine. He had served in Saint Domingue as royal physician-botanist (*Médecin Botaniste*) from 1749 through 1751, and he was the author of a

96. See Poissonnier (1783) and notice, SRdM-H&M, vol. 4 (1780-1781), Histoire, pp. iv, 239.

97. AN-MAR D^3 41, fol. 29.

98. SRdM, "Manuscrits #7," p. 146 ("Séance du vendredy 30 avril 1779"); SRdM, "Manuscrits #8," p. 378 ("Séance du Mardi 24 Avril 1781"). For another medical report filtering into to Poissonnier from the navy, see AN-MAR G 102 (dossier 2).

99. On these exchanges, see relevant entries in BCMNHN, Ms. 1327: "Journal des Envois de Plantes, Arbres et Graines faits au Jardin du Roy par ses Correspondants"; Ms. 314: Papiers de Thouin, 1778-1791; Ms. 848: "Lettres adressées par Laurent."

100. See letter dated "7 avril 1770" and related undated correspondence, AN-COL E 89; copies PA-DB (Poissonnier).

101. See, for example, ministerial document dated "le 1.er 9bre 1778." AN-MAR C^7 344 (Verdelet), where Poissonnier indicates his approval of an appointment of one Verdelet as "Chirurgien Aide Major."

102. AN-MAR C^7 253 (Poissonnier Desperrieres)

Traité des fièvres de St. Domingue (1763), produced on government orders, and a *Traité des maladies des gens de mer* (1767). Both of these appeared in second editions in 1780 from royal presses. The appointment was obviously nepotistic, but its bureaucratic impact was large and important.

The navy's elaborate medical structures grew incrementally but revealingly, with the addition in 1783 of a special medical school at Brest, the École de Médecine Pratique. The École de Médecine Pratique was created especially to introduce recent graduates of traditional medical schools (that is, graduate M.D.s) to diseases particular to the colonies and port cities, and to train naval doctors for service in naval hospitals in France and overseas in the colonies.[103] The navy ministry was centrally involved in planning this school and bringing it into being.[104] In the few short years of its existence the École de Médecine Pratique at Brest became so much the established entrée into naval medical positions in the metropolis and abroad that other doctors complained that the school cut them off from these lucrative career opportunities.[105] An equivalent proposal of 1785 to establish an *École Pratique pour les Chirurgiens Navigateurs de la Marine* seems not to have gone anywhere, but the initiative is indicative of the navy's need and drive to provide itself with qualified medical and surgical personnel.[106]

[Medal: PRIX POUR LES CHIRURGIENS DE LA MARINE DU ROI FONDÉ EN 1768]

Other Experts. Of course, the Royal Navy deployed a small army of construction engineers to build and maintain port facilities and to superintend the steady production of naval vessels. The gradual professionalization of this staff

103. Boudet, pp. 34-36; Hamet, pp. 329-30; SRdM, Archives, 132 dr. 59 ("Médecins de la marine à Brest"). SRdM ms. 132 dr. 59 contains important documentary material; see also AN-MAR C2 111, fol. 101.

104. SRdM, Archives, 132 dr. 59; SHMB, 1L10/120 r°, (Série L-1L: Règlements..., 1677-1785), 22 May 1783.

105. See letter of Mallet de la Brossière dated Paris, "le 4 octobre 1787," SRdM, Archives, 109 dr. 4.

106. AN-MAR G 147, #2.

meant that the French built the finest ships of the era.[107] And at the highest levels, positions such as that of chief engineer of naval ports and arsenals (*Ingénieur en Chef des Ports et Arsenaux de la Marine*), held for many years by Marc Blondeau (1742-1793), are of interest because of their inevitable connections with the rest of the Colonial Machine. Antoine Groignard (1727-1799), for example, rose through the ranks from being a student at Duhamel's École du Génie Maritime to becoming chief naval engineer (*Ingénieur Général de la Marine et Ordonnateur Civil*). Groignard was elected an ordinary member of the Marine Academy at Brest, and he wrote technical essays that won prizes from the Académie des Sciences in 1761 and 1765. He was paid a total of 25,400 livres for his services in 1789![108] Jean-Joseph Verguin (1701-1777) was a naval engineer at Toulon who accompanied La Condamine, Bouguer, and Godin to Peru in the 1730s, becoming a correspondent of the Paris Academy and a member of the Marine Academy at Brest.

Elsewhere in the bureaucracy of the navy, individuals here and there held other technical posts worthy of note. The leading French royal clockmakers, Ferdinand Berthoud (1766-1807) and his nephew Pierre-Louis Berthoud (1767-1813), for example, were both *Horloger de la Marine*. The famous astronomer-physicist and correspondent of the Académie des Sciences, Abbé Ruggiero Giuseppi Boscovich (1711-1787), served as *Astronome opticien de la Marine* and *Directeur d'Optique* from 1773 until his death in 1787 with a salary of 2000 livres; he was succeeded by the abbé Rochon of the Academy of Sciences.[109] Extending his power, Poissonnier had Louis-François Rigaut (also spelled Rigaud) appointed as *Médecin Naturaliste Physicien et Chimiste de la Marine* in 1764 with 2400 livres; Rigaut served until 1777 when his position was judged "useless."[110] Around 1784 the navy nevertheless went on to employ another official chemist, a certain M. Gautier, *Officier Chimiste*.[111] The navy and the Dépôt des Cartes et Plans also employed official artists and painters.[112]

As is apparent, then, a not insubstantial number of scientific, medical, and technical experts found a home and a livelihood within the huge metropolitan bureaucracy of the Marine Royale, making the French Navy a key and multi-faceted institution within the overall framework of the Colonial Machine.

107. Pritchard (1987a), p. 126, and (1987b) makes this point; on naval engineers, see also Neuville (1882), pp. 86-89ff; on the contemporary literature regarding naval construction, see Demarcq (2005).

108. AN-MAR C[7] 132 (Groignard).

109. AN-MAR C[7] 38 (Boscovich).

110. AN-MAR C[7] 276 (Rigaud). Rigaud was able to keep a pension of 1000 livres. Rigaud was also a correspondent of the Academy of Sciences.

111. SHMB, Série A, Sous-série 3A, 3A 91, #200, 4 July 1778.

112. Chassagne, p. 287.

The Académie Royale des Sciences (1666)

[Jeton: Académie Royale des Sciences]

The Académie Royale des Sciences (1666-1793) has long been recognized as the centerpiece of the contemporary French scientific establishment, and in many senses it represents the scientific core of the contemporary Colonial Machine.[113] The Academy has been well studied to date, but no one has fully investigated the extraordinary extent to which the Academy was oriented to the overseas and to scientific matters pertaining to the colonies.[114]

Prior to its reform in 1699, the Académie Royale des Sciences was of small size, closed character, and committed to a collective, Baconian approach to making new knowledge.[115] The seventeenth-century Academy is sometimes thought to be more insular than it in fact was. Two major sources of information belie the common view: the records of the Academy's meetings that show information and objects drifting in from overseas; and the dramatic series of expeditions sponsored by the Academy from its beginnings. In early 1667, for example, just weeks after the Academy's founding the previous December, academicians began discussions of geodesy, longitude, and observations to be made in Madagascar.[116] And already in 1668 the Academy sent one Delavoye sailing in the Atlantic to perform tests of Huygens's pendulum clocks as marine chronometers.[117] The voyage by the academician Jean Picard in 1671-72 to Denmark to

113. The literature surrounding the Académie Royale des Sciences (1666-1793) is highly developed. For an entrée into that literature and the history of the Academy in general, see McClellan (2007), McClellan (2000a); Éric Brian in Brian and Demeulenaere-Douyère, eds. (1996), pp. 397-407, provides a complete bibliography to 1996. See also Gillispie (1980), pp. 81-99; Stroup; Brian and Demeulenaere-Douyère, eds. (2002); Broc (1975), pp. 16-20; Dhombres (2003a). The Ur-source on the Academy remains Hahn (1971).

114. Lacroix (1938) is the notable exception.

115. On the early Academy, see esp. Stroup.

116. See Auzout report ("Toutes les methodes pour mesurer la grandeur de la terre..."), PA-PV 2 (1667-Avril 1668 [Registre de Mathématiques]), pp. 33-35; "Memoire des Instrumens & autres choses necessaires dont il faudra fournir ceux qui iront à Magadascar [sic]" and "Memoire des Observations qu'il faudra faire à Magadascar [sic]," pp. 37-58, and report for meeting ("Ce 11e de Janvier 1667"), p. 155.

117. Olmsted (1942), p. 117; Iliffe, p. 621. On this and related voyages, see further below, Part IIA.

calibrate Tycho Brahe's historic observations at Uranibourg, and Jean Richer's path-breaking trip to Cayenne in 1672-73 to conduct astronomical and geodetic work for the Academy likewise evidence this outward orientation. So too does an early work published by the Academy, the *Recueil d'observations faites en plusieurs voyages par ordre de sa Majesté, pour perfectionner l'astronomie et la géographie* (1693). Notably also, in this early period, through common members and other ties, the Academy affected early and strong connections with the Jardin du Roi and with another new institution of high importance to the Colonial Machine, the *Observatoire Royal.*

The Academy underwent a major reform in 1699. The Letters Patents of 1699, the institution's first formal set of rules and procedures, fundamentally restructured its membership, orientation, and function in the contemporary world of science. With minor modifications the letters patents of 1699 governed the Academy's operations until its closure in 1793. They also formalized the authority of the Academy as a technical consultant to the crown. That is, inventors seeking patents from the crown now had to pass through the Academy and gain its approbation. This power made it not only a center for the natural sciences, but also for contemporary technology and, particularly for our purposes, naval technologies.

Continuing in the tradition established in the seventeenth century, the Academy is well known for sponsoring an impressive series of scientific expeditions throughout the eighteenth century: the valiant journeys to Lapland and Peru to measure the shape of the Earth in the 1730s, the transit of Venus expeditions of 1761 and 1769, and the epic voyages of discovery by Bougainville in the 1760s and La Pérouse in the 1780s, to mention only the most dramatic which the Academy sponsored or with which it was associated as part of the Colonial Machine. As notable, even heroic, as these overseas adventures were, and as indicative as they are of an outward-looking Academy, they have tended to obscure the more mundane ways in which the ordinary experience of navigation and colonization was deeply integrated into the life of the institution from the beginning. The grand, Academy-sponsored expeditions will require further attention, but in the meantime the more ordinary aspects of the Academy's functioning in the Colonial Machine can be highlighted here.

Overseas Correspondents. The Academy's colonial connection was established first and foremost through individuals it elected as correspondents. As the Academy's permanent secretary, Grandjean de Fouchy, put it in 1760, expeditions were one thing, but "reports produced by intelligent observers living on the spot are the more important ones. It is perhaps only by dint of their aid that we will ever arrive at certain knowledge of the world around us."[118]

118. HARS 1760, pp. 16-17. See similar remarks by Fontenelle in 1733; *Histoire et mémoires de l'Académie Royale des Sciences de 1666 à 1699, Tome II (1686-1699)* (Paris: chez Gabriel Martin, Jean-Baptiste Coignard fils, and Hippolyte-Louis Guérin, 1733), pp. 1-2, where he writes: "L'Académie...a adopté des Correspondants qui ont appris par Elle à interroger la Nature à propos, & à regarder les choses avec des yeux de philosophes."

And so it was that the Academy systematically elected a subset of correspondents located in the colonies.

The Academy counted roughly seventy correspondents with colonial or naval connections.[119] Out of a total of 357 official correspondents in the period 1699-1793, the number of these correspondents may seem small, and it puts the colonial dimension of the Academy in perspective.[120] These correspondents represent about 20 percent of the total number of correspondents; crucially for us, most of them held some other official position in the Colonial Machine: royal doctors and royal botanists in particular, along with a handful of jurists, colonial administrators, high-ranking naval officers, and Jesuits in Asia. In Canada, for example, Michel Sarrazin (1659-1734) held the post of médecin du roi at Quebec from 1697 to 1734, and he was elected in the first batch of the Academy's correspondents in 1699. His successor as médecin du roi from 1735 through 1759, Jean-François Gaultier, was the above-mentioned correspondent of Duhamel du Monceau from 1745. In the 1730s, Réaumur and his correspondent on Île de France, J.-F. Charpentier de Cossigny, a military engineer and employee of the Compagnie des Indes, undertook and published a series of concurrent weather observations in the Indian Ocean colony and in France.[121] The Academy similarly elected a series of correspondents in the colony of Cayenne in South America, including Pierre Barrère (1690-1755), royal Médecin-Botaniste in 1725 and Jacques-François Artur (1708-1779), royal doctor in Cayenne from 1735 to 1770 and correspondent from 1753. Another of the initial set of correspondents of 1699 was Marc de Vaux de la Martinière, médecin du roi on Martinique from 1691 until his death in 1716; la Martinière was followed both as royal doctor and Academy correspondent in Martinique by François-Antoine Le Dran (1690-1724). Similarly, over on Guadeloupe, the royal doctor-botanist Jean-Baptiste Lignon (1667-1729) was a correspondent from 1699, and he was succeeded by Jean-André Peyssonnel (1694-1759), royal doctor-botanist on Guadeloupe from 1729 and correspondent from 1723. Jacques-Alexandre Barbotteau was superintending royal botanist for the Windward Islands, member of the *Conseil Supérieur* of Guadeloupe, and Duhamel's correspondent from 1776. The Academy elected the intendant at Guadeloupe and Martinique, François-Joseph Foulquier (1744-1789), correspondent of J.-D. Cassini IV in 1781.[122] We will encounter these and other correspondents of the Academy in subsequent pages, but the connections mentioned here seem typical of the overlapping ties estab-

119. See Biographical Appendix.

120. McClellan (1981), p. 547.

121. On these individuals, see Institut de France, *Index biographique*. For works published by academicians using materials supplied by their correspondents: (Sarrazin by Tournefort) MARS 1704, pp. 48-65; (Gaultier by Duhamel) MARS 1747, pp. 466-88 and MARS 1750, pp. 309-10; (Cossigny by Réaumur) MARS 1733, pp. 417-37; MARS 1734, pp. 553-63; MARS 1735, pp. 545-76; and MARS 1738, pp. 387-403.

122. On these men, see entries in Lacroix (1938), vol. 3.

lished by the Academy to the colonies and to the rest of the Colonial Machine through the position of correspondent. The unaffiliated colonist had little to do with the Academy or the Colonial Machine in general.

To this list of official correspondents we need to add another two dozen or so unofficial ones who maintained occasional contact with the Academy from the colonies or who had not been formally elected before the Academy closed in 1793. These include one Hocquart who sent samples from a rain of ashes in Quebec in 1734; the Jesuit, Crossat, who dispatched astronomical observations from French Guiana in the 1720s; the royal engineer Frézier writing from Saint Domingue in the 1720s; Kerbiquet Lunven, an officer of the East India Company, who provided geographical data to the Academy in the 1740s; the Jesuit apothecary in Martinique, Father Yon, corresponding on botanical matters in the early years of the century; a certain Fougeroux who wrote to the Academy concerning volcanic activity on Guadeloupe in the 1770s; the Saint Domingue colonist Dubuisson, in touch with the Academy about bamboo in the 1770s; François de Neufchâteau, the royal prosecutor in Saint Domingue, and Michel René Hilliard D'Auberteuil, the former royal prosecutor in Martinique and Grenada, both of whom offered their services to the Academy.[123] The obvious point is that the Académie Royale des Sciences became *the* center not only for its official correspondents but for any colonial who had anything of scientific interest to report to the Colonial Machine.

The Academy did not use its correspondents passively to collect information and convey it back to the center, but in a not obvious fashion it trained its correspondents and diffused techniques, systems, and procedures outward that allowed the Academy to define and structure ever more precisely scientific activity on the ground in the colonies. The Academy's correspondent in Canada, Gaultier, for example, complained of the unreliability of his rapporteurs in the field and his inability to get better or more samples to Paris because his correspondents did not or could not follow instructions.[124] And thus, visible across the board in the Academy's dealings with correspondents and in sending out expeditions, detailed instructions flowed concerning how to collect, present, and transport samples needed for quiet study in the metropolis; notices concerned observing procedures and systems of classifications, techniques for

123. For these examples, see variously HARS and PA-PV. Regarding Neufchâteau and D'Auberteuil, see Neufchâteau letter to "Monsieur" dated "À Paris le 1er aoust 1783" and D'Auberteuil letter, "Mémoire A Messieurs de l'Academie Royale des Sciences de Paris" signed "Ce 1er septembre 1783, hilliard D'auberteuil," PA-PS (15 November 1783) and PA-PV (1783), fol. 210 ("Samedi 15 Novembre 1783"). For the Hocquart (also Hocquard) example, see "Lettre du Ministre annonçant l'envoic à l'Académie des Sciences de la letter de M. Hoquant sur la pluie de cendres tombée à Quebec," AN-COL F^3 12, 165 and subsequent correspondence AN-COL B^{60} (1734), fol. 7 (26 January 1734) and B^{61} (1734), fol. 516 (13 April 1734). Also Regourd (2000), pp. 342-48.

124. Gaultier makes these complaints repeatedly in letters to Réaumur in early 1750s; PA-DB (Gaultier).

chemical analysis of plants or minerals, or methods that specified how to use thermometers, pendulums or optical apparatus.[125] These rules became more and more precise and complex, and from this perspective the Academy of Sciences was a pedagogical institution in the Colonial Machine, diffusing expert know-how of scientific observation and the ordering of results.

Academicians with Overseas Connections. In addition to colonial correspondents, the Academy had a complementary set of resident academicians with significant colonial and/or naval connections of their own. Many of these resident academicians, particularly those in botany and natural history, were linked closely with the Jardin du Roi and with the colonial and overseas worlds, notably Buffon, Thouin, the physician Le Monnier, Tessier, d'Angiviller, and other members of the botanical scientific establishment. The honorary academician, the Jesuit Thomas Gouye (1650-1725), deserves to be singled out here. From 1699, when he was the first person appointed by Louis XIV as an honorary member of the Academy, until his death in 1725, Gouye held a leadership position within the Academy. He regularly served as an annual vice president and president, and he provided a broad institutional liaison between the Academy and Catholic missionaries in India, China, and Canada, with whom he corresponded officially on behalf of the Academy.[126] Veterans of the Academy's expeditions, such as Maupertuis, La Condamine, the abbé Lacaille, the abbé Rochon, Father Pingré, and others, likewise added to the overseas orientation of the Academy. Other academicians had substantial experience living in the colonies. Michel Adanson (1727-1806) served a botanical tour of duty with the Compagnie des Indes in Senegal from 1749 to 1753 and was a correspondent before he returned to Paris and rose through the ranks from adjunct in 1759 to *pensionnaire* in botany in 1782. The associate astronomer Guillaume J.-H.-J.-B. Le Gentil de la Galaisière (1725-1792) wandered across the Indian Ocean and Southeast Asia from 1760 to 1771.

A number of other academicians brought with them strong ties to the royal navy, many representing the elite of navy brass. These were of a type labeled by Michel Vergé-Francheschi as "learned officer" (*l'officier-savant*).[127] Jean-Charles de Borda (1733-1799), for example, a mathematician and naval captain, rose to the rank of *pensionnaire* within the Academy; in 1784 Borda succeeded Duhamel du Monceau as inspector of naval constructions and head of the *École des Élèves Ingénieurs Constructeurs de la Marine.*[128] Gabriel de

125. Regourd (2005a); see also Rousseau (1966) as a case in point; see also Collini and Vannoni for other examples, including E.-F. Turgot's *Mémoire instructif.* Melançon (2007), p. 267, gives an example of such instructions given by La Galissonnière for his Canadian correspondents.

126. See entries in Halleux et al. On Gouye's contact with Jesuits in Canada, see correspondence in AN-COL C^{13A} 1 (1678-1706); see also inventory by Menier et al.

127. Vergé-Francheschi (1999); see also his (1991).

128. On Borda, see especially volume edited by E. Neuville; Taillemite (2008), pp. 176-83, and (2002), pp. 58-59; and AN-MAR C^7 37 (Borda). Borda did not, however, replace Duhamel du Monceau as inspector general of the navy.

Bory (1720-1801), a free associate of the Academy, was a navy commodore (*Chef d'escadre*), as was the marquis de Chabert (1724-1805), who also became inspector general of the navy's Dépôt des Cartes et Plans.[129] Bernard Renau d'Éliçagaray (1652-1719), an honorary, served as a naval captain and chief engineer of the navy from 1691; J.-B-H. du Trousset de Valincour (1653-1730) was an honorary and secretary-general of the navy. In this connection one can also mention André-François Boureau-Deslandes (1690-1757) who was born in India and who became an adjunct in geometry in the Academy as well as commissioner-general of the Navy from 1736. He was the author of, among other works, *Essai sur la marine et sur le commerce*, 1743.[130] Most of these officers were also academicians of the Académie Royale de Marine. The connection between the Académie des Sciences and the naval establishment was more or less a formal one, as the superintending minister of the Academy in Paris observed somewhat awkwardly in 1758 apropos of the election of the marquis de Chabert:

> It is customary that the Academy have a naval officer among its Free Associates, and because naval matters pertain to the subjects with which the Academy is most concerned, the interests of the Navy and those of the Academy concur to anticipate a vacancy among the class of Free Associates in order to admit an officer of the Navy, there not being one among them at present.[131]

The elite of colonial administration likewise figured among the members of the Academy, including specifically honorary members and free associates – Roland-Michel Barrin de La Galissonnière (1693-1756), lieutenant general of the navy and at one point governor-general of Canada; A.-L Rouillé, the count de Jouy (1689-1761), minister of the navy and the colonies from 1749 through 1754; and, of course, the count de La Luzerne (1737-1799).

Along these lines, we should also signal first royal geographers (*premiers géographes du roi*) who from 1730 were de jure members of the Academy with the title of *adjoint-géographe*, notably the succession of Philippe Buache (1700-1773), J.-B. d'Anville (1697-1782), and J.-N. Buache de la Neuville

129. Taillemite (2002), p. 59. On Bory and the men mentioned here, see also personnel files, AN-MAR C^7, and other material, PA-DB.

130. On Deslandes, see Taillemite (2002), p. 135; Laurent; PA-DB (Deslandes).

131. "M. de Chabert Lieutenant des Vaisseaux du Roy désire Monsieur d'être admis à l'Accademie en qualité d'Associé libre. Comme il est d'usage que l'Accademie ait parmy ses Associés libres un Offer. de Marine, et qu'en effet c'est à elle que sont relatifs plusieurs objets dont elle s'occupe le plus, l'Interest de la Marine et celui de l'Accademie concourent à anticiper le moment d'une Place vacante dans la Classe des Associés libres, pour y admettre un Off.er de Marine n'y en ayant point à présent." Saint Florentin to de Fouchy, "A Vlles. Le 25 Aoust 1758," PA-PS (30 August 1758).

(1741-1825), who also succeeded one another as premier géographe du roi.[132] Although they and their geographer and mapmaking colleagues in the Academy, particularly Guillaume and Joseph-Nicolas Delisle, were stay-at-home *géographes de cabinet*, their collective geographical interests spread wide and reinforced other orientations of the Academy outward to the colonies and the overseas world. Finally in this connection, a handful of key academicians exercised major scientific and administrative positions in other institutions and thus solidified the Academy's ties to the rest of the Colonial Machine. We have already mentioned Duhamel du Monceau, inspector-general of the navy; and Pierre-Isaac Poissonnier, director-general and inspector of naval and colonial medicine; to this list we can add the count de Buffon, *pensionnaire* and treasurer at the Academy of Sciences and Intendant at the Jardin du Roi for over fifty years; and the Cassinis, who ruled for twice that long at the Observatoire Royal.

A Collection Center for Overseas Knowledge and Naval Innovation. In a word, then, important facets of the Academy's membership and structure oriented it toward matters colonial and to the rest of the scientific-technical bureaucracy of French colonial administration. The Academy's naval and colonial dimensions also manifest themselves plainly in a drumbeat of letters and reports that came before its meetings. A survey of the Academy's minutes (its *Procès verbaux*) reveals this steady stream of items flowing from the colonies and pertaining to colonial expansion. In 1770, for example, – to pick a representative year, – the Academy received samples and reports concerning: salted meat for navy rations, rubber from South America, marine chronometers and related instruments, a coconut from the Indian Ocean, the natural history of Louisiana, dwarfs in Madagascar, meteorological data from China, North American plants and animals, a mule from Saint Domingue, another coconut, minerals from Madagascar, unusual ostrich eggs, a report on navigation in the Indian Ocean, sea biscuits, a comet observed in Saint Domingue, a machine to keep the sweat of slaves from ruining sugar production, a cometary observation from Île de France, the discovery of a new island off Madagascar, the cement of pre-Columbian Indians, and astronomical observations from Mexico.[133] A partial list for 1788 includes meteorological reports from North America, Saint Domingue, Île de France, and China; papers on shipping colonial spices, on the mineralogy of Guiana, and on cultivating wheat in Madagascar; the observation of a comet taken at Île Bourbon; commissioners' reports on sugar cane and indigo; several reports on navigating the Indian Ocean; samples of indigo from a colonial plantation on Île de France, cacti from Peru, and the trunk of a palm tree offered by the minister – all this, not

132. On the positions of premier géographe du roi and adjoint-géographe (from 1785 *associé-géographe*) in the Académie des Sciences, see Pedley (2005), p. 199; Broc (1975), pp. 481-82.

133. PA-PV 89 (1770), passim.

to mention the visit on Wednesday, September 5, 1788, of ambassadors from India, for whom Lavoisier performed chemical experiments.[134]

Similarly, when one examines the list of referees and referees' reports commissioned by the Academy throughout the whole of the eighteenth century, this drumbeat of maritime and colonial topics sounds strongly and consistently over the years. Regarding only matters naval, for example, one reads of machines, projects, proposals, and reports concerning finding longitude at sea, navigation methods, ventilation and fire prevention aboard ships, capstans, anchors, life preservers, maps, compasses, maritime clocks (including pendulum clocks and a water clock, not to mention exquisite spring-loaded machines), locks and shipbuilding, oars and means of rowing and towing ships, techniques and machines for masting ships, rigging for spars, measuring the speed of ships, gauging the capacity of ships, tar and cordage, shipboard distillation equipment, means of floating sunken or grounded ships, barrels for holding fresh water, naval varnishes, tillers, dredging equipment, cranes, pumps (including fire pumps), cladding ships, machines to off-load ships and boats, naval artillery, botanical shipments, staunching leaks, and so on.[135] The list makes the point.

This colonial-naval dimension of the Academy likewise shows up in its publications. The *Histoire* section of its annual volume, the *Histoire et Mémoires*, regularly announced notable episodes recorded in the minutes, reporting variously, for example, on coffee on Île Bourbon; on an eclipse observed in Louisiana; on bamboo, earthquakes, and hurricanes in Saint Domingue; and on an electric eel in Surinam.[136] A series of formal papers published by academicians in the *Mémoires* likewise concerned colonially inspired topics, such as, to pick just two, Philippe Buache's 1764 "Observations géographiques sur les Isles de France et de Bourbon, comparées l'une avec l'autre" and Abbé Tessier's 1789 "Mémoire sur l'importation et les progrès des arbres à épicerie, dans les colonies françoises."[137] And similarly, the Academy's *Savants Étrangers* series likewise incorporated a string of papers reflecting this dimension of the Academy's identity; in this connection we may mention d'Après de Mannevillette's 1768 "Mémoire sur la navigation de France aux Indes," Adanson's 1755 "Latitude de Podor," and Pierre Sonnerat's 1776 "Description du Cocos de l'Île de Praslin, vulgairement appelé Cocos de mer."[138]

134. PA-PV 107 (1788), passim; the pochettes de séance preserved for each session likewise contain much material with a colonial-naval slant. See also the personal notes taken by Fougeroux de Bonderoy of Academy meetings: APS Manuscripts, Fougeroux de Bondaroy Papers B, F8245, "Séances de l'Académie Royale des Sciences, 1786-1789," e.g., meeting for Saturday, 29 April 1786 where the discussion concerned a (failed) experiment in transporting fruits from Cap François.

135. See PA-PV-110.

136. See HARS 1716, p. 34; 1730, pp. 104-105; 1752, pp. 16-17; 1760, pp. 21-23; 1769, pp. 77-78.

137. MARS 1764, pp. 1-6; MARS 1789, pp. 585-96.

138. SE 2 (1755), pp. 605-6; SE 5 (1768), pp. 190-232; SE 7 (1776), pp. 263-66.

Prizes. The Academy's series of prize contests deserves attention here as another dimension of its institutional role in driving research connected with colonization and the world overseas. Its series of prizes established in 1714 for determining longitude at sea – the famous Meslay prize – is preeminent among these. In 1714 Rouillé de Meslay bequeathed 125,000 livres to Academy for two prizes.[139] The first, on the motion of the planets, on light, and on movement has attracted more attention from mainstream historians of science. The second, to determine "more exactly the altitudes and degrees of longitude at sea and discoveries useful to navigation and overseas voyages" is less well studied. The navigation prize contest was nominally biannual and came with a substantial 2000 livres award attached to it. Contests began in 1719, with the first award made in 1720. Subject questions for the prize asked about keeping pendulum movements and water and sand clocks steady at sea (1720 and 1725), keeping time at sea (1747, 1769, and 1773), and compass variations (1731, 1743), the latter subject at one point was thought to bear on the problem of longitude. Questions about minimizing the rolling and pitching of ships (1755, repeated in 1757 and 1759) and about ballast and loading vessels (1761 and 1765) had implications for the longitude problem, too. Indicative of the Academy as part of the "research engine" of the Colonial Machine, other, more practically oriented questions asked about the best means of masting ships (1727), determining latitude (1729), the best kinds of anchors (1733), the best ways to construct a windlass (1740), determining ocean currents (1751), supplementing the power of wind by oars or other means (1753), the best way to make magnetic needles (1777), as well as other, more straightforwardly technological questions. Some surprisingly highbrow men of science won these decidedly technological and practically oriented contests, including Johann and Daniel Bernoulli and Leonard and J. A. Euler, all of whom were regular contestants and multiple prize winners. But so, too, were a certain *sous-constructeur* Cauchot, winner of the 1753 prize on improvements to ship construction, and two *Ingénieurs-Constructeurs de la Marine Royale* winners in 1761 and 1765.[140] With the marine chronometer established as a reliable means of determining longitude in the 1770s, the Meslay prize of the Paris Academy petered out with a question for 1787 about maritime insurance.

But the Meslay prize was not the only one administered by the Academy that dealt with longitude and navigation. In 1716 the regent, Philippe d'Orléans, offered 100,000 livres for a solution of the longitude problem; the prize was never awarded. In 1785 Baron de Bernstorff established a prize on the theory of the movement of ships; the Academy proposed the prize for 1787, but it received no submissions and the funding was withdrawn. The abbé Raynal, whose work *Histoire philosophique et politique...des deux Indes* (original edition

139. See Maindron, pp. 14-22, 44, 46, 48-49.

140. Llinares (1996), p. 46; DDP I:21; Filliozat (1993), p. 153. On the general point, see also DDP I:53.

1770) prompted a famous debate on contemporary colonial policy, funded a prize at the Academy in 1788 "for practical navigation" ("relatif à la Navigation pratique"); the Academy proposed a question for 1792 about measuring parallax and made an award in 1791. The Raynal prize for 1793 on the best method to determine latitude was never awarded.[141]

A Center of Expertise and Action. Finally in this connection, the Academy is also well known for serving as a technical consultant for the crown. It judged submissions for patent monopolies, for example. Not a few of these concerned matters nautical, such as the 1731 technique of M. Delandes, the naval controller at Brest, for drying wood for naval construction, or the diving bell invented in 1774 by one Sieur Freminet.[142] Indicative of its larger role in the Colonial Machine, in 1772 the Paris Academy forwarded a patent application on rowing ships to the Académie de Marine at Brest for further study and development.[143] Also in 1772, on behalf of the Academy, Le Monnier, Cassini, and Bory responded to the request of the minister regarding plans for a new set of dikes in Brittany.[144] In addition, on more than one occasion delegations of academicians traveled to port cities at the request of the navy ministry to make inspections or report on improvement projects. In 1782, for example, Condorcet, Bory, Bézout, and Bossut journeyed to Le Havre to evaluate proposals to expand the harbor there. In 1783 Borda, whom Étienne Taillemite calls "a kind of technical consultant to the minister for questions of naval construction" ("une sorte de conseiller technique du ministre pour les questions de constructions"), the abbé Rochon, and two other naval personnel ventured to Saint-Malo to evaluate the possibility of a new royal naval installation there.[145] In 1785-86 Borda and Maraldi were called in by the minister to consult on jetties, docks, and locks to protect the port of Dunkerque.[146] In this connection, one might signal the ministerial project to install lightning rods on the navy's powder magazines at Brest in 1784-1786. Le Roy of the Academy and an assistant ended up installing 121 rods on various government buildings and another 20 rods on

141. See previous notes and APS Manuscripts, Fougeroux de Bondaroy Papers, B F8245, "Séances de l'Académie Royale des Sciences, 1786-1789" session du "Vendredi 9 Mai [1788] extraordinairement...M. labbé Raynal propose a l'acad.e un prix dont la fond sera de 24,000# et qu'il desireroit avoir pour objet principalement la navigation pratique..." See also Fougeroux's notes for the subsequent meeting of "le Samedi 14 fevrier 1789." The original commissioners to propose and judge this prize were Borda, de Bory, Bougainville, Le Monnier, and the abbé Rochon. See also list of largely 'naval' prize subjects complied by Delisle, AN-MAR G 141.

142. See AN-MAR G61, fols. 88-89 for the Delandes case; AN-MAR G235, #52, 65, 66, 67, 73, and 82 and below for Freminet.

143. See Sartine letter dated "Versailles le 10 7.bre 1779" in SHMV, Ms. 93.

144. AN-MAR D^2 58, fols. 241-2.

145. Taillemite (1999); see also Taillemite (2008), p. 178, where Borda is labeled "engineer."

146. On these examples, see AN-MAR D^2 9, #41, 45, 47, 72, 73; AN-MAR D^2 22, fols. 276-325; AN-MAR D^2 5, fols. 307-321.

vessels of the Marine Royale harbored at Brest, at a cost of almost 40,000 livres.[147]

Matters naval and colonial, therefore, constitute a defining element of the Academy of Sciences. They percolate through the history of the Academy and, indeed, through the other institutions of the Colonial Machine. This background noise, so to speak, concerning the colonies and the world overseas has generally been overlooked in the literature surrounding the Academy, as it has in studies of other contemporary French scientific and technical institutions. And so, again, a principal objective of the present work is to draw attention to this outward, extra territorial, and distinctly modern feature of the Academy and French scientific establishment in the Old Regime.

Beyond all that, the Academy of Sciences launched its own institutional initiatives for systematic contact with the colonies. The first of these concerns the mission of the Montpellier trained physician, Michel Isambert, sent by the Academy to Martinique in 1716 specifically to be an observer on the spot "to cultivate useful plants and to undertake a correspondence with the Company on his activities related to botany and natural history."[148] A striking feature of this episode is that Isambert was elected as a correspondent of the Academy as a whole, and not attached to a single academician, as was universally the norm otherwise. Isambert proceeded to Martinique with a cargo of coffee plants; bees; silkworms; and mulberry, olive, walnut, and pistachio trees. Unfortunately, he died shortly after his arrival. Isambert's demise brought to an abrupt end this effort at implanting a permanent Academy representative in the colonies. The death of the coffee plants also signaled the failure of the first effort to bring coffee production to the French Caribbean.

All was not lost, however. Jean-Baptiste Lignon (1667-1729), a correspondent of Tournefort and the Academy from 1699 and a royal botanist previously installed on Guadeloupe, retrieved what he could from Isambert's plants and incorporated them into the royal garden on Guadeloupe. The abbé Bignon, Academy honorary and its superintending administrator encouraged Lignon, writing: "Coffee and the other plants are very important, but honeybees and silkworms are even more so, and you cannot apply yourself too much to their

147. On this complicated episode, see letter from "Le Maréchal de Castries" dated "Versailles Le 7 7.bre 1784" in SHMV. Ms. 93 and materials SHMB, Série A: Sous-série 1A, Corespondance (1783-1785), pp. 286-87 ("Le 20 8.bre 1784") and p. 312 ("Le 17 9bre. 1784"). Also SHMB, Série A, Sous-série 3A, Registres 3A4 à 3A10, esp. 3A 94, 1786, #58: "Rapport" dated "A Brest le 13 Mai 1786." And, SHMB, Procès-verbaux de délibérations du Conseil de Marine de 1772 à l'an II (1793-1794) 20 mai 1786.

148. "L'Académie a choisi, de l'agrément de Mgr le duc d'Orléans, M. Isambert, docteur en médecine de la Faculté de Montpellier et apotiquaire de S.A.R. pour aller aux isles de l'Amérique, dont il a bien voulu entreprendre le voyage. Il y cultivera des plantes utiles qui ne peuvent venir icy et entretiendra correspondance avec la Compagnie, sur ce qu'il fera en botanique se rapportant à l'histoire naturelle"; PA-PV 35 (1716), fols. 97-97v ("Le Samedi 14 Mars 1716"). On Isambert, see also Lacroix (1938), vol. 3, pp. 23-27.

care….Utility consists in multiplication."[149] (Here Bignon sounds themes we will hear again.) For his efforts, the regent awarded Lignon 1000 livres and a gold chain and medal.[150]

It was perhaps the demise of the Isambert mission that prompted Bernard-Laurent d'Hauterive, the public prosecutor in Fort-Royal, Martinique, to volunteer his services as the Academy's agent in the Antilles. In a remarkable document dated November 30, 1716, Hauterive laid out an impressive and detailed research agenda "concerning geometry, astronomy, mechanics, anatomy, chemistry and botany" to be pursued in the tropics. In essence, he proposed to develop locally an extension of the Academy itself in Martinique: "If it pleases the Academy, I will involve some capable people we have here to assist me in this appealing plan and by their expertise to help me in the discoveries and the shipments that would be appropriate to make to the Academy, noting our correspondents for each particular science."[151]

Hauterive apparently found a supporter in Bignon and the local intendant in Martinique.[152] Through the offices of Réaumur and Dortous de Mairan, a *pensionnaire* in geometry, the Academy sent Hauterive a questionnaire, and Hauterive's responses duly appeared in the Academy's *Histoire et Mémoires* regarding the tides in Martinique, racial mixtures, a green stone from the Orinoco River used to prevent epilepsy, a plant to ward off snakes, a metal amalgam specific for migraines, and the vanilla plant that Hauterive was successfully growing on his own plantation.[153] Hauterive sent drawings of American plants and animals "and a lot of the most curious materials he was

149. Bignon letter to Lignon, dated 11 January 1717, quoted in Lacroix, vol. 3, p. 27.

150. See Regourd (2000), pp. 322-27; Lacroix (1938), vol. 3, p. 23; and ALS "Lignon Le Jeune" to Regent dated "de la Guadeloupe le 27e avril 1717," PA-DB (Lignon); Lignon's elder brother, himself named botanist to Guadeloupe at one point, disputed this award.

151. "...Je veux seulement faire observer a Messieurs de L'Academie des Sçiençes que le pais leur pouvoit Estre Utile...; et sils souhaitent en faire quelque usage,...Je me feray un honneur singulier en donner mon tems et mes peines pour tascher de Satisfaire a ce qu'ils souhaiteront la dessus...J'Engageray si cela fait plaisir a l'Academie quelques habiles gens que nous avons Icy a me seconder dans un si beau dessein, et a m'aider de leurs lumieres pour les decouverte et les Envoys qu'il Conviendroit luy faire, en vous Indiquant les Correspondants a qui nous nous addressions pour Chaque sçience en Particulier...," Ms. "Memoire de la Martinique pour Messieurs de l'Academie royalle des Sçiençes" dated and signed "À la Martinique Le 30e Novembre 1716, De hauterive," PA-PS (1717). See also Hauterive letter of 13 April 1717 in same folder. See also Regourd (2000), pp. 348-51 and 716-20 for further documentation. Lacroix (1938), vol. 3, p. 35.

152. See letter from the Intendant of Martinique, Blondel de Jouvancourt, dated 22 August 1723 that praises Hauterive's "talents de botaniste" and reveals the protection of Bignon; AN-COL C^{8A} 32, fol. 194.

153. HARS 1724 (1726), pp. 17-19. See also Gasc and Laissus, p. 33; Regourd (2000), pp. 348-51. Regarding tides, Hauterive was authoritative through century; see Fleurieu (1773), p. 396: "La briéveté de mon séjour à la Martinique, ne m'a permis de faire aucune observation sur les marées de l'île; mais je ne puis me dispenser d'inviter les Marins à vérifier celles qu'y fit M. de Hauterive, anciennement Procureur général du Conseil Supérieur de la *Martinique*, & qu'il avait adressées, en 1724, à l'Académie Royale des Sciences, dont il était Correspondant." See also Hauterive materials, PA-DB (Réaumur).

able to collect." He was also responsible for the early introduction of cinnamon into Martinique, which he likewise forwarded to the Academy, via the Intendant and the Ministry of the Navy and the Colonies.[154] Hauterive was elected a correspondent of de Mairan and the Academy in 1724.[155] His connection to the Academy seems to stop at that point, but he shows up again later that same decade as the agent in Martinique responsible for sending botanical specimens to the royal garden in Nantes.

Others of the Academy's correspondents and various of its expeditions kept the institution in touch with the colonial world over the course of the eighteenth century. This orientation may be said to have culminated with the formal "association" forged in 1789 between the Academy of Sciences in Paris and the Cercle des Philadelphes in colonial Saint Domingue. Essentially a French provincial learned society set in the tropics, the Cercle des Philadelphes originated in Saint Domingue in 1784, and it received letters patent in 1789 transforming it into the Société Royale des Sciences et des Arts du Cap François. The Cercle des Philadelphes is independently a colonial institution of considerable note in this narrative, and has been mentioned already. Here we can remark simply that the link forged between the Paris Academy of Sciences and this outpost on the colonial frontier establishes the high-water mark of the Academy's colonial connections as well as of the Colonial Machine as a whole. Beyond that, enough has been said to make plain the Academy's orientation toward the sea and the colonies and its position as a central institution of the Colonial Machine.

The Jardin du Roi and Metropolitan Botanical Gardens

> *...il est de la grandeur du Roy de concourir à la perfection de la botanique et de réunir dans ses jardins les plantes les plus rares; il est de Sa bonté de procurer à nos colonies les arbres et plantes qui peuvent y étre utiles soit pour le commerce soit pour la nourriture des habitants et surtout des esclaves.*
>
> – Anonymous ministerial secretary, 1777[156]

154. See letter to this effect from the indendant, Charles Bénard, to the ministry dated 21 November 1720, AN-COL, C^{8A} 27, fols. 384-85, and Lacroix (1938), vol. 3, p. 35. See also report by Father Laval from Martinique in 1720: "Mr. d'Haute Rive Procureur général du conseil de Martinique me donna ce jour là une petite branche de canelle que j'ay encor, laquelle est d'un arbre planté dans son jardin. elle a le même goust que celle des Indes orientales, aussi l'arbre en est il venu. on voit par là que les cannelliers viendroient fort bien en ce pais là. il me dit qu'il preparoit diverses Plantes et autres curiositez du Pais pour les envoyer à l'académie, avec laquelle il etoit en commerce; il m'en auroit chargé si nous eussions resté plus longtems au mouillage du fort Royal," OdP, B 5, 2 [8*] "Observations en Amérique...," p. 364.

155. PA-PV 43 (1724), fol. 255 ("Le Samedi 2 Décembre"). It is not clear why Hauterive was elected a *correspondant* after having been in contact with the Academy for so many years.

156. Anonymous [1777] "Mémoire pour le jardin et la serre chaude du Roy etablis à lorient," AN-MAR D^2 29, fol. 79.

An extensive system of state botanical gardens and horticultural stations dotted Old Regime France and its colonies. No other nation at the time came close to rivaling the state-funded botanical infrastructure or worldwide botanical outreach of contemporary France. The network of official French gardens and the professional staff that ran them collectively constituted a formidable part of organized science and the Colonial Machine of the Old Regime. In a subsequent section we examine gardens in the colonies and the role of royal botanists (*botanistes du roi*) overseas. Here, we focus on the substantial and fundamental botanical infrastructures that arose in France itself, beginning with the Jardin du Roi.

The Jardin du Roi. The magnificent Jardin du Roi, or Jardin des Plantes, on the left bank of the Seine in Paris stood as the centerpiece of the system of contemporary French botanical gardens and nurseries.[157] The Jardin des Plantes, or Jardin du Roi, emerged in the 1620s and 1630s as a chartered royal garden for medicinal plants. A royal edict of 1626 under Louis XIII launched the institution, and another of 1635 created it definitively. The Jardin du Roi was thus the first expert institution of the Colonial Machine, but in its early years it remained very much an apothecary's garden associated with the court and the Paris medical faculty. Its organization was not settled and it was not a player in the overseas game until the reforms launched by Louis XIV and Colbert after 1660 that brought the Colonial Machine into being. Nevertheless, it is telling for us that around 1640 the Dominican missionary Jean-Baptiste Du Tertre made a point to stop at the Jardin du Roi before heading to the islands.[158] It is no less telling that from an early date the Jardin du Roi likewise came to possess beds reserved for "plants from the Indies."[159]

The Jardin du Roi developed into a very different institution by the end of the seventeenth century and the beginning of the eighteenth. Colbert took over the administration of the Jardin in 1671, placing it under the auspices of the Bâtiments du Roi. The professorship in botany was revitalized that year and the important position of head gardener instituted the year following; a professorship in chemistry followed in 1695 and another in anatomy in 1727. The appointment in 1693 of Guy-Crescent Fagon (1638-1718) as *Intendant Gen-*

157. On the Jardin du Roi, see Letouzey; Spary (2000); Daugeron (2007) and (2009); see also Laissus (1964), (2003), (2005); and Jardine et al. Mathieu sets the botanical scene in sixteenth- and seventeenth-century France and, pp. 135-54, details the creation and early years of the Jardin du Roi. Laissus (1981) and Barthélemy, pp. 223-52, examine naturalist voyagers associated with the Jardin; and in emphasizing the colonial and acclimatizing aspects of the Jardin, the study by Kury especially merits consulting. An up-to-date and straightforward institutional history of the Jardin du Roi is still a desideratum.

158. "M. Robin, l'un des plus habiles hommes de son siècle pour la botanique, qui gouvernait le jardin du roi en 1640, m'en fit voir un petit, haut de deux pieds, qui lui avait été apporté des îles, auquel il avait donné le nom de laurier aromatique," Du Tertre, vol. 2, p. 173. Vespasien Robin was sub-demonstrator of plants at the Jardin from 1635 to 1662.

159. Dorst, p. 595; Mathieu, p. 227.

eral of the Jardin du Roi marked the transformation of the institution into a full-blown scientific center devoted to the study of botany and the cultivation of plant specimens. Fagon served as first royal doctor to Louis XIV and honorary at the Académie des Sciences from 1699, and he was an early, activist administrator at the Jardin du Roi. In particular, for our purposes, Fagon encouraged the cultivation of plants from overseas and lent the sponsorship of the Jardin to the scientific voyages of Fathers Charles Plumier and Louis Feuillée to the Antilles and South America and that of his protégé, Tournefort, to the Levant.[160]

By the end of the eighteenth century the Jardin du Roi flourished as a world center, if not *the* world center for botanical research. The Jardin du Roi was a well-endowed state institution with a large, paid staff of professors, demonstrators, curators, and gardeners. Georges-Louis-Marie Leclerc, the count de Buffon and author of the renowned *Histoire naturelle*, ruled as Intendant of the institution from 1739 until his death in 1788.[161] *Pensionnaire* and treasurer at the Académie des Sciences and one of the immortals of the Académie Française, Buffon was another of the great scientific administrators whose hands pulled the levers of the Colonial Machine; yet judging from residues in the archives, he seems unexpectedly remote from it. More important for our story was Louis-Guillaume Le Monnier (1717-1799), professor of botany at the Jardin from 1759 to 1786. (He was known as Le Monnier le Médecin, to distinguish him from his brother, Pierre-Charles Le Monnier l'Astronome.) In addition, Le Monnier was first ordinary physician to Louis XV and then Louis XVI, adjunct member of the Académie des Sciences from 1743 and *pensionnaire* from 1758, and also in charge of the royal gardens at the Trianon.[162] André Thouin (1747-1824) was another key personage of the Colonial Machine at the Jardin.[163] He was born at the Jardin and succeeded his father as Chief Gardener there in 1764 at the age of seventeen. Even as the Jardin du Roi gave way to the Muséum d'Histoire Naturelle in 1793 Thouin continued in ever-growing importance in the position of chief gardener until his death in 1824. Other scientific positions at the Jardin du Roi included two chairs in botany, two in chemistry, and other posts in natural history, anatomy, surgery, and pharmacy. The holders of these chairs constituted a virtual *Who's Who* of eighteenth-century botany and natural history: Joseph Pitton de Tournefort (1656-1708); Sébastien Vaillant (1669-1722); and the Jussieu dynasty of Antoine (1686-1758), Bernard (1699-1777), and Antoine-Laurent (1748-1836), to

160. Laissus (1964), p. 291; Roger Williams, p. 78. On these voyages, see further, below, Part IIC. The head of the Jardin, Fagon, expressed strong interest in coffee and quinine; Dorst, p. 595; Allorge, pp. 127-30.

161. On Buffon, see Gillispie (1980), pp. 143-51; Roger Williams, pp. 45-55; Roger; Allorge, pp. 236-49. See also www.buffon.cnrs.fr.

162. Plantefol (1973); Roger Williams (2003), pp. 8-9.

163. On Thouin, see, among others, Letouzey; Spary (2000); Allorge, pp. 361-80.

name only the most major figures. And, needless to say, each of these men had ties to other institutions of the Colonial Machine, notably the Academy of Sciences.[164]

The Jardin du Roi was also a teaching institution with free public courses offered in botany, anatomy, and chemistry. For these purposes it possessed an amphitheater with 800 to 900 seats. Early on, the senior botanists and gardeners at the Jardin trained *Garçons-Jardiniers* who apprenticed there before moving on to jobs in the many private and royal gardens elsewhere in France.[165] These recruits were so sought after that a formal *Ecole Pratique de Jardinage* within the Jardin du Roi was proposed in 1785 to fill this need.[166] The Jardin was an expensive operation; in 1775, for example, costs for upkeep alone, exclusive of salaries, amounted to 27,890 livres, a sum that included expenses for plumbing, masonry, cabinetry, weekly time sheets for laborers, livery, and other materials and supplies.[167]

Almost by definition the Jardin du Roi was a center of botanical collections with a worldwide reach. Six thousand different species of plant were cultivated on the grounds of the Jardin du Roi, many in hothouses especially intended for foreign and tropical plants.[168] In the period 1772 through 1792 Thouin recorded over 41,000 botanical specimens sent *to* the Jardin from all over the world, and 164,000 *from* the Jardin to correspondents.[169] In his botany course in 1782, A.-L de Jussieu, for example, presented plants from Canada, Labrador, Pennsylvania, Maryland, Virginia, the Carolinas, New Jersey, Florida, Mississippi, Bermuda, Louisiana, the Caribbean, Saint Domingue, Guadeloupe, Guiana, Mexico, Peru, Chile, Buenos Aires, Patagonia, the Falkland Islands, Africa, Arabia, Abyssinia, Crete, Ethiopia, Cape of Good Hope, Senegal, India, Bengal, Madagascar, the Malabar Coast, Pondicherry, Ceylon, Île de France, China, Siberia, and the South Seas.[170] The most recent historian of the Jardin, E. C. Spary, has mapped Thouin's correspondents; and, taking

164. See list of positions in Laissus (1964) and Spary (2000), pp. 259-61; on the Jussieux, see Lacroix (1938), vol. 4, pp. 99-181; on Vaillant, see Roger Williams, chapt. 1; Chassagne, pp. 223-26.

165. Note the reference ("la modification de l'Écolle de botanique du Jardin du roy qui s'est commencée à l'automne 1773..."), BCMNHN, Ms. 2134.

166. ALS "Thouïn" to Angiviller dated "A Paris ce 5 Aoust 1785," AN O[1] 2113A: "Pépinieres d'Amérique. Correspondance Générale pour la Mission du S. Michaux, Botaniste du Roi, Année 1785." Nothing came of this proposal.

167. AN O[1] 2124: Jardin Royal des Plantes, Folder marked: "Jardin du Roi Année 1775."

168. Duhamel du Monceau put the number of plants cultivated at the Jardin du Roi at 8,000-10,000; see undated Ms. "De Lutilité quon peut Retirer d'une pepinière Royalle," APS-MS, 580/D881 #37.

169. "Etat Général des Envois de Plantes, Arbres, Graines &c. faits au Jardin National par ses Correspondants avec ceux faits par le dit jardin à ses Correspondants," BCMNHN, Ms. 1327.

170. APS-MS, A.-L. de Jussieu, Ms., "Catalogue des Plantes démontrées en 1782 au Jardin du Roy." Here and elsewhere the geographical origin of a plant is taken from its botanical name (e.g., *Tsuga canadensis*), and caution has to be exercised as to an actual locale.

Thouin as a surrogate for the Jardin as a whole, her results show the national, European, and global reach of the institution.[171] Thouin's own records indicate that in 1790, for example, he received sixty shipments of plants for the Jardin, forty-two of which (70 percent) came from outside of Europe.[172] Of the plants received in these forty-two shipments, approximately 31 percent came from the Antilles, Louisiana, and Cayenne, 26 percent from North America, 20 percent from China and the East Indies, 8 percent from Africa, 8 percent from Spanish America, and 7 percent from the Pacific. In 1790 Thouin planted specimens sent to him by correspondents in Île de France, Martinique, Saint Domingue, Cayenne, Madagascar, the Cape of Good Hope, Île Bourbon, Florida, Louisiana, China, India, Botany Bay, Manila, and the East Indies.[173]

Year after year botanical samples arrived at the Jardin, and year after year Thouin shipped out seeds to his correspondents worldwide. Many of these correspondents were private individuals, but most were officially affiliated with one institution or another. Obviously, material infrastructures and the rest of the Colonial Machine had to be mobilized in order to effect the Jardin's international outreach and connections. In 1749, for example, the Ministry of the Navy duly informed the intendant in Canada that it had transmitted to the Jardin du Roi seeds collected in Canada for the Jardin by the royal doctor, Gaultier.[174] In 1759, to pick another example, the royal doctor in Saint Domingue, de Pas, sent to Louis-Jean-Marie Daubenton (1716-1800), the keeper of the natural history cabinet at the Jardin du Roi, two barrels of pickled animals and another of shells that were sent aboard a navy frigate and forwarded by naval authorities in Brest.[175] A 1773 shipment from Île de France came in thirty-four cases and contained a variety of botanical and natural history specimens.[176] In 1783 a box from Saint Domingue arrived with thirty different types of seed, forwarded through Le Havre.[177]

Beginning in 1773, the international outreach of the Jardin du Roi was reinforced institutionally by Buffon through the creation of the position of correspondent of the King's Cabinet (*correspondant du Cabinet du Roi*). The

171. Spary (2000), figures 3-5. See also BCMNHN, Ms. 314 ("Etat de la Correspondance de M. A. Thouin") and Ms. 1327 ("Journal des Envois de Plantes, Arbres et Graines faits au Jardin du Roy par ses Correspondants, avec ceux faits par le Jardinier dudit Jardin du Roy aux Correspondants d'icelui").

172. BCMNHN, Ms. 1327, entry for 1790.

173. See BCMNHN, Ms. 691, "Catalogues des graines semées au Muséum pendant les années 1790-1801," fols. 2-18.

174. AN-COL B[89] (1749), fol. 25 (18 April 1749).

175. SHMB, Série E-1E: Intendant de la Marine et successeurs, 1672-1827, E-1E.157, fol. 139, 28 July 1759.

176. AN-MAR G 101, Dossier 4, Histoire naturelle, fols. 177-178, Ms. "Etat des Effets d'histoire Naturelle Embarqués sur la vaisseau du Roy la Victoire Commandé par M. de Joannie."

177. ALS "Mistral" dated "Au havre le 28 Aoust 1783" to "Monsieur," AN O[1] 2112: Pépinières, 1783-1787.

Cabinet du Roi, or *Cabinet d'Histoire Naturelle* itself emerged in 1729 as the repository of drugs and natural history collections, and the Cabinet du Roi early on came to house artifacts from the world overseas.[178] After several extensions and reorganizations and with an enlarged staff, the Cabinet du Roi, under Daubenton's direction, opened to the public weekly on certain days and had its collections enriched through purchases and donations. Some of the best pieces came from Bonnier de Mosson, a famous collector, but also from Adanson's and Réaumur's collections, among others. Buffon's initiative regarding the *correspondants* of the Cabinet du Roi of 1773 can be traced back to an earlier one of 1757 to tap the Colonial Machine to supply the Cabinet du Roi. A request in June of that year filtered down from the Bâtiments du Roi through the Ministry of the Navy and Colonies to naval officials and officers in Brest, ordering naval captains "whose destination is for other parts of the world beyond Europe [to collect] all the natural curiosities that they can gather easily and 'without cost' for the Cabinet du Roi."[179] Nothing seems to have come of this plan, doubtless because of the outbreak of the Seven Years War, but Buffon went on to formalize this activity by establishing his own circle of correspondents. Correspondents of the Cabinet du Roi were given certificates and on occasion at least supplied with reference books.[180] The correspondents of the Cabinet du Roi have not yet been systematically reconstituted or studied in detail; we do not know how many there were or where they were located. A small file in the French national archives holds fifteen commissions of these *correspondants*, including those of Sonnerat on Île de France (1773), Sonnini de Manoncourt in Guiana (1775), André Michaux in Persia and then North America (1779), Jean-André Mongez with La Pérouse (1785), and one Roussillon in Senegal (1788).[181] The network so created bolstered and further institutionalized the international profile of the Jardin du Roi.[182] Among other examples, one can note the shipment of plants and seeds for the Cabinet du Roi

178. On the Cabinet du Jardin du Roi, see among others, Laissus (1964), pp. 294-98, and Lamy, esp. pp. 39-58.

179. Our emphasis. "M. Le M.is de Marigny Directeur général des batimens du Roy qui m'a prié de luy procurer par la voye des Commandants des vaisseaux du Roy dont la destination est pour les autres parties du monde que l'Europe, toutes les curiosités naturelles qu'ils pourront ramasser sans peine et sans depense pour le Cabinet de Sa Majesté: je vous prie de faire remettre à l'avenir une copie de cette note à tous les Commandants des Vaisseaux dont la destination pourra les mettre à portée de trouver quelque chose d'utile en ce genre," SHMB, Série A: Sous-série 1A, 1757, fol. 274, "Versailles le [?] Juin 1757" and copy fol. 275.

180. Note the 2623 livres paid to the printer Panckouche "pour les livres d'histoire naturelle qui ont été envoyés et donnés aux correspondants du Cabinet...," AN O[1] 2124, folder marked "Jardin du Roi Année 1775."

181. AN AJ[15] 510 (items 364-382), "Brevets de Correspondants du Jardin et du Cabinet du Roi."

182. Laissus (1964) discusses these correspondents briefly, pp. 297-99. See also Thésée, p. 22; Kury (1995), p. 101.

sent in 1788 by a certain Crosnier le Jeune, correspondent of the Cabinet du Roi from outside Cap François in Saint Domingue.[183]

In 1768, Abbé Expilly described the Cabinet in his *Dictionnaire des Gaules et de la France*:

> The room preceding the Natural History gallery is furnished with beautiful large cabinets containing mainly anatomical items. A large desk in the middle offers an elegant display of selected shells. You go next into a splendid gallery…loaded with all sorts of the weapons, gear, and attire of savages, fruit from the Indies, reptiles, quadrupeds, amphibians, fish, snakes, and so on. The walls are all around neatly and richly hung with the most precious specimens belonging to all three kingdoms: animals, metals, salts, stones, talcs, shells, bezoars, saps, gums, etc. all of them in phials or jars artfully displayed on the shelves of large cabinets with small compartments in the lower parts containing all sorts of fossils, every class of precious stones, topazes, jaspers, agates, jades, cornelians, stones from Florence, pebbles from Egypt, and other marbles, alabasters, crystals, and so on.…Other armoires are crammed with foreign woods, fruits and seeds and with ores and petrifactions, insects, and animal fragments in the lower compartments. All twenty-two of those large armoires are surmounted and crowned, some with Indian feathers and dress, others with various artifacts from the seas…the Cabinet du Roi could be said to be the treasure of nature and the triumph of good taste.[184]

The image here suggested is that of an "abstract of the universe" where nature's productions are placed side by side with traces left by faraway human civilizations, much like what Louis-Sébastien Mercier proposed as an ideal cabinet, which appears in his utopia, *L'An 2440*, celebrating the perfect harmony of an ordered and controlled world absorbed at one glance in its entirety.[185] The Cabinet, as described by Expilly, benefiting from the protection of a monarch and a vast network of correspondents, presided over the collection process on a planetary scale.

Trianon and Rambouillet. As weighty as it was, the Jardin du Roi was only one component of a larger system of French botanical gardens. Administratively, in fact, it was not the top component, falling as it did under the Bâti-

183. ALS "Mistray" dated "Au havre le 26 Juin 1788" to "M. Le C.te Dangeviller," AN O[1] 2113A "Pépinières d'Amérique, Année 1788."

184. Cited in Hamy (1890), n. 2, p. 32.

185. Louis-Sébastien Mercier (1786), cited in Kury (1995), p. 86. The *Encyclopédie* in the article "Cabinet," repeats a similar definition: "Un Cabinet d'histoire naturelle est donc un abrégé de la nature entière"; cited in Schnapper, p. 10.

ments du Roi, the agency that ran all the royal gardens and nurseries. We turn to these establishments next, but before doing so we should point out that certain administrators in the Bâtiments du Roi itself assumed leadership roles in the botanical establishment, most notably Charles-Claude Flahaut de la Billarderie, count d'Angiviller (1730-1809), director-general of the Bâtiments du Roi under Louis XVI. D'Angiviller was installed as an associate academician in chemistry at the Académie des Sciences in 1772, becoming a veteran *pensionnaire* in 1779; and he served as a conduit from the Bâtiments du Roi to the Academy.[186] D'Angiviller took a very active interest in expanding the botanical outreach of his office.[187] He was the maestro who initiated an important botanical overture to China and who sent the royal botanist, André Michaux, on a ten-year mission to North America from 1785 to 1796. D'Angiviller was so involved in botany that at one point he maneuvered to succeed Buffon at the Jardin du Roi. Working alongside d'Angiviller, the abbé Pierre-Charles Nolin served as controller general of French nurseries from 1773 to 1792.

[Jeton: Rambouillet, 1712]

The royal Trianon and Rambouillet gardens appear in this context. The Grand and Petit Trianon gardens, next door to Versailles, were put on a new footing beginning in 1750 with the father and son team of Claude and Antoine Richard as head gardeners.[188] The gardens and nurseries at the Trianon complex came to include plants from Persia, India, the Cape of Good Hope, China, Virginia, Pennsylvania, Bermuda, Africa, and Mauritania.[189] These gardens were notable, not only as major botanical installations, but also for nurturing

186. See entry, Institut de France and PA-PV-110, La Billardière, 15 May 1790 ("Ouvrage de Botanique sur les plantes des philippines et principalement de L'Isle de Java"). This d'Angiviller is often confused in the literature with his older brother, August-Charles-César de Flahaut de La Billarderie, the Marquis d'Angiviller (1724-1793).

187. Gillispie, p. 20. See further Part IIC, and, for example, d'Angiviller's letter to the king's agent in Macao, De Guignes, asking for botanical samples and information, dated "Vers.les 7 X.bre 1785," AN O[1] 1292, #18.

188. For background, see Lamy; Nolhac, esp. chapts. 2-3; Roger Williams, chapt. 3.

189. At least according to an inventory prepared by the "Commission de la recherche des Plantes dans les Jardins de la liste Civile et dans ceux des Émigrés," 1792-1793; BCMNHN, Ms. 306. On Marie Antoinette and the Trianon, see also Roger Williams, pp. 156-57.

gardeners and botanists who later filled positions elsewhere in the French botanical establishment. The Trianon gardens might have developed as a major botanical and horticultural station, but Louis XVI gave the Petit Trianon as a gift to his bride, Marie Antoinette, who transformed the property into an English-style pleasure garden, and the Trianon thus lost its usefulness both as a scientific garden and as a nursery.

To make up for this loss, Louis XVI acquired the property at nearby Rambouillet in 1784 with the specific goal of turning it into a station for experimental horticulture and the propagation particularly of American species of trees.[190] Thus, the Jardin du Roi in Paris was the "scientific" garden, and Rambouillet became the "industrial" garden for acclimatizing and propagating plants.[191] Duhamel du Monceau made this point explicitly in his memoir on the usefulness of royal nurseries:

> One might say that the Jardin du Roi could fulfill these functions, but we assert the opposite. The efforts it takes to tend eight or ten thousand plants more than suffice for the gardeners there, and the space is hardly enough for one or two trees of each species, such that, for example, two cedar trees are all that are needed for the Jardin du Roi, but one cannot thereby distribute them to other gardens which, however, should be the purpose of the royal nurseries.[192]

The abbé Henri-Alexandre Tessier (1741-1837) managed the royal gardens at Rambouillet. He was appointed by Louis XVI who liked to visit the gardens and to converse with its manager.[193] Tessier joined the Academy of Sciences in 1783 and was promoted to associate in natural history in 1785. He collabo-

190. Spary (2000), pp. 98-99, 133; Guillaumin and Chaudun, pp. 123-26. See related papers by Thouin submitted to the Academy of Sciences; PA-PV-110, Thouin, 8 March 1786 ("M.re sur l'utilité d'une pepinière" and "M.re sur la culture des plantes de différentes contrées qu'on doit placer dans un Jardin de Botanique"). Well before these more sustained efforts, in 1775 d'Angiviller and other officials were already using the nurseries at Choisy and Nemours as sites for the importation of foreign trees – Turgot writes of 2000 Canadian poplars being shipped to Nemours that year; see ALS "hazon" dated "Choisy 13 X.bre 1774" to "Monsieur" and ALS "Turgot" dated "Paris Le 22. Novembre 1775" to "M. Le Comte D'angiviller," both AN O[1] 2110.

191. The Director-General of the Bâtiments du Roi, under which all these gardens fell, was clear on this point. See d'Angiviller note appended to ALS "Bachelier et Saubisson" dated "havre le 19 Aoust 1789" to "M.r Dangiviller a Versailles"; where he writes, "il falloit envoyer les caisses en question au jardin du roy. car rien de ce qu'elles contiennent ne regard les Bat. ni Rambouillet." AN O[1] 2113A: "Pépinières 1789."

192. "On dira peut estre que le jardin Royal peut remplir ces vues, mais nous pouvons assurer le contraire. les soins qu'il faut pour elever 8 ou 10 milles plantes qu'on y cultive sont plus que suffisants pour les jardiniers et le terrein suffit a peine pour avoir un ou deux arbres de chaques Espesse de sorte qu'il y a par examp. deux Cedres au jardin du roy qui sont tout ce qu'il y faut mais qu'on ne peut communiquer a d'autres ce qui doit cependant estre lobjet des pepinieres R. lles," APS-MS, 580/D881 #37.

193. On Tessier, see Silvestre.

rated with André Thouin at the Jardin du Roi, and he was known as a promoter of colonial botanical exchanges.[194] Tessier likewise forged contacts with the Société Royale de Médecine.[195]

Our concern is with the scientific dimension of these gardens, but we should not lose sight of their larger social role in the Old Regime. Under d'Angiviller and Nolin, a whole series of royal nurseries annually supplied trees and ornamental bushes and flowers for royal properties and for persons of rank who requested them.[196] Eleven different nurseries dotted the Paris/Versailles region in 1763; they employed fifty gardeners, and their operations cost over 100,000 livres a year (31,000 livres for the third quarter of 1763 alone). In 1779, officials put forward a plan to establish a new nursery at Vincennes.[197] Requests from princesses, countesses, dukes, marquis(es), and other nobles and notables for their private gardens inundated the administrators of the royal nurseries.[198] Madame de Victoire wanted 5500 trees of various species; the Marquis de Mouchy wanted 600 to 700 Italian poplars.[199] Early in the century authorities shipped tulip trees from Louisiana for the Princess de Conti.[200] Royal nurseries handed out approximately 600,000 plants and trees in 1751 alone. The royal facility in Toulon had the special function of supplying early flowers from the south.[201] Nolin became annoyed by these requests: "It is hardly possible for us to deal with creating schools for botany and trees. It's the bane of fashion...These sorts of fantasies cost a lot of time, and otherwise I think it would be more appropriate to propagate these new species in order to put them at Rambouillet than to scatter them in the gardens of the amateurs of today who will soon grow tired of taking care of them."[202]

194. See Tessier (1789). On the extensive contact between Tessier and Thouin, see entry in BCMNHN, Ms. 2310.

195. See Tessier communication regarding clove plants, SRdM, Ms. 7, p. 250 ("Séance du vendredi 30 Juillet 1779").

196. Bouchenot-Déchin provides an entry into the royal gardens of the Bourbon kings. See also Velut for the larger contemporary social context. Gillispie (1980), pp. 333-35, provides additional background. On royal nurseries themselves, see AN O^1 2101 and other folders in the series O^1 and inventory by Curzon.

197. AN O^1 2111: Pépinières, 1778-1782, Draft letter in d'Angiviller's hand to "M. de sartine" dated "Vers.les le 24 aoust 1779."

198. "Demandes faites à Monsieur le D.eur Général concernant les L'épiniéres, Année 1784. et 86. 86.87.88.89"; AN O^1 2113A "Pépinières 1789."

199. ALS "Nolin" dated "a paris Le 24 8.bre 1788" to "Monsieur Le Comte"; AN O^1 2113A "Pépinières de 1788."

200. See letters AN-COL C^{13A} 21 (1736), fol. 278 and C^{13A} 22 (1737), fols. 128, 155.

201. See AN O^1 2124: Jardin de Toulon, 1683-1761.

202. "Il n'est guere possible que nous nous occupions de former des Ecoles De botanique et darbres. C'est La maladie a La mode...Ces sortes de fantaisies font perdre un Temps considerable, d'ailleurs je crois quil seroit plus convenable de multiplier ces especes nouvelles pour les placer a rambouillet que de Les Eparpiller dans Les Jardins des amateurs actuel qui bientôt n'en ferons plus de cas," ALS "L'abbé Nolin" dated "a paris Le 28 9bre 1788" to Angiviller, AN O^1 2113A "Pépinières de 1788."

Rare fruits represented a particular form of currency at court that also entangled the Colonial Machine. In 1670 Colbert prodded the Compagnie des Indes for a systematic investigation of American animals, fruits, and flowers still unknown in Europe, but he especially wanted pineapples for domestication and Louis XIV's table.[203] Over a century later, as we noted in the Introduction, Cossigny *fils* was pleased to send a new species of pineapple from Île de France for degustation by the Louis XVI, Marie-Antoinette, and the Royal Family. The pleasure of the king and an almost fetishistic concern over pineapples is seen no less in a letter of 1738 from the Intendant of the Jardin du Roi, Dufay, to the royal doctor in Saint Domingue, Pouppé-Desportes: "I encourage you always to send us samples of pineapples because the King [Louis XV] really loves them, and they grow marvelously. You only have to send small ears, pulled-up small plants, or the fruit itself wrapped up in moss in a barrel, they all come back nicely."[204] Two barrels of green oranges were requested of authorities in Canada in 1686 for the dauphin, and in 1765 an elaborate ceremony was arranged for the presentation of ripe Chilean strawberries to the king.[205] The details are worthy of being reported. One Duchesne, Trudaine de Montigny's secretary and the possessor of the plant, wrote to court authorities:

> With the permission you have given us, I am leaving tomorrow morning at 5:00 for Compiegne with my son, and we are bringing with us the strawberry plant from Chile you saw. The largest of the four strawberries it carries is almost ripe and the plant will be suitable to be presented to His Majesty on Sunday the 7th at the latest.
>
> As this strawberry plant has only given fruit twice at the Jardin du Roi since 1716, Monsieur de Jussieu strongly desires that Mlle Basseporte [the royal miniaturist] paint these strawberries on their stalk, which has not yet been done, for deposit at the Cabinet du Roi at the Royal Library. I think it would be good if the Count de Saint Florentin [secretary of state for the king's household] could be alerted, and the [uniden-

203. "Surtout envoyez-moy de l'ananas," Colbert letter of 21 June 1670, quoted in McClellan and Regourd (2000), p. 41.

204. "je vous recommande toujours de nous envoyer des encheilletons d'ananas parce que le Roy les aime fort et qu'ils veuillissent a merveilles, il n'y a qu'a envoyer des orilletons [oreilletons], des petits pieds arrachez et des couronnes enveloppez de mousse dans un tonneau, cela reprend a merveilles," ALS, Dufay to Pouppé-Desportes, "a Paris ce 20e 9bre 1738," PA-DB (Pouppé-Desportes).

205. For the former, see AN-COL B^{12} (1686), fol. 65 (26 April 1686); Roger Williams, p. 37, points to interests in and the popularity of strawberries, as does Allorge, pp. 172-73.

> tified] Marquis, being with the King, could slip him a word about it and get him to be in the King's quarters at the time of the presentation.[206]

The *Ménagerie du Roi* deserves to be touched on here, although it was more of a symbolic mirror to the power and glory of the king than a scientific repository or a vital part of the Colonial Machine.[207] Designed by the architect of Versailles, Le Vau, the elaborate Royal Menagerie there dated to the 1660s and came to house animals from around the world, at various points including lions, tigers, leopards, elephants, camels, a rhinoceros, gazelles, zebras, seals, porcupines, monkeys, baboons, and many species of rare birds, including ostriches, toucans, and a condor from South America. Getting these animals from their exotic points of origin to Versailles was no easy task and brought into motion the wheels and gears of the Colonial Machine.[208] Academicians from the Académie des Sciences regularly dissected deceased animals from the Ménagerie, but Buffon did not think of the institution as especially useful for the study of natural history. The administration of the Ménagerie fell under the Bâtiments du Roi, but it seems not to have had a scientific staff of its own. Nevertheless, the Ménagerie, in filling the needs of the monarchy, found its niche in the Colonial Machine.

These pressures to service the whims of the monarchy notwithstanding, in the 1780s with d'Angiviller and Nolin orchestrating things from the Bâtiments du Roi, with Thouin and Daubenton at the Jardin du Roi, and with Tessier at Rambouillet, the men and institutions were in place for one of the great chapters in the history of the Colonial Machine and of French botany in the eighteenth century – the effort to fortify French silviculture and alleviate the problem of the shortage of naval timber by importing seedlings from America and propagating them at Rambouillet. For this purpose, d'Angiviller sent the naturalist André Michaux to the new United States in 1785, where Michaux established official royal gardens in New Jersey in 1786 and in South Carolina

206. "Suivant la permission que vous nous avez données je pars demain à 5 heures du matin pour Compiegne avec mon fils et nous emportons avec nous le fraisier du Chili que vous avez vu. Le plus grosse des quatres fraises qu'il porte est presque mûre et sera en etat d'etre presentée sur son pied a Sa Majesté le Dimanches [sic] 7 au plutard. Comme ce fraisier n'a porté fruit que deux fois au jardin du Roi depuis 1716. M. de Jussieu desireroit fort que M.lle Basseporte put peindre ces fraises sur leur pied, pour la donner ensuite au cabinet du Roi a la Bibliotheque où ce fraisier n'est pas encore peint. Je crois qu'il seroit bon que M. Le Comte de S. florentin en fut prevenu, et M. Le Marquis se trouvant chez le Roi, pouroit lui en toucher un mot et l'engager a setrouver chez le Roi a l'heure de la presentation…," AN O^1 1293, #317, ALS "Duchesne" to "Monsieur" and dated "V.lles 4 Juillet 1765."

207. On the Ménagerie du Roi, see the wonderful study by Roberts, esp. chapt. 2; and J. B. Lacroix (1978).

208. Roberts makes this point clearly and documents several cases of what it took to transport animals, pp. 52-60; for a Dutch example, see Harold Cook, p. 326. See also ministerial correspondence related to importing animals from Canada AN-COL B^{36} (1714), fol. 27v (27 January 1714); B^{47} (II) (1724), fol. 1222 (20 June 1724); B^{48} (1725), fols. 853, 877 (29 May 1725 and 5 June 1725); B^{49} (1726), fol. 672v (11 May 1726). See also below, Part IIC.

in 1787. This episode is so telling of the nature and activities of the Colonial Machine that it receives detailed, separate treatment in a subsequent part (IIC) of this volume.

Other botanical gardens dotted the provinces of Old Regime France, in part as an outgrowth of the 1707 edict of Louis XIV that each medical faculty was to have a professor of botany and an associated botanical garden.[209] One Revolutionary document lists thirty-two state/official gardens in contemporary France; another puts that number at fifty-two, counting colonial outposts.[210] The botanical garden of the Duc d'Orléans, established at Saint-Cloud in 1756 with Pierre Gasser as its gardener, deserves notice in this regard.[211] Provincial academies and learned societies were also great promoters of botanical gardens.[212] But four government gardens dotting the Atlantic coast – those in Brest, Lorient, Nantes, and La Rochelle – require further examination because they were the chief points of botanical contact between the Paris region and the overseas colonies.

[Map: Major French Atlantic Ports]

209. Velut, p. 69; Spary (2003), p. 16.

210. BCMNHN, Ms. 308, "Projet de Decret" and "Projets pour la conservation et L'Etablissement des Jardins...." See also, Spary (2000), fig. 3.

211. For the Saint-Cloud garden, see reference, BCMNHN, Ms. 797.

212. On the provincial academies, see Roche (1978). For a case study of an academy garden, see Bret (1999b).

Brest. Strategically situated at the Atlantic entrance to the English Channel, the port of Brest on the western tip of Brittany was a major installation of the French Royal Navy. In addition to the infrastructure one would expect at such a military port, Brest was also the site of a large naval hospital and medical school. Substantial and ongoing botanical and medical links connected Brest with the establishment in Paris. Brest was likewise the home of the Royal Academy of Naval Sciences (the Académie Royale de Marine) and a school to train naval cadets. It was also a manufacturing center for making compasses and other nautical scientific equipment.

[Medal: Port of Brest, TUTELA CLASSIUM OCEANI, 1681]

Not least in this connection, a royal *Jardin Botanique* came to exist at Brest.[213] The Brest garden originated in 1694 as an apothecary garden; it was upgraded in 1740, then closed between 1760 and 1768, but was renewed and significantly expanded after that. Backed by Poissonnier-Desperrières, Étienne Chardon de Courcelles pushed the significant expansion of the Brest garden; Courcelles was chief royal physician (*premier médecin du roi*) at Brest and director of the anatomy school there for many years.[214] Antoine Laurent was the head gardener from 1771 through at least 1815, with emoluments of 1200 livres. Laurent trained at the Petit Trianon and then at the Jardin du Roi, and he maintained very close ties with Thouin in Paris. The Jardin Botanique possessed several large greenhouses, one fifty feet in length, and it benefited from the services of convict labor from the galleys stationed at Brest. The local intendant endowed the garden with a modest natural history cabinet containing birds from Cayenne and the Amazon region. From their travels, naval officers and medical personnel occasionally brought back plants and seeds to Brest;

213. These details are taken particularly from BCMNHN, Ms. 848, "Lettres adressées par Laurent, directeur du Jardin botanique de Brest, à Thouin." and BCMNHN, Ms. 1327, "Journal des Envois de Plantes, Arbres et Graines faits au Jardin du Roy par ses Correspondants." Laurent's correspondence has now been published by Roussel et al., who likewise provide details concerning all the port gardens of concern here, notably pp, 58-69, regarding Brest. See also Huard et al. (1967); Allain, pp. 134, 136-37; and Guéguen, passim and here, p. 379.

214. DDP IV: 29.

and, despite Laurent's complaints about the irregular and unsystematic nature of this traffic, the garden thereby received plants from the Canary Islands; from Martinique, Guadeloupe, and elsewhere in the Caribbean; and from New England, Guiana, the Cape of Good Hope and elsewhere in Africa, India, the Indian Ocean, China, and Australia (Botany Bay).[215] Notably, Laurent established contact with royal government gardens on Île de France in the Indian Ocean. Not unexpectedly, the Brest Jardin Botanique sent botanical samples back to the Jardin du Roi and to the Trianon gardens, notably from India and the Americas. In the 1780s the Brest Jardin served as a transit garden to rehabilitate trees and plants after their long sea voyages, particularly from North America and before being shipped on through Le Havre to Paris.[216] Thouin reciprocally enriched the Brest garden with specimens from his collections in Paris.

Lorient. Down the Atlantic coast of Brittany from Brest, the new town of Lorient was laid out in 1666 as the home port for the Compagnie des Indes, the French East India Company. The Compagnie des Indes conducted an enormous trade with India, China, and Canada while that province was in French hands. Naturally, this trade came to include botanical and natural history specimens in addition to commercial items such as tea, rhubarb, and beaver pelts for European markets, and American ginseng destined for the Chinese market. The Compagnie was instrumental in forging and maintaining commercial and botanical links to China; at one point it imported *Cannabis sativa* from China, doubtless as a resource for the all-important cordage industry. The interests of the Compagnie des Indes in botany, pharmacy, and materia medica were substantial, given that the Compagnie had its own hospital at Lorient and sent out its famous East Indiamen with staffs of surgeons and pharmaceutical supplies.[217]

Not surprisingly, therefore, the king came to underwrite a royal botanical garden at Lorient.[218] Dr. François Gallois (or Galloys) (1719-1779), a royal

215. In addition to sources sited in previous notes, see also BCMNHN, Ms. 691, "Catalogues des graines semées au Muséum pendant les années 1790-1801" and BCMNHN, Ms. 1410: "Catalogues de jardins botaniques (Brest)." See also ALS "Thouïn" to "Monsieur le Comte" dated "A Paris ce 8 7.bre 1786" that discusses a valuable shipment of trees from India arriving at Brest, AN O[1] 2113A.

216. Thouin recorded 154 communications from Laurent from 1771 through 1803, BCMNHN, Ms. 2310. See also Laurent entries, BCMNHN, Ms. 1327 (e.g., "1786: de M. Laurent de Brest reçu 6 Esp. de Gr. D'Arbres arrivantes d'Amerique nord") and Ms. 308. See also Laurent correspondence with d'Angiviller at the Bâtiments du Roi related to transshipping plants through the Brest garden, AN O[1] 2112: Pépinières, 1783-1787. This correspondence has now been published by Roussel et al.

217. On these points see Haudrère (1989), pp. 591-94; Romieux (1987); SHML, 1P 266, Liasse 3, #44 and further below, this Part.

218. Allain touches on this garden, pp. 131-32, 139-41; Roussel et al., pp. 45-51, have gone a long way towards remedying the fact that the garden of the Compagnie des Indes at Lorient has yet to be investigated by historians. See manuscript sources AN-COL C[7] 114 (Galloys), materials in AN-MAR D[2] 29, #77-95, and undated historical note by Gallois, BCMNHN, Ms. 357. See also mention of this garden in the minutes of the contemporary Comité d'Agriculture in Foville and Pigeonneau, eds., pp. 104, 107; other avenues of inquiry are to be found in Legrand and Marec, passim.

doctor in Lorient, headed the garden from its instauration under Louis XV (probably in 1768) until Gallois' death in 1779, with annual appointments of 2000 livres.[219] The garden of the Compagnie des Indes in Lorient was established especially to be an entrepôt to acclimatize plants received from China and India and destined for the Jardin du Roi and the Trianon. The idea was that plants could not be shipped directly from Asia to Paris, but needed to be tended immediately on their arrival in port in France. As one commentator, probably Gallois himself, remarked in 1777: "India and China hold treasures in trees and plants that can be advantageously procured for France and for America. To achieve this end the directors of the Jardin du Roi [in Paris] recognized the necessity of a station in Lorient where trees can be revived after their long trip and prepared for transport to Paris and to America."[220] Gallois built an orangerie and a hothouse for tropical plants and planted the alleys of the garden with bamboo.[221] He corresponded with Thouin, regularly sending him and Jussieu Chinese, Indian, and American seeds and plants via Lorient for the Jardin du Roi in Paris.[222] A shipment of 1777 to Thouin, for example, included plants from China (three orange/grapefruit trees, a hybrid lemon tree, a species of mandarin orange, a species of creeper, a pomegranate), from Île de France (a rose-apple, a jujube tree, a cactus), from India (an acacia), from Cayenne (a rose), and from Peru (the Quito orange). The accompanying consignment of seeds of twenty-three types of plants from India prompted Thouin to respond to Gallois, "Continue, Sir, to enrich an establishment that will soon owe you a part of its fame."[223] Gallois also acted as the agent of the Paris botanists, Le Monnier in particular, in shipping European plants outward to the colonies.[224]

More than implicit in these developments was the idea that economically valuable plants could be shipped from the East Indies through Lorient in

219. On Gallois, see his personnel file AN C[7] 114; and SHML, 1P 284 Liasse 104, #73: letter from "Les Sindics" dated "A Paris le 5 Juin 1765."

220. "...l'inde et la chine renferment des trésors en arbres et plantes qu'il est très avantageux de procurer à la france et à l'amerique; pour y parvienr il à [sic] été reconnu par Mrs. les Directeurs du jardin des plantes qu'il faloit un entrepot à lorient où on put, après les longues traversées ranimer les arbres et les mettre en etat d'etre transportés à paris et à l'amerique"; AN, MAR D[2] 29, #79 "Mémoire pour le jardin et la serre chaude du Roy etablis à lorient." The 1777 date comes from a supporting letter of August 25, 1777 from A.-L. de Jussieu attached to this piece.

221. For a contemporary report on the state of the garden at Lorient see BCMNHN, Ms. 1995, letters from Cossigny to Le Monnier dated "L'Orient le 5 May 1775" and "L'Orient le 15 May 75."

222. See entry BCMNHN, Ms. 314. See also various entries in BCMNHN, Ms. 1327. See also mention of shipments from Cossigny on the Île de France through Gallois and Lorient to Le Monnier in Paris, BCMNHN, Ms. 1995.

223. "...continués, Monsieur, d'enrichir un établissement qui vous devera bientot une partie de sa célébrité...," AN-MAR D[2] 29, #83, "Copie de la lettre de M. Thouin premier jardinier du jardin du Roy écritte à M. Gallois le 19 août 1777."

224. BCMNHN, Ms. 1995: Letter from Cossigny to Le Monnier dated "Au Port Loüis Isle de france le 20 avril 1770."

France and thence onward to the West Indies, and vice versa. In other words, authorities envisioned the garden at Lorient as a staging ground for a vast exchange of economically useful plants back and forth from South Asia to the Americas. A disinterested internal memorandum of 1777 within the ministry of the navy made the case for this type of exchange:

> The main goal of the garden in question is to procure rare plants for the Jardin du Roi in Paris and to restore those that suffered on their crossing and that would perish if sent [directly] from the Orient to Paris. We are assured that, through Mister Gallois's care, many useful plants from China are doing very well in France.
>
> In addition to those plants that can be raised and propagated in France, once received and cared for in the greenhouses of Lorient, this station can serve to grow plants from India that we wish to introduce into the American colonies. Many shipments have already been made to the colonies, and Mr. Gallois awaits the breadfruit tree that should be sent to him from the Philippines. If this tree can be introduced and multiplied in our sugar islands, that will be a great resource for feeding slaves. To thus furnish our colonies with plants from India, they must be received and tended at Lorient on their arrival by sea. One cannot forget that all the coffee trees in the French colonies in America derived from a [single] tree raised at the Jardin du Roi in Paris.[225]

Briefly during the 1770s, the Lorient garden operated as just this kind of transshipment station that redistributed colonial botanical products. In 1772, on government orders, Gallois expedited onions, mangoes, "Hottentot bread," and trees called "vacois" from the Indian Ocean to Guadeloupe, Martinique, and Saint Domingue in the Caribbean.[226] Another dispatch to officials in Guade-

225. "Le Jardin dont il est question a pour objet principal de procurer au Jardin du Roi à Paris des Plantes rares et de remettre en bon Etat celles qui ont souffert dans le trajet et qui periroient si, dans leur état de Langueur, elles étoient envoyées de l'Orient à Paris. Mr. Gonet assure que par les soins du Sr. Gallois il y a plusieurs plantes utiles venues de la Chine qui reussissent très bien en france. Independanment de celles qui peuvent être élevées et multipliées en france, après avoir été recues et soignées dans les serres de L'Orient, cet Entrepôt sert encore pour élever les plantes de L'Inde qu'on cherche à introduire dans les Colonies de L'Amériqque. Il a deja été fait plusieurs envois dans nos Colonies et le s.Gallois attend L'Arbre à pain qui doit lui être envoyé des Manilles. Si cet arbre peut être introduit et se multiplier dans nos Isles à Sucre, il sera de la plus grande ressource pour la nourriture des Esclaves. Pour parvenir a procurer ainsi à nos Colonies les plantes de L'inde, il faut qu'elles puissent être recues et soignées à L'Orient à leur arrivée de la mer. On ne peut oublier que tous les arbres a caffé des Colonies françoises de l'Amérique provienent d'un Arbre élevé à Paris au Jardin du Roi," AN-MAR D^2 29, #86, internal ministry memorandum dated "11 Juillet 1777." Allain, pp. 96-97, has noted this function for the gardens at Lorient and Brest.

226. "...manguiers, des vacois, des vavanquers des oignons [?], pain des hotentots," AN-MAR D^2 29, #79, undated, unsigned "Mémoire pour le jardin et la serre chaude du Roy etablis à lorient."

loupe in 1777 prompted thanks and the request that "the richest present you can procure for the colonies would be the breadfruit tree."[227] Similarly, on Île de France, the chief royal gardener noted his pleasure in exchanges via Lorient, with plants from his Indian Ocean garden being sent to the Antilles and his receiving in turn plants from the Antilles via Gallois in Lorient.[228] Authorities with the Compagnie des Indes were especially concerned that vessels arriving from China offload botanical cargo quickly and transfer it immediately to the Lorient garden, "because one day of delay can ruin everything."[229]

After Gallois's death, the garden at Lorient passed for a while into private hands, those of a Monsieur Dodun, a former tax farmer in Lorient, "on the condition that the gardener will send to the Jardin du Roi in Paris all the plants from China and from India that he will be able to cultivate in the Lorient garden."[230] The American War doubtless interfered with the transshipment of plants from the East and West Indies, but Dodun apparently let the garden fall into desuetude; and later in the 1780s the minister approved its conversion to a storage area for naval timber.[231] That was unfortunate, of course, but by then, as we will see in Part II, the ministry had set in motion a better plan for transferring plants, not least the precious breadfruit, from one continent to another.

Nantes. Still further down the Atlantic coast, at the mouth of the Loire, the botanical garden in the port city of Nantes occupied a special place among contemporary botanical gardens in France.[232] Nantes was a major port for the extensive French slave trade and so had a strong orientation toward Africa and the French colonies in the Americas. The local medical school furnished

227. "le plus richye present que vous puissés procurer à nos colonies ce seroit l'arbre de fruit à pain," AN-MAR D^2 29, #8, "Extrait d'une lettre de la Guadeloupe écrite à M. Galloys le 3 janvier 1777 par M. Vatable major d'un bataillon."

228. See BCMNHN, Ms. 1995, Ms. titled: "Etat de ce que contient la Caisse marquée No. 1. *Graines pour le roi*... a l'adresse de Mr. Galloys directeur des serres du roi et medecin de la marine à L'orient," with marginal date, "A L'isle de france le 25 juin 1778." Also, BCMNHN, Ms. 303, "Recensement de tout ce que renferme Le Jardin du roi, le Montplaisir; remit par Mr. De Ceré... au Jardin du Roi, Le Monplaisir, Isle de france, le 21 Juin 1785." This latter document refers to an earlier time ("cy-devant") for the exchanges with Gallois in Lorient, probably the mid to late 1770s. The colonist Cossigny de Palma likewise made many shipments through Lorient and Gallois' garden; see, for example, ALS, Cossigny to Le Monnier dated "Palma le 29 Mars 1777," BCMNHN, Ms. 1995, "Lettres de Cossigny à Le Monnier."

229. "...parce qu'un jour de retard peut tout faire perir," SHML, 1P 284, Liasse 104, #72: "Copie de la lettre ecritte à la Comp.e par M. De la Verdy dattée de Versailles le 3 juin 1765"; see also #73: ALS, "Les Sindics" to Buisson, dated "A Paris le 5 Juin 1765."

230. "...sous la Condition qu'il Enverra au jardin royal de paris toutes les plantes de Chine et de l'inde qu'il aura pü cultiver dans celuy de l'orient...," AN C^7 114, Poissonnier letter to Monseigneur dated "A Paris ce 26 9bre 1779."

231. AN-MAR D^2 29, #88, dated "A Lorient le 29 Janvier 1787."

232. On the Nantes and its garden, see Auvigne, pp. 7-12 and illustrations; Roussel et al., pp. 51-55; see also Kernéis et al.; Laissus (1964), p. 293n; Juhé-Beaulaton, pp. 271-73; Guéguen, pp. 377-79; Allain, pp. 133-36; McClellan (1992), p. 149.

apothecaries and surgeons for slave ships and for the colonies in general, and it was not for nothing that in 1750 a medical chair was established for "diseases of seamen."[233] It was not unnatural either that Nantes should be chosen as the major botanical entrepôt for plants from the American colonies.[234] The Nantes botanical garden originated in 1688 as a typical apothecary garden under control of the university and the town. In 1709 the royal government took over the Nantes garden for the specific purpose of acclimating plants from overseas. Letters Patent in 1719 and 1726 confirmed this role and made the royal garden at Nantes a formal subsidiary of the Jardin du Roi in Paris. As the 1726 royal edict ordered:

> The garden will take the title of Royal Botanical Garden and will be annexed and subordinated to His Majesty's Jardin du Roi in Paris. Under the supervision of a Director to be named, the garden at Nantes will serve as a depository for raising and maintaining plants arriving in port from foreign countries....To this end His Majesty orders that all royally supplied surgeons departing from Nantes for overseas voyages bring back from wherever they make land four plants or shrubs rooted in their native soil and all seeds that they can collect, properly labeled and boxed. They will take care to keep specimens moist and to deliver them to the Director of the said garden at Nantes on their arrival. His Majesty further orders captains of said vessels to see to this matter on pain of a ten-livre fine for each infraction.[235]

In the early years following its royal charter, the botanical garden at Nantes actively fulfilled its role as a collection center for colonial plants, and its work was vigorously promoted and enforced at the highest levels of government. Navy secretary, the count Maurepas ("who has taken this business to heart," remarked the regional governor), pressured officials in Martinique and Saint

233. Roussel et al., p. 55.

234. See previous note and AM-Nantes, DD 48. #7. This document is an 8-page printed "Mémoire pour le jardin royal et botanique de Nantes." Anonymous and undated, it is by François Bonamy and dates from 1757-1758.

235. "...ordonne que ledit Jardin prendra la quallité de Jardin Royal des plantes et Sera annexé et subordonné au Jardin des plantes de sa Majesté du fauxbourg saint victor de Paris auquel il servira dentrepot pour la culture et pour l'entretien des plantes qui viendront des pais Estrangers dans le port de Nantes sous l'inspection du Directeur qui sera nommé pour le Jardin de la dite ville...Ordonne sa Majesté a cet effet a tous Chirurgiens qui se pourvoiront a Nantes des Remedes pour des voyages de long cours de raporter des païs ou les vaisseaux tomberont quatre plantes ou arbustes dans leur propre terre, et toutes les graines qu'ils pouront ramasser avec leurs Etiquettes qu'ils auront soin de Rénfermer dans des Boettes et de les garantir de l'humidité et de les remettre au Directeur dudit Jardin de Nantes a leur arrivée. Ordonne sa Majesté aux Capitaines desdites Vaisseaux d'y tenir la main a peine de dix livres d'amande pour Châque contravention...," AM-Nantes DD 49 #4, undated fair copy of regulations of 1726; see related details in Allain, pp. 20-21; Auvigne, pp. 9-10.

Domingue to comply with the regulations to send plants and seeds to the Nantes garden.[236] Maurepas himself rewarded captains who took care of plants in their charge, and he reprimanded those who did not.[237] The provincial governor-general, the Maréchal d'Estrées, coordinated with Maurepas and likewise exerted himself for the success of the venture. D'Estrées collected reports from ships' captains, badgered officials in the Caribbean, and held out hope that the garden at Nantes would one day "top all the other gardens in France."[238] The mayor of Nantes, Gérard Mellier, was also actively involved in promoting the effort. Forms were printed up with lists of plants to be collected from along the coasts of Africa and from the Caribbean. Apparently to remind departing captains of their botanical duty, these forms were to be signed by two port officials at Nantes with the name of the ship in question.[239]

Officials in Martinique seem to have been especially effective in delivering specimens from the Caribbean to Nantes and thence to Paris.[240] The early success of the Martinique-Nantes connection was greatly reinforced by the presence and support of Bernard-Laurent d'Hauterive, the enthusiastic local royal prosecutor in Martinique we encountered above, who in 1716 articulated a multifaceted research program in conjunction with the Academy of Sciences in Paris. A few captains did in fact transport plants from Martinique to Nantes in these years under seal from d'Hauterive, but difficulties presented themselves. Some captains were simply uncooperative, despite the threat of fines. Many plants died on the Atlantic crossing. They did not always come with labels or instructions that would allow the gardeners at Nantes to know what they were or how best to tend them. And, the scale of shipments was always small.[241] These were problems that continually undermined the success of intercontinental botanical exchanges throughout most of the eighteenth century.

If the Nantes garden was not the immediate success hoped for, over the years it continued to receive botanical specimens from overseas. Before the loss of Canada in 1763, the Nantes garden obtained many plants from that region, at least in part because the French governor of Canada from 1747

236. See Maurepas letters, AM-Nantes, DD 49, #10, #44, #48; DD 52 #11. The quote, "qui a pris cette affaire à cœur," is from the letter cited in a following note.

237. See remarks to this effect in ALS, Dufay to Pouppé-Desportes, "a Paris ce 20e 9bre 1738," PA-DB (Pouppé-Desportes).

238. "...que ce jardin des plantes de Nantes l'emporte sur les autres jardins du Royaume," "Lettre de M. le Marechal d'Estrées [Gouverneur de Nantes] à M. Mellier Maire. A Paris, le 14 Avril 1727", in "Mémoire pour le jardin royal et botanique de Nantes," AM-Nantes, DD 48. #7, pp. 2-3. See also AM-Nantes, DD 49, #31, #36.

239. AM-Nantes, DD 50, #1, printed 2pp. sheet for Africa; #5 and #8 for America. Apparently the same order for ships captains to bring back plants and seeds was issued for Marseilles and ships sailing to the Levant; see "Mémoire pour le jardin royal et botanique de Nantes," AM-Nantes, DD 48. #7, pp. 2-3.

240. AM-Nantes, DD 49, #3, #5, #7 ff. contain receipts for plants from Martinique.

241. AM-Nantes, DD 49, #3, #5, #6, #7, #16, #20, #32, #36, #40, #41, #45, #47, mostly list plants arriving in Nantes, are all suggestive of these problems.

through 1749, the Marquis de La Galissionnière (1693-1756), was from the Nantes region. La Galissionnière's chateau was just ten miles outside of Nantes; he embellished his property with many Canadian and other colonial specimens, and there was a hearty exchange with the Nantes garden.[242]

In the early years Antoine de Jussieu provided a strong liaison between the royal garden at Nantes and the Jardin du Roi in Paris. Under Maurepas' auspices and impulse, Jussieu sent instructions and plants for the Nantes garden via the mayor of Nantes, and received specimens back for the Jardin du Roi from the same.[243] Dr. François Bonamy (1710-1786) superintended the Nantes garden from 1730 until his death fifty-six years later. Bonamy was professor of botany and doctor regent at the University of Nantes; personal physician to the Marquis de la Galissionnière; friend of Réaumur, Duhamel, and the Jussieux; correspondent of the Royal Society of Medicine in Paris; and author of a *Florae Nannentensis Prodromus* (1782). Naturally, Bonamy was in contact with Thouin at the Jardin du Roi, and vice versa.[244]

Assisted by a talented Capuchin apothecary, Bonamy witnessed the steady growth of the Nantes garden, and he left it a flourishing establishment at his death. He developed strong botanical contacts with Guadeloupe and the Saint Domingue colony; and at least to judge by their botanical names, the garden came to harbor plants from the Americas, Africa, India and China, although an effort to secure and import true vanilla seems to have come to naught.[245] Bonamy complained about the garden's shortcomings, notably the lack of a hothouse, but he likewise transformed the Nantes garden into a center of correspondence with provincial gardens in Caen, Rouen, Montpellier, Angers, and elsewhere throughout France.[246]

Rochefort. Finally in this connection, to complement the establishment in Brest, in 1666 Louis XIV created a new port for the Royal Navy at Rochefort on the Charente River. Rochefort came to possess a comparatively substantial naval hospital and school for naval medical personnel. A garden emerged, associated with the hospital, in the 1690s; but the first beds for a formal botanical garden were dug in 1738, and the Rochefort royal garden was officially inaugurated in 1741 with part of its mission to be an acclimatizing garden for American plants.[247] The garden was situated at the port along the river to

242. "Mémoire pour le jardin royal et botanique de Nantes," AM-Nantes, DD48. #7, pp. 4-5. See also AM-Nantes, DD 49, #6, undated, unsigned 24 pp. ms. labeled "Jardin des Plantes etrangeres." See also Kernéis et al., pp. 64-66.

243. AM-Nantes DD 49 #4 which is a five-page catalogue of 104 different plants. See also DD 56bis, #1 and #4, Jussieu letter to Mellier, "a Paris ce 22 Mars 1727."

244. See entries BCMNHN, Ms. 314; Ms. 1327; Roger Williams, p. 113.

245. AM-Nantes, DD 49, #50 and #4. On the vanilla episode, see DD 49, #11, #15, #27. See also Kernéis et al., p. 62.

246. See Bonamy's remarks to this effect, "Mémoire pour le jardin royal et botanique de Nantes," AM-Nantes, DD 48, #7, pp. 4-5; Auvigne, p. 10.

247. On the Rochefort garden, see Sardet (1993) and (2001); Roussel et al., pp. 39-41; see also Bitaubé; Dejussieu; Allain, pp. 137-38.

receive plants easily, and it was equipped with a crane especially to off-load plants from ships. Already by 1732, cases of seeds and plants began to arrive annually in the Rochefort garden: pineapples from Saint Domingue, Louisiana tulip trees, and plants from Guiana. Botanical specimens ultimately came in from North America, the Caribbean, Guiana, Senegal, the Mascarene Islands, India, and the Pacific. A census of 1794 counted 700 species in the garden, which was equipped with an orangerie and two hot houses, one for tropical plants and one for temperate plants.[248] In the 1770s and 1780s the king allocated the substantial sum of 1000 livres a year for the garden's upkeep.[249] Succeeding his father as chief royal physician at Rochefort, Gaspard Cochon Dupuis was named professor of the botanical garden at Rochefort, and he taught a public course in botany.[250] Until his death in 1788 he, too, was in contact with Thouin, and, as would be expected, botanical specimens flowed back and forth between Rochefort and the Jardin du Roi in Paris.[251]

[Medal: Port of Rochefort, URBE ET NAVALI FUNDATIS, 1666]

Although there was no official botanical garden there, the French port of Le Havre in Normandy on the English Channel needs to be noted in this regard. Transporting botanical specimens overland from the Atlantic ports to Paris proved difficult, and in the end it was least inconvenient for ships to put in at Le Havre and for specimens to be shipped to Paris up the Seine River. The botanical garden created in 1756 in the nearby town and staging point of Rouen, in Normandy, might also be included in this list of royally supported gardens with a colonial slant. Rouen was not a coastal town, but it did connect the port of Le Havre to Paris via the Seine. The *Académie des Sciences, Belles-*

248. BCMNHN, Ms. 1427: "Catalogues de jardins botaniques...Rochefort."

249. See documents related to garden in AN MAR C^7 70 (Gaspard Cochon Dupuis). At one point the ministry made Duhamel du Monceau inspector of the Rochefort garden; Allard (1970), p. 54.

250. See previous note and MAR C^7 70 (Jean Cochon Dupuis).

251. See entries BCMNHN, Ms. 314; Ms. 1327.

Lettres et Arts (1744) managed the Rouen garden through its superintendent (*Intendant du Jardin*) and the royal professor of botany, in addition to the gardener, Varin, who was in contact with Thouin in Paris. The Rouen garden, known as the *Jardin de l'Académie*, was endowed with a hothouse and functioned as an entrepôt for plants on their way to the Jardin du Roi.[252] In addition, royal gardens at Bordeaux, Marseilles (from 1733), Toulon (from 1680), Hyères, and elsewhere likewise became involved in transshipments of colonial products.[253] Thouin and the Jardin du Roi had links to other provincial gardens, and doubtless these, too, had their "colonial" dimension.[254] Just before the Revolution, a plan was underway for creating yet another transplantation garden in Bayonne near Bordeaux.[255] Plans to establish transplantation gardens and nurseries on Corsica never seem to have materialized, but these plans fit more with largely "overseas" gardens examined below.[256]

A complementary set of gardens did arise overseas in the colonies. In addition to a number of sometimes substantial gardens that grew up around private plantations, hospitals, and government residences in the colonies, official royal (or company) gardens made their appearance on Île de France (1735, 1748), Île Bourbon (1769), Guadeloupe (1707, revived 1775), Louisiana (1728), Port-au-Prince (1777), Cayenne (1778), New Jersey (1786), and South Carolina (1787). A subsequent section examines these colonial gardens and the functioning of the unsurpassed intercontinental system of French botanical gardens that arose with them. But enough has been said here to make it plain that metropolitan France was strongly endowed with an unrivaled set of botanical gardens, particularly along the Atlantic coast, that formed an element of the highest importance to the structure of the Colonial Machine and that provided an impressive infrastructure for promoting systematic exchanges of plants from around the world.

252. See references to the "Jardin académique" in BM/Rouen, Ms. B4/1, Registre no. 1 (1744-1763) and Ms. B4/2, Registre no. 2 (1764-An XII), e.g., p. 277, 1778 *Hortus Regius Academiae Rostomanensis*. See also references in Burckard and Boulier; BCMNHN, Ms. 1428, "Collection de catalogues de jardins botaniques... Rouen."

253. Juhé-Beaulation, pp. 273-74; Huard et al., pp. 115-20. On the Marseilles garden see also AN-MAR D^2 40, fols. 226, 524, 526; on the Toulon garden see AN-MAR D^2 40, fols. 118-236. Roussel et al., pp. 41-44, make clear that Toulon possessed two royal gardens; one dating from the seventeenth century which provided bulbs and flowers for the court, and the garden that fell under the aegis of the Ministry of the Navy; the latter arose in 1769 and was upgraded in 1786.

254. See list of these 'corresponding' gardens in section headed "Distribution de Graines...tant d'Après les Catalogues des Jardins des Correspondants, que d'après leur listes de demandes...," in BCMNHN, Ms. 1327: "Journal des Envois de Plantes, Arbres et Graines faits au Jardin du Roy par ses Correspondants, avec ceux faits par le Jardinier dudit Jardin du Roy aux Correspondants d'icelui."

255. BCMNHN, Ms. 357: Papiers de L. G. Le Monnier, "Dossier André Michaux."

256. See, for example, ALS, De Reine "a Mgr. Le Marechal prince de soubise," dated "a Versailles le 5 jan. 1773," BCMNHN, Ms. 357: Papiers de L. G. Le Monnier, Dossier de Reine.

The Observatoire Royal (1667)

[Medal: Observatoire Royal, SIC ITUR AD ASTRA, 1667]

Determining the precise location of the colonies and general technical problems concerning longitude, geodesy, cartography, navigation, and meteorology were of central concern to colonial authorities. In addition, contemporary astronomy placed an increasing emphasis on coordinated observations worldwide. The resulting astronomical component of the Colonial Machine was therefore substantial, and it revolved around the Observatoire Royal in Paris.[257]

A Home for Official Astronomy. Louis XIV founded the Observatoire in 1667 as a specialized component of the Academy of Sciences. The facility was complete by 1673. Jean-Dominique Cassini (I) (1625-1712) came from Italy to head the establishment. Administratively, the Observatoire fell variously under the aegis of the Maison du Roi and the Bâtiments du Roi, but J. D. Cassini and his son (Jacques Cassini [Cassini II], 1677-1756), nephew (Jacques-Philippe Maraldi, 1665-1729), grandson (Cassini de Thury, 1714-1784), great-grandson (Jean-Dominique [Cassini IV], 1748-1845), and great-nephew (Jean-Dominique Maraldi, 1709-1788) ran the Observatoire virtually as a family fiefdom through the eighteenth century. In addition to these leaders, a lesser staff of astronomers, astronomer-trainees, calculators, guards, custodians, and hangers-on saw to the operations of the Observatoire. Although it never became a direct extension of the Academy of Sciences, the Observatoire cannot be considered separately from the Academy either. The major astronomers at the Observatoire were all *pensionnaires* of the Academy, for example, and the Academy served as the publication outlet for the French astronomical estab-

257. The Observatoire, too, is another institution in need of a comprehensive modern study. In the meantime, see Gillispie (1980), pp. 99-130; Hahn (1964b), p. 654; Pelletier (2002), passim and pp., 44-45; McClellan (2003b), pp. 98-99; McClellan (2007), pp. 717, 730.

lishment.[258] The Observatoire was reformed and upgraded in 1785 with new instruments and the institution of a formal class of astronomer-trainees.[259]

The Cassinis made their own observations at the Observatoire, but the institution was intrinsically dependent on observations elsewhere, and it quickly became an international center for the collection and coordination of the observations of other astronomers and other astronomical stations.[260] The Observatoire was oriented first and foremost to other observatories and specialist astronomers elsewhere in Europe, but of necessity overseas connections came into play, including connections to the colonies.[261]

An example can suggest the importance of the overseas for the Observatoire and the importance of the Observatoire as the center for scientific work undertaken in the colonies. On the 20th of August, 1690, the exiled king of England, Charles II, paid a ceremonial visit to the Observatoire. His hosts, the academicians of the Academy of Sciences, pointed out to him the great map of the world chiseled into the floor.

> We showed his Majesty the places that had been determined by the direct observations of members of the Academy sent out on the orders of the King – those by Picard, de la Hire, Richer, Varin, du Glos and Deshayes in Denmark, on the coasts of France, in Cayenne, at Cape Verde, in the Antilles and those by Jesuit Fathers and royal mathematicians at the Cape of Good Hope, Siam [and China].[262]

The quotation suggests the extent to which Academy and the Observatoire had already cast their net worldwide by the 1690s. For the expeditions listed and the ones that followed, the Cassinis and the Observatoire institutionally provided detailed astronomical instructions.[263] In return, they received data and reports back from explorers on mission. In one notable case, the royal

258. Pelletier (2002), chapt. 3; McClellan (2001), pp. 26-29.

259. See OdP, D 5 38: "Observatoire de Paris. Sur l'établissement de 1784" and OdP, D 5 40: "Observatoire de Paris. Correspondance avec les ministres."

260. Bigourdan (1895) makes this point, p. F46n [sic]; see also OdP, D 5 40: "Observatoire de Paris. Correspondance avec les ministres."

261. See 1214-page document, OdP, B 4 1: "Observations astronomiques" and register of 1046 pages, OdP, B 4 4: "J. Cassini II. Mémoires sur différents points d'astronomie."

262. "On montra à Sa Majesté les endroits qu'on avoit établis par les observations immediates de Messieurs de l'Academie par l'ordre du Roy, par messieurs Picard, de la Hire, Richer, Varin, du Glos & des Hayes en Denemark sur les Costes de France en Cayenne, au Cap Vert, aux Antilles, et par les Peres Jesuites Mathematiciens du Roy au Cap de Bonne Esperance & à Siam...," PA-PV 13 (1690), fols. 20v-34 ("Le 20e d'Aoust 1690"), here fol. 25; the reference to China is interpolated from fol. 23. See also Jacob, pp. 131-32.

263. OdP, B 2.7 lists many of these sets of instructions, e.g., "Projet pour les observations geographiques dressé par Mr. Cassini pour les missionaires 29 juin 1681," "Instruction particuliere pour le voyage des canaries," "pro expeditione brasiliana," "Instruction pour les observations quil faudra faire dans l'isle de St. Thomé," and so on.

engineer Pierre Baron prepared for his posting to Louisiana in 1728 by training at the Observatoire, and he duly sent astronomical observations back to Cassini II in Paris.[264]

In addition, especially in the arena of determining longitude, the demands of contemporary astronomy were such that astronomers at the Observatoire necessarily had to coordinate with observers in the colonies in simultaneous observations of celestial phenomena. For example, in 1685 J.-D. Cassini and Philippe de la Hire in Paris determined the longitude of Quebec by observing a lunar eclipse in conjunction with Deshayes on the spot in Canada.[265] (Notably, at this early date of 1685, the astronomical arm of the Colonial Machine incorporated missionary observations of this same eclipse from Goa[266]). Valuable reports such as these did not go simply to the Observatoire, but filtered through the entire Colonial Machine. Father Feuillée, for example, sent his observations from America to the Duc d'Orleans (the regent) who transmitted them to the abbé Jean-Paul Bignon (honorary academician, royal librarian, and director of the *Journal des Sçavants*) who then passed them on to Cassini II, to the cartographer Guillaume Delisle, and to the Academy as a whole.[267] In Cayenne, the Jesuit de Crossat was the first to observe the comet of 1723; his observations filtered through local administrators to Father Gouye and Maraldi in Paris, and were ultimately printed in the *Mémoires* of the Paris Academy.[268] Other observers in the colonies sent in observational dribs and drabs to the central depot and calculating center of the Observatoire. The comet of 1742, observed by the Jesuit Boutin from Saint Domingue, might be mentioned in this regard.[269] As part of the failed effort to establish a colony at Kourou in South America, M. Béhague and his successor, M. de Fiedmond, were sent to Cayenne in 1762 to superintend astronomical observations.[270]

Other Observatories at Home and Abroad. Outside of Paris, the Jesuits staffed the Navy's first observatory in Marseilles.[271] Father Laval founded the naval observatory at the Jesuit college there in 1702, which was elevated to the full status of an *Observatoire Royal de la Marine* in 1749; the Jesuit Esprit Pézenas (1692-1776) ran the observatory and taught hydrography at Marseilles from

264. Langlois (2003), pp. 152-53.

265. See report of this observation by J.-N. Delisle and further coordinated observations with J.-B. de Chabert in 1750 to confirm Quebec's position in Delisle (1751).

266. Communicated by Thévenot, OdP, B 2.7, p. 195.

267. See report, OdP, B 5, 3 [9*], "Nombreuse collection d'observations...," p. 31n.

268. OdP, A 4 dossier 11 (comète de 1723), pièce B (journal des observations de la comète de 1723). Maraldi (1723), here pp. 375-76.

269. OdP, A 4 6, dossier 4 ("Relation Observations de la comète de 1742, tirée de différentes gazettes"), pièce B.

270. Ms. Code de Cayenne, vol. 4, 1761-1767, pp. 25, 155, 161, 363, 375, in AN-MAR G 237.

271. On these observatories, see Débarbat and Dumont (1990), p. 19; Stephan; Neuville (1882), pp. 53-54.

the 1729 through 1762. From 1763, the secular Guillaume de Saint-Jacques de Silvabelle (1722-1810) succeeded Pézenas as director of the naval observatory (*Directeur de l'Observatoire de la Marine*); Saint-Jacques de Silvabelle was not a *correspondant* of the science academy in Paris, but he published three papers in its *Savants Étrangers* series, and from 1781 he was assisted by Pons-Joseph Bernard (1748-1816), who ultimately *was* elected a correspondent of the Academy of Sciences attached to Méchain.[272] Formal control of the Marseilles observatory passed to the *Académie des Belles-Lettres, Sciences et Arts* of Marseilles in 1781. The Jesuits and the navy combined to establish another observatory at the naval school in Toulon in 1718. As a reward for his scientific travels through the colonies, in 1714 Louis XIV granted Father Louis Feuillée permission to maintain a second naval observatory at the Minim monastery at Marseilles.[273] This observatory operated until the mid-1730s and maintained relations with the Observatoire and the rest of the naval establishment.

The Académie Royale de Marine at Brest (1752)

[Jeton: Académie Royale de Marine, 1778]

The Académie de Marine, or Academy of Naval Sciences, at Brest was another notable cog in the Colonial Machine.[274] Founded in 1749, it began in

272. AN-MAR C7 293 (Silvabelle); on these developments, see also Neuville (1882), pp. 58-62.

273. Feuillée was aided in this endeavor by his nephew, Father Sigalloux, who was elected a *correspondant* of the Académie des Sciences in 1729. Hahn (1964b), p. 656; Froeschlé (1990), p. 29; Lacroix (1938), vol. 3, p. 17; Bigourdan (1923), pp. 8-17. On Feuillée, see further below, this part, "Overseas Missions and Missionaries."

274. Vergé-Franceschi (1996) provides a solid overview. The special issue of *Neptunia* [226, #2 (2002)], devoted to the 250th anniversary of the founding of the Académie de Marine, is likewise a valuable source, as is the anniversary volume issued by today's Académie de Marine, but the Académie Royale de Marine at Brest still needs a full and updated historical treatment. The series of articles from the 1870s by Doneaud du Plan (DDP) remains the starting point. See also, Levot; Konvitz, pp. 75-77; McClellan (1992), pp. 124-25; McClellan (1985), pp. 99, 265; Balcou, esp. 125-207; Henwood (1987), Broc (1975), pp. 280-81 underscores the importance of the Académie de Marine. For manuscript sources, see SHMV, "Fonds de l'Académie de Marine," Ms. 64-110; AN-MAR G 93 "Pièces diverses sur l'Académie royale de marine" and BnF, Mss. N.A.F. 9412 and 9490.

classic fashion as a nominally private society by a group of ranking naval officers posted to Brest. Duhamel de Monceau was behind the idea, as was the minister at the time, Rouillé. The Brest Academy received a formal charter from the crown in 1752.[275] This *Règlement* of 1752 dictated that the Academy was to concern itself particularly with the different parts of mathematics and *physique* pertaining to navigation. But at the same time, "Although those parts of mathematics having a direct connection to nautical matters are the principal subject of the Academy's work, academicians are nonetheless invited to extend their researches to everything that might be useful or interesting in other parts of mathematics and physics, as well as the arts and natural history."[276]

An Academy of the Seas and Overseas. Administratively, the Academy was a dependency of the Ministry of the Navy and the Colonies. Organizationally, its structure mirrored that of any other academy, but with honorary, ordinary, and associate members all chosen from navy personnel; free associates and correspondents were not necessarily navy men. Sébastien-François Bigot, Viscount de Morogues (1705-1781), who was ultimately a navy lieutenant general and inspector-general of naval artillery paid 12,000 livres per annum, served as the Director of the Academy until his death.[277] About seventy-five individuals constituted the membership at any one time with 196 members over the lifetime of the institution.[278] The first ordinary academicians included two commissioners-general of the navy, eight captains, three naval commissioners, nine naval lieutenants, one ensign, three royal shipbuilders, and four naval engineers.

The Marine Academy was a weak institution in its first years. Initially, there were jokes and resistance from within the navy, thinking the Academy to be a remedial school for officers.[279] The Marine Academy pushed for formal Letters Patent early on, but nothing came of the initiative because of the Seven Years War, which took a great toll on the vitality of the institution.[280] After the war, however, it became an invigorated organization, graced with letters patent in 1769 as the Académie Royale de Marine. The membership and structure remained essentially the same.[281] Its funding, cut back to 3000 livres in the

275. On materials relative to the founding of the Académie de Marine in 1752, see SHMB, Série A: Sous-série 1A-4 Correspondance entre la Cour et le Commandant de la Marine à Brest.

276. Règlement of 1752, article 25, quoted in DDP I:9; see also Académie Royale de Marine (1773), front matter. For Academy charters and other legal privileges, see SHMV, Ms. 081 ("Registre des séances de l'Académie royale de Marine, 1752-1765") and Ms. 099 ("Règlements pour l'établissement d'une Académie de marine au port de Brest").

277. Taillemite (2002), pp. 49-50; Levot; AN-MAR C^7 30 (Bigot de Morogues); PA-DB (Morogues).

278. Académie de Marine, pp. 39-49, lists the members through 1793.

279. SHMV, Ms. 87, letter "a M. Pellerin" dated "1er Mars 1754."

280. On these initiatives see SHMV, Ms. 87, Letter to "Monseigneur" dated "Le 1er Mars 1754," letter "a M. Pellerin" dated "1er Mars 1754."

281. Compare 1769 Règlement in Académie Royale de Marine (1773), frontmatter, article 2.

Seven Years War, was handsomely restored to 6000 livres a year.[282] The Marine Academy was put under direct protection of the secretary of state for the navy.[283] A new set of Règlements followed in 1774 that added a class of six foreign associates drawn from others of the world's navies.[284]

The Marine Academy strengthened and matured in the second half of the eighteenth century, and it evidences the coming-into-being of secondary royal institutions rounding out the full structure of the Colonial Machine. Despite interruptions occasioned by the Seven Years War and then the American War, the Académie de Marine pursued its own substantial program of work and research. The Academy met weekly for two hours. The range of its concerns included naval architecture and construction; naval medicine and the health of crews; artillery, navigation, and the sciences applied to navigation (physics, mathematics, hydraulics, nautical astronomy).[285] The Brest Academy became a center of research on marine chronometers, potable drinking water, sea rations, lightning rods, ships' ventilation, sail design, dry dock construction, and more. The work of the Marine Academy circulated to the navy ports of Rochefort and Toulon and also throughout the rest of the Colonial Machine.[286] Although it functioned as a typical Old Regime academy, the Marine Academy did not hold public meetings or sponsor prize contests. Using funds from the ministry, the Academy had jetons struck in 1769 and 1778. Two portraits of Louis XVI hung on the walls of its assembly room in Brest.[287]

The names of the most prominent ordinary, honorary and free-associate members are or will become familiar to readers: Duhamel du Monceau (inspector general of the navy), Poissonnier (inspector and director of naval and colonial medicine), de La Galissonnière (commodore and former governor-general of Canada), Bory (commodore and former governor of the Lesser Antilles), Bellin (*Ingénieur de la Marine, au Dépôt des Cartes et Plans, à la Cour*), d'Après de Mannevillette (ship's captain and cartographer of the Compagnie des Indes), the count de Fleurieu (naval captain and *Directeur des Ports et Arsenaux* with a compensation of at 27,000 livres), and the Jesuit Pézenas of the Marseilles naval observatory.[288]

Brest and Paris. The connection between the Brest Academy and the Academy of Sciences in Paris deserves to be explored. The links between the two

282. On these points see Sartine letter to the count d'Orvilliers dated "Versailles le 22 X.bre 1777" in SHMV, Ms. 93. Académie Royale de Marine (1773), p. iv, puts funding at 4000 livres.

283. See new Règlement of 1769 reprinted in Académie de Marine (1773), pp. xi-xx; DDP II:3-10.

284. "Copie d'une lettre du Ministre a l'Académie Royale de Marine à Versailles le 13 mars 1774," AN MAR G93, fol. 58; see related letter, fol. 59.

285. Vergé-Franceschi (1996), p. 13; DDP I:19.

286. SHMV, Ms. 87, Letter to "Monseigneur" dated "Le 1er Mars 1754."

287. DDP VI:64.

288. DDP I:12; Académie Royale de Marine (1773), pp. i-x; Académie de Marine, pp. 39-49.

institutions were reinforced through members in common, of course. In addition to the above-mentioned correspondents, Bézout, Borda, Bouguer, Chabert, Lalande, the astronomer Le Monnier, J.-B. Le Roy, Pingré, and R.-B. Sabatier were academicians from Paris who were also elected to the Brest Academy. (Pierre Bouguer, *pensionnaire* in astronomy at the Paris Academy, was obviously flattered by his election as a correspondent of the Brest Academy[289]). In turn, Bigot de Morogues, Étienne Chardon de Courcelles (1705-1775, director of the medical school at Brest), and Antoine-Jean-Marie, count Thévenard (1733-1815, vice admiral), became correspondents of the Paris institution, the first named publishing on the ventilation of ships in its *Savants Étrangers* series. The abbé Rochon started out as the librarian and keeper of maps of the Marine Academy and was elected a correspondent of the Paris Academy in 1767; but beginning in 1771 Rochon became an associate in mechanics in his own right in Paris, rising to the rank of *pensionnaire* in 1783.[290]

But more than that, in 1771, coincident with its elevation to the status of a *royal* academy, a formal "affiliation" united the Brest Academy with the Academy of Sciences in Paris.[291] The move was put forward by the minister of the navy and approved by the Academy of Sciences. As the internal report in Paris noted, "Whatever advantages the correspondence between a few members of the two Companies has produced to date, there will be even greater ones to be hoped for if this correspondence could be between the two Companies themselves."[292] Once some minor matters of precedence were ironed out, the affiliation became official with the secretaries of the two institutions charged to carry it out.

This affiliation was not between equals, however. The Paris Academy was the dominant organization, the Brest Academy the subservient one. In many senses the Brest Academy was an extension of the senior institution in Paris, devoted specifically to matters naval. An early historian of the Brest Academy, Charliat, called it the "laboratory" for the Paris institution.[293] For example, at one point the Paris Academy proposed electrical experiments to be carried out

289. "Je suis infiniment flaté de me vois attaché à l'Academie...," ALS, Pierre Bouguer, dated "à Paris le 19 Sept.bre 1752," PA-Fonds Bertrand, Carton 5.

290. On Rochon, see Taillemite (2002), p. 458; Collet (1996); Guy Jacob (1990), esp. pp. 44-48; Lacroix (1938), vol. 4, pp. 15-24; Fauque; and materials in PA-DB (Rochon).

291. DDP II:59-65; McClellan (1985), pp. 99, 184, 340n; Vergé-Franceschi (1996), p. 13; Académie de Marine, pp. 15-17; AN-MAR G93, fols. 49-51v.

292. "Que quelques avantages qu'ait déjà produits une correspondance entre quelques membres des deux Compagnies, il y en aurait encore de plus grands à espérer si cette correspondance pouvait avoit lieu entre les deux Compagnies mêmes," "Extrait des Registres de l'Académie Royale des Sciences du 13 fevrier 1771," AN-MAR G 93. For other materials related to this affiliation, see DDP II:59-65; SHMV Ms 71, fols. 206-209; PA-PS (13 February 1771); PA-PV 90 (1771), fols. 52-54 ("Mercredy 13. Fevrier 1771"); PA-PV-110, Marine, Académie de, 13 February 1771 ("Sur un projet d'affiliation proposée entre L'académie de marine et des Sciences"), marked "Agréé par l'académie/ Maillebois, Trudaine, Duhamel, De Thury, De Bory & Bezout."

293. Charliat, p. 76.

by the junior institution in Brest.[294] In 1777 the minister wrote to the naval commander at Brest that the Academy of Sciences had persuaded the king of the necessity of systematic observations of the tides. On the king's orders, the minister sent the Paris Academy's published instructions on how to make these observations; he requested that they be passed on to the Brest Academy, and the Brest Academy duly chose observers to report to Paris.[295] In another instance, in 1775, the Brest Academy itself petitioned the mother academy in Paris regarding making available papers from a prize contest on the magnet.[296] In another odd case in 1772, the two institutions functioned as one. The naval lieutenant de Charnières submitted his work on determining longitudes at sea to the Academy of Sciences in Paris, but as a member of the Marine Academy he claimed the right to be evaluated by that institution as well. In his Solomon-like decision, the minister, de Boynes, ordered that the Brest Academy appoint Bory and Chabert as its commissioners, the same men named by the Paris Academy, the same report serving both institutions.[297]

Expertise and Publications. Like the Paris Academy, the Brest institution was called on by the government for technical advice, as, for example, when it gave its opinion on a new compass or when the minister approached it regarding a new sort of mortar.[298] At one point it was asked to review one Pereire's secret of moving ships in calm water.[299] At another, the ministry approached the Academy about port construction, "The Academy will do something agreeable to the King in dealing with this important matter and in giving its advice on something pertaining to theory and to practice."[300] In 1782 a certain Romme, professor of mathematics at the officer training school at Rochefort, submitted treatises on sails and masts to the navy board at Rochefort (the *Conseil de la Marine de Rochefort*) which approved them. Nevertheless, the minister of the navy wrote to the Academy in Brest requesting that commissioners be appointed to review Romme's work.[301] They approved,

294. SHMV, Ms. 105, vol. I.

295. DDP IV:72 and Sartine letter to "M. Le C.te d'Orvilliers, Commandant la Marine au Port de Brest, en date du 25 9.bre 1777" in SHMV, Ms. 93.

296. Copy of Academy letter dated "28 Avril 1775," "À Messieurs de l'Académie Royale des Sciences" in SHMV Ms. 092. De Fouchy replied that the contest had been put off. On related connections between the Brest and Paris academies, see DDP V:90-91. Chapuis, pp. 154-58.

297. See de Boynes letter to "M. de Rosnevet L.t de V.au Sécretaire de l'Acad. de Marine à Brest," dated "À Versailles le 14 mars 1772," SHMV, Ms. 88.

298. DDP IV:20 The Academy called for tests it would perform; see also DDP III:76; SHMB, Série A, Sous-série 3A, 3A 92, #64, s.d.

299. DDP V:16.

300. "l'Académie fera une chose agreable au Roy en s'occupant de cette matiere importante et en donnant son avis sur un objet qui tient à lá theorie et à lá pratique," ministerial letter dated "A Versailles le 18 Janvier 1771," SHMV, Ms. 88.

301. DDP V:86-87. For related efforts regarding sails, see SHMV, Ms. 93, letter to minister read 13 January 1774 and letter dated "à Versailles le 31 janvier 1774."

but it is revealing that the Marine Academy in Brest trumped navy authorities in Rochefort as the body for overarching technical judgments.

The crown granted the Academy its own imprimatur, and perhaps the major institutional thrust of the Académie de Marine was its monumental *Dictionnaire de Marine*, a compendium of matters nautical in the encyclopedic spirit of the age. For decades the Academy worked on its dictionary project and assigned topics to various of its members and correspondents, who likewise responded variously. This activity mobilized a nontrivial segment of the Colonial Machine over several decades. Many traces remain in the archives, including reports that the king himself was paying for the engravings and the printing of the Academy's *Dictionnaire* on royal presses in Paris or Versailles.[302] At another point, the minister offered the services of the engravers of the Dépôt des Cartes and permission to print the work at Brest.[303] Turgot granted postal cover for all communications related to the project.[304] Lalande volunteered his services for all articles having to do with astronomy.[305] But the great work never appeared directly under the imprint of the Academy.[306]

In 1773 its first and only volume of *Mémoires* (600 pages, 10 plates) did appear, a collection of 20 comparatively theoretical papers devoted mostly to mathematics, astronomy, and physics.[307] The explicitly programmatic statement heading the *Mémoires* shows the concern of these academicians to unite theory and practice:

> [The Academy de Marine is] principally composed of members who intrinsically join theory and practice....We have never better understood to what extent theory is necessary for practice in navigation, and never have we given ourselves over with more ardor and greater success to applying one to the other....Although the Academy's works are intended to be useful to navigators, they are not limited to things of immediate utility. The Academy will more surely attain the goal it has proposed for itself by directing its research towards other parts of these sciences that are sometimes seemingly foreign to nautical matters, but

302. DDP III:95; Levot, pp. 71-72.

303. SHMV, Ms. 88, Boynes lettter dated "Versailles le 11.e Avril 1774."

304. SHMV, Ms. 88, Turgot letter dated "Compiégne le 11.e aoust 1774."

305. SHMV, Ms. 89, Lalande letter dated "À Paris le 13 Janvier 1771."

306. On the dictionary project, see Tromparent; DDP I:22-23; IV:64; Académie de Marine, p. 18; SHMV, Ms. 66, fols. 70-78, and Mss. 103-104 ("Vocabulaire des mots pour le Dictionnaire de marine"); AN-MAR G 101, folder 5, "Projets de Dictionnaire des termes de marine." In a letter dated "À Paris le 13 Janvier 1771" (SHMV, Ms. 89), the astronomer Lalande compared the project to the *Dictionnaire* of the Académie Française. All was not lost, however, because much of this material did appear in the *Encyclopédie méthodique, Marine*, published in four volumes between 1783 and 1787, of which Blondeau of the Marine Academy was one of two editors. Also, Tucoo-Chala (1987).

307. 1000 copies were printed at a cost of 5,295 livres; DDP II:56-57.

> the advancement of which has a necessary connection to advancing the naval arts. New knowledge in physical astronomy of the motion of the moon can lead to a complete solution of the problem of longitude. The discovery of the true principles of hydrodynamics will shed new light on loading ships and on their construction. In a word, there is almost no branch of the physical-mathematical sciences that cannot be advantageously applied to nautical matters. Even those that cannot directly serve to advance the art of navigation should not be excluded from the work of the Academy. All parts of mathematics reinforce one another. One cannot neglect one without harming the progress of those one cultivates.[308]

The Academy placed itself in the tradition of literary societies dotting France, yet it sharply differentiated itself from them:

> But of all these literary societies, there is perhaps none whose works can be more useful than those of an academy of naval sciences. The subjects encompassed by navigation are so important, so extensive, and connected in so many ways to the mechanical arts, to *physique*, and to the exact sciences. They impact the power of nations and the happiness of humanity so strongly that it is astonishing that the advantages associated with the creation of a Company equipped to cultivate knowledge in this domain has not been felt sooner.[309]

This same preface emphasizes the importance of communicating the Academy's research; tellingly, it tips its hat to the talent of young officers of the nobility, "the Navy's ornament and hope."[310] The question of theory and practice in the Academy might be framed in another, more sober way, too, that of the minister who in 1772 said of the Académie de Marine that "its primary objective is to produce things useful for the Navy."[311]

With ministry backing, one of the Academy's most active members, the professor of mathematics and hydrography at the naval college at Brest, Etienne-

308. Académie de Marine (1773), pp. v, vii, ix-x. Already in 1752 at the inaugural meeting of the Academy Bigot de Morogues emphasized the importance of uniting theory and practice: "...théorie sans l'expérience ne navigue et n'opère sans danger que dans le cabinet, mais que l'expérience sans la théorie est longue, incertaine, dispendieuse, que ce n'est le plus souvent qu'un tâtonnement aveugle qui retarde le progrès...," quoted in DDP I:16.

309. Académie Royale de Marine (1773), p. ii.

310. "Nous sommes en partie redevables de ce nouveau degré de lumiere aux soins qu'on a donné à l'institution d'une jeune noblesse, l'ornement & l'espoir de la Marine...," Académie de Marinc (1773), p. v. On the necessity of communicating its work, see p. viii.

311. "...que son premier objet est de produire des choses utiles au Service de la Marine," de Boynes letter dated "Versailles le 26 Avril 1772," SHMV, Ms. 88.

Nicolas Blondeau (†1783), published fifteen numbers of the *Journal de Marine, ou Bibliothèque Raisonnée de la Science du Navigateur* from 1778 through 1783; in a sense this was also an Academy publication, not least because the Academy itself censored the Journal.[312] Blondeau's *Journal de Marine* was distributed widely throughout the Colonial Machine, but according to Lalande, there were not enough competent officers in the navy to sustain the publication.[313] The Marine Academy published a translation of the English *Nautical Almanack* in 1772 and 1773, as well as lesser, practical brochures concerning compasses and rescuing the drowning.[314]

The Academy provided the imprimatur for several works, including, among others, Fleurieu's *Voyage par ordre du Roi en 1768-1769 sur l'Isis pour éprouver en mer les horloges marines de Berthoud* (1773), Bouguer's *De la manœuvre des vaisseaux* (1757), and Pingré's *L'État du ciel à l'usage de la Marine* (1753), the latter dedicated to the Academy.[315] The Academy also superintended the production and publication of the second edition of d'Après de la Mannevillette's *Neptune Oriental* (1775), the major cartographical work concerning the Indian Ocean in the eighteenth century.[316]

The creation of a specialist library that naval officers and ultimately the public could consult formed a part of the institutional mission of the Académie de Marine for which it received extra funding in its early years.[317] The head of the *Bibliothèque du Roi* in Paris and free associate of the Brest Academy, one Lefebvre, regularly sent cases of books to help build this library.[318] The

312. On this and related publications discussed here, see Junges (2005) and (2002); DDP IV:60; V:9; AN-MAR C^7 32 (Blondeau). See also ministerial correspondence related to the *Journal de Marine* in letters dated, "Versailles le 24 Avril 1776," "25 7bre 1776," and "À Versailles le 30 9bre 1776," in SHMV, Ms. 93; Neuville (1882), pp. 45-48 adds valuable details.

313. For distribution to the Société Royale de Médecine, see SRdM, Ms. 7, p. 195r ("Séance du vendredi 11 Juin 1779"). On the lack of support, see DDP V:6: "Malheureusement, ajoute Lalande, la marine ne fournissait pas alors assez de personnes instruites et curieuses pour que le journal pût se soutenir."

314. The initiative to translate the English *Nautical Almanack* may well have come from d'Après de Mannevillette; see d'Après de Mannevillette letter dated "À Kx.gars le 30.e Avril 1771," SHMV, Ms. 89. Lalande sheds light on this episode in a letter dated "À Paris le 12 Juin 1771," SHMV, Ms. 89. The minister was not pleased with the fact that Paris longitude was not used; see de Boynes letter dated "Versailles le 26 Avril 1772," SHMV, Ms. 88.

315. DDP: I:17; III:69; see also letter "a M. Pingré Ch.ne Regulier à S.te Genevieve" dated "Du 24 aoust 1753," SHMV, Ms. 87. For more on the Acacemy's imprimatur and other, limited powers of the institution, see McClellan (2003c), p. 14.

316. DDP III:74 and copy of Academy letter to d'Après de Mannevillette, dated "5 7.bre 1775," in SHMV, Ms. 92. The original edition dates from 1745, paid for by the Compagnie des Indes.

317. SHMV, Ms. 108. On the Library being open to the public, see de Boynes letter to Goimpy, dated "Versailles le 8 7bre 1771" in SHMV, Ms. 88 and Castries letter dated "Versailles le 28 7.bre 1782" in SHMV, Ms. 93. De Boynes approved that the Academy library be opened on Mondays, Wednesdays, and Saturdays of each week to military officers, members of the administration, and engineers, physicians, and surgeons of the navy.

318. DDP I:21 and Letter "a M.r Le febvre" dated "Du 20 avril 1753," SHMV, Ms. 87. For plans to create naval libraries in Rochefort and Toulon similar to the Academy's library in Brest, see reference in ms. "Acquisition du cabinet de Mrs. de L'isle pour le depôt des Cartes, plans et Journeaux de la Marine"; AN-MAR C7 186 (Lisle).

Academy issued catalogues of its collections in 1781 (80 pages) and again in 1788 (216 pages). The latter listed 1888 works, 100 by its members, mostly concerning science, technology, and accounts of voyages.[319] The library's collections came to include globes and astronomical instruments; these the Academy solicited early on from the minister with Duhamel du Monceau and de la Galissonnière as its agents.[320] The abbé Rochon held the post of librarian and guard of astronomical instruments (*Garde de cette Bibliothèque et des Instruments astronomiques*) with appointments of 1200 livres; the library was otherwise attended by the Academy's amanuensis, a certain *Sieur* Vincent.[321] The ministry annually sent upwards of 100 copies of the *Connaissance des Temps* for the library and for distribution to Brest academicians and qualified mariners capable of making scientific observations.[322] Soon, however, over 200 copies were not enough for all the ships' captains, colonial agents, and Academy members who could use the *Connaissance des Temps*.[323]

Fabrication and Distribution of Scientific Instruments. As an unexpected dimension of its activities, the Marine Academy came to have responsibility for manufacture of nautical instruments for the Royal Navy. The navy operated a workshop in Brest (*La Cadrannerie*) that made compasses and related instruments for its vessels. In what seems to have been a power grab, in 1774 the Academy complained about the quality of the compasses produced in this workshop, and the minister charged the Académie de Marine with supervising the department, and so emerged the full-fledged *Atelier des Boussoles* producing high-quality compasses and other nautical instruments for the Marine Royale. The Academy attended to this duty assiduously for the next twenty years.[324] In 1777 the Academy appointed its secretary, Marguerie, as director of the Atelier des Boussoles, but he died the following year in combat off Grenada.[325] Until his own death in 1783, the above-mentioned Blondeau took over and supervised the Atelier's production.[326] In February of 1782, for

319. Académie Royale de Marine (1781); DDP V:56; VI:29.

320. See letters [from Bigot de Morogues] to "M.r Duhamel" dated "du 3 May [1754]" and to "Mgr." dated "Du 30 Janvier 1753," SHMV, Ms. 87. See also Niderlinder.

321. See de Boynes letter dated "fontainebleau le 24 octobre 1773," SHMV, Ms. 88.

322. See de Sartines letter, dated "Versailles le 23 Décembre 1774," to M. Le C.te de Breugnon (of Brest Academy), SHMV, Ms. 88. See also Sartine letter dated "Versailles le 3 8bre 1775," SHMV, Ms. 93; Chassagne, pp. 154-56.

323. Letter dated "9e 7bre 1775" from Le Begue to Sartine, SHMV, Ms. 93.

324. DDP III:73-74 and annual reports following; SHMV, Mss. 101-102, "Ouvrages exécutés à l'atelier des boussoles (1778-1785)," and compass-related materials, AN-MAR G 99. See also correspondence relating to atelier, SHMV, Ms. 88, e.g., de Boynes letter dated "Versailles le 21 Janvier 1774." For evidence of the Académie superintending the Atelier, see, e.g., undated copy of Academy letter "À M. Sage, de l'Académie Royale des Sciences...à Paris" and Sage reply, "Paris ce 30 Juillet 1776" in SHMV, Ms. 092 "Académie R. de Marine. Lettres et Réponses, par l'académie."

325. DDP IV:67; Levot, p. 77.

326. AN-MAR C^7 32 (Blondeau).

example, the workshop employed seven artisans at various levels of skills producing and repairing a variety of compasses, compass needles, jewel settings, sand clocks, and ship's barometers.[327] The navy supplied instructions for the use and maintenance of this equipment.[328] The Academy and the Atelier became quite involved in compass technology, displaying special concern for nonferrous metals and solders used in the constructions of pivots and housings.[329] The quality of instruments improved under Academy control, and it did valuable work in this area by publicizing the discovery that two compasses on the same deck – a redundancy favored by the navy until then – could interfere with one another. This discovery was publicized in a printed notice from the ministry distributed through the bureaucracy; the Bordeaux port director further distributed it to and through the chambers of commerce of Bordeaux and Bayonne; administrators in Corsica had it translated into Italian.[330]

A word needs to be added about barometers produced in the Atelier. The thought was that changes in barometric pressure indicated changes in the weather. Lavoisier provided Blondeau with a barometer for trials, and Blondeau wrote in the *Journal de Marine* advising officers on the use of barometers to predict upcoming storms at sea.[331] But barometers were ordinarily too fragile to carry aboard ship, notably because of the effects of cannon fire, and so Blondeau perfected one using an iron tube.[332] Blondeau's design was strongly endorsed by the Marine Academy and tested by ministerial order on a voyage to Saint Domingue.[333] The Atelier de Boussoles subsequently manufactured

327. AN-MAR D^2 25, fol. 441 (and fols. 443-46 for other months). The total salaries paid for the month came to 354 livres. See other materials related to the operations of the Atelier de Boussoles, SHMV, Ms. 93, e.g., Academy letter to minister dated "6 X.bre 1776" and response to D'Orvilliers ("Lieutenant Général des Armées Navales Commandant de la Marine du Port de Brest") dated "23 X.bre 1776" approving the Academy's continuing superintendence of Atelier des Boussoles.

328. SHMB, Série A, Sous-série 3A, 3A 93, #83, 5 December 1784: "Instruction pour les Pilotes du Roi concernant la conservation et l'usage des boussoles et des barreaux aimantés."

329. See revealing details in SHMV, Ms. 88, Sartine letter dated "Versailles le 31 Janvier 1775." See also earlier exchanges between Chabert and d'Après de Mannevillette in letters dated "À Paris le 20.e Janvier 1771" and "a hennebon le 12.e fevrier 1771," both SHMV, Ms. 89. The dispute concerned the magnetic properties of brass ("cuivre jaune") versus copper ("cuivre rouge"). The Academy, the Atelier de Boussoles, and the instrument maker for the Compagnie des Indes rejected using brass because it deflected the magnetic needle. Chabert admitted that brass had 'ferringiferous' particles, but questioned whether this makes any difference; compass housings being circular and the metal ordinarily near compasses aboard ship (e.g., cannons) having greater effects.

330. See SHMV, Ms. 89, La Porte letter dated "À Bordeaux le 11 Juin 1771." See also d'Après de Mannevillette letter in same collection, dated "À Kx.gars le 6 Juillet 1771," and letter from Reigner du Tillet dated "À Bastia en corse le 4 aoust 1771."

331. Giroux, p. 154; Junges (2002), p. 15.

332. AN-MAR, C^7 32 (Blondeau); AN-MAR G 99 Dossier 1: "Baromètre de Blondeau professeur d'hydrographie à Brest." See also SHMV, Ms. 93, letter from Le Ch.r de la Coudraye to the minister dated "Brest le 24 X.bre 1779."

333. AN-MAR G 99 Dossier 1, fols. 12-14 ("Rapport sur les Barometres marins de M. Blondeau").

Blondeau's barometers, using tubes supplied by the royal manufactory at Tulle, and the Atelier was enlisted to provide every royal navy ship with a Blondeau barometer.[334]

In the 1780s the Marine Academy became involved with installing lightning rods on ships.[335] The idea harks back to the 1750s and Benjamin Franklin's *Observations and Experiments on Electricity*, but this particular episode seems to have bubbled up from a suggestion offered by a naval officer to the *Société Royale des Sciences* in Montpellier concerning arming ships with lightning rods. The minister directed the Marine Academy to investigate, but the Academy was uncertain over possible consequences and looked to the Académie des Sciences in Paris for guidance. In the meantime, on orders from the minister, Le Roy from the Paris Academy went to Brest to install lightning rods on naval installations; the Brest Academy was to consult on how best to install rods on ships. Le Roy attended a special meeting of the Marine Academy and read a paper on the necessity of arming ships with lightning rods. The Marine Academy appointed commissioners, and the minister pressed for a report. When it finally came, the Brest commissioners were hardly as convinced as Le Roy of the utility and safety of lightning rods, and they called for further tests, in this case saying that theory and practice were two different things![336] But it did not matter because, as mentioned, assisted by a certain Billiaud employed by the Paris Academy, Le Roy went on to outfit ships and installations in Brest and then moved on to Lorient and Rochefort to do the same thing.[337]

Particularly as it applied to navigation, astronomy was a science of special interest to the Academy. The Academy acted as a center for circulating astronomical information, and in a few cases it served to coordinate simultaneous astronomical observations. In this connection it interacted with the Cassinis, Delisle, and the astronomy community in Paris.[338] In a draft set of new letters-patent in 1775, the Academy proposed to build and run a new observatory for the navy at Brest.[339] The Academy pressed for the observatory, and the minister agreed in principle.[340] In 1783 the Academy purchased land as a site for

334. DDP IV:29-30, 70, and letter to "Monseigneur," dated and signed "Brest le 25 fevrier 1782, du Val le Roy, sous-secretaire de l'academie Royale de Marine," AN-MAR G 93, fols. 94-95.

335. See DDP V:45-47, VI:6-9, 21-22, and materials in AN-MAR G 100. See also ministerial/ Academy correspondence related to lightning rods: SHMV, Ms. 93 (e.g., de Castries letters dated "Versailles Le 11 Avril 1781" and "Versailles Le 7 7.bre 1784"). See also, SHMB, Série A: Sous-série 1A: Corespondance for 1783-1785, p. 312, Le 17 9bre. 1784.

336. SHMB, Série A, Sous-série 3A, 3A 94, 1786, #58: "Rapport de la Commission Nommée Par Le Conseil de Marine pour constater La dépense qu'a pu occasionner La Construction d'un Paratonnere…," dated "A Brest le 13 Mai 1786."

337. See DDP VI:21; Le Roy.

338. DDP VI:66.

339. AN-MAR G 93, "Pièces diverses sur l'Académie royale de marine," fols. 62-73: Ms. "Projet d'un nouveau Règlement pour l'Académie Royale de Marine avec des observations en marges," Article 67. Dated and signed, "A Brest le 22 7bre 1775, De Marguery Secretaire de l'Académie R. de Marine." Also, DDP IV:30-42.

340. DDP IV:70, and Academy letter to the minister dated "3 Mai 1784" and reply from de Castries dated "A Versailles le 15 aoust 1784," SHMV, Ms. 93.

the observatory, but unfortunately, astronomy seemed a bit beyond the organizational and technical reach of the institution.[341] At one point, for example, the Academy had to admit to Cassini de Thury that it lacked competent observers, and the efforts to establish an observatory came to naught.[342] A naval observatory at Brest arose only in Year V of the Republic (1797).[343]

In moves that weakened the Academy as a center for astronomy, the minister siphoned moneys from its budget to pay for astronomical instruments for use by other observers, notably Father Pézenas in Marseilles.[344] In 1757 Academy funds went for an 1800-livre quadrant that Duhamel somehow appropriated.[345] The Academy kept an inventory of its instruments, but the lending of instruments and authority over them were constant problems.[346] The Academy consented to lending instruments to accompany the abbé Rochon on an expedition to the Indian Ocean in 1767-1770, but it cautioned the Minister that the Academy was in no position to provide instruments for the many ordinary sailings of naval vessels and "could do so only for those who aim to make astronomical observations."[347] Indicative of the power relations involved, the minister warned the Academy away from thinking that it could decide on how to assign instruments belonging to it.[348] Rochon's instruments were eventually returned to the Brest Academy, but only after the minister intervened and prevented the intendant on Île de France, Pierre Poivre, from retaining them there.[349] A related squabble arose over the papers and observations of the astronomer Véron, who died on Île de France while voyaging with Bougain-

341. DDP V: 68.

342. DDP III:73: "Le 26 août, Blondeau lut une lettre de Cassini de Thury, invitant l'Académie à faire des observations correspondantes à celles qu'il faisait lui-même à l'occasion des réfractions. Nous ne voyons pas qu'il ait été donné suite à cette lettre. See related materials," AN-MAR C^7 32 (Blondeau).

343. DDP V:68; AN-MAR D^2 25, fol. 437 for report on the prospective site of the observatory.

344. Pézenas himself commissioned a telescope made in London for 7,300 livres paid on the accounts of the Académie de Marine. See litany of complaints from the Academy: "À Mg.r de Berryer Ministre" dated "Du 12 Xbre 1760," "A Mg.r Le Duc de Choiseul" dated "Du 7 Juin 1762," and "à Mg.r Le Duc de Choiseul" dated "du 6 aoust 1764," in SHMV, Ms. 87. See duplicates and related material, AN-MAR G 235 Divers, esp. #41.

345. DDP I:65.

346. SHMV, Ms 108, # 10: "Mémoire des instruments et autres effets appartenant à l'Académie." See also SHMV, Ms. 87, letters to "Duhamel de Monceau" dated "5 Mars 1756" and to "M.r le Ch.er de la Motte Baracé" dated "23e Janvier 1756," with reference to academicians privately ordering instruments without the Academy's approval.

347. Paraphrased by DDP IV:46-47.

348. See Sartine letters dated "Versailles le 12 Décembre 1775" and "31 Mars 1776" and related materials in SHMV, Ms. 93.

349. "...je vous prie d'en prévenir M.rs de l'Académie afin de la tranquilliser sur le sort des Instruments dont il s'agit," wrote Minister de Boynes in a letter dated "À Versailles le 9 X.bre 1773," SHMV, Ms. 88; see related letter in the same collection dated "À fontainebleau le 9.e 9.bre 1774" and ministerial correspondence dated "versailles le 21 Janvier 1774" in SHMV, Ms. 93. For correspondence regarding Academy instruments left by Rochon on Île de France, see Academy letter "À M. de Tromelin à L'Isle de france" and "À M. de la Brillane, Com.dt à l'Isle de France," dated "16 Mars 1776" and from "M. Le Ch.er de Tromelin" dated "À l'Isle de france le 1.er aoust 1776": in SHMV, Ms. 92; letter from "Retail" to "Monsieur Petit lieutenant de Port Secretaire de l'Académie de marine à Brest" dated "À L'orient le 3 9.bre 1770," SHMV, Ms. 89.

ville; Le Gentil de la Galaisière of the Paris Academy grabbed them, but d'Après de Mannevillette asserted the rights of the Brest Marine Academy.[350]

French and Colonial Connections. The Marine Academy enjoyed connections to other institutions, notably the navy's Dépôt des Cartes et Plans. The chevalier d'Oisy, an ordinary member of the Brest Academy and assistant director of the Dépôt, served as a link between the two institutions. Chabert, academician at Brest and Paris and then director of the Dépôt des Cartes et Plans backed the connection.[351] In 1775, coincident with the successful use of marine chronometers to solve the problem of longitude at sea and for cartography generally, the minister Sartines approved the Academy's plan to create in Brest a duplicate repository of the Dépôt des Cartes et Plans in Paris; and he ordered that all maps and publications from the Dépôt be sent annually to the Marine Academy.[352] The Académie de Marine was put in formal charge of this new Dépôt.[353] Books and maps made their way to Brest as part of this project, including works by Bellin, de Mannevillette, Chabert, and Lalande.[354] The Academy's Dépôt became an operational center complementing the navy's repositories in Paris and in Lorient.[355] Sartines himself remarked on the positive results he saw from this connection between science and the navy: "In effect it is through the officers of the Navy that Geography and Navigation are going to make rapid progress that one can rightly expect from the backing that astronomical observations receive today from the use of marine chronometers for the perfection of maps."[356]

At one point, the Academy proposed a formal "correspondence" with the Dépôt des Cartes et Plans in Paris and individuals in various ports.[357] Whether anything came of this initiative remains uncertain, but it signals the importance of the Academy-Dépôt connection. The Academy similarly had ties to the Royal Society of Medicine (the *Société Royale de Médecine*). The inspector

350. See d'Après de Mannevillette letter dated "À Kx.gars le 6 Juillet 1771," SMHV, Ms. 89.

351. See Academy letter to Chabert, dated "10 février 1775" and reply by Chabert dated "Paris le 27 Mars 1775" in SHMV, Ms. 92.

352. Sartine letter dated "À Versailles le 3 8.bre 1775," SHMV, Ms. 88 and copy, Ms. 93.

353. See Sartine letter to the count d'Orvilliers dated "Versailles le 22 X.bre 1777" and letter of 26 Avril 1777 from M. Le C.te D'Orvilliers from Brest in SHMV, Ms. 93.

354. DDP, IV:5, 21; Sartine letter dated "Versailles le 14 mars 1775," SHMV, Ms. 88 and copy, Ms. 93.

355. See SHMB, Série A, Sous-série 3A, Conseil de Marine, Registres 3A4 à 3A10: Procès-verbaux de délibérations du Conseil de Marine de 1772 à l'an II (1793-1794), 4 August 1787 and 13 October 1787.

356. "C'est en effet par les Officiers de la Marine que la Géographie et la Navigation doivent faire les progrès rapides qu'on a droit d'attendre du Secours important que les observations astronomiques reçoivent aujourd'hui de l'usage des horloges marines pour la perfection des Cartes," Sartine letter dated "Versailles le 14 Mars 1775," SHMV, Ms. 93.

357. AN MAR G93, fols. 62-73: Ms. "Projet d'un nouveau Règlement pour l'Académie Royale de Marine avec des observations en marges," Article 65.

and director-general of medicine for the Navy, the omnipresent Poissonnier, was a key member of both institutions, of course. The Brest Academy received the publications of the medical society in Paris, and the Academy requested that Royal Navy surgeons submit to the Academy duplicates of the reports they were required to send to Poissonnier.[358] Regarding the *Académie Royale d'Architecture* (1681), the abbé C.-É.-L. Camus was a *pensionnaire* at the Academy of Sciences, an honorary of the Brest Academy, and also secretary of the architecture academy, and there seems to have been some link between these institutions concerning naval architecture.[359] Along these lines, the Brest Academy elected Julien-David Le Roy (1724-1803) of the Academy of Inscriptions and Belles-Lettres on the basis of his work on ancient navies.[360] Connections to the provincial academies in Montpellier and in Dijon and the *Società Italiana* in Verona, Italy round out the list of the institutional contacts of the Marine Academy at Brest.

By its very nature, the Académie de Marine is an institution of note for our Colonial Machine, but it had its own colonial orientations as well. Partly, this aspect of the Academy's identity came from members who lived in or traveled to the colonies. Jean-René-Antoine, the marquis Verdun de la Crenne (1741-1805) might be cited in this regard.[361] Associate member of the Brest Academy in 1771 and full member in 1777, Verdun de la Crenne had a long and active career in the navy, rising to vice admiral and ultimately commodore; in 1785 he was in charge of the navy's operations in the Caribbean. He took a great interest in scientific navigation and cartography and captained two voyages (1768, 1771) to test marine chronometers, the results of which were published by the government in 1778 in two volumes as *Voyage en diverses parties de l'Europe, de l'Afrique et de l'Amérique*. Like Verdun de la Crenne, virtually all of the leaders of the great hydrographic and chronometric voyages in the second half of the century were members of the Academy and reported to it, including the chevalier de la Cardonnie, who observed the passage of Mercury from Saint Domingue in 1752. Lesser-known members of the Marine Academy at Brest include Du Dresnay des Roches, ship's captain and governor-general of Île de France and Île Bourbon for the Compagnie des Indes; and Daniel-Marc-Antoine Chardon, intendant at Saint Lucia in 1763-1764 and author of an *Essai sur la colonie de Sainte-Lucie* (1779). The Academy concerned itself in detail with maps and mapping of sea routes to the colonies, and the great cartographers Bellin and d'Après de Mannevillette had many dealings with the Academy. The Brest Academy likewise received occasional

358. DDP, IV:20.

359. See SHMB, Série A, Sous-série 3A, Conseil de Marine, Registres 3A4 à 3A10: 12 September 1786, 23 September 1786, and materials in AN MAR G145.

360. DDP VI:41, 44.

361. On Verdun de La Crenne, see Taillemite (2002), pp. 525-26.

reports from the colonies, including ones concerning monstrous births from Île de France, careening at Île de France, mapping Madagascar, the port at Quebec, native canoes in the Philippines, and astronomical observations from Gorée and from China.[362]

The colonial connection of the Brest Academy is manifest most directly in its dealings with its correspondent on the island of Saint Lucia in the Caribbean, the royal doctor Jean-Baptiste Cassan (1768-1803?). A graduate of the medical faculty at Toulouse, Cassan spent an additional three years training, at government expense, at the above-mentioned École de Médecine Pratique at Brest; after that, in 1786, he was posted as royal doctor to Saint Lucia, "having been charged by the government," he wrote, "to make every observation I can concerning natural history during my stay in the New World."[363] Cassan became a formal correspondent of the Royal Society of Medicine in Paris and the Cercle des Philadelphes in Saint Domingue, and he was active with the Academy of Sciences, the Royal Society of Agriculture in Paris, and with the Academy in Bordeaux.[364] His name will come up again, but here his close connections to the Académie de Marine in Brest need emphasis. Before he left for the New World, Cassan read a paper to the Brest Academy on sea rations and scurvy. This paper was well received in the Academy, and Cassan was duly elected a correspondent.[365] Other papers followed from his colonial outpost: on fossil bones, on volcanic activity on Saint Lucia, on flour-eating insects, and on hurricanes in the Antilles; and Cassan also submitted two meteorological logbooks that were published in Rozier's *Journal de Physique*. He likewise presented to the Brest Academy his 1790 book, *Considérations sur les rapports qui doivent exister entre les colonies et la métropole*.[366] With the approval of the minister, in 1790 the Brest Academy elected Cassan as a full-fledged associate member.[367] Cassan's case was exceptional, but indicative of the Marine Academy as an institution oriented toward the colonies.

362. DDP I:27, 53; III:54; V: 7, 33. See also SHMV, Ms. 87 passim, d'Après de Mannevillette papers in Académie Royale de Marine (1773), and SHMV, Ms. 361: "Mémoire sur le Sénégal."

363. "Ayant été chargé par le Gouvernement de faire pendant mon séjour dans le nouveau monde toutes les observations dont je serais capable sur l'histoire naturelle…," "Mémoire sur les ossements fossiles, lu à la séance du 22 juin 1786," SHMV, Ms. 78, p. 342. On Cassan, see Regourd (2000).

364. Regourd (2000). Cassan was elected a correspondent of the Académie de Marine in 1786 and a full-fledged associate member in 1790. For his submission to the Academy of Sciences in Paris, see PA-PV-110, Cassan, 10 February 1790 ("Plusieurs mémoires de géographie, de physique et d'histoire naturelle").

365. Regourd (2000); DDP VI:38-39, 44. See also Cassan's papers SHMV, Ms. 78, pp. 316-34; pp. 335-36; Ms. 79, fols. 35-37.

366. On these points, see Regourd (2000); DDP: VI:74, 83; Cassan (1790).

367. DDP VI:83; CAOM EE 405 (Cassan).

The Société Royale de Médecine (1778)

[Jeton: Société Royale de Médecine]

The Société Royale de Médecine was among the most vital of French learned societies in the last quarter of the eighteenth century.[368] Nominally, the Société grew out of a commission on epidemics established by Turgot in 1776 in the wake of a cattle plague in 1774-1776. This commission was charged to build and maintain a network of provincial correspondents for the exchange of medical and veterinary information, but in fact the move was aimed to launch a national learned medical society over the objections and particularism of the Paris medical faculty. In 1778, under the leadership of Félix Vicq d'Azyr (1748-1794), this commission received letters patent, transforming it into the Société Royale de Médecine. Modeled directly after the Académie des Sciences and with a substantial budget, the Société de Médecine quickly developed strong colonial connections, and it immediately assumed an absolutely central position within the larger Colonial Machine.[369]

The Société Royale de Médecine has attracted the attention of historians, but its colonial dimension has not been noticed until now. Given the specialist concerns of the institution and its particular mission to develop a network of correspondents, however, its colonial connections come as no surprise. Although initially the Société focused on recruiting correspondents within metropolitan France, already from 1777 it began to elect a not insubstantial cadre of correspondents from the colonies. Thirty-two official correspondents from the colonies and the navy are listed in the Biographical Appendix. When one adds an additional twenty-eight men overseas who maintained active contact with the Société, but who were not, apparently, official correspondents, the colonial face of the institution becomes even more apparent. These corre-

368. Although the need exists for a full modern study, the literature on the Société Royale de Médecine is not insubstantial; see Gillispie (1980), pp. 194-203; Goubert; Desaive et al.; Hannaway (1972); Hannaway (1974); Ganière, chapt. 2; Simonetta.

369. Ganière, p. 30, puts the initial budget at 6000 livres, later raised to 20,000 livres.

spondents hailed mostly from the colonial centers in the Antilles and South America; the Indian Ocean colonies seem underrepresented. With the exception of two token foreign doctors from Chile and Brazil, the correspondents were all French and mostly all physicians and surgeons trained in France. As might be expected, correspondents of the Société were also drawn primarily from the ranks of royal doctors or official naturalists posted to the colonies.[370]

A deeper prosopographical analysis is not possible here, but we may signal a few especially noteworthy correspondents. One is J.-B. Cassan, again, the royal physician from Saint Lucia whose connection to the Marine Academy at Brest we just encountered. Elected a correspondent in 1790, Cassan sent substantial and well-received papers to the Paris medical society on tropical topics: climatological influences, characteristic nosologies, a pharmacopeia, and meteorological reports.[371] Bertin, the surgeon and correspondent from Guadeloupe, was in regular contact with the Société, notably with his 1783 paper on stomach diseases of slaves.[372] The royal doctor and head of the royal botanical garden in Port-au-Prince in Saint Domingue, Joubert de la Motte, maintained regular contact with the Société Royale de Médecine, which elected him a national associate (*associé régnicole*) in 1787.[373] The royal veterinarian from Guadeloupe, Hapel Lachênaie (1760-1808), was another notable colonial correspondent, elected in 1789.[374] A protégé of the chemist and academician Fourcroy, Hapel Lachênaie taught chemistry at the veterinary school of Maisons-Alfort before being sent to Guadeloupe "on the orders of government" in 1784. He was in close contact with Vizq d'Azyr through whom he obtained the post of royal chemist-apothecary and his election to the Société de Médicine. Hapel Lachênaie was in contact with the Société from 1785, over the years sending it memoirs on ants, corn, mineral waters, epizootics, meteor-

370. It is worth pointing out that potential correspondents had to submit papers to the Société which were reviewed; not everyone passed muster, as in the case of the Chevalier de St. George whose 1788 paper on skin diseases in Saint Domingue was judged "incomplete and insufficient to merit the title [of correspondent] solicited by the author," SRdM Ms. #11, fol. 221r ("Séance du Vendredi 24 Octobre 1788"); fol. 234r ("Séance du Vendredi 7 Novembre 1788"); fol. 235r ("Séance du Vendredi 14 Novembre 1788"). Conversely, especially valuable correspondents gain spots on a waiting list for election as national associates (associé régnicoles); colonial correspondents Dazille, Joubert, and Mallet de la Brossière earned this promotion.

371. See Regourd (2000); Simonetta, p. 232, and above.

372. SRdM, Ms. 10, p. 34 ("Séance du mardi 9 decembre 1783") and follow up, p. 192 ("Séance du Vendredi 18 Juin 1784"). For other Bertin contact with the Société Royale de Médecine, see also SRdM, Ms. 9, pp. 153, 155, 157, 161, 187, 199, 366; Yaqubi, p. 3.

373. On Joubert's contact with the Paris medical society, see, for example, SRdM, Ms. 10, p. 73 ("Séance Du vendredi 23 J.vier 1784") and p. 76 ("Séance Du mardi 27 janvier 1784"). See also SRdM, Archives, 109 dr. 3, #111, letter dated "Paris le 22 mars 1784." See also McClellan (1992), pp. 155-56.

374. On Hapel Lachênaie, see Demeulenaere-Douyère (2000); Gerbaud; Lacroix (1938), pp. 117-24. See also materials CAOM EE 1095 (Hapel); PA-DB (Hapel). On his contact with the Société Royale de Médecine, see SRdM, Ms. 10, fol. 355; Ms. 11, fol. 222v, 265v, 266, 270, 272, 285; Ms. 11bis, fol. 92, 97, 108; SRdM, Archives, 175 dr. 1, #1-4; and paper SRdM-H&M, vol. 4 (1780-1781), pp. 325-33.

ology, and the medical topology of Guadeloupe; the last of these won a prize from the Société.

Among the unofficial correspondents of the Société, a number of names crop up. These include Bertrand Bajon, the royal surgeon from Cayenne; and the royal engineer Cossigny (fils) from Île de France, who sent papers on worms and ipecac.[375] Others are Joseph Gauché, administrator of the royal spa at Port-à-Piment in Saint Domingue; the surveyor on Guadeloupe, Badier, who sent quinine samples for testing; and the Jesuit Duchoisel in Pondicherry, who sent memoirs on rubies.[376] The inspector of colonial artillery in the Antilles, the count Du Puget d'Orval, sent quinine samples.[377] The noted colonial lawyer and jurist, Médéric-Louis-Élie Moreau de Saint-Méry (1750-1819), sent his *Loix et constitutions des colonies françoises de l'Amérique Sous-le-Vent* (1784-1790) to the Society and a medical remedy of his own composition for their review.[378]

One correspondent, Charles Arthaud (1748-1793?), needs to be singled out as the medical Society's most active contact in the colonies. Trained at the medical faculty at Nancy, Arthaud began a private medical practice in Saint Domingue in the 1770s and worked his way up in the colonial medical establishment, becoming chief médecin du roi in 1783. The Société Royale de Médecine in Paris elected Arthaud a correspondent in 1777, and he regularly communicated with the Society from that date until at least the end of 1791.[379] Arthaud submitted papers and reports on scurvy, leprosy, tetanus, elephantiasis, rabies, dysentery, epizootics, birth defects, the benefits of nursing, hospital administration, medical topography, disease outbreaks, menstruation, magnetism, albinos, anthropology, natural history, poisons, and case histories of various sorts. But more than that, Arthaud served as the link between the Société Royale de Médecine in France and the scientific society founded in Saint Domingue in 1784, the Cercle des Philadelphes. Arthaud was the moving force behind this key colonial institution of science and medicine, and as its permanent secretary he made it a point to forward the Cercle's every announcement and every publication to the Société Royale in France. In sending the prospectus for the Cercle in 1785, Arthaud asked that "the Société use its credit to help

375. SRdM, Archives, 168, dr. 4, #18-21.

376. SRdM, Ms. 11, fol. 239r ("Séance du Vendredi 5 Décembre 1788") and fol. 253r ("Séance du Vendredi 13 fevrier 1789"); Ms. 125 dr. 15, #30-32 (missing).

377. SRdM, Ms. 11, fol. 280r ("Séance du Mardi 26 Mai 1789").

378. SRdM, Ms. 10, p. 385 ("Séance du Vendredi 18 mars 1785") and Ms. 11bis, fol. 81v ("Séance du Mardi 25 Janvier 1791").

379. On Arthaud, see Hamy (1908); McClellan (1992), pp. 137-38; Taillemite (1958), p. x; DPF, p. 1445. SRdM, Archives, 136 dr. 1, contains forty submissions from Arthaud in this period; the Procès-Verbaux of the Société de Médecine indicate an even more substantial contact. At one point Arthaud requested the Société name him "correspondant en chef," SRdM Ms. #7, p. 236 ("Séance du mardi 13 Juillet 1779").

its establishment."[380] Although nothing formal resulted, Vicq d'Azyr did write back to Arthaud to express "in the name of the Company" its satisfaction over the coming-into-being of the Cercle des Philadelphes in Saint Domingue.[381] Some of the Cercle's work, for example on tetanus, was clearly undertaken in harmony with similar projects of the Société Royale. When one reexamines the list of the correspondents of the Société Royale, this link between the Société and the Cercle des Philadelphes becomes even more apparent, in that many of the official and occasional correspondents of the Société were also key members of the Cercle des Philadelphes, including notably Simeon Worlock, Lefebvre-Deshayes, Jean Gélin, Joseph Gauché, and especially Moreau de Saint-Méry. In establishing ties between the Antilles and the metropolis in general and the Société Royale de Médecine and the Cercle des Philadelphes in particular, Arthaud and Moreau de Saint-Méry were influential figures in the Colonial Machine.[382]

In addition to input from its formal and informal correspondents, the Société Royale de Médecine received occasional reports from Louisiana, Saint Domingue, Martinique, Guadeloupe, Saint Lucia, Guatemala, South America, Africa, Ceylon (Sri Lanka) and the Indian Ocean colonies, and from the Jesuits in China.[383]

But these contacts hardly represent the full extent of the institution's colonial orientation or its place in the larger Colonial Machine. The Société's imprimatur, which it used to approve a number of works on colonial medicine, illustrates the point.[384] The program of prizes sponsored by the Société and the awards it presented to colonials make even more plain the centrality of colonial medicine to the mission of the institution. For example, the Société proposed a series of prizes on how to maintain the health of troops in different seasons; and in its program of 1782 it asked questions about diseases prevalent

380. SRdM, Ms. 10, p. 341 ("Séance du Mardi 8 fevrier 1785...une Lettre de M. Arthaud med. à Cap [sic] avec un Prospectus du Cercle des Philadelphes qui y est établie. il desire que la société emploie son Credit pour cet établissement").

381. "J'ai transmis au cercle des phyladelphes les deux lettres dans lesquelles vous exprimés, au nom de votre compagnie, la satisfaction qu'elle a éprouvé de son établissement," Arthaud letter dated "au Cap, le 10 septembre 1785," SRdM, Archives, 136 dr. 1, #26.

382. Arthaud and Moreau de Saint-Méry were also brothers-in-law. For other links between the Société Royale de Médecine and the Cercle de Philadelphes, see SRdM, Ms. 11, fol. 113r ("Séance du Mardi 24 Juillet 1787"), where Arthaud materials are passed on to the Société through the Ministry of the Navy; and, SRdM, Ms. 11, fol. 205r ("Séance du Mardi 22 Juillet 1788"), where a winner of a prize competition sponsored by the Cercle des Philadelphes received permission to publish his winning essay with the imprimatur of the Société Royale de Médecine.

383. SRdM, Mss. 7-11, passim and Ms. 11, fol. 115r ("Séance du Mardi 31 Juillet 1787... M. Saillant a fait la lecture de l'extrait d'une lettre qui a été écrite de Pekin par le S. Amyot concernant l'etat actuel de la médecine parmi les Chinois").

384. See, for example, SRdM, Ms. 10, p. 433 ("Séance du Mardi 24 may 1785...M. Andry a lu un rapport fait avec Mrs Poissonnier, Geoffroy et Desperrieres sur un ouvrage de M Dazile intitulé observations Générales sur les maladies des pays Chauds.... d'après la Lecture du rapport La Compagnie a pensé que cet ouvrage meritoit son approbation et dêtre imprimé sous son privilege").

in the summer and "the means to prevent them or diminish their effects in very hot countries, such as the Greater and Lesser Antilles?"[385] The Society explicitly sent its program to the colonies, and military surgeons were especially invited to make submissions.[386] Its prize for 1785 asked about the advantages and disadvantages of cinchona in treating intermittent fevers. In a like vein, given its commitment to environmental causes of disease, compiling medical-topographical surveys of the various regions of France represented one of the principal goals of the Société, and it ended up awarding special medals to a number of its colonial correspondents for the work they submitted regarding overseas locales.[387] Lefebvre-Deshayes was given an award for his detailed observations on the mineral waters of Saint Domingue, marine worms, and albinos.[388] Bertin won a gold medal for his survey of Guadeloupe, as did Simeon Worlock for his study of an animal plague in Saint Domingue in 1780. The Société Royale crowned Mallet de la Brossière for his medical topology of various locales in Saint Domingue; Hapel Lachênaie, as mentioned, for his study of Pointe-à-Pitre in Guadeloupe; and a certain Dr. Ramel on the topography of the *comptoir* at Calle on the African coast.[389]

The independent colonial face of the Société Royale de Médecine is already obvious. What needs emphasis here are the ways in which the institution plugged into the rest of the Colonial Machine to become the scientific center for colonial medicine. Partly this status was achieved through its resident membership, notably Pierre-Isaac Poissonnier, whom we have encountered many times already as the director-general and inspector for naval and colonial medicine. Poissonnier was very active in the Société and an important intermediary with various correspondents, and not surprisingly a theme running through the Société's records concerns the salubrity of ports and the health of seamen. Poissonnier regularly spoke officially for his department, and he served on many committees appointed by the Société, writing many reports. In the Société Royale de Médecine, Poissonnier was backed up by his brother, Antoine Poissonnier-Desperrières, whom

385. SRdM-H&M, vol. 4 (17810-1781), Histoire, pp. 2-3.

386. "On en enverra les Programmes dans les colonies," SRdM, Ms. 9, p. 70 ("Séance du Mardi 15 février 1782") and p. 88 ("Séance Publique du Mardi 19 fevrier 1782"). The records of the Société do not seem to indicate a winner of this contest, but Delandine, p. 160, indicates that a certain M. Thion de la Chaume, physician to the military hospitals, won the 400 livre award. For another submission see, SRdM, Ms. 10, p. 8 ("Séance du mardi 28 octobre 1783").

387. See summary remarks regarding 226 medical-topographical reports received, SRdM, Ms. 11, fol. 264v, ("Séance publique du Mardi 3 Mars 1789").

388. For these awards, see SRdM-H&M, vol. 5 (1782-1783), pp. 8, 10-11; vol. 7 (1784-1785), p. 11-12; vol. 8 (1786), p. 12; vol. 9 (1790), p. xv; vol. 10 (an VI, 1798), p. iv; SRdM, Ms #10, p. 244 ("Séance publique du Mardi 31 Août 1784"), p. 358 ("Séance publique du Mardi 15 fevrier 1785"), and p. 402/502 ("Séance Publique du Mardi 30 août 1785"). See also *Journal de Physique* (1784), p. 393.

389. Another correspondent in Saint Domingue, Dr. Ycard, also won one of these awards, but for his description of locales in Languedoc in France.

he had appointed as adjunct administrator for colonial medicine in his administration. The two men embodied connections between the Société Royale de Médecine and the rest of the colonial medical establishment within the Navy.

The elder Poissonnier was merely one of a number of members of the Société Royale de Médecine who were simultaneously academicians of the Académie Royale des Sciences, and this human connection makes plain that the Société de Médecine functioned, in effect, as an arm of the Académie des Sciences for matters medical, much like the Académie de Marine did in its areas of specialization. Vicq d'Azyr, for example, permanent secretary of the Société, was associate in anatomy at the Académie. Joseph-Marie-François de Lassone, another first doctor to Louis XVI, was prominent at the Society of Medicine and *pensionnaire vétéran* at the Academy of Sciences. Other common members included the anatomists R.-B. Sabatier and Daubenton, the botanists/zoologists A.-L. de Jussieu, Tessier, and Mathieu Tillet, and the chemists Fourcroy, Pierre-Joseph Macquer and Lavoisier, all of whom brought their specialist expertise to bear on scientific and medical questions that arose within the Société de Médecine.[390]

The respective roles of the two institutions seem to be have been recognized by the players at the time. The Société Royale regularly forwarded materials, such as astronomical observations, it felt more appropriate for consideration by the science academy.[391] Just as regularly, the Académie des Sciences sent over materials it deemed more strictly medical in nature.[392] And then, a division of labor is evident in their interactions. For example, from Saint Domingue in 1789 the peripatetic naturalist Palisot de Beauvois sent samples of local cinchona explicitly to the Paris Academy for chemical analysis and to the Société for study of cinchona's therapeutic properties.[393] In a letter to the Société from Île de France in 1780, Cossigny de Palma wrote about a memoir he had sent concerning a vegetable oil: "I refer you to the paper in question for its medicinal properties. I think that this paper is more for the Société Royale de Médecine than for the Académie des Sciences, but you will be the judge of how best to use it."[394] Similarly, Charles Arthaud's shipment of wild cochineal in 1787-1788

390. At one point the Société referred to its "Membres Botanistes"; SRdM, Ms. 11bis, fol. 181r ("Séance du Mardi 22 9bre 1791").

391. See SRdM, Ms. 7, p. 330 ("Séance du du vendredi 15 8bre 1779"). See also the paper that "a été remis a l'Académie des sciences Comme concernant plus directement cet Comp. que la Société Royale de med," SRdM, Ms. 7, p. 113 ("Séance du Vendredy 26 mars 1779").

392. See SRdM, Ms. 7, p. 266 ("Séance du mardi 17 aoust 1779").

393. See de Beauvois letter dated "Du Cap François le 1er juin 1789..." in PA-DB (Palisot de Beauvois). See also SRdM, Ms. 7, p. 266 ("Séance du mardi 17 aoust 1779") and Ms. 11, fol. 159r ("Séance du Mardi 22 Janvier 1788"); McClellan (1992), p. 161.

394. "Je vous renvoie au Mémoire en question sur ses propriétés médicinales. Je crois cet écrit plutôt fait pour la société Royale de Médecine que pour l'Académie des Sciences. Vous en ferés l'usage qui vous plaira," ALS, Cossigny to Le Monnier dated "Palma le 20 Xbre 1780," BCM-NHN, Ms. 1995.

went to both institutions, one sample for chemical analysis, the other for medical. About this shipment the Société Royale de Médecine remarked pointedly: "Trials concerning cochineal for dyeing in no way concern the Société Royale de Médecine and are absolutely the department of the Académie des Sciences which we know already is aware of this matter."[395]

The Société Royale de Médecine had contact with the Académie Royale de Marine and the Société Royale d'Agriculture, but more than anything else its ties to central government cemented the place of the Société Royale de Médecine as a core institution of the Colonial Machine.[396] The Société quickly became *the* institution tapped by government authorities for expert advice concerning medicine in general and colonial medicine in particular.[397] More specifically, every minister of the navy and the colonies from 1776 regularly had the Société de Médecine undertake large-scale investigations across a range of medical issues of practical concern to the ministry: proper rations for seamen, regimens in naval hospitals, treatments for tetanus and elephantiasis, maintaining potable water aboard ship, comparing Peruvian cinchona and colonial substitutes, and like topics. In many lesser instances as well, the Société served as a technical consultant to the Ministry: De Castries asked for its views on Crespy's treatment of illnesses in Senegal; La Luzerne received a report on maintaining the health of slaves on the Middle Passage.[398] Not just the navy minister, but the whole colonial apparatus worked with the Société de Médecine, as is clear when La Luzerne, while governor-general in Saint Domingue, sent colonial botanicals back to de Castries at the ministry at Versailles. The latter in turn passed them on to the Société Royale to examine their medical properties.[399] By the same token, the Société sent reports up the chain of com-

395. "L'essai de la cochenille à teinture ne regarde aucunement la S. R. et est absolument du département de l'Académie des Sciences que nous savons avoir déjà eu connoissance de cet objet," marginal note to Arthaud letter dated "au Cap, le 29 novembre 1787," SRdM, Archives, 136 dr. 1, #32. See also, SRdM, Ms. 11, fol. 159r ("Séance du Mardi 22 Janvier 1788").

396. On the connection with the agriculture society, see, for example, SRdM, Ms. 11, fol. 207r ("Séance du Mardi 5 Août 1788"). We can note the contact between the Société Royale de Médecine and the Académie de Marine at Brest through the person of Blondeau; see SRdM, Ms. 7, p. 195r ("Séance du vendredi 15 Juin 1779...[sic] Le redacteur du Journal de Marine de Brest a envoyé à la Compagnie un Cahier de son Journal") and p. 207 ("Séance du vendredi 25 Juin 1779"), where he is referred to by name as secretary of the Marine Academy at Brest. The Société elected Jean-Baptiste Herlin, chief naval physician at Brest, as a correspondent 1770 and an ordinary member in 1776; see DDP: V:24.

397. Hahn (1971), p. 103 notes this role of the Société as a "technical consultant" to the crown.

398. For the former case, see SRdM, Archives 140 dr. 22; SRdM Ms. #11, fol. 53v ("Séance du Mardi 17 8bre 1786") and fol. 55v ("Séance du Vendredi 27 Octobre 1786"). The Société forwarded a positive report to the minister; for the latter, SRdM, Ms #11, fol. 160v ("Séance du Mardi 29 Janvier 1788").

399. SRdM, Ms. 11, fol. 72v ("Séance du Mardi 30 Janvier 1787") and Ms. 11, fol. 113r ("Séance du Mardi 24 Juillet 1787"). For an example of the Société working with administrators in Cayenne, see SRdM, Ms. 11, fol. 224v ("Séance du Vendredi 3 Octobre 1778").

mand when it felt the ministry needed to know of its findings.[400]

The Société Royale de Médecine meshed with central government not only through the ministry of the navy and the colonies. The Société Royale originated under the controller general's office, and its permanent secretary held an ex officio appointment within that bureau.[401] In the Ministry of War (the *Ministère de la Guerre*), the Société's permanent secretary served on a standing committee overseeing military hospitals.[402] And, on at least one occasion the Société de Médecine had dealings with the Ministry of Foreign Affairs. While not all of these interactions had to do with the colonial world per se, they do illustrate the interlocking directorates that characterized the Colonial Machine and the ways in which the Société Royale de Médecine became inextricably woven into the administration of medical affairs and the colonies of Old Regime France. That the Société took a great interest in chocolate as a medicine may or may not reflect its colonial orientation.[403]

The Académie Royale de Chirurgie (1731)

[Jeton: Académie Royale de Chirurgie, 1741]

Like the Académie de Marine at Brest or the Société Royale de Médecine, the Académie Royale de Chirurgie was a secondary specialized institution

400. See, for example, SRdM, Ms. 11, fol 31v ("Séance du Mardi 11 Juillet 1786...M. Mauduyt a lu un rapport redigé avec M. Desperrieres sur un Mémoire de M. DelaBorde concernant l'insalubrité de la ville de Cayenne a l'occasion d'un terrein desseché à la proximité de la Ville. Ils ont pensé qu'il étoit apropos de Communiquer ce Mémoire au Ministre").

401. SRdM, Ms. 10, p. 386/486 ("Séance du Mardi 2 août 1785"). See also SRdM, Ms. 11, fol. 201v ("Séance du Mardi 8 Juillet 1788") where Vicq d'Azyr records the Société's contact with the Controle Général regarding its experiments to treat insanity; this initiative also put the Société in touch the the Lieutenant de Police of Paris; on connections to the Paris police, see also SRdM, Ms. 11, fol. 201v ("Séance du Mardi 8 Juillet 1788").

402. SRdM, Ms. 11, fol. 192r ("Séance du Vendredi 6 Juin 1788").

403. See SRdM, PV, passim; including, Ms. #10, pp. 52-3 for "un chocolat aphrodesiaque"; see also Boucher, p. 347n.

within the Colonial Machine.[404] Created by royal fiat as the Sociéte Académique des Chirurgiens de Paris 1731 and receiving formal letters patent in 1748, the Académie Royale de Chirurgie antedated the Société Royale de Médecine by almost half a century, but its precocious emergence has to do more with the flexibility of the Parisian surgeons' guilds and the ability of court surgeons to organize such an Académie, in contrast to the Parisian medical faculty which into the 1770s resisted the organization of medicine into a learned society. With a budget of 7000 livres, the Académie de Chirurgie was a successful organization, holding meetings, offering prizes, and publishing a series of *Mémoires*.[405] Yet the surgeons' academy did not effect quite the same strong colonial ties that characterized the royal medical society or the Marine Academy.[406] Perhaps this circumstance had to do with the more guildlike orientation of the Académie de Chirurgie; perhaps also the explanation is to be found in the fact that, independent of the court or Paris institutions for surgeons, the Ministry of the Navy formally controlled the training and governance of naval surgeons and royal surgeons posted to the colonies. Nevertheless, how could the Académie de Chirurgie not be touched by the overseas and the colonial that so permeated French scientific and official life? And so, as may be surmised on first principles, the Académie de Chirurgie formed part of the Colonial Machine and had a noticeable colonial component of its own, especially from the 1770s on.

Like other Parisian learned societies, the Académie de Chirurgie meshed with the rest of officialdom. The First Royal Surgeon, for example, was de jure the president of the Academy. In particular the Académie de Chirurgie had strong links to the Académie des Sciences and its anatomy section. The first royal surgeon, Georges Mareschal (1658-1736), founded the Académie de Chirurgie and was its first president, followed by François Gigot de La Peyronie (1678-1747) who was a free associate (*associé libre*) at the Académie des Sciences. Jean-Louis Petit (1674-1750, known as Petit le Chirurgien) was the first director of the surgery academy and a *pensionnaire* in anatomy in the science academy; Sauveur-François Morand (1697-1773) was another surgeon *pensionnaire* in anatomy, and Raphaël-Bienvenu Sabatier (1732-1811) was an associate in anatomy in the science academy.[407] The Académie de Chirurgie also elected a number of naval and port surgeons, and it seems to have played some role in

404. On the Académie Royale de Chirurgie, see Gillispie (1980), pp. 205-206; Gelfand (1980), esp. pp. 63-67; Chassagne, pp. 190-93; Ganière, chapt. 3; theses by Dordain and by Yaqubi, esp. chapt. 1.

405. The surgeons academy published five volumes of *Mémoires* between 1743 and 1774 and granted its imprimatur to other works after that date. Ganière, pp. 42-43, discusses the financial arrangements.

406. Dordain emphasized how difficult it is to compile a list of non-resident members of the Académie de Chirurgie, noting, p. 117: "Le nombre des praticiens d'outre-mer, correspondants de l'Académie royale des Sciences et de la Société royale de Médecine est important. Moins nombreux sont les correspondants de l'Académie royale de Chirurgie."

407. On R.-B. Sabatier and his connection to the Académie des Sciences through the Académie de Chirurgie, see Hahn (1971), pp. 105-106.

making nominations of surgeons to these positions.[408] And, not a little of the work of the Académie de Chirurgie concerned topics pertaining to the navy and the colonies, notably the health of troops and the sad litany of colonial diseases: tetanus (in Batavia, Cayenne, and Brazil), yellow fever, and yaws.

The Académie de Chirurgie was committed to the dissemination of surgical information, and to this end it appointed a commissioner for correspondence. In addition, the Académie elected at least four correspondents with strong ties to the colonies, including, once again, the royal physician and motive force behind the Cercle des Philadelphes, Charles Arthaud. Pierre Campet (1723-1801), chief surgeon at the government hospital at Cayenne from 1754 to 1772, sent reports on tetanus to the Académie.[409] Even more active were two individuals whom we have likewise encountered previously. One was Bertin, a physician and surgeon in Guadeloupe and correspondent of the Société Royale de Médecine. Bertin sent the Académie papers on diseases of the island and on the influence of Guadeloupean air on surgical treatments.[410] The other was Bertrand Bajon, chief royal surgeon of the military hospital at Cayenne from 1764 to 1776 and correspondent of the Paris Academy of Sciences.[411] Bajon perfected an operation in Cayenne for cleft palate which the Académie de Chirurgie published in its *Mémoires* and for which it awarded him a gold medal. Bajon also wrote to the surgical Académie on tetanus and liver disease. He was otherwise heavily engaged elsewhere in the Colonial Machine, and from this point of view, the notable point may be the place of the Académie de Chirurgie in Bajon's world, rather than the other way around. In any event, the Colonial Machine was enhanced by both.

The Société Royale d'Agriculture (1761)

[Medal: EX UTILITATE DECUS, 1785]

408. See Dordain, p. 19; Yaqubi, p. 333.

409. Dordain, pp. 117-119; Yaqubi, p. 333; Chaïa (1958), p. viii; Campet (1802).

410. Dordain, p. 19; Yaqubi, p. 333.

411. Lacroix (1938), vol. 3, pp. 71-75; Institut de France; for Bajon papers submitted to the Academy, see PA-PV-110, Bajon, 8 September 1774 ("Sur le Paragona"), 3 September 1777 ("Recueil de Mémoire pour servir à l'Histoire de cayenne et de la Guiane") and 25 February 1778 ("Second Vol. D'un Ouvrage sur l'hist.re naturelle et la médecine").

The Société d'Agriculture for the Paris region (*Généralité de Paris*) arose in 1761 as part of the government-sponsored initiative to install local improvement societies across the kingdom.[412] The Paris Société d'Agriculture got off to a good start with a membership that included Buffon, Bernard de Jussieu, and Duhamel de Monceau, and with a first volume of *Mémoires* that appeared in 1761.[413] But the institution soon languished for lack of funds and was a dead letter after 1769. It was not until the 1780s that, reanimated by the local intendant, the organization began to flourish. It elected an expanded, national membership; in 1785 it launched its trimestrial *Mémoires d'Agriculture, d'Économie Rurale et Domestique*, and in 1788 the revitalized institution received letters patent transforming it into the Société Royale d'Agriculture. In charging it to be a clearing house for all French agricultural societies, the royal charter expanded the institutional horizon of the Paris Société d'Agriculture and centralized its authority as the national society for agricultural science and practice.[414] As a feature of these changed circumstances, the Société Royale roundly embraced the colonies and colonial agriculture and so integrated itself fully into the Colonial Machine.

A Strong Colonial Dimension. This integration proceeded in the first instance through local members who held overlapping positions in other metropolitan learned societies. The energetic permanent secretary of the Société d'Agriculture from 1785, Pierre Auguste Broussonet (1761-1807), joined the Academy of Sciences that same year as an associate in anatomy. Notably, in its reorganization of 1785, the botanical section of the Académie des Sciences became the section on botany and agriculture. More familiar than Broussonet, perhaps, are other members of the Société d'Agriculture who were concurrently academicians at the science academy and who likewise were affiliated with other core institutions of the Colonial Machine: Borda d'Oro, Daubenton, Desmarest, Fougeroux de Bondaroy, Fourcroy, Lavoisier, Poissonnier, Tessier, Thouin, Tillet, Vicq d'Azyr. From this perspective the Société d'Agriculture needs to be thought of as a specialist spin-off of the science academy in the same way that the Société Royale de Médecine, the Académie de Marine, and the Académie de Chirurgie represented specialist extensions of the mother institution of science. More than that, the Société d'Agriculture elected a number of its own correspondents from the colonies, many of whom are also now familiar: the jurist Moreau de Saint-Méry from Saint Domingue; Céré

412. On the Société Royale d'Agriculture, see esp. Regourd (1998) and Gillispie, 368-387; see also Passy and Boulaine (1990a), (1990b). Justin provides background. Archival sources are largely lacking because of the 1910 Paris flood, but some papers are conserved in AN H^1 1501, H^1 1503, H^1 1504, H^1 1510, and F^{10} 222. See also the personal minutes for 1789 taken by Fougeroux de Bondaroy, APS-MS, B D87 Group 26: Notebook labeled "1789 Société royale d'agriculture."

413. Société Royale d'Agriculture de Paris, pp. 8-9 and 24-26; Passy, p. 118.

414. See "Règlement fait par le Roi, du 30 mai 1788," in *Mémoires d'agriculture, d'économie rurale et domestique*, summer trimester, 1788, pp. i-xvii. Boulaine (1990b), p. 54.

(government botanist on Île de France), Badier from Guadeloupe, Hypolite Nectoux (royal botanist in Port-au-Prince), Jean-Baptiste Leblond (royal physician-naturalist in Cayenne), Joseph Martin (royal botanist in Cayenne), one Villard from New Orleans, Lescallier (naval second in command in Guiana), Jean-Baptiste Auvray (president of the Société Royale des Sciences et Arts in Cap François, the Cercle des Philadelphes), and one Fréon (a judge on Île de France, elected prior to 1781).[415] Some of these correspondents were in foreign countries, as, for example Saint-Jean de Crèvecœur. The abbé Raynal, author of the famous critique of colonial administration, the *Histoire philosophique et politique des établissements et du commerce des Européens dans les deux Indes*, should also figure on this list of correspondents of the Société with strong colonial connections.

A number of unofficial correspondents augmented the colonial component of the Société d'Agriculture: J. F. Dutrône La Couture (1749-1814), a physician, formerly of Saint Domingue and author of the *Précis sur la canne* (1790); François de Neufchâteau, the royal prosecutor in Cap François; Duchemin de l'Étang (former royal physician in Saint Domingue), one Cailleau (a royal clerk on Île de France); and, once again, J. B. Cassan, the royal physician on Saint Lucia.[416] In addition, eight or so property owners in the Antilles may have been associated with the Société d'Agriculture.[417]

Fougeroux de Bondaroy's private minutes of the meetings of the Société Royale d'Agriculture confirm its overseas dimension, with topics that included for 1789: sugar production, mules in the colonies, animal fodder, cotton cultivation in Saint Domingue, storing seeds on Île Bourbon, bananas, cloves from Île de France, sorghum from Africa, sea biscuits, pecans from Illinois country, and quinoa from Brazil.[418] Broussonet functioned as an intermediary in forwarding seeds from the colonies to Thouin and the Jardin du Roi for planting.[419] The colonial dimension manifests itself further in papers submitted by correspondents and published in the *Mémoires* of the Société. Topics included rice grown on Île de France, animal forage from Africa and the colonies, preserving grain in the colonies, the soils of French Guiana, the juice of sugar cane, the transport of plants from the colonies, and acclimatizing exogenous

415. Article XX of the "Règlement" of 1788 required that prospective correspondents present at least two memoirs on agriculture or rural economy to the Society.

416. See Regourd (1998); Passy; and the Society's *Trimestres*. See also otherwise unidentified 1784 document where François de Neufchateau and a certain Becht from Cap François are proposed as correspondents, as well as Schilling in Surinam, Clark in Jamaica, Dombey on mission in Peru, Michaux on mission in Basra, one Desfontaines in Tunisia, and Koenig in India (Tranquebar), BCMNHN, Ms. 308, Dossier: "Nottes et Projets de Mémoires Pour L'Academie et la Sociètè."

417. Regourd (1998); Debien (1953), p. 254.

418. APS-MS, B D87 Group 26.

419. BCMNHN, Ms. 314 and Ms. 691, "Catalogues des graines semées au Muséum pendant les années 1790-1801," fol. 131.

cultivars in France.[420] Several of these papers and their authors deserve special notice. Jean-Baptiste Cassan, for example, published a lengthy article on agricultural production in Saint Lucia, with remarks "on the feeding of slaves, on commercial agricultural commodities and products of the Lesser Antilles, and on new cultivations introduced into the colony and how to make them succeed."[421] As part of an initiative concerning cotton production in the colonies, undertaken by the Société d'Agriculture with the support of the Ministry of the Marine, Cassan likewise volunteered a thirteen-point program to grow experimental plots of cotton on Saint Lucia.[422] As part of the same initiative in 1788 the colonist Barthélemy de Badier from Guadeloupe presented the results of his own experimental cultivation of eighteen species of cotton raised in test beds on that island; these tests were likewise supported by the ministry.[423] Badier also published on bananas and yams in the *Mémoires* of the Société d'Agriculture, and he himself enjoyed further connections to the Académie des Sciences, the Société Royale de Médecine, the Jardin du Roi, and the office of the controller-general.[424] Because of Badier's connections and activity, André Thouin mistakenly thought that Badier was an official royal botanist, rather than a simple colonist.[425] These multifaceted connections to the Colonial Machine make Badier, like Cassan, Bertrand Bajon, Charles Arthaud, and others, a colonial of particular note and analytical importance.

Moreau de Saint-Méry: A Colonial Pivot. No colonial, however, was of greater importance to the Société Royale d'Agriculture or better illustrates the role of key individuals pulling the levers of the Colonial Machine than does M.-L.-E. Moreau de Saint-Méry. A true colonial – a Creole born in Martinique in 1750, Moreau de Saint-Méry studied law in Paris and thereafter practiced in Saint Domingue where he was ultimately appointed a judge on the *Conseil Supérieur* of Cap François.[426] In the 1780s the Ministry of the Navy co-opted him back to Paris for a project to compile a compendium of colonial jurispru-

420. Regourd (1998) gives a complete list of colonial memoires published by the Société Royale d'Agriculture. About 10 percent of the *Mémoires d'agriculture* of the Society were devoted to topics pertaining to overseas and the colonies.

421. Cassan (1790); see also Debien (1966).

422. See "Plan d'expériences proposé par M. Cassan" and Broussonet letter dated "de Paris le 30 septembre 1790," both CAOM EE 405 (7) (Cassan, Jean-Baptiste); Regourd (2002) These documents also mention Cassan's contact with the Académie des Sciences.

423. CAOM E 14 (Badier) and APS-MS B/D87, Group 26: "du jeudi 5 fevrier 1789...rapport sur des Cotons cultivés aux Antilles – le memoire par Mr. de Badier – qui compte 18 especes cultivés dans ses jardins – a la guadeloupe."

424. See SRdM, PV, passim, and references in BCMNHN, Ms. 314 and Ms. 1327. See also Badier (1788), where he writes, "En 1778 j'en fis fabriquer des chapeaux que j'ai eu l'honneur de présenter avec le Coton à l'Académie des Sciences & à la Chambre du Commerce, qui reconnurent qu'il méritait la préférence sur celui qu'on est dans l'usage de cultiver, ainsi qu'il est constaté dans la lettre que me fit l'honneur de m'écrire à ce sujet M. Necker."

425. Thouin, BCMNHN, Ms. 314, labels de Badier an "observateur naturaliste pour le roy."

426. See McClellan and Regourd (2006) and its sources.

dence. Moreau's six fat volumes of *Loix et constitutions des colonies françoises de l'Amérique Sous-le-Vent* (1784-1790) resulted, for which he received upwards of 12,000 livres, a very hefty sum for this kind of technical work. Various departments of government and the entire colonial administration from the king and ministers in Versailles down to obscure agents in the backwater coasts of Saint Domingue lent their aid and offices to Moreau in this great project. The case illustrates that technical expertise deployed for colonization was not limited to science, medicine or technology, but that the law needs to be counted in this connection, too.

Moreau de Saint-Méry quickly earned the reputation as the leading authority on colonial legal and policy affairs, but he had his own scientific interests as well, and these came to the fore in his dealings with the Société Royale d'Agriculture which elected him a correspondent in 1787 and then upgraded him to an ordinary (resident) member in 1791. (At a certain moment he even filled in as secretary at the Société's meetings).[427] In 1788 and 1789 Moreau published six papers with a colonial slant in the *Mémoires* of the Société, and more than anyone he was responsible for the Société's expanded interests in colonial agronomy in these years. His first paper described a kind of cotton grown in Saint Domingue, which precipitated experiments by the Société's commissioners showing the superiority of Saint Domingue cotton over that grown in Guiana.[428] Other articles concerned potatoes, wooden locks used by slaves, making wine from oranges, and "animals useful in the colonies of France" (the latter article fifty-three pages long).[429] Moreau de Saint-Méry was no doubt responsible for the election of Badier, Nectoux, and Auvray to the Société d'Agriculture and for the publication of Badier's and Cassan's papers in the *Mémoires* of the Société. Moreau was a non-trivial intermediary between the colonies and French agriculture. His paper on cotton, for example, sought to expand production in the Antilles, whereas the one on the potato dealt with acclimatizing foreign plants in France.[430]

Moreau de Saint-Méry's other affiliations with the Colonial Machine will come as no surprise. He regularly sent materials to the Société Royale de Médecine, a connection reinforced by that of his brother-in-law, Charles Arthaud, who reported on medical experiments they undertook together.

427. Passy, pp. 296, 374, 391.

428. The interests of the Société in colonial cotton, seen in its dealings with Cassan and de Badier, are to be noted again. See Moreau de Saint-Méry (1788), and DPF, p. 1263. The reviewers for Moreau's paper were Louis-Paul Abeille and two members of the Academy of Sciences, Nicolas Desmarets and the head gardener at the Jardin du Roi, André Thouin.

429. See list in Regourd (1998). In his personal minutes of the meetings of the Société d'Agriculture, Fougeroux de Bondaroy (who was also a member of the Académie Royale des Sciences), noted that Moreau de Saint-Méry's observations "sont d'une grande utilité" and his examples "sont rares," APS-MS, B/D87, Group 26, meetings "du jeudi 30 avril 1789" and "du jeudi 2e juillet 1789."

430. It was not an accident, perhaps, that Parmentier was vice director of the Société d'agriculture when Moreau's paper on the potato appeared.

Moreau seems to have been involved in the nautical dictionary project of the Académie de Marine; and independent of the Société d'Agriculture, he was in contact with André Thouin and the Jardin du Roi. Moreau was also a leading member of the Cercle des Philadelphes in Saint Domingue, serving as the primary liaison of the colonial scientific society with institutions in the metropolis, not least the Ministry of the Marine and the Académie des Sciences.[431] Moreau was one of the architects of the association between the Cercle des Philadelphes and the Academy of Sciences, and in another of his works, the *Description de la Partie française de l'Isle de Saint-Domingue*, he boasted of having effected a formal correspondence between the Cercle des Philadelphes and the Société Royale d'Agriculture.[432] Given his strong links to Freemasonry and the contemporary *Musée* (or free-school) movement in France, Moreau's relations with the Colonial Machine per se were not unalloyed. He was not someone like Duhamel de Monceau or André Thouin who spent their entire careers within the confines of the Colonial Machine. Yet these ambiguities aside, the sum of Moreau's ties to the Colonial Machine obviously reinforced those he had with the Société d'Agriculture, and, reciprocally, they reinforced the place of the Société d'Agriculture within the larger framework of the Colonial Machine.

Prizes. Certain of its prize contests and awards also reinforced the colonially oriented character of the Société Royale d'Agriculture. In 1787, for example, the Société proposed a 600-livre prize on acclimatizing plants, and in 1789 it sponsored a contest for 300 livres on increasing cotton production.[433] The Société compiled a list of forty-four possible awards for the successful cultivation of foreign trees in France. In 1788 in a public meeting it presented such an award to the bishop of Apt, Michel Éon de Cely, for being the first person to cultivate the guava plant and the Chinese tallow tree (Suif, *Croton Sebiferum*) in the open air in France, thus "enriching our southern provinces with a new fruit and a tree useful to the arts."[434] The abbé Raynal was likewise a laureate of the Société for his support for French farmers.[435] In 1790 the Société awarded a gold medal as its "prize for rural virtue" to Jean Jasmin (Aloou-Kinson), a freed slave in Saint Domingue who founded a hospital for the poor and abandoned newborns.[436]

431. On these points, see McClellan and Regourd (2006).

432. DPF, p. 348. The election of the President of the Cercle, Auvray, to the Société d'agriculture in 1791 was a direct result of this affiliation.

433. Société Royale d'Agriculture, *Mémoires d'agriculture*, *Trimestre d'automne* 1789, pp. xix, xxiii and *Trimestre d'automne* 1791, p. xxv. The acclimatization prize was not awarded because the entries were judged too theoretical.

434. Société Royale d'Agriculture, *Mémoires d'agriculture*, *Trimestre d'automne* 1788, p. vii.

435. Société Royale d'Agriculture, *Mémoires d'agriculture*, *Trimestre d'hiver* 1789, p. xi.

436. When the Cercle des Philadelphes tried to make a similar award to Jasmin, local authorities intervened to prevent it; DPF, 411-16; Regourd (1998); McClellan (1992), p. 92; McClellan and Regourd (2006).

More notably for our purposes, in 1788 the Société d'Agriculture awarded a gold medal to the royal botanist on Île de France, Céré, for his work propagating spice plants so precious "for Agriculture and Commerce."[437] Similarly, at the end of 1789 it crowned the gardener-botanist Joseph Martin for his quasi-heroic voyage that succeeded in transshipping spice plants and the breadfruit tree from the Indian Ocean to the Americas – plants "whose propagation can contribute to the wealth of the nation and the happiness of humanity."[438]

We will have more to say later about Céré and Martin and their botanical activities. For now, their examples testify to the pivotal place and role of the Société d'Agriculture in the history of colonial botany and agronomy in the Old Regime.[439]

Related Agricultural Institutions. Other organizations are worthy of attention in this connection. One, the *Feuillle du cultivateur*, an unofficial publication that succeeded the *Mémoires d'Agriculture* of the Société Royale d'Agriculture from 1790, published a number of articles with a colonial slant through 1793.[440] More notably, four *Chambres d'Agriculture* appeared in the colonies themselves in the early 1760s, one in Martinique, one in Guadeloupe, and two in Saint Domingue.[441] A Chambre d'Agriculture was at least suggested for Louisiana before the loss of the colony in 1763.[442] Another Chambre followed for Cayenne in 1771, and overtures were made to the minister of the navy in 1773 to establish one on Île de France.[443] Pondicherry seems also to have been endowed with a Chambre d'Agriculture.[444] The Caribbean Chambres d'Agriculture had a paid representative in France who connected the colonial branches with the national organization of Chambres d'Agriculture in France.[445]

These colonial Chambres d'Agriculture were not always the sleepy groups contemporaries thought them to be, as Moreau de Saint-Méry, who was assist-

437. Société Royale d'Agriculture, *Mémoires d'agriculture, Trimestre d'automne* 1788, p. vi. In a special ceremony held on Île de France in August, 1789 the local Governor and Intendant made this award to Céré; they conveyed to Céré the particular satisfaction of the Minister of the Navy, La Luzerne. See BCMNHN, Ms. 47, Printed notice: "Agriculture. Le 17 Août 1789, en l'Hôtel de l'Intendance."

438. Société Royale d'Agriculture, *Mémoires d'agriculture, Trimestre d'automne* 1789, p. xiii.

439. See below, Part IIC.

440. Regourd (1998); Janrot; Boulaine (1990b). Subjects concerned oranges, pumpkins, okra, and corn; see *Feuille du Cultivateur*, vol. 1, pp. 497-98 (14 December 1791), vol. 2, p. 193 (20 June 1792), vol. 3, pp. 60 (16 February 1793), 104 (27 March 1793), 131-32 (20 April 1793 and 5 October 1793).

441. Regourd (1998); Regourd (2000), pp. 521-23; Durand-Morlard, passim; Banks, pp. 206-207; AN-COL A7 (1758-1760) and A8 (1761-1763); AD/Martinique, Série B, Registre 9, fols. 151-151v; registre 13, fol. 180ff.

442. AN-COL C^{13B} 1, fol. 341.

443. Ms. Code de Cayenne, vol. 5, p. 503 (12 February 1771), in AN-MAR G 237; Crépin, p. 46.

444. See reference CAOM SOM (III P. 461-9).

445. The legal records of the Chambre d'Agriculture in Martinique provide clear insight into the organization of these colonial institutions; see AD/Martinique, Série B, Registre 9, passim; AD/Martinique, Série B, Registre 13 Durand-Molard, vol. 2, pp. 85-91, 130-33, 186, 570-71; vol. 3, pp. 74-75. See also AN-COL C^{8A} 62, fol. 400 and materials, AN-COL F^3 124-26.

ant secretary of the Chambre d'Agriculture in Cap François from 1783 to 1788, repeatedly sought to deny.[446] The Cap François Chambre d'Agriculture, for example, was in touch with the Société Royale de Médecine in 1783 regarding making bread from sweet potatoes.[447] And, in Moreau's collections one can find papers read to colonial Chambres d'Agriculture about a variety of agricultural, commercial, and civic matters: the *mapou* tree, the cochineal insect, agricultural mills powered by steam engines, colonial roads, coastal and foreign shipping, colonial botanical gardens, debtors, and provisioning colonies with slaves, mules, horses, salted cod and beef. *En passant*, these papers evidence the serious interest of colonial authorities, notably the navy minister, in the activities of the Chambres d'Agriculture.[448]

Finally in this connection, we need to evoke the ephemeral *Comité Consultatif d'Agriculture*, an informal assemblage of scientific experts (Lavoisier, Poissonnier, Tillet, Darcet, and a few others) that met in 1785-1787 under the auspices of the office of the Controller-General.[449] Organized after the terrible drought and famine of 1784-1785, this Comité advised the government on actions to improve agriculture and the food supply, and it met nearly seventy times between 1785 and 1787 before being absorbed into the reformed Société Royale d'Agriculture in 1788. Lavoisier proposed that the Comité tap the Académie des Sciences for expertise outside of its immediate competence and that it use the *Journal de Physique* as a vehicle to collect and disseminate information. The Comité accepted Poissonnier's suggestion that the Comité enlist the network of the Société Royale de Médecine for the same ends, and it contacted Vicq d'Azyr to this effect; the Comité ended up mobilizing some 900 agricultural correspondents. The colonies never figured as a priority, but even so, matters concerning colonial agronomy did come before the Comité. The Comité seems to have been especially struck by the successful introduction of the cochineal insect into Saint Domingue by Nicolas-Joseph Thiery de Menonville and the initiative of the Cercle des Philadelphes to publish his manuscript – to the point where, not only did they subscribe to the publication, but the Comité proposed awarding gold medals to other "Hercules" like Thiery de Menonville, should it receive proper funding. Duchemin de l'Étang's recommendation of guinea grass as a fodder from Saint Domingue resulted in overtures to Thouin and inquiries about the plant grown in the Jardin du Roi. Thouin kept the Comité informed of the labors of André Michaux in North America and related efforts of the abbé Nolin to propagate North American

446. DPF pp. 490-491, 1030, and subsequent.

447. SRdM, Ms., #9, pp 350, 379 ("Séance du Jeudi 2 Janvier 1783 ...Séance du mardi 11 fevrier 1783")

448. AN-COL F^3 124-126. See also AN-MAR G 101, dossier 4, fols. 162-70. The *Mapou* tree produced a kind of cotton thought suitable to replace felt made from beaver fur; the Academy of Sciences became involved in testing this product.

449. See Foville and Pigeonneau, who published the minutes of this committee. See also Regourd (2008a), p. 142; Boulaine (1990a), pp. 215-216; Passy, pp. 202-208, 372.

trees in France. The plan to introduce martins and shrikes into France from Île Bourbon and frigate birds from the Americas to control insect pests received some attention, as did suggestions for improving methods of sugar refining in Saint Domingue.[450] In addition, this Comité studied the creation of a royal garden at Hyères as a new depot for acclimatizing exotic plants. A component of the larger body of institutionalized science and agriculture at the end of the Old Regime, the Comité consultatif figures as a small and ephemeral, but notable wheel in the Colonial Machine.

The Compagnies des Indes

In providing a reliable link to the Indian Ocean and Asia, the series of Compagnies des Indes served the Colonial Machine first and foremost as a shipping company.[451] For a fee, the Compagnie transported goods under seal for individuals, as it did, for example, in sending boxes of seeds, instruments, machine models, and books from Denis de Boiseler to his brother on Île de France in 1766, or in expediting a box of medical supplies the same year to the surgeon Tenebre, also on Île de France.[452] The Compagnie transported French missionaries and their effects for free, and it did the same regularly for scientists sent out by the king.[453] The Lacaille astronomical expedition to the Cape of Good Hope and the Mascarenes in the 1750s, for example, was effected entirely through the infrastructure of the Compagnie des Indes.[454] So, too, was Le Gentil de la Galaisière's decade-long sojourn through South Asia in the 1760s, as was the mission of the royal physician and naturalist Philibert Commerson (1727-1773) to Madagascar and Île de France.[455]

450. See previous note and Delaleu (1777), vol. 1, pp. 306-309, 327-29, 386-87; vol. 2, pp. 132-134.

451. On the Compagnies des Indes, see at the beginning of this Part, "The Sponsoring Authorities."

452. See relevant correspondence, SHML, 1P 284 Liasse 107, #31, #38, #49, and #64. For another example of paid medical shipments, see letter, Syndics to Buisson, "à Paris le 3 Juin 1769," SHML, 1P 284 Liasse 112, #80.

453. See, for example, ALS "Les Syndics et Directeurs de la Comp.e des Indes" to "M. La Vigne Buisson à Lorient," dated "à Paris le 2 aoust 1766," SHML, 1P 284 Liasse 107 #16.

454. Legrand and Marec, p. xi; "Note des Caisses Laissées dans les Magasins de La Compagnie appartenantes a M. De la Caille," SHML, 1P 266, Liasse 1, #17 [1754].

455. On the Compagnie and the Le Gentil expedition, see Legrand and Marec, p. xvi and SHML, 1P 279, Liasse 22, Correspondance Godeheu, especially #81, ALS "Boutin" to "M.r [Godeheu] dygoville" dated "a Paris le 7 mars 1760," wherein the Compagnie instructed its agent that "M Le Gentil de l'academie des sciences doit se rendre incessament Monsieur a L'Orient pour passer dans l'Inde où vous sçavez qu'il est envoyé par order du Roy pour y faire des observations astronomiques, quoi que je prévoye que vous aurez de la peine a le faire embarquer commodement sur le Massiac je vous prie de faire de votre mieux pour luy procurer touttes les facilités qui dépendront de vous et de faire Embarquer avec luy tous ses effets. vous voudrez bien aussy le recommander de ma part au Capitaine du Vasseau sur le quel il embarquera." Note in different hand at top: "Re[çu] le 20 que Je l'ai placé sur le Berrer, et qu'il y a une chambre." Regarding Commerson and the Compagnie des Indes, see correspondence, SHML, 1P 284, Liasse 112, #86 and Liasse 113, #2.

A Shipping Company for the Colonial Machine. The Compagnie des Indes served as an intermediary between metropolitan institutions, notably the Jardin du Roi, and correspondents abroad. Early in the colonization of Île de France, for example, Mahé de la Bourdonnais sent botanical specimens back for Dufay at the Jardin du Roi via the Compagnie des Indes.[456] Réaumur and de Jussieu received natural history specimens in 1754; the chemist Macquer acquired four packages of geological samples in 1760, compliments of the Compagnie des Indes.[457] The Compagnie forwarded to Buffon preserved birds for the Cabinet du Roi from Lanux, correspondent of the Académie des Sciences on Île Bourbon.[458] (Notably, at one point the Compagnie did not charge freight for natural history samples, "given the small value of these things"!)[459] The loss of three tea bushes sent by the Compagnie from China in 1766 was "even more grievous," said one of its directors, "since for a long time we have wanted to be able to place this tree in the Jardin du Roy."[460]

Particularly with regard to botanical shipments from China headed for the Jardin du Roi, the role of the Compagnie's own botanical garden in Lorient is not to be overlooked.[461] The count d'Angiviller at the Bâtiments du Roi was especially eager to forge links with the Compagnie des Indes and its contacts in China.[462] Then, the episode over the tiger sent aboard the *Ajax* by the governor of Île de France for the royal menagerie bordered on the comical, the Compagnie "wishing to be rid of the animal soon."[463] This traffic, borne on

456. ALS, "De fulvy, Le 6 juin 1738" to "M Despremesnil, à L'orient," SHML, 1P 278 Liasse 1.

457. Legrand and Marec, p. xi; SHML, 1P 266, Liasse 1, #1 ("Nottes des articles destinés pour M. de Réaumur, expediés à l'adresse de la Comp.ie des Indes Le 11 8bre 1754") and #29 ("Caisse plate pour M. de Reaumur; Item en carré pour idem; idem pour M. de Jussieux"). Re materials for Macquer, see 1P 279, Liasse 22, #85 (ALS, Boutin, "Ce 10 may 1760") and #84 (ALS "Boutin" dated "A Paris ce 23 Avril 1760").

458. SHML, 1P 284, Liasse 113, #60. Syndics to Buisson, "a Paris Le 4 X.bre 1769" +1P 284, Liasse 113, #61. "Copie de la lettre ecrite a la Compagnie par M. Daubenton Le 2 Décembre 1769."

459. "...vû le peu de valeur de ces objets," SHML, 1P 284, Liasse 109, #72 (letter dated "le 7 9.bre 1767"); see also #53 (letter dated "30 7.bre 1767").

460. "...cette perte est d'autant plus facheuse que depuis longtems on desire de pouvoir placer cet arbre dans le Jardin du Roy," ALS, "Les Syndics et Directeurs de la Comp.e des Indes" to "M. La Vigne Buisson à Lorient," dated "à Paris le 5 Juillet 1766," SHML, IP 284, Liasse 107, #3.

461. See instructions for the chief gardner of the Lorient garden concerning botanical shipments from China ("Copie de la lettre ecritte à la Comp.e par M. De la Verdy dattée de Versailles le 3 juin 1765"), SHML 1P 284, Liasse 104, #72 and #73 (ALS, "Les Sindics" to Buisson *et alia* dated "A Paris le 5 Juin 1765").

462. See draft in d'Angiviller's hand to "Mons. Les administrateurs de la Comp.e des indes," 4 pp. "Memoire sur les graines et plantes qu'on desireroit se procurer de la Chine soit pour Rambouillet soit pour le jardin du Roi," and other material related to these initiatives, AN O[1] 2112: Pépinières, 1783-1787. Angiviller offered to pay the Compagnie for any expenses involved in shipping plants and other specimens back to France.

463. "...La Compagnie, ayant informé le Ministre de l'arrivée à Lorient du Tigre destiné pour la Ménagerie du Roy, il est à présumer qu'il aura donné des ordres pour le faire rendre à sa destination, nous souhaitons que vous puissiés bientôt en être débarassé," SHML, 1P 284, Liasse 104, #64 (ALS, "Les sindics," dated "A Paris le 22 may 1765" to "M. Lavigne Buisson A Lorient") and related correspondence #58 (ALS "Les Sindics," dated "Paris le 13 may 1765") and #61 ("Copie de la lettre de M. Le Controlleur General ecrite à la Compagnie le 21 May 1765"). See also Roberts for related episodes.

the bottoms of the Compagnie's ships, also went in the other direction, as, for example, with the shipment of cows and sheep from Brittany to Île de France in 1736 or, even more tellingly, of fruit trees sent to Île de France by the Société Royale d'Agriculture in 1765.[464]

A Pool of Experts and Specialists. Beyond this relatively passive role, the Compagnie des Indes, like the crown, gathered unto itself a body of scientific and technical experts to facilitate its aims and operations. This finding is especially evident in the substantial cadre of medical and surgical personnel, about 500 in total, attached to the Compagnie.[465] Each of the Compagnie's ships carried at least one staff surgeon, often with an assistant or two, and the Compagnie stationed other surgeons at its entrepôts overseas: eight to ten in India, one in Canton, one in Juda (Whydah, Benin), six in Senegal from 1725, two, then five on Île Bourbon, four on Île de France, and a couple in Louisiana.[466] Residues in the archives shed a little light on these men. One Coulet, for example, chief Company surgeon in torrid Senegal, died in 1752 leaving effects worth 1500 livres; these included a silver sword, a silver watch, a porcelain coffee service, a box of wigs, a silver tobacco canister, a violin, a beaver-fur hat, a taffeta parasol, and an unspecified box of books valued at fifty-one livres.[467] The Compagnie also posted apothecaries to these locales. From 1724 it maintained a huge apothecary division in Lorient, separate from the hospital facility and headed by a chief apothecary; this bureau was responsible for outfitting each of the company's ships with considerable surgical and pharmaceutical supplies.[468] In addition to infirmaries maintained in its overseas outposts, the Compagnie also ran its own substantial hospital in Lorient from 1731 to 1759; this institution functioned on a par with those run by the Royal Navy to treat seamen and others ill in the Compagnie's employ.[469] A Madame Thierssan received 2150 livres as chief hospital administrator in 1742, and she was complemented by a physician, a barber-surgeon, an apothecary and his assistant, and a building engineer and his assistant, in addition to a nursing staff. Annual

464. Legrand and Marec, p. xiii; ALS "Ce 4 fev. 1736, De Fulvy" to "M. D'Espremesnil, à L'orient," SHML, 1P 278, Liasse 1; ALS, "Les sindics et Directeurs de la comp. des Indes," dated, "à Paris le 2 mars 1765" to "M. Le Vigne Buisson à Lorient," 1P 284, Liasse 104.

465. On medical infrastructures of the Compagnie des Indes, see Haudrère (1989), pp. 340, 591-94. See also Yannick Romieux (1987), pp. 19-33; Yves Romieux (1996); Chaligne, chapt. 4; SHML, 1P 266, Liasse 3, #44.

466. See previous note and SHML, 1P 277 Liasse 3, "Répertoire du Livre D'engagements."

467. "Inventaire et vente des effets du feu Monsieur Coulet, Chirurgien Major du Senegal mort le 3 Octobre 1752," SHML, 1P 274 Laisse 8, #66. Another inventory lists five surgery books ("État général des ouvriers et officiers mariniers, Port Louis, Isle de France, ce 1er fevrier 1758"), 1P 272 Liasse 11. The extreme poverty of ordinary seamen, some of whom left estates valued at only a few livres, contrasts sharply with the state, station, and level of education of top company surgeons.

468. Yves Romieu, pp. 120-21.

469. Chaligne, chapt. 2; see also "Copie de la lettre écrite par M. de Maurepas a M. le Controleur général le 25 aoust 1737," SHML, 1P 278, Liasse 2.

operating expenses topped 100,000 livres.[470] Finally, in this connection, the medical-surgical arm of the Compagnie des Indes blended with that of the medical establishment of the Marine Royale. The navy's own facilities trained many of the surgeons of the Compagnie, and the connection between these two elements of the Colonial Machine can be seen in the order of 1769 from the directors of the Compagnie that ships' captains and senior surgeons read and report on Poissonnier's multivolume study of illnesses of seamen "so that we can utilize the means that seem the most effective to cure these kinds of diseases."[471]

In promoting cartography and mapping the Indian Ocean in particular, the Compagnie des Indes made its most positive scientific contribution as an agent and element of the Colonial Machine. We have already encountered Jean-Baptiste-Nicolas-Denis d'Après de Mannevillette (1707-1780).[472] D'Après de Mannevillette's father was a captain for the Compagnie, and the son entered the service of the Compagnie at the age of twelve, sailing over the years to the Caribbean, the coasts of Africa, India, and China as he ascended in rank to become a full captain in the Compagnie's fleet in 1752. D'Après de Mannevillette became inspector general for the Compagnie des Indes, and he was active in Company affairs.[473] Even as the eighteenth century unfolded, the Indian Ocean was not well charted, and in 1745 he published his *Neptune Oriental* and a navigator's manual (*Routier des côtes des Indes orientales*) for sailing to the Indian Ocean.[474] The *Neptune Oriental* appeared under the auspices of the Compagnie des Indes and with the approval of the Paris Academy of Sciences, and was duly presented to the king.[475] D'Après de Mannevillette was naturally a member of the Académie de Marine, and he was elected a correspondent of the Académie des Sciences at the early date of 1743. The Parisian institution

470. "Tableau" of personnel, 1742; SHML, 1P 305, Liasse 69, Déliberations de la Compagnie, Registre 1739-1755, fol. 44v.; see also fol. 5 ("Hopital de La Compagnie des Indes"), 1P 266, Liasse 3 ("Hôpital de la Compagnie") and 1P 294, Liasse 279 ("Hôpital de la Compagnie à Lorient").

471. "Vous voudrés bien faire distribuer aux principaux chirurgiens majors et aux Cap.es des Vaisseaux de la compagnie les 12 Volumes sur les maladies des gens de mer par M. Poissonnier, que nous vous avons cy devant adressés; les engages a lire cet ouvrage, a vous en dire leur sentimens afin que d'après L'Examen qui en sera fait, on puisse Employer les moyens qui paroitront les plus convenables pour la guerison de ces sortes de maladies," Syndics to Buisson, dated "a Paris le 6 Mars 1769," SHML, 1P 284, Liasse 112, #35. On the connection with the Marine Royale, see further, Taillemite (1996), p. 320; Llinares (1996), pp. 45-47.

472. On d'Après de Mannevillette, see Filliozat (1993) and (2005), 137-40; Lacroix (1938), vol. 3, pp. 167-68; Chassagne, pp. 239-41; Taillemite (2002), p. 16; AN-MAR, C^7 6 (Après de Mannevillette); PA-DB (d'Après de Mannevillette).

473. For d'Après's influence in the Compagnie, see, for example, ALS, "Les Sindics et Directeurs de la Comp.ie des Indes," dated "A Paris, le 30 janv. 1765" to "M. Delavigne Buisson, A Lorient," SHML, 1P 284, Liasse 104.

474. See d'Après de Mannevillette (1745a), (1745b) and (1768); see also Andrew Cook, p. 174; Chassagne, pp. 239-42; DDP I:29. For background, see Le Monnier (1742).

475. Andrew Cook, p. 174; Filliozat (1993), p. 230.

backed him in a priority dispute and published two substantial pieces of his in its *Savants Étrangers*. One was his "Relation d'un voyage aux isles de France et Bourbon" of 1763, the other his general instructions of 1768 on navigation in the Indian Ocean, which the Ministry of the Marine reprinted as a separate pamphlet for distribution by the Compagnie des Indes.[476] The Compagnie created its own "Dépôt des Cartes à l'Orient" for d'Après de Mannevillette, which he subsequently maintained in Lorient under the auspices of the navy and as an off-shoot of its main Dépôt des Cartes et Plans. D'Après de Mannevillette was ennobled by Louis XV, and he became another of the key human pivots around which turned the wheels and gears of the Colonial Machine. The Compagnie des Indes made his career in science, and his career exemplifies the place of the Compagnie des Indes within the structure of the Colonial Machine.

Specialists employed by the Compagnie des Indes represented virtually every craft and trade in the Old Regime France. How much the Compagnie underwrote the botanical garden at Lorient is unclear – the staff seems to have been on the payroll of the Marine Royale, but the Compagnie did employ gardeners at seemingly all of its overseas stations. Two company gardeners toiled on Île de France in 1737 and 1738, for example, for 400 livres; one Jean Cassagne worked as a Company gardener in Senegal from 1733 through at least 1744, for 300 livres.[477] Doubtless many of these tended simple kitchen gardens, but their potential for science should not be overlooked. (One recalls that Michel Adanson, for example, worked as a clerk for the Compagnie in Senegal and did substantial botanical research while in Africa).

Company engineers are of special relevance to this discussion. A "Conseil de construction de la Compagnie" oversaw the Compagnie's works from at least the early decades of the eighteenth century, and the Compagnie kept forty-four engineers on its rolls or 4 percent of its officer corps.[478] Five were deployed to Louisiana when that colony was administered by the Compagnie, two to Senegal, and thirty-seven to Asia and the Mascarenes. Thirteen were royal engineers (*ingénieurs du roi*) detached to the Compagnie's service, yet another indication of the melding of the Compagnie with royal officialdom. Volunteers and civilian surveyors complemented the paid engineering staff.[479] The Compagnie's chief engineer, a certain de Saint Pierre, was paid a hardly insubstantial 3548 livres in 1740.[480] The Compagnie employed its own instru-

476. Filliozat (1993), pp. 100-101, 170-226; see further Bibliography 1 and Polak (1976/1983), #127-134bis for a list of d'Après' works.

477. See SHML 1P 274, Liasse 3, #2, #5, #9, and #11 ("Estat des sommes qui sont dûës par la Compagnie dans sa concession du senegal, aux ouvriers non-classez..."); 1P 274, Liasse 7, #63 ("Etat des retenues faites aux gens classés de la concession du senegal pour l'année 1743 à 1744"); 1P 277, Liasse 3, "Répertoire du Livre D'engagements."

478. Llinares (1996), p. 43; Conseil de Construction de la Compagnie des Indes, AN-MAR D1 29, 1729-1812.

479. Haudrère (1989), pp. 339-40, 760-761.

480. SHML, 1P 305, Liasse 69, fol. 17, "Gratification au sieur de St. Pierre Ingenieur en chef de la Compagnie."

ment maker at Lorient, one M. Retail (*ingénieur d'instrumens de mathématiques et d'astronomie*) to make compasses for the Company for at least two decades from about 1750; d'Après de Mannevillette praised Retail as "an excellent artisan; he and Sieur Canivet in Paris are the only two men of the sort in whom we can have confidence."[481] The Compagnie in Senegal employed at least two chemists to exploit the gold mines at Galam in the 1720s and 1730s. The first, Jean Louis Le Grand, was employed by the Compagnie in Senegal until his death in December 1730; a certain Pélays replaced Le Grand as *essaieur de mines* and chemist, but in 1734 this Pélays was murdered by slaves, porters and others in his party on a trip to the interior, and as a result the Company had to buy its gold from the natives.[482] Also standing out among such scientific and technical specialists employed by the Compagnie are an English interpreter in Senegal and Mme Arnald and her two daughters engaged in the Compagnie's service to establish silk production on Île de France and Île Bourbon.[483]

The disbanding of the Compagnie des Indes after the Seven Years War eliminated it formally as an institution of the Colonial Machine, both as a sponsoring authority and as itself a center of expertise. Yet the men, women, and institutions of the Compagnie did not disappear overnight. They were absorbed by the central government, and it may be argued that the Colonial Machine became strengthened by this move. The reestablishment of the Compagnie in 1785 heralded its return as part of the Colonial Machine and a patron of science and expertise.

Overseas Missions and Missionaries

[Jeton: French Clergy, Donation to the French Navy, 1782]

481. "Excellent artiste et qu'on peut même regarder ainsi que le Sieur Canivet à paris, comme Les deux seul hommes de confiance que nous ayons en france en ce genre," letter dated "a hennebon le 12.e fevrier 1771," SHMV, Ms. 89.

482. Froidevaux (1899), pp. 97-100; "Copie d'un mémoire présenté à M.r La Brue le 15 Fevrier 1752," in PA-DB (Adanson); SHML, 1P 274, Liasse 3, #32, "Etat des Employés Decede a la concession du Senegal."

483. Legrand and Marec, p. v; SHML 1P 274, Liasse 3, #32 ("Etat des Employés Decede a la concession du Senegal"); see also 1P, 272 Liasse 11 ("État général des ouvriers et officiers mariniers, Port Louis, Isle de France, ce 1er fevrier 1758").

The Catholic Church was an integral element of the Colonial Machine. Every French voyage and every overseas settlement had its priest, and the French Catholic Church, largely conceived, was a powerful force in overseas expansion and colonization.[484] Like the Compagnie des Indes, various Catholic orders incorporated their own experts, notably medical personnel, but Catholic religious played a variety of different roles that need to be differentiated. Even before the Colonial Machine took form, Catholic missionaries led the way as the earliest scientific rapporteurs from the world overseas. Later, missionaries functioned in situ as collectors and agents for the Colonial Machine as well as for their own international networks, including Rome. Other priests, like Father Louis Feuillée, traveled as expert botanists and astronomers on royally sponsored expeditions with no religious mission; still other individuals, such as the abbé Nolin, operated as functionaries within the Colonial Machine, where their scientific specialties trumped their religious orientations. These latter groups evidence an important trend in the history of the Colonial Machine, wherein the French government and the instrumentalities of the Colonial Machine co-opted individual Catholic religious personnel and to a large degree swamped the influence of the Church in the Colonial Machine as the eighteenth century unfolded.

Early missionary voyagers. Catholic missionaries predated the strong, direct involvement of the state and government in overseas expansion, and in a way they represent the first element in the coming-into-being of the Colonial Machine.[485] For example, in 1635 Richelieu sent the Dominican priest, Raymond Breton (1609-1679), with the original party colonizing Guadeloupe. Breton returned to France to publish his *Dictionnaire caraïbe* and *Grammaire caraïbe* in the 1660s.[486] Another early missionary, the Jesuit Jacques Bouton (1592-1658), returned from his station in the Caribbean to publish in 1640 his *Relation de l'establissement des François depuis l'an 1635 en l'Isle de la Martinique l'une des Antilles de l'Amérique. Des mœurs des Sauvages; de la situation, & des autres singularitez de l'Isle*. Likewise Richelieu's agent, the Dominican priest Jean-Baptiste (Jacques) Du Tertre (1610-1687), sojourned in the French West Indies for sixteen years in the 1640s and 1650s, returning to Paris to publish his four-volume *Histoire générale des Antilles* (1654, 1667-1671), the most complete and precise compilation of his day on the history and natural history of the young French colonies of the West Indies.[487] Dominicans like Du Tertre

484. On the French religious orders, see above, Prologue and this Part, "The Sponsoring Authorities". The starting point for French scientific missionaries overseas is Fournier; see also typed ms., "Inventaire sommaire de Colonies," AN-F^{5A} Missions religieuses (28 articles).

485. See Boucher and Deslandres for the larger background; Pizzorusso (1995) examines the missionaries of the French West Indies.

486. Regourd (2000), pp. 194-98; McClellan (1992), pp. 111-12; and Ms. "Relation de l'isle de la Guadeloupe faite par les missionnaires dominicains à leur Général en 1647," BnF, Ms fr. 24974, cited by Petitjean Roget in *Annales des Antilles*, 1963, n° 11, p. 22.

487. On Bouton and Du Tertre, see Fournier, pp. 35-40; Regourd (2000), pp. 98-99; McClellan (1992), p. 112; also Broc (1975), p. 79.

stand out in the first phase of French colonization prior to the age of Louis XIV.[488] Father Mathias Du Puis figures in this group, having spent six years (1644-1650) in Guadeloupe and returning to issue his *Relation de l'établissement d'une colonie française dans la Guadeloupe* in 1652.[489]

The Dominican priest Jean-Baptiste Labat (1666-1738) is another well-known example and the quintessential missionary in this regard.[490] Labat gave up a position as professor of philosophy and mathematics at Nancy to volunteer for service in the islands. He spent the years 1694-1706 traveling throughout the colonial and pirate worlds of the French Antilles. Mostly he stayed on Martinique, where he ran a sugar plantation for his order. He returned to France to publish his *Nouveau voyage aux Isles de l'Amérique* in 1722 and an expanded edition in 1724. French missionaries led early efforts at reporting the botanical riches of the New World, and in his botanical sections Labat, like his naturalist brothers, showed special interest in economically useful plants, presenting virtual monographs on sugar, cotton, cacao, ginger, vanilla, and other commodity products. As an applied botanist, the merry Father Labat seems to have specialized in making wines and distilling rum!

Serving the Colonial Machine. Officialdom and the Colonial Machine came to recruit Catholic priests into their direct service. This was a key development. Previously, missionaries semi-independent of the state represented the vanguard of French colonial expansion and the original scientific informants of the world overseas. As the Colonial Machine gathered momentum, the state co-opted scientifically inclined religious individuals, thus bringing the sponsoring authorities of state and church closer together in the activities of the Colonial Machine. Indicative of these new ties, the Minim priest Charles Plumier (1646-1704) made three government-sponsored trips to the West Indies between 1687 and 1697, on the first of which he was accompanied by a former Minim, the physician Surian.[491] The foremost naturalist of his day to study the flora and fauna

488. See previous notes and Mandonnet.

489. Fournier, pp. 35, 40; Regourd (2000), pp. 172-80, 200-218; Pizzorusso (2001). In this period religious personnel were the most active agents collecting colonial materials, but traveling merchants and rich collectors, if sometimes less easy to identify by historians, are not to be neglected as sources of information for French scientific networks. The manuscript descriptions of plants and animals, known as the Anonyme de Carpentras, is relevant here; see Moreau, ed. (1994); see also the published *Histoire et voyage des Indes Occidentales, et de plusieurs autres régions maritimes & esloignées, divisé en deux livres, par Guillaume Coppier Lyonnois* (Lyon: chez Jean Huguetan, 1645). See also letter relating to documents received from a French pirate who had traveled to the Strait of Magellan; AN-COL B[18] (1697), fol. 461 (20 March 1697). This kind of material deserves further study.

490. On Labat, see Regourd (2000), pp. 328-34; Boucher, passim and pp. 254-55; Fournier, pp. 63-67, and Broc (1975), pp. 79-84.

491. See Fournier, pp. 53-59; Regourd (2000), pp. 282-290; Jacquet; Allorge, pp. 144-59; McClellan (1992), p. 113; BCMNHN, Mss. 1-37, "Œuvres de botanique du Père Charles Plumier." Surian performed chemical experiments while Plumier botanized; see AN-COL B[13] (1687), fol. 25 (30 June 1687). Hrodej settles the chronology of Plumier's three trips, often mistaken in the historiography.

of the colonies, Plumier was a disciple of Joseph Tournefort, professor of botany at the Jardin du Roi. On royal orders, Michel Bégon, former intendant in Saint Domingue, recruited Plumier for his first voyage, and Plumier reported to Guy-Cresent Fagon, the head of the Jardin du Roi. Louis XIV appointed Plumier as Botaniste du Roi and gave him a pension, and the state paid for the engraving and printing of Plumier's handsome works devoted to American botany, including his *Nova Plantarum Americanum* (1703) and a posthumous *Tractatus de Filicibus Americanis* (1705). Plumier died in 1704 on his fourth voyage to Peru. He left behind some 6000 drawings of botanical and zoological specimens and twenty-two volumes of unpublished manuscripts at the Jardin du Roi, including eight volumes for an unpublished *Botanicon americanum*.

The career of the Minim priest Louis Feuillée (1660-1732) similarly speaks volumes about the government tapping religious experts and about the Academy, the Observatoire, and extraterritorial astronomy in the early decades of the eighteenth century.[492] Feuillée was named a correspondent of J.-D. Cassini in the Academy of Sciences in 1699, and Cassini sent Feuillée on a voyage to the Levant in 1699-1701 to determine longitudes of ports there. As an emissary of the Observatoire and the Academy, Feuillée traveled in the Caribbean from 1703 through 1706, and he then undertook a voyage to South America in 1707-1711. For the latter, Louis XIV charged him as *mathématicien du roi* to "transport himself to the Indies and America to make every observation necessary for the perfection of the sciences and arts, the exactitude of geography, and to improve the reliability of navigation."[493] Feuillée's astronomical work has been hailed as "the most advanced work by any astronomer in the Americas to date."[494] Coordinating with Feuillée, for example, Jacques Cassini II used simultaneous observations of the satellites of Jupiter in 1703 and 1704 to establish the coordinates of Martinique.[495] Feuillée returned to France to publish in 1714 his *Journal des observations physiques, mathématiques et botaniques faites par l'ordre du Roy sur les Côtes Orientales de l'Amérique Meridionale, & dans les Indes Occidentales*. This work reported his astronomical and geodesical observations and included many maps, and his travel reports also had a botanical/natural history dimension to them.[496] Although primarily an astronomer, Feuillée undertook his voyages as agent of Fagon and the Jardin du Roi. Indeed, Feuillée styled himself

492. On Feuillée, see Bourgeois (1970); Lacroix (1938), vol. 3, pp. 14-21; Fournier, pp. 60-63; Lescure, p. 61; Cañizares-Esguerra (2001), pp. 15-17; Kellman, pp. 177-261; Guerra (1971); Allorge, pp. 175-77; McClellan (1992), p. 119; Regourd (2000), 294-98; see also undated biographical notice by Raoul Eymar, "Le Pere Louis Feuillee, Minime (1660-1732)," PA-DB (Feuillée).

493. Lacroix (1938) quotes Feuillée's charge, vol. 3, p. 17.

494. Stehlé, p. 88.

495. See Cassini (1704); McClellan (1992), p. 119.

496. McClellan (1992), pp. 113-114.

"mathematician and botanist of His Majesty."[497] Feuillée's case well illustrates the ways in which the Colonial Machine co-opted talent in religious orders.

The importance of missionary travelers, even those sponsored by the government, lessened in the eighteenth century. Regarding many priests associated with the Colonial Machine, their religious affiliations seem largely coincidental in comparison with the concerns of their scientific avocations. Fathers Plumier and Feuillée already illustrate the point, but it is no less plain in the cases of the many abbés who populated the Colonial Machine: Abbé Rochon (*pensionnaire* in mechanics at the Académie des Sciences, traveler to the Indian Ocean, and librarian at the Académie de Marine), Abbé Nolin (director-general of the king's nurseries), Abbé Boscovitch (naval astronomer and optician), the abbé Pézenas (naval astronomer at Marseilles), Abbé de Lacaille, Abbé Pingré, Abbé Cotte, Abbé Tessier (director of the royal nurseries at Rambouillet), Abbé Rozier (publisher of the *Journal de Physique*), Abbé Nouet (priest with the Puységur expedition to the West Indies and later an astronomer at the Observatoire), not to mention the abbé Bignon, royal librarian and administrative superintendent of the Académie des Sciences.[498]

Missionaries in situ. For the most part, the cadres of priests and nuns in the colonies, even those with some expertise, merely added to the weight of French colonization in their respective locales. But a handful of other religious in situ were more active agents of the Colonial Machine. In Canada, the Jesuit Joseph-Pierre de Bonnecamps (also Bonnécamps, 1707-1790), was royal professor of hydrography and mathematics at Quebec from 1741 through the fall of that city in 1759. In 1749 La Galissonnière sent him to accompany a French expedition down the Ohio River Valley, detailing its flora and fauna, making astronomical observations, and producing a map of the region. Bonnecamps published the account of his voyage, and his map found its way to the Dépôt des Cartes et Plans of the Marine.[499] And, through the 1750s Bonnecamps was in contact with Delisle and La Galissonnière over cartographical issues, the former of whom shared Bonnecamps's researches with the royal geographer Buache.[500] Similarly in Canada through the 1730s and 1740s naval authorities vigorously encouraged a certain Sieur Gosselin, the missionary of the Missions Étrangères at Yamasqua, to collect botanical specimens for the Jardin du Roi in Paris, at one point paying him an incentive of 150 livres for this work.[501]

497. "Mathematicien et Botaniste de Sa Maiesté," letter dated "de Marseille le 24e Novembre 1716" to "Monseigneur Votre Altesse Royale," PA-PS (24 November 1716).

498. On the little-known Nouet in particular, see AN MAR C^7 228 (Nouet) and Bret (1993).

499. Audet, pp. 30-35; Lortie, pp. 15-17. Bonnecamps' report is reprinted in Thwaites, vol. 59 (1899), pp. 149-199; see also AN-COL B^{89} (1749), fol. 12 (11 April 1749) regarding the ministry providing Bonnecamps with scientific instruments.

500. See AN-COL, B^{108} passim; AN MAR 2JJ 66, 68; for Bonnecamp's observations received at the Observatoire, see OdP, A 7, 6 [64^3] "Observations météorologiques."

501. See ministerial correspondence that regularly exhorted Gosselin to keep up his collecting for the Jardin du Roi: AN-COL B^{63} (1735), fol. 507 (24 May 1735); B^{64} (1736), fol. 435 (26 April 1736); B^{66} fol. 32v. (13 May 1738); B^{72} fol. 13 (8 April 1741); B^{76} fol. 24 (15 April 1743) and fol. 86 (21 May 1743); and B^{78} fol. 61 (17 April 1744), where the payment of 150 livres is noted.

In Louisiana, Father Le Maire of the Missions Étrangères and a missionary at Île Dauphine from 1706, wrote a "Relation de la Louïsiane," which is mostly a description of the botanical riches of the region.[502] In 1716 Le Maire passed on his account to his fellow priest, Jean Bobé (1654-1735), who was one of the directors of the Compagnie des Indes and who was passionate about Louisiana. Bobé himself published maps of the region, and he distributed Le Maire's material to the rest of the Colonial Machine, including Le Maire's maps that made it to the high council of the Marine.[503] Le Maire himself was also in contact with A.-T. Danty d'Isnard, professor of botany at the Jardin du Roi and associate botanist at the Académie des Sciences. In the 1710s, from the wilds of that part of Louisiana, now Kansas, Father Gabriel Marest, missionary of the Immaculate Conception, forwarded hydrographical, agricultural, natural history and mineralogical information to the authorities in France.[504] Later, in 1765 the Académie des Sciences christened the Franciscan Father Archange, who was on the verge of departing for Louisiana, a correspondent; and in 1768, as control in the colony passed to Spain, Archange was able to send a packet of scientific memoirs to the astronomer Le Monnier back in France.[505]

On Martinique, at the turn of the eighteenth century, the Jesuit Yon, the Jesuit mission's apothecary, sent botanical samples (Martinique "tea") back to Lémery at the Jardin du Roi.[506] Yon's colleague and contemporary on Martinique, the Jesuit Adrien Le Breton passed on a substantial collection of seeds to the Jussieux at the Jardin du Roi and to Gouye and the Académie des Sciences.[507] Yon's and Le Breton's materials were duly noted in the *Histoire* of the Académie for 1702, 1703 and 1704. This same Le Breton also left substantial manuscripts dealing with Caribbean botany.[508] Another Franciscan on Martinique, Father Christophle, contributed his observations of the transit of Venus of 1769, and at the end of the century Maxime Duval, the priest-apothecary on nearby Saint Lucia, offered his chemical analysis of the local mineral waters.[509]

502. BCMNHN, Ms 948: "Relation de la Louïsiane par Lemaire"; see also Duprat, p. 232; Langlois (2003), p. 345.

503. DPF, 223; Petto, pp. 103-105; Dawson, pp. 109-116; Broc (1975), p. 127; see also re Lemaire, AN-COL C^{13A} 4, p. 205 (29 August 1716); re Bobé, see AN-COL C^{13A} 6, fol. 295 (1722).

504. AN-COL C^{13A} 2, p. 775 (9 November 1712).

505. AN-COL C^{13A} 45, fol. 223 (18 January 1768). Archange was never officially confirmed as the Academy's correspondent because it had limited the number of its *correspondants* to 100; see PA-PV 84 (1765), fol. 135 v ("Samedi 16 Mars 1765").

506. HARS 1702, p. 49.

507. HARS 1703, p 57; HARS 1706, p. 42; Lapierre, pp. 27-33. Breton apparently returned to France and was seemingly used as a kind of consultant to the Jardin du Roi regarding matters American; see BCMNHN, Ms. 668.

508. BCMNHN, Mss. 63, 667-68, 937-39; Lapierre.

509. MARS 1772, pp. 421-25; letter of La Borie, dated 8 juin 1786, AN-COL C^{10} C 3, fol. 37-49v; for background, see Kacy.

In general these missionaries were not as bereft of intellectual resources as might be thought a priori. The Jesuits on Martinique in particular maintained a substantial library at the mission home at Saint Pierre. As the missionary Jean Mongin wrote already in 1679:

> We are not deprived of the satisfaction of having the best and most recent books. We have a library that is very well stocked because we are regularly sent the best of what is published [in France.] The provinces [in France] are very much closer to the source of these wonderful things, but they do not receive them any sooner or more regularly than we do here....The *Journal des Savants* are all sent to us three or four times a year at the least, accompanied by all sorts of new imprints.[510]

Missionaries on Saint Domingue were steadfast agents. Through Gouye and in conjunction with de La Hire at the Observatoire, Jesuit Father Boutin observed lunar eclipses in 1706 and 1707 to establish the longitude of ports in Saint Domingue.[511] (Father Feuillée was observing the 1706 eclipse at the same time in Martinique). Boutin was apparently still sending astronomical observations back to France as late as 1742.[512] The first missionary naturalist to be permanently stationed on the island was the Jesuit Jean-Baptiste Le Pers (1675-1735). Le Pers came to Saint Domingue in 1709 and remained there as curate at Limonade in the northern part of the island until his death three decades later. Le Pers took up a systematic studies of the island's botany, having read the famous botanist and academician Tournefort, and he compiled a large manuscript of eighteen years' worth of notes and observations.[513] In 1730 a Jesuit colleague of Le Pers, P.-F. Xavier de Charlevoix (1682-1761), published this manuscript in France in two quarto volumes with government support as the *Histoire de l'Isle espagnole ou de S. Domingue*. The Dominican, J.-B.-M. Nicolson (1734-1773) served as Dominican superior in Léogane, and his *Essai sur l'histoire naturelle de St. Domingue* was posthumously published on royal command in 1776.[514] Later still in the century, Brother Gerard, superior of the

510. Quoted in Chatillon, p. 63. Traveling to the islands in 1703-1706, Father Feuillée confirmed the availability of printed matter in Martinique; Feuillée, p. 179; see also Regourd (2000), pp. 192-93; Petitjean Roget, p. 68; Gelfand (1987), p. 94, mentions the substantial library of the Jesuits in Quebec.

511. See La Hire's papers, MARS 1706, pp. 481-82; MARS 1707, pp. 381-82; Cassini II (1708), pp. 15-16; see also HARS 1706, p. 113; HARS 1707, pp. 82-83, and Guillaume Delisle, pp. 251-52 ; see also portrait of Boutin in letter of the Jesuit Margat, dated "Au Cap, ce 20 juillet 1743," in the *Lettres édifiantes et curieuses* reprinted in Reichler, pp. 277-90, esp. p. 288.

512. *Mémoires de Trévoux* (1742), p. 1676; OdP, Ms A 4 6, dossier 4 ("Relation Observations de la comète de 1742, tirée de différentes gazettes").

513. On Fathers Le Pers and Charlevoix, see Fournier, pp. 48-52; DPF, p. 270; McClellan (1992), p. 112.

514. Fournier, pp. 68-72; McClellan (1992), p. 114.

Brothers of Charity in Cap François, was sending seeds back to d'Angiviller and the Jardin du Roi.[515] And finally in this connection, Abbé de la Haye, curé of the remote parish of Dondon in the mountains of northern Saint Domingue, stands out as a serious botanist and botanical illustrator and the prototype of religious agents in the colonies in the service of the Colonial Machine. De la Haye befriended Palisot de Beauvois, the correspondent of the Académie des Sciences botanizing in Saint Domingue in the 1780s, and de Beauvois praised de la Haye as "the only man in the whole colony from whom one can get both exact information and decent specimens."[516] De la Haye provided the illustrations for Thiery de Menonville's 1787 *Traité de Nopal*, published by the Cercle des Philadelphes. Himself a colonial associate of the Cercle des Philadelphes, de la Haye produced and illustrated his own substantial botanical manuscript, *Florindie, ou Histoire phisico-économique des végétaux de la Torride*. Through de Beauvois and A.-L. de Jussieu in France, the Cercle des Philadelphes launched plans to publish de la Haye's work, and La Luzerne subscribed for ten copies, but 1789 intervened.[517] Beyond that, de la Haye sent descriptions and illustrations of colonial plants to the Académie des Sciences, and both with de Beauvois and on his own, he communicated with Thouin at the Jardin du Roi from 1783 apparently through 1801.[518] He experimented à la Parmentier with potato flour, and he sent meteorological reports back to the minister in France as well.[519] De la Haye lived and worked at the far reaches of the French colonial world in the eighteenth century. His ministry and his science encapsulate the essential place and role of religious missionaries in the Colonial Machine.

Many religious personnel were specialized health-care providers, and running local hospitals and medical facilities was one way local priests and nuns fulfilled their religious callings and attended to the faithful. The best example is the substantial hospital founded in 1698 at Cap François in Saint Domingue by the Brothers of Charity.[520] The Charity hospital was an official royal hospital that received letters patent in 1719. Located on a large property just outside of town, it had twelve medical wards and treated upwards of 800 patients at any one time. The corps of royal doctors and surgeons was nominally

515. BCMNHN, Ms. 1327: (1790) ("Du Pere Gerard Superieur de la Charité. Recu 36 Especes de Graines récoltées a S.t Domingue"). See also BCMNHN, Ms. 691, fols. 10-10v.

516. ALS, Palisot de Beauvois to Jussieu, dated "Du Cap le 15xbre 1788...," PA-DB (Palisot de Beauvois), quoted in McClellan (1992), p. 351; see also pp. 114, 161.

517. McClellan (1992), pp. 246-47 and notes; the Revolution put a stop to this project. De la Haye's manuscript can be found SHMV, "Manuscrits de la Bibliothèque historique de la Marine 0100. XVIII Florindie...par de la Haye."

518. McClellan (1992) p. 267; BCMNHN, Ms. 1289 ("Lettre à l'Académie des Sciences, avec la description de quelques plantes de Saint-Domingue, par l'abbé de La Haye"), and other contact of De la Haye with Thouin at the Jardin du Roi/Muséum d'Histoire Naturelle, BCMNHN, Ms. 314 and Ms. 2310.

519. See reference, Moreau de Saint-Méry (1789), p. 50, and meteorological report, "Signé Delahaye Curé du Dondon," AN-MAR G 101, Dossier 3, fols. 77-84.

520. McClellan (1992), pp. 92-94.

responsible for medical treatment, but the Brothers of Charity ran the hospital and provided much of the medical treatment. Not unexpectedly, tension arose between the official medical establishment and the priests who were themselves expert medical providers.

Government officials coordinated with religious authorities to see that colonial outposts were provided with staff to attend to the spiritual and health needs of colonists, as, for example, when the navy ministry solicited the Brothers of Charity for priests and medical attendants for the settlement at Louisburg at the mouth of the Saint Lawrence River.[521] In this context we need to highlight French nuns, who were particularly present in the system of colonial hospitals.[522] On Île de France, for example, the *Sœurs Hospitalières de Chartres* held the directorship of the royal hospital.[523] Their counterparts from Dieppe served in the hôtel-dieu in Quebec.[524] In Louisiana the Ursuline order ran the hospital in New Orleans and other "grey sisters" were at least requested for Mobile.[525] Nuns specializing in pharmacy and midwifery figure among this personnel.

The Jesuits... in Asia. The Jesuit order constituted a key, if multifaceted, element in the Colonial Machine. The Jesuits' stranglehold on teaching in France and in Canada has been noted. Many of the early missionaries just surveyed as well as those on station in the Americas were members of this order, including in Cayenne.[526] The Jesuits also published two journals that independently disseminated information from the colonial world to readers in France: the *Lettres édifiantes et curieuses* (34 vols., 1702-1776) and the *Mémoires de Trévoux* (265 vols., 1701-1767).[527] The latter was a more general journal of science and learning, and among its regular reports one reads the observations of the royal doctor in Canada on local waters; the Jesuit Berthier's astronomical observations of the 1761 transit of Venus; A.-F. Boureau-Deslandes, of the Academy of Sciences and the Ministry of the Marine, on fresh water aboard

521. AN-COL B^{106} (1757), fol. 161 (14 December 1757).

522. For background, see Dufourcq.

523. AN MAR C^{7} 161 (Lajust).

524. See letters patent "pour permettre l'établissement des religieuses hospitalières de Dieppe à l'Hôtel-Dieu de Québec," AN COL C^{11A} 1, fol. 117 (April, 1639); for a quick overview of this point and bibliographical references, see Gelfand (1987), 85-87.

525. See Clark, passim and pp. 43-46, and ministerial correspondence, AN COL C^{13A} 23, fol. 109 (12 March 1738), fol. 119 (15 March 1738), C^{13A} 28, fols. 342-45 (31 December 1744), C^{13A} 35, fol. 367 (21 April 1751).

526. For Jesuits and science, see Romano (2000, 2002); for more on the Jesuits in Cayenne, see AN-MAR G62, fol 210vff; AN-MAR G101, fol. 184.

527. See entries in Sgard. Lénardon provides an entrée to the contents of the *Mémoires de Trévoux*. Several contemporary and modern editions of the *Lettres édifiantes et curieuses* have appeared, including the massive bilingual compendium of Thwaites; see also editions, selections, and introductions by Desbarats; Reichler; and Vissière and Vissière (1979) and (1993) with bibliographic information, p. 394.

ships; and so forth.[528] The *Lettres édifiantes* concentrated more on frontline evangelical triumphs among the heathens, but not without attention to God's work in nature; as the editors put it, "Our Fathers have not omitted material they thought should be agreeable to the learned and curious. Observations are included that can be useful for geography and astronomy and that will serve to rectify our maps and make navigation more secure every day, along with exact and singular descriptions of rarities they have encountered in the lands where Providence has led them."[529] Thus, to pick two examples from among hundreds from the *Lettres édifiantes*, the Jesuit Père Taillandier reported from Cap François in 1707 on the similarity of bushes in Saint Domingue to the tea plant in China, a similarity he hoped to investigate further when he arrived on the other side of the world in the East. In 1730 Father Papin wrote in on medical practice in India.[530]

All that notwithstanding, the role of the Jesuits in the Colonial Machine was greatest in China and India.[531] The Jesuit outposts there loom large in the landscape of this account for two reasons: first, existing as they did at the outer limit of the contemporary French colonial effort, the Jesuit missions reinforced patterns of trade by the Compagnie des Indes and life in the *comptoirs* in India and the formal colonies on Île de France and Île Bourbon; and secondly, the Jesuits were nontrivial scientific rapporteurs imbedded in the Colonial Machine.

The Jesuits reached China at the turn of the seventeenth century, but the separate French Jesuit mission began with six Jesuit mathématiciens du roi sent out by Louis XIV via Siam (Thailand) in 1685.[532] These emissaries and their succes-

528. *Mémoires de Trévoux* (1735), pp. 956-59; (1761), pp. 1720-1726; (1730), pp. 409-423.

529. "Nos Pères n'ont pas omis ce qu'ils ont crû devoir estre agreable aux Sçavans & aux Curieux. Il y a des observations qui pourront estre utiles pour la Géographie & l'Astronomie, & qui serviront à rectifier nos Cartes, & à rendre la navigation plus seure de jour en jour, avec des descriptions exactes & curieuses de ce qu'ils ont rencontré de rare dans les païs où la Providence les a conduits, Lettres édifiantes et curieuses Écrites des Missions Étrangeres, par quelques Missionnaires de la Compagnie de Jesus," vol. 3 (Paris: Chez Jean Barbou, 1713), pp. ii-iii.

530. "Lettre du P. Taillandier, missionnaire de la Compagnie de Jésus, au P. Willard de la même compagnie, À Pondichéry, le 20 février 1711," *Lettres édifiantes et curieuses*, vol. 11 (1715), pp. 92-179, here pp. 99-101, reprinted in Reichler, pp. 215-42, here, p. 217; "Lettre du Père Papin au Père le Gobin, sur les Arts & sur la Médecine des Indiens," vol. 19 (1730), pp. 418ff; see also Desbarats, p. 239.

531. The standard reference work is Pfister (1932-1934). Hsia (2009) is now the starting point for any consideration of this topic. Harris (2005) provides an important review of the literature of Jesuit scientific activity overseas in the early modern period generally. He notes, p. 72, "At its peak around 1750, the Society operated more than 500 colleges and universities in Europe, a hundred more in overseas colonies (mostly in Spanish America), and roughly 270 mission stations scattered around the globe. The Society, in other words, presided over one of the most extensive and complex institutional networks of the *ancien régime*." See also the special section of the *Annales Histoire, Sciences Sociales* presented by Jami (2004) and her introduction, pp. 697-99; Vissière and Vissière (1979), pp. 13-27, provide additional background, as do Li; Duteil; and Jami (2008).

532. Landry-Deron provides the authoritative account; see also retrospective letter of 1703 Father Fontaney transcribed in Vissière and Vissière (1979), pp. 117-43; Duteil, 41-44; Harris, p. 74, who notes: "For the next half century, French Jesuits dominated – and altered the character of – the mathematical sciences in the China missions."

sors were not lone agents driven to the Asian wilderness by the spirit of Saint Ignatius; they were, rather, official parties sponsored by French authorities and connected to the Colonial Machine. For example, prior to his departure for China the Jesuit Antoine de Beauvollier, known as Father Barnabé, received his official appointment from Louis XIV as mathématicien du roi in 1688: "We wish that in this capacity he transport himself to India and to China for the sake of the arts and sciences in order to make all the observations necessary for the perfection of geography and to put navigation on an ever more sure footing."[533] Once in China, French Jesuits maintained close contact with the scientific establishment back in France, notably with the Observatoire and the Académie des Sciences, which directed their research activities.[534] We have already quoted the 1686 letter from the Jesuit astronomer Jean de Fontaney that called the Jesuit mission in Beijing an outpost of the Academy in Paris. In that same letter he went on to express his hope that "we can do something in this great Empire for the glory of God and for the perfection of the sciences. I say the same thing to the members of the Academy, to whom and on their account I have the special joy of seeing myself united. Everything we do in China is their work because it will be done according to the instructions we have taken from the Academy."[535]

A succession of French Jesuits maintained their own house in Beijing.[536] The Académie des Sciences elected a number of these as its correspondents: Joachim Bouvet (1656-1730), Antoine Chomel (1668-1702), Jean de Fontaney (1643-1710), Jean-François Gerbillon (1654-1707), Jean-Alexis de Gollet (1664-1741) Father Saint-Leu (1668-1700), Claude de Visdelou (1656-1737), Antoine Gaubil (1689-1759), and Pierre Noël Le Chéron d'Incarville (1706-1757). These Asian correspondents connected to the Academy especially through the Jesuit honorary member Thomas Gouye. Working through Réaumur, a certain Father Orry, although never officially attached to the Académie, seems to have

533. "voulons qu'en cette qualité il puisse se transporter aux Indes et a la Chine, pour y faire toutes les observations necessaires pour la perfection et la curiosité des arts et des sciences l'exactitude de la Geographie, et establir de plus en plus la sureté de la navigation...," letters patent, dated 12 fevrier 1688; AN-MAR C^7 23 (Beauvollier).

534. On data sent back to the Observatoire and the Académie from the Jesuit "mathématiciens du roi," see "Extrait des registres de l'Académie des sciences," OdP, B 2.7, pp. 200, 286; "Observations Astronomiques faites a Lounau dans le Royaume de Siam l'an 1687, par les PP. Jesuites envoyes a la Chine," OdP, B 5, 2 [8*], pp. 749-54; Hsai; Duteil, pp. 278-333, also provides many details on Jesuit scientific activity in China.

535. "...nous puissions faire quelque chose dans ce grande Empire pour la gloire de Dieu et pour la perfection des sciences. Je dis la mesme chose a Messrs. de l'academie, a la quelle j'ay une joye tres particuliere de me voir uni a cause d'eux. Tout ce que nous ferons a la Chine est leur ouvrage, par ce qu'il sera fait selon les instructions que nous avons tirées de l'academie...," Fontaney letter dated "A Siam le 6e May 1686," OdP, B 4.10. On later direct contact between the Academy and Jesuits in Beijing, see letters of 1730, 1735, and 1740 from Father Parennin to Dortous de Mairan at the Academy, reprinted in Vissière and Vissière (1979), pp. 358-98.

536. Location #81 on the map of Beijing, RS/L, L&P III, #322a; see also *Phil Trans* 50 (1758), pp. 704-26; Vissière and Vissière (1979), p. 58.

succeeded Gouye in this role.[537] The role of the astronomer and academician Joseph-Nicolas Delisle as a conduit linking missionaries in Asia with the scientific establishment in France has already cropped up. The royal observatory in Marseilles and Jesuit observers in China likewise maintained contacts.[538] Duhamel de Monceau provided botanical instructions for missionaries to China.[539] The count Angiviller, head of the Bâtiments du Roi, played this role later in the eighteenth century with regard to matters botanical and natural historical, writing that "one can usefully employ French or Chinese missionaries there who are salaried by the Emperor. They need the Compagnie des Indes for their correspondence with Europe, and it is likely that they will be anxious to fulfill the requests of its agents in China."[540] As much as the Jesuits in Asia served as informants to savants in Europe, so too did the latter, particularly Delisle, insure that the Jesuits in Asia were adequately supplied with the latest scientific publications from Europe; the link between Paris and Beijing has to be seen as a two-way street, albeit a long and winding one.[541]

Of the many French missionaries in China in the eighteenth century, two are exemplary in melding science and service to God, Antoine Gaubil and Pierre d'Incarville. Gaubil arrived in China in 1722 and remained there for the next thirty-six years, never to return to Europe. On his way he stopped at Île Bourbon to conduct declination experiments, and he went on to be the best astronomer and historian among the Jesuits in China in the eighteenth century.[542] His initial interests were in chronology, and he became caught up in controversies over the chronology of Chinese empires. He later turned to Chinese astronomy.[543] He became an official correspondent of Delisle and the Académie des

537. See his letter to his counterparts in China, dated "à Paris le 21 déc 1722," wherein he writes, "...Mr.s de L'académie des Sciences avec lesquels je me suis lié cette année plus etroitement pour entretenir entre vous et eux une correspondance reglée. Ils peuvent nous faire honneur et Mr. de Reaumur entre autres qui a tout credit auprès de Mr. le Duc d'Orléans et dans cette Académie peut nous faire beaucoup de bien," in PA-DB (Gaubil).

538. See Gaubil letter to Delisle, dated "à Peking Ce 2 7br 1756," AN-MAR 2JJ 68.

539. APS-MS, Duhamel du Monceau/Fougeroux de Bondaroy papers B D87 Group 24, ms in Duhamel's hand: "avis donnes a deux pretres missionaires partants pour la chine."

540. "on pourroit y employer utilement les missionnaires soit francois soit chinois qui sont entretenus par l'empereur. Ils ont besoin de la compagnie des indes pour leur correspondance avec l'Europe, et il est probable qu'ils s'empresseront de remplir les demandes des chef de son administration à la chine...," 4pp. "Memoire sur les graines et plantes qu'on desireroit se procurer de la Chine soit pour Rambouillet soit pour le jardin du Roi," AN O[1] 2112: Pépinières, 1783-1787.

541. Note remarks by Father Gaubil in 1732: "J'ai la Geographie réformée de Riccioli, les voiages ou observ. de MM. de L'Academie, de l'Imprimerie Roiale 1693, les Mem. de l'Acad. jusqu'en 1725 avec les 3 recueils du P. Gouie 1688, 1692, 1693, les transact. de Luthorp jusqu'en 1700, les actes de Leipsick jusqu'en 1724, l'ancien Journal du P. Feuillée," "Copie de la lettre du P. Gaubil au P. Souciet de Pekin le 7e Juin 1732, reçue à Petersbourg le 9 Nov. 1733 N. St." AN-MAR 2JJ 62.

542. On Gaubil, see Fournier, p. 76; von Collani; Pernet; article by Henri Cordier in the *Catholic Encyclopedia*; and materials in PA-DB (Gaubil). Renée Simon published Gaubil's letters from Beijing.

543. "What I have thus done in chronology, I am desirous of doing with regard to astronomy, or, at least, furnishing materials ranged in order upon that science," "Extracts of two Letters from Father Gaubil, of the Society of Jesus, at Peking in China," RS/L, L&P II, #385; see also *Phil. Trans.* 48 (1753), pp. 309-317.

Sciences only in 1750, but he had been in active contact with Delisle from at least 1731, passing on observations of the satellites of Jupiter, of lunar and solar eclipses, of conjunctions and other astronomical phenomena from China to Delisle and to Cassini, Maraldi, and de Mairan at the Académie des Sciences. Delisle reciprocated with astronomical and cartographical information from France and Russia.[544] Gaubil translated many works on Chinese astronomy and chronology, and he supervised the production of maps of China, Tibet, Tartary, Korea and other parts of Asia, including the best to that point on the islands of Japan.[545] Gaubil's science and scholarship appeared in print in France in a series of volumes edited by Étienne Souciet (*Observations mathématiques, astronomiques, géographiques...aux Indes et à la Chine*, 1729-1732), in the Jesuit *Lettres édifiantes et curieuses*, and in the *Connaissance des temps*. Gaubil was likewise elected a corresponding member of the Imperial Academy at Saint Petersburg and Fellow of the Royal Society of London. For a while at least, the Jesuit Michel Benoist took up Gaubil's mantle after his death as the contact person in China for French astronomy.

Although a generation younger, Pierre d'Incarville was Gaubil's contemporary equivalent for Chinese botany, d'Incarville having been labeled "the first professional botanist in China."[546] He joined the Jesuits in 1727 and spent the years 1730-1739 at their mission in Quebec, where he taught grammar and rhetoric and where he observed similarities between Canadian and Chinese ginseng, an observation with significant economic consequences for the Compagnie des Indes. On his return to France d'Incarville studied botany in Paris with Bernard de Jussieu and chemistry with Hellot and Geoffroy of the Académie des Sciences. He wrote about his preparation:

> When my superiors had destined me for the Missions, I thought that the knowledge of some medicinal Plants would be of use, in order to make up some galenical Medicines occasionaly, for our poor Christians, who sometimes have not the necessary assistance. With this Intent I applied to Mons. Geoffroy, and beg'd leave to come to his house, to learn some Compositions. He received me in the most obliging manner, and would himself be my master. As soon as he know that I was to go to China, he advised me to learn a little Natural History; to wch I was easily persuaded, from a natural Propensity....M. Geoffroy procured me the acquaintance of M. Bernard de Jussieu, who was so good to instruct me

544. See Delisle correspondence, OdP B 1, 10 [150] and AN-MAR 2JJ 62-66, 68. On Gaubil contact with Paris Academy of Sciences, see, for example, PA-PV 69 (1750), p. 314 ("Samedi, 22e Août 1750").

545. See previous notes and Comentale; Szczesniak; Duteil, pp. 294-97, and list of Gaubil's publications, pp. 361-62.

546. Bernard-Maître, p. 338 (p. 5 of off-print). On Incarville, see Bernard-Maître; Fournier, pp. 76, 79-83; Lacroix (1938), vol. 4, pp. 138-40; Allorge, pp. 249-50; PA-DB (Incarville).

> in Botany: thus instead of some usefull Plants which I aim'ed to knowing, I found myself embark'd in Botany by Principles; because M. de Jussieu intended that I should extend my views beyond Charity to our poor Christians, & be of Service to the Botanic Garden. Such are the sources of the little knowledge I have acquired in Natural History and Botany....[547]

D'Incarville moved on to China in 1740, where he came to serve as director of the imperial gardens in Beijing with access to the emperor, all of which increased his abilities to send Chinese botanicals back to France.[548] D'Incarville and Gaubil proposed themselves as correspondents of the Académie des Sciences in 1750, a proposition warmly welcomed, with d'Incarville becoming the correspondent of Geoffroy.[549] D'Incarville compiled a French-Chinese dictionary of nearly 1200 manuscript pages, and he served as a conduit for passing on technical information to French authorities concerning the silk industry, cotton manufacture, dyeing, fireworks, candle making, and other mechanical and fine arts.[550] He published three miscellaneous pieces in the Académie's *Savants Étrangers* series, and in the 1740s and 1750s he continued in close and sustained contact with Bernard de Jussieu, his brother Antoine, the academician Geoffroy, and Controller General Machault. D'Incarville was instrumental in getting Chinese plants, notably the golden rain tree (*Koelreuteria paniculata*) and the pagoda tree (*Sophora japonica*) to the Jardin du Roi in Paris.[551] Botanical exchanges went both ways, with d'Incarville receiving plants and pictures of plants from France for presentation to the emperor of China, who seemed especially to enjoy colorful flowers. D'Incarville sent substantial herbaria of his own, copies of historical Chinese herbaria, pharmaceutical collections, and a few natural history specimens to the Jardin du Roi and academicians (notably Hellot) in France via the infrastructure of the Compagnie des Indes. In addition, he, like Gaubil, made use of the slow-moving annual caravan between Beijing and Moscow to send his communications back to Europe. D'Incarville undertook a variety of assignments, including raising silkworms, for Machaut (controller-general and keeper of the seals) and Trudaine (*Intendant des Finances* and founder of the *École des Ponts et*

547. RS/L, L&P II, #153: Ms. headed: "A Letter from Father d'Incarville to Dr. Mortimer Secr. R.S. Dated at Peking, Nov. 11: 1748." Contemporary translation.

548. In this connection see in particular Incarville letter [to B. de Jussieu] "à Peking ce 27 8bre 1753," PA-DB (Incarville).

549. PA-PV, 69 (1750), p. 393 ("Samedi, 22e Août 1750"): "sur La Proposition faite à L'Académie des P.P. Gaubil et D'Incarville Jesuites Missionnaires à la Chine pour ses Corresp.es elle les a agréé L'un et L'autre et m'a chargé de leur en expédier des Lettres."

550. Incarville letter to Père Fiteau, 13 octobre 1754, reprinted in Bernard-Maître, pp. 40-42.

551. See previous notes and Spongberg. On d'Incarville's publications, see Halleux et al. The original *Sophora japonica* is still alive and can be seen today in the Jardin des Plantes in Paris.

Chaussées) back in France. The library that he and Father Gaubil collected for the French mission in Beijing included the *Mémoires* of the Académie des Sciences and of the Académie des Inscriptions et Belles-lettres, Gouye's publications, ten volumes of the *Encyclopédie*, various of the *Phil. Trans.* of the Royal Society of London, the *Description des arts et métiers*, the *Acta Eruditorum*, the *Mémoires de Trévoux*, and Feuillée's *Journals*. D'Incarville died at the age of fifty-one, two years before Gaubil. He was aware of the secular and the sacred in his science, commenting: "I work on this in my free time. Everything is good in God's eye when it refers back to Him. And then, I render our Mission a service in meriting the protection of the King and his ministers." He closed another letter to Jussieu with the prayer, "May God grant the fruition of the projects I conceive for His glory and the advantage of our France."[552]

Missionary outposts in Pondicherry in India complemented those in China.[553] As a mid-eighteenth-century manuscript described Pondicherry: "This city is the foremost of the *comptoirs* of the Compagnie des Indes....The Compagnie has a magnificent garden there that serves as a public space. The Jesuits have a college where they teach reading, writing and mathematics. The Missions Étrangères have a House there, the Franciscans a monastery, and the Hindus two pagodas"[554] At one point, working through the elder Cossigny, the correspondent of the Académie des Sciences on Île de France, the Franciscans in Pondicherry collected meteorological data for Réaumur.[555] As early as 1733, Indian rulers brought Jesuits to their courts, in part because of their astronomical expertise.[556] The Jesuit Gaston-Laurent Cœurdoux (1691-1779) was France's most notable priest-scientist in India in the eighteenth century. In addition to being an Orientalist and the linguist who remarked on the kinship of Sanskrit with European languages, Cœurdoux maintained a regular scientific correspondence with Delisle in the 1750s and 1760s. Cœurdoux sent Delisle his own and others' observations of the 1753 passage of Mercury, the return of Halley's comet in 1759, the 1761 transit of Venus, and other eclipse and cometary observations.[557] Cœurdoux thought of himself more as a cartographer, and he worked closely with Delisle and d'Après de Mannevillette, sup-

552. Incarville letter to Père Fiteau, 13 October 1754, quoted in Bernard-Maître, pp. 40-42; and Incarville letter to B. de Jussieu, "à Peking ce 17 novemb. 1742," ["Dieu veuille que les projets que je forme pour sa gloire, et l'avantage de nôtre France, ayent lieu"], PA-DB (Incarville).

553. On missionaries in India, see Vincent, chapt. iv.

554. "Cette ville est Le premier des comptoirs de la compagnie des indes et son plus bel Etablissement dans L'inde. ...La Compagnie y a un jardin magnifique qui sert de promenade publique; Les Jesuites ont un College ou ils enseignent a Lire, a Ecrire et Les mathematiques. Les missions Etrangères y ont une maison; Les Capucins un Convent et Les gentils deux Pagodes...," BCM-NHN, Ms. 1765, Ms. "Table Geographique et historique des Indes," entry for "Pondichery, ville dasie."

555. Réaumur (1736).

556. Vincent, p. 100.

557. See Delisle-Cœurdoux correspondence, OdP, Ms. 1029 and AN-MAR 2JJ 66-68; see also Murr.

plying up-to-date information for improving European maps of India, notably those of the royal geographer, J.-B. d'Anville. Delisle reciprocated by sending maps and other scientific materials to Cœurdoux in India. Cœurdoux completed the circle through contacts with his fellow Jesuits in China and back in France. Like d'Incarville, his compatriot in China, Cœurdoux was a French patriot who detested the English.

For almost a century until the dissolution of the order in 1773, Jesuits manned a strong outpost for the Colonial Machine in the East. Working overseas in science with such constancy, application, and competence was, of course, part of their global policy, which was to gain political support from the king of France as well as from the Chinese emperor to promote their evangelical goals. As a result, they served the Colonial Machine and were served by it.

Tertiary Institutions – Secular Connections to the East

The Colonial Machine was a vast and complicated affair. Roughly speaking, one set of institutions – the Académie des Sciences, the Observatoire, the Jardin du Roi, and the Marine Royale – constituted its core. A secondary group – the Académie de Marine, the Société de Médecine, the Dépôt des Cartes et Plans, the Société d'Agriculture – complemented these primary units. A third set of tertiary institutions, such as the Cercle des Philadelphes or the Atelier de Boussoles, rounded out the whole of the Colonial Machine. This layered structure is especially apparent in considering further connections of the Colonial Machine to China and India, where secular connections reinforced those effected by the Jesuits and other missionaries in the East and where another tertiary institution, the *Académie Royale des Inscriptions et Belles-Lettres*, complemented the role of the predominant science academy and the Observatoire back in the motherland.

[Jeton: Académie Royale des Inscriptions et Belles-Lettres]

Founded by Colbert in 1663, the Académie Royale des Inscriptions et Belles-Lettres was the French learned society devoted to the humanities, notably history, numismatics, and philology. The Académie des Inscriptions et

Belles-Lettres came to have a distinctly "Orientalist" dimension. It became part of the Colonial Machine and was entangled in the East because the erudition of French missionaries in Asia in the areas of languages, chronology, geography, and the astronomical traditions of China and India inevitably drew the attention of the literary savants and academicians in Paris, not least because of the need to reconcile the chronologies of these great civilizations with traditional Christian views.[558] For example, also relying on and citing the work of Father Gaubil, the permanent secretary of the Académie, Nicolas Fréret (1688-1749), stirred up an academic controversy by publishing on Chinese chronology in the *Mémoires* of the Académie des Inscriptions. This work pitted Fréret and the Académie des Inscriptions against Father Étienne Souciet (1671-1744) and the Jesuits.[559] The missionary in Nanjing, Amiot, wrote on Chinese music and was made a correspondent of the Académie des Inscriptions.[560] The royal cartographer, J.-B. d'Anville, an associate academician, presented his maps of India and the Americas and his historical maps to the Académie des Inscriptions. Delisle tapped the Académie des Inscriptions for help in translating Chinese materials.[561] De Guignes *fils*, the French consul in Canton, offered a Chinese planisphere to the Académie which elected him a correspondent and sent him a list of questions to research concerning the history of China.[562] Joseph de Guignes *père*, professor of Syriac at the Collège Royal and *pensionnaire* of the Académie, had serious interests in these questions. This was evidenced by one of his papers, among others: "Historical Research on the Establishment of the Indian Religion in Tartary, Tibet, and China and on the Fundamental Books of This Religion That Were Translated from Indian into Chinese."[563] The Academy examined a map of the Ganges

558. See Godlewska, p. 29; Broc (1975), p. 20. See also Institut de France, Archives de l'Institut, Registres des Délibérations et des Assemblées de l'Académie Royale des Inscriptions, passim. On orientalism in Louis XIV's France, see Dew (2009).

559. Delisle's letter to Gaubil, dated "de Paris le 24 Décembr. 1758" (AN-MAR 2JJ 66, #66), sheds light on the internal politics of the Académie des Inscriptions at the time, notably the suggestion that Gaubil was counseled to drop his work on Chinese chronology in order not to embarrass Fréret. See also Delisle letter to Gaubil, dated "à Paris le 27 Octobre 1754," that discusses Gaubil's connection with the Académie des Inscriptions; AN-MAR 2JJ 68: #10. See also PA-DB (Gaubil).

560. Institut de France, Archives de l'Institut, Cote: A: 68, "Registre des Assemblées & Délibérations de l'Académie Royale des Inscriptions & Belles Lettres pendant l'Année 1785," pp. 45, 50 ("Fevrier 1785, Du Vendredi 18"); see also Cote A 67, "Régistre des Assemblées & Délibérations de l'Académie Royale des Inscriptions & Belles Lettres pendant l'Année 1780," p. 7 ("Janvier 1780, Du Vendredi 14") for earlier Amiot connection to the Académie des Inscriptions.

561. See reference in Delisle letter "au P. Hallerstein Jesuite missionaire à Pekin," dated "à Paris le 27 Octob. 1754," AN-MAR 2JJ 68.

562. Institut de France, Archives de l'Institut, Cote: A: 68 "Registre des Assemblées & Délibérations de l'Académie Royale des Inscriptions & Belles Lettres pendant l'Année 1783," p. 13 ("Janvier 1783, Du Vendredi 24"); p. 151 ("Decembre 1783, Du Mardi 9"); and pp. 157-158 ("Decembre 1783 Du Vendredi 19").

563. "Recherches historiques sur l'établissement de la religion Indienne dans la Tartari, le Tibet & la Chine, & sur les livres fondamentaux de cette réligion, qui ont été traduits de l'Indien en Chinois." See reference, Institut de France, Archives de l'Institut, Cote: A: 66, "Registre des Assemblées & Délibérations de l'Académie Royale des Inscriptions & Belles Lettres pour l'Année 1776," p. 83 ("Juin 1776, Du Mardi 25").

River drawn by Father Tieffentaller, the papal emissary to India, and it heard from a certain Mr. Bhitsingh, an agent of the Compagnie des Indes in Bengal, regarding Japanese history.[564] Although peripheral to the main body of our Colonial Machine, these traces in the Académie des Inscriptions reveal the extent to which the colonial and overseas worlds penetrated the fabric of Old Regime institutions and mentalities.

The initiatives launched by Henri-Léonard-Jean-Baptiste Bertin (1720-1792) with the Jesuits and with secular figures in China cemented contacts with the East.[565] Bertin was controller-general from 1759 to 1763 and then state minister for agriculture, commerce, mining and manufacturing. He was an honorary member of both the Académie des Sciences and the Académie des Inscriptions, and a key member of the Société Royale d'Agriculture; he employed a special secretary to keep track of his correspondence with China, and through his high office botanical specimens regularly arrived from Asia via the Compagnie des Indes for the Société d'Agriculture and for the Jardin du Roi.[566] Bertin sent mathematical instruments to the Chinese scholars Ko and Yang in Canton via the agent of the Compagnie des Indes in Canton.[567]

The above-mentioned consul general of France in Canton, Chrétien-Louis-Joseph de Guignes (1759-1845), should also be highlighted in this context. Elected a correspondent in 1783, de Guignes had strong connections to the Académie des Sciences. De Guignes himself translated and submitted Chinese cometary observations to the science academy by a Chinese astronomer, a certain "*Natuon-lin* Sav. Chinois."[568] And he sent meteorological observations back to the Academy from China, published in part by Le Monnier in the *Mémoires* of the Académie des Sciences.[569] De Guignes also sent anthropo-

564. On these last examples, see Institut de France, Archives de l'Institut, Cote: A: 66, "Registre des Assemblées & Délibérations de l'Académie Royale des Inscriptions & Belles Lettres pour l'Année 1776," p. 115 ("Aoust 1776, Du Mardi 20"); Cote 69, "Registre des Assemblées & Délibérations de l'Académie Royale des Inscriptions & Belles Lettres pendant l'Année 1786," p. 165 ("Novembre 1786, Du Vendredi 17").

565. See BCMNHN, Mss. 1515-1526, "Correspondances des Jésuites missionnaires en Chine avec H.L.J.B. Bertin (1744-1798)."

566. See relevant entries, BCMNHN, Mss. 314, 357, 1327; Foville and Pigeonneau, p. 91; Hamy (1890), p. 29.

567. See letter dated "A Paris le 10 mars 1771" to "M. de la vigne buisson, Cap.ne du Port a lorient," signed "Caveux," SHML, 1P 288, Liasse 179, dossier 37, #33.

568. PA-PV-110, Guignes, 22 March 1782 ("Traduction d'un *Catalogue des Comètes observées à la Chine* extraites des livres Chinois, par *Natuon-lin* Sav. Chinois"). See also SE, vol. 10 (1785), addendum, pp. 1-76, and PA-PV-110 3 August 1785 ("Diverses Observations astronomiques et physiques envoyées de canton").

569. MARS 1789, pp. 597-600. Also, PA-PV-110, Guignes, 12 August 1786 ("Observations Météorologiques envoyées de Canton"), 1 August 1787 ("Observations Météorologiques faites à Macao proche Canton, année 1786"), 6 December 1788 ("Observations météorologiques faites à La Chine précédans [sic] la plus grande partie des années 1786 et 1787") and 21 July 1792 ("Observations météorologiques faites à Canton en Chine pendant l'année 1791"); PA-PV 107 (1788), fols. 266v ("Mercredy 19. Novembre 1788"), 276-277v ("Samedy 6. Décembre 1788); and further, APS Manuscripts, Fougeroux de Bondaroy Papers, B F8245, "Séances de l'Académie Royale des Sciences, 1786-1789," "du Samedi 6 décembre [1788.]"

metric data back to Tenon of the Académie and seeds to Thouin at the Jardin du Roi.[570] Coincident with the Michaux mission to America, d'Angiviller at the Bâtiments du Roi launched a systematic effort to enlist the Compagnie des Indes and the Jesuits in China to further botanical researches, and as part of this effort he recruited de Guignes to ship trees from China back to France.[571]

Colonial interests permeated the institutions of Old Regime France, and colonial residues can be found in institutions beyond the main ones examined here. At the *École des Mines*, for example, founded in 1783, mineral samples arrived from Madagascar, Île de France, Île Bourbon, Cayenne, Guadeloupe, and Saint Domingue, as well as the rest of the world.[572] The Royal Mint struck a special series of coinage for the colonies, including for Île de France, Île Bourbon, and Cayenne.[573] Founded in 1766, the royal veterinary school at Alfort responded to a request of the minister of the navy and the colonies, Sartine, and as a result in 1774 began to select veterinary students destined for the colonies.[574] The royal manufactories at Les Gobelins and Sèvres played roles in tests of dyeing materials from the colonies and attempts to duplicate Chinese porcelain.[575] The Académie Française elected in 1779 the Creole playwright and musicologist, Michel-Paul-Guy de Chabanon (1730-1792) who hailed from Saint Domingue and was already a member of the Académie des Inscriptions et Belles-Lettres. Further research would doubtless reveal more of a colonial face to the Académie Française, too.[576]

[Colonial Currency]

570. BCMNHN, Ms. 1327, entry for 1789, and de Guignes letter to Tenon "de l'academie des sciences," dated "Quanton Ce 28 janvier 1785," PA-DB (de Guignes).

571. See material to this effect AN O^1 2112: Pépinières, 1783-1787 and 4 pp. Ms. "Notte relative a la reponse que M.r Le C.te D'Angiviller doit faire à M.r De Guigne sur les Envois de Graines et de Plantes en nature qu'il a faits de la Chine en 1786." And other documents related to this matter, AN O^1 2113A: "Pépinieres d'Amérique. Correspondance Générale pour la Mission du S. Michaux, Botaniste du Roi, Année 1786." And AN O^1 1292, #18. ministerial AL dated "Vers.les 7 X.bre 1785."

572. See Darnis, and Archives de la Monnaie (Paris), Série AA. D.-1-5: Création École des Mines, Dossier labeled "Musée de la Monnaie, Collection minéralogique" and Dossier labeled "Musée de la Monnaie, Etat du Cabinet de Minéralogie après le décès de M.r Sage."

573. See Zay; Archives de la Monnaie (Paris), Série H 1-3, Dossiers anciens des colonies françaises.

574. Demeulanaere (2000a), p. 123; Gerbaud, pp. 20-25 and passim.

575. For the latter, see AL draft in d'Angiviller's hand to "M. de Montigny" in Canton dated "Vers.lles. 5 mars 1786," AN O^1 2112: Pépinières, 1783-1787.

576. On Chabanon, see Guertin and Quetin; McClellan (1992), p. 237.

The Test Case of the Cabinet de Physique et d'Optique

Not every French institution of science and learning in the Old Regime constituted part of the Colonial Machine to an equal degree, and the dividing lines between the central institutions and extraneous, less connected, or even antagonistic elements help clarify the boundaries of the institutional and administrative structure we have been exploring. Part III, which concludes this study, systematically examines limits and alternatives to the Colonial Machine and the centrifugal forces that developed in contradistinction to the hegemony of the Machine as directed from Versailles. Here, a small, but revealing example – a test case – can illustrate the boundary in question: to wit, the king's *Cabinet de Physique et d'Optique* at the château of La Muette in Passy, just outside of Paris. A royal scientific entity created in 1759 under, once again, the administration of the Bâtiments du Roi, the Cabinet de Physique housed the personal scientific instruments of Louis XV, notably a state-of-the-art telescope and the full armamentarium of scientific instruments of the day. The Cabinet de Physique represented quite an expense; there was the building at La Muette, a custodian, two assistants, a domestic, and the head of the Cabinet, the Benedictine Father Nicolas Noël. Livery was provided and at least occasionally, if not permanently, the Cabinet also employed a carpenter, a forger, a clockmaker, a lathe operator, a technician, and a polisher. The minister approved an annual budget of 21,800 livres for the Cabinet, including 6000 livres for Noël himself.[577] The instrument collection grew rapidly, and the Cabinet gives every evidence of being a serious institution of science that one would naturally expect to be integrated into the Colonial Machine.

But into the 1780s the Cabinet de Physique was a sham institution. It was created by Mme de Pompadour for her protégé, Noël! Previously, Noël had been a clockmaker and occasionally sold scientific instruments to Louis XV, who was "amused" to look through his telescope; and Mme de Pompadour took advantage of this, as well as other of the king's weaknesses, to set Noël up handsomely. As an internal ministry document later put it, Noël showed off the king's instruments "to persons of quality," but was not interested in pursuing science. "He goes so far as to have the ridiculous pretension to compare himself to the celebrated Tycho Brahe, but the truth is Abbé Noël never enjoyed the slightest consideration in the learned world."[578] Tellingly for us, Noël rebuffed efforts by the navy astronomer, Méchain, to make the Cabinet de Physique a genuine outpost for science and naval astronomy. In an effort to

577. This story emerges from documents in AN O[1] 1584: Cabinet de physique et d'optique du Roi, 1755-1792; see also Fauque, pp. 10-15.

578. "L'Abbé Noel pretend etre seul en etat de conduire et d'entretenir Le Cabinet. Il va jusqu'à la pretention ridicule de se comparer au Celebre Ticho Brahé. La verité est que l'Abbé Noel n'a jamais joui de la moindre consideration dans le monde sçavant...," undated, unsigned "Memoire" AN O[1]1584, #383.

save money and to wring some productive use out of the Cabinet, in 1777 d'Angiviller – the head of the Bâtiments du Roi – moved to install "really learned men" to run it: notably Abbé Rochon for the astronomical and optical equipment and Le Roy for the rest so that the Cabinet might, finally, "become a resource for those who cultivate the sciences, that they might make *physique* and astronomy useful, as well as to entertain your Majesty."[579] Incredibly, Noël successfully resisted his ouster, and it was only after his death in 1783 that Rochon and Le Roy were able to amalgamate the Cabinet de Physique into the Colonial Machine. Rochon and Le Roy were otherwise well attached to the Colonial Machine, and Rochon argued to expand the facility so that it could be of greater service to the community of astronomers in the Paris region. In the end, the Colonial Machine assimilated the Cabinet de Physique. Indicative of its new status, Rochon began to build telescope mirrors at the Cabinet using platinum sent by the royal naturalist, J.-B. Leblond, from South America, via the superintendent of the Bâtiments du Roi, and using other platinum obtained from the explorer Joseph Dombey who had been sent at great expense by the crown to South America.[580] The example of the Cabinet de Physique is a small one, but it illustrates not only that the Colonial Machine was an assemblage of institutions in action, but that not all institutions were part of that assemblage to the same degree or at the same time, and that boundaries defined the Colonial Machine of the Old Regime.

* * *

As is clear already, its various parts interacted, they shared overlapping members, they undertook collective projects, and they interconnected bureaucratically to the point where the whole Colonial Machine became something greater than the sum of its parts. The institutional infrastructure examined in this Part I gave France and the Bourbon monarchy an unprecedented and unchallenged ability to exploit science, medicine, and expert knowledge for colonial development and simultaneously to use the nation's colonial experience to advance science and medicine. To repeat a point made earlier, this bureaucratic assemblage was not an abstract entity, reified as the Colonial Machine, but was real insofar as it existed in action. What exactly the French Colonial Machine accomplished in these regards is the subject of Part II of this study.

579. "...attachant enfin des hommes vrayment sçavans, il devenoit un moyen de faire du bien à ceux qui cultivent les sçiences, qu'ils pouvoit trouver à l'utilité de la Physique de l'Astronomie, ainsi qu'à l'Amusement de Votre Majesté," undated, unsigned "Memoire sur le Cabinet de Physique et d'Optique de la Muette," AN O[1]-1584, #384.

580. See previous notes and Smeaton, p. 112.

Part II

The Colonial Machine in Action
A – The Conquest of Space

Il y a des sciences...qui, pour leur progrès ont un besoin nécessaire du pouvoir des Souverains; ce sont toutes celles qui exigent de plus grandes dépenses que n'en peuvent faire les particuliers, ou des expériences qui dans l'ordre ordinaire ne seroient pas pratiquables.

– Maupertuis, *Lettre sur le progrès des sciences* (1752)

Creating and maintaining colonies in the early modern period depended first and foremost on mastery of the sea and overcoming the great distances and practical difficulties involved in sailing the world's oceans. Like other colonial powers, the French state needed reliable knowledge of the position of its colonies and how to travel there and back. Mastery of the seas and colonial space was quickly seen to depend on mastery of geographical and astronomical space. As a result the state turned to a constellation of sciences – astronomy, geodesy, cartography – with the aim of securing such mastery, and it established the set of institutions we have seen to address the core problems of navigation and cartography, notably the Académie des Sciences, the Observatoire, the Académie de Marine, the Dépôt des Cartes et Plans de la Marine, and cadres of royal geographers and cartographers. French officialdom also installed royal engineers and surveyors in the colonies to chart local space, and these agents formed part of the Colonial Machine and added to formal efforts to map the colonial and overseas worlds. In addition, the infrastructures of the Colonial Machine served as a gigantic research engine tackling the many practical problems of long-distance sailing. And finally, France, like the other great maritime powers, faced the overriding problem of finding longitude at sea; the Colonial Machine was central to – indeed, identical with – the lesser-known story of extensive French work on this problem. This second part of this book examines these nautical and navigational aspects of the Colonial Machine in action.

Cartography and Navigation

La Navigation seroit encore dans son enfance, si L'Astronomie ne L'avoit perfectionner; et Le Commerce des pays etrangers n'avoit pas apporté en Europe des sommes immenses, Si L'Astronomie n'avoit pas traçé Les Chemins aux Navigateurs et posé sur Les Cartes Les Situations des Lieux d'ou L'on a tiré ces grandes richesses.

– Father Louis Feuillée (1719)[1]

Personne n'ignore la liaison intime que la Géographie et l'astronomie ont avec la navigation et la Marine, car on ne peut construire de bonnes cartes marines, dont on est obligé de se servir dans la navigation, sans connoitre exactement la situation des cotes...d'ou l'on voit qu'independement de l'utilité de la géographie et de l'astronomie en elles mêmes, combien elles peuvent servir au progrés de la navigation et de la marine.

– J.-N. Delisle (1750s)[2]

...La Sûreté [et] le salut de la Marine et du Commerce dépendent principalement de l'exactitude et de la fidelité qui doivent être mises dans la Construction, l'impression et le débit des Cartes qui indiquent aux navigateurs les routes qu'ils doivent suivre, et les dangers qu'ils doivent éviter.

– Anonymous (1780s)[3]

The success of the colonial endeavor depended on the strength of French cartography, and there was considerable pressure throughout this period to create accurate and serviceable maps.[4] The world of contemporary cartography was a complicated one. Manuscript maps continued to circulate as they had in the seventeenth century, and entrepreneurial mapmakers and printers published

1. ALS Feuillée, dated "de Marseille Le 15 Avril 1719" to the Duc d'Orléans, "Ms 2003 Bibliothèque de l'Institut"; copy PA-DB (Feuillée).

2. AN-MAR C^7 186 (Lisle), Ms. "Projet d'ajouter au Dépôt des journeaux [sic], Plans, et cartes de la marine."

3. Anonymous memoir; AN-MAR G 234, #20.

4. The literature surrounding eighteenth-century French cartography is now highly developed; for entrées see Broc (1975); Chapuis; Konvitz; Pedley (1992); Pedley (2005); Pelletier (1990); Pelletier (2002); and Petto, esp. ch. 3. Although focusing on a slightly later period, Godlewska offers important insights. Dawson is another key source; Neuville (1882), pp. 71-85 provides overlooked background.

maps and stole from one another regularly.[5] Our concern is with one, narrower slice of that world, the official astronomical and cartographical endeavors sponsored by the French state within the Colonial Machine. The Colonial Machine became intensively occupied with astronomy, cartography, and navigation, and sizeable segments of the Colonial Machine were devoted to the collection of cartographical information and the production of maps. The great hydrographical and cartographical landmarks, such as d'Après de Mannevillette's *Neptune Oriental* (1745) or Bellin's *Neptune François* (1753), not to mention the mapmakers and map-making capabilities that produced them, were glorious gifts of the Colonial Machine.

Early Astronomical and Geodesic Voyaging. From the 1660s on a steady stream of scientific voyages sailed under the flag of the Colonial Machine. The earliest expeditions of the Académie des Sciences tested Huygens's pendulum clock as a marine chronometer. The first expedition set out in 1668, with one Delavoye sailing the Atlantic with Huygens's clocks. That test was a failure, and a second voyage followed in 1669 from Toulon to Crete and back. The results seemed better, and a third trip left for Canada in 1670 with Jean Richer (1630-1696) aboard to test clocks and Jean Deshayes (†1706) to test a method of determining longitude using lunar distances.[6]

Jean Richer's return voyage to the Americas in 1672 changed the direction of geodesic research.[7] Richer was an original *élève astronome* of the Académie des Sciences in 1666. In what has been labeled the "first modern scientific expedition," the Academy sent Richer to Cayenne in South America to make astronomical and geophysical observations.[8] There he undertook simultaneous observations with Cassini at the Observatoire to determine the longitude of Cayenne, and he made the discovery that a pendulum of a standard length beats more slowly near the Earth's equator than near the poles.[9] At first, Richer's pendulum results were thought to be in error, but voyages following Richer's timed the beat of a standard pendulum in various parts of the world, and by the 1690s it gradually became clear that the phenomenon was real and needed explaining.[10]

5. On contemporary maps and the map trade, see Pedley (2005), esp. chapt. 4; on cartography in the colonies, see Regourd (2000), pp. 393-419; re 'private' maps: see collection of manuscripts of the Service hydrographique (Dépôt des Cartes et Plans) (SHMV). This collection makes the point that maps were being made constantly.

6. Olmsted (1942), p. 117; Broughton, p. 176; Iliffe, p. 621; Dew (2008), pp. 60-64.

7. On Richer, see Kellman, pp. 101-111; Regourd (2000), pp. 272-77; Olmsted (1942) and (1960); Broughton; Lévy; Lacroix (1938), vol. 3, pp. 12-14.

8. Olmsted (1942), p. 128; see also Rosen, and "Commission pour le sieur Richer, ingénieur et professeur de mathématiques, qui part pour les Indes Occidentales pour y faire des observations astronomiques," AN-COL B[2] (1670), fol. 12 (26 February 1670), and related ministerial correspondence, fol. 17 (10 March 1670). Richer became a full academician in 1679.

9. See Richer (1729). Pelletier (2002), p. 86, and McClellan (1992), pp. 118-19, tell this story.

10. By 1690 the shape of the earth controversy and its link to Richer's results surfaced in the Académie des Sciences; see PA-PV 13 (17 décembre 1689-31 août 1693), fol. 31 ("Le Samedy 27e d'Aout 1690").

More than that, the problem became central to the controversy between Cartesians and Newtonians over the shape of the earth, and Richer's discovery ultimately prompted the great Academy-sponsored expeditions to the northern polar region and the equator in the 1730s to settle the matter. Richer naturally reported back to the Academy and to the Observatoire.[11] In extraordinary language that shows the link in the minds of the actors between astronomy and European expansion, the Academy itself noted Richer's voyage: "This was apparently the first time that careful and exact observations were made in the New World. Precision astronomy thus took possession of it."[12]

Jean Deshayes, who had sailed with Richer to Canada in 1670, undertook three more state-sponsored voyages to the Americas in the decades that followed. In 1682 he sailed to the Cape Verde islands, Martinique, and Guadeloupe.[13] In 1685 he returned to map the Saint Lawrence River, and in 1699-1700 under the command of the naval captain and Academy honorary, Bernard Renau d'Éliçagaray, he once more ventured to Cayenne and the Antilles. In toto he observed in Canada, Africa, Cayenne, Saint Domingue, Martinique, Guadeloupe, Grenada, and Saint Christopher.[14] The Academy carefully crafted detailed instructions directing Deshayes' astronomical and geographical observations.[15] The top academicians at the Academy taught Deshayes how to conduct geodesic surveys, and the map he produced of the Saint Lawrence, based on 300 survey triangles, "was the first chart of the St. Lawrence to be published and was so good that it became the standard for the best part of a century."[16] Deshayes ended his career in Canada as *hydrographe du roi* stationed in Quebec.[17]

In Part I we enumerated the state-sponsored voyages of Fathers Plumier and Feuillée. Pierre-François-Xavier de Charlevoix, S.J. (1682-1761), continued this

11. OdP, Ms. A 2 1, A 2 5; PA-PV 10 (15 novembre 1679- juin 1683 [Registre de Physique.]), fol. 45v ("Le Mercredy 14e d'Aoust 1680"); Olmsted (1960).

12. "c'était apparemment pour la première fois que l'on observait avec soin & avec justesse dans le Nouveau Monde. L'Astronomie exacte en prit alors possession," "Année 1672. Mathématiques. Astronomie," in *Histoire de l'Académie Royale des Sciences*, vol. I: *depuis son établissement en 1666 jusqu'à 1686*, (Paris, chez Martin, Coignard fils et Guerin, rue St-Jacques, 1732), p. 158.

13. On this expedition with Varin and De Glos, also designated as observers "of the Academy," by Cassini; see Regourd (2000), pp. 277-82; Dew (2010).

14. See Dehayes journals and observations, OdP, Ms. B 5 2 [*8] and Deshayes correspondence with Cassini, OdP, Ms. B 4.9bis. On his contact with the Académie des Sciences, see, for example, PA-PV 19 (1700), fols. 301-304v ("Le Vendredi 30 Juillet 1700"); BM/Aix Ms #232 (759). On the Deshayes and Renau expedition, see Regourd (2000), pp. 291-294; see also Dew (2010).

15. See OdP, Ms. D 1 11, and "Instruction générale pour les Observations Géographiques & Astronomiques à faire dans les Voyages," included in "Voyages au Cap Verd en Afrique et aux Isles de l'Amérique," *Mémoires de l'Académie Royale des Sciences de 1666 à 1699* (Paris: Par la Compagnie des Libraires, 1729, vol. VII), pp. 432-38.

16. Broughton, p. 197; see also ministerial traffic relative to this map and the Academy's involvement, AN-COL B^{11} (1685), fol. 123 (16 May 1685) and fol. 128v (17 June 1685).

17. Broughton, pp. 196-197, 206; Chartrand, et al., pp, 25-6; Audet, pp. 26-27.

tradition of the priest-scientist in 1719-1722 on a government-sponsored voyage in search of the "Western Sea" (the Pacific) and the Northwest Passage. Charlevoix traveled down the Saint Lawrence to Quebec. From there, with two canoes and a party of eight and goods for trading, Charlevoix followed in LaSalle's footsteps, venturing through upper Canada, out to Illinois country, and then down the Mississippi to the newly founded burg of New Orleans before returning to France via Saint Domingue.[18] On his return the ministry granted Charlevoix complete access to materials in the Dépôt des Cartes et Plans for writing up the account of his travels.[19] In 1744 his three-volume *Histoire et description générale de la Nouvelle France avec le Journal Historique d'un Voyage fait par ordre du roi dans l'Amérique septentrionale* finally appeared. This work included several maps by Bellin, including Bellin's "Carte de la Partie Orientale de la Nouvelle France ou de Canada."[20] (Charlevoix also served the Colonial Machine by publishing Father Le Pers' history of Saint Domingue, and he may have returned to Canada as royal hydrographer in the 1740s.)

The Jesuit, Antoine de Laval (1664-1728), is also representative of the role of priest-scientist in the Colonial Machine.[21] Backed by Abbé Bignon, he founded the naval observatory in Marseilles in 1702 and was professor of hydrography at the naval school at Toulon from 1718 until his death. Laval was in contact with Cassini, Maraldi, and Delisle in Paris; and in 1720, appointed as royal mathematician, he undertook an astronomical expedition to Louisiana on behalf of Cassini and the Académie des Sciences. He followed up on instructions from Cassini II, sending back his observations from Martinique, Saint Domingue, and Louisiana. In 1720 he also published his *Voyage en Louisiane fait par ordre du roi en 1720* in 1728, along with a notable map of the Mississippi and the Gulf of Mexico, and he remarked that he was the first astronomer from the Academy to visit the Americas since Feuillée in 1712.[22] Even as a man of faith, Laval's commitment to science was strong:

18. On Charlevoix, see Broc (1975), pp. 93-98, 132, 155-56; Rousseau (1957), p. 152; Robbins, p. 43; Chinard (1957a), pp. 267-68; Banks, pp. 93-94; see also AN-COL B^{42} (1720), fol. 448 (17 June 1720); on the French quest for the Northwest Passage, see Havard and Vidal, pp. 135-41.

19. See Broc (1975), p. 83 and ministerial correspondence related to this matter, AN-COL B^{53} (1729), fol. 33 (17 April 1729), fol. 75v. (15 September 1729), and AN-COL B^{56} (1732), fol. 28 (1 April 1732).

20. Petto, pp. 106-113; Mathieu, p. 112.

21. On Laval and his scientific work, see especially Débarat and Dumont (1990); see also R. Taton, pp. 431, 656; Langlois (2003), pp. 62-63, 153-54; Banks, p. 193; Kellman, 285-98; Laval, (1728); AN-MAR C^7 170 (Laval). See also observational reports, OdP, Ms. B 5 1: "Observations...par les PP. Laval et Feuillée"; B 5, 1 [7*] "Observations ...à la Louisiane, à la Martinique..."; B 5, 2 [8*] "Observations en Amérique..."; B 5, 3 [9*] "Nombreuse collection d'observations..."; PA-PV 39 (1720), fol. 22 ("Le Mercredi 17 Janvier"); Bibliothèque de l'Institut, Ms. 2721, Recueil de cartes; see also G. Delisle (1726).

22. Laval, "Observations au fort royal de La Martinique, au Cap francois a St. Domingue, a L'isle Dauphine sur la cote de la louisiane" in OdP, B 5, 2 [8*] "Observations en Amérique...," p. 363.

> In the litigation we have against nature, the more documents we file against her the more surely we will win our suit. It may take two or three generations of plodding, but in the end with patience and time we or our nephews [sic!] will prevail. It's too bad that we didn't initiate these proceedings two hundred years ago, but our ancestors didn't possess our method. Metaphysical verbiage was more to their taste, but fortunately for the sciences tastes have changed.[23]

Field Observatories. Eighteenth-century observatories were not always the formal affairs we tend to think, yet, with one exception, one is hard-pressed to speak of colonial observatories in the same way that we recognize the existence of colonial botanical gardens. The exception concerns the Imperial Astronomical Bureau in Beijing run by the Jesuits. China was not a European colony by any means, yet, as discussed further on, missionary outposts there and elsewhere merged with the rest of the Colonial Machine. Father Gouye transmitted astronomical observations from Jesuits in China to the astronomers of the Academy of Sciences, a role taken up by J. N. Delisle in his extensive correspondence with missionaries in China and India.[24] Writing to Cassini I from Siam in 1686, the Jesuit astronomer, Fontaney, made explicit the connection between Beijing and Paris: "I am causing you considerable trouble, but remember, Sir, it is for God and for the Royal Academy of China, an outpost of the Academy of Paris, which is so honored to count you among its members."[25]

In general, however, unlike the botanical enterprise, which became deeply rooted in the colonies, astronomy remained more centered in the metropolis with only feeble institutional extensions to the colonies. Father Bonnecamps installed an observing post on the roof of the Jesuit college in Quebec, but little

23. "surtout dans le Proces que nous avons avec la Nature; plus nous aurons des pieces à produire contre elle, plus surement nous gaignerons nôtre Proces. il en sera peutetre de celuy cy, comme de plusieurs qui se perpetuent deux ou trois générations en certaines cours; mais enfin avec la patience et le tems nous en viendrons à bout nous, ou nos neveux. c'est dommage qu'on n'aye pas commencé il y a deux cent ans l'instruction de ce proces. nos devanciers n'avoient pas cette espece de Procedure. un verbiage de metaphysique leur plaisoit davantage. heureusement pour les sciences ce gout a changé," OdP, Ms. B 5, 1 [7*] "Comparaison des Variations observées en 1720...."

24. See below and OdP, B 2.7, "Extrait d'une partie des observations faites a la Chine par Les R. P. jesuittes francois donné a Lacademie par le R. P. Gouie."

25. "Voilà bien de la peine que je vous donne: mais souvenez vous, Messieur, que c'es pour Dieu, et pour l'Academie Royale de la Chine [sic!], un detachement de celle de Paris, qui se trouve si honnorée de vous compter entre ses membres," OdP, B 4.10, Fontaney letter to Cassini, dated "A Siam le 6e May 1686." Fontenay was named an official correspondent of Gouye in 1699. Writing from Beijing to Delisle in Paris in 1732, Father Gaubil was considerably less sanguin about the state of the Imperial observatory, remarking that "Quant donc on parle en Europe d'un observatoire Imperiale à Pekin, il faut bien se garder de penser à quelque Comparaison aux observatoires de Paris, Graenvik [sic], Petersbourg, Berlin, &c," Gaubil letter to Delisle dated "à Pekin Ce 28 May 1732," AN-MAR 2JJ 62.

came of this for lack of instruments.[26] Proposals to build formal observatories in Canada and in Louisiana were rejected outright.[27] Astronomers sent out from the Observatoire and the Academy, however, did set up temporary observatories in the colonies. Father Laval, for example, was troubled that cattle as well as people had access to the "observatory" he set up for himself in a wooden house on Dauphine Island near New Orleans.[28] Likewise in Louisiana, in 1730 the engineer, Baron, installed his "observatory" on the terrace of the governor's house. Jean Richer wrote about his sojourn in Cayenne in 1672: "I found a place where two millstones lay for the last eight years. I had the natives build a little house around these in their fashion, twenty-four feet long and eighteen feet wide, covered with palm branches and leaves and closed in with the bark of trees. This served as my observatory while I was on the island."[29] Seventeen years later, Deshayes made observations from the same spot, noting the holes where Richer had placed his octants.[30] In Martinique in 1769 Fleurieu made his geodesic observations from the artillery post at the south end of the docks in Saint Pierre.[31] And as part of the same expedition Father Pingré set himself up in Saint Domingue for the 1769 transit of Venus.

> The site we judged the best for establishing an observatory is located on the northern outskirts of Cap François. North latitude 19° 57' 3"; as for its longitude I cannot say anything, except that the site is 4 hours and 58 or 59 minutes west of the meridian of the Observatoire royal in Paris. It is a rectangular pavilion with only a single room on the ground floor and open to every wind by a very large number of crossing passages. It is called the "Red House," and it is situated on a small hill. A gorge separating two high mountains allows the sun to be observed from there in June and July until six in the evening.[32]

26. Audet, p. 32; Pighetti, pp. 167-68.

27. See ms. by B. Boivin, "Notice biographique sur J. F. Gaultier 1708-1756)," p. 2 in PA-DB (Gaultier), and AN-COL C^{13A} 13, fol. 3 (25 November 1731).

28. "Observations au fort royal de La Martinique, au Cap francois a St. Domingue, a L'isle Dauphine sur la cote de la louisiane," OdP, B 5, 2 [8*] "Observations en Amérique...", p. 377

29. "Arrivant à Caïenne, je trouvai un endroit, où depuis huit années il y avait sur terre deux meules de moulin, auprès desquelles je fis bâtir par les Sauvages une petite maison à leur manière, de vingt-quatre pieds de long sur dix-huit de large, couverte de branches & de feuïlles de palmiers, & fermée par les costez avec des écorces d'arbres, laquelle m'a servi d'Observatoire pendant que j'ay eté dans cette Isle," Richer, p. 279.

30. OdP, B 5, 2 [8*] "Observations en Amerique par Mr. Deshayes," pp. 325-356; OdP, B 2.7, pp. 323-25 ("On voit encore le trou dans lequel [Richer] avoit placé son octans..."); see also "Lettre de Mr. des Hayes," PA-PV 19 (1700), fols. 301-304v ("Le Vendredi 30 Juillet 1700").

31. Fleurieu, p. 99: "Je chargeai M. le Chevalier d'*Isle* de se rendre au *Fort-Royal*, pour instruire de notre arrivée M. le Comte d'*Ennery*, Gouverneur Général des îles du Vent, & lui demander son agrément pour établir un observatoire à *Saint-Pierre* sur la plate-forme de la batterie de *Sainte-Marthe* située à la partie méridionale de la rade."

32. Pingré (1769), p. 513; see also Pingré (1770), pp. 503-504.

In India, at Pondicherry, the peripatetic Le Gentil de la Galaisière seems to have created a somewhat more substantial astronomical station, apparently with the concurrence of the local French governor-general, Law de Lauriston, and the neighboring English. As Le Gentil wrote to the Academy in 1768, "A kind of observatory was erected for me on the ruins and remnants of Mr. de Lally's chambers in the middle of the remains of the citadel and on the foundations of this superb structure raised in an earlier day by Mr. Dupleix." In the same letter Le Gentil sent his observational data of the satellites of Jupiter and a lunar eclipse that he hoped would better establish the longitude of Pondicherry.[33]

From Canton in 1754 d'Après de Mannevillette described his makeshift observatory:

> To usefully employ the time I had to stay there, I had a small house built on the Isle of Wam-houen nearby our ships, where I could conveniently observe and keep my instruments. Just outside the house I had a large flat stone set on which I traced a meridian, and I put my octant there. This instrument has a radius of twenty-two inches with two telescopes with two convex lenses, one placed along the radius, the other perpendicularly. This instrument was built by Mister Canivet, designer of mathematical instruments in Paris, and it has been checked by M. abbé de La Caille and myself from time to time.[34]

Back in the New World, the new intendant at Guadeloupe, Foulquier de la Bastide, apparently had plans to establish a permanent observatory in that colony. Before he left for the Caribbean, Foulquier was elected a correspondent of the Academy attached to Cassini de Thury. He was accompanied by an astronomical assistant, a certain Tondu.[35] He wrote to Cassini from Basse

33. "Sur les ruines et les debris de la chambre de Mr. de Lally...on m'a elevé une espece d'observatoire qui se trouve comme au milieu des ruines de la citadelle; et sur les fondements de ce superbe edifice Elevé jadis par mr. dupleix," Le Gentil letter to de Fouchy, dated "a pondichery le 9 de 9bre 1768," PA-DB (Le Gentil).

34. "Pour employer utilement le tems que de devois y séjourner, je fis bâtir, sur l'Isle de Wamhouen où étoient les bancasals de nos vaisseaux, une petite maison d'où je pouvois commodément observer, & y loger mes instrumens. Je fis mettre au dehors & sans entourage, une grande pierre plate, solidement assise, sur laquelle je traçai une méridienne : j'y plaçai mon octant. Cet instrument est de vingt-deux pouces de rayon, garni de deux lunettes à deux verres convexes; l'une placée le long du rayon, & l'autre perpendiculaire. Il a été construit par le sieur Canivet, Ingénieur d'instrumens de Mathématiques à Paris, & vérifié par M. l'Abbé de la Caille & par moi-même de tems en tems," d'Après de Mannevillette (1773), p. 295. In conjunction with Lacaille d'Après used these same instruments to establish the longitudes of Rio de Janerio, the Cape of Good Hope, Île de France, and Île Bourbon.

35. On Foulquier, see Lacroix (1938), vol. 3, pp. 83-90, Regourd (2000), pp. 411-12. Tondu himself sent reports back from Martinique to the Paris Academy of Sciences; see PA-PV-110, Tondu, 19 June 1784 ("Observations faites sur les marées en 1783 à la Guadeloupe").

Terre, Guadeloupe on 11 May 1782, "Finally my dear correspondent, we have all arrived. None of our instruments was broken, which is surprising, given the risks they ran. You will see really interesting things. We have limitless tools, and nothing will escape our observations."[36] The following year, writing of his "zeal and activity demanded by the work we have undertaken for you and for the sciences in general," Foulquier outlined an elaborate program of research he anticipated making in the Gulf of Mexico, asking only that Academy have the minister purchase a needed quadrant.[37] Foulquier and Tondu made cometary and planetary observations from the intendancy of Guadeloupe, which they sent back to Cassini, Messier, and Méchain in Paris.[38] Tondu, ill, returned to France, and the climate began to destroy Foulquier's instruments ("My poor compasses are busted. This country is really hell for instruments").[39] Still, before he transferred over as intendant at Martinique, Foulquier kept recording tidal data (à la Hauterive) and held to his dream of a general survey of the Gulf of Mexico. In Martinique, as we will see further, Foulquier was also active in fostering botanical shipments to the mother country, but nothing further came of this attempt to establish an astronomical station in the colonies. The attempt and its failure trace the limits of the astronomical arm of the Colonial Machine.

Longitude Observations. Maps are only as accurate as the information on which they are based, and until late in the seventeenth century cartographers had to rely on notoriously unreliable mariners' reports to estimate the longitude of positions on the Earth. But the Observatoire and the Colonial Machine offered the capability of using much more accurate astronomical techniques for fixing the longitude of points on land. (The determination of longitude at sea is a related, but more complicated problem, given the unsteady platform of a

36. "Enfin Mon tres amable Correspondant nous sommes tous arrivés....Nous n'avons Eu Rien de Cassé dans nos instruments et apres les Hazards quils ont Couru la chose Est surprenante nous travaillons a force pour Etablir l'observatoire et vous verres vraiement des choses interessantes...," Foulquier letter dated "Basse Terre Guadeloupe 11 mai 1782," Foulquier correspondence, OdP, B 4.10.

37. "...vos lettres et les marques de votre amitié sont l'excitateur le plus assuré pour Entretenir le Zelle et Lactivité necessaires aux travaux que nous avons Entrepris pour vous et pour les Sciences En General! Jai Concu le projet denvoyer mesurer La largeur de L'isthme de panama veriffier si la hauteur des deux mers est Egalle visitter toutes ces Cottes jusque au Golphe de mexique et a l'embrouchure du Mississipy En lever Les Cartes faire des observations sur les marées, sur les boussoles, le thermometre, le barometre, emporter autant d'objets d'Histoire naturelle quil Sera possible et revenir par le Golphe de la floride ou le Canal de bahama, mais jaurois besoin pour Cella dun Grand quart de Cercle...; Tachez sil vous Est possible par vous même ou par lacademie de determiner le ministre a achepter et a menvoyer Ce quart de Cercle....je voudrois faire Executer mon projet dans le Cours de L'année prochaine....Je vous prie dassurer Lacademie et touts les messieurs qui Composent Ce Corps vraiement Celebre de mon respectueux attachement....P.s. Comme je scais que vous aimez les bonbons je vous en adresserai par le premier navire qui partira de ces Colonies pour le Havre," letter dated "Guadeloupe 11 8bre 1783," Foulquier correspondence, OdP, B 4.10.

38. See reports in Cassini IV, p. 666; Messier, p. 321; Méchain, p. 365.

39. "mes pauvres boussolles sont en deroute Ce pays Cy est vraiement perfide pour Les instruments...," letter dated "Guadeloupe 21 Janvier 1783," Foulquier correspondence, OdP, B 4.10.

moving ship for astronomical observations.) But on land, using "simultaneous" astronomical observations of Jupiter's satellites, eclipses, or occultations of stars, observers in Paris and in a overseas locale could determine the difference in time and hence the longitudinal separation between Paris and the station in question.

Variously sponsored by the Paris Academy, the Marine Academy, the Observatoire, the Royal Navy, and the Compagnie des Indes, observers in France and in the colonies successfully employed these techniques to determine points of longitude and the geographical location of French overseas territories for decades prior to the development of the marine chronometer in the 1760s.[40] We have already seen several examples: Richer in South America in 1672 determined the longitude of Cayenne in conjunction with J.-D. Cassini in Paris; the same Cassini and Philippe de La Hire ascertained the longitude of Quebec in 1685 by observing a lunar eclipse in conjunction with Deshayes in Canada; Jacques Cassini II and Father Feuillée used simultaneous observations of the satellites of Jupiter in 1703 and 1704 for the longitude of Martinique; La Hire and the Jesuit Father Boutin performed like observations in 1706 and 1707 to pin down the position of ports in Saint Domingue.[41] Sailing in the Indian Ocean in 1740, d'Après de la Mannevillette worked with the academician P.-C. Le Monnier in Paris to determine the longitude of Île Bourbon. The effort to determine the longitude of Foulpointe, Madagascar, in 1757, for example, coordinated observations of Jupiter's satellites by d'Après de Mannevillette in Madagascar and other observers in Paris, Toulouse, Béziers, Marseilles, and Rouen.[42] Efforts were likewise made in the 1750s to fix the longitude of Macao in China and that of Oyapock in Guiana.[43] This kind of work could get complicated, as in 1755 when J.-N. Delisle deduced the longitude of Quebec by comparing observations taken in Quebec and on Île Bourbon in the Indian Ocean. Delisle, who received his Canadian data through de la Galissonnière, knew the longitude of the Indian Ocean possession measured from Paris ("derived from several exact, corresponding observations made separately in Europe and on Île Bourbon by Mr. d'Après de Mannevillette and the abbé de La Caille"). The data gave the longitudinal distance between Quebec and Île Bourbon, and hence he could derive the separation between Paris and Quebec. (He ended up thinking that the longitude of Quebec was still not fully estab-

40. See below and materials concerning French work on the problem of longitude, AN-MAR G 94-96.

41. See La Hire's papers, MARS 1706, pp. 481-82; MARS 1707, pp. 381-82. On the impressive series of longitude observations taken in Canada by Catholic missionaries in the seventeenth century, see Broughton, pp. 188-208.

42. Académie Royale de Marine (1773), p. 257.

43. Regarding the longitude of Macao, see the correspondence of the Jesuit Chanseaume, OdP, Ms. 1029, B 1-6 (XI) 1750-51; on Oyapock, see PA- PV, vol 69 (1750), p. 191 ("Samedi 30eme May").

lished.)[44] As the secretary of the Academy of Sciences remarked about this kind of activity in 1745: "Points of longitude determined astronomically are to Geography what medals and monuments are to History."[45]

The results of efforts to determine the longitude of geographical points around the globe showed up in the publications of the astronomers belonging to the Académie des Sciences and the Observatoire Royal. By 1734 the *Ephemerides* published by Philippe Desplaces (and other astronomers before and after him) gave a list of more than 300 precisely determined geographical coordinates in Africa, America, and Asia, including 161 in China, thanks to the industry of Jesuit missionaries.[46] The *Connoissance des Temps*, published under the auspices of the Observatoire Royal from 1679 for the use of the Marine Royale and mariners generally, provided such lists every year. In 1702, in addition to almost 100 points in France and Europe, fourteen places from overseas are to be noted: Beijing, Macao, Canton, Goa, Malacca, Cape of Good Hope, Cape Verde, Hierro Island, Olinde in Brazil, Quebec, and some others. Cayenne and Martinique appeared for the first time in 1706. Asterisks indicated that those coordinates, as well as about thirty others on the list were the result of astronomical observations "made by the Academy." Pondicherry and Chandernagor appeared in 1748, Louisbourg (Canada) in 1754, Île Bourbon and Île de France in 1757, and so on. In 1788, the list in the *Connoissance des Temps* counted more than 800 places, including 367 out of Europe.[47]

A steady stream of papers appeared in the *Mémoires* and the *Savants Étrangers* series of the Paris Academy on colonial longitudes and related matters. The Canary Islands became a special focus of attention in this connection, in part because virtually every vessel leaving France for America or the Indian Ocean stopped at or passed by them and because, following ancient tradition, a royal ordinance of 1634 set the French prime meridian on Hierro (Ferro) Island in the Canaries.[48] For similar reasons exact knowledge of the Iberian coasts was essential for safe passage to and from French ports, and therefore Louis XV sent the naval captain, Gabriel de Bory (1720-1801), on two expeditions to determine the longitude and latitude of the major headlands of the Iberian peninsula; the first expedition sailed to Spain in 1751, the second to Portugal in 1753.[49] Bory trained in astronomy at an early stage in his naval

44. "concluë de plusieurs observations exactes et correspondantes faites en Europe et dans l'Isle de Bourbon par Mrs. D'aprez de Mannevillette et Mr. l'Abbé de la Caille separement," ALS, Delisle "au Reverend Pere Bonnecamps de la Compagnie de Jesus à Quebec," dated: "a Paris le 1 mars 1755," AN-MAR 2JJ 68.

45. HARS 1745, p. 77.

46. Desplaces, vol. 3, pp. xx-xxviii.

47. *Connaissance des Temps.*

48. In this connection see papers by J.-D Maraldi (1742) and Lacaille (1746); for larger background, see Débarbat and Lerner.

49. Taillemite (2002), p. 59; AN-MAR C^7 38 (Bory); and official correspondence related to these voyages, SHMB, Série A: Sous-série 1A: 1-A5, fol. 167 (29 June 1753), fol. 257 (23 September 1753); 1-A6, fol. 67 (7 March 1754).

career, and he became a free associate of the Paris Academy and member of the Marine Academy at Brest. He also published reports of these voyages in the Academy's *Mémoires* and worked closely with the Marine Academy in planning his missions.[50] He determined latitudes by measuring the height of the pole star and longitudes through simultaneous observations of Jupiter's satellites, a solar eclipse, and occultation of stars by the moon. Gabriel de Bory was the quintessential "scientific" officer of the French Navy at the time, and another remarkable representative of the Colonial Machine.

Lapland and Peru. In the 1730s French expeditions journeyed to Lapland and Peru to measure the shape of the Earth. These were the two most important scientific voyages in the first half of the eighteenth century.[51] At issue was perhaps the most compelling scientific question of the day: the exact shape of the Earth and deciding between the physics and worldviews of Isaac Newton and René Descartes. Using triangulation techniques well-developed by the Cassinis and French cartography, each party – one toward the northern pole, the other at the equator – surveyed a degree of arc along the meridian of the Earth's surface, and the results ultimately showed that the Earth was indeed flattened at the poles, as predicted by Newton, and not at the equator, as the Cartesians would have it. The expeditions to Lapland and Peru were launched under the authority of the Académie des Sciences, but they required the full resources of the Colonial Machine for their success, tapping as they did the Observatoire, the Ministry of the Navy, the Marine Royale itself, and the Ministry of Foreign Affairs.

[Medal: MENSOR ORBIS TERRARUM, 1744]

The established academician and *pensionnaire*, Pierre-Louis Maupertuis led the mission to the north. He was accompanied by three young scientific asso-

50. See papers by Bory (1768), (1771), and (1772, Part 2).

51. On these expeditions and their scientific importance, see Terrall (2002), chapt. 4; Godlewska, pp. 49-54; Broc (1975), pp. 37-42, 116-21; Pelletier (2002), pp. 84-88; McClellan (1985), pp. 200-202; (1992), pp. 118-22; Allorge, pp. 269-98; Lacroix (1938), vol. 3, pp. 55-60; Chassagne, pp. 167-90; and most recently, Safier (2008a), (2008b).

ciates: the astronomers and geometers Alexis Clairaut, Abbé C.-É.-L. Camus, and P.-C. Le Monnier. The former two were already *associés* in the Academy, the latter was elected an adjunct member on his return. The priest was Réginald Outhier (1694-1774), a correspondent of Cassini and an accomplished astronomer himself. Joined by Anders Celsius in Sweden and given ample logistical and financial support by the French and Swedish governments, this group had a comparatively easy time of it, leaving France in May of 1736 and returning by August of 1737. Already in 1738, Maupertuis was in print with his *Figure de la terre...faites par ordre du Roy au cercle polaire*, published by the government. It is a stretch to consider the expedition to Lapland as part of the *colonial* Machine, and although the analytical ambiguities of the case highlight the fluidity of the Machine's boundaries, the episode deployed the same institutional structures that we have been studying and that unambiguously underwrote the complementary mission to South America.

The Academy's *pensionnaire* astronomer Louis Godin (1704-1760) nominally headed the star-crossed odyssey to the equator. The *pensionnaire* astronomer, Pierre Bouguer (1698-1758), and the then-associate geometer, Charles-Marie de La Condamine (1701-1772), were senior members of the scientific staff. A young Joseph de Jussieu, then unaffiliated with the Academy but associated with it through his more famous brothers, was aboard as physician and naturalist. An engineer, two geographer-surveyors, a draftsman, a clockmaker, a surgeon, and six domestics rounded out the party. Departing France in 1735, they stopped in Martinique and Saint Domingue before meeting up with their Spanish scientific collaborators, Antonio de Ulloa and Jorge Juan y Santacilia. The expedition then proceeded overland to Quito where, over the following two years, its members made the requisite measurements. Unlike their counterparts in the north, this group met with great difficulties. Promised letters of exchange were not forthcoming, and Godin, La Condamine, and Bouguer all feuded during the expedition and afterward, refusing to share their data.[52] La Condamine did not return to Paris until 1745, having heroically descended the Amazon and gaining Cayenne, where he repeated Richer's pendulum experiments. Bouguer did not make it back until 1744. Godin stayed on in Lima as professor of mathematics and did not return to Europe until 1751. Despite these travails, the results of the expedition to Peru supported those from the North in resolving the shape-of-the-earth controversy, and they added considerable findings to the geography and natural history of South America. While in Cayenne, for example, La Condamine's party identified samples of rubber, and La Condamine returned with some 200 botanical and natural historical samples for Buffon and the Jardin du Roi.[53]

52. McClellan (2003c), pp. 51-53 tells this story.
53. Safier (2008b), esp. p. 251.

Chabert and La Cardonnie. A less well-known but equally revealing scientific voyage followed in 1750-1751 – the cartographical mission to Canada of Joseph-Bertrand de Cogolin, the marquis de Chabert (1724-1805).[54] The goal was to gather data to improve maps of Louisbourg and the Saint Lawrence. In 1750 Chabert, a royal ensign (*Enseigne des Vaisseaux du Roy*), was sent as an agent of the Colonial Machine. La Galissonnière, then director of the Dépôt des Cartes et Plans, was involved in planning this expedition. Chabert received guidance and instructions from Delisle and remained part of his circle ever after.[55] Chabert received official orders from the Ministry that also alerted local authorities in Canada.[56] The result was Chabert's book of 1753, *Voyage fait par ordre du Roi en 1750 et 1751, dans l'Amérique septentrionale, pour rectifier les cartes des côtes de l'Acadie, de l'Isle royale & de L'isle de Terre-Neuve; et pour en fixer les principaux points par des observations Astronomiques*.[57] The title says it all about the mission! Submitted to and approved by the Marine Academy at Brest, Chabert's volume was favorably received there when it appeared. (La Galissonnière and Bouguer wrote the report.) On his return, Chabert was the only ensign named as ordinary academician of Marine Academy at its foundation in 1752, and he remained active in the Brest institution. Thereafter he rose in the ranks of the navy to lieutenant and ultimately commodore (*chef d'escadre*); in 1758 he was elected a free associate (*associé libre*) of the Academy of Sciences, and he was active in the life of the Paris Academy for the next three decades.[58] 1758 also brought his appointment as inspector-general of the Dépôt des Cartes et Plans, succeeding La Galissonnière – a post Chabert held to the end of the Old Regime. Chabert knew Poissonnier, and in 1766 he commanded the ship that tested Poissonnier's desalinization apparatus.[59] Sailing twice during the American War, Chabert conducted late sea trials of the marine chronometer, and later, in the 1780s, he and André Thouin at the Jardin du Roi exchanged letters and botanical specimens.[60] Chabert's voyage to Canada in the 1750s is noteworthy for French cartography and hydrography, but from our perspective it is also one episode in a long career that made Chabert truly a man of the Colonial Machine.

54. Taillemite (2002), pp. 92-93; Lamontagne (1964a).

55. See Delisle (1751).

56. AN-COL B[91] (1750), fol. 21 (19 May 1750), B[92] fol. 126 (19 May 1750) and further ministerial correspondence, B[92] fol. 14 (31 May 1751).

57. See DDP I: 28, 54; report HARS 1753, pp. 242-56; and Chabert's original observations and journal, OdP, Ms. B 5, 3 [9*].

58. See PA-PV passim, e.g. PA-PV 107 (1788), fol. 254v-57 ("Mercredy 5. Septembre 1788"), and commissions on which Chabert served, PA-PV-110, passim.

59. AN-MAR D[3] 41: "Mémoires et Projets," fols. 195, 197, 209ff.

60. See entries in BCMNHN, Ms. 314 and Ms. 1327.

Hard on the heels of the Chabert expedition came the less well known voyage of the chevalier Boutier de la Cardonnie to the Antilles to collect hydrological and cartographical data and to observe the transit of Mercury of 6 May 1753. The scientific object of the observation was to help determine the Earth-Sun distance. (This fundamental value in astronomy – the astronomical unit, AU – was still not accurately known by the mid-eighteenth century). The Mercury transit also set the stage for the later and more successful transit of Venus observations in the 1760s.[61] But even this "pure science" objective was tied to cartography and navigation, at least in the eyes of J.-N. Delisle, who was a major promoter of the observations of the Mercury and Venus transits: "Given that these kinds of eclipses last only momentarily," wrote Delisle, "one can use them to determine the longitude of places [which is] one more way to perfect geography and navigation."[62] For Delisle, in 1752, the longitude of Cap François in Saint Domingue was still not known with certainty, and so the ministry sent the naval ensign Jacques Boutier de La Cardonnie (1727-1791) to Saint Domingue to observe the transit of Mercury and to make longitude observations, a task for which he studied at the Observatoire.[63] La Cardonnie was also called upon to map the treacherous islands north of Saint Domingue that every returning voyage had to navigate.[64] La Cardonnie sailed carrying instruments from the Academy of Sciences he received via Duhamel du Monceau. La Cardonnie was elected an ordinary member of the Marine Academy, and he, like Chabert, went on to have an outstanding naval career, rising through the ranks to become a commodore himself.

Regarding the 1753 Mercury transit, corresponding observations were to be made by Jesuits in Canada with de la Galissonnière's help.[65] Delisle sought additional observations in India, China, and Cayenne for the 1753 Mercury transit.[66] Data from China collected during that transit proved sufficiently accurate to be able to determine the longitude of Beijing vis-à-vis Paris.[67] But uncertainties remained regarding the astronomical unit.

61. Woolf, pp. 41-53.

62. "Ces sortes declipses ettant mommentanées; on peut encore s'en servire...pour determiner la position des lieux en longitude...un moyen de plus pour perfectionner la geographie et assurer la navigation," letter "A Monseigneur De Rouillé. Passage de Mercure sur le Soleil du 6 may 1753," AN-MAR 2JJ 69.

63. Taillemite (2002), pp. 67-68; see also Delisle letter to La Cardonnie, dated "Paris le 14 Décembre 1752," OdP, Ms. 1029, B 1-7 (XII, #30), and La Cardonnie to Delisle, dated "rochefort, 9 xbre 1752," B 1-7 (XII, #32). Also, Académie de Marine letter to "M.r Duhamel," dated "du 7 Xbre 1753," SHMV, Ms. 87.

64. DPF, pp. 1410, 1413.

65. Delisle-Galissonnière correspondence, OdP, Ms. 1029, and Ms. "Passage de Mercure sur le soleil du 6 may 1753 a observer dans le Canada," AN-MAR 2JJ 69.

66. See Delisle letter to the Jesuit Neuville ("Neuvialle"), dated "13 octobre 1752," OdP, Ms. 1029, and Delisle letter to Bellin, "De Reims Le 30 octobre 1752," OdP, Ms. 1029, B 1-7(XII, #6).

67. "conclure assez bien la difference de Longitude de Pekin et de Paris," draft letter in Delisle's hand "au R. P. Bondier missionaire de la C. de J. à Chandernagor," dated "de Paris le 24 Janvier 1756," AN-MAR 2JJ 68.

The Special Role of J.-N. Delisle. As the La Cardonnie expedition illustrates, the astronomical and cartographical arms of the Colonial Machine were considerably enhanced by the initiatives and indefatigable energy of the astronomer Joseph-Nicolas Delisle (1688-1768).[68] His remarkable career began in Paris as a naval geographer. Delisle worked his way up through the ranks of the Academy of Sciences from adjunct in 1714 to associate academician in 1716. He developed an extensive correspondence with astronomers, and he assiduously collected historical astronomical documents and data. In 1725 he accepted a position as imperial astronomer at the Imperial Academy of Sciences in Saint Petersburg, where for twenty years he orchestrated astronomical activity in Russia, in France, and throughout Europe. In 1747 he returned to France, and in the arrangement discussed above, he became *astronome-géographe* attached to the Dépôt des Cartes et Plans. Delisle resumed work as an active *pensionnaire vétéran* in the Academy of Sciences in Paris, added the professorship of mathematics at the Collège Royal, and continued to encourage astronomical work and research until his death, notably as a key promoter of the transits of Venus observations of the 1760s.

In one facet of his role as an intermediary, Delisle kept up a correspondence with Jesuits in China and India and with agents of the Compagnie des Indes for almost forty years, from the late 1720s through the 1760s.[69] Throughout, he displayed a constant interest in astronomy and geography, proposing a program to improve the cartography of Asia by enlisting the network of Jesuit missions and missionaries. In 1752 he wrote from the Dépôt des Cartes et Plans to the Jesuit, Bondier, at Chandenagor on the Ganges: "The new studies that have appeared since the death of my brother [Guillaume Delisle, in 1726] can improve the maps of all these countries [of India and China], and this especially by tapping Jesuit missionaries in China and the East and West Indies. This is what I will try and do."[70] The next year he sent Bondier a package with the latest European maps of the East:

> ...all these maps, I say, along with the *Atlas oriental* of Mr. d'Après de Mannevillette, make up the best and most recent that we have on the East

68. On Delisle see Chapin (1971); Jaquel; Dawson; Woolf, passim, presents much valuable information concerning the life and times of Delisle. See also PA-DB (Delisle). A major scientific biography of Delisle would be a contribution to the historiography of eighteenth-century science.

69. See Delisle correspondence, OdP and AN-MAR, e.g., Delisle letters "Aux R.R.P.P. Missionaires Jesuites, à Pekin, de Petersbourg le 20 Sept. 1731 N. St." and "Aux R.R.P.P. Missionaires Jesuites à Pekin de Petersbourg le 25 Juillet 1732 N.St," AN-MAR 2JJ 62. Be it noted that Delisle's network extended to Canada; see correspondence with Father Bonnecamp, AN-MAR 2JJ 68.

70. "On pourroit bien reformer les Cartes de tous ces pais par les nouveaux memoires que l'on a depuis la mort de mon frere, et surtout en se servant de ceux des missionnaires Jesuites de La chine et de ceux des Indes; c'est ce que je tacherai de faire," Delisle to Bondier, dated "à Paris 13 Octobre 1752," OdP, Ms. 1029, B 1-6 (XI #253).

> Indies, and they can serve as the bases for new research by missionaries....It seems to me that to keep you up to date with new maps of the countries you travel through is the best and easiest way to enable you to enrich geography with your own observations, and this can be as useful to the curious in Europe as to missionaries abroad....I am going to advise the Jesuit superior for missions, I say, that the best maps that come out in Europe be sent to all the missions, enjoining the missionaries there to carry on a correspondence over geography with a skillful and zealous geographer in Europe [presumably Delisle himself!].[71]

Delisle's proposal was not for naught. His correspondence with Father Gaubil in China, for example, is replete with cartographical exchanges regarding China, Tibet, Tartary, Japan, Korea, and Cochinchina.[72] His communications with Father Cœurdoux in India were similar in nature regarding the cartography of various regions of India (especially northern India) and Southeast Asia. Delisle likewise corresponded over matters astronomical and cartographical with the Jesuits Chanseaume and Neuville in Macao.

The maps of India and Asia produced by the royal geographer Jean-Baptiste Bourguignon d'Anville (1697-1782) became a special focus of discussion, and the case reveals the importance of these local agents.[73] Delisle queried Gaubil closely about d'Anville's maps of Asia: "You would do me a favor to indicate what you would find to redo because I am always hoping for new memoirs and clarifications over this [frontier] that you can provide me better than anyone else. I am also awaiting your judgment on the other maps of Asia by Mr. d'Anville and especially if you would have corrections to make to the part touching on the Indies."[74] At first Delisle was not able to send the d'Anville

71. "...toutes ces Cartes dis je, jointes avec L'Atlas oriental de Mr. d'Aprez de Manevillette composent tout ce que l'on a de meilleur et de plus resçent sur les Indes orientales, qui doit servir de fondement aux nouvelles recherches des missionnaires,....Il me paroit que c'est le meilleur moien et le plus facile pour vous mettre en état d'enrichir la Geographie de vos remarques que de vous envoier toujours les nouvelles cartes des Pais que vous parcourez ce qui peut être aussi utile aux missionnaires qu'aux curieux D'Europe....Je vais luy conseiller [the Jesuit superior for missions], dis je, que l'on envoie dans toutes les missions les meilleurs cartes qui auront parû en Europe dans lesquels seront repandus les missionnaires, leur enjoignent qu'ils aient correspondance au sujet de la Geographie avec quelque habile et zelé Geographe d'Europe," Delisle to Bondier, dated "Paris le 30 Octobre 1753," OdP, Ms 1029, B 1-7 (XII #237).

72. Delisle-Gaubil correspondence, AN-MAR 2JJ 62; see, for example, undated Gaubil letter to "Mr. Lange," marked by Delisle, "reçu le 9 Nov. 1733."

73. On d'Anville, see Broc (1975), pp. 31-36; Petto, pp. 77-81; PA-DB (Anville). Hostetler (2001, 2009) fully discusses cultural contacts between European and Chinese maps and mapmakers in early modern times.

74. "Vous me feriez plaisir de me marquer ce que vous y trouverez à redire; car j'attens toujours de nouveaux memoires et eclaircissemens sur cette [illegible: frontière?] que vous pouvez me fournir mieux que tout autre. J'attends egalement vôtre jugement sur les autres cartes de l'Asie de Mr. Danville et surtout si vous aurez quelques corrections à faire à la partie qui regarde les Indes." AN-MAR 2JJ 68, Draft, "Au P. Gaubil à Pekin." Dated: "à Paris le 27 Octobre 1754."

map of India (*Carte de l'Inde*, 1752) because it was commissioned by the Compagnie des Indes for its exclusive use. Soon, however, the map made its way to India and China, and Delisle was anxious to get responses from his missionary correspondents on the spot. Regarding this map, Delisle wrote to Bondier in India the same day he wrote to Gaubil in Beijing requesting "...new reflections and corrections that you, Reverend Father, are in a better position to make than anyone else; so I am asking you to please apply yourself to this end and for the advancement of geography inform me of all the remarks you have to make."[75] D'Anville was a favorite of the Jesuits, however, and they were reluctant to criticize his work. Cœurdoux, whom Delisle likewise pressed for responses to the d'Anville map, wrote back saying that "I esteem Mr. d'Anville so highly that it pains me to not always agree with him…and I pray you not to use my comments without the knowledge of this able geographer whom I esteem too highly to wish to cause him pain."[76] Whether Delisle complied with Cœurdoux's request is unknown, but Delisle did write back thanking his Jesuit agent especially for Cœurdoux's "memoirs and geographical observations that contain your remarks and fairly substantial corrections to the geographical maps of the Indies of my brother, and Mr. d'Anville, and the charts of Mr. d'Après."[77] In the letter quoted above, Delisle asked specifically for Bondier to distribute maps to those who can make corrections and for Bondier to serve as his contact. What is apparent in these exchanges is that local observers were better equipped and absolutely necessary for providing feedback to Delisle and the big-shot "géographes de cabinet" back in the metropolis.[78]

75. "…de nouvelles reflexions et corrections que vous etes plus en etat de faire que tout autre, Mon Reverend Pere; cest à quoy je vous prie de vouloir bien vous emploier particulierement, et de me faire par de toutes les remarques que vous ferez pour l'avancement de la Geographie," ALS "au P. Bondier à Chandernagor, &," dated "à Paris le 27 Octobre 1754," AN-MAR 2JJ 68.

76. "j'estime si fort Mr. Danville que je me scay mauvais gré de n'etre pas toujours de son avis…je vous ay prié en meme temps de n'en point faire d'usage sans la participation de cet habile Geographe, que j'estime trop pour luy vouloir faire de la peine," Cœurdoux to Delisle, dated: "a Pondicheri ce 29 octobre 1759," MAR 2JJ 66 #179.

77. "[vos] memoires et observations Geographiques consistent dans les remarques et corrections assez importantes sur les Cartes Geographiques des Indes de mon frere & de Mr. Danville et du routier de Mrs. Daprez," draft Delisle letter to Cœurdoux in Pondicherry, dated "de Paris le 8 decemb. 1757," AN-MAR 2JJ 69. See also Delisle's Ms. "Remarques sur la carte de Mr. Danville,"AN-MAR 2JJ 69.

78. Delisle also promoted the putative discovery of the Northwest Passage by an elusive Spanish Admiral, de Fuente, in the 1640s, reports of which first appeared in London in 1708. Delisle's map (*Carte des nouvelles découvertes au N. de la mer du Sud*) of 1753 spurred interest by the Académie de Marine at Brest and by the Jesuits in China and Canada. The navy ministry approved an expedition for Delisle himself to go in search of the Northwest Passage, but at age 65 Delisle chose to remain at home and a *géographe de cabinet.* See Petto, pp. 113-19; AN-MAR C^7 186 (Lisle); DDP I:30; ALS, Bonnecamp to Delisle, dated "a Quebec ce 30.eme Oct. 1754," AN-MAR 2JJ 68; Ms. trans. of Gaubil letter to Cromwell Mortimer, dated "Peking 30 Octob. 1751," RS/L, L&P II, #422; see also Ms. "Extract of a letter of Monsr. De Guignes, Interpreter to his most Xtian Majesty for the Oriental Languages. To Mr. Birch, dated at Paris June 1, 1752 NS," RS/L, L&P II, #306.

The Transits of Venus. With J.-N. Delisle as *magister ludi*, organized science everywhere in Europe mobilized to observe the transits of Venus across the face of the sun in 1761 and 1769.[79] These observations represent the biggest "Big Science" efforts to date in history, and their high profile exerts a distorting historiographical influence on our tale. The scientific object was the same as the transit of Mercury of 1753, that is, to use coordinated observations to calculate the astronomical unit, or the Earth-Sun distance. The very nature of the exercise required observers scattered as widely as possible *and* the return and processing of data at a central locale. Many expeditions fanned out around the globe to observe these rare events. In 1761, astronomers made 120 observations at 60 points in Europe and around the world; 177 observers at 77 stations took data in 1769. It was a great accomplishment of eighteenth-century science that European astronomers were thus able to calculate the astronomical unit for the first time with confidence and to a reasonable degree of accuracy.

The French and the Académie des Sciences were leading participants, but the efforts to observe the transits of Venus mobilized virtually every European scientific society, observatory, and active astronomer in the 1760s. The Swedes and the Danes were major observers both in 1761 and 1769. The Bolognese academy was responsible for six stations in 1761 and participated again in 1769. With French help, the Russians sent observers to Siberia in 1761; and thirteen Russian-based astronomers recorded the event in 1769. For the 1761 transit the British government and the Royal Society of London sent Charles Mason and Jeremiah Dixon to the Cape of Good Hope, and Nevil Maskelyne to Saint Helena. The British and the Royal Society of London led the way for 1769. There were seven observers from Greenwich, including Maskelyne, and others from Oxford and elsewhere in Britain. Dixon went to Hammerfest, Jardine was at Gibraltar, and reports came in from India. But by far the premier Venus transit observation – by the British or anyone else – was that of Captain James Cook during his first voyage to the Pacific, where the party observed the transit at Tahiti. In America, the American Philosophical Society in Philadelphia, with financial support from the Pennsylvania Assembly, sponsored three local observations in 1769.

The French were major participants in the transit of Venus observations. Twenty-nine observers in France itself reported in 1761, and twenty-six in 1769. These included multiple parties at the Observatoire, at the naval observatory at the Hôtel de Clugny, and even at the Cabinet de Physique et d'Optique. In addition, the transits occurred in June in both years, and several academicians (Le Monnier, La Condamine, Lacaille, Bailly) observed them at their country estates. The Seven Years War, being fought in 1761, prevented the full participation of competent personnel in the navy; but three observations (by Blondeau, Le Roy,

79. Woolf remains the definitive treatment of the transits of Venus; see also McClellan (1985), pp. 208-20; Iliffe, pp. 622-24. On Delisle's role, see AN-MAR 2JJ 67, "Passage de Venus sur le Soleil 6 Juin 1761."

and Verdun) were reported from naval headquarters at Brest for 1769, and d'Après de Mannevillette reported from Kergars, his property in Brittany.

These metropolitan observations testify to the strength of French astronomy in the third quarter of the eighteenth century, but the expeditions launched by the French in 1761 and 1769 show the real power of the Colonial Machine to advance astronomy. Not counting Cassini de Thury who went to Vienna for the 1761 transit, the French sent Alexandre-Guy Pingré (1711-1796) to Île Rodrigues in the Indian Ocean, Jean Chappe d'Auteroche (1728-1769) to Tobolsk in Siberia, and Le Gentil de la Galaisière (1725-1792) to Pondicherry in India. For 1769, Chappe observed from Baja California, Pingré went to Saint Domingue, and Le Gentil was to observe again from the Indian Ocean.[80] The Académie des Sciences planned and organized these expeditions, but needless to say, their success depended on deploying the full armamentarium of the Colonial Machine, including the Académie des Sciences, the Observatoire, the Ministry of War (for 1761), the Ministry of the Navy, and the Ministry of Foreign Affairs, the navy itself, and the Compagnie des Indes. Other observations filtered in from Martinique and Île de France. French Jesuits in China likewise participated, and the Académie served as the central clearing house for processing data and disseminating results. Transit observations were connected to the work of establishing longitudes on points of land. As might be expected, these trips returned not just astronomical and longitude data, but also botanical and natural history information and samples; and to this extent Buffon and the Jardin du Roi need to be included in the story of the Venus transits.[81]

The eleven and one-half year odyssey of Guillaume J.-H.-J.-B. Le Gentil de la Galaisière, from 1760 to 1771, in the Indian Ocean was also tied to the transits of Venus.[82] An adjunct astronomer, Le Gentil was sent by the Academy to observe the 1761 transit from Pondicherry. Despite war with Britain, his expedition proceeded with the approval of the minister of war, the secretary of state, the minister of the navy, and the control general. He sailed from Brest aboard the *Le Berryer*, but the entrepôt at Pondicherry had fallen into the hands of the English by the time he arrived in the Indian Ocean. To his further misfortune, he was stuck at sea on 6 June 1761, the day of the transit. Unable to observe the 1761 transit with scientific rigor, Le Gentil remained in the Indies to observe the transit of 1769, returning to France only in 1771. When not ill, he sailed on the ships of the Compagnie des Indes and sent reports back to Paris concerning the longitude and cartography of Île de France, Île Bourbon, Île Rodrigues, Pondicherry, Mada-

80. On the Pingré expedition, see Pingré (2004) and the original of his journal, SHMV, SH 158 2537; Lacroix (1938), vol. 3, pp. 177-84.

81. On this point, see Woolf, pp. 102-103, 109, 113-114, and "Chappe d'Auteroche. Journal du voyage en Californie," OdP, Ms. C5 20-24.

82. See Chapin (1973); Woolf, pp. 58-60, 126-30; Allorge, pp. 405-415; Lacroix (1938), vol. 3, pp. 169-76; PA-DB (Le Gentil).

gascar, and Manila. He returned to publish his two-volume *Voyages dans les mers d'Inde* in 1779-81.[83]

Mapping the Indian Ocean. The Le Gentil adventure is indicative of the huge investment made by the Colonial Machine in exploring and mapping France's Indian Ocean possessions and in developing navigational aides for sailing to and from the Indian Ocean. The merest, remote specks in the great body of water of the Indian Ocean – Île de France, Île Bourbon, and their dependency, Île Rodrigues – were the key transit stations for French mercantile voyages traveling to India and China. Accurate knowledge of the location of these islands and how to get to them, beyond them, and then back to France was obviously essential for such trade. With Compagnie des Indes involvement through the 1760s, followed by direct government support, for half a century at least the Colonial Machine diligently attended to mapping and charting the Indian Ocean.

The Pingré observation of the transit of Venus in 1761 on Île Rodrigues was a related episode in the continuing French efforts to chart the Indian Ocean.[84] The paper by royal geographer Philippe Buache, "Observations Géographiques sur les Isles de France et de Bourbon, Comparées l'une avec l'autre," appearing in the *Mémoires* of the Paris Academy in 1764, also evidences this interest in mapping those colonies. We have seen J.-N. Delisle and his network of Jesuit missionary cartographers and natural historians who were active from the 1730s through the 1760s. We have likewise encountered the cartographer of the Compagnie des Indes, d'Après de Mannevillette, and his *Neptune Oriental*, his compendium of maps of the Indian Ocean originally published in 1745 and underwritten by the Compagnie des Indes. In 1766 the Compagnie reprinted d'Après de Mannevillette's *Neptune* and distributed it throughout its organization.[85] It was at this point (in 1765) that Mannevillette and the Compagnie published his *Mémoire sur la Navigation de France aux Indes*, a work based on his widely-circulated manuscript of 1758 that likewise appeared in the *Savants Étrangers* series of the Academy of Sciences.[86] The full title of a related piece of his ("Account of a Voyage to Île de France and Île Bourbon with Many Astronomical Observations for Finding Longitude at Sea as Well as for Determining the Geographic Position of These Islands"), published by the Académie des Sciences in 1763, evidences the focus on the Indian Ocean at this time.[87] Readers

83. Le Gentil de la Galaisière (1768), (1771), (1777), and (1779-1781).

84. See Pingré (1761) and HARS 1761, pp. 107-111; "Lettre Ecrite à M. de fouchy par M. Pingré, datée au Port-Louis de L'isle de france le 19 7bre 1761," AR-MAR 2JJ 67. According to Pingré, even the *latitudes* of the Indian Ocean possessions were still not known with sufficient accuracy.

85. See letter from "B.au des archives" to "Mr. delavigne buisson a lorient," reception dated "Le 10 Xbre 1766," SHML, 1P 288, Liasse 180.

86. See 1758 ms. "Instructions Pour aller au Indes Oriental," AN O^1 597, #33; d'Après de Mannevillette (1765) and (1768).

87. D'Après de Mannevillette (1763).

will recall that in 1762 the Compagnie created its own Dépôt des Cartes et Plans in Lorient for d'Après de Mannevillette to house his library and collections so that he could work on a revised edition of his *Neptune Oriental*. The crown maintained this facility as a subsidiary to the Navy's main Dépôt des Cartes et Plans in Paris once it formally took over from the Compagnie des Indes in 1770; and, as we have seen, d'Après remained as *Inspecteur et Garde du Dépôt des Cartes, Plans et Journaux* concernant la Navigation de l'Inde a l'Orient with a salary that rose to 4000 livres in 1775.[88] The ministry provided him with housing and a 10,000-livre bonus for the plates of the 1775 edition of his *Neptune Oriental*.[89]

The 1775 *Neptune Oriental* was an entirely new edition and a major undertaking, in which the Académie Royale de Marine played a leading role. D'Après de Mannevillette was already an important member in the Marine Academy, and in 1773 he wrote to it with his plan for a new edition of his *Neptune Oriental* with 56 maps.[90] The Marine Academy wrote back with its approval and the wish that de Mannevillette's work be printed at Brest under its supervision.[91] The Marine Academy was in touch with le chevalier d'Oisy, the Director of the Dépôt des Cartes et Plans over this project. The ministry subscribed for 400 copies.[92] Published in Paris and Brest, the work, with 60 plates, appeared in 1775, and was dedicated to the king. Separately published with the maps came an *Instructions sur la navigation des Indes orientales pour servir au Neptune oriental*. A supplement to the new *Neptune Oriental*, was issued in 1781 after the death of d'Après de Mannevillette.

D'Après de Mannevillette had a difficult personality. He did not get along with Bellin, accusing Bellin of plagiarizing his *Neptune Oriental*, a view upheld by a commission of the Paris Academy of Sciences.[93] Adanson, who sailed to Senegal with d'Après, said he was crazy.[94] And, d'Après' temper became an issue in the Rochon-Grenier dispute, to which we turn momentarily. On the other hand, d'Après de Mannevillette's reputation was more secure in Asia, where active sailors depended on his work, than in Europe. Father Cœurdoux reported that "this

88. Filliozat (1993), pp. 92-102; Chapuis, pp. 249-52; and ministerial note dated "24 avril 1773," AN-MAR C^7 6 (Après de Mannevillette). De Mannevillette was allocated an assistant, Benoit Grognard du Justin (1742-1814); see AN-MAR C^7 132 (Grognard du Justin). The Dépôt at Lorient folded into the Paris one with d'Après de Mannevillette's death in 1780.

89. On these points see previous note and ALS, Turgot to Sartines, dated "Versailles 31 Juillet 1775," AN-MAR D^2 29, fol. 135).

90. See Mannevillette materials, SHMV, Ms. 064-75: Registres des procès-verbaux des séances de l'Académie royale de marine, ainsi que des Mémoires qui y ont été lus (1769-1787), passim, esp. 072-074; see also copy of Academy letter to D'Après de Mannevillette, dated "5 7.bre 1775," SHMV. Ms. 092.

91. DDP III:74.

92. As noted by Cossigny in a letter to Le Monnier, dated "Paris le 28 Mars 1775," BCMNHN, Ms. 1995.

93. Filliozat (1993), pp. 100-101; PA-PV 69 (1750), p. 313 ("Samedi 4eme Juillet 1750").

94. ALS, Adanson to [Bernard de Jussieu] dated "Au Senégal le 15 aoust 1749," in PA-DB (Adanson).

skilled geographer,...is not treated with equal fairness in France as abroad," and even before the second edition of the *Neptune Oriental* appeared, Le Gentil de la Galaisière declared flatly that "The maps of Mr. d'Après are superior to all the other maps [of the Portuguese or the Dutch]."[95]

D'Après de Mannevillette was earlier involved in another major episode in the story of eighteenth-century French cartography in the Indian Ocean. This was the trip he captained in 1750-1753 on the orders of the king, the Paris Academy, and the Compagnie des Indes. The main goal of this mission was to establish more accurate longitude positions in the Indian Ocean. Once there, d'Après de Mannevillette used the lunar observation method for refining the established longitudes of the Cape of Good Hope and Île de France. He again undertook coordinated astronomical observations with the astronomer Le Monnier, and he followed up with a set of maps for the area in 1763.[96] Another part of d'Après' assignment for the voyage in 1750 was to transport Abbé Lacaille to the Cape of Good Hope for Lacaille's and the Academy's own program of astronomical observations from the tip of southern Africa.

With ministerial approval and with the material support of the Compagnie des Indes (which Lacaille praised for its "zeal for Navigation which in large measure depends on that for Astronomy"), the Academy sent Lacaille, an associate astronomer, to the Cape of Good Hope to map the stars of the southern hemisphere.[97] Determining the exact geographical position of the Cape was also one of the principal aims of the voyage, and, armed with letters from the Prince of Orange and the Dutch East India Company, Lacaille fulfilled this duty with exactitude, including producing a map of the Cape area. Because his stay at the Cape became prolonged, Lacaille was also able to undertake a set of coordinated astronomical observations with Lalande in Berlin. In 1753, at the request of the Compagnie des Indes, the king ordered Lacaille to move on to Île de France and Île Bourbon to fix their "exact geographical location." Lacaille had seen d'Après de Mannevillette's recent data and thought them perfectly satisfactory, but there was not time to get his orders countermanded. And so, he spent 1753-54 in the Indian Ocean.[98] The planet Jupiter was then in conjunction with the sun, so longitude observations using Jupiter's satellites proved impractical. Nevertheless, Lacaille made good use of his time on Île de France and Île Bourbon, including three tough

95. "...j'ay vu seulement en passant Mr. D'aprés au quel on ne rend pas en france la meme justice que chez les etrangers. Le mauvais etat de sa santé et de la mienne, joint au prochain depart de cet habile Geographe, me mettront apparemment hors d'état de le voir davantage," ALS Cœurdoux to Delisle, dated "a Pondicheri ce 18 aout 1758," AN- MAR 2JJ 66; Le Gentil de la Galaisière (1779-1781), vol. 1, p. 619.

96. See d'Après de Mannevillette (1763); and P.C. Le Monnier (1751).

97. "...zèle pour la Navigation, qui dépend en grande partie de celui de l'Astronomie," La Caille (1751), p. 521. On La Caille, see also Woolf, pp. 35-38; Lacroix (1938), vol. 3, pp. 161-65.

98. On La Caille's unhappiness about this development, see "Copie dune lettre de Mr. de la Caille à Mr. Duhamel lue à l'academ. le 23 juin 1753," OdP, Ms. 1029, B 1-7 (XII, item #154); "Journal historique de mon voyage," OdP, Ms. C 3, 26.

months in the bush in the rain with a small team of soldiers and slave porters.[99] He reported longitude observations where possible, other astronomical observations, declination and pendulum experiments, local cartographical work, and meteorological observations, including an annotated list of monsoons for the period 1753 to 1754.[100] While on this trip Lacaille befriended Lanux *père*, Conseiller of the Conseil Supérieur on Île Bourbon; in 1754 the Académie des Sciences elected this Lanux a correspondent of Réaumur.[101]

Lacaille's contact with Lanux, father and son, illustrates a significant feature of extraterritorial astronomical and geographical research undertaken by the Colonial Machine in the eighteenth century – the overwhelming dependence on astronomical emissaries from France. In providing regular astronomical observations from Île Bourbon, the Lanuxes were the exception proving this rule. Father and son supplied astronomers of the Paris Academy with cometary observations, in particular from Bourbon, from the 1750s through the 1780s.[102] The observations of the younger Lanux of the comet of 1788 were instrumental in determining its orbit, and the Académie des Sciences duly appreciated his contributions and their regrettably exceptional nature:

> It is very advantageous for the astronomy of comets that so zealous an observer as M. de Lanux lives in the southern hemisphere under such a serene sky. He already observed the comet of 1784 more than five weeks before it appeared on our horizon. He is in a position to discover many more that are not visible to us, and we must hope that he will occupy himself with finding the one we are waiting the return of this year. We encourage him in this endeavor even more because this comet should be seen first in the southern hemisphere.[103]

99. For an early report on how difficult it was to penetrate these forests, see letter "A M.rs les directeurs généraux de la Compagnie des Indes," dated "a l'Isle de Bourbon le 20 Xbre 1731," in Lougnon (1934), p. 148.

100. See La Caille (1754a) and (1754b) and HARS 1754, pp 110-16.

101. See report in Cossigny letter, dated "Isle de France le 24e mars 1755," reprinted in Jean-François Cossigny (1939-1940), pp. 312-13; on Lanux père, see Lacroix (1938), vol. 3, pp. 185-90.

102. See PA-PV 90 (1771), fols. 257-58 ("Mercredy 18 Décembre"); PA-DB (Lanux), notably copy of Lanux senior letter to Pingré, "À l'Isle de Bourbon, le 3 9bre 1765; "Observations astronomiques et lettres adressées de l'Ile Bourbon au P. Pingré par M. de la Nux père et fils 1755-1784," Bibliothèque Ste. Geneviève, Ms. 1085.

103. "Il est très avantageux pour l'Astronomie des Cométes, qu'il se trouve dans l'hemisphère austral, et sous un beau Ciel, un observateur aussi Zélé que M. De la Nux; il avait déja observé la Cométe de 1784. plus de 5. Semaines avant qu'elle parut sur notre horison; il est à porté d'en découvrir plusieurs qui ne peuvent être visibles pour nous, et on doit esperer qu'il s'occupera de la recherche de celle dont on attend le retour cet [sic] année ci. Nous l'y éxhortons d'autant plus que cette Cométe doit commencer à paraitre dans l'hémisphère austral…," PA-PV 107 (1788), fols. 209-210v ("Mercredy 6 Aoust 1788").

The Rochon-Grenier Dispute The mapping voyage undertaken by Abbé Alexis-Marie Rochon and Ensign Jacques-Raymond Grenier of the Royal Navy in 1767-1770 underscores the importance royal authorities placed on reliable navigation in the Indian Ocean. The dispute that subsequently arose between Rochon and Grenier allows us to see better the inner workings of the Colonial Machine and how secular authorities reacted when experts disagreed.

The purpose of the Rochon-Grenier voyage was to chart the isles and banks to the north of the Mascarene Islands and to ascertain the shortest route to India from Île de France and Île Bourbon.[104] The traditional route took about 100 days and swept in a great circle westward and northward along the African coast before turning east to India. Sailing directly north to India offered the possibility of shaving 2000 miles (800 lieues) and 25 to 30 days off the trip from Île de France to India. In addition to surveying this route, Rochon was instructed to make corrections to d'Après de Mannevillette's maps and to find the source of the true coconut, the *coco de mer.*[105]

Naturally, the Ministry of the Navy and the Colonies sponsored the expedition. Rochon was a good choice. He was an accomplished astronomer, keeper of scientific instruments of the Brest Marine Academy, and an author of a work on finding longitude at sea. He sailed as an official naval astronomer (*astronome de la Marine*) attached to the Dépôt des Cartes et Plans, and on the occasion of this voyage he was elected a correspondent of the Academy of Sciences attached to Cassini de Thury. He shared additional connections to the Academy through Duhamel du Monceau and Bézout. The chevalier Grenier captained the corvette *L'Heure-du-Berger*, which arrived in the Indian Ocean in early 1769.[106] He and Rochon performed longitude and cartographical work in May through October 1769, stopping at Île de France, the Seychelles, the Maldives, the Malabar Coast, Ceylon, Pondicherry, and Diego Garcia. Rochon observed the 1769 transit of Venus while in the Indian Ocean, returning home with Grenier in early 1770.

At first, Rochon reported the happy discovery with Grenier that the direct route north was safe and preferable to the traditional one.[107] But quickly, a heated dispute arose between Rochon and Grenier over the precise route and whether sailors had previously used, but abandoned, the northward track because of the dangers of shipwrecks on reefs and islands. (Hence the rationale for charting these dangers and determining their longitudes in the first place.)

The dispute played itself out first within the Marine Academy at Brest, where Grenier presented his findings.[108] The Brest academicians wholeheartedly

104. On Rochon and this episode, see Fauque; Guy Jacob, pp. 44-45; Lacroix (1938), vol. 4, pp. 15-24; Collet; Allorge, pp. 403-405; PA-DB (Rochon); AN-MAR G 91 fols 214ff.

105. Lionnet; Lacroix (1938), vol. 4, pp. 16-17.

106. DDP II:43.

107. ALS Rochon to Monseigneur dated "a l'orient le 8 avril 1770," AN-MAR G 91 fol. 203.

108. DDP II:44.

approved Grenier's report and sent it on to the minister. In a twist, the minister, de Terray, consulted Rochon who disagreed vehemently with Grenier's views, saying that the proposed route was not new and was too dangerous. The minister then sent the whole matter back to the Marine Academy asking for new commissioners and further adjudication. He seemed upset that two experts who had been on the same voyage would disagree: "The difference of opinion of two informed persons who are themselves witnesses seemed to His Majesty to merit the most serious attention and the most thoughtful study. He thought, Sirs, to submit the matter to you."[109] A second panel at the Brest Academy thus took up the matter again. It seemed to have been influenced by an intemperate report from d'Après de Mannevillette, who took umbrage at Rochon's criticisms of his maps. The report of this second panel was likewise favorable to Grenier and rejected Rochon's criticisms as unfounded.[110]

In the meantime, Rochon sent his own observations to the Marine Academy and to the Academy of Sciences in Paris.[111] In Paris the committee report by Cassini de Thury, Maraldi, and Bory was much more positive, approving of Rochon's views and authorizing publication in the *Savants Étrangers* series.[112] Given this fundamental institutional disagreement, the new Minister of the Navy, de Boynes, brought the whole matter back before the Paris Academy. Grenier was afraid that de Boynes and insiders in the Paris Academy were partial to Rochon and that the Paris Academy did not have the expertise to judge the matter![113] At issue, again, was how to deal with conflicting experts; and in de Boynes' view the Paris Academy was the ultimate scientific arbiter:

> Although the king has the greatest confidence in the wisdom of his Académie de Marine and in that of Mr. de Mannevillette, he has nonetheless thought that so important a question cannot be examined closely enough and that one cannot choose better judges in a dispute that divides knowledgeable people who are already involved than a Company as enlightened as the Académie des Sciences whose indefatigable zeal is ceaselessly occupied with everything that can contribute to the progress of useful knowledge. His Majesty therefore desires that the

109. "Cette contrariété d'avis de deux personnes instruites et qui ont vû par elles mêmes, a paru à Sa Majesté mériter la plus sérieuse attention, et l'Examen le plus réfléchi. elle a crû, Messieurs, de vous soumettre [la] question," ministerial letter, Terry to "M.rs de L'académie R.ale de Marine à Brest," dated "À Versailles Le 18 mars, 1771," SHMV, Ms. 88; see also DDP III:16-17.

110. SHMV, Ms. 071 (vol. VIII), pp. 289-316; see also related memoirs and reports, pp. 175-87, 189-202; see also, Ms. 073 (vol. X) for d'Après and Rochon materials related to this matter.

111. "Mémoire de l'abbé Rochon sur voyage dans la mer des Indes," SHMV, Ms. 068 (vol. 5) fols. 51-129; and DDP II:42-43.

112. This affair plays out in PA-PV 89 (1770); see also PA-PS (18 July 1770).

113. See letter of Le Chevalier Grenier, dated "A Paris ce 10 Juillet 1771," SHMV, Ms. 89.

Académie des Sciences charge itself to examine the memoir of the chevalier Grenier and the observations of the Abbé Rochon.[114]

De Boynes submitted a substantial dossier with all relevant materials, and the Academy duly appointed Le Monnier, Bory, and Father Pingré to investigate. Pingré wrote the report, which blasted d'Après de Mannevillette's partisanship toward Grenier and exhorted him to return to his natural character of candor and honesty.[115] Nevertheless, while praising Rochon's goodwill and citizenship, the report concluded that "even so, his notion is not solidly based" and that Grenier's route possesses "an invaluable advantage."[116] The official minutes of the Academy for 6 July 1771 indicate only that the commissioners gave their report. In Lavoisier's personal minutes, however, one learns, less diplomatically, that "Messrs. Le Monnier, Bory, and Pingré read their report on the dispute between Mr. Grenier and the Abbé Rochon on a new route for going more directly to the Coromandel coast and concluded ~~in favor of Mr. Grenier~~ [sic] that the route proposed by Mr. Grenier is new and secure and that the objections of the Abbé Rochon are devoid of any substance."[117]

As much as the matter seemed settled in France, it remained a life-and-death issue for sailors in the Indian Ocean; and to eliminate any lingering uncertainty the ministry sent Rochon back to the Indian Ocean accompanying Yves-Joseph Kerguelen (1734-1797) on the latter's first voyage in search of a southern continent in the waters of the Antarctic. Rochon proved a quarrelsome fellow, and Kerguelen expelled him from the expedition in 1771 on Île de France.[118] Kerguelen then wrote from Île de France in 1772 that "the route indicated by

114. "Quoique le roi ait la plus grande confiance dans les lumieres de son Academie de Marine et dans celles du Sr. de Mannevilette, cependant il a pensé qu'une question aussi importante ne pouvoit être examinée avec trop d'attention, et qu'on ne pouvoit choisir de meilleur juges, dans une contestation qui divise deux Personnes instruites, et qui ont vu par elles mêmes, qu'une Compagnie aussi éclairée que l'Academie des Sciences, dont le zele Infatigable est sans cesse occupé de tout ce qui peut contribuer aux progrés des connoissances utiles. Sa Majesté désire donc que l'Academie des Sciences se charge de l'examen du Mémoire du Ch.er Grenier, et des observations de l'Abbé Rochon," ALS "de Boynes" to "M. de fouchi." dated "A Marly le 17 juin 1771," PA-PS (26 June 1771).

115. "Rapport d'une commission de l'Académie des sciences pour examiner un écrit de l'abbé Rochon...," Bibliothèque de Sainte-Geneviève: Ms 2540; copy, PA-DB (Pingré).

116. "quant même son sentiment ne seroit pas appuyé sur des fondemens bien solides... un avantage bien précieux," in "Rapport," cited previous note.

117. "MM. lemonnier de Borry et Pingré lisent leur rapport sur la contestation entre M. grenier et M. l'abbe rochon sur une nouvelle route pour aller plus directement a la Cote de Coromandel ~~on~~ et Conclud ~~en faveur de M. Grenier~~ que la route propose par M. Grenier est nouvelle et sure et que les objections de M. labbé rochon sont destituees de tout fondement;" PA-PS (1771), "Plumitif par Lavoisier," "Seance du 6 juillet"; Lavoisier's deletions. See also PA-PV 90 (1771), fol. 165v ("Samedi 6. Juillet. 1771").

118. On Kerguelen, see Brossard; Taillemite (2002), p. 274. Rochon and Kerguelen played out their own dispute within the confines of the Académie des Sciences; see PA-PV 91 (1772), fol. 281 ("Mercredy 5 Aout 1772"), fol. 284 ("Samedy 8 Aout 1772"), and fol. 340 ("Mercredy 18 Novembre 1772").

Grenier is not only practical, but even preferable to that followed until now."[119]

Rochon did not return to France until 1773, but when he did, he did not let the matter rest. He brought the question back before the Marine Academy at Brest, and he asked the Paris Academy likewise to revisit its decision.[120] In Paris, the Academy appointed an expanded committee that included the original rapporteurs Bory, Le Monnier, and Pingré, complemented by Duhamel and the veteran of the Indian Ocean, Le Gentil de la Galaisière. Bory wrote the report. Once again it absolved Rochon from the calumnies leveled against him by d'Après de Mannevillette, but otherwise the group avoided the issue by saying that vacations were at hand and that further input would ultimately settle the shortest and safest route of travel in the Indian Ocean.[121] In point of fact, however, some of the documents originally submitted to the Academy by de Boynes had been tampered with; the committee sought originals from the Marine Academy at Brest and from Rochon, but seems to have let the matter drop rather than pursue who might have tampered with the documents or why.[122] In a private letter Duhamel wrote: "There, Sir, is the decision of the Academy on the subject of the dispute between the Chevalier Grenier and the Abbé Rochon. You will see that the Academy is taking no part and is seeking only further clarifications."[123]

Rochon's status in the Colonial Machine was uncertain for a while after that. During his mission, he was attached to the Dépôt des Cartes et Plans; afterward, he was forced to return to his position as librarian of the Marine Academy at Brest. When he finally regained Paris in 1783, on the publication of his book about his voyages, he was elected an associate academician (1780), then full *pensionnaire* in the Academy (1783).[124] Ironically, as late as 1788 uncertainties remained over what was the safest route from Île de France to

119. Quoted in DDP III:51.

120. DDP III:52, 101; PA-PV 92 (1773), fol. 141 ("Mercredy 23 Juin 1773"), fol. 148 ("Samedy, 26 Juin 1773"); PA-PS (1773), "Plumitifs," 23 June 1773; "Liste des Commissions données en 1773," 26 June 1773.

121. PA-PV 92 (1773), fols. 187-88 ("Samedy 4 Septembre 1773"); PA-PV-110, Rochon, 4 September 1773 ("Sur une contestation élevée entre Lui et M.r le Chev. Grenier relativem.t aux différentes routes que les Vaisseaux peuvent tenir pour passer, dans l'arrière Saison, de l'Isle de france à Pondichery"), marked "L'acad.e reprendra ce travail après la rentrée. Bory."

122. See the 28 page untitled ms. signed "Tel est mon sentiment a l'observatoire Royal le dix d'aoust mil sept cent soixante et treize. Legentil," PA-PS ("Samedi 4 Septembre 1773").

123. "Voila, Monsieur, l'arreté de l'Académie au sujet de la contestation de M. le Chevalier Grenier et de M. l'abbé Rochon. vous verez que l'Academie ne prend aucune part et ne cherche qua avoir de plus grands eclaircissements." He continued: "il n'en est pas de meme pour ce qui regarde la lettre de M. Daprès qui est un tissu d'invectives revoltantes sans procurer rien de certain sur le fond de la question," ALS "Duhamel du Monceau" to unspecified recipient, dated "A Paris ce 6 7bre 1773," AN-MAR G 91 fol. 214.

124. See Rochon (1783); see also Rochon (1791). Regarding his administrative troubles, see copy of letter to Rochon, dated "versailles le 21 Janvier 1774," SHMV, Ms. 093, and other Rochon materials, AR-MAR G91 (dossier 3).

India.[125] For us, the moral is not so much who was right, but what the story reveals of the complex mechanisms for adjudicating the dispute and what they show us of the internal workings of the Colonial Machine in action.

Voyages of Discovery: Bougainville and La Pérouse. The great voyages of discovery in the second half of the eighteenth century can easily bedazzle. Then and now, for example, the three heroic circumnavigations spanning 1768-1779 by Captain Cook in particular focused all eyes on valiant bands of Europeans wending their way to parts unknown and bringing back wonders to civilization. The related French circumnavigations led by Bougainville in 1766-1769, what actually preceded Cook's, and then by the count de La Pérouse in 1785-1788 likewise elicit this awe. The fact that "Science" accompanied these voyages made them even more magnificent to contemporaries and to us. Two analytical problems undermine this view. One, although they were circumnavigational, the voyages of Bougainville and La Pérouse were hardly the first or only French "scientific" voyages of the seventeenth or eighteenth centuries; from Richer's trip to South America in the 1670s, through the voyages to Lapland and Peru in the 1730s, to Grenier and Rochon in the Indian Ocean in the 1770s, we have already encountered many state-sponsored scientific expeditions. Secondly, the great voyages of discovery did not take place in a vacuum. They were not, somehow, vague efflorescences of the Age of Enlightenment. Rather, the expeditions of Bougainville and La Pérouse need to be seen concretely as manifestations of the Colonial Machine in action.

In 1766-1769 Louis-Antoine de Bougainville (1729-1811) led the first French expedition to sail around the world.[126] It was the first circumnavigation with professional naturalists and geographers on board. Bougainville had distinguished himself with Montcalm in Canada, and in 1764 he had transported displaced Acadians to the Falkland Islands to establish a French colony there. The first part of Bougainville's voyage in 1766 had a distinctly diplomatic cast to it, as he returned to the Falklands to hand control back to the Spanish, who had objected to the implantation of a French colony on its territory. But the rest of Bougainville's mission was entirely a scientific voyage of discovery. Officers and crew numbered 330, sailing aboard the royal frigate, *La Boudeuse*, accompanied by the storeship, *L'Étoile*. The scientific staff included the royal botanist and naturalist Philibert Commerson (appointed by dint of his associations with Bernard de Jussieu, Le Monnier, and Poissonnier); the naval astron-

125. See "M.re sur des Isles ou bancs reconnus dans la Mer des Indes," by one Fortin, PA-PV-110, Fortin, 5 September 1788, and PA-PV 107 (1788), fol. 229 ("Mercredy 20. Aout 1788") and report, fols. 254v-257 ("Mercredy 5. Septembre. 1788").

126. On the Bougainville expedition, see first and foremost the modern critical edition of Bougainville (2001) edited by Bideaux and Faessel, esp. their introduction, pp. 5-44, and complete bibliography, pp. 461-72; see also, Taillemite (2008), pp. 125-33, (2002), pp. 63-64, (1997), (1977); Mollat du Jourdin and Taillemite; Dunmore (1965); Allorge, pp. 329-380; Lacroix (1938), vol. 4, pp. 60-61; Iliffe, p. 625; see also Gasc and Laissus, pp. 86, 92.

omer Pierre-Antoine Véron (1736-1770), accompanied by an assistant, the chevalier du Bouchage; a royal engineer and cartographer, Charles Routier de Romainville; and two surgeons. A writer/historian completed the party. Commerson (and presumably the others) was paid 200 livres a month and ate at the captain's table.[127] The expedition did considerable longitude work, especially in the Strait of Magellan, where Véron, a protégé of Lalande, established his station on the "Île de l'Observatoire." A later lunar eclipse helped fix other points of longitude at the Solomon Islands. Véron's use of lunar distances in the Philippines determined the width of the Pacific for the first time with some accuracy.[128] In the Pacific during March 1768, Poissonnier's distillation machine was used successfully to supplement the water ration; and while in the neighborhood of Batavia, trading with locals suggested to Bougainville the possibility of French involvement in the spice trade. Commerson spent his time collecting specimens of local flora. Among the new plants he found in Brazil was a violet-colored flowering climber that Commerson named "Bougainvillea" after the expedition's leader. In his botanizing, Commerson was aided by his not inexpert "valet," Jeanne Baret, a woman disguised as a man; her true identity was not revealed until she went ashore with Commerson in Tahiti, where the islanders immediately saw through her disguise. (She thus became the first woman to circumnavigate the globe.)[129] This expedition made only a short stay of nine days at Tahiti, but it had a profound effect on the European imagination in bringing back reports of Tahiti as a utopian paradise inhabited by the "Noble Savage" living outside the confines of European social and sexual oppression.

The expedition came to an effective end when it put in at Île de France in November 1768. This was because the scientific cadre elected to stay in the Indian Ocean. Commerson, who was chronically seasick, wanted to botanize there and work with the botanically inclined intendant, Pierre Poivre, on a natural history of the Mascarenes and Madagascar. (Commerson died in 1773 on Île de France.) The astronomer, Véron, was aware of the impending transit of Venus and wanted to go to Pondicherry to observe the phenomenon. (He died on Île de France in 1770.) Romainville, who had drawn the charts for the voyage, also decided to remain on the island. Bougainville regained France, was awarded 50,000 livres for the success of his adventure, and became vice admiral in the navy. He was justly proud that only nine men died from illness during his trip around the world. Bougainville's *Voyage autour du monde par la frégate du roi la Boudeuse et la flûte l'Etoile* appeared in 1771-72. Unfortu-

127. Commerson was granted an additional 1200 livres for books and equipment; see Ly-Tio-Fan (1976), p. 54, and Thouin letter "a M.r Dombey Docteur en Medecine et Botaniste du Roy," BCMNHN, Ms. 222.

128. Further on Véron, see PA-DB (Véron); AN-MAR C[7] 344 (Véron). For other Véron astronomical work, see HARS 1772 (Pt. 1), p. 81, and Lalande (1772).

129. Jolinom, pp. 78-89.

nately, the fact that Commerson and Véron had not returned to France undermined the worth of Bougainville's account because most of the scientific material remained in the Indian Ocean or filtered back to France only over the succeeding years. Nevertheless, Bougainville's trip opened the Pacific to French exploration. The Ministry of Foreign Affairs, the Royal Navy, the Ministry of the Navy, the Jardin du Roi (via the botanist Le Monnier), and the Observatoire were all involved in the Bougainville expedition. Yet the Académie des Sciences seems not to have been a direct participant. Bougainville's late election to the Académie des Sciences in 1789 as a free associate may testify to how the Académie saw his expedition in retrospect.

Just shy of two decades elapsed between the departures of the great French expeditions to the Pacific of Bougainville in 1766 and La Pérouse in 1785. As we have indicated, many French ships charged with scientific missions sailed in the interim, and the French Colonial Machine was active on all fronts in this period working to improve navigation and cartography. The La Pérouse expedition should therefore not be singled out as more exceptional than it was or as simply a product of the Age of Enlightenment. That said, the doomed voyage of Jean-François de Galaup, Count de la Pérouse (1741-1788) was a classic, perhaps *the* classic eighteenth-century voyage of scientific discovery and a quintessential manifestation of the Colonial Machine in action.[130] The very names of the two royal frigates commanded by La Pérouse, *L'Astrolabe* and *La Boussole* (*The Astrolabe* and *The Compass*), signal the high scientific character of the undertaking. This undertaking enlisted every element of the Colonial Machine, beginning with Louis XVI who took a deep personal interest in organizing the expedition.[131] With the king, the Minister of the Navy de Castries coordinated with La Pérouse and another veteran naval captain, the count de Fleurieu, to plan it. Fleurieu was an academician at the Académie de Marine, and he brought that organization into the planning. Bougainville participated. The Jardin du Roi was mobilized for its input.[132] Lavoisier and the Académie des Sciences were tapped for the scientific components of the voyage, while the Société Royale de Médecine prepared questions of "medicine and allied sciences" to be sent with the expedition.[133] The Royal Navy itself

130. On the La Pérouse and d'Entrecasteaux expeditions, see Taillemite (2008), pp. 184ff, (2002), pp. 303-304; Richard (1986); Gaziello (1984); Bravo (1999); Dunmore (1985); Melguen (2005); Roger Williams (2003); Duprat, p. 233; Allorge, pp. 487-98; Iliffe, p. 638.

131. In this connection see the over 300-page collection of royal documents, "Projets, instructions, mémoires et autres pièces, relatifs au voyage de découvertes, ordonné par le Roi, sous la conduite de M. de La Pérouse," Bibliothèque Mazarine, Ms. 1546; on the royal instructions for the La Pérouse expedition, see also Roger Williams (2003), pp. 15-31, cf. cover illustration.

132. BCMNHN, Ms. 1928, "Documents relatifs à l'expédition de La Pérouse autour du monde, commencée en 1785"; for Thouin's instructions, see Roger Williams (2003), pp. 33-50.

133. See Huard and Zobel; SRdM, Ms. 10, pp. 438, 336/446, 368/468, and 371/471, meetings for May, June, and July of 1785. See Richard (2005) for books sent with this and the d'Entrecasteaux expeditions.

took charge of outfitting and seeing to the seaworthiness of the 500-ton *L'Astrolabe* and the 500-ton *La Boussole.*[134]

[1785 Medal Commemorating the Voyage of La Pérouse]

By the time he was appointed at the age of forty-four to lead the expedition to the Pacific, La Pérouse was already a navy veteran and a veteran of the colonial seas, having spent 1772-1777 based on Île de France. With a total crew of 220, the *Astrolabe* and *La Boussole* were well-equipped and well-staffed floating laboratories of science.[135] That staff included two astronomers, Joseph Le Paute d'Agelet (1751-1788?), adjunct in astronomy at the Académie des Sciences and professor of mathematics at the École Militaire, and Louis Monge, also a professor at the École Militaire and brother of the associate academician in the physical sciences, Gaspard Monge. (Louis Monge had to disembark at Tenerife for health reasons.) The abbé Jean-André Mongez (1750-1788?), physical scientist and editor of the famed *Journal de Physique* from 1780 to 1785, brought his expertise. A bevy of naturalists and botanists were aboard the two ships: Joseph de la Martinière (physical scientist and naturalist), Jean-Paul de Lamanon, (mineralogist and meteorologist), one Dufresne (naturalist and brother of the government official, who debarked at Macao), Father Receveur (naturalist and chaplain, who died at Botany Bay), and Nicolas Collignon (1761-1788?), gardener and botanist. Two engineers added their expertise; the royal engineer and cartographer, de Monneron, and the Ingénieur Géographe, Bernizet, assisted by Pierre Guéry, watchmaker and armorer. In addition to the

134. See "rapport de la commission chargée de la visite des vivres pour l'expédition des frégates *L'Astrolabe* et *La Boussole*," SHMB, Série A, Sous-série 3A, Conseil de Marine, Registres 3A4 à 3A10: Procès-verbaux de délibérations du Conseil de Marine de 1772 à l'an II (1793-1794), 23 juillet 1785.

135. Bruno Latour famously makes much of these ships as "floating laboratories" reporting back to the "centers of calculation" in France. Latour uses the stop at Sakhalin Island by the La Pérouse expedition as a case study of the "extraction" of local knowledge and its transmutation into universal "scientific" knowledge by and at the "centers of calculation." For a critical analysis and an entrée into the surprisingly rich literature surrounding this episode, see Bravo (1999).

surgical staff of four, three artists rounded out the scientific complement of the La Pérouse expedition: Duché de Vancy (figures and landscapes), Guillaume Prévost (who did botanical drawings only), and Jean-Louis Robert Prévost (nephew of the former, natural history/zoology). The diplomat and Russian interpreter, Jean-Baptiste-Barthélemy de Lesseps (1766-1834), likewise sailed with La Pérouse, traveling back across Asia with news and reports of the expedition after a meeting with the Russians in Kamchatka in 1787.

The Académie des Sciences and the Académie de Marine provided a large complement of instruments. The Dépôt des Cartes et Plans de la Marine furnished charts and maps. Thouin, from the Jardin du Roi, sent seeds and plants for dissemination in the Pacific, and the expedition possessed a substantial scientific library that included the *Encyclopédie*. The Poissonnier device to distill seawater went aboard, too. The La Pérouse expedition was completely state-sponsored and a grand manifestation of the Colonial Machine in operation. The budget was a staggering 1,000,000 livres.[136]

The expedition left Brest in 1785, stopping in Brazil and rounding Cape Horn before turning for Chile, Easter Island and the Sandwich Islands. It headed north to Alaska and then down the coast of North America to California. At that point La Pérouse crossed the Pacific to Macao and then went on to Manila, Taiwan, and Sakhalin Island. He then turned south to Samoa and Tonga. In 1788 the expedition put in at the incipient British station at Botany Bay in Australia. It left Botany Bay in March 1788 and was never seen again. Modern marine archeology has revealed that the *Astrolabe* and *La Boussole* crashed on the reefs of the Pacific island of Vanikoro. The survivors made camp on the island, but their ultimate fate is still unknown.

Despite its tragic ending and although much was lost, the La Pérouse expedition returned valuable scientific information because La Pérouse and others sent back letters and reports at every opportunity and port of call. De la Martinière and Collignon shipped seeds and plants to Thouin and the Jardin du Roi; shipments were made from Cartagena and Chile.[137] Mongez reported to the Académie des Sciences, and the consignment heroically carried by de Lesseps overland across Asia and back to France was a rich one. The explorations in and around Sakhalin Island were especially productive, including longitude data returned from another "Île de l'Observatoire" set up there.

Worry mounted, of course, when La Pérouse failed to show up as expected at Île de France in June of 1789. In 1791, in the midst of the French Revolution, the National Assembly voted with patriotic fervor to send an expedition in search of La Pérouse but also aimed at gathering scientific data. This patriotic move was prompted by de Fleurieu, who had become minister of the navy,

136. Melguen, p. 50.

137. See ALS "Thouïn" to "Monsieur le Comte," dated "A Paris ce 8 7.bre 1786," AN O[1] 2113A "Pépinieres d'Amérique. Correspondance Générale pour la Mission du S. Michaux, Botaniste du Roi, Année 1786."

by the Académie des Sciences, and by the Société d'Histoire Naturelle; and it had the support of Louis XVI, then, but for only a short time longer, king of the French. Antoine de Bruny, chevalier d'Entrecasteaux (1739-1793), was selected to command the expedition.[138] Previously, d'Entrecasteaux headed the French Navy in the East Indies and was governor of Île de France and Île Bourbon from 1787 to 1790. D'Entrecasteaux sailed on the frigates *La Recherche* and *L'Espérance* with a scientific party that included the distinguished hydrographical engineer Charles-François Beautemps-Beaupré (1766-1854).[139] The expedition explored extensively in the waters around Australia in 1792-1793, where d'Entrecasteaux himself died of scurvy. This expedition ended in disasters of its own. News of the increasing radical turns of the Revolution in France prompted dissention among the crew, and the two ships commanded by royalists surrendered to the Dutch, then at war with republican France. The British later captured the scientific journals and documents of the d'Entrecasteaux expedition, and these were not returned to France until 1802. For us, the voyage of d'Entrecasteaux is little more than a footnote to the La Pérouse expedition, and the revolutionary circumstances under which it sailed speak to the turbulent end of the Colonial Machine itself.

Confronting Other Astronomies. Although the French in the Western Hemisphere took great interest in the botanical and medical knowledge of Native Americans, they paid little heed to native astronomy or cartography. In his *Relation de la Guadeloupe* (1652), for example, Father Mathias Dupuis set the tone when he wrote: "They have some rough sense of the stars, but the fables they mix with the truth remove all certitude from them."[140] Still, astronomers must have been surprised that natives (slaves?) in Saint Domingue observed the comet of 1742 a month before any European observer.[141]

The situation in the East was altogether different. There, the French encountered high civilizations fully the equivalent of their own, with rich technical traditions in astronomy and cartography, and they quickly recognized the value of this indigenous knowledge and the necessity of mastering it for their own purposes.

The records of the Observatoire contain a host of manuscripts related to Indian and Chinese astronomy, geography, and chronology.[142] The Jesuit Éti-

138. See previous notes and Chapuis, pp. 327-88; Richard (1984, 1986); Taillemite (2002), pp. 170-71.

139. On Beautemps-Beaupré and his role in modern geography, see the fully detailed Chapuis.

140. "Ils ont quelque grossiere connoissance des Astres: mais les fables qu'ils meslent avec la verité en ostent toute la certitude," Dupuis (1652), p. 197. Father Le Breton circa 1700 had the same idea of aboriginal knowledge of mathematics, writing "Les Caraïbes ignorent totalement le bel art du calcul," in Lapierre, p. 99.

141. "Relations et Observations de la comète de 1742, tirée de différentes gazettes," OdP, Ms. A 4 6, dossier 4, pièce B.

142. Cf, "Astronomie orientale, chinoise et indienne," OdP, Ms. A 1, 12; "Mémoires pour servir à l'histoire de l'Astronomie et de la Chronologie chinoises," OdP, Ms. B 1, 11-13.

enne Souciet (1671-1744) compiled an important, but controversial, collection of "mathematical, astronomical, geographical, chronological and physical observations taken from early Chinese books or newly made in India and China by the Fathers of the Company of Jesus," which appeared in three volumes between 1729 and 1732. The second and third volumes of this work in particular contained a history and treatise of Chinese astronomy by Father Gaubil.[143] Not surprisingly, J.-N. Delisle played a pivotal role in what amounted to a massive effort to absorb Eastern astronomy and chronology. From Cœurdoux in India, for example, he solicited "ancient as well as modern" cartographical and geographical information.[144] And, as early as 1722 to a correspondent in Amsterdam he wrote about Chinese cometary records that: "We need first of all to ascertain the true dates and circumstances surrounding these observations in the selfsame books of the Chinese, and in a word to study their astronomy in order to extend and make better use of what they make of astronomy."[145] In 1734 Delisle wrote revealingly to the Jesuit Kögler in Beijing:

> But I would wish not only to have modern maps drawn or corrected by your hand, but if it were possible also to have copies of earlier maps made by the Chinese themselves in different eras and especially this old map that I learned you have in your possession which was made a long time before the arrival of the Jesuits in China. Father Gaubil informs us that the Chinese for a long time have known the declinations of the stars fairly well and that they have made catalogues wherein they have recorded changes that have arisen over the course of time. One might perhaps extract as much utility from these records as one can from the former positions of the fixed stars observed by Timocharis and Hipparchus and reported by Ptolemy....I hope that this consideration will lead you, my Reverend Fathers, to contribute to enlarging this library by a collection as complete as possible of antique as well as new Chinese books dealing with astronomy.[146]

In China, the Jesuit Gaubil was instrumental in transmitting Chinese materials back to France, reporting to Delisle in 1732 that "in recent years I have sent to Messrs. Cassini and Maraldi in France long memoirs on the astronomy of the ancient Chinese, on astronomical observations, and on the calculations

143. See Souciet (1729-1732); Duteil, pp. 283-302.

144. "tant anciens que modernes qui ne seroient pas venus a ma connoissance...," Delisle to Cœurdoux, "A Paris le 13 Octobre 1752...," OdP, Ms. 1029, B 1-6 (XI, # 252).

145. ...il faut auparavant s'assurer des veritables dates & circonstances de ces observ. dans les livres même des Chinois, & en un mot étudier leur astronomie; pour les pouvoir entendre, & faire un meilleur usage de ce qu'ils rapportent d'Astronomie," Delisle letter to Nicolas Struyk in Amsterdam, dated "De Paris le 20 Avril 1722," OdP, Ms. 1029, B 1-2 (II, #36).

of Chinese astronomers. I attached long dissertations on what one can reasonably take as fixed and certain regarding Chinese chronology."[147] And Gaubil continued to send materials on ancient Chinese astronomy to de Mairan and the Academy of Sciences at least through the late 1750s.[148] Apparently, the Chinese maintained an official bureau of maps, or at least Gaubil so inferred in writing to Delisle in 1758.[149] To which Delisle responded: "You gave me real pleasure in sending me what you were able to obtain from the mandarin in charge of maps regarding the map by the Jesuits Spinhu and Rocha of the areas to the west and north of Hanoi."[150]

In India, Father Cœurdoux performed the same function as Gaubil in China. He himself reported regarding Indian geography:

> I have slowly but surely gathered up memoirs on the geography of these districts either by means of missionaries or by that of Indians. As the latter are accustomed to speak colloquially using "north," "east," etc., they have a facility to know the location of different places that the people of Europe do not have. I have already amassed a considerable quantity of these memoirs, but once I have collected a few more that I am hoping for, I will try to unite them into a single map.[151]

146. "Mais je ne souhaiterois pas seulement avoir les cartes modernes dressées ou rectifiées par vos soins; mais, s'il étoit possible, aussi, des copies des cartes antérieure & faites par les Chinois mêmes, en differens tems, & sur tout de cette ancienne carte que j'ai appris que vous aviez entre les mains, qui a été faite longtems avant l'arrivée des Jesuites à la Chine. Comme le P. Gaubil a mandé que les Chinois avoient depuis longtems assez bien connu les declinaisons des ét. & qu'ils en avoient fait des catalogues où ils avoient marqué les changemens qui y étoient arrivés par la suite des tems, L'on en pourroit peut être tirer autant d'utilité que l'on peut faire des anciennes situations des ét. fixes observées par Tymochyaris & Hipparque, & rapportées par Ptolemée.... j'espere qu'en cette consideration vous voudrez bien Mes R.R.P.P contribuer à l'augmentation de cette biblioteque [sic] [à Paris] par un recueil aussi complet qu'il sera possible des livres Chinois tant anciens que nouveaux qui traitent d'Astronomie," Delisle letter "au R. P. Kögler le Dec 1734 avec une addition du 15 Janv. 1735,"AN-MAR 2JJ 63.

147. "Ces années Passées j'envoyai En France à Mrs. Cassini et Maraldi de longs Ecrits sur L'astronomie des Anciens Chinois, sur les observations astronomiques et sur les calculs des astronomes Chinois. Je joignis de longues dissertations sur Ce qu'on peut raisonablement fonder de fixe et de certain en fait de chronologie chinoise," Gaubil to Delisle, dated "A Pekin Ce 15 May 1732,"AN-MAR 2JJ 62.

148. See Gaubil letter to Mairan, dated "de Pekin le 20 9bre. 1754," AN-MAR 2JJ 68.

149. "...C'est par un mandarin du tribunal des Cartes que j'ai Eu les points de la Carte faitte par les RP Spina et Rocha de quelques lieux à ouest et nord de la ville de *hanoi*," Gaubil letter to Delisle, dated "a Peking le 20 9bre 1758," AN-MAR 2JJ 66.

150. "Vous m'avez fait bien de plaisir de menvoier ce que vous avez pu obtenir d'un mandarin des cartes sur les points de la Carte faite par les R.P. Spinhu et Rocha de quelques lieux à ouest et nor de la ville de Hanoî," Delisle to Gaubil, dated "a Paris le 30 decembre 1759," AN-MAR 2JJ 66.

151. "Je ramasse peu a peu des memoires sur la geographie de ces cantons tant par le moyen des Missionnaires que par celuy des Indiens. comme ils sont accoutumez a parler dans l'usage de la vie par Nord, par est &c. cela leur donne une facilité pour scavoir la situation des differens lieux, que le peuple na pas en Europe. j'en ay deja ramassé une quantité considerable que je tacherai de reunir en une seule carte, quand j'en auray recueilli encore quelques uns que j'espere," Cœurdoux to Delisle, dated "a Pondicheri ce 14 Octobre 1755," AN-MAR 2JJ 66.

Later Cœurdoux wrote that he had "started something on chain measures of India, on an old Indian method of calculating the months, and on the geography of certain districts of India."[152] To which Delisle replied: "You will judge, Sir, how much these treatises will give me pleasure and will be useful in serving as the basis for the geography of India and for Indian astronomy, of which I have already put together a fair bit of information. It remains only to compare this with Oriental and European astronomy."[153]

The actors involved seemed to recognize that knowledge in the East was equal or sometimes superior to what Europeans knew. The Jesuit Amiot, for example, wrote from Beijing in 1756 about a mapping expedition sent out by the emperor with two Jesuits and "two monks who are pretty good mathematicians for this country."[154] Writing from Canton in 1776, the botanist and Academy correspondent Pierre Sonnerat insisted that the prevalent view of the inferiority of the Chinese was wrong; having procured a Chinese encyclopedia, he said it showed their high level of sophistication, and "based on this work one can fully grasp the genius and character of the Chinese."[155] Cœurdoux earlier wrote from India regarding an eclipse of 1763: "Basing our belief on Lacaille's ephemerides, we had established that the eclipse would not be visible in the Indies. An Indian calculator upheld the contrary opinion, and the calculation being made, we found to our shame that the Indian was right."[156]

This interest in the astronomy, cartography, and chronology of China and India was not limited to the Jesuits and the Delisle circle, and such interest did not diminish as the eighteenth century progressed. French scientists who traveled to India and China in the second half of the century were likewise impressed and displayed serious interest in indigenous astronomical and geographical traditions. While in Asia for the 1761 Venus transit, for example, Pingré took daily lessons from an Indian calculator on Indian methods of pre-

152. "...j'ay quelque chose de commencé sur les mesures chinoises, de L'Inde sur une ancienne methode Indienne pour calculer le Commencement [?] des mois, &c. sur la Geographie de certains cantons de l'Inde...," Cœurdoux to Delisle, dated "a Pondicheri ce 27 septembre 1757," AN-MAR 2JJ 66.

153. "vous juger Mr. combien ces traitez me pourroient faire plaisir et pourroient estre utile pour servire de fondement à la Geographie de l'Inde et à l'Astronomie Indienne sur la quelle jay déja rassemblé bien des connoissances quil ne reste que de comparer avec l'Astronomie Orientale et Europeene," Delisle to Cœurdoux, dated "à Paris le 16 Fevrier 1759," AN-MAR 2JJ 66.

154. "...avec les deux Jesuites il y a deux Lamas assez bons mathematiciens pour Ce pais," extract of "une lettre du R. P. Amiot au P. de la Tour [dated] de Pekin le 24 nov. 1756," AN-MAR 2JJ 69.

155. ...d'apres cet ouvrage on peut réelement connoitre le genie et le caractere du chinois, comme d'apres le Notre il est aisé de juger de la nation francoise et à quel point elle a porté les sciences et les arts...," ALS Sonnerat to Adanson, dated "de Canton le 29 decembre 1776," HI, AD 237.

156. "Nous avions assuré, sur la foy des Ephémerides de M. de la Caille, qu'elle ne seroit pas visible aux Indes. Un calculateur Indien a soutenu le contraire, et le calcul fait, on a trouvé à notre honte que l'Indien avoit raison," Cœurdoux to Delisle, "De Trinquebar, 28 Août 1762- 27 Sept 1764," OdP, Ms. 1029, B 1-7 (XV, #119).

dicting eclipses.[157] Sonnerat traveled extensively in South Asia in the 1770s as an agent of the Colonial Machine; his 1782 work, *Voyage aux Indes orientales et à la Chine, fait par ordre du Roi, depuis 1774 jusqu'en 1781*, includes chapters on Indian arts and crafts, Indian medicine, and Indian astronomy.[158] The astronomer and longstanding academician in Paris, Jean-Sylvain Bailly (1736-1793), took a deep interest in the subject of non-Western astronomy, publishing his *Traité de l'astronomie indienne et orientale* in 1787. The internal report on Bailly's book in the Academy of Sciences, signed by Le Gentil and Laplace, noted: "The object of this work is to collect everything that remains of the ancient astronomy of the Orient, to explain the astronomical methods of the Indians, and to compare the different tables that these astronomers use.... We think this work should interest philosophers and astronomers, and in every way it merits the approbation of the Academy and publication under its privilege."[159] And so it was.

The initiatives undertaken through the Colonial Machine regarding cartography and navigation were manifold, successful, and ongoing throughout the later seventeenth and eighteenth centuries. This mechanism that organized and captured networks of informants, from missionaries to appointed explorers, was indeed an efficient tool for French overseas expansion.

Local Actors

On the ground in the colonies a host of engineers and surveyors labored to master and rationalize local spaces.[160] They planned new towns. They organized munition stores and erected local defenses. They saw to fountains and water supply. They built irrigation systems. They laid out roads and oversaw their construction. They spanned rivers and streams. They drew maps. They surveyed property and maintained cadastral records.

Government Engineers. The work of surveying and mapping local space was essential for colonial development, and cadres of royal engineers and surveyors appeared in the colonies. Even if they did not require the highest scientific qualifications, government engineers in the colonies bolstered the critical mass of the Colonial Machine.

157. See Ms. "L.S. Astronomie indienne" in PA-DB (Pingré) and original in the Bibliothèque Sainte-Geneviève, Paris.

158. Sonnerat (1782); Ly-Tio-Fane (1975, 1976).

159. "l'objet de cet ouvrage est de recueillir tout ce qui nous reste de l'ancienne astronomie orientale, d'expliquer les methodes astronomiques des indiens, & de comparer entre elles les differentes tables dont ces astronomes font usage....nous pensons qu'il doit interesser les philosophes & les astronomes, & qu'il merite a tous egards l'approbation de l'academie, & d'estre imprimer sous son privilege," report PA-PS (23 December 1785).

160. Pedley (2005), pp. 19-22, 199 surveys surveyors and surveying; Langlois (2003), p. 223, notes the limited literature on engineers in the colonies.

The story of colonial engineers and surveyors is tied to that of French military engineers in France in the period. Royal engineers were detached to the colonies and depended on the Secretary of State for the Navy until 1690, then from 1691 on the Department of Fortifications for land and coastal places, supervised by its director, Michel Le Pelletier de Souzy, and by the Commissioner of Fortifications, Sébastien Le Pestre de Vauban.[161] The training and intellectual backgrounds of engineers varied widely. Their scientific culture was based on drawing and math instruction provided in colleges or military academies, or by independent masters.[162] Their training culminated with the study – sometimes with the help of a master, sometimes not – of manuals dealing with architecture and fortifications published from the early seventeenth century onward. Quite frequently their fathers were architects or master masons, and they benefited from experience gained from working with them in the field.[163] Their qualification as royal engineers required a license that was not granted to them without letters of recommendation and until their aptitude had been duly examined, a process favoring experience and competence gained in the field.[164]

By the eighteenth century in France the army had its own, well-developed engineering corps.[165] Military engineering and expertise in fortifications and artillery were integrated into the French Army by the sixteenth century; and in 1696 Vauban created a separate corps of military geographers-surveyors or engineer-geographers (*ingénieurs géographes* in French) that ultimately became integrated into the ranks of the officer corps of the army.[166] The creation in 1748 of the royal engineering school at Mézières institutionalized the training of ingénieurs-géographes and military engineers of all sorts.[167] The army's presence in the colonies was weaker than in France, but especially in times of war colonial fortifications were a matter of high importance, and the army organized a special council to oversee them.[168] Especially, too, in the great colonial wars of the period – the Seven Years War and the American War – the French Army and the French Navy melded into a single fighting force.[169] That said, the royal engineers who came to staff positions in the colonies by and large seem to have come from the world of the army and not the navy. The Marine Royale possessed its own engineering cadres that supervised

161. Blanchard, pp. 60-80.

162. Hahn (1964a), pp. 513-16; Hahn, (1964c).

163. Blanchard, pp. 109-111.

164. Vérin (1993), pp. 181-95 and pp. 243-333.

165. For background, see Vérin (1993); Picon.

166. On these points see Pedley (2005), p. 23; Berthaut; Dainville, p. ix.

167. Taton (1964), pp. 559-615.

168. See Archives de l'Armée de Terre, Vincennes, Série A, passim; e.g., "procès-verbaux des séances du Conseil des fortifications des colonies."

169. Lesueur makes this point.

port operations and ship construction, and rivalries seem to have broken out between the sets of military and naval engineers.[170] But on the ground in the colonies, engineering staffs formed part of local government and reported to local intendants and governors-general and not to the navy. The circumstances regarding local engineering cadres and the Navy were similar to what we will see between local medical communities and medical personnel of the Royal Navy.

The history of engineers in the French West Indies began in the mid-1660s at the moment when Colbert took direct control of those colonies. Indeed, in 1666, Nicolas-François Blondel (1618-1686), an engineer and teacher of mathematics at the Collège Royal, author of several papers on fortifications in Provence and along the Atlantic coast (Rochefort in particular), was sent to the West Indies on a mission to inspect fortifications and to assess the defensive power of the colonies recently acquired by the king. Blondel was answerable directly to Colbert, who was then in charge of the corps of engineers.[171]

Feeling strongly supported and aware of the importance of his mission, Blondel visited the islands as a proud commissioner in the king's service. He made inspections, wrote reports, drew plans, and launched some works while bringing others to a halt. On Tortue, he launched La Tour, a fortification requested by Governor d'Ogeron in April, 1667. On Guadeloupe and Martinique, he drew fortress plans, initiated projects, and toured existing fortifications. He stood as the champion of a new generation of fortification engineers and architects of the type Colbert wanted to promote. Blondel was well trained in mathematics. He knew about ballistics, geography, cartography, and drawing. He exemplified the implantation on the colonial scene of a new generation of competent specialists. These engineering experts played a decisive part in the creation of colonial territories by the monarchy, thanks to their fieldwork but also thanks to the instruments of evaluation they provided for the ministry. Blondel's cartography of the main islands, Guadeloupe, Martinique and Grenada, as well as minor ones, remained a point of reference at least until the beginning of the eighteenth century and served for the rest of the navy and colonial officialdom.[172]

Engineers were posted to several colonial strongholds in order to improve them and to set up new coastal batteries. One Gemosat was appointed to the West Indies islands as a whole in 1672.[173] In the 1670s and 1680s, at governor

170. Contamine, pp. 471, 474.

171. Regourd (2000) pp. 300-301 and Le Blanc.

172. DPF, p. 695, and Pérotin-Dumon, pp. 118-119. AN-COL E 35; AN C^{8B} 1, fol. 22: "Mémoire sur les fortifications de la Martinique, de la Guadeloupe et de Saint-Christophe, envoyé par M. Baas, gouverneur général, à la suite de la mission accomplie par l'ingénieur Blondel." See also CAOM DFC VI Guadeloupe 1, 3 and 5 ; and BnF Cartes et Plans SH Pf. 155 #3, Pf. 156 #2, and Pf. 157bis #21.

173. AN-COL B 4, fol. 12 (25 January 1672), "pouvoir d'ingénieur pour Gemosat, envoyé aux Antilles pour y travailler aux fortifications."

Blénac's request, Descombes (also spelled Decombes) created the first city plan of the new capital (Fort Royal) on Martinique, and he conducted various improvements on the fortification of Tobago and Saint Christopher.[174] Payen, who had been at work on Martinique as early as 1680, sent memoirs from there about fortifications, and he was succeeded in 1692 by a certain Caylus (or Cailus), who had seemingly worked on the famous Canal du Midi in 1691 but did not give full satisfaction on his colonial labors.[175]

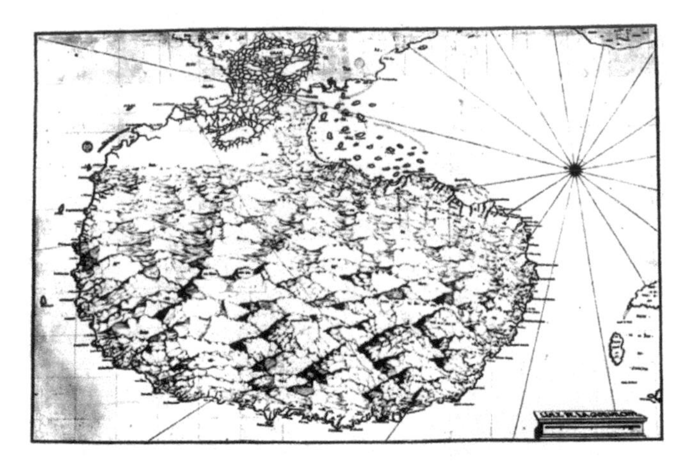

[Ms. Map of Guadeloupe by Blondel (1667)]

The number of engineers in the French West Indies seems to have been kept to a minimum under Louis XIV. Apparently there never was but one engineer per island and repeated demands by the administrators in this regard testify to the scarcity of staff; the governor-general and the intendant requested the appointment of a second chief engineer for Guadeloupe in 1697; and again in 1715 the engineer for Saint Domingue himself asked for help, claiming that two engineers were indispensable for these strategic outposts. At first, colonial engineers had indeed to hop from island to island, like Binois (or Binoist), who worked on fortifications in Saint Christopher in 1700, then on Guadeloupe around 1703, and finally in Grenada where he filled several posts in different

174. AN-COL B 6, fol.35 (1681), "Projet de la ville à construire au cul de sac de la Martinique avec la contrescarpe du Fort-Royal, dressé par le sieur Decombe, ingénieur du Roy." BnF Cartes et Plans SH Pf. 154 #21 piece 3D: "Plan du fort de la Basse terre, 1678"; also Blanchard (1981), p. 177.

175. AN-COL Colonies C^{8A} 1, fols. 396-417 (October and November 1680). On Caylus: AN-COL Colonies B 14, fol. 249 (22 June 1691) and fol. 264 (12 September 1691). Also AN-COL C^{8A} 12, fol. 80 (8 January 1692) and AN-COL C^{8A} 15, fols. 425-42 (1705). Caylus was back in France in 1722: AN-COL C^{8A} 30, fols. 358-59.

grades from 1706 until his death in 1735. In 1738 there were still but three engineers in these islands: one chief engineer and two seconds to assist him.[176]

Payen's brother, Marc Payen, present in Saint Domingue from 1687, also worked as a government engineer in charge of fortifications. He was licensed in 1694 (apparently the first licensed engineer in Saint Domingue) after having produced several designs for forts and installations.[177] He was replaced by one Grivelé in 1698, who was himself replaced after his death in 1699 by Fleury. Alas, Fleury died as soon as he took his post.[178] Then La Broue and a "second chief engineer," Henault, were appointed to Saint Domingue in 1700 in order to fortify the southern region of the island.[179]

A few years later Frézier, a talented engineer who had at that time already published an account of a journey to South America, operated in Saint Domingue as director of fortifications. He drew manuscript maps (1720-21) and later (1724) printed maps addressed to the navy minister and the marquis d'Asfeld, director-general of fortifications. Those maps were the most complete and precise ever drawn at that date, as Guillaume Delisle, first royal geographer, himself testified.[180] Louis Floixel Cantel, the chevalier d'Ancteville (1738-1785), started his career as a military engineer in the royal artillery corps and then moved to Saint Domingue as a royal engineer in 1762, ultimately becoming the chief engineer at Cap François and the Môle Saint Nicolas until his death in 1785.[181] By 1775, royal engineers were in place in each of the three administrative divisions of Saint Domingue.[182]

Requests for engineering staff were regularly issued for Guiana, where there was considerable work to do.[183] The engineer Paquine was at work there around 1690 and was seemingly the first to be appointed to the post.[184] The

176. AN-COL C^{8A} 10, fol. 20-39 (May 1697); C^{8A} 12, fols. 208-17 (30 June 1700); C^{8A} 15, fols. 43-46 (21 September 1703); C^{8A} 16, fols. 15-17 (20 June 1706); C^{8A} 20, fols. 457-61 (July 1715), and C^{8A} 49, fols. 10-13 (February 1738).

177. DPF p. 590. See also AN-COL C^{8A} 4, fol. 330 (28 July 1687); AN-COL B 18, fol. 11, "Brevet d'ingénieur à St-Domingue, pour le Sr Payen" (1st January 1694): "Le Roy... voulant faire choix d'une personne expérimentée au fait du genie pour faire les fonctions d'ingénieur à Saint-Domingue et sachant que le Sr Payen a les qualitez nécessaires pour s'en bien acquitter et qu'il y a fait les fonctions d'ingénieur dans le service de terre pendant plusieurs années avec approbation, Sa Majesté l'a retenu et ordonné, retient et ordonne Ingénieur à Saint-Domingue... etc.". Payen died in 1697: AN-COL C^{8A} 10, fols. 2-29. See Also BnF Cartes et Plan, SH Pf. 149 #4, piece 4 D and SH Pf. 149 #5, piece 1.

178. AN-COL B 21, fol. 14 (15 January 1698, to Ducasse); B 21, fol. 351 (9 May 1699) and B 21, fol. 472 (6 January 1700).

179. AN-COL B 21, fol. 516 (13 January 1700); B 21, fol. 518 (20 January 1700), and B 21, fol. 530 (1st February 1700).

180. On Frézier, see Pinon; Regourd (2000), pp. 297 and 304.

181. DPF, p. 1444; AN-COL E 4 (Ancteville).

182. DPF, pp. 483, 590, 1167.

183. See correspondence, AN-COL C^{14} 10, fols. 7-15.

184. AN-COL C^{14} 2, fols. 75-80 (1st August 1690) and C^{14} 2, fols. 111-52.

minister sent François Fresneau (1701-1770) to Guiana in 1732 as chief royal engineer in Cayenne. Fresneau served there for over two decades making local maps, but was also engaged in many other things. In 1734, for example, he submitted to the administrators an ambitious project for opening a road through the forest in order to reach the cocoa trees that had been discovered in Haut-Camopi in 1727. He also performed botanical research and chemical experiments, notably regarding the rubber plant, following up on La Condamine's passage through Cayenne.[185]

On the Mascarene Islands in the Indian Ocean, the Compagnie des Indes maintained engineering cadres, sometimes bringing personnel over from India to staff positions.[186] Early on, the Compagnie pressed hard for cadastral maps, and it prepared model cadastral maps for use in the Indian Ocean colonies.[187] In 1743 the engineer Belval received 2000 livres and sat on the Conseil Supérieur on Île de France.[188] The French government naturally maintained royal engineering cadres once it took over from the Compagnie des Indes.[189]

In Canada, side by side with the series of royal hydrographers, the crown appointed a parallel staff of royal engineers, surveyors, and related technical personnel.[190] The first *ingénieur du roi* for Canada was named in 1678, with powers comparable to those of naval engineers working in ports in the kingdom.[191] In 1703 officials spelled out the duties of one Lhermite as military engineer in Canada to include fortifications, munitions supply, water supply, and mapmaking.[192] For a certain Dubois Berthelot de Beaucours, in 1712, the post of engineer in Canada came with the military Cross of Saint-Louis, indicative of the importance of the position and its holder.[193] At one point a certain Gédéon de Catalogne served as the second engineer-in-residence in Mon-

185. AN-COL C^{14} 16, fols. 26-30 (9 September 1734, Cayenne). La Condamine (1751), p. 323; HARS (1751), pp. 17-22; Chaïa (1977), p. 25. Safier (2008b) p.251.

186. Ms. Code de l'Ile de France, vol. 1, in AN-MAR G 237; see also letter "A M.rs les directeurs généraux de la Compagnie des Indes," dated "20e Xbre 1731," in Lougnon (1934), p. 167; and "Extrait du registre général des Déliberations de la Comp.ie des Indes. Du 27 juin 1741 [and] Du 26 juin 1742" in Lougnon (1940), pp. 32, 48.

187. Lougnon (1933), p. xxxv; see also letter from the Compagnie des Indes to "M.rs du Conseil Supérieur de l'Isle de Bourbon," dated "A Paris ce 23 Xbre 1730," In Lougnon (1934), p. 105; and "A Mrs. du Conseil Supérieur de l'Isle de Bourbon. A Paris le 17 9bre 1732," in Lougnon (1933), pp. 96-99.

188. "Extrait du Registre général des déliberations de la Compagnie des Indes. Du 19e juin 1743...Ordres au Conseil de l'Isle de France," in Lougnon (1940), p. 103.

189. "Copie de la lettre de M. le duc de Praslin, Ecritte à la Compagnie, En datte de Versailles le 2 Aoust 1766," SHML, IP 284, Liasse 107, #17.

190. On royal hydrographers, see above Part I. On engineers in Canada, Chartrand (2005, and 2008); Palomino; Lafrance with Charbonneau.

191. Chartrand et al., p. 22; see also Banks, pp. 90-91.

192. AN-COL B^7 (1703), fol. 293 (17 March 1703).

193. AN-COL B^{34} (1712), fol. 333 (21 June 1712).

treal.[194] Succeeding Robert de Villeneuve (1685-1693), Jacques Levasseur de Néré (1694-1712), and Josué Dubois Berthelot de Beaucours (1712-1715), Gaspard Chaussegros de Léry became chief engineer of Canada from 1716 to 1756. At the strategic place of Louisbourg, succeeding Jacques de Lhermite in 1715, Josué Dubois Berthelot de Beaucours became chief engineer, and Étienne Verrier had the post from 1725 to 1745. The Ministry appointed Louis-Joseph Franquet as chief civil engineer (*ingénieur en chef de terre*) at Louisbourg, and as director of fortifications for Canada (1750-1757).[195] A certain Sarrebource de Pontleroy then occupied the position of chief engineer and was instructed at Louisbourg in his duties by Franquet.[196]

The need for engineers and surveyors seemed especially pressing in Louisiana, where an official engineering corps of the Compagnie des Indes likewise existed from 1720, when four engineers sent by the king arrived to develop the colony: Le Blond de la Tour, Pauger, Franquet de Chaville (Louis-Joseph Franquet's brother), and Boispinel.[197] A chief inspector and surveyor supervised concessions of land to colonists in Louisiana.[198] In 1724 the chief engineer received the seemingly high salary of 3600 livres a year.[199] Accredited by the Paris Academy of Sciences, Pierre Baron worked as an engineer in Louisiana from 1728 to 1730, as well as making astronomical observations and working on economic botany.[200] In 1731 Broutin became chief engineer in Louisiana, and designed the first fortifications for New Orleans.[201] Still, the colony apparently lacked competent surveyors and engineers, and local administrators regularly begged authorities in France for the same.[202]

As we see, a hierarchical organization of royal engineers arose in the colonies. Nominally, but always with staffing gaps, that hierarchy included a director of fortifications, a chief engineer, a second chief engineer, ordinary engineers, and inspectors. Salaries were not very high but still allowed government engineers to cut a good figure in the colonial society of the time. In 1742

194. AN-COL B^{34} (1712), fol. 344 (24 June 1712).

195. AN-COL B^{91} (1750), fol. 60 (14 June 1750); B^{99} fol. 14 (1 May 1754); Chartrand (2005), p. 9; Charbonneau with Lafrance; Lafrance with Charbonneau.

196. AN-COL B ^{105} (1757), fol. 12 (9 April 1757).

197. See Langlois (1999a and 1999b); also AN-COL B^{43} (1722), fol. 274 (19 December 1722), listing the staff and their emoluments.

198. "grand-voyer et arpenteur" AN-COL C^{13A} 32 fol. 12 (20 juin 1748).

199. AN-COL B^{47} (1724), fol. 1239 (9 May 1724).

200. On Baron, see Langlois, pp. 152-57; AN-COL C^{13A} 12 (1729-1730), fols. 191, 412, 419; C^{13A} 13 (1731), fol. 3 ("faisant fonction d'ingénieur") and fol. 57; see also SHMV, Ms. 0073, XVIII, vol. 26 ("Recueil d'Ordonnances... 1707-1779").

201. Chartrand (2005), p. 9.

202. See, for example, AN-COL C^{13A} 1 [Louisiana] (1678-1706) fol. 47; C^{13A} 9 [Louisiana] (1725-1726) fol. 221 (1st September); C^{13A} 36 (1752) fol. 252 (12 September 1752). For more on engineers and surveyors in Louisiana, see C^{13A} 10 [Louisiana] (1726-1727) fols. 276, 284, 287; C^{13A} 25 (1740), fols. 9, 102.

the salary of 800 livres for the royal engineer in Martinique was judged insufficient, and by 1763 it had risen to 1500 livres.[203] One, Amelot, received 3000 livres as chief engineer in 1761.[204]

It is clear that those engineers who possessed both theoretical and practical knowledge in science and technology were experts valued by the administration as the competent, devoted, and subservient auxiliaries of colonial development. More often than not, engineers found themselves by necessity compelled to assume, mostly singlehandedly, the functions of cartographers, explorers, architects, project managers, and supervisors of the ongoing works. At the same time they functioned as experts in fortifications and hydraulic works, having building plans to draw up, coasts and rivers to sound, and eventually mines to explore.

Auxiliary Forces: Surveyors, Architects, and Officers. Other technicians helped engineers locally and also formed part of the Colonial Machine. For building and construction, the services of some talented master mason or some more or less qualified private architect stranded in the tropics could be tapped by the administration. Thus in Martinique in 1719, one Ju, a self-appointed architect, was requesting a regular salary from the administrators in compensation for his services.[205] The odd specialist could occasionally be swept up in the Colonial Machine, as was the case of a Portuguese Franciscan set ashore by pirates at Cayenne, who put to use his much-needed knowledge in mineralogy; in 1720, the governor of French Guiana enticed him to stay on – for money – to conduct an exploration for mineral resources in the lands situated along the Approuague River.[206]

But engineers did receive auxiliary help essentially in the field of cartography.[207] Some surveyors – following the example of Thimotée Petit, the surveyor general for Martinique appointed in 1690 (and who died in 1723) – not only surveyed and kept records of plantations and their boundaries, but sometimes made more general plans and maps.[208] With a smattering of math and geometry necessary to make exact surveys and basic angular measurements, surveyors mostly relied on their competence in drawing. Thus, it is no surprise that, as young Joseph-Nicolas Delisle contemplated applying for a post as royal surveyor on Martinique in 1707, he first turned to a drawing master to progress further.[209] The basic theoretical knowledge necessary for surveyors

203. AN-COL C^{8A} 54, fol. 52: Champigny and Lacroix (24 April 1742).

204. AN-COL E 4 (Amelot).

205. AN-COL C^{8A} 26, fols. 296-300 (2 August 1719).

206. AN-COL C^{14} 12, fols. 32-34 (12 February 1720).

207. See, for example, the royal commission for le S. Chasservent as "Arpenteur du Roi pour l'île de la Martinique," AD/Martinique, Série B, Registre 9, fol. 230.

208. AN-COL C^{8A} 19, fols. 87-134 (May 1713); C^{8A} 21, fols. 300-305 (May 1716) and C^{8A} 32, fols. 30-31 (August 1723).

209. "Éloge de M Joseph-Nicolas De l'Isle," *HARS*, 1770 (1768), p. 169.

was found in didactical treatises such as *La praticque et usage d'arpentage* by Claude Flamand (Montbéliard, 1661), *L'école des arpenteurs* by Philippe de la Hire (Paris, 1689), or *L'art de dessiner proprement les plans, profils, etc.*, by Henri Gautier (published in Paris in 1697). Texts like these flourished in the eighteenth century, and *L'art de lever les plans* by Dupain de Montesson, first published in 1764 and regularly reprinted into the 1820s, was the most complete. One learnt the rest through practice in the field.[210]

Royal surveyors were charged with maintaining a general cadastral register of all surveying work.[211] Just as royal doctors and surgeons in the colonies licensed private practitioners, local administrators, royal engineers, and royal surveyors closely monitored and credentialed individuals who worked as private surveyors.[212] Newcomers were examined in geometry, calculating techniques, and drawing to insure a minimal competence.[213] Local surveyors plotted property lines, and local ordinances regulated their qualifications and pay. The chief surveyor on Martinique, for example, received 42 livres a day for his work.[214] On Île de France, the royal surveyor received 36 livres a day for his work and 20 sols per page for his reports; private surveyors got 30 livres a day, and chainmen (*portes-chaines*) 7 livres and 10 sols.[215] From Quebec to the Mascarenes via the West Indies and Guiana, these forgotten actors in colonial conquest were in many respects the first practitioners of colonial cartography, even if their technical abilities were often called into question.[216]

Another set of specialists called on to assist government civil engineers in their work was made up of artillery or navy officers. These experts could draw maps and coastal profiles, carry out exploration missions, sound rivers, and build fortification works. Before there were any civil engineers appointed – that is, until the end of the seventeenth century – military officers had perforce to carry out themselves the construction of essential fortifications and the first cartographical surveys needed by squadrons for mooring.

Indeed, some military officers were well prepared for scientific work in the colonies by their training in France. Such was dispensed in the many private schools that opened in France in the seventeenth century (in Paris, Caen, Rouen,

210. Regourd (2000), pp. 405-406.

211. "Ordonnance de MM les Général et Intendant, sur le dépôt des Plans et Procès-Verbaux d'Arpentage, du 1er janvier 1772," Durand-Molard, vol. 3, pp. 98-100.

212. Durand-Molard, vol. 2, pp. 257-58.

213. "Ordonnance du Roi, portant Réglement pour la reception des Arpenteurs aux îles du Vent de l'Amérique," Durand-Molard, vol. 3, pp. 162-64.

214. "Ordonnance de MM les Général et Intendant portant Règlement et Tarif général de tous les Émoluments attribués à divers fonctionnaires publics. Du 30 avril 1771," Durand-Molard, vol. 3, pp. 38-67ff, here p. 65.

215. "Ordonnance...concernant les Arpenteurs... Du vingt-un Novembre 1774," Delaleu, vol. 1, pp. 372-73; "Ordonnance...qui fixe les droits & honoraires des Arpenteurs de l'Isle de France. Du 17 Juillet 1777," Delaleu, Supplément 1, pp. 239-41.

216. Boudreau, pp. 17-57.

Lille, Angers, etc.) or as part of the training of young noble officers (called *cadets*) under the authority of a mathematics master.[217] The results of that instruction are visible today in the numerous memoirs, maps, and coastal surveys stored in the archives of the navy.[218] A striking example is the case of one Gabaret de Lérondière (who signed himself La Rondière on occasion), an army captain in Cayenne early in the eighteenth century who served as an engineer without being licensed or properly paid for the job. He explored inland zones, sounded various rivers, drew several maps, and made contacts with Amerindian tribes, like the Arouas and the Palicours, from whom he endeavored to obtain information about certain spices and out-of-the-way mines. He sent numerous manuscript maps of his explorations to the navy administration back in France, and his work helped later in drawing improved maps of the region.[219]

The Special Set of Engineer-Geographers. We have to draw particular attention to the corps of the ingénieurs-géographes (engineer-geographers) who appeared in the colonies after the Seven Years War. The context harks back to the 1730s and the grand project to map the kingdom, undertaken on royal authority by the Cassini family, notably in the 1740s and 1750s by César François Cassini de Thury (1714-1784). This "geometrical" large-scale map of France was a formidable enterprise that rested on the techniques of triangulation. It required extensive and meticulous surveys and a perfect mastery of the rules of trigonometry. This huge work resulted in a network of 3,000 precisely determined geographical points in France, and the effort continued until the end of the Old Regime with the first of 182 sheets published in 1769.[220]

The initial successes of Cassini's map were widely known at Versailles and in military and geographical circles. The example of mapping the motherland induced Choiseul, the minister, to think about mapping the American colonies in a similar way in order to provide a reliable instrument to tacticians in charge of colonial defense after the disastrous Seven Years War. For the Colonial Machine, Choiseul co-opted these ingénieurs-géographes, most of them trained in the Cassini project and previously in military service in Europe, and assigned them to posts and work in the colonies.

The royal engineer-geographers, a kind of super-surveyors in royal service, thereafter wrote a distinctive chapter in the history of cartography in the colonies.[221] During the Seven Years War, the number of these engineer-geogra-

217. Hahn 1964c, p. 514.

218. AN-COL C^{8A} 13, fols. 305-308 (February 1701). Hundreds of accounts are in BnF Cartes et Plans, SH; CAOM DFC; AN-COL C^7, C^8, C^9, C^{14} and AN-MAR JJ various series.

219. AN-COL C^{14} 6, fols. 31-39 (February 1711); C^{14} 9, fols. 198-204 (September 1716); C^{14} 9, fols. 285ff. (December 1716) and C^{14} 10, fols. 33-37 (June 1717).

220. Pelletier (1990, 2002) and Konvitz. The last sheet of this gigantic map was published in 1817. See also De Reynal de Saint-Michel.

221. On the ingénieur-géographes du Roi, see Glénisson; also Regourd (2000), pp. 401-409; SHMV, Ms. 0141. XIX, #33; Berthaut, vol. 1, p. 31 and pp. 37-39; Rinchenbach, ed., pp. 7-8. According to Pedley (2005), pp. 206, 238, the salary for these individuals was 1800 livres.

phers rose to forty, and after the war the decision was made to deploy them in the colonies to draw up good local maps. They were sent to Saint Pierre and Miquelon, to Martinique, Guadeloupe, Saint Lucia, Saint Domingue, Guiana, and even to the Falklands, in connection with the ephemeral colony of Acadians founded by Bougainville we mentioned previously. In 1770s three royal engineer-geographers led by René Moreau du Temple produced a large and exquisite map of Martinique (2.7 by 4.85 meters) based on triangulation work there in 1764-1766.[222] Their colleagues on Guadeloupe created a similarly impressive map of that island.

But though the Colonial Machine provided a strong support, the story was less successful in other colonies. The sixteen ingénieur-géographes sent to Saint Domingue fared less well, with only four remaining in the colony after a short while; and nothing of substance resulted from their work, in part because the financial burden – projected at over 200,000 livres – fell on local government authorities.[223]

The engineer-geographers in charge of drawing the map of Saint Lucia and Guiana were not so lucky as their colleagues on Guadeloupe or Martinique. One of them, Morancy, fell ill soon after the operations started and his map was never completed. Finally, in the mid-1780s, Lefort de Latour, surveyor general for Saint Lucia, undertook to draw a large-scale map. But it was far less ambitious than the project of the engineer-geographers as it consisted in putting together the surveying plans available in the colony without any reference to original triangulation work. In Guiana, Joseph-Charles Dessingy, who had won fame for the cartography of the right bank of the Rhine and in Canada, arrived in 1762 and served there until 1786, drawing a number of topographical maps in Cayenne and elsewhere.[224] In July 1763, Simon Mentelle joined him; Buache and Lalande in Paris had been Mentelle's masters, and he had also been trained as part of the Cassini map project group. This very active engineer-geographer made numerous maps of the South American colony, first in connection with the Kourou colonization project, then as requested by administrators or exploration missions which he accompanied deep inland. Their colleague, François-Étienne Haumont, died in Cayenne as early as 1767, having mapped the right bank of the Kourou River. Other engineer-geographers came to Guiana one after the other without apparently being able to apply the exacting principles of the reformed cartography recommended by the Académie des Sciences and by the minister. Triangulation work on an essentially coastal territory almost entirely covered in thick forests proved hardly feasible. It was not until 1777 that Dessingy managed to produce a map based

222. This manuscript map is discussed and reproduced in Moreau du Temple (1998). See also Regourd (2000), p. 404.

223. Glénisson, vol. 1, pp. 146-47; Regourd (2000), p. 408.

224. Berthaut, p.31.

on the triangulation of the isle of Cayenne.[225] Finally, a royal order in 1784 suppressed the title of "Ingénieur-Géographe" for the colonies.[226]

Fashioning the Colonial Landscape and Mastering Water. If mapping was an essential element for the conquest of space orchestrated by the Colonial Machine, implanting and maintaining colonies was also a strategic challenge. The ability of specialists in the field to subjugate the environment to the needs of colonial prosperity in eighteenth-century colonies, impressed observers. Moreau de Saint-Méry, referring to an alluvial embankment that had led to regaining sixty-three acres of land from the sea for a coastal habitation on Saint Domingue, expressed his awe: "Now you can see a piece of land where small boats and launches used to sail eight or ten years ago yielding an abundant crop, man having become through his industry master of nature and a sort of creator."[227] The "industry" attached to colonists rested mainly on the shoulders of contractors and architects coming from a wide range of origins, and most of all on the engineers who gradually shaped colonial space from the end of seventeenth century.

In towns, engineers and their expert associates drew up plans, built ramparts and wharves, created shaded promenades, and installed public fountains.[228] Road surveyors had the responsibility of organizing the maintenance of roads and existing communication routes, which they did at the head of teams of slaves requisitioned from neighboring plantations. Engineers often intervened to decide on the laying out of pathways, roads, and highways, as did Fresneau in the forests of Guiana. Some individuals took it upon themselves to draw up their own projects before submitting them to the administrative authorities, who then consulted local engineers and also other elements of the Colonial Machine. Thus, for example, two drafts for building bridges in Saint Domingue are kept in the archives of the Académie des Sciences: one for a suspension bridge (submitted in 1789); the other for an iron bridge, submitted by the plantation-owner Barré de Saint-Venant the following year.[229] Some engineers from the *Corps des Mines*, owing to their specific competences, were even sent to Guiana in the late 1780s to investigate veins of iron and platinum ore – engineers like Chapel and Guilot-Duhamel *fils*, the latter being in the first class of students at the *École des Mines*, founded in 1783.[230]

Among the most spectacular projects in the colonies of eighteenth-century France were the irrigation systems installed in Saint Domingue, especially in

225. Chaïa (1979b).

226. Berthaut, p. 115.

227. DPF, pp. 196-197.

228. Regourd (2000) pp. 386-389. D'Orgeix and Vidal eds.

229. PA-PS 29 May 1789 and 10 February 1790.

230. AN-COL C^{14} 60 fol. 327 (1786); C^{14} 63 fol. 92 (1789); C^{14} 63 fols. 227-231 (1787) and C^{14} 91, fol. 1.

the Cul-de-Sac plain around Port-au-Prince and along the Artibonite Plain to the north. Most of these, be it noted, were undertaken by private plantation owners or private syndicates that had vested interests in installing hydraulic works to feed agricultural production.[231] But the Colonial Machine made itself felt here, too, notably from 1783 in the form of an official hydraulic-engineer (*ingénieur-hydraulicien*) appointed by the colonial administrators to plan and approve irrigation projects and to adjudicate disputes over water rights. One Jean-Joseph Verret, an engineer who had been in Saint Domingue since 1772, held this post from 1783 until the collapse of the colony; and he superintended the planning and execution of many irrigation works in the interim. Not unexpectedly, Verret and his rival hydraulic engineers in the colony, Jean Trembley and Jean-Joseph Bertrand Saint-Ouen, became associates of the Cercle des Philadelphes of Cap François.

The complicated story of irrigating the Artibonite Plain is most revealing.[232] Sometimes on private initiative, sometime on public, but always with the approval of the local intendant and governor-general – and with the ministry back in France being kept informed – projects to irrigate the Artibonite Plain were attempted in 1744, 1749, 1751, and 1753, but nothing resulted because of the strength and capriciousness of the Artibonite River. In 1755 a syndicate finally implanted an effective system, but it failed in 1761. Another attempt followed in 1766. In 1768 the colonial government rewarded a local planter, M. Bertrand *père*, for his exemplary efforts to irrigate his own plantation. In the early 1770s the local governor-general at the time, the count de Nolivos, and the ministry exchanged memoranda on the subject. In 1773 a new plan was launched, this time with the backing of the highest levels of government in France. The ministry accepted the proposal of the colonist and royal engineer, the chevalier Auguste-François-Gabriel de Courrejolles, and forwarded it to the Académie des Sciences which gave its stamp of approval. Jean-Rodolphe Perronet (1708-1794), director of the royal department of bridges and roads (the *Ponts et Chaussées*) and free associate of the Paris Academy, consulted on the plans; and Trudaine de Montigny (1733-1777), intendant in the Finance Ministry and Academy honorary, released two engineers, Varaigne and Dausse, and their 8000-livre salaries, from the Ponts et Chaussées to evaluate the proposal on site. Ultimately, these experts reported back that the original proposal by de Courrejolles was impractical and too expensive, and they supported an alternative by two local engineers at half the price, which still came to nearly 3.5 million livres.[233] Capping this significant planning effort, the king himself offered 1.2 million livres as a subvention for

231. On these points, see McClellan (1992), pp. 71-74; Regourd (2000), pp. 386-92; DPF, passim.

232. For these details, see DPF, pp. 819-43.

233. DPF, 827; Regourd (2000), p. 391; Bruneau-Latouche, p. 53; CAOM E 4 (1 pièce).

the enterprise. Funding for this project was thus three or four times greater than that spent on the La Pérouse expedition!

The Artibonite River flooded again in 1780 and wiped out what had been constructed to date. Three plans vied for approval to be tried again. Sojourning in France, Verret submitted his to the Académie des Sciences. A local plantation owner in Saint Domingue, Jean Trembley, pushed for funding for his plan from the colonial administrators. The third proposal – the successful one – came from Jean-Joseph Bertrand de Saint-Ouen, son of the planter rewarded for his example in 1768. In 1781 authorities appointed Bertrand de Saint-Ouen as royal commissioner in charge of irrigating the Artibonite plain. Bertrand de Saint-Ouen and the others all proposed using steam engines to pump water for irrigation; the government spent 72,000 livres for a Watt-style steam engine built in France by the Perrier brothers, and brought over an engineer from France to superintend the operation of the machine. On 11 November 1786, Bertrand de Saint-Ouen ceremonially inaugurated the pumping system in the presence of the count de La Luzerne, then the colony's governor. Unfortunately, unstable soil near the riverbank literally undermined the usefulness of this steam engine. Bertrand de Saint-Ouen died in 1787, and the latest project to irrigate the Artibonite died with him. The Colonial Machine itself did not last much longer.

Experimental Spaces

The French Navy faced huge technical challenges in the problem of survival for extensive periods at sea, and the navy and the Colonial Machine, in general, functioned as a giant research engine and laboratory to address these challenges.[234] The historian of the naval archives, Étienne Taillemite, writes of the "scientific role of the Navy as an incubator of inventions," and he emphasizes that these activities represent a new, peacetime function for the eighteenth-century French Navy that had not been studied to date.[235] Despite resistance, commitments to traditional ways of doing things, and an inbred suspicion of "machines," the navy indeed became a vibrant center for research and innova-

234. For the variety of inventions and projects entertained by the navy, in general see AN-MAR, Série G passim and partial inventory by Henrat, et al.; see also AN-MAR Série D (Matériel) and AN Série O[1] 1293. "Inventions: mémoires et correspondances." Llinares (2005) takes up some of this material.

235. "...ce rôle scientifique de la Marine laboratoire d'inventions." Taillemite (1991), p. 123. Allard (1970), too, p. 94, makes somewhat the same point, writing "...des liens très étroits unissaient l'Académie des Sciences et le Ministère de la Marine; l'une devenant le « laboratoire » de l'autre. Une intégration des ressources de la société française se réalisait." Llinares (2005) has recently highlighted this function of the French navy and provides new details.

tion.[236] Later in this Part II we will examine the problem of finding one's longitude at sea, a problem that preoccupied the Colonial Machine for decades, and sections in the next part take up medical research undertaken by the navy along with the rest of the Colonial Machine. Here, our focus is on a range of other, perhaps less dramatic but equally telling issues affecting long-distance sailing. Here, as elsewhere, these activities demonstrate the reality of the Colonial Machine in action.

Preserving Fresh Water at Sea. The lines from Samuel Taylor Coleridge's 1798 poem "The Rime of the Ancient Mariner" ("Water, water, every where,/ Nor any drop to drink") capture the issues, and the necessity of preserving drinking water at sea. The problem was that, stored in barrels, fresh water quickly became corrupted and putrefied. It grew dark and foul; it stank; larvae and insects roiled in it. Beer and wine could serve as partial substitutes, but ultimately ships crews needed fresh water to sail long distances. As a result, across the seventeenth and eighteenth centuries there were a host of proposals and trials to somehow develop a reliable means of preserving potable water at sea.[237] These proposals were always taken up at the highest levels of government, and the Colonial Machine became deeply involved in the matter. Colbert, for example, received so many proposals that he seemingly became fed up with them.[238] The Académie des Sciences was early and perpetually involved in this work.[239] The Société Royale de Médecine received more than one proposal, and the navy itself was besieged with others.[240]

The range of solutions offered testifies to the pressing nature of the problem and to the sustained attention it received. In 1737, for example, Cossigny *père*

236. On this resistance, see letter from André-François Boureau Deslandes, commissioner-general of the navy, to Réaumur, dated "Ce 28eme d'Avril [1724?]," "Nous ne sçaurriez croire, Monsieur, combien on est prévenu dans la Marine contre tout ce qui s'appelle Machines et Machinistes. On ne manque jamais de les faire échouer. Il est vrai aussi qu'à regarder les embarras de la navigation et le peu d'habilité des Officiers majors et mariniers, on ne peut avoir dans un vaisseau des choses trop simples et trops faciles. En général, tout ce qui s'y peut faire à mains d'homme, est préférable, coûtât-il plus de dépense et plus de tems....Depuis 1716 que je sers dans la Marine, j'ai remarqué que toutes les Machines qu'on y a voulu introduire, n'étoient que des imitations palliées de l'Hemisphere nautique, qu'on trouve dans l'Hidrographie du Pere Fournier," PA-DB (Deslandes).

237. See, for example, AN-MAR D³ 42 (Vivres), Section #1, "Moyens de purifier et de conserver l'eau douce," containing materials covering 1703-1799! Allain, pp. 28-30, points to this problem.

238. "Extrait de la Lettre de M. Colbert à M. de Pomponne," dated 4 Juillet 1670," AN-MAR D³ 40, fol. 3.

239. PA-PV-110, passim; PA-PV, passim, e.g., vol. 92 (1773), fols. 203-204 ("En l'Assemblée extraordinaire tenuë à Versailles le 12 Septembre pour la presentation du Volume de 1770"); PA-PS, passim, e.g., 17 November 1773, 23 April 1774. See also, AN-MAR D³ 40 fols. 9-23; D³ 42 fols. 68, 77, the latter ms. dated at top "24 Aoust 1773," "Machine hydraulique propre à préserver les gens de mer des maladies occasionnées par la corruption de l'eau."

240. See previous notes. Regarding the Société Royale de Médecine, see immediately below and, for example, SRdM, Ms. 7, p. 207 ("Séance du vendredi 25 Juin 1779"). For the navy, see SHMB, Série A, Sous-série 3A, Conseil de Marine, Registres 3A4 à 3A10: Procès-verbaux de délibérations du Conseil de Marine de 1772 à l'an II (1793-1794), passim.

suggested retrieving water from great depths in the sea on the presumption that fresh water might lie below salt water.[241] In the 1750s he proposed rinsing barrels with sulfur, a proposal tested in ten trials from 1753 to 1759 in the Indian Ocean, using the resources of the Compagnie des Indes.[242] Cossigny *fils* later thought that filtering water offered a solution, a solution the Académie des Sciences publicized.[243] In the period from 1780 to 1784 a surgeon-major of the navy, one Bouebe, proposed a machine to aerate and filter water, a machine that had the advantage of doubling as a fire pump. Bouebe's machine was successfully tested on a voyage to America and was approved by the Paris Academy; the naval command at Brest took a keen interest in his machine and ordered it installed on all royal ships.[244] At one point the navy tried lead-lined tanks for storing sweet water, but for reasons that are not clear (taste? weight?) did not follow this up.[245] Many proposals were rejected or clearly far-fetched, including that of one Pierre Boyer who in 1746 informed the minister "that his studies of Alchemy have led him to discover the way…to make water brought aboard ships for long sea voyages incorruptible, always pure, good to drink, and even very effective against scurvy which often affects sailors, and this by means of an inexpensive salt."[246]

But it was the suggestion made in 1777 by one La Peyre, a surgeon-major of the royal troops on Île de France and Île Bourbon, that seems to have drawn the most attention, and the case best reveals this side of the Colonial Machine in action. La Peyre's idea was simple: to prepare special casks impregnated with lime. He made his proposal to the Société Royale de Médecine and received a positive report signed by Poissonnier, Macquer, Poissonnier-Desperrières, d'Arcet, and Bucquet – upon which the minister of the navy and the colonies, de Sartine, gave his approval for formal testing.[247] The test was con-

241. DDP I:26.

242. "Essai sur la maniere de conserver à la mer l'eau potable dans les voyages de long cours," 22p. ms., AN-MAR D^3 42, fols. 27-37; see also PA-PV 79 (1760), fol. 288 ("Samedi 7 Juin 1760").

243. See Charpentier de Cossigny (1760) and (1774).

244. See AN-MAR, C^7 40 (Bouebe); AN Marine D^3 42, fols 210-39. SHMB, Série A, Sous-série 3A, Conseil de Marine, Registres 3A4 à 3A10: Procès-verbaux de délibérations du Conseil de Marine de 1772 à l'an II (1793-1794), 26 August 1780, 23 September 1780; SHMB, Série A, Sous-série 3A, Conseil de Marine, Pièces étudiées par le Conseil de Marine, 1745-1792, 3A 91, 1783, #1, 18 January 1783.

245. See letter dated 6 August 1739, SHMB, Série A: 1A. Porcelain tanks were rejected as too heavy.

246. "Le S.r Pierre Boyer...prend la liberté de vous exposer que son application dans l'Alchimie lui a decouvert la facon d'extraire le Mercure du plomb et de rendre l'Eau qu'on embarque dans les Vaisseaux pour les Voyages de long cours incorruptible toujours pure, et bonne a boire, meme tres efficace contre le Scorbut qui attaque souvent les Marins, et cela par le moyen, d'un Sel qui coute peu...," AN-MAR D^3 42, fols. 25-26, "Juin 1746." For a proposal rejected by the Academy of Sciences, see PA-PV-110, Joyeuse, 30 April 1777 ("M.re Contenant L'exposition des moyens de conserver l'Eau douce et de l'empêcher de se Corrompre, tans sur terre que sur les Vaisseaux"), marked "N. approuvé, Macquer, Darcy, Bory."

247. "Premier rapport sur le procédé de M. La Peyre pour prévenir la corruption de l'eau dans les voyages de long cours," SRdM-H&M, vol. 1 (1776). pp. 348-51

ducted aboard the royal vessel *Le Flamand*, sailing from Lorient to Île de France.[248] La Peyre prepared 300 barrels to provide drinking water on the voyage, water that proved perfectly potable according to certificates submitted by the captain and other notables aboard. Two barrels were set aside under seal for testing on Île de France. One had putrefied and turned black; the other was not corrupted but had a bitter taste. La Peyre explained that the first was an untreated test barrel and that too much lime had been placed in the second.[249]

Because of these slightly ambiguous results, the Société Royale de Médecine proposed to the minister that further trials be conducted. These were conducted in 1785 with the chemist Fourcroy in charge of the tests. Notably for our purposes, Fourcroy approved La Peyre's method unreservedly, saying that the original reports back from Île de France were tainted because "the required checks were made by administrators *and not by chemists*."[250] He concluded that La Peyre should be sent to all French ports to instruct artisans in how to construct and prepare barrels and that he should be given a reward for his efforts. The medical society continued to supervise further tests of La Peyre's method for preserving fresh water at sea through at least 1787.[251] The Marine Academy at Brest, which itself had been variously involved with preserving fresh water at sea since the 1750s, was also party to these and related trials.[252]

Desalinization and Distillation. Removing salt from seawater recommended itself to all concerned with provisioning ships with fresh water.[253] Beyond the obvious utility of this approach, our sources reveal two unexpected themes affecting the thinking of those involved. One, desalinization was seen to be especially valuable in the slave trade. As the physician-inventor Gautier from Nantes put the matter to the Regent in 1716, "Those engaged in the slave trade will have more space aboard and can transport a greater number of slaves. Sailors and slaves will no longer be obliged to drink corrupted water that

248. On these and related details, see SRdM-H&M, vol. 7 (1784-1785), p. 280; SRdM, Ms. 7, p. 183 ("Séance du mardi 1er Juin 1779"); Ms. #10, p. 253 ("Séance du Vendredi 3 7bre 1784"), p. 257 ("Séance du Mardi 14 7bre 1784"), and p. 262 ("Séance du Mardi 21 Septembre 1784"). See also La Peyre materials, SRdM, Archives.

249. In fact La Peyre suggests that sabotage may have been involved; see La Peyre letter, dated "Isle de france le 5 9bre 1778," in SRdM, Archives (La Peyre).

250. "l'examen ordonné a été fait en administrateurs et non en Chymistes." Our emphasis. "Affaire de M. La Peyre conservation de l'eau douce en Mer," "donné en 9bre 1785," SRdM, Archives (La Peyre).

251. SRdM, Ms. 11, fol. 145r ("Séance du Vendredi 23 Novembre 1787") and fol. 146r ("Séance du Mardi 27 Novembre 1787").

252. DDP I:44, V:96-97, VI:37-38; SHMV, Ms. 87, letter to "Monseigneur" dated "Le 8. Fevrier 1754."

253. See AN-MAR D^3 40, "Divers procédés pour dessaler l'eau de mer: 1670-1800," including historical review, fol. 91, by Poissonnier in a letter dated "a Paris ce 26 Xbre 1762." See also AN-MAR D^3 41, "Mémoires et Projets. Vivres. Distillation de l'eau de mer," and AN-MAR C7 276 (Rigaud).

causes illnesses and the usual high level of mortality."[254] Secondly, the matter was thought to be so important as to be maintained as a state secret! As Minister for Foreign Affairs, the count de Vergennes, suggested to the Minister of the Navy, de Castries, in 1781: "It would be difficult and even inappropriate to make a secret of something of this nature, but you will perhaps also conclude that the best thing would be to prevent other Powers from mastering this discovery that they themselves might keep secret at least for a while."[255]

As might be expected, inventors proffered dozens of desalinization processes all through the late seventeenth and eighteenth centuries. Some involved precipitating salts; some filtering; one proposed to extract fresh water from sea ice.[256] But by far, most proposals were for distillation machines of one sort or another. As might be expected, all the relevant elements of the Colonial Machine became involved in evaluating these proposals. As early as 1669, for example, the Paris Academy of Sciences performed experiments in its laboratory on rue Vivienne to test a proposal from one Othon from the provincial academy at Caen.[257] And the Academy's contact with the ministry and involvement in desalinization work continued from there through the 1790s.[258] The ministry regularly ordered land and sea trials of these machines at Le Havre, Bayonne, Lorient, Dieppe, Saint Malo, Marseilles, Nantes, Cherbourg, Rouen, and Paris (where sea water had to be imported!); and the king just as regularly underwrote the expenses and rewarded inventors with pensions and gratuities for this kind of work.[259] Among the many proposals considered and

254. "Ceux qui font Le commerce des negres auront plus d'espace et en pouront transporter une plus grande quantité. Les equipages et Les negres ne seront plus obligés de boire des eaux corrompues qui causent des maladies et des mortalités trop ordinaires...," Gaultier memoir to Regent dated "à Nantes le 8e 10bre 1716," AN-MAR D^3 40, fol. 63. The same thought is expressed, fol. 121, "Mémoire pour servir de moyen de dessaler l'eau de mer et de la rendre potable, par Le. Sr. Chervain habitan de Mok a quartier de Limonade, dépendance du Cap."

255. "...Il seroit difficile et même peu convenable de faire un secret d'une chose de cette nature, mais aussi vous jugerez peutêtre que le mieux seroit d'empêcher que d'autres Puissances ne se rendissent maitresses de cette découverte qu'elles pourroient cacher au moins pendant quelque tems," ALS, de Vergennes to de Castries, dated "à Versailles le 21 aoust 1781," AN-MAR D^3 40, fol. 141. See also AN-MAR D^3 41, fol. 104 for another reference to desalinization as a 'secret'.

256. On the latter, see "Analyse et observations sur l'eau provenant des glaces formées sur la mer Par M. Rigault," dated and signed, "a calais le 23 janvier 1768. rigaut physicien et chymiste de la marine," PA-PS ("Mercredi 10 février 1768") and report by Poissonnier and Macquer, PA-PS ("Samedi 6 août 1768").

257. PA-PV 6 (5 janvier 1669-14 décembre 1669. [Registre de Physique]), fol. 189 ("Du Samedy 7.e Decembre 1669"); see further, PA-PV 13 (17 décembre 1689-31 août 1693), fol. 3v ("Le Mecredy 22eme de fevrier 1690").

258. See PA-PV-110, passim, e.g., De Feuquieres, 6 August 1681 ("Expérience sur l'Eau de mer adoucie par précipitation"); Gauthier, 1 September 1717 ("Modèle d'une Machine pour dessaler l'Eau de La Mer"); Dezicourt, 20 June 1764 ("Sur une Méthode pour dessaler l'Eau de La Mer"); Moreau, 24 March 1772 ("Sur la manière de conserver l'eau douce à La Mer"); Rigaud, 7 March 1778 ("Observations faites en 1764 sur l'eau qu'on embarque et les changements qu'elle Eprouve sur mer dans une longue traversée"), and so on.

259. AN-MAR D^3 40, passim and ALS, Duhamel du Monceau to "Monseigneur," dated "a Paris ce 20 Juin 1764," and related materials, fols. 29, 35ff.

tests made, it suffices to highlight two, one from the first half of the eighteenth century, one from the second half.

The first concerns a proposal by the above-mentioned Gautier (or Gauthier), a *docteur-régent* of the University of Nantes.[260] Gautier proposed his distillation machine to the regent in 1716, who forwarded the proposal to Abbé Bignon and the Academy of Sciences. The Academy found Gautier's idea "new and very ingenious."[261] In 1717 the governing council for the navy ordered initial tests of Gautier's machine at Lorient. The machine cost 1684 livres to build. The chief physician, surgeon, and apothecary of the Compagnie des Indes at Lorient performed a battery of chemical tests and had workers drink the resulting distilled water. Apparently, the solder used in manufacturing the machine created an unpleasant aftertaste, and although Gautier's water was more "gray" than ordinary water, the results proved satisfactory. Two sea trials followed in 1719 aboard the royal ships *L'Entreprenant* and *Le Toulouze*, one of which went from Toulon to Alexandria. The crown was prepared to pay Gautier an amazing 15,000 livres for his invention. Whether it did so remains unknown. The Academy preserved Gautier's machine among its collections.

As inspector-general for naval and colonial medicine, the omnipresent P.-I. Poissonnier, was regularly involved in evaluating desalinization machines, and in 1763 he proposed one of his own.[262] Whether because his was a superior proposal or because he was a powerful administrator, Poissonnier's desalinization machine received the full and respectful attention of the Colonial Machine.

The first sea trial of Poissonnier's machine took place in 1763 aboard *Les Six Corps* sailing from Lorient to Brest with Poissonnier himself aboard. Further tests followed on land and on *Le Brillant* sailing from Brest in January, February, and May of 1764. Poissonnier personally reported the results to the Academy of Sciences, where he was a free associate.[263] Poissonnier invited the members of the Academy's commission to his home for lunch and to judge the machine, which doubtless resulted in a positive report. The academician and officer Chabert commanded the *Hirondelle* equipped with a Poissonnier machine in 1766; the royal frigate *L'Enjoué* sailed to Miquelon in 1768; and further tests by the navy followed into the 1770s and 1780s.[264] Apparently, all these tests were successful;

260. On this episode, see AN Marine D[3] 40, fols. 47-76, and historical review by Poissonnier, fol. 91.

261. "nouvelle et fort ingénieuse," PA-PV-110, Gauthier, 1 September 1717.

262. On this episode, see AN-MAR D[3] 41: "Mémoires et Projets. Vivres. Distillation de l'eau de mer: appareil de Poissonnier et de Rigaut. 1763-1769."

263. PA-PV 84 (1765), fol. 244v ("Mercredi 22 Mai 1765").

264. SHMB, Série A: Sous-série 3A, Conseil de Marine, Registres 3A4 à 3A10: Procès-verbaux de délibérations du Conseil de Marine de 1772 à l'an II (1793-1794), 19 September 1772. In this same series, see also related naval tests of preserving fresh water, 11 September 1779, 26 September 1779, 23 October 1779, 3 June 1780, 12 August 1780.

already well paid, Poissonnier received an additional pension of 6000 livres from the king for his contributions. In 1764 the king (read minister) mandated that all ships of the Royal Navy headed for the high seas be equipped with Poissonnier machines, and royal authority likewise instructed French chambers of commerce to work with Poissonnier to see that the machines were installed on merchant ships, too.[265] The Compagnie des Indes similarly adopted Poissonnier machines for all its vessels. It installed Poissonnier's machine on *Le Berryet*, destined for China in 1765, and subsequent vessels sent out from Lorient were likewise so equipped.[266] Private slave ships sailing to Africa and the Caribbean also began carrying the Poissonnier machine.

Poissonnier had his detractors, and in some quarters authorities were suspicious that desalinization machines would be used to avoid the salt tax![267] Nevertheless, there was general praise for the machine. For example, the *Journal de Médecine, Chirurgie et Pharmacie Militaire* for 1786 wrote about "the sound and ingenious process of Mr. Poissonnier. We have this intelligent physician to thank for the construction of a distillation device that unites reliability with economy of men and fuel, and in procuring a cornucopia of sweet water for ships undertaking long distance voyages, he has given them a gift of inestimable value."[268]

The problem with Poissonnier's and all these distillation machines was that they used a great amount of charcoal. The navy was clearly concerned about this use and conducted separate tests in 1769, monitoring charcoal use for distillation.[269] Writing about this matter, the historian of the Marine Academy, Doneaud du Plan, concluded that the Poissonnier machine used too much charcoal to make it worthwhile.[270] In a revealing effort to apply science to this shortcoming inherent in all desalinization machines, in 1783 the royal engineer and then correspondent of the Académie des Sciences, Jean-Baptiste-Marie-

265. "Projet de Lettre aux Chambres de Commerce" (1764), AN-MAR D[3] 41, fol. 104.

266. Legrand, p. xxiii. See also letter from Compagnie Syndics to the Duc de Choiseul, dated "a Paris le 19 May 1765," AN-MAR D[3] 41, fol. 131 and fol. 128 ("Journal des Experiences faites avec la Machine a distiller N.o 3 fabriquée a Brest en Mars 1765 pour servir sur le fregat la Terpsicore") and related letter fol. 136. See also seven-page printed "Instruction pour le service de la machine a dessaller l'eau de mer," AN O[1] 597 #34, and Compagnie letter, dated "Le 19 jan.er 1764," indicating the placement of machines on all Compagnie ships; SHML, 1P 288, Liasse 179, 37, #2.

267. AN-MAR D[3] 40, fols. 77, 79, 81.

268. "...le procédé solide & ingénieux de M. Poissonnier: c'est à ce Médecin intelligent que l'on doit la construction d'un vaisseau distillatoire qui réunit la sûreté à l'économie des hommes & du combustible, & qui, en procurant aux Vaisseaux qui font des voyages de long cours, l'abondance & la bonté de l'eau douce, leur a fait un présent véritablement inestimable." *Journal de médecine, chirurgie et pharmacie militaire, Publié par ordre du Roi. Fait et Rédigé par M. Dehorne*, vol. 5 (1786), p. 262.

269. "Epreuves faites...pour constater le produit de distillation de l'eau et en faire la comparaison avec la consommation de charbon de terre...," SHMB, Série A, Sous-série 3A 120, Commissions Spéciales, #35, 14 May 1769. A six-to-one ratio was felt to be satisfactory.

270. DDP I:41.

Charles Meusnier (1754-1793), proposed that distillation take place at reduced atmospheric pressure and hence lower temperatures that would result in less use of fuel.[271] Despite its limitations, Poissonnier's machine was widely adopted. Whether because it was effective or because Poissonnier himself occupied such a powerful position within the Colonial Machine is hard to judge.

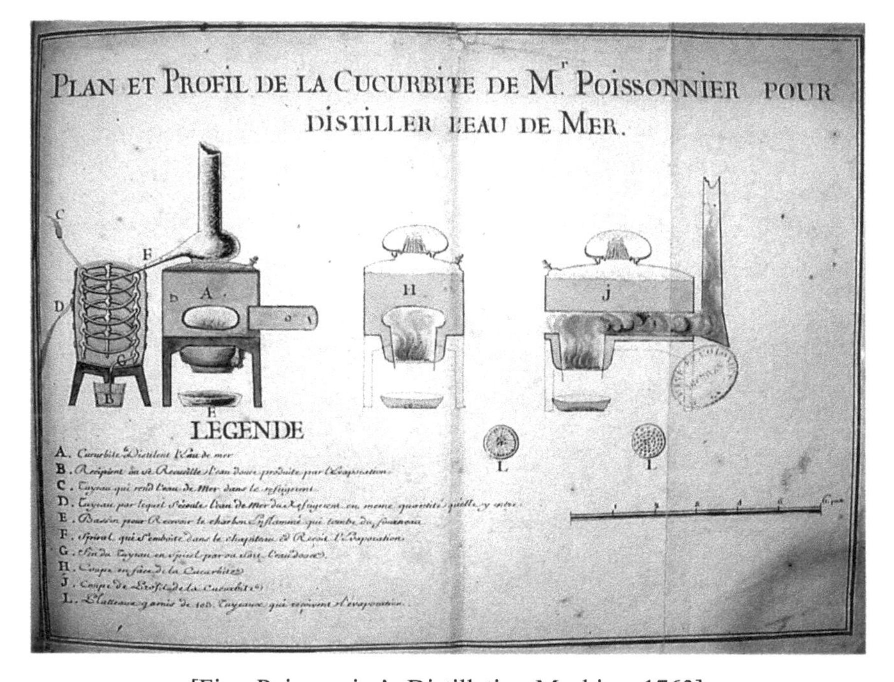

[Fig.: Poissonnier's Distillation Machine, 1763]

Preserving Flour and Meat. For voyages that could last months without putting in for supplies, preserving foodstuffs aboard ships proved a related, general problem of ongoing concern to the Colonial Machine.[272] Looking back on his work of 1753, the *Treatise on the Conservation of Grains and Espe-*

271. "Mémoire Relatif à une nouvelle Maniére de dessaler L'eau de la Mer, avec une chaleur très Médiocre et une oeconomie considérable decombustible; Executée à Cherbourg pendant Les années 1781, 1782 et Le commencement de 1783. Par Le Sr. Meusnier Lieutenant en 1er au Corps royal du Genie et Correspondant de L'Academie royale des Sciences," AN-MAR D[3] 40, "Divers procédés pour dessaler l'eau de mer: 1670-1800," fol. 142. Meusnier went on to become a resident associate of the Academy of Sciences.

272. See especially AN-MAR D[3] 43-45, "Approvisionnements et subsistances"; also, AN-MAR G 145: "Copie d'un mémoire indiquant le moyen de conserver l'eau et le biscuit à la mer, fut-ce pendant 18 mois...par le S.r Raby de Kepeach."

cially Wheat (*Traité de la conservation des grains et en particulier du froment*), for example, Duhamel wrote in 1774:

> Finding myself at Brest at the arrival of a ship loaded with grain that had become fouled with moisture and a bad odor, I saw all the trouble taken to rehabilitate this grain, and this gave me the idea to work in this area. One sees the number of experiments I made on this subject in the three duodecimo volumes I had printed. I will limit myself to saying that I came to preserve grain for twelve years left alone and absolutely without rotting or insects. It was the same with flour I sent to Saint Domingue that was sent back to me at the end of five years and with which I had a very good bread made.[273]

As might be expected, throughout the eighteenth century inventors made a whole series of proposals to preserve foodstuffs. The same Bouebe who (as mentioned previously) proposed a machine to filter seawater came up with a "varnish" (*vernis*) to preserve sea biscuits. Bouebe received a positive report from Bory, Cadet, and Poissonnier at the Academy, and his process proved successful on a voyage to India.[274] The Académie de Marine naturally received its share of such schemes.[275] No less involved were the Société Royale d'Agriculture, the Société Royale de Médecine, and the Chambre d'Agriculture of Cap François in Saint Domingue, all of which tested the sea biscuits developed by the noted agronomist, Antoine-Augustin Parmentier (1737-1813).[276] A similar set of proposals and tests concerned methods of seasoning and preserving meat for use on long voyages; the relevant players again included the Compagnie des Indes, the Académie des Sciences, the Société Royale d'Agriculture, and the Société

273. "M'etant trouvé a Brest à l'arrivé d'un vaisseau chargé de grains qui avoient contracté de l'humidité et une mauvaise odeur, je vis toutes les peines qu'on se donnoit pour retablir ces grains, ce qui me fit naitre l'idée de travailler sur cet objet; et l'on voit dans trois volumes in 12 que j'ai fait imprimer le nombre d'experiences que j'ai fait à ce sujet. je me bornerai a dire que je suis parvenu a conserver des grains pendant 12 ans sans les remuer et absolument exempts de fermentation et d'insects: il en a été de même des farines que j'ai envoyées à S. Domingue qui m'ont été renvoyées au bout de cinq ans et avec les quelles j'ai fait faire de très bon pain," "Etat de mes services depuis 1730 que j'ai eu la Commission d'Inspecteur général de la Marine," signed and dated, "Duhamel du Monceau a Paris ce 13 Septembre 1774," AN-MAR C^7 93 (Duhamel du Monceau).

274. See ministerial documents AN-MAR C^7 40 (Bouebe).

275. See, for example, "La fabrication des farines propres à l'usage de la marine," SHMV, Ms. 0194 XVIII.

276. Llinares (2005), pp. 171-74; Anonymous (1931). On the involvement of the Société d'Agriculture, see APS-MS, B/D87, Group 26, entry for "jeudi 29. janvier 1789." For the Société de Médecine, see SRdM, Ms. 8, p. 356 ("Séance du Vendredi 23 Mars 1781"). For the Cap Chambre d'Agriculture, see reference, SRdM, Ms. 9, p. 350 ("Séance du Jeudi 2 Janvier 1783") and p. 379 ("Séance Du mardi 11 fevrier 1783").

Royale de Médecine.[277] Not surprisingly, the name of P.-I. Poissonnier shows up repeatedly. At another point, the Compagnie des Indes seems to have been involved in testing the preservation of eggs on long voyages.[278] Indicative of the importance of these issues, the apothecary Jean-Baptiste Meunsier (1735-1782) received an astounding reward of 12,000 livres from the king and a lifetime pension of 1000 livres for his recipe for bouillon cubes, certified as incorruptible by Poissonnier.[279]

Thévenard's Experiments on the Resistance of Fluids, 1768-1769. The problems and the research we have examined thus far were obviously practical in orientation. The state had practical interests, so it suffices for our analysis to say that the state employed scientific and technical experts to solve practical problems associated with its colonial and overseas initiatives. By the same token, the various disciplines of contemporary science and medicine struggled with their own theoretical and research questions. These were not necessarily congruent with the practical interests of the state, and we have seen instances – such as pendulum experiments – where pure science questions piggybacked on top of other activities undertaken by the Colonial Machine. Yet the distinction is not all or nothing, and in one instance the practical interests of the state fitted admirably with cutting-edge theoretical work, in this case in hydrodynamics, and the result was a remarkable series of experiments sponsored by the Colonial Machine to test hydrodynamical theory.

A grand tradition of theoretical and mathematical research in mechanics and hydrodynamics unfolded in the seventeenth and eighteenth centuries. The issues involved were central to Cartesian physics with its aetherial vortices, and Newton devoted the entire Book II of the *Principia* to the physics of bodies moving in fluids. Deep into Book II, in fact, after discussing the shape of a body that would offer minimum resistance in moving through a fluid, Newton disingenuously added the footnote, "Indeed, I think that this proposition will be of some use for the construction of ships."[280] How best to design ships and to sail them was the practical spur that motivated official interest, but those subjects were also topics of vigorous, high-level, and ongoing scientific debate. For example,

277. For the Compagnie des Indes, see letter to "Messieurs les Directeurs g.aux de la Compagnie des Indes," dated "Au Port Louis isle de France ce 10 avril 1745," in Lougnon (1940), p. 213. For the Société d'Agriculture, see "1778, Viande dessechée pour la Conserver," APS-MS, Duhamel du Monceau/Fougeroux de Bondaroy Collection, B/D87, group 34. For the Académie des Sciences, see PA-PV 89 (1770), fol. 2 ("Mercredi, 10 Janvier 1770"), and Ms., "Expériences faites en 1772 pour dessecher la chair de boeuf et la rendre susceptible de la conserver sans se corrompre," PA-DB (Rigaut), with marginal note: "MM Macquer et Fougeroux [commissioners]; Lu le 28 Janvier 1778." For the Société de Médecine, see SRdM, Ms. 7, p. 146 ("Séance du vendredy 30 avril 1779"). For the Académie de Marine, see DDP IV:49-50 and Sartine letter dated "Versailles le 23 février 1776," SHMV, Ms. 093.

278. ALS "Ce 4 fev. 1736, De Fulvy" to "M. D'Espremesnil, à L'orient," SHML, 1P 278, Liasse 1.

279. AN-MAR C^7 207 (Meunsier).

280. Newton, *Principia*, Book II, Section 7, Prop. 34, Theorem 28 scholium.

in 1689 the navy captain, military engineer, and later Academy honorary, Bernard Renau d'Éliçagaray (1652-1719), published his *De la Théorie de la Manœuvre des Vaisseaux*, a work that sparked a baroque series of criticisms by Huygens and replies by Renau.[281] Johann Bernouilli entered the debate in 1714 with his *Essay d'une nouvelle théorie de la manœuvre des vaisseaux*. In 1749 Leonard Euler contributed his two-volume work on constructing and steering ships, *Scientia Navalis seu Tractatus de Construendis ac Dirigendis Navibus*, published in Saint Petersburg. From France, Pierre Bouguer added his *De la manœuvre des vaisseaux, ou traité de mécanique et de dynamique* in 1757, and others across the eighteenth century investigated hydrodynamics at a high level of theoretical and mathematical sophistication.[282] Most issues concerning the theory of fluids were still not settled by the 1760s.

In 1768 Antoine-Jean-Marie, the count Thévenard (1733-1815), then a captain with the Compagnie des Indes, proposed a set of experiments to test hydrodynamical theory.[283] The knowledge gained, Thévenard promised, "will indicate new methods for calculating that will prove definitive in practice."[284] The Minister of the Navy and Bézout and Borda from the Academy of Sciences unanimously approved the proposal. With the backing of the Academy and support from the Compagnie des Indes, the ministry provided 5000 livres in subventions and ordered the work to proceed.

On the property of the Compagnie des Indes in Lorient, Thévenard and company built an experimental canal, 180 feet long, 18 feet wide, and 10 feet deep, to which they added dikes, locks, and scaffolding. D'Après de Mannevillette and one Choquet were on hand to observe from the Académie de Marine. "My zeal is boundless, as is my desire to usefully employ it for the State," declared Thévenard on the occasion.[285] For three days in October 1769, the experimenters conducted tests under ideal conditions. Sixty different experiments analyzed seventeen differently shaped test objects placed in a constant flow in the canal. Choquet reported that "the results of these experiments destroy the theory founded on the calculations of Mr. Newton…and on this score many singular and satisfying things emerged that can be very useful in the construction of ships."[286] De Mannevillette was in agreement that the

281. See Renau (1694); on Renau, see Bagneux and Taillemite (2002), p. 443.

282. Vérin (2000), pp. 162-64; Levot, pp. 67-69, and Boudriot (2000) add important background.

283. These details taken from AN-MAR G 101, Dossier 1, "Expériences de Thévenard sur la résistance des fluides. 1768-1769." On Thévenard, see Taillemite (2002), pp. 499-500.

284. "Cette connoissance fera indiquer de nouvelles mèthodes de Calculs, sures pour la Pratique," AN-MAR G 101, Dossier 1, Thévenard "Mémoire," fol. 8.

285. "Mon Zele est sans bornes, ainsi que mon desir de l'employer utilement pour L'Etat...," ALS "Thevenard" to "Monseigneur," dated "A Lorient le 22 aoust 1769,"AN-MAR G 101, Dossier 1, fol. 23.

286. "les résultats de ces expèriences détruisent la Théorie fondée sur les calculs de M. Newton,…et sur ce point il s'est fait des choses singulières et satisfaisantes, et qui peuvent être très utiles à la construction des vaisseaux...," ALS, "Choquet" to "Monseigneur" dated "au Port Louis le 18 Octobre 1769," AN-MAR G 101, Dossier 1, fol. 27.

experiments "clearly prove the falsity of the principles adopted until now by all the learned about resistance, and it is not surprising that ships constructed in consequence are far from having the good qualities that one expected from them."[287] Whether there was more to this story, we do not know, but the conjuncture of theory and practice in the Colonial Machine could not be more remarkably evident.

Other Experimental Research. The relentless quest for utility that drove this research becomes powerfully clear when one considers the entire range of nautically related problems taken up by the Colonial Machine. Across the eighteenth century various elements of the Colonial Machine became involved in projects and tests concerning dry docks, diving bells and diving suits, ships' tonnage, fans and ventilators for ships, devices to measure the speed of ships under sail, anchors, pulleys, mills, pumps, rigging, machines to raise sunken ships, machines to work under water, improved oars, bathtubs, cannon borers, worm damage, gunpowder stocks, compasses, and so on and so forth.[288]

The belief that unhealthy air caused alarming illnesses among sailors prompted sustained work on ventilating ships. Our now familiar naval captain and academician, Gabriel de Bory, was involved from the 1750s, and he continued to work into the 1780s, publishing his paper on mechanisms to purify air in ships in the *Mémoires* of the Academy of Sciences for 1781.[289] Bigot de Morogues of the Académie de Marine at Brest offered his paper in the *Savants Étrangers* series of the Paris Academy in 1750 on the corruption of air in ships, and the Marine Academy likewise continued discussion of ventilators for ships through the 1780s.[290] The Compagnie des Indes installed fans on its ships and printed up instructions for their use by its personnel.[291] On the other hand, the Société Royale de Médecine recommended larger scuttles for the natural passage of air.[292]

Regarding gauging the tonnage of ships, uniformity of measurement was key. Certified experts (*jaugeurs jurés*) performed these measures based in part on instructions from the Academy of Sciences.[293] In 1721 and again in 1724

287. "prouvent avec évidence la fausseté des principes adoptés jusquici par tous les scavans sur cette mesme résistance & il n'est pas étonnant que les vaisseaux contruits en consequence nayent pas a beaucoup près les bonnes qualites qu'on en attendoit," ALS "D'aprés de Mannevillette" to "Monsieur," dated "a Lorient le 13.e 8bre 1769," AN-MAR G 101, Dossier 1, fol. 31. See related work by Thévenard, "Calcul raisonné de la force d'un Appareil pour tirer un Vaisseau à terre," Marine Academy *Mémoires* (1773), pp. 364-76.

288. It is impractical to document every case here, but for an entrée to these and related projects, see further AN-MAR G 145 and G 146.

289. Bory (1781); for his work in the 1750s, see later report, *Collection Académique, Partie Française* vol. 16 (1787), pp. 107-112.

290. DDP VI:37.

291. See seven-page printed instructions to captains of vessels of the Compagnie des Indes, AN O^1 597, #8.

292. See Poissonnier remarks to this effect, SRdM, Ms. 7, pp. 352-3 ("Séance du mardi 9 9bre 1779").

293. "Instruction pour le Jaugeage des Vaisseaux approuvée par l'academie," AN-MAR G 61, vol. 1, fol. 155.

Jean-Baptiste Dortous de Mairan (1678-1771), *pensionnaire* in geometry in the Academy, published papers and detailed instructions in the Academy's *Mémoires* on measuring the capacity of ships.[294] Father Pézenas followed up in 1749 with his *Théorie et pratique du jaugeage des navires et des tonneaux des navires et de leurs segments*. Another article by Pézenas published in the *Savants Étrangers* of the Academy in 1750, "Solution to Kepler's Problem on the Proportions of Segments of a Cask Cut Parallel to Its Axis," emphasizes the theoretical dimension of this problem.[295]

In a like spirit, experts concluded that dry docks needed to be built in the shape of a cycloidal curve.[296] Regarding anchors, in 1736 the Ministry of the Navy ordered the Academy of Sciences to examine the optimal shape for anchors, but this work evoked a revealing objection from a certain Lamothe, who wrote to the minister: "Although I do not at all doubt that this academy is sovereign over geometrical calculations, I nonetheless believe that a great sailor, a good mechanic, and a good forger would be better equipped to decide these questions than men who spend their lives in their offices and who carry on discussions among themselves that are beyond the reach of ordinary artisans."[297] The minister responded by inviting Lamothe to submit his thoughts to the Academy.[298] Reprising the matter two decades later, Lamothe was probably right in concluding that "all these studies have hardly resulted in any change in the form of anchors."[299]

The several cases of diving bells and diving suits presented to the Academy of Sciences in Paris would seem entirely straightforward, except for the revealing instance in 1774 regarding the diving apparatus of one Sieur Freminet.[300] After tests in the River Seine the Academy initially rejected Freminet's invention as impractical and dangerous, but Freminet had powerful backers in the dukes d'Orleans and de Chartres, who prevailed upon the minister to have the Academy revisit its judgment. As instructed, the Academy added two commis-

294. Dortous de Mairan (1721) and (1724).

295. Pézenas (1750); see also Pézenas (1749); Stephan; Boistel (2005), pp. 119-22. A second edition of Pézenas' 1749 work appeared in 1778.

296. Ms. "Mémoire où on Détermine la forme la plus avantageuse que doivent avoir les Chantiers sur les quels on Construit les Vaisseaux," AN-MAR G 141.

297. "Quoyque Je ne doüte point que Cette accademie possede Les plus grandes Lumieres sur tous Les Calculs Geometriques, Je suis Cependant dans La Croyance qu'un grand Marin, un bon Mecanicien, et un bon forgeron pourroient Etre plus propres à décider Ces questions que des hommes qui passent Leur Vie dans des Cabinets, et qui Entre Eüx tiennent des propos qui ne sont point à La portée des artistes ordinaires...," ALS, Lamothe "A Monseigneur Le Comte de Maurepas Ministre de la Marine," dated "10 Octobre 1736," AN-MAR G 145.

298. ALS Maurepas to "M. de La Mothe," dated "Versailles le 14 Octobre 1736,"AN-MAR G 145.

299. "...que tous ces ouvrages n'ont point donné Lieü à auçuns changements sur La forme des ancres," ALS, Lamothe "A Monsieur Ruis Embito Commissaire General, et ordonnateur au Department de Rochefort," dated "a Bordeaux ce 5e septembre 1757," AN-MAR G 145.

300. On this episode, see AN-MAR G 235, #52, #65, #66, #67, #73, #82.

sioners with naval experience (Chabert and Bory), but it still refused to approve Freminet's machine. Nevertheless, despite acknowledging the Academy's authority, the next year the king granted Freminet a royal patent and exclusive salvage rights for fifteen years using his machine. (This was probably what was really at stake.) As the royal patent disingenuously noted:

> For some fifteen years Mr. Freminet has applied his diligence and his work to arrive at constructing a machine that he presented to us in our Council, the design backed up by with certificates of experiments first done in the Seine River under the supervision of commissioners named by our Academy of Sciences....As a result of these first experiments, our Academy of Sciences, *the true judge of new discoveries*, recognized Mr. Freminet as the sole inventor of this ingenious device...without, however, approving the machine because one can dive in the Seine only to a depth of fifteen feet, and everyone recognizes the danger of descending to great depths with such a machine.[301]

That Freminet was able, through his powerful connections, to override the judgment of the Academy is a telling exception proving the rule that the Academy decided such matters.

Another obscure case likewise reveals cracks in the Colonial Machine. This one concerns tests in 1782 of a "loch," or a device to measure the speed of a ship under sail devised by Jean-Baptiste Degaulle, professor of hydrography at Le Havre.[302] Degaulle submitted his machine to the Academy of Sciences where, after requesting modifications, the Academy panel (consisting of Le Monnier, Bézout, Bory, and Borda) approved the invention. The results were passed down to the Académie de Marine, which in turn appointed its own commissioners. At first the commissioners at Brest thought it sufficient to accept the report of their Parisian counterparts, but they suspected that the

301. ...[Freminet] a depuis environ quinze ans mis son application et son Travail pour parvenir à construire la Machine dont il nous a fait présenter dans notre Conseil le Dessin appuyé des Certificats des expériences qui en ont été faites dabord dans la Riviere de Seine sous les yeux des Commissaires nommés par notre Académie des Sçiences....Qu'il résulte des premieres expériences que notre Académie des Sçiences, vrai Juge des nouvelles Découvertes, et qui a reconnu comme entierement appartenante au S. freminet l'invention ingénieuse...sans néanmoins approuver la Machine, à cause qu'on n'avoit pu plonger dans la Seine qu'à quinze pieds de profondeur, et que l'on sait tous les danger [sic] qu'il y auroit de descendre à une grande profondeur avec une telle machine...," "Privilege exclusif accordé au S. freminet auteur d'une nouvelle machine à plonger," "Donné à Versailles Le [blank space] jour du mois de Janvier, L'An de Grace Mil sept cent soixante quinze et de notre Regne le premier," AN-MAR G 235, #82. Our emphasis. On a case in 1765 where the Duc de Praslin directed Paris Academy to evaluate a diving bell used in Marseilles and approved by the Chambre de Commerce there, see PA-PV 84 (1765), fol. 4 ("Mercredy 16. janvier 1765"), and negative report, fols. 50-53 ("Mercredi 30. janvier 1763").

302. AN-MAR G 99, dossier 2, fols. 20-27; on Degaulle, see Taillemite (2002), pp. 129-30; Neuville (1882), pp. 34-35.

speed-measuring device was not as accurate as its inventor claimed.[303] As would be typical, the Minister then authorized the Brest Academy to run a series of tests, but the results proved so unsatisfactory that the Brest Academy protested to Borda over his approval of the invention before any experiments were tried.[304]

Work to improve gunpowder fits into this analytical category. Working with both Le Monnier and the count d'Angiviller in France, Cossigny on Île de France conducted experiments at the royal powder plant to improve gunpowder production for the French in the Indian Ocean, notably by trying different woods as sources of charcoal.[305] At another point, an explosion occurred at the gunpowder manufactory on Île de France. Fortunately, as the Compagnie des Indes wrote back to the local Conseil, "the damage was minimal and none of the slave workers or white artisans were wounded."[306]

In and of themselves and for whatever else they might tell us, these cases are minor, but they and still others we might explore demonstrate once again, if any such demonstration be necessary, the relentless pressure to be useful that was felt throughout the Colonial Machine. Nowhere was this pressure more keenly felt than in efforts to solve the nagging problem of finding longitude at sea.

The Problem of Longitude at Sea – Version Française

While astronomical techniques for determining longitude worked well on colonial terra firma, the solution aboard ship lay elsewhere, and by the 1760s the problem was not so much determining longitude in general but determining longitude at sea. This more narrow concern was famously solved by the invention of a reliable marine chronometer by the British clockmaker John Harrison (1693-1776), who in 1765 was awarded the prize for his accomplishment by the British Board of Longitude. Harrison and his clock have been the subject of several fine historical studies, and there is no need to summarize those accounts here.[307] Less well known is the story, stretching over a century, of efforts to solve the problem by the French and the Colonial Machine.

303. DDP V:70-71.

304. DDP V:85-86.

305. See ALS, Cossigny to Le Monnier dated "A L'Isle de France le 20 9bre 1779," and ALS, Cossigny to Le Monnier dated "A L'Isle de France le 16 Avril 1780," both, BCMNHN, Ms. 1995. See also "Résultats des Expériences qui ont été faites au Moulin à Poudre du Roi," dated "le 10 Mars 1781," BCMNHN, Ms. 357, and ALS Cossigny to "M. le C.te D'Angivillers" dated "A L'Ile de France le 15 X.bre 1783, AN O[1] 1292.

306. "Au surplus il est heureux que les avaries ayent été médiocres, qu'aucun des Noirs et des Blancs frabricateurs n'ait été blessé," "Copie de la lettre écrite par la compagnie à M.rs du Conseil supérieur de l'isle de France dattée à Paris le 28 février 1749," in Lougnon (1949), p. 134.

307. Andrewes, ed.; Sobel; Despoix, esp. pp. 19-72; Haudrère (1997), pp. 111-20 provides useful background, as does Iliffe, pp. 634-36.

The Académie des Sciences was a prime agent in the French longitude story. We have seen the Academy's role in sending out observers such as Richer and Deshayes in the seventeenth century to determine points of longitude on land, and we have seen how the Paris Academy directed longitude research through its administration of the prize established by the parlementarian Rouillé de Meslay. We previously described this long series of contests organized by the Academy, most pertaining directly or indirectly to the longitude problem. The astronomers of the Observatoire played a role in these matters. Moreover, from the 1660s on the Academy adjudicated an unending stream of longitude proposals brought before it directly by inventors or as ordered by ministerial authority, initially Colbert.[308] The Academy was variously supportive of some and dismissive of others. In some ways this stream of longitude proposals had almost the character of those concerning squaring the circle or trisecting angles.[309]

Declination and other Work before the Chronometer. Ultimately, there were two solutions to the problem of longitude at sea – the invention of the marine chronometer and the perfection of a technique involving lunar observations. These methods, especially the chronometer, came into play in the late 1760s; but before that achievement, experts pursued any avenue that seemed promising. One with apparent potential involved compass variations and observed differences between magnetic north and true north.[310] Based on work published by the Englishman, Edmund Halley, in 1701, the idea was that the observed differences between true and magnetic north might provide a standard basis for determining one's position at sea. When it was discovered that compass readings themselves varied at any one spot over time, the hope was that *variations* in these readings might be regular and so offer another entrée to solving the problem.

Investigations of compass variation proceeded in a virtually unbroken series of observations across the eighteenth century, undertaken both for their intrinsic scientific interest and for solving their potential for the longitude problem. In the first decade of the century, J.-D. Cassini I published three notable papers on the subject, and "declination" observations repeatedly show up in the expeditions and longitude research reported here previously.[311] The voyage to Louisi-

308. See above Part I. On early longitude projects and Colbert's involvement, see, for example, PA-PV 3 (11 avril 1668-mars 1669 [Registre de Mathematiques]), fol. 261 ("du mercredy 20e feburier 1669"); PA-PV 5 (3 avril 1669-18 décembre 1669 [Registre de Mathematiques]), fols. 184-190 ("du mercredy 16e octobre 1669"). Strikingly in this early longitude work, Academy operated in what is usually thought of as its eighteenth-century mode of judging projects.

309. See PA-PV-110, passim.

310. On this work see Filliozat (1993), 146-50; Chapuis, pp. 58-62; AN-MAR G 99; AN-MAR G 101 dossier 2; for background, see Iliffe, p. 619.

311. See Cassini I (1705a), (1705b), and (1708). Cassini's and French use of the word, "declination," to distinguish true and magnetic north should not be confused with true declination or the "dipping" of the compass needle between the equator and either of the poles; see below.

ana in 1720 by Father Laval, for example, incorporated an effort to establish tables and navigational aids based on compass variations; Laval concluded there was no certainty regarding annual or periodic variations of the compass needle, and he expressed the hope that the Paris Academy might take up the matter as part of a longer-term research endeavor.[312] Father Gaubil sent compass variation observations from Île Bourbon on his way to China in 1722.[313] Father Feuillée reported his magnetic variations to the Academy in November of 1731; earlier, in June of that year, Meynier's discussion of observations at sea garnered praise from an Academy committee headed by Jacques Cassini II, Academy *pensionnaire* and leader at the Observatoire.[314] In 1734, backed by Maurepas, the Minister of the Navy and the Colonies, the Paris Academy started a systematic effort to observe compass variations in the major ports of France. Reviewing this work the next year, the Academy's commissioners Maraldi and Buache "judged that these observations have not yet attained the degree of perfection of which they are capable."[315] Other observations filtered in from Île Bourbon and Pondicherry in the 1730s.[316] In his *Lettre sur le progrès des sciences* (1752), Maupertuis supported the compass variation solution, acknowledging that the direction of the compass needle varies over time at any one spot; but he held that the *rate* of such variation is probably constant and hence "in some fashion can provide what we lack concerning knowledge of longitudes at sea."[317]

312. See Débarbat and Dumont (1990), pp. 24-25; Laval's "Réflexions sur les observations de la Variation dans le voyage de la louisiane en 1720" and his "Comparaison des Variations observées en 1720 avec la Variation dans les mêmes Parages determiné pour 1700 par les Courbes décrittes par Mr. Halley," OdP, Ms. B 5, 1 [7*], pp. 43-48, 49-58. The Academy was already involved in this work; see PA-PV, 29bis (1710), fol. 270v-281 ("Le Samedi 2 Aoust 1710"), and OdP, Ms. B 2.7, "Extrait des registres de l'Académie des sciences," pp. 336-38.

313. Ms. copy of letter from Gaubil to "mon Reverend Pere le R. P. Souciet de la Comp. de Jésus à Paris," dated "à Poulo Condor ce 23 février 1722," in PA-DB (Gaubil).

314. See PA-PS (November 1731); PA-PV-110, Meynier, 13 June 1731 ("Sur un Ecrit concernant l'Observation de la déclinaison de l'aiguille en Mer").

315. "Jugé que ces observations n'ont pas encore atteint le degré de perfection dont elle [sic] sont susceptibles," PA-PV-110, Maurepas (Le Comte de), 30 Avril 1735 ("Rap: Sur des Observations demandées sur la Déclinaison et l'inclinaison de l'aiguille aimantée en conséquence de l'instruction approuvée par l'académie le 4 Août 1734 et envoyée dans les différents Ports du Royaume par M.r de Maurepas").

316. See Cossigny letters, dated "A St-Paul de l'isle de Bourbon, le 20e décembre 1732" and "A L'Orien en Bretagne," reprinted in Cossigny (1939-1940), pp. 186-192 and pp. 233-34. For the Pondicherry reports, see PA-PV 59 (1740), fols. 152-53 ("Le Mercredi 13. Juillet 1740"). See also, PA-PV-110, 1734, Quereineuf, 23 June 1734 ("Sur une Méthode pour trouver en mer La variation de L'aiguille aimantée").

317. See his *Lettre* appended to Maupertuis (1997), pp. 124-26. On related projects submitted to the Academy in this period, see PA-PV-110, La Croix, 2 September 1744 ("Ecrit touchant la recherche des Longitudes par l'inclinaison des aiguilles aimantées"); Mandillo, 12 February 1746 ("M.re *Sur la manière de trouver La Latitude et la Longitude au point de Midi par le réglement magnétique*"). Pedley (2005), p. 207, signals 144 ("very rare") manuscript charts and maps displaying magnetic variation.

The interest in compass variations to determine longitude did not diminish in the second half of the century. Even as the marine chronometer emerged as the effective solution to the problem, in 1767 the marquis de Courtanvaux, someone heavily involved in chronometric work, reported more or less optimistically to the Academy:

> Astronomy offers several ways to discover longitude at sea. Until now we owe the physical sciences only one such method leading to the same end. That's the variation of the magnetic needle. There are anchorages where this variation alone suffices to know very closely the longitude of a vessel. But beyond the fact that the method is not generally applicable, it suffers the further default of being subject to a degree of uncertainty. The declination of the compass varies not only from one place to another, it is even more variable in the same place. We have not really discovered the law behind this double variation.[318]

That uncertainty notwithstanding, in 1776 the meteorologist Louis Cotte along with Marc Blondeau of the Marine Academy at Brest was still making observations.[319] In 1777 Le Gentil chimed in from the Indian Ocean.[320] In 1778 P.-C. Le Monnier, who had been actively involved in this research since the 1750s, published an updated map of geomagnetic lines in his substantial *Lois du Magnétisme comparées aux observations et aux expériences, dans les différentes parties du globe terrestre, pour perfectionner la théorie générale de l'aimant, et indiquer par-là les courbes magnétiques qu'on cherche à la mer, sur les cartes réduites* (The Laws of Magnetism Compared with Observations and Experiments in Different Parts of the Terrestrial Globe to Perfect the General Theory of the Magnet and to Indicate Thereby the Magnetical Lines Sought at Sea on Reduced Maps), two volumes published at government expense.[321] Later, in 1781, Father Hallerstein contributed his observations from China.[322] In 1786

318. "Tels sont les secours que l'Astronomie peut offrir pour la découverte des longitudes sur mer. Jusqu'à présent nous ne sommes redevables à la Physique que d'une seule méthode, qui puisse tendre au même but; C'est celle de la Variation de l'aiguille aimantée. Il est des parages où cette variation seule suffit pour connoître à très peu près la longitude du Vaisseau. Mais outre que la Méthode n'est pas générale, elle a encore le defaut d'être sujette à quelque incertitude; vu que la déclinaison de l'aiguille ne varie pas seulement d'un lieu un autre; elle est de plus inconstante dans le même lieu; et l'on n'a pas encore découvert bien décisivement la loi de cette double variation," Courtanvaux ms. memoir, "Verification de quelques instrumens destinés à la determination des longitudes en mer," AN-MAR G98, dossier 2, fols 87-98, with note at top: "Lû Le 14 9bre 1767. A La rentrée de L'academie."

319. See Cotte letter, dated "De Montmorency 19 Janvier 1776" in PA-DB (Cotte).

320. Le Gentil de la Galaisière (1777).

321. P.-C. Le Monnier (1776-1778); see also AN-MAR G 99, fol. 93; PA-PV-110, Lemonnier, 11 March 1778 ("Ouvrage Intitulé, *Lois du Magnétisme* Seconde partie, qui contient les nouvelles recherches sur la Situation Géographique de L'Equateur, &c").

322. PA-PV-110, Hallerstein, 17 March 1781 ("Observations de la déclinaison de L'aiman, faites en Chine").

Cotte reported on a concerted international effort centered on the Observatoire Royal using observers at Brest and La Haye, and in London at the Royal Society to monitor variations of terrestrial magnetism. A daily cycle of needle variation was apparently discovered.[323] By this time, of course, studies of compass variation were undertaken for understanding terrestrial magnetism itself and not to solve the problem of longitude at sea, but the investment in compass-based work illustrates the overriding importance of the longitude problem to eighteenth-century science and to the Colonial Machine.

Compass work measuring true versus magnetic north was not the same as investigating geophysical declination, or the "dipping," of the compass needle between the equator and either of the poles. The former was more astronomical in character and practical. About the latter, we will only say that the true declination, or dipping, compass appeared in France in 1777. The same Degaulle, whose "loch" so disappointed, made the instrument, and the Academy of Sciences approved it in 1777. After sea trials conducted by the Académie de Marine, Degaulle received 1200 livres and was given permission to sell his instrument to the English.[324]

Other, astronomical approaches likewise promised to help sailors fix their longitude. In 1767 and 1768 Abbé Rochon proposed a method using a suspended telescope to make shipboard observations of the satellites of Jupiter.[325] The Academy of Sciences (Bory, Maraldi, Cassini de Thury as commissioners) approved, and the minister added a test of Rochon's method to Rochon's voyage to the Indian Ocean with the chevalier Grenier in 1768-1770. The Académie de Marine was also involved, according to Rochon. The results seemed positive, but Rochon's method ended up a dead letter because of the difficulties of observing the satellites of Jupiter at sea and ultimately because of the greater ease and reliability of the marine chronometer.

The method of lunar distances represented another astronomical approach to solving the problem. Although complicated in practice, the idea was simple: Comparing the measured angular distance between the moon and a fixed star with a standard table in principle allowed one to ascertain the longitudinal distance between Paris and a remote locale.[326] There were several problems with this method: the creation of accurate tables, the difficulties of observing aboard ship, moonlight obscuring the very stars one needed to observe, and the complexity of the required calculations. The exact tables of the German Tobias Mayer and the observational work of the Royal Astronomer in England, Nevil Maskelyne, who pushed hard for the astronomical solution, eventually demon-

323. See Cotte letter to unknown recipient, dated "Laon, 18 février 1786," OdP, Ms. B 4.9bis.

324. AN-MAR G 99, dossier 3, fols. 32-78.

325. AN-MAR G 91, fols. 188-205.

326. Filliozat (1993), pp. 133-9, gives a more detailed technical exposition of the method of lunar distances, as does Chapuis, pp. 67-72, 83.

strated the practical feasibility of the method of lunar distances in 1765. Mayer shared the British Board of Longitude prize with Harrison.[327] In the meantime, French astronomers also applied themselves to perfecting the method of lunar distances.

D'Après de Mannevillette was making longitude measurements at sea using lunar distances as early as 1751, and he later deplored the lack of competent personnel to make these recondite observations.[328] In the late 1760s and early 1770s Father Pézenas was also deeply involved in preparing tables and arguing for the method of lunar distances. The Paris Academy entertained Pézenas' proposals, but ultimately rejected them as too complicated for the ordinary sailor or naval officer and too likely to lead to errors.[329] Another effort using lunar distance was the *mégamètre*, a large, octantlike device capable of accurately measuring large angles, invented by naval lieutenant Charles-François-Phillippe de Charnières (1740-1780).[330] In 1766 Charnières presented his invention to the Academy, which gave a positive report. The ministry sponsored a test voyage to Guadeloupe and Saint Domingue in 1767-1768; using the mégamètre, the ship's party arrived within a margin of error of seven leagues, making the cutoff limit of ten. Charnières published his report on the expedition, *Expériences sur les longitudes, Faites à la Mer en 1767 & 1768*, at government expense in 1768. The Observatoire and the crown combined to publish Charnières' *Théorie et pratique des longitudes en mer* (Theory and Practice of Longitudes at Sea) in 1772.[331] (There was a squabble between the Academy of Sciences in Paris and the Marine Academy in Brest over providing the imprimatur for this tiny treatise.[332]) The Academy again approved: "The method employed by Mr. De Charnières is founded on clear and evident principles, just those desired by the Academy."[333] Charnières was awarded gratuities of 1000 and 1200 livres to cover his costs, but an item in the minutes of the meetings of the Academy for 24 January 1770 tells a less upbeat story: "Mr. Le Monnier read a letter in which Mr. Pingré informs him from Cap François that the mégamètre sent along was not worth anything; upon which we resolved not to speak of it in the report."[334] The final solution to the problem of longitude

327. Forbes.

328. See d'Après de Mannevillette letter, dated "À Kx.gars le 30.e Avril 1771," SHMV, Ms. 89; Filliozat (1993), p. 139.

329. AN-MAR G 91, fol. 53 ff; Boistel (2002).

330. Taillemite (2002), p. 98; on this episode see AN-MAR G 91, dossier 2, fols. 44-165.

331. See SHMV, Ms. 0251 1766.

332. See De Boynes letter to "M. de Rosnevet L.t de V.au Sécretaire de l'Acad. de Marine à Brest," dated "À Versailles le 14 mars 1772," SHMV, Ms. 88.

333. "...la methode Emploiée par M. De Charnieres est fondée sur des Principes clairs, Evidents Et tels que L'academie les desire," "Extrait des Registres De l'Academie Royale des Sciences, Du 24 Février 1768," AN G 91, fol. 69.

334. "M. Lemonnier a lu une lettre par laquelle M. Pingré lui mandait du cap français que le mégamètre qu'on avoit envoïé ne valait rien du tout sur quoi il a été résolu de n'en point parler dans le rapport," PA-PV 89 (1770), fol. 11v ("Mercredy 24 Janvier 1770").

at sea was not the mégamètre, but the development of a reliable marine chronometer.

The French Marine Chronometer. Less well known, the French side of the chronometer story closely parallels the English one.[335] The French themselves possessed a sophisticated technical tradition of watch- and clockmaking, and royal largesse supported a bevy of royal clockmakers. The French, notably in the person of the royal clockmaker attached to the navy, Ferdinand Berthoud (1766-1807), followed not far behind the English in developing comparable French clocks that could serve as marine chronometers. [336]

The French did not invent or reinvent the chronometer, however, but received the necessary technical secrets from Harrison himself, albeit not easily. The story begins in 1763 with a proposal to send French academicians to England to check out Harrison's clock. Because the Academy was on vacation at the time, an ad hoc group of academicians met to discuss the trip. They nominated the senior academician, Abbé Charles-Étienne-Louis Camus (1699-1768), and Berthoud to go. The goal was to learn what they could of Harrison's clock and for Berthoud to "make a model in order to construct a similar clock as soon as possible after his return and to test it beginning this summer on ships of the Marine Royale."[337] Harrison apparently stonewalled them, and the French emissaries returned empty-handed. On the promise that Harrison would reveal his secret in return for 4000 pounds sterling, Berthoud undertook a second trip to England in early 1766, for which he studied English![338] With the ambassador from Saxony, the count de Brühl, acting as an intermediary and hosting a series of dinners for the two, Harrison finally instructed Berthoud in all the secrets of his clocks for a mere 500 pounds, which Berthoud paid directly himself, "...the sum seemed to me too modest to doubt that the Minister would not consent to it."[339]

A pamphlet, *The Principles of Mr. Harrison's Time-Keeper*, appeared in London in 1767 and was immediately translated by the Jesuit Pézenas and disseminated throughout the Colonial Machine.[340] Asked by the Minister to comment on the translation, in a letter dated 24 October 1767 the naval officer and

335. On this story, see materials in AN-MAR G 97 and especially G AN-MAR 98, "Montre marine du Sr. Harrison." Boistel (2003, 2005) provides detailed background.

336. On F. Berthoud, see Cardinale.

337. "...en faire un modèle pour en construire une pareille aussitôt après son retour et la mettre dès cet été en expérience sur les Vaisseaux du Roy," ALS, de Fouchy to "Monseigneur," dated "à Paris Le 4 avril 1763," AN-MAR G 98, dossier 1, fol. 6. See also Cardinale, p. 25.

338. ALS, Ferdinand Berthoud to "Monseigneur," dated "Paris Le 26e Xbre 1765," AN-MAR G 98, dossier 1, fol. 9.

339. "...la somme m'a parûe trop modique pour pouvoir doûter Monseigneur que vous n'y consentiez...," ALS Ferdinand Berthoud to "Monseigneur," dated "a Paris le 14 Mars 1766," AN-MAR G 98, dossier 1, fol. 12.

340. AN-MAR G 98, dossier 1, fols. 19-71, and Pézenas' letters, fols. 72-74.

Brest Academy member, the count de Fleurieu, praised it highly, going on to say:

> If the English have prevailed over us in developing marine chronometers, My Lord, the success of those of Mr. Berthoud made on your order and that are ready for delivery will concede nothing to Mr. Harrison's English clock. I do not hesitate to say that the construction of French clocks is far superior to that of the London maestro....If success matches our hopes (and I daresay they are well founded), France will enjoy sooner and at less cost than England the fruits of this important discovery.[341]

The chronometric solution to the problem of longitude may have emerged from the world of the crafts and technology, but it took French astronomers and the Royal Navy to test and certify its utility. Four French sea trials ensued.[342] The first occurred in 1767 aboard the *Aurore*, sailing in northern European waters. At a public meeting of the Academy of Sciences in November of 1767, the honorary academician, François-César Le Tellier, the marquis de Courtanvaux (1718-1781) proposed this expedition. He had a frigate built especially for it at his own expense. Aboard were Courtanvaux himself, the astronomer and free associate of the Academy A.-G. Pingré, the naval astronomer C.-J. Messier, and the artist Ozanne – the latter two detached from the Dépôt des Cartes et Plans for the purposes of the trip. Sailing from Le Havre along the coasts of France and the Low Countries to Amsterdam and back, Berthoud's clocks performed well, but the trip was not long enough nor outside temperate waters to be thought a real test.[343] A second, more substantial trial followed in 1768 on the *Enjouée*, captained by M. de Trongoly, with J.-D. Cassini IV and the royal clockmaker Le Roy sailing from Le Havre to Saint Pierre and Miquelon in the North Atlantic and then south to

341. "Si nous avons été primés par les Anglais dans l'essai des horloges marines, je crois, Monseigneur, que le succes de celles que le S.r Berthoud construit par vôtre ordre, et qui sont prêtes à êtres livrées, ne le cèdera en rien à celui de Mr. Harrison et sa montre angloise. je ne crains pas même d'avancer que la construction des horloges françoises est de beaucoup supérieur à celle de l'artiste de Londres... Si le succès repond à nos espérances (et j'ose dire qu'elles sont bien fondées) la France joüira plutôt, et à beaucoup moins de frais que l'Angleterre, du fruit de cette importance découverte...," ALS Fleurieu to "Monseigneur," dated "Paris le 24 8bre 1767," AN-MAR G 98, dossier 1, fol. 75.

342. See AN-MAR G 98, dossier 2 ("Epreuves des montres marines du Sr. Le Roy") and dossier 3 ("Nouvelle épreuve dans les mers d'Europe, d'Afrique et d'Amérique des montres des S.rs Berthoud et Le Roy"); see also AN-MAR G 234, #56, #57; AN-MAR G 235 passim; McClellan (1992), pp. 122-23.

343. See Chassagne, pp. 183-86; Courtanvaux (1768); and Courtanvaux memoir, "Vérification de quelques instruments destinés à la détermination des longitudes sur mer," AN-MAR G 98, dossier 2, fols, 87-98; copy AN O[1] 597: #30. See also anonymous memoir, "Projet de Campagne pour la frégate destinée à l'épreuve des differens moyens de déterminer les Longitudes à la mer," AN-MAR G 235, #39.

the Canary Islands before returning to Le Havre. The ship was a veritable floating laboratory, the objective being to test a variety of clocks and instruments, including the clocks of Berthoud's rival, Le Roy. (Le Roy's clocks performed poorly.[344]) A third trial ensued immediately thereafter in 1768-69 and was designed explicitly to test Berthoud's chronometers against Le Roy's. Accompanied by the academician Pingré, the count de Fleurieu captained the *Isis*, sailing to Saint Domingue and back.[345] Prior to the voyage Fleurieu studied clockmaking with Berthoud, and Berthoud's chronometers seem to have proved superior.[346] A fourth voyage followed in 1771 with Pingré and Borda repeating the voyage to Saint Domingue aboard the *Flore*.[347] The 1771 voyage of Kerguelen to the Southern Ocean might be considered in this light, too, as he and Rochon took Berthoud's clock #6 along with them; Kerguelen's 1772 voyage took Berthoud's clock #8.[348]

The Académie des Sciences and the Marine Academy at Brest were both heavily involved in all of these trials. The two academies planned, supervised, and provided the personnel for the expeditions. The Marine Academy worked closely with Berthoud.[349] The navy and the Marine Academy distributed to competent officers printed instructions on how to determine longitude at sea.[350] Reports were regularly highlighted at the twice-annual public meetings of the Academy in Paris in this period.[351] The story of French chronometers is a triumphant chapter in the history of the Colonial Machine.

The results of the chronometer tests of the 1760s and 1770s demonstrated the perfection of the instrument for determining longitude at sea, and henceforth French navigators regularly employed such clocks in sailing back and forth to the colonies. The end of the longitude story, at least as far as the French were concerned, is already plain in Pingré's paper that appeared in the *Mémoires* of the Academy for 1770: "Précis of a Voyage to America or Geographical Essay on the Position of Several Islands and Other Places in the Atlantic Ocean, Accompanied by a few Observations concerning Navigation."[352] Marking this same transition,

344. See Cassini's "Journal de l'épreuve des montres marines de Le Roy en 1768," OdP, C 5 25-26; see also relevant materials, AN-MAR G 98.

345. "Rapport de Joseph-Bernard, marquis de Chabert, et de Bezout sur le voyage de Fleurieu à bord de l'*Isis* et l'épreuve des horloges de Berthoud," SHMV AM Ms. 73, pp. 166-77. See also Chapuis, pp. 74-82; Doublet (1910b); Taillemite (2002), pp. 185-86.

346. Regarding Fleurieu's study of clock making, see various references, AN-Marine G 98, dossier 1.

347. Taillemite (2008), p. 178.

348. Brossard.

349. SHMV, Ms. 064, vol. I ("Registres des procès-verbaux"), fols. 133-50.

350. See Duhamel du Monceau letter, dated "À Paris ce 5 Avril 1770," SHMV, Ms. 89.

351. See for example, "Opèrations faites sur Mer, Pour la détermination des Longitudes et autres objets concernant la Navigation. Par MM. de Verdun, Chevalier de Borda et Pingré. Lu à la rentrée de l'Académie le 21 avril 1773," AN-Marine G 98, dossier 3, fols. 205-215.

352. Pingré (1770).

the marquis de Chabert's "Memoir on the Usage of Maritime Chronometers relative to Navigation and Especially to Geography to Determine the Difference in Longitude of Some Points of the Antillean Islands and the Coasts of North America with Fort-Royal in Martinique or with Cap-François in Saint Domingue," was read at the Easter public meeting of the Academy in 1783 and appeared in its *Mémoires* for that year.[353]

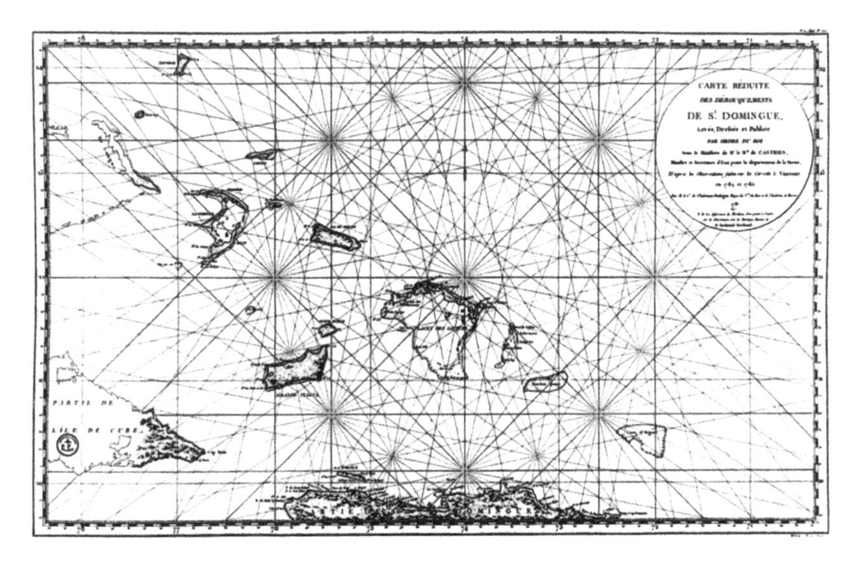

[Printed Map, "Débouquements de Saint-Domingue"
by Chastenet de Puységur (1787)]

The wording in the title of Chabert's paper ("...Especially to Geography ...") makes explicit that the status of the chronometer changed in the 1780s. The chronometer became a precision instrument, not just for determining longitude at sea but for cartography, because tiny longitudinal distances could now be measured using the chronometer with the results integrated into much higher quality maps. Already with Pingré's second voyage to the West Indies in 1771, the chronometer came into use as a tool for mapmaking.[354] And so, the history of the problem of determining longitude at sea comes full circle with the history of eighteenth-century cartography. Given the perceived inaccuracies of existing maps and with this new technology, in 1784-85 French authorities sent A.-H.-A. de Chastenet, the count Puységur, to Saint Domingue to draw up a new set of maps of the colony

353. Chabert (1783).

354. Navy minister Sartines was explicit about this new use of chronometers in his letter dated "Versailles le 14 mars 1775," SHMV, Ms. 88.

and its waters using these advanced chronometric techniques.[355] Assisted by an astronomer-chaplain, Dom Nouet, detached from the Observatoire, Puységur spent fourteen months in and around Saint Domingue taking observations and recording data.[356] He returned to France in 1785, and in 1787 the government published his *Le pilote de l'Isle Saint-Domingue* and his *Détail sur la navigation aux côtes de Saint-Domingue*, which were highly praised in reports by Fleurieu and Borda in 1787.[357] Adolphe Cabon, who has written on this matter, compares Puységur's work with the extensive cartographical efforts of the U.S. Marines after they invaded Haiti in 1915. Cabon says the Marines had better instruments, spent more money, and took longer (five years), but produced no better maps of Haiti's coasts than did Puységur in the eighteenth century.[358] All in all, a century of scientific investments to master longitude and cartography paid off handsomely for the French.

* * *

Thanks to the Colonial Machine, from the 1660s and the era of Colbert and Louis XIV to the 1780s and the French Revolution, the French achieved impressive intellectual and technical mastery of the seas and unparalleled access to the world abroad. At the outset of the period, the French sailed more or less blindly to fragile outposts overseas; by the end they had precision instruments and exquisite maps that reliably told them where they were at sea and how to navigate safely back and forth from home ports to the colonies. They knew the world better, and thanks to the research efforts undertaken by arms of the Colonial Machine, the French had better ships with better provisions and had developed better processes for sailing their great tall ships. The successful expansion of French colonies up to 1763 and the extraordinary maturation of the remaining ones after that date depended on these technical and intellectual triumphs.

The importance of the Colonial Machine as the state agency facilitating the conquest of space was not limited to improving navigational charts or solving the problem of longitude at sea, however. A related domain concerned local space and, as we have seen, success came here, too, in the form of cadres of government experts and specialist agents of the Colonial Machine implanted on the ground in the colonies themselves. These men gained intellectual and technical control over local colonial space, and so gave the colonizing metropolis yet additional powers for colonial conquest.

355. Regourd (2000b); Chapuis, pp. 228-29; McClellan (1992), pp. 124-26; see also archival materials, SHMV, Ms. 074, vol. XI; AN-MAR C7 61 (Chastenet de Puységur).

356. On Dom Nouet, see AN-MAR C7 228 (Nouet); Bret (1993).

357. SHM AM Ms. 74: Comptes-rendus des séances de l'Académie (tome XI), pp. 443-46.

358. Cabon, pp. 65-68; McClellan (1992), p. 126.

But contemporary French colonization faced issues that extended well beyond these fundamental necessities. Matters medical likewise loomed large when it came to preserving the health of sailors at sea, say, or the health of colonists and slaves in far-off and medically challenging environments beyond the shores of the metropolis. In this connection, the Colonial Machine proved no less an essential element to the success of the burgeoning colonial enterprise of the French.

PART II

THE COLONIAL MACHINE IN ACTION
B – LIVING AND DYING IN THE COLONIES

If the French encounter with the world overseas during the Old Regime depended on the work of French astronomers and cartographers, the point applies with equal force to French physicians and surgeons and to the structures of French medicine that underpinned contemporary colonization and long-distance sailing.[1] The centrality of matters medical to French expansion in the late seventeenth and eighteenth centuries cannot be overemphasized. Medicine was a direct instrument of colonization and overseas expansion. A prime concern was the health of crews aboard ships of the Marine Royale and the Compagnie des Indes, not to mention slavers and their notoriously precious cargoes. Another concern was the well-being of colonists adjusting to foreign climes, where three out of ten French immigrants to the colonies were not expected to survive and where one out of nine slaves perished annually.[2] Disease and unhealthy conditions threatened the French overseas enterprise at every turn, and it is no surprise that the Colonial Machine developed substantial medical structures and devoted substantial energy to combating disease and researching cures. To this extent, the story of the Colonial Machine is as much the story of medicine as it is of natural science. Put another way, science piggybacked upon colonial medicine. Colonial authorities enlisted experts of all sorts, and, as it happened, the medical and the scientific converged in the service of colonial France. In this section of Part II we examine the medical component of the Colonial Machine and the broad front along which the Colonial Machine operated to promote the health and welfare of colonists and sailors. In so doing, a number of themes reassert themselves. The institutionalized and

1. On French naval and colonial medicine, see Pluchon (1985); Homer; McClellan (1992), chapt. 9; Brau; and Hannaway; Quindlan (2005) provides a overview of contemporary colonial-medical thought.

2. At least as applies to immigrants and slaves in the Antilles; see remark, SRdM-H&M, vol. 10 (1798), pp. 463-64; McClellan (1992), p. 52.

state-sponsored character of the Colonial Machine becomes more manifest. The relentless pressure for useful outcomes is once again plain – pressure that drove the research engine of the Colonial Machine in medicine. A new note, or at least one not sounded so strongly up to this point, is that colonial administrators and physicians sought out native medicines and medical practices and so looked across the social-cultural divides separating themselves from the world of the "other."

Colonial Medical Structures Revisited

We have already encountered the highly developed character of organized medicine in place in France to address issues surrounding colonial medicine. The Royal Navy led the way in this regard with its cadres of naval physicians and surgeons maintained at crown expense; with navy hospitals at Brest, Rochefort, and Toulon; and with the three navy medical schools associated with these hospitals.[3] The creation of the post of director general and inspector of naval and colonial medicine in 1764, held by the redoubtable Pierre-Issac Poissonnier, was a milestone in the institutionalization of naval and colonial medicine. Poissonnier's subsequent establishment of the position of inspector of naval and colonial hospitals for his brother in 1775 reinforced this bureaucratizing tendency. And then, the particularly colonial orientation of naval medicine saw its apogee in 1783 with the creation of the École Pratique de Médecine, the school explicitly for postgraduate training of physicians for colonial service.

The Compagnie des Indes is not to be overlooked in this connection, having its own substantial staff of doctors, surgeons, and apothecaries and maintaining its own large hospital at Lorient. Then, from 1778 on the Société Royale de Médecine in Paris was another major element of contemporary organized medicine and a key player in the story of colonial medicine. Much of what follows in this Part deals with the Société Royale de Médecine as the scientific center for promoting and coordinating research in colonial medicine. The Académie de Chirurgie had its colonial facet, and we should not leave out royal court physicians who often served as intermediaries at the Société de Médecine or at the Académie des Sciences. Then, regarding medicaments and drugs in particular, at a certain point the medical and the botanical merged within the Colonial Machine, and the infrastructures of French botany, such as the Jardin du Roi, also come into play in considering colonial medicine.

3. On naval medical personnel, see Pluchon (1985) and AN-MAR G 102 ("Personnel médical, 1678-1788"). See also SHMB, Série L-1L ("Règlements, ordonnances, édits...enregistrés par le contrôleur de la Marine, 1677-1785"), where, for example, at 1L8/207 r° one finds a list of navy of medical staff for the port of Brest in 1767.

Royal Physicians in the Colonies and Their Duties. Royal medical facilities and royal medical personnel installed in the colonies constituted an extension of this core of men and institutions in metropolitan France, ready and able to attack problems of colonial medicine. Médecins du Roi, royal doctors stationed in the colonies, have already made a scattered appearance in this book, but their role as institutionalized agents of the Colonial Machine needs to be brought into sharper focus. These were highly regulated, official positions, to be distinguished from those of royal physicians and surgeons in the Marine Royale.[4] Their holders, supported by the state, ranked among the privileged *entretenus* whose daily bread was assured. They were the chief medical officers in colonial hospitals and in various military and civilian medical facilities, performing a complicated set of medical, administrative, and legal functions. Many of the médecins du roi stationed in the colonies became correspondents of the Académie Royale des Sciences or the Société Royale de Médecine. While, on the one hand, médecins du roi organized and supervised the delivery of medical services on the ground in the colonies, they also functioned as expert agents for the Colonial Machine in reporting back to the metropolis about a wide range of scientific subjects.

Two royal doctors in French Canada can serve as examples of the role and extensive functions sometimes played by médecins du roi. Michel Sarrazin (1659-1734) (whose name was also spelled as Sarrasin) arrived in Canada in 1685 as a naval surgeon.[5] By 1691 he had risen through the ranks to become surgeon major. He then went back to France to get an M.D. from the medical faculty at Reims, during which time he studied botany with Tournefort. He returned to Canada in 1697, and in 1699 the Académie des Sciences elected him one of its first correspondents, attached to Tournefort. In Canada, Sarrazin was médecin du roi and from 1707 *Médecin des hôpitaux*, a post ultimately paying only a modest 800 livres. Beginning in 1717 he sat on Quebec's governing *Conseil Souverain.*[6]

Sarrazin conscientiously treated the ill – military, civil, and Native American – in Canada for nearly five decades, becoming sick many times himself; but he came to be valued more as a scientific/botanical agent than for his work as a physician and medical administrator. Already in 1698 the minister wrote to the intendant in Canada that "Mr. Sarrazin is charged to collect in Canada the plants, fruits and other things particular to the country that can be useful in

4. See, for example,"Réglement du roi pour l'instruction des médecins destinés à servir dans les hôpitaux militaires de la marine tant en Europe que dans les colonies," SHMB, Série L-1L ("Règlements, ordonnances, édits... enregistrés par le contrôleur de la Marine, 1677-1785"), 1L10/ 128 r° (1783); on official and unofficial medical practitioners in France Canada, see Gelfand (1987), pp. 79-90.

5. On Sarrazin, see Vallée; Rousseau (1957), pp. 152-55; Rousseau (1969); Lortie, pp. 9-11; Théodores (1959); Pighetti, pp. 248-49; Litalien; Kernéis et al., p. 65; Gasc and Laissus, p. 26; Allorge, pp. 162-64; Lacroix (1938), vol. 4, pp. 113-16; PA-DB (Sarrazin).

6. On these points, see previous note and relevant ministerial correspondence, AN-COL B^{16} (1691), fol. 33 (16 March 1691), B^{22} fol. 90v (23 April 1700), B^{29} fol. 118 (30 June 1707).

the royal gardens. You are to see that the cases and boxes of Mr. Sarrasin are embarked on vessels of the royal navy."[7] Commanders at the various French forts scattered about North America were likewise ordered to send seeds and plants to Sarrazin.[8] Sarrazin similarly relied on government forestry agents to bring back interesting samples.[9] He was a protégé of the abbé Bignon, and in 1717 Bignon persuaded the *Conseil de Marine* to award Sarrazin 500 livres for his "zeal in his research on plants."[10] In 1724, this supreme royal naval council again rewarded Sarrazin for "his application not only to his functions but more for his important scientific observations."[11] It did so to the same effect again in 1728.[12] Sarrazin was in contact with Réaumur, becoming Réaumur's official correspondent at the Academy of Sciences after Tournefort's death.[13] Sarrazin became connected to several botanists at the Jardin du Roi, notably Sébastien Vaillant (1669-1722) to whom for over twenty years he sent botanical specimens that formed a significant part of Vaillant's and later the Jardin's herbarium. Based on Sarrazin's notes, in 1706 Vaillant compiled a 200-page manuscript of Canadian plants, and Antoine de Jussieu prepared yet another.[14] Among the more notable botanical items Sarrazin sent to Vaillant and the Jardin du Roi in 1717 was Canadian ginseng.[15] Sarrazin communicated with François Chicoyneau (1672-1752), chief royal physician, over mineral waters.[16] Through Tournefort and Réaumur, he submitted material to the Académie des Sciences, and he came to specialize in anatomical and what we would call zoological work, publishing substantial and well-illustrated papers in the Academy's *Histoire et Mémoires* on the water rat, porcupines, beavers, seals, and the wolverine.[17]

7. Quoted in Rousseau (1969), p. 622.

8. AN-COL B[48] (1725), fols. 879-80 (5 June 1725).

9. See ALS Sarrazin [to Réaumur?], dated "a quebec ce 15 8bre 1722," PA-DB (Sarrazin).

10. Per Taillemite (1969) and AN-COL B[39] (1717), fol. 112 (30 January 1717); see also B[41] fol. 30v (8 February 1719), and ALS Sarrazin, dated "a quebec ce 15 8bre 1722," PA-DB (Sarrazin).

11. Per Taillemite (1969) and AN-COL B[47] (1724)(II), fol. 1114 (30 May 1724); see also other Sarrazin letters in this series.

12. AN-COL B[52] (1728), fol. 542 (24 May 1728).

13. AN-COL B[51] (1728), fol. 481v (4 May 1728).

14. "Plantes envoyees de Canada par M.r Sarrazin Conseiller du Conseil supreme et Medecin du Roy en Canada," BCMNHN, Ms. 944. See also ms. by B. Boivin, "Notice biographique sur J. F. Gaultier (1708-1756), Naturaliste-médecin du Roi au Canada," p. 15 in PA-DB (Gaultier). D'Anty Isnard wrote that Vaillant made a point of adding the herbaria of Sarrazin, Plumier, and Tournefort to his own botanical researches; see ALS, dated "a Paris le 11 Avril 1722," RS/L, Miscellaneous manuscripts, Sh. 244-275.

15. See ALS to Bignon, dated "a quebec le 5 9bre 1717," in PA-DB (Sarrazin); Rousseau, p. 623. On reports of Canadian ginseng in the Academy, see also PA-PV 36 (1717), fols. 309v-311v ("Le Samedi 11 Décembre 1717).

16. AN-COL B[58] (1733), fol. 431 (14 April 1733) and fol. 470 (12 May 1733).

17. See bibliography, references in Halleux et al., and HARS, 1713, p. 12-14; HARS 1718, p. 32; PA-PV 39 (1720), fol. 28 ("Le Samedi 27 Janvier 1720"); 49 (1730), fol. 199 ("Le Samedi 19.e Aoust 1730").

Sarrazin's successor, after his death in 1734, was Jean-François Gaultier (1708-1756).[18] Actually, the position of médecin du roi in Canada was held open for Sarrazin's son, then studying medicine in France, but the younger Sarrazin himself died in 1739, so Gaultier was appointed médecin du roi in Quebec in 1741 at 800 livres and an occasional 500 livres gratuity. From then until his own death in 1756, Gaultier was active scientifically and a part of the Colonial Machine functioning in Canada. He performed the duties of médecin du roi, which included vetting new surgeons in the colony. Almost ex officio, he became a correspondent of the Paris Academy in 1745; he was also attached to and dealt closely with Duhamel du Monceau, who received a 400-page manuscript from Gaultier on American trees and plants.[19] Zoological specimens went to Réaumur, to whom Gaultier sent fishes, a caribou pickled in a barrel, and a porcupine.[20] Gaultier's botanical shipments made it to the Jardin du Roi, to and through Bernard de Jussieu. He also kept substantial meteorological and nosological logs in the 1740s, which Duhamel published in the *Mémoires* of the Academy.[21] Additional Gaultier work went to Duhamel, who published it in some of his volumes.[22] Gaultier himself published on the sugar maple in the Academy's *Savants Étrangers*.[23] He worked closely with La Galissonière when the latter was governor in Canada and prepared instructions for collecting specimens that were distributed throughout French Canada. He contributed to Guettard's mineralogical work, sending him at least four shipments of mineralogical samples.[24] He, too, sat on the governing Conseil of Quebec, and he was in contact with Delisle, promising annual botanical-meteorological reports.[25] Gaultier

18. On Gaultier (also Gaulthier), see Lamontagne (1964b); Wien; Vallée (1930); Pighetti, p. 167; Lortie, pp. 13-14; Rousseau (1957), pp. 155-57; Bonnault, pp. 172-73; see also PA-DB (Gaultier), which contains the ms. by B. Boivin, "Notice biographique sur J. F. Gaultier (1708-1756), Naturaliste-médecin du Roi au Canada."

19. BnF, N.A.F. 3345, "État de la dépense à faire pour la gravure et l'impression des six volumes de l'Histoire des Plantes de l'Amérique." See also letter, B. Boivin to Murphy Smith, dated "Ottawa, Ontario, January 23, 1973," APS manuscripts, Duhamel du Monceau, Table of Contents Folder.

20. On these points, see various letters from Gaultier to Réaumur, PA-DB (Gaultier), e.g., letter dated "le 30 7bre 1750 a quebec."

21. See Slonosky, and "Observations Botanico-Metheorologique faittes en Canada par M. Gautier," SHMV, Ms. 499; "Journal des observations Meteorologiques, Etc., de Mr. Gauthier à Kebec depuis le 1 octobre 1747 jusqu'au 1 octobre 1748," OdP, Ms. A 7, 6 [64^3]; "Documents botaniques et météorologiques (Canada)," APS-MS, Duhamel de Monceau-Fougeroux de Bondaroy Collection of Botanical Manuscripts, B/D87, Group 25; see also MARS 1744, pp. 135-55; 1745, pp. 194-229; 1746, pp. 88-97; 1747, pp. 466-488; 1750, pp. 309-10.

22. APS-MS, Duhamel de Monceau-Fougeroux de Bondaroy collection of manuscripts, B/D87 Group 20.

23. "Histoire du Sucre d'Erable," SE, vol. 2 (1750), pp. 378-92.

24. Lamontagne (1965); ms. by B. Boivin, "Notice biographique sur J. F. Gaultier 1708-1756), Naturaliste-médecin du Roi au Canada," p. 22 in PA-DB (Gaultier).

25. AN-COL B^{81} (1745), fol. 60 (5 May 1745); "Journal des observations Meteorologiques, Etc., de Mr. Gauthier à Kebec depuis le 1 octobre 1747 jusqu'au 1 octobre 1748," fol. 1r, OdP, Ms. A 7, 6 [64^3].

escorted the traveling Swedish naturalist, Pehr Kalm, around Quebec for two months in 1749. Authorities named one Lebeau to succeed Gaultier in 1756 as médecin du roi at Quebec and charged him to continue this kind of work, but the fall of Canada cut off that story.

In Louisiana, Louis Prat (M.D., Montpellier, 1719) served as médecin du roi from 1725 with a salary of 2000 livres. He was replaced by his brother, Jean Prat (M.D., Montpellier, 1731), from 1735 to 1746.[26] In the Caribbean, Marc de Vaux de La Martinière was the initial médecin du roi on Martinique, appointed in 1691 with a stipend of 1500 livres.[27] He was elected in the first group of correspondents of the Academy of Sciences in 1699, attached to Lemery. De Vaux de La Martinière established administrative control over medical practice in the Martinique colony and ruled for twenty-five years. He was an acquaintance of Plumier, knew and met Labat, and was in contact with Fagon at the Jardin du Roi. He died in 1716 on Martinique, replaced by François-Antoine Le Dran (1690-1724), who was médecin du roi from 1716 to 1721 and who was likewise named a correspondent of the Academy in Paris in 1717.[28] Le Dran was succeeded by Jean-Baptiste Fauste Alliot de Mussay, who in turn became a correspondent of the science academy, leaving for Martinique in 1721 and dying in Saint Domingue in 1730.[29] J.-B.-René Pouppé-Desportes (1704-1748) was médecin du roi for a long while in Cap-François in Saint Domingue and Academy correspondent, first of Dufay in 1738 and then of Bernard de Jussieu in 1745.[30] Pouppé-Desportes sent materials to Dufay back at the Jardin du Roi through official channels.[31] We have already encountered Charles Arthaud as a major correspondent of the Société Royale de Médecine, and he was also the first royal doctor in Cap François. In Cayenne, Raymond Laborde was chief médecin du roi from 1770 through late 1786 and evidently was a correspondent of the Société Royale de Médecine.

Regulations spelled out the functions of the chief royal doctor.[32] On Martinique he had to visit the wards of the facility in Saint Pierre and superintend medical treatment of the state's sick. Part of his duties after 1770 included reporting to P.-I. Poissonnier on the illnesses treated and "discoveries to be had in medicine, pharmacy and botany, as well as in things relative to natural his-

26. Leroy; Langlois (2003), pp. 344-45; Lamontagne (1962b).

27. See materials PA-DB (La Martinière); Lacroix (1938), vol. 3, pp. 23-24.

28. See materials PA-DB (Le Dran); Lacroix (1938), vol. 3, p. 24.

29. Lacroix (1938), vol. 3, p. 24.

30. On Pouppé-Desportes see Lacroix (1938), vol. 3, pp. 52-53, vol. 4, p. 138; and correspondence in PA-DB (Pouppé-Desportes).

31. ALS, Dufay to Pouppé-Desportes, dated "a Paris ce 20e 9bre 1738," PA-DB (Pouppé-Desportes).

32. For the Martinique case, see "Règlement du Roi, concernant l'Administration générale de la Colonie de la Martinique, du 24 mars 1763," in Durand-Molard, vol. 2, pp. 169-173, and registration of the rules concerning the médecin du roi and the chirurgien-major by the Conseil Souverain of Martinique, AD/Martinique, Série B, registre 9, fol. 229.

tory."[33] For appearances in court, or opening cadavers, or while traveling on official business, chief royal doctors in Martinique received twenty-four livres a day; chief royal surgeons got eighteen.[34]

Colonial Hospitals. Colonial hospitals arose all over the colonies. In Canada, the hospital in Montreal was founded in 1688, the one in Quebec in 1692, and in Louisbourg in 1722. Hospices, hôtels-dieu, were also be to be found in Quebec (1639), Montreal (1642), and Trois-Rivières (1694).[35] The government hospital in Saint-Pierre, Martinique, added weight to the Colonial Machine. Cayenne had its well-staffed government hospital.[36] On the islands of Île de France and Île Bourbon, the government took over hospitals from the Compagnie des Indes.[37] The one on Île de France had beds for 400 to 500 patients.[38]An entire system of official colonial hospitals emerged in Saint Domingue. The centerpiece turned out to be the secular, government-run establishment in Port-au-Prince that dated only from 1751 but that had a large medical staff and treated thousands of patients a year.[39] The charity hospital in Cap-François, founded in 1698 and chartered in 1719, was located just outside of town, with 800 to a 1000 beds in several pavilions for different classes of patients. Superintended by the médecin du roi at Cap-François, the medical staff residing at the Cap hospital included physicians, surgeons, and pharmacists. The Brothers of Charity, some with medical training, otherwise ran the establishment. The hospital in Léogane received letters patent in 1719, but remained a small hospital of only fifty beds. In 1766 another small hospital arose in Les Cayes in Saint Domingue. A hospice for Native Americans operated in Montréal.[40] A separate hospital for women arose in Martinique.[41] Among the more interesting colonial hospitals in the French colonies in the

33. "des découvertes qu'il pourra faire tant dans la médecine la pharmacie et la botanique que dans les objets Relatifs à l'histoire naturelle," AD/Martinique, Série B, Registre 13, fols. 23-23v. This act registered the *brevet* of Sr. Laguarigue as médecin du roi, dated "Versailles, 2 juillet 1771."

34. "Ordonnance de MM les Général et Intendant portant Règlement et Tarif général de tous les Émoluments attribués à divers fonctionnaires publics. Du 30 avril 1771," in Durand-Molard, vol. 3, pp. 38-67.

35. Gelfand (1987), pp. 85-88; Havard and Vidal, pp. 181-82; "Lettres patentes du roi portant établissement d'un hôpital général à Québec," AN-COL F^3 vol. 7, fol. 93 (March, 1692); see also AN-COL B^6 (1675), fol. 93 (26 March 1675) ordering payment of 2000 livres for medicaments and the upkeep of the hospital at Quebec.

36. Ms. Code de Cayenne, vol. 1 (1579-1711), pp. 291, 379-382; vol. 3 (1740-1760), p. 15; vol. 4 (1761-1767), p. 173; vol. 5 (1768-1777), pp. 123-4, 854, in AN-MAR G 237.

37. Ms. Code de l'Ile de France, vol. 1, p. 215, in AN-MAR G 237.

38. See Ms. "Table Geographique et historique des Indes," under entry "France (Les isles de)...," BCMNHN, Ms. 1765, "Relation des Indes orientales."

39. See McClellan (1992), pp. 91-94 for these details.

40. "Articles accordés entre Mgr. L'Abbé de [Queylus?] et les R.R. Mères hospitalières de Québec pour la fondation d'un hopital des sauvages invalides à Montréal," AN-COL F^3 3, p. 97.

41. AD/Martinique, Série B, Registre 6, fol. 138v; Durand-Molard, vol. 1, pp. 538-42: Regulations dated 3 mars 1750.

eighteenth century was the one in New Orleans, run first for the Compagnie des Indes and then for the government by the Ursuline order of nuns from 1727.[42] All of these hospitals owned slaves and were maintained by slave labor, but the New Orleans hospital seems unusual in reserving a position, albeit a lowly one, for a slave pharmacist; in 1744 this was a certain Jean-Baptiste, aged about thirty.[43] Another government medical facility operated in Louisiana on nearby Île Dauphine.[44]

Leper colonies constitute another medical institution of note. Authorities established a leper colony at Île Mère in Cayenne in 1777 and forced the relocation there of lepers and victims of elephantiasis.[45] Much earlier in the century – at least from the 1720s – authorities instituted a similar colony on the island of La Désirade off Guadeloupe to extirpate what was called "this public Cancer."[46] The royal medical staff on Guadeloupe was charged with containing leprosy and had police powers to force lepers onto the island. A special tax was assessed on slave owners to cover the costs of medical visitations by the royal doctor and surgeon to identify cases of leprosy in the colony.[47] In 1728 the médecin du roi, André Peyssonnel, now posted to Guadeloupe, identified 187 leprous slaves, 117 of them already in the leper colony at La Désirade.[48] Peyssonnel had misgivings but sent 60 lepers to La Désirade in 1728.[49] On this or another visit 22 whites, 6 mixed-race individuals, and 97 slaves were declared contaminated with leprosy and immediately packed off to La Désirade.[50] Slaves apparently escaped the island and returned to their masters, so legislation was introduced to the effect that lepers could be shot by anyone if found off La Désirade, and boat captains were to be shot if they transported lepers off the

42. AN-COL C[13A] 28 (1743-1744), fols. 342-45 (31 December 1744); see also Clark, passim and pp. 43-46, 201-205; Langlois (2003), p. 347, notes a new hospital and pharmacy built in New Orleans in 1733-1734.

43. AN-COL C[13A] 28 (1743-1744) fol. 343v (31 December 1744), Article 11.e: "Sa Majesté laissera aussy a lhopital le Negre nommé Baptiste qui Continuera de Travailler a la Pharmacie, et luy fera delivrer de Ses magasins pour Sa Subsistance, une ration en pain et lard seulement, Sur le même pied quelle se délivre aux Troupes Et aux autres Rationnaires, et au cas de mort dud. Baptiste Sa Majesté le remplacera par un autre Negre Capable ou par un garcon apoticaire pour travailler a la Pharmacie." See also fol. 346 and ms, "Negres Negresses Negrillons et Negrittes appartenant au Roy attachez au service de Lhopital," "Savoir: Jean Baptiste Apoticaire agé denviron trent ans."

44. AN-COL C[13A] 5 (1717-1720), fol. 341 (18 October 1719).

45. See ms. Code de Cayenne, vol. 7, 1783-1803, pp. 71, 161, 189, 357, 363, 549, 687; also ms. Cayenne, Annales du Conseil 1713-1780, pp. 660, 857, both in AN-MAR G 237.

46. "...l'extirpation de ce Cancer public," André Peyssonnel, "Procés Verbal et Resultat de la Seconde Visite des l'épreux fait a la Guadeloupe en 1748," RS/L, L&P III, #214b.

47. André Peyssonnel, "An Account of a Visitation of the leprous Persons in the Isle of Guadeloupe," RS/L, L&P III, #214, pp. 41-42; see also Yaqubi, p. 333.

48. Ms. "Traité de la Lepre à la Guadeloupe par M. Peyssonel," with ministerial date of "25 May 1728," AN-MAR G 102, dossier 3, #102. See same dossier for related material on this subject.

49. AN-MAR G 102, dossier 3, #116.

50. Durand-Molard, vol. 1, p. 314.

island. Authorities paid for six months of upkeep for new arrivals, and they imported a few sheep and cows for their sustenance. No exports were allowed off the island, and the quarantined leper colony otherwise managed itself.[51] Authorities in Martinique and Saint Lucia, too, conducted systematic searches for lepers, and they also sent their lepers to La Désirade. In 1788 the leper colony there consisted of 44 blacks and a few children with 95 more lepers slated for deportation to La Désirade.[52]

Surgeons, Midwives, and Others. Alongside the first médecin du roi in these various locales stood a chief royal surgeon, the chirurgien du roi. These two government medical officials generally controlled private medical practice in the colonies, notably in licensing private physicians and especially surgeons, who had to report to the médecin du roi every six months.[53] The private practice of medicine and surgery in the colonies never amounted to that much, but it did reach an apogee late in Saint Domingue's history, where the guild structure of the motherland replicated itself in the colonies; this structure was poles apart from the officialdom of the Colonial Machine represented by the médecin du roi and the chirurgien du roi. These latter officials were sometimes complemented by a second royal doctor or second royal surgeon, as well as by other medical specialists working in government service, notably royal pharmacists, veterinarians, and midwives.[54] Thomas-Luc-Augustin Hapel Lachênaie (1760-1808) is the paradigmatic royal veterinarian. Lachênaie was veterinarian and chief pharmacist at the naval hospital at Guadeloupe. He was trained in the national veterinary school at Alfort, and was a correspondent of the Société Royale de Médecine, to which he submitted memoirs. He worked on epizootics, illnesses of slaves, and sugar production, and he set up a chemical lab to analyze drugs.[55] Jean Gelin, also a correspondent of the Société Royale de Médecine, played a similar role as royal veterinarian on Saint Domingue.[56]

The government developed and supported a cadre of midwives in the colonies. In 1722 Madelaine Bouchette, 39, received 400 livres a year as midwife in Quebec; later Mme Braquenard, the royal midwife at Quebec, petitioned the government to create two schools to spread midwifery through the territory occupied by the colony.[57] Marie Grisot was the royal midwife in Louisiana in

51. On these points see Durand-Molard, vol. 1, pp. 311-15.

52. "Arrêt du Conseil Souverain sur les mesures à prendre contre la maladie de la Lèpre. Du 10 novembre 1786," Durant-Molard, vol. 3, pp. 715-16; see also AN-MAR G 102, dossier 3, #120 and #121. For miscellaneous leprosy materials in naval archives, see further AN-MAR G 102 (dossier 3).

53. For the rules regarding private medical practice in the Indian Ocean colonies, see ms. Code de l'Ile de France, vol. 2, in AN-MAR G 237; Delaleu, vol. 1, pp. 336-39.

54. Homer, pp. 114-200 gives a general overview of medical personnel in Saint Domingue.

55. Demeulenaere-Douyère (2000a); Gerbaud, pp. 86-91; PA-DB (Hapel), and above Part I.

56. McClellan (1992), p. 234.

57. Gelfand (1987), p. 84; AN-COL B^{101} (1755), fol. 1 (2 March 1755); see also B^{89} fol. 24 (18 April 1749); Havard and Vidal, p. 182, provide other examples from Canada.

1709; Mme Faguier was the royal midwife in New Orleans in the 1750s.[58] The Compagnie des Indes paid for the training of "demoiselle Tugeret" as a midwife, then to be sent to Île de France in 1750.[59] In Saint Domingue, the colony's surgeons trained local midwives, and a board consisting of all physicians examined midwife candidates and licensed them. In 1764 the government sent an additional group of paid, licensed midwives from France to practice in the colony, but they apparently rose above their station and were suppressed in 1773. In that year a physician-obstetrician was appointed and given exclusive rights to examine midwives, and the position was upgraded in 1787. Courses on midwifery included one using a mechanical doll, and midwives advertised with the approbation of physicians. The law did not permit slaves and free people of color to act as midwives or medically in any fashion, but the Conseil Supérieur of Cap François made a legal exception for the experienced mulatto midwife in Cap François, the widow Cottin.[60]

Missionaries and others in various Catholic orders constituted a nontrivial element of the delivery of health care in the French colonies in the eighteenth century, but the aforementioned official, government agents of the Colonial Machine ran the show. Comparatively little medical science issued from the medical religious. From Pondicherry, the Jesuit Duchoisel submitted a paper on rabies to the Société Royale de Médecine.[61] Another medical report (this one about birth defects) filtered in from a second Jesuit in India.[62] From Guadeloupe, Father Valentin Delaporte of the Frères Precheurs also sent a piece on indigenous herbs to the Société Royale de Médecine, but it was rejected.[63]

Medical Literature. We might point out the not insubstantial medical literature of the day, much of it printed on royal presses, that dealt with colonial medical topics. Much of this literature has been mentioned in passing. Duhamel's piece, *Moyens de conserver la santé aux équipages des vaisseaux* (Means to Conserve the Health of Ships' Crews, 1759), is an early example.[64] Poissonnier-Desperrières's *Fièvres de St. Domingue* (Fevers of Saint Domingue, 1763) and his *Traité des maladies des gens de mer* (Treatise on the Illnesses of Seamen, 1767) appeared in new editions in 1780 at government expense – the former with an expanded section on the diseases of sailors.[65] Pouppé-Desportes' *Histoire des*

58. AN-CPL C^{13A} 39 (1755-1757), fol. 241 (28 July 1756).

59. SHML, 1P 305, Liasse 69. fol. 85v. (4 November 1750).

60. These points are lifted verbatim from McClellan (1992), pp. 134-35; see also Homer, pp. 127-28.

61. SRdM, Archives, 125 dr. 15, #30-#32.

62. SRdM, Archives, 182 dr. 4.

63. SRdM, Archives, 97 dr. 72.

64. See Duhamel de Monceau (1759) and 1786 review in *Collection Académique, Partie Française* (1786), pp. 84-85; see also Villiers.

65. See Bibliography and references, SRdM, Ms. 8, p. 202 ("Séance du Vendredi 15 7bre 1780") and SRdM, Archives, 201 dr. 9, report "delivré au louvre le 27 juin 1786."

Maladies de S. Domingue (History of Illnesses of Saint Domingue, 1770) was of a piece with this literature. One of the senior Poissonnier's agents, the physician Jean-Barthélemy Dazille (1733?-1812), was especially active and productive, publishing on illnesses affecting slaves (1776), on tropical diseases in general (1785), and on tetanus (1788). He compiled the most significant record of work on colonial and tropical medicine of anyone in the eighteenth century. Dazille was a regional associate of the Société Royale de Médecine, not a mere correspondent, and all of his works appeared with the approval and approbation of the Société Royale de Médecine, with the one on tropical diseases being distributed to members of the Société Royale.[66] On royal order, Jean-Jacques Paulet's two-volume *Recherches historiques & physiques sur les maladies épizootiques* (Historical and Physical Research on Epizootic Diseases) appeared in 1775 in response to epizootic outbreaks in the south of France and in Guadeloupe. Bertin, the médecin du roi on Guadeloupe, wrote a two-volume *Mémoire sur les maladies de la Guadeloupe* in 1778-1780 and *Des Moyens de conserver la santé des blancs et des nègres aux Antilles* (Means of Preserving the Health of Whites and Slaves in the Antilles) in 1785.[67] In 1787 J.-F. Lafosse, a physician in Saint Domingue, offered his medical handbook for colonists. These medical works merely add to the body of expert literature issuing from both private presses and those of the Colonial Machine, and it evidences the weight of the colonial medical establishment.

The Medical Research Engine

The Colonial Machine was a gigantic research engine that deployed experts to develop solutions to the problems faced in French colonial expansion in the Old Regime. This dimension of the Colonial Machine shows itself in many areas. In the previous Part, we encountered work in the broad subjects of cartography and navigation, as well as more narrow research into, for example, preserving fresh water at sea. In the following Part IIC, we will examine the substantial agronomic and botanical research undertaken by the Colonial Machine. Here, our focus is on medical research carried out under its auspices. Needless to point out, all of this work and research – the Colonial Machine as a research enterprise – was of a piece and facets of the Colonial Machine in single-minded action.

66. On Dazille and the Société Royale de Médecine, see SRdM, Ms. 10, p. 433 ("Séance du Mardi 24 may 1785") and Ms. 11, fol. 57r ("Séance du Mardi 7 Novembre 1786"), fol. 60r ("Séance du Vendredi 24 Novembre 1786"), fol. 104r ("Séance du Mardi 19 Juin 1787"), and fol. 108r ("Séance du Mardi 3 Juillet 1787"); see related materials, SRdM, Archives, 118 dr. 113, #7; 126 dr. 13, # 28. Of note is that Dazille submitted his book as a member of the Société Royale and de Castries submitted it as part of official channels through the Royal Navy.

67. See Bertin (1778-1780, and 1786); Quindlan, p. 108, and reference SRdM-H&M, vol. 5 (vol. for 1782/1783, published in 1785), Histoire, p. v; Guerra (1994) provides a complete bibliography for the West Indies. Romieux (2005), sketches books used in the naval medical service.

The Société Royale de Médecine stands out as a great cog in the Colonial Machine and central to the story of medical research focused on the colonies and the world overseas. The Société displayed great concern with the health of sailors, for example, making comparisons of crews that did and did not get sick. Poissonnier and the navy kept careful statistics on the navy's ill, using printed forms to collect data.[68] The Société paid attention to medical conditions in ports and, working through the ministry, regularly sent agents to ports to check on illnesses.[69] The Société had its own network of medical correspondents in the colonies, and its call for medical topologies stimulated responses from the colonies, as did its prize questions, such as the one on scurvy.[70] The Société Royale de Médecine renews our attention here as the center for a number of major medical research projects. We look at four episodes of medical research in the Colonial Machine – projects concerning elephantiasis, quinine, tetanus, and rations for sailors – and we conclude with an overview of some lesser but still relevant instances of medical research carried out under the aegis of the Colonial Machine. These cases are noteworthy for the history of colonial medicine, and they shed a further light on the functioning of the Colonial Machine as a research engine.

Mal Rouge. Mal rouge (the "red disease") was the name the French had for the parasitic worm disorder of elephantiasis, also known as Cayenne leprosy.[71] The administrators in Cayenne triggered the episode in question in 1784 by writing to the Minister of the Navy and the Colonies, de Castries. Mal Rouge "reigned" in Cayenne, and the administrators were worried that the disease might be contagious and, if so, whether to constrain the movements of affected colonists.[72] De Castries mobilized the Colonial Machine in response. In September 1784 he directed the query to the Société Royale de Médecine asking for its advice on ways to halt the effects of the contagion and cure those affected by it. De Castries also ordered a pamphlet to be prepared (a *mémoire instructif*) on the disease "to be made public throughout the colonies and in the

68. Printed forms in SRdM, Archives, 143 dr. 17 ("Administration medicale de la marine"); see also SRdM, PV passim, e.g., Ms. #7, p. 333 ("Séance du Mardi 19 8bre 1779") where sailors from Brest and the Charente are compared, and pp. 352-53 ("Séance du mardi 9 9bre 1779"). The Société argued that the navy should provide uniforms for sailors, like the army did.

69. E.g., the trip of De La Porte to Brest and La Rochelle in 1780, SRdM, Ms. 8, p. 38 ("Séance du Mardi premier Fevrier 1780") and p. 241 ("Séance du Mardi 31 8bre 1780"). See many references to these activities in SRdM, Ms. 7 and 8, passim.

70. On the Société's prizes and calls for topologies, see above Part I (Société Royale de Médecine), pp. 125ff; see also SRdM-H&M, vol. 3 (1779), Histoire, p. 5, for the scurvy prize, and on its medical topologies, see SRdM-H&M, vol. 10 (an VI, 1798), p. iv.

71. On this episode, see especially folders, SRdM, Archives, 160 dr. 1, #1-#6 ("Mal rouge ou éléphantiasis de Cayenne") and 191 d 31 (13 items on quinine). See also Société minutes for this period, SRdM Ms. #10, passim. Mosquitoes transmit the parasitic worms causing elephantiasis, a tropical skin and lymphatic disease causing swelling of the extremities.

72. These points are recapitulated in de Castries' letter to Vicq d'Azyr, dated "A Versailles le 17 mai 1787," SRdM, Archives, 191 dr. 31.

ports of France, where persons afflicted with mal rouge might debark every day."[73] The Société was happy to oblige the minister and immediately formed a six-man investigating committee headed by Poissonnier.[74] Two months later in November 1784, the committee submitted its report. The minister was delighted with this new proof of the zeal of the institution for the public good and quickly had the report printed on royal order.[75] The brochure, *Rapport des commissaires de la Société royale de médecine, sur le mal rouge de Cayenne, ou éléphantiasis* (Report of the Commissioners of the Royal Society of Medicine on the Red Disease of Cayenne or Elephantiasis) appeared in January of 1785.[76] De Castries then distributed it throughout the Colonial Machine. By September 1785 the administrators in Martinique acknowledged receipt of copies of the Société Royale's memoir on elephantiasis, "which we immediately distributed to all doctors and surgeons on this island" along with orders to keep track of the disease and to send reports back.[77]

On Martinique the médecin du roi, the chirurgien du roi, and the apothicaire du roi all responded to the governor and the intendant, who in turn forwarded their reports to the minister at Versailles.[78] The chirurgien du roi observed that "the disease rampant in Cayenne, at Pointe-à-Pitre on Guadeloupe, and at Marie-Galante is very rare on Martinique."[79] The count Du Puget d'Orval, then on his tour as inspector general of colonial artillery, regretted not receiv-

73. "…pour être rendu public dans toutes les Colonies, et dans les ports de france, ou il peut debarquer journellement des personnes attaquées du Mal Rouge," ALS, de Castries to Vicq d'Azir [sic], dated "Versailles le 18 7bre 1784," SRdM, Archives, 160 dr. 1 (Bajon).

74. SRdM, Ms. 10, p. 263 ("Séance du Mardi 21 Septembre 1784").

75. SRdM, Ms. 10, pp. 291-92 ("Séance du Mardi 16 Novembre 1784…Séance du Vendredi 19 Novembre 1784...M. De Chamseru a continué et terminé la lecture des recherches quil a faites avec M. Poissonnier, Desperrieres, Andry, Coquereau et Thouret sur la maladie appellé mal rouge ou Elephantiasis au sujet de laquelle M. Le Mal. de Castries à [sic] consulté la société"); see also SRdM Ms #10, p. 319 ("Séance du Mardi 11 Janvier 1785...J'ai lu une lettre de M. le M.chal de Castries qui annonce à la Compagnie la reception du mémoire qu'il l'avoit priée de rediger sur le mal rouge ou Elephantiasis. il apprend qu'il va le faire imprimer pour le rendre public, tant dans les Colonies que dans les ports de france et il temoigne à la société sa réconnoissance de cette nouvelle preuve de son Zèle pour ce qui interesse le bien public").

76. SRdM Ms #10, p. 319 ("Séance du Mardi 11 Janvier 1785) and de Castries letter, p. 324; see also SRdM-H&M, vol. 5 (1782-1783, published in 1785), p. 244.

77. "Nous avons reçu la lettre que vous nous avez fait l'honneur de nous écrire le 16 juin dernier et les exemplaires du mémoire sur le mal rouge qui l'accompagnaient. Nous en avons sur le champs fait la distribution à tous les médecins et chirurgiens de cette isle, en leur mandant voir l'intendance à ce sujet, sitôt qu'ils nous aurons envoyé leurs observations," Laborie letter of 20 September 1785 from Martinique, CAOM C[10] 3 (8), fol. 16.

78. "Nous avons l'honneur de vous envoyer, en exécution de votre dépêche du 16 juin dernier les mémoires du médecin, chirurgien et apoticaire du Roi à Saint-Pierre en réponse à la communication du mémoire que vous nous aviez adressé sur la maladie Rouge. C'est tout ce qui nous a été transmis d'observations jusqu'à ce moment," letter from de Damas and de Viévigné, dated 5 December 1786 from Martinique, CAOM C[10] 3 (6), fol. 61.

79. "…cette maladie qui règne à Cayenne, à la Pointe-à-Pitre isle Guadeloupe, et à Marie-Galante, est très rare à la Martinique," report by "[Fermond?], chirurgien du roi inspecteur," CAOM C[10] 3 (6), fol. 62.

ing the Société's memoir prior to his departure for Martinique, but reported that he had "verified on the ground in Martinique everything that has been sent to the Société Royale on Mal Rouge, especially that of the médecin du roi, Mr. Delaborde."[80] The minister collected all of these and other colonial reports for the Société Royale de Médecine.[81]

The Société Royale de Médecine received encomiums from several intendants in the French provinces to whom it sent its memoir, but elephantiasis was probably not much on the minds of these administrators.[82] A fuller response came from Saint Domingue, where the ministerial directive and the medical memoir of the Société Royale passed through the Saint Domingue administrators to Charles Arthaud, the chief médecin du roi at Cap François and his colleague, the chief royal surgeon, J. Cosme d'Angerville. These medical officials did their jobs in further distributing what amounted to an elephantiasis survey in the colonies; but using their additional base in the newly founded Cercle des Philadelphes, Arthaud and company produced a separate local circular, "Avis à MM. les Médecins & Chirurgiens de la dependance du Cap" (Alert to the Physicians and Surgeons in the Cap Region), which issued from the royal press at Cap François in 1785.[83] On his own and as médecin du roi for Cap François, Arthaud was an official correspondent of the Société Royale de Médecine in Paris, and he wrote to Vicq d'Azyr concerning the mal rouge project, promising to send the collected results from Saint Domingue directly to the Société Royale.[84]

The mal rouge initiative had most direct impact in Cayenne itself, where the royal surgeon and surgeon-major, Bajon and Remy, became involved. As mentioned, authorities had installed a leper colony on Île Mère at Cayenne, and local health inspectors became empowered to search for persons afflicted with

80. "J'ai été bien faché de n'avoir pas reçu avant mon départ ce que M. Vicq D'azyr m'avoit promis pour Cayenne mais j'ai verifié sur les lieux tout ce qu'on avoit écrit a la Société Royale sur la maladie rouge, et que j'ai trouvé la plus grande exactitude dans les Memoires de Mr. Delaborde [former "premier médecin du Roi en cette colonie"], qui repasse en france et mérite d'être accueilli par les Corps savans," undated ms."Extrait d'une Lettre de Mr. Dupuget en date du 10 avril 1785, au fort royal de la Martinique." SRdM, Archives, 160 dr. 1 (Bajon); see also 160 d 3 (Dupuget).

81. "J'ai reçu, M. avec la lettre de Mrs de Damas et de Viévigné, les observations de divers officiers de santé sur le mémoire de la Société royale de médecine relativement au mal rouge. Je les transmettrai à cette société avec celles que j'attends des autres colonies," ministerial correspondence dated Versailles, 11 February 1786, CAOM C^{10} 3 (6), fol. 60.

82. "...Plusieurs Lettres D'Intendants de Province qui font à la société leurs remerciemens de l'attention qu'elle a eue de Leur envoyer le rapport sur ... le mal rouge," SRdM, Ms. 10, p. 378/478 ("Séance du Vendredi 22 Juillet 1785").

83. Cercle des Philadelphes (1785); copy in AN-MAR G 102, dossier 3, #119.

84. "Monsieur et très honnoré confrère: Nous avons reçu par la voie de M.M. nos administrateurs le rapport de la Société sur le mal rouge de Cayenne. C'est pour suivre les intentions du ministre, satisfaire à nos devoirs et répondre à l'invitation de la S. R. que nous vous envoyons les observations que nous avons recueilliez," Arthaud letter, dated "au Cap, le 29 décembre 1785," SRdM, Archives, 136 dr. 1, #27.

Mal Rouge and to send them to the leper colony. Slaves with elephantiasis were to be reported within forty-eight hours. (Apparently authorities became overzealous in searching out victims.)[85] Remy, in Cayenne, offered an updated memoir on elephantiasis in 1789 that La Luzerne as minister sent to the Société Royale de Médecine for publication and dissemination.[86] At another point, given positive reports that a change in climate helped with mal rouge, La Luzerne proposed to the Société Royale that some of the sick with mal rouge in Cayenne be sent to the leper colony at Marseilles for experimental treatments by correspondents of the Society "to make this illness disappear."[87] In the meantime he recommended that the Société conduct tests with aloe for the treatment of mal rouge.

The Cinchona Project. The colonial-medical arm of the Colonial Machine that was centered on the Société Royale de Médecine from its creation, similarly worked to develop French sources of "Peruvian bark" as a specific against intermittent fevers, or what we know as malaria. Intermittent fevers often proved deadly, and Peruvian bark (containing quinine) was the most wonderful of wonder drugs before penicillin.[88] Joseph de Jussieu described the tree when he passed through South America in 1737.[89] Linnaeus named the plant *Cinchona officinalis* in 1742, but the secret of cinchona bark was known from the 1640s when the Countess de Chinchon, after whom it was named, the wife of the viceroy in Peru, learned of its medicinal properties from the natives. The Spanish controlled the Andean sources of the bark and profited greatly from their monopoly. The French had every incentive to be attentive to alternative sources.

In the 1720s Sarrazin suggested that cinchona might be found in Canada because its latitude north putatively matched Spanish sources at latitudes

85. See reference to inquiry by the Société Royale ("Elle desireroit en connoître la teneur, vû que l'execution paroît avoir donné lieu, de l'aveu des officiers de santé, a une recherche trop rigoureuse d'individus non attaquès du mal Rouge, pour etre cependant comme les autres rélegués dans l'islot qui sert de lazaret"), undated, untitled ms., SRdM, Archives, 160 dr. 1; see related ALS, de Castries to Vicq d'Azyr dated "A Versailles le 17 mai 1787," in same folder.

86. SRdM, Ms. 11, fol. 288r ("Séance du Mardi 7 Juillet 1789...J'ai fait la lecture d'une Lettre de M. Le Comte De la Luzerne; il annonce qu'il a reçu le Rapport sur le Mémoire de M. Remy, Médecin de Cayenne relatif au mal rouge et il invite la Société a publier le plustôt possible la nouvelle instruction qu'elle s'est proposé de donner sur cette maladie").

87. "...J'ai fait la lecture d'une Lettre de M. Le Comte De la Luzerne concernant les reglemens relatifs aux personnes attaquées du mal Rouge; il dit que d'après les renseignemens qu'il a reçu des administrateurs de Cayenne, il est à peu près reconnu que le changement de Climat et surtout le passage d'un pais chaud à un pays froid est salutaire à cette maladie. Il desire avoir l'avis de la société sur le projet qu'il a conçu de faire venir au Lazaret de Marseille quelques sujets atteints du mal Rouge pour y être traités par les Correspondans de la Société qu'il engage d'ailleurs à combiner tous les Mémoires qui lui ont été adressés sur cette maladie et a rediger des instructions fixes d'après lesquelles les officiers de Santé feroient des expériences pour parvenir à faire disparoitre cette maladie," SRdM, Ms. 11, fol. 222r ("Séance du Mardi 23 Septembre 1788").

88. Allorge, pp. 213-218, provides the background here, as does Bleichmar, pp. 242-45.

89. See Jussieu (1936); Safier (2008b), pp. 252-54.

south.[90] In 1747 the médecin du roi on Saint Domingue, Pouppé-Desportes, sent Saint Domingue cinchona to Jussieu at the Jardin du Roi, but without the plant attracting any special attention. Ultimately, three species of cinchona-like plants were discovered on Saint Domingue, and another was found by Badier growing on Guadeloupe and Martinique, *quinquina-piton*.[91] Badier's report in the 1770s seems to have sparked interest in researching the efficacy of Caribbean cinchona shrubs as alternatives to genuine Peruvian bark. Badier was the first to bring to France samples of these plants along with reports on treatments undertaken with a colleague on Guadeloupe, a certain Mallet.[92] From its formal chartering in 1778, the Société Royale de Médecine showed sustained interest in cinchona, with many reports submitted to it, so much so that febrifuges (fever reducers) became to the Société Royale de Médecine what longitude proposals were to the science academy.[93] On Saint Domingue the state-employed administrator of the Eaux de Boynes spa at Port-à-Piment, Joseph Gauché, tried native cinchona and wrote up his description of the virtues of the plant.[94] Governor La Luzerne in Saint Domingue sent this document to the minister in Versailles along with eight pounds of bark for consideration by the Société Royale de Médecine, if it so pleased the minister.[95] Minister de Castries asked for speedy consideration at the Société Royale. Gauché's memoir was read in camera, and the Société appointed a four-man committee to make trials with the cinchona sent by La Luzerne. This was in effect a separate cinchona committee of the Société Royale and was headed by Jussieu, who noted the cinchona plant in the herbarium of the Jardin du Roi, and by the chemist Fourcroy. The minutes of the meetings of the Société where these matters were discussed record that "several other Members of the Society volunteered to

90. "...on y trouveroit peutestre le Kinkina, et Lypecacuanha, parcequon parcouriroit les mesmes hauteurs dans la partie du nord, que celle ou les Espagnols les recueïllent dans la partie du sud," see undated [probably 1709], unsigned Sarrazin ms. in PA-DB (Sarrazin). For a later case where ecological analogy lead to a putative discovery, see SRdM-H&M, vol. 4 (1780-1781), Histoire, p. 343.

91. Mallet (1779).

92. SRdM, Ms., #11, fol. 269v ("Séance du Vendredi 27 Mars 1789") and fol. 286v ("Séance du Mardi 30 Juin 1789"). In all likelihood, this Mallet was Mallet de Brossière, the naval physician experienced in the colonial world and Regional Member (Associé Régnicole) of the Société Royale de Médecine.

93. See SRdM, PV, passim, e.g., SRdM, Ms. 7, p. 21 ("du mardy 15 xbre 1778"), p. 24 ("Du vendredi 18 Decembre 1778"), p. 77 ("Du Vendredi 19 fev. 1779"), p. 79 ("Séance Publique du Mardy 23 fev. 1779"), and so forth.

94. "Description d'un quinquina indigène à Saint-Domingue," BCMNHN, Ms. 1275. The Eaux de Boynes spa was incorporated as a government facility in 1772. Gauché was its paid director. Gauché performed chemical analyses of the waters at Port-à-Piment and was active and published in the Cercle des Philadelphes. Charles Arthaud served as medical inspector of the Eaux de Boynes facility; see McClellan (1992), pp. 142-44; DPF, p. 1492.

95. SRdM, Ms. 11, fol. 103r ("Séance du Vendredi 15 Juin 1787") NB remark in these minutes: "M. De Jussieu a dit avoir en herbier cette plante qui lui a été envoiée par M. Desportes [in 1747]." See also SRdM, Ms. 11, fol. 280r ("Séance du Mardi 26 Mai 1789") and fol. 281r ("Séance du Vendredi 29 Mai 1789").

perform trials with this cinchona and to report their observations."[96] Indicative of the integrated nature of the Colonial Machine, the Société Royale de Médecine received Saint Domingue cinchona from yet other sources, d'Angiviller at the Bâtiments du Roi and the abbé Tessier at Rambouillet.[97] This cinchona ultimately came from the naturalist traveling in Saint Domingue, Palisot de Beauvois, who expedited it to d'Angiviller with the recommendation that chemical dyeing trials be conducted at the Académie des Sciences and medical ones by the Société Royale de Médecine.[98]

Botanists transferred the newly discovered cinchona plants to the Jardin du Roi at Port-au-Prince in Saint Domingue, where they prospered. Like Gauché, Hypolite Nectoux, the royal botanist in charge of that garden, wrote on *quinquina indigène*, and the administrators asked Nectoux for samples to send to Paris for tests.[99] The médecin du roi in Port-au-Prince, Peyré, submitted his observations on the medical virtues of Saint Domingue cinchona "made at the royal military hospital at Port-au-Prince."[100] A military and naval surgeon major stationed in Port-au-Prince, Chézé, contributed his report.[101] Du Puget d'Orval sent samples of cinchona from Martinique.[102] Once he became minister, the former governor-general of Saint Domingue, La Luzerne, pushed cinchona and the Société Royale hard.[103] He ordered new shipments of cinchona for the Société Royale from the administrators in Saint Domingue.[104] The plant propagated well, and La Luzerne hoped to get living plants and seeds directly from Peru.[105] The chemist and member of the Société Royale Antoine-François Fourcroy was deeply involved in these tests.[106] The botanical network and establishment was brought to bear on the cinchona project, as the royal

96. "plusieurs autres Membres de la Société se sont offerts pour en éssaier et en donner ensuite leurs observations," SRdM, Ms. 11, fol. 113r ("Séance du Mardi 24 Juillet 1787"); see also, fol. 72v ("Séance du Mardi 30 Janvier 1787").

97. SRdM, Ms. 11, fol. 113r ("Séance du Mardi 24 Juillet 1787") and fol. 155r ("Séance du Vendredi 4 Janvier 1788").

98. See ALS Palisot de Beauvois to Jussieu, dated "Du Cap François le 1er juin 1789" in PA-DB (Palisot de Beauvois). Palisot suggested usefulness of cinchona for preserving libraries and against ants, possibly in response to prize contests of the Cercle des Philadelphes and the Conseil Souverain of Martinique; see below Part IIC (Ecological Cause and Effect) and Part III (Obstacles to Success).

99. McClellan (1992), p. 161 and n.

100. "...faites à l'hôpital royal et militaire de Port-au-Prince," see SRdM, Archives, 191 dr. 31 (Peyré), fol. 281v.

101. SRdM, Ms. 11, fol. 281v [sic] ("Séance du Mardi 9 Juin 1789").

102. SRdM, Ms. 11, fol. 280r ("Séance du Mardi 26 Mai 1789).

103. SRdM, Ms. 11bis, fol. 28v ("Séance du Mardi 22 7bre 1789") and fol. 30r ("Séance du Vendredi 25 7bre 1789").

104. SRdM, Ms. 11, fol. 201r ("Séance du Mardi 8 Juillet 1788").

105. SRdM, Ms. 11, fol. 160v ("Séance du Mardi 29 Janvier 1788").

106. SRdM, Ms. 11bis, fol. 50v ("Séance du Mardi 10 Novembre 1789"); fol. 51v ("Séance du Vendredi 13 9bre 1789"), fol. 53r ("Séance du Mardi 17 Novembre 1789"), and fol. 181r ("Séance du Mardi 22 9bre 1791").

botanists Leblond and L.-C. Richard were sent to search for true cinchona plants in Cayenne and Guiana. Coordinating the Colonial Machine in action, La Luzerne (at the Navy Ministry) notified d'Angiviller (at the Bâtiments du Roi) that he was mobilizing the local Intendant and Governor General in Cayenne to give Leblond every support.[107]

The Société Royale carefully compiled information on the various types of cinchona it received and their mostly positive medical effects, and the royal medical society kept working on cinchona through 1791, as the Old Regime collapsed around it.[108] The outcome of all this experimental work is not altogether clear, but Thouin at the Jardin du Roi in Paris wrote to Nectoux at the Jardin du Roi in Port-au-Prince that an unnamed but famous chemist performed the chemical analysis and an "expert" physician conducted the clinical trials. Medical authorities passed the ambivalent word back that Saint Domingue cinchona could replace genuine Peruvian bark in certain respects but that it proved inferior in others.[109] The Cercle des Philadelphes became involved, using its connections with the provincial academy in Rouen to test Saint Domingue cinchona as a dyeing agent. In the end, La Luzerne and company did not break the Spanish monopoly or successfully develop production of cinchona for medical use. The French had greater success with other, more strictly botanical undertakings having to do with timber and spices, but the cinchona story provides its own snapshot of the Colonial Machine mobilized for research and application.

Tetanus. Much the same story repeated itself with regard to official attention paid to lockjaw, or the disease of tetanus, the disease we now attribute to bacterial toxins.[110] State Minister Maurepas consulted the new Société Royale de Médecine over a treatment for tetanus in 1779, and the succeeding minister of the navy, de Castries, came back to the Société in 1785 with a similar request. This was backed up with letters from the médecin du roi in Cayenne, Laborde, regarding a particular treatment.[111] Headed by Poissonnier, Geoffroy, and Poissonnier-Desperrières, the same commissioners who reported in 1779 did so again in 1785.[112] The Société undertook to study the disease systematically, and one of the members of the commission, Carrère, polished up the report that

107. ALS "La Luzerne" to "M.r Le C.te d'Angiviller," dated "A versailles le 19 Juin 1788," AN O[1] 1292, #20. On Richard's involvement, see "Lettre du ministre à Fitz Maurice et Lescallier, du 30 décembre 1785," AN C[14] 58, fol. 273, reprinted in Jandin, p. 72.

108. SRdM, Ms., #11, fol. 286v ("Séance du Mardi 30 Juin 1789...La société m'a chargé de porter sur ses plumitifs le résumé suivt. concernant les différentes espèces de quinquina qui ont été soumises à son examen") and fol. 181r ("Séance du Mardi 22 9bre 1791").

109. These details are lifted from McClellan (1992), p. 161; see also Thouin letter of December, 1789 in "Correspondance et papiers du naturaliste Hypolite Nectoux," BnF, Mss., n.a.f. 9545, fols. 18-20, 38.

110. Quindlan (2005), discusses concern over tetanus, pp. 113-15.

111. See report, SRdM, Ms. 10, p. 369 ("Séance du Mardi 1er mars 1785").

112. See previous note and SRdM, Ms #10, p. 372 ("Séance du Vendredi 4 mars 1785").

the Ministry printed in 1786 under the title of *Projet d'instruction sur une maladie convulsive, fréquente dans les colonies de l'Amérique, connue sous le nom de tétanos* (Information Gathering regarding the Convulsive Disease Common in the American Colonies, Known as Tetanus).[113] The title makes clear that the Société Royale was attuned to the colonies and that gathering information about tetanus was something it could do and do well. The pamphlet was widely distributed through ordinary channels to the rest of the Colonial Machine.[114] The Société received substantial responses from physicians in France and in the colonies.[115] Bertrand Bajon, royal surgeon-major at the military hospital at Cayenne and correspondent of the Académie des Sciences, was an active agent funneling tetanus information to the Société Royale de Médecine.[116] The 1788 book *Observations sur le Tétanos* by J.-B. Dazille fits with this same effort; Dazille was regional associate of the Société, former médecin du roi in Saint Domingue, and national associate of the Cercle des Philadelphes, and the work appeared with the formal approbation of the royal medical society.[117]

Receipt of the Société's tetanus circular in Saint Domingue prompted unexpected developments. The authorities approached Arthaud about tetanus in 1785, more as the colony's first royal physician than as president of the new Cercle des Philadelphes. But Arthaud seized the opportunity for the Cercle. He compiled the papers the Cercle had already received on the subject, adding his own observations and a summary overview. The Cercle approved the assembled manuscript, the administrators added their approbation and permission to print the volume in the colony, and the Cercle's 104-page *Dissertation et Observations sur le Tétanos* (Dissertation and Observations on Tetanus) appeared with a 1786 imprint.[118] This expeditiously produced volume, essentially a first volume of memoirs of the Cercle des Philadelphes, provided considerable visibility for it, at least in the matter of tetanus. Arthaud communicated all of these materials to the Société Royale in Paris.[119] The minister encouraged the Société to keep up the good work and update its tetanus brochure.

113. See SRdM (1786) and notice of this pamphlet, SRdM-H&M, vol. 5 (1782-1783, published 1787), p. 244, and SRdM, Ms. 10, p. 432/532 ("Séance du Lundi 31 Octobre 1785"). See Carrère's reports, SRdM, Ms. 10, p. 376/476 ("Séance du Vendredi 15 Juillet 1785") and p. 378/478 ("Séance du Mardi 19 Juillet 1785").

114. See SRdM, Ms. 11, fol. 2v ("Séance du Mardi 14 Mars 1786").

115. See SRdM, Archives, 144 dr. 46 ("Mémoires sur le Tétanos").

116. See SRdM, Ms. 10, 11, passim; e.g., Ms. 11, fol. 61v ("Séance du Vendredi 1 Décembre 1786").

117. See Dazille (1788) and SRdM, Archives, 118 dr. 113.

118. These words are taken from McClellan (1992), pp. 228-29 and 329.n54.

119. SRdM, Ms. 11, fol 86r ("Séance du Mardi 27 Mars 1787...Le Ministre desire que la Société s'occuppe s'il y a lieu d'un nouveau Mémoire sur cet objet..."), fol. 106v ("Séance du Mardi 26 Juin 1787"), and fol. 107r ("Séance du Jeudi 28 Juin 1787").

Rations for Seamen. The Colonial Machine was similarly stirred to action in official quests to prevent scurvy and to rationalize shipboard rations.[120] The Société Royale de Médecine again played a central role, but the Marine Royale was also involved, and this is a cautionary tale because it shows areas where things did not go smoothly and where we see tension and conflict within the Colonial Machine, notably in this instance between the Société Royale de Médecine and the Royal Navy.

The work in naval medicine by the English physician James Lind, F.R.S. (1716-1794), forms a key part of the background here.[121] The first edition of Lind's *Treatise on Scurvy*, appearing in 1753, reported a rapid beneficial effect of citrus juice on the symptoms of scurvy. Lind's *Scurvy* was translated into French by 1756.[122] Lind's *Essay on the Most Effectual Means of Preserving the Health of Seamen* appeared first in 1757 and was also translated into French.[123] His *Essay on Diseases Incidental to Europeans in Hot Climates* came out in 1768, and the 1785 translation into French was based on the third English edition of 1777.[124] The French knew about Lind and were struggling like Lind and the Royal Navy in Britain to improve administrative and medical practices in the Marine Royale in France. Scurvy was still a grave danger for long-distance crews. Bering, d'Entrecasteaux, and hundreds of other sailors died from it over the eighteenth century. Distributing lots of malt, sauerkraut, and a juice concentrate called "rob," Cook had success against scurvy in his voyages of 1768, 1772, and 1776, but until then no one had a remedy for scurvy at sea, and even then what exactly was the best antiscorbutic practice and the lag in the spread of these practices meant that outbreaks and deaths from scurvy continued into the 1780s. Ships' provisions and diet for sailors was an obvious concern for all.

In France the Colonial Machine sprang into action over a vegetable diet for sailors proposed by Antoine Poissonnier-Desperrières sometime around 1768 or 1769. The gist of his proposal was to eliminate salted meat from the rations the state provided its sailors.[125] Salted meat was presumed to cause scurvy, and Poissonnier-Desperrières suggested substituting a vegetable diet of cooked dried beans, rice, lentils and like grains and legumes. By this time Poissonnier-Desperrières was an experienced hand in colonial medicine, although not yet elevated to the post of inspector of colonial hospitals. But his elder brother was already installed as inspector general of naval and colonial medicine, and the

120. Villiers (2000) provides important background here, as does Pritchard (1987a), pp. 178-83; see also Gelfand (1987), p. 78; Allorge, pp. 529-30.

121. On Lind, see Roddis; Harvie details this story from the English point of view, touching on the French case, pp. 152-53; see also Iliffe, pp. 636-38.

122. Lind (1753), (1756), and (1953).

123. Lind (1757) and (1758).

124. See Lind (1768) and (1785).

125. See Poissonnier-Desperrières (1771).

Colonial Machine was pressed into action. A test voyage aboard the navy frigate *La Belle Poule* sailed to Saint Domingue in 1770-71 to try the rations, but the test seems not to have been a success.[126] In September and October of 1771 an experiment was tried at the naval hospital at Brest dividing patients with scurvy into three groups and feeding them three different regimens: the Poissonnier-Desperrières vegetable diet, ordinary sea rations, or ordinary hospital rations. The last had a lot of bread and wine, and patients in this group did better.[127] Poissonnier-Desperrières was undaunted, responding to critics and reprinting his 1771 pamphlet "on the advantages to be had in completely changing the grub for seamen" in 1772.[128] It says something that this later document was printed on the presses of the French *Army* and not the French Navy, apparently because the Marine Royale and the medical wing of the Navy did not like being told what to feed sailors by the fancy physicians at the Société Royale de Médecine.

Poissonnier-Desperrières came under attack from Etienne Chardon de Courcelles (1705-1775), the chief naval physician at Brest and director of the medical school there.[129] Cochon-Dupuy *fils* from the navy's facilities at Rochefort chimed in against Poissonnier-Desperrières. The naval ensign who had sailed on the test voyage, the chevalier de La Coudraye, reported negatively and privately to the Marine Academy at Brest.[130] Desperrières's 1772 edition of his *Essai* responded to Courcelles and de La Coudraye.[131] The Brest Academy appointed commissioners to investigate, but equivocated, pleading summer vacation, and when the matter came back before it, the Académie took a moderate stance vis-à-vis de Courcelles' proposals, saying that any novelties needed to be introduced slowly. In the meantime, Poissonnier-Despierrières and/or his powerful elder brother apparently intervened to attack de Courcelles behind the scenes beyond any norm of civility. The Marine Academy found this behavior offensive and reported it to the minister.[132]

126. See DDP III:25-26 and report by the Chevalier De la Coudraye, "Observations sur le mémoire de M. Poissonnier Desperrières" in Poissonnier-Desperrières (1772), pp. 17-31.

127. "Observations sur le Memoire de Monsieur Desperrieres" by Chardon de Courcelles, AN-MAR C^7 60 (Chardon de Courcelles).

128. Poissonnier-Desperrières (1772); see copies, AN-MAR C^7 253 (Poissonnier Desperrières) and AN-MAR D^3 37. See also Yaqubi, pp. 13-33.

129. "Observations sur le Memoire de Monsieur Desperrieres, concernant les avantages qu'il y aurait à changer absolument la Nourriture des gens de mer," AN-MAR C^7 60 (Chardon de Courcelles).

130. On these points see DDP III:25-26; on François-Célestin de Loynes, chevalier de La Coudraye (1743-1817), see Taillemite (2002), pp. 286-87.

131. See documents, AN-MAR C^7 253 (Poissonnier Desperrières); Poissonnier-Desperrières (1772), pp. 35-60 contains his "Mémoire en réponse à M. de la Coudraye, Enseigne de Vaisseau, Sur le Régime Végétal."

132. DDP III:60.

The elder Poissonnier's memoir on hospital rations and pharmaceutical standards made it through channels in 1777 without provoking any resistance, but the American War had just broken out.[133] The matter popped up again forcefully in 1783, after the war, when Minister de Castries formally approached the Société Royale de Médecine about "how best to organize the rations of seamen in their states of health and sickness."[134] Given that the diet supposedly should not contain meat, de Castries went on to ask what amount of salted meat or fish as well as legumes and beverages would be ideal in rations for sailors. He inquired about practices in other nations, and what the most famous navigators had written. He also charged the Société Royale with a parallel task, to determine the best rations for various classes of patients in navy hospitals.[135] The Société leaped into action, appointing a large committee, first of eight members, headed by Lavoisier and Poissonnier, that then expanded to eleven with the addition of Poissonnier-Desperrières and two others.[136] This expansion undoubtedly reflected political maneuvering in the Société. This group in essence formed a separate standing committee of the Société Royale on nutrition, and issues surrounding government rations for sailors and hospitals occupied the Société on and off for years. In responding to de Castries, however, the committee and the Société Royale stalled, suggesting an exciting series of prize contests that would spark useful contributions.[137] The minister rejected prize contests out of hand as taking too long and not being sufficiently responsive to immediate needs.[138] The Société seemed to be picking a fight with the navy; it invited all of its members to take up the research questions posed by the minister and to consult naval ordinances that specified navy rations.[139]

In the meantime the naval command at Brest and the Brest Marine Academy unleashed their ammunition against the Poissonnier brothers and the Société Royale de Médecine. The Brest Marine Academy invited royal naval surgeons to submit duplicate reports to the Academy as well as to Poisson-

133. SHMB, Série A, Sous-série 3A, Conseil de Marine, Registres 3A4 à 3A10: Procès-verbaux de délibérations du Conseil de Marine de 1772 à l'an II (1793-1794), entry for 19 July 1777.

134. "...déterminer la meilleur maniere de composer La nation des gens de mer tout dans l'etat de santé que dans l'etat de maladie," SRdM, Ms. 9, p. 582 ("Séance du mardi 30 septembre 1783").

135. "...on demande quelle pouroit être la composition de la ration d'hopital la plus generalement appropriée," SRdM, Ms. 9, p. 582 ("Séance du mardi 30 septembre 1783").

136. SRdM, Ms. 9, p. 585 ("Séance du mardi 30 septembre 1783"). See published "Rapport sur plusieurs Questions proposées à la Société Royale de Médecine, par M. le Maréchal de Castries, Ministre de la Marine, relativement à la nourriture des Gens de Mer," SRdM-H&M, vol. 7 (1784-1785), pp. 221-305 and copy at AN-MAR D^3 37.

137. The prize proposal suggestion is found in the minutes of the Société, SRdM, Ms. 10, p. 73 ("Séance Du vendredi 23 J.vier 1784").

138. "Lettre du ministre Continue dans Le plumitif Du 10 fevrier 1784," SRdM, Ms. 10, pp. 90-91.

139. SRdM, Ms. 9, p. 585 ("Séance du mardi 30 septembre 1783").

nier.[140] The naval command at Brest appointed its own twelve-member committee that reported negatively on the proposal of the Société Royale de Médecine regarding rations for seamen.[141] The navy diplomatically said that it would stick to the new regime as closely as possible and would explore new provisioning contracts.[142] It even proposed a new set of experiments to adjudge the rations suggested by the Société Royale.[143] But the new rations obviously upset the status quo. In the final analysis, navy officials were not convinced scientifically, and they were particularly opposed to eliminating salted meat. The switch seemed impractical; other things needed to be considered to match the success of Captain Cook, they said, adding that wartime always presented a different situation.[144] They wanted a wartime test to be confident.

The squabble continued for a while. Antoine-Chaumont Sabatier (1740-1798), a navy staff physician at Brest, an associate of the Marine Academy, and also a correspondent of the Société Royale, pronounced against Poissonnier-Desperrières.[145] In 1785 he proposed his own memoir on conserving the health of crews aboard ship.[146] Needless to say, when Sabatier submitted his paper, the Société Royale de Médecine rejected his views in favor of those of Poissonnier-Desperrières.[147] The médecin du roi on Saint Lucia, Cassan, chimed in from the colonies against Poissonnier-Desperrières in 1786, imploring the Académie de Marine to stand firm against the vegetable diet.[148] Surprise, surprise! Sabatier's report to the Marine Academy on Cassan's stance was positive.[149]

140. DDP IV:20.

141. See set of documents related to this matter from March and April, 1784, SHMB Série A, Sous-série 3A 93, 1784, #1-#5.

142. "Envoy de l'opinion de quelques off.rs sur le nouveau regime proposé par la societé Rle de medecine, pour les gens de mer... le 24 mars 1784," SHMB Série A, Sous-série 1A, 1-A117, p. 138.

143. "Essai de la Nouvelle Ration projettée," dated "A Brest Le 31 Mars 1784" and signed by 12 Commissioners, in SHMB Série A, Sous-série 3A 93, 1784, #5.

144. See previous note and "Rapport de la Commission nommée par le Conseil de Marine pour éxaminer le projet de Composition de la Ration des Gens de Mer à bord des Vaisseaux du Roi, d'après l'avis de la Societé Royale de médecine... A Brest le 22 mars 1784," SHMB Série A, Sous-série 3A 93, 1784, #1. See also #2, #3, and #4 and copy of this report, SHMV, Ms. 0240 1784.

145. DDP V:21.

146. DDP VI:26-27 and undated ms. "Mémoire sur la conservation des équipages," SRdM, Archives, 201 dr. 9.

147. See report signed Geoffroy, Jeanroi, Coquereau, Halléz [?], "delivré au louvre le 27 juin 1786," SRdM, Archives, 201 dr. 9.

148. Cassan ms, "Examen de la question si les viandes salées sont une cause puissante du scorbut des marins," SHMV Ms. 78, pp. 316-35; DDP VI:38-39.

149. "Rapport du Mémoire de Mr Cassan, lu à la séance du 6 avril 1786," SHMV Ms. 78, pp. 335-36.

It was a standoff between organized medicine in Paris and at the court as centered on the Poissonniers and the Société Royale de Médecine on the one hand and navy officers, physicians and surgeons, and the Académie Royale de Marine at Brest on the other. Sea rations did not change. The episode seems little more than a dust-up over turf, but the turf involved and the conflict among the parties highlight the structure of the Colonial Machine and the limitations to its good working order.

Mineral Waters, Medical Electricity, and Smallpox. The medical research arm of the Colonial Machine became involved in a broad range of investigations where matters medical impinged on the success of the colonial endeavor.

Interest in local mineral waters was another medically related theme that runs through the archives of the Colonial Machine. Reports on mineral waters were filed regularly.[150] In 1772 the French monarchy created a standing committee on mineral waters (*Commission Royale des Remèdes Particuliers des Eaux Minérales*) that systematically collected data in France and in the colonies. Upon its creation in 1778, the Société Royale de Médecine absorbed this commission on mineral waters, and this effort meshed naturally with the other information-gathering activities of the Société de Médecine and was reinforced by a ministerial order of 1787 for investigations of mineral waters.[151] The Société printed and distributed a circular, and the results poured in.[152] The first (and only) volume of *Mémoires* of the Cercle des Philadelphes of 1788 was essentially a response to this solicitation; the volume contained fourteen reports on mineral water sources in the French Antilles, including the royal Eaux de Boynes spa on Saint Domingue.[153] The colonist Lefebvre-Deshayes sent his forty-three-page report to the Société de Médecine on thermal sources near the Jérémie River in Saint Domingue, for which he received a gold medal from the Société.[154] Arthaud also contributed a paper on the waters at the military hospital at Cap François and their curative properties.[155] From Pointe-à-Pitre on Guadeloupe, Hapel Lachênaie wrote in about a local hot spring.[156]

150. See, for example, communications to and from Sarrazin in 1733 regarding a source he discovered at Trois-Rivières in Canada; AN-COL, B^{58}, fols. 431, 454v.

151. "Premier registre des arrêtés du comité et des délibérations de la Société royale de médecine, au sujet des eaux minérales," SRdM, Archives, 94 dr. 3; see also Pascal.

152. SRdM, Archives, cartons 88-95, and cartons 111-15; see also Muller.

153. McClellan (1992), pp. 142-44, 243-45; Homer, 261-306.

154. "Essai analytique sur eaux thermales minérales dites de la Grande-anse ou du bras-gauche de la grande rivière de Jérémie," SRdM, Archives, 89 dr. 30; SRdM, Ms. 10, p. 136 ("Séance du Mardi 23 mars 1784") and p. 358 ("Séance publique du Mardi 15 fevrier 1785"); SRdM-H&M, vol. 5 (1789), p. 10.

155. SRdM, Ms. 11, fol 216v ("Séance du Vendredi 29 Août 1788").

156. SRdM, Ms. 11, fol. 222v ("Séance du Vendredi 26 Septembre 1788"). See further SRdM, Ms. 11bis, fol. 92r ("Séance du Mardi 1er Mars 1791"). There may have been an inspector of mineral waters on Guadeloupe; see AN-COL, E 348 (Regnaudot).

Similarly, an *Analyse des eaux minérales de la Martinique et de Sainte-Lucie, d'après les ordres du Roi* appeared under royal authority.[157]

Using electrical discharges from static electrical machines to treat medical conditions, notably paralysis, became something of a fad in French medicine of the 1780s, and the medical use of electricity can be found within the Colonial Machine.[158] The Société Royale de Médecine naturally expressed interest in the topic.[159] A significant effort in the 1780s extended the practice of medical electricity under the aegis of the Académie de Marine.[160] The minister ordered instruction in medical electricity to be given to surgery students at Brest, and the Academy itself proposed a public course in the subject.[161] The Academy had its own electrical equipment that was authorized for use to treat patients in the royal naval hospital at Brest.[162] There seemed particular interest in medical electricity in the colonies.[163] Arthaud at Cap François used electrical equipment to effect cures; Dubourg of the Cercle des Philadelphes similarly experimented with medical electricity, although he complained that the humid conditions of the tropics made manipulating the instruments difficult. One Thomin, of the Chambre d'Agriculture of Port-au-Prince, was similarly involved. Related work took place on Martinique, and Cassan on Saint Lucia likewise tried experiments, saying that the apparent scarcity of the electrical fluid mitigated effects of medical electricity there.[164] The electric eel from Cayenne may be counted here. Bajon, the surgeon major at the military hospital at Cayenne and Academy correspondent, reported that shocks from eels cured gout.[165] Eels from South America were therefore imported with great care through ministerial channels to the Société Royale de Médecine![166]

One might have expected the Colonial Machine to have become involved in inoculation against smallpox in a major way. The technique of inoculation was introduced from Turkey into Europe in the 1720s, and inoculations were tried in Saint Domingue as early as 1745, nearly two decades before the Paris medical faculty approved the practice, and three decades before it became widely

157. See undated, unsigned 54 p. pamphlet, BnF, call number Te163 1105.

158. For an overview, see Delbourgo, pp. 201-77.

159. SRdM, Archives, 182 dr. 3.

160. See ministerial documents AN-MAR C^7 87 (Diard. Horloger et électricien à Brest).

161. Copy of Academy letter to minister dated "23 Janvier 1784," in SHMV, Ms. 093, "Académie R. de Marine. Lettres et Réponses au Ministre."

162. DDP IV:20.

163. See DPF, pp. 509-10; McClellan (1992), pp. 173-74.

164. DPF, p. 510; Cassan (1803b), p. 29; Homer, p. 112; Regourd (2002).

165. SRdM-H&M, vol. 4 (1780-1781), Histoire, pp. 344-45; SRdM, Ms. 9, p. 114 ("Séance du Mardi 5 Mars 1782") and related report, p. 122. See also Richer.

166. SRdM, Ms. 11, fol. 45v ("Séance du Mardi 5 Septembre 1786"); see also "Expériences sur le gymnotus electricus, avec une description de cet animal," by the Swiss engineer, Samuel Guisan (1740-1801), working for the French in Cayenne, dated "Cayenne, le 1er août 1790," BCMNHN, Ms. 355; on the electric eel story in Europe, see Delbourgo, pp. 165-99.

accepted in France.[167] Siméon Worlock (brother-in-law of the English inoculator, Daniel Sutton), a correspondent of the Société Royale de Médecine and a resident associate of the Cercle des Philadelphes, was in Saint Domingue from the 1770s, inoculating plantation slaves by the thousands! Joubert de la Motte held an M.D. from Angers and in due course became botaniste du roi at Port-au-Prince, correspondent of the Société Royale de Médecine, and colonial associate of the Cercle des Philadelphes; he, too, vigorously promoted inoculation in Saint Domingue. But, inoculation remained controversial, for the slightly risky procedure still had its critics, and smallpox epidemics continued to occur in France and in the colonies through 1789 and beyond.

The Société Royale de Médecine took only a little notice of inoculation.[168] But, within the Académie des Sciences a battle royal waged over whether to endorse the practice.[169] The topic hardly shows up in the archives of the Académie de Marine or in ministerial records. All in all, the comparative lack of interest of the Colonial Machine in contemporary research or application of smallpox inoculation is unexpected and difficult to explain. The connection between slavery and the early and widespread use of inoculation in Saint Domingue should be clear because at very little cost inoculation protected and preserved the valuable capital property of slave owners. Yet inoculation remained underground or at least on the sidelines where individual colonists and slave owners privately took advantage of the lifesaving practice without the involvement of the Colonial Machine. This surprising finding helps us define the limits of the Colonial Machine.

Finally, in this connection, the record of the Colonial Machine shows a diffuse but wide-ranging interest in ethnology, anthropology, and archaeology, particularly of the American cultures it encountered. Bobé reported on "the customs and mores of the Canadian Indians."[170] An Indian canoe, arrows, and a quiver made it to France from Canada.[171] Based on hatchets, arrows, and other artifacts he received from Canada and the Antilles, Jussieu suggested parallels between the peoples of the Old and New Worlds, and so launched the field of comparative archaeology. Fontenelle, presenting Jussieu's memoir in the *Histoire et Mémoires* of the Académie des Sciences in 1725, eloquently summed up the far-reaching consequences of his discovery:

167. On inoculation in Saint Domingue, see McClellan (1992), pp. 144-45; Homer, pp. 315-16; see also Weaver, pp. 51-54; DPF, passim. On the general background to contemporary inoculation, see Peter.

168. E.g., the late entry, "...M. Belin a fait la lecture d'un Mémoire sur l'inoculation," SRdM Ms. #11, fol. 135v ("Séance du Vendredi 5 Octobre 1787"). Peter indicates some further attention of the Société Royale de Médecine to inoculation, but likewise underscores the caution and low level of interest in inoculation on the part of the institution.

169. On this episode, see McClellan (2003c), pp. 69-71.

170. "Détails sur les moeurs et coutumes des Sauvages du Canada," AN-COL F^3 2, fols. 441-44.

171. AN-COL B^{48} (1725), fol. 282 (23 January 1725); Regourd (2000), pp. 215-17, Feest.

> The origin of these stones is very evident and very certain as soon as you see similar ones carved by the Savages of America and used to chop wood or to arm their arrows....Our continent was inhabited by Savages in ancient times, & the same wants, the same shortage of iron inspired them to the same industriousness....If fossil stones are monuments left after big physical revolutions, these ones are a monument to another great revolution that can be called *moral*, and the comparison between the New World and the Old one is the means to prove both types of revolutions.[172]

La Condamine brought back grave goods from Peru that underscored those similarities.[173] Charles Arthaud and the Cercle des Philadelphes in Saint Domingue took a special interest in the remnants of indigenous American culture on Saint Domingue. Arthaud produced a piece "On the Native Inhabitants, Their Arts, Industry, and Means of Subsistence," wherein he took pains to depict the aboriginal, and by his day extinct, Arawaks as "Noble Savages" at peace in their natural surroundings.[174] As a result of Arthaud's work and that of other Cercle associates, the Cercle's *cabinet* became a minor repository of Indian artifacts. J.-B. Auvray, a founding member of the Cercle des Philadelphes, donated an Indian head recovered from a grave; Baudry des Lozières, likewise a founder of the Cercle, contributed some stone axes, and the collection came to include miscellaneous pieces of Indian pottery, jewelry, several stone fetishes, a stoneware mortar, other tools and utensils, and doubtless other artifacts. The count d'Ingrandes, a colonial associate of the Cercle from the southern part of the colony, wrote on the customs and rituals of the natives of the island.[175]

Racial differences and research on skin color provoked sustained interest and inquiry and touched on larger contemporary debates concerning the origins and connections of various varieties of humans.[176] The Cercle des Philadelphes issued a questionnaire that included questions about different groups from Africa, and the Société Royale de Médecine was fascinated with an African albino in France in 1788 and other cases of albinism elsewhere in the colonies.[177] Reports came in all through the eighteenth century concerning differ-

172. "Sur les pierres de foudre, les Yeux de Serpent, & les Crapaudines," HARS 1723 (1725), p.17.

173. Hamy (1870), pp. 22-24, 31.

174. *Recherches sur la Constitution des naturels du pays, sur leurs arts, leur industrie et les moyens de leur subsistance*; Arthaud (1786). On these points, see also McClellan (1992), pp. 268-69.

175. See reference in index entry, DPF, p. 1500.

176. On these points see Cohen; Greene, chapts. 6-8; and McClellan (1992), pp. 241-42.

177. SRdM, Ms. 11, fol. 234v ("Séance du Mardi 11 Novembre 1788"); SRdM 191, dr. 17, "Dissertation sur les nègres blancs, ou albinos" by Lefebvre Deshayes from Saint Domingue; and Arthaud (1789a).

ences in skin color and possible explanations.[178] Many other papers evidence the growing interest among colonial physicians in the anatomy of the world's peoples. For instance, Arthaud published his "Dissertation sur la conformation de la tête des Caraïbes" (Essay on the Shape of the Heads of Caribs) in 1789 in the *Journal de Physique*, which opened the way for comparative anthropometry that was destined to flourish in the nineteenth century.[179]

Indeed, physicians availed themselves fully of the memoirs and samples addressed to them by correspondents in the colonies and developed techniques of comparative anatomy. For instance, Vicq d'Azyr, permanent secretary of the Société Royale de Médecine, after dissecting the throat of a Red Howler monkey, or *Stentor seniculus*, sent to him by Malouet (naval ordonnateur in Guiana), established some hitherto unnoticed differences between the disposition of the larynx in monkeys and in humans, thus taking a step forward toward understanding the formation of the human voice.[180]

Such subjects, however important they might be, were nevertheless not considered of primary interest by administrative circles in terms of immediate utility and profitability. That is why they were given little attention in the cogwheels of the administration and why they found their way more easily to the attention of provincial academies. On the contrary, meteorology retained all the attention of the Colonial Machine.

Meteorology and the Tides

Individually, the historical vignettes we have encountered to this point do not amount to much in themselves, but collectively they add up to define the reality of the Colonial Machine of Old Regime France and demonstrate the breadth, depth, and regularity of its activities. The efforts expended in and around the Colonial Machine to conduct meteorological research in the eighteenth century only add to the weight of this observation.

Climate conditions encountered by the French overseas varied tremendously, of course, from the seasonally frozen north in Canada, to the perpetually tropical jungles in South America, and on to monsoon-swept South Asia. Changing weather conditions at sea meant life or death for ships and crews. Obviously, success in overseas endeavors depended on reliable knowledge of the weather and its seasonal changes. Meteorology became an even more cru-

178. See, for example, Barrère (1741a) entitled *Dissertation sur la cause physique de la couleur des nègres, de la qualité de leurs cheveux, et de la dégénération de l'un et de l'autre*; "Dissertation sur la couleur noire des Éthiopiens (1709)," by François Bellet; BM/Bordeaux, Ms. 828 XIII, dossier 16. Quindlan (2005), pp. 111-12, discusses these matters.

179. Arthaud (1789b).

180. Chaïa (1980), pp. 11-12.

cial area of study, given the dominant medical philosophy of the Société Royale de Médecine in the 1770s and 1780s that posited environmental causes for disease. It is no surprise, therefore, to discover the Colonial Machine mobilized in organized efforts to master matters meteorological. Several elements of this mobilization are perhaps not so obvious. One, local observers were not merely desirable but essential for collecting meteorological data. Secondly, observations needed to be sustained in order to be useful, which meant that transient observers (such as the astronomers of the Colonial Machine or botanical explorers) were of less value than persons established in situ. Then, reports had to be funneled into a central reporting station if anything more than local knowledge was to be had. And finally, instruments and measurement scales (notably temperature and barometric pressure, but also wind strength, descriptions of conditions, etc.) had to be standardized in order to harmonize reports coming in from scattered locales.[181] These parameters at once define the meteorological activities of the Colonial Machine and at the same time explain its comparatively limited success in this arena.

Gathering Data. The story of meteorology and the Colonial Machine centers largely on the massive effort of Father Louis Cotte and the Société Royale de Médecine to collect and publish weather data from France, elsewhere in Europe, and overseas beginning in the 1770s. But that campaign was merely the last and largest in a series of meteorological activities dating back at least to the early decades of the eighteenth century. In 1717 in Saint Domingue, for example, the engineer son of the academician Joseph Sauveur collected temperature and barometric observations using equipment that had been calibrated in the sub-basements of the Observatoire.[182] Beginning in the 1730s, J.-N. Delisle pushed coordinated meteorological observations just as he did astronomical and cartographical ones. He sent standardized thermometers and his "brief instructions on how to make meteorological observations" to his correspondents in China and India, and he promised barometers.[183] Delisle wrote about this initiative to Gaubil in Beijing:

> Because Missionaries often travel in the different provinces of China or might have correspondents there, I have thought that they could procure for us journals of meteorological observations made in far-away places. If we compare these with those maintained regularly in Europe, all this, I say, can only lead us one day to have a universal history of changes

181. On the use of instruments for comparative observations, see Cotte (1774), pp. 279-280; Bourguet and Licoppe (1997, 2002); Regourd (2005a).

182. PA-PS (24 April 1717).

183. "Instructions abrégées sur la manière de faire les observations météorologiques," OdP, Ms. A 7 4.

> of the atmosphere across the whole extent of the Earth where these observations might be made.[184]

Boudier, the Jesuit missionary in Chandernagor, provided ten years of temperature and barometric readings from 1740 to 1750 which Delisle passed on to Réaumur in Paris for comparison.[185] At least one Jesuit in China, Amiot, followed up on Delisle's call and sent back to France meteorological data from China for 1757 to 1772.[186] In the 1740s authorities sent Réaumur thermometers and other instruments to Canada for use by the royal doctor Gaultier and by the Jesuit professor of mathematics at Quebec, Father Bonnecamps.[187] Gaultier was in sustained contact with Delisle, forwarding him a total of seven years' worth of botanical-meteorological reports in the 1740s and 1750s – some undertaken in collaboration with Bonnecamps.[188] From the 1740s through the 1770s, weather journals regularly filtered into the Academy of Sciences from the Antilles and Guiana, such as those published by Jean-Baptiste Thibaut de Chanvalon in 1751.[189]

In the Indian Ocean, Cossigny *père*, using the new Réaumur alcohol thermometer, sent five years' worth of observations over the period 1732-1739[190] The abbé de Lacaille collected data during his voyage to the Indian Ocean in 1751 and 1752. In the 1760s a certain Grésil (or Gresey) likewise recorded and sent to the Paris Academy meteorological observations from Île de France.[191] The academician Le Gentil de la Galaisière supplied eighteen months of meteorological data from Manila and the Indian Ocean during his prolonged stay in those parts in the 1760s and early 1770s.[192] One Cazaud sent observations

184. "…car j'ai consideré que comme les Missionaires font souvent des voiages dans les differentes provinces de la Chine, ou peuvent y avoir des correspondances, cela nous pourroit procurer des Journaux d'observ. meteorologiques dans des lieux fort éloignés de nous, les quels étant comparés avec ceux que l'on tient regulierement en Europe, tout cela, dis-je, ne peut pas manquer de nous donner un jour l'histoire universelle des changemens de l'air dans toute l'étendue de la terre où ces observ. auront été faites," Delisle "au R. P. Gaubil à Peking. Petersbourg le 7 Decembre 1739 N.S.," AN-MAR 2JJ 65.

185. ALS "au P. Boudier à Chandernagor, &," dated: "à Paris le 27 Octobre 1754," AN- MAR 2JJ 68; Pueyo, p. 366.

186. Pueyo, p. 366.

187. AN-COL B[78] (1744), fol. 41 (30 March 1744); B[81] fol. 23 (12 April 1745) and fol. 59 (5 May 1745).

188. See Wien; Slonosky; and trove of Gaultier materials, "Observations météorologiques," OdP, A 7, 6 [64[3]]; Pueyo, p. 367.

189. In 1740, for example, the Académie Royale des Sciences received a year's worth of barometric and thermometric observations from Martinique, and in 1751 Thibaut de Chanvallon sent his from Saint Pierre, Martinique; PA-PS (1740) and (1751). For observations from Cayenne, see PA-PV-110, Macaye, 25 January 1772; Chaïa (1968) gives details on Artur's work for Réaumur since 1736.

190. Pueyo, pp. 364-65.

191. PA-PV-110, Grésil, 1 June 1763 ("Observations Météorologiques faites à l'isle de france") and 3 September 1768 ("Observations météorologiques faites à l'Isle de france").

192. Pueyo, p. 365.

from Grenada for 1772-73.[193] In this same period other occasional, but substantial meteorological observations arrived from the Cape of Good Hope, Île de France, Île Bourbon, Madagascar, and Senegal, the last provided by David, director of the Compagnie des Indes in Senegal.[194]

Heures des Observations	Thermometre au soleil. Matin. Midi.	Thermometre à l'ombre. Matin. Midi.	Vents.	Barometre.	Lune.	Pluie, Tonnerre, &c.	Aiguille aimantée.
... M	......	20 ...	Est.	28 4½	22 Dern. quart.	8 P. ...	......
... M	27	......	..	28 5½	..	......	......
... S	23	25½ ..	Est.	28 5	..	......	......
... S	......	21 ...	..	28 5½	..	......	......
... M	......	20 ...	Est.	28 4½	23 [illegible]	......	......
... M	30	......	..	28 5½	...	P.	......
... S	31	26 ...	S. E.	28 4½	..	P.	......
... S	......	19½ ..	..	28 4½	..	6½ P. ..	......
... M	......	19½ ..	Est.	28 5½	24	P.	......
... S	32	25½ ..	Est.	28 4	..	38 P. T. ..	......
... S	......	19½ ..	Est.	28 4½	..	P.	......
... M	......	20 ...	Est.	28 4	25 [illegible]	37½ ...	......
... M	28	......	Est.	28 4½	..	P.	......
... S	32	26½ ..	Est.	28 4½	..	P.	......
... S	......	20½ ..	..	28 5	..	P.	......
... M	......	19 ...	Est.	28 4	26	A.	......
... M	27	......	Est.	28 4½	..	P. ...	......
... S	33½ ...	26½ ..	Est.	28 4½	..	P.	......
... S	......	21 ...	Est.	28 [illegible]	..	......	......
... M	......	19½ ..	Est.	28 4	27	P.	......
... M	26	......	Est.	28 4½	..	P.	......
... S	32	25½ ..	Est.	28 4	..	8 P. ...	......
... S	......	20	Est.	28 4½	..	E.	......

Jours du Mois. Août 1751.	VARIATIONS DU TEMS.
23	Vent médiocre : petite pluie à quelques reprises : ciel chargé de beaucoup de nuages.
....	Vent un peu fort : quelques nuages : le thermometre du soleil a été observé à deux heures & demie du soir.
....	Le vent a diminué à quatre heures, & est devenu calme ; il a repris un peu ce soir : ciel serein : quelques nuages à l'est.
24	Vent très-petit.
....	Vent fort, mais inégal : quelques gouttes de pluie à deux ou trois fois.
....	Grand vent : beaucoup de nuages chargés à l'est : à trois heures grande pluie d'un quart d'heure.
....	Vent petit : ciel serein : quelques nuages ; il a plu trois ou quatre fois assez abondamment ; mesuré 6 pouces ½ d'eau.
25	Gros tems cette nuit, pour parler comme les Marins, avec pluie fréquente & abondante qui continue, & petit vent : ciel tout pris : mesuré 38 pouces d'eau vers les neuf heures du matin.
....	Pluie toujours abondante ; le ciel est presque toujours entierement couvert : vers les cinq heures du soir deux coups de tonnerre sourds.
....	Vent médiocre : ciel à demi-couvert : nuages à l'est. La pluie a cessé à l'entrée de la nuit.
26	Vent petit ; beaucoup de nuages à l'est : mesuré 37 pouces ½ d'eau.
....	Grand vent : petite pluie très-souvent depuis ce matin ; beaucoup de nuages à l'est.
....	Petite pluie fréquente toujours : nuages à l'est.
27	Vent petit : nuages à l'est & au nord-est. Arc-en-ciel à 8 heures du matin.
....	Calme : beaucoup de nuages : petite pluie à quelques reprises depuis 7 h. ½.
....	Le vent qui avoit été petit ou médiocre, a renforcé après midi ; un peu de pluie à [illegible] heures : nuages autour de l'horison.
....	Vent fort ; ciel serein par-tout.
28	Calme, quoiqu'aux nuages le vent soit fort : il regne du sud & de l'est : il a plu ce matin.
....	Ciel chargé : il a plu légerement.
....	Vent médiocre ou calme : petite pluie de tems-en-tems : mesuré 8 pouc. d'eau.
....	Vent fort : nuages chargés à l'est : éclairs vifs & fréquens du nord.

[Meteorological Table for Martinique, 1751 by Thibault de Chanvalon]

Cotte and the Société Royale de Médecine. By the last quarter of the eighteenth century agents of the Colonial Machine had produced a substantial amount of meteorological data, but the problem remained of how to coordinate this and like data. The Colonial Machine found its meteorologist in Louis Cotte (1740-1815), the Oratian priest who was curate at Montmorency and passionate about meteorology. Cotte was elected a correspondent at the Académie des Sciences in 1769, and he became associated with the Société Royale de Médecine and the Société Royale d'Agriculture when these bodies came into being. Cotte also had his own wide correspondence, co-opted into the Colonial Machine by the Société Royale de Médecine.[195] His *Traité de*

193. Pueyo, p. 368.

194. Pueyo, pp. 366-67.

195. Pueyo, p. 369; Regourd (2005a), pp. 41-42 tells this story, too; see also SRdM, Archives, 122 dr. 14, 149 d39, and 167 dr. 5. Further on Cotte and his work, see Taylor; and Cotte (1774) and (1778).

météorologie of 1774, published on royal presses, launched a collective research effort, describing meteorological instruments, printing tables for data collection, and providing instructions on making meteorological observations.[196] In his *Météorologie*, Cotte pointed back to previous work at the Observatoire and abroad from the 1660s, mentioning Delisle, the astronomer Lalande's correspondents, Asian missionaries, and the botanical-meteorological observations of Duhamel du Monceau.[197]

But a larger, more sustained meteorological project took off when Cotte hooked up with the Société Royale de Médecine after that institution fully came on the scene in 1778. The program embraced by the Société Royale de Médecine in conjunction with Cotte was Baconian to the core and was intended to help solve the problem of centralizing the reporting of meteorological data.[198] The Société and Cotte would be the central agents. The project began in 1778 and continued through 1793 and the closing of the Société Royale. The environmentalism of the Société Royale de Médecine privileged its commitment to investigating the weather and climatological influences on disease. The Société printed 1000 copies of a large data-collection form drawn up by Cotte and grandly labeled "Correspondence of the Société Royale de Médecine."[199] Each form had columns for a daily weather record for a one-month period with entries (morning, noon, and night!) for temperature, barometric pressure, humidity, winds, rain amounts, and general weather and seasonal conditions, including agricultural conditions. The Société went on to "invite observers to use the method published by Father Cotte that the Royal Society has sent to its members and correspondents."[200] (Incidentally, the Compagnie des Indes and presumably the Royal Navy used similar printed sheets for recording maritime conditions.[201])

The Société distributed its form widely, and the data poured in. Participation from the colonies was strong, with Guadeloupe well represented. The médecin du roi on Guadeloupe, Vergnies, vigorously participated in the program.[202] A local physician, Le Gaux, and a colleague sent data to the Société Royale.[203]

196. Cotte (1774).

197. Cotte (1774), pp. xviii-xxii.

198. Desaive et al; McClellan (1992), pp. 163-64.

199. On the 1000 copies, see reference, SRdM, Ms. 8, p. 329 ("Séance du Vendredi 16 f.r 1781"). For the form itself, see Desaive et al., following p. 28 and copy, SRdM, Archives, 176 dr 4, #8 (Madier). See also similar sheets from 1763 in Thibault de Chanvalon; see also Regourd (2000), p 738.

200. "On invite les Observateurs à se servir de la méthode publiée par le P. Cotte, & que la Société Royale a adressé à ses membres & à ses correspondants," SRdM, Archives, 176 dr. 4, #8 (Madier). The form notes that "la Société n'exige pas que tous les détails de cette Table soient remplis."

201. See documents SHML, 1P 296, Liasse 5.

202. SRdM. Ms, 10, p. 306 ("Séance du Vendredi 17 décembre 1784"). See further SRdM, Mss. 10 and 11, passim; SRdM, Archives, 149 dr. 39.

203. SRdM, Ms. 9, p. 418 ("Séance Du mardi 1er avril 1783)" and SRdM, Archives, 191 dr. 18.

Hapel Lachênaie, the pharmacist and veterinarian at the navy hospital on Guadeloupe, contributed.[204] In Saint Domingue, Charles Arthaud, however, resisted the project of Cotte and the Société Royale to collect weather data, and the Cercle des Philadelphes did not participate.[205] But reports from Saint Domingue did flow in to Cotte and the Société Royale de Médecine. The Cap-François physician and inoculator, Simeon Worlock, for example, collected data from Saint Domingue inclusively from 1783 through 1786.[206] The abbé de la Haye reported on winds in Saint Domingue.[207] The private colonist and owner of a remote sugar plantation, Étienne Lefebvre Deshayes (1732-1786) was one notable rapporteur from Saint Domingue as part of this project. He recorded and sent seven years' worth of data to the Société Royale.[208] Deshayes was an official correspondent of the Société Royale de Médecine, an in-colony associate of the Cercle des Philadelphes, and one of Buffon's correspondents through the Cabinet du Roi and the Jardin du Roi. Deshayes was also connected to the Colonial Machine through the astronomer Jérôme de Lalande, who from 1760 to 1775 directed the *Connaissance des temps*; and it was Lalande who transmitted Deshayes' observations to Cotte.[209] A rich merchant from Bordeaux, one Laffont de Ladebat, acted as intermediary between observers in Saint Domingue and Cotte. He had instruments sent to Saint Domingue and had his agent on his plantation in Saint Domingue make and report back meteorological observations.[210]

Even though meteorology in Saint Domingue and the project of the Société Royale de Médecine lacked the backing of the Cercle des Philadelphes, the project received a local boost from Charles Mozard, the editor and publisher of the colonial newspaper, the *Affiches Américaines*. In 1784 Mozard took on the job of providing a central meteorological clearinghouse in Saint Domingue and began to promote colonial meteorological investigations. Mozard invited colonists to send him their reports for publication in the *Affiches Américaines*, and he persuaded the Intendant to provide postal cover for the activity.[211] Beginning in the later part of 1784 meteorological observations started to flow to Mozard and the *Affiches Américaines*, with regular reports from Port-au-Prince, Cap François, Léogane, Les Cayes, and elsewhere in the colony. Mozard's enthusiastic claim in 1787 that "meteorological observations are

204. SRdM, Ms. 11bis, fol. 92r ("Séance du Mardi 1er Mars 1791") and SRdM, Archives, 146 dr. 7.

205. For the story of meteorology in Saint Domingue, see McClellan (1992), pp. 163-68.

206. SRdM, Ms. 11bis, fol. 70v ("Séance du Mardi 29 Decembre 1789").

207. AN-MAR G 101, dossier 3, fols. 77-84.

208. SRdM, Archives, 136, dr. 1, #16; 89, dr. 30; SRdM, Ms. 10, p. 136 ("Séance du Mardi 23 mars 1784"). On Lefebvre Deshayes, see further McClellan (1992), pp. 271-72, 334; DPF, 1513; Pueyo, p. 368.

209. SRdM-H&M, vol. 1 (1779), Histoire, p. 153.

210. SRdM-H&M, vol. 1 (1779), Histoire, p. 153.

211. McClellan (1992), p. 165.

being made everywhere in the colony" exaggerates the situation, but several dozen stations seem to have reported.[212] To guide meteorological work in Saint Domingue, in 1788 Mozard printed extracts from the work of the Italian theorist Guiseppe Toaldo. At the newspaper and press office in Port-au-Prince Mozard himself sold mercury and alcohol thermometers and blank tables for recording meteorological observations.

Reports drifted in from elsewhere in the Antilles and from Île de France.[213] Data from some thirty extra-European observers were directly incorporated into Cotte's work, evidence of a higher level of activity generally surrounding meteorology at the time. Meteorological data was duly published in successive volumes of the *Histoire et Mémoires* of the Société Royale de Médecine. Cotte himself summarized and published more data in the two tomes of his government-produced *Mémoires sur la Météorologie* of 1788.

On tiny Saint Lucia, the médecin du roi Cassan was a formal correspondent of the Société Royale de Médecine and in touch with it over meteorology. Cassan also was associated with the Marine Academy at Brest and by 1790 with the *Société Royale des Sciences et Arts* of Cap François, which had formally succeeded the Cercle des Philadelphes. He shared his meteorological findings with these institutions as well.[214] Cassan wrote up his observations on the meteorology of the torrid zone that were approved by the Académie des Sciences and printed in Rozier's *Observations sur la physique* for April of 1790, after which other matters intervened to disrupt research in meteorology.[215] One footnote on Cassan as a meteorologist is that he had at least one of his instruments, a Saussure hygrometer, made on the spot in Saint Lucia.[216] Clearly, this approach could not be the norm in systematic and large-scale meteorological research.

Standardized Réaumur thermometers reached the Indian Ocean, and meteorological observations from Île de France made it to the Société Royale in Paris.[217] Jean Baptiste Lislet (later known as Lislet-Geoffroy, 1755-1836) was a geographer and correspondent of the Académie des Sciences, born on Île Bourbon and living on Île de France. Using French-made instruments, Lislet kept detailed meteorological logs on Île de France that he sent to the Academy

212. Mozard quoted in McClellan (1992), p. 166.

213. See SRdM-H&M, vol. 1 (1779), p. 153; vol. 5 (1782-1783), pp. 246ff; vol. 8 (1790), pp. 84ff; SRdM, Archives, 36 dr. 1, #27; 122 dr. 14; 194 dr. 29; 167 dr. 5 no. 1 (Dufour); 191 dr. 18; McClellan (1992), p. 334; SRdM, Michel Beaurepaire, Ms. "Société Royale de Médecine. Extrait du fichier de la S.R.M. détenu par la Bibliothèque de l'Académie Nationale de Médecine."

214. SHMV, Ms. 79 ("Mémoires des correspondants"), vol. 4, fols. 35-37.

215. Printed "Observations météorologiques faites sous la zone torride par M. Cassan," AN-MAR G 101 Dossier 3 fols. 101-109; Regourd (2002).

216. "un hygromètre composé suivant la méthode de M. de Saussure, que j'ai fait faire sous mes yeux par un ouvrier très-intelligent," Cassan (1803a), pp. 162-63.

217. SRdM-H&M, vol. 8 (1786), Histoire, p. 84.

in Paris regularly from 1786 through 1789. The Academy approved Lislet's submissions for publication and made him a correspondent.[218] The Academy seemed pleased to get Lislet's meteorological reports and wanted to encourage him, writing at the end of one committee report: "We hardly know any sustained meteorological observations on Isle de France as extensive or as complete as those of Mr. Lislet of which we just gave an account. We therefore think that his work will be very useful to meteorologists, that he merits the approbation of the Academy, and it is to be wished that he continues with the same zeal and the same assiduity."[219]

Meteorological reports made it all the way from China, notably from the French consul in Canton, de Guignes. De Guignes produced a sustained series of meteorological reports from Macao and Canton covering the years 1786-1791, approved for publication by the Academy of Sciences.[220] The Colonial Machine took notice of monsoons in the Indian Ocean; Lieutaud, a surgeon with the Compagnie des Indes reported on them from Pondicherry.[221] The science academy approved the translation of another work on Indian Ocean monsoons and monsoon patterns, and Le Monnier pushed for the publication of monsoon data.[222]

Tides. Within the scope of meteorology, we might cite research done on tides under the auspices of the Colonial Machine. The Académie des Sciences undertook an early study of tides beginning in 1716 on orders from the abbé Bignon and the regent. The Academy printed a circular and tapped into the network of royal professors of hydrography to conduct observations. Another episode unfolded in 1734-35, again on ministerial initiative. The Académie des Sciences developed explicit instructions for its correspondents for measuring

218. PA-PV-110, Lislet, 5 September 1786 ("Observations Météorologiques faites au port Louis de L'Ile de france"), 6 August 1788 ("Recueil d'observations météorologiques faites à l'Isle de france en 1786 et 1787"), and 4 March 1789 ("Journal d'une Expédition relative à l'hydrographie accompagné d'observ.ons météorologiques faites à l'isle de france, depuis le 16 Juillet Jusqu'à la fin de L'année 1787"); see also PA-PV 107 (1788), fols. 69 ("Samedy 5 Avril 1788"), 211 ("Mercredy 6. Aoust 1788), and 295v ("Mardy 23 Décembre 1788").

219. "...nous ne connaissons point de suites d'observations météorologiques dans l'Isle de France, aussi complettes et aussi étenduës que celles de M. Lislet dont nous venons de rendre compte. Nous pensons donc que son travail sera très utile aux météorologistes, qu'il mérite l'approbation de l'Academie, et qu'il est à desirer qu'il le continue avec le même zèle et la même asiduité," PA-PV 107 (1788), fol. 210 ("Mercredy 6 Aoust 1788").

220. PA-PV-110, Guignes, 12 August 1786 ("Observations Météorologiques envoyées de Canton"), 1 August 1787 ("Observations Météorologiques faites à Macao proche Canton, année 1786"), 6 December 1788 ("Observations météorologiques faites à La Chine précédans [sic] la plus grande partie des années 1786 et 1787") and 21 July 1792 ("Observations météorologiques faites à Canton en Chine pendant l'année 1791"); and AN-MAR G 101, dossier 3, fol. 88.

221. "memoire contenant tout ce que j'ai remarqués pendant mon sejour de deux mois a pondichery," marked in margin: "Par Mr. Lieutaud Chirurgien pour la Compagnie des indes," BCM-NHN, Ms. 293.

222. PA-PV-110, Forest, 28 January 1786 ("Mém.re sur les moussons de L'inde /traduction"); ALS Le Monnier to "Monseigneur" dated "a Paris ce 16 aoust 1786," AN-MAR G 101, dossier 3, fol. 87.

high and low tide. The Academicians also asked for meteorological data (winds, barometer, thermometer) and for information about any iron mines or deposits in the area. The Academy was not entirely thrilled with what it got back, but praised the hydrographer at Bordeaux, one Montequt, "who worked with great care and intelligence."[223] D'Après de Mannevillette supervised a major investigation of tides in 1768 that produced 300 folios of results.[224] Lalande was in contact with the Brest Academy in 1771 over new systematic studies of tidal periods, "the originals of those done six years ago having been lost and what appears in the *Mémoires* of the Academy being insufficient."[225] Another ministerially inspired directive by the science academy in Paris in 1777 for a tidal monitoring station filtered down to the Académie de Marine at Brest and to the commander of the navy port at Brest.[226] A tide chart is for local uses, albeit important ones, but there was some thought that mastering tidal periods on a larger scale might be an aid to navigation.[227]

The greatest usefulness extracted from all of this meteorological data may have been by recent historians who have been able to reconstruct meteorological conditions of pre-Revolutionary France.[228] Still, the Colonial Machine became substantially involved in advancing meteorology, and meteorological research helped define the Colonial Machine, not least for introducing standards of precision and exactitude in meteorology.[229]

Mesmerism in Saint Domingue

The story of the German physician and physical scientist Franz Anton Mesmer (1734-1815), famous for discovering "animal magnetism," is well known and well documented.[230] Suffice it to say here that Mesmer arrived in Paris in

223. "M.r Montequt Hydrographe à Bordeaux qui paroit travailler avec beaucoup de soins et d'intelligence," PA-PV 54 (1735), fols. 105r-106v ("Samedi 30 Avril 1735"), whence the details for this episode are taken. See earlier PA-PV 53 (1734), fol. 231v ("Mercredi, 4 August 1734"), and PA-PV-110, Maurepas (Le Comte de), 30 April 1735 ("Rap: Sur des Observations demandées sur la Déclinaison et l'inclinaison de l'aiguille aimantée en conséquence de l'instruction approuvée par l'académie le 4 Août 1734 et envoyée dans les différents Ports du Royaume par M.r de Maurepas").

224. SHMV, Manuscrits de la Bibliothèque du Dépôt des cartes et plans de la marine, Ms. 55-57 666.

225. "...les originaux de celles qui furent faites il y a six ans sont perdus, et ce qu'on en a mis dans les mémoires de l'Academie [des sciences] n'est pas suffisant," Lalande letter, dated "À Paris le 13 Janvier 1771," SHMV, Ms. 89.

226. DDP IV:72

227. See ms., "Memoire sur l'Observation du flux et du reflux de la mer, Dressé par ordre de l'Accademie royale des Sciences" (1779), AN-MAR G 141.

228. For what one group of researchers did so successfully with this data, see Desaive et al. The twelve volumes of *Ephemerides* of the *Societas Meteorologicæ Palatinæ* of Mannheim, Germany, published between 1783 and 1795 offer the same prospect; see McClellan (1985), pp. 220-27.

229. This point is made by Bourguet and Licoppe (1997), p. 1120; Regourd (2005a).

230. On Mesmer, see Gillispie (1980), pp. 261-89; Darnton (1968); Riskin, pp. 189-226; Meheust; Peter (1999).

1778 with a firm belief in a completely orthodox, Newtonian ethereal fluid. Mesmer collected this ethereal fluid in specially prepared tubs containing iron filings and bottles of "mesmerized" water. Ropes and iron bars connected to his tubs, and through them Mesmer could channel his fluid through the human body to provoke "crises" and to cure a multitude of ailments. Over the next years Mesmer made a name for himself in Paris through a flamboyant medical practice that involved a variety of manipulations of animal magnetism and that provoked miracle cures. Mesmer fell foul of the scientific and medical establishments in France and Saint Domingue, however, because he insisted that his discoveries be kept secret among initiates and because he profited from an extravagant and monopolistic medical practice. Mesmer suffered numerous run-ins with the Académie des Sciences, the Société Royale de Médecine, the Paris medical faculty, and the court itself. Questions of public safety and morals were raised, and in 1784 Louis XVI famously appointed two commissions to investigate the scientific and medical claims underlying mesmerism. These royal commissions were, naturally, peopled with academicians of the Académie Royale des Sciences and the Société Royale de Médecine, and members included Lavoisier and Benjamin Franklin. Here Mesmer ran up against the powerful scientific and medical establishment in France that we have seen under other guises as core elements of the Colonial Machine.

Spread to Saint Domingue. The interest for us lies in the revealing and consequential impact of the great wave of mesmerism that crashed down on Saint Domingue in 1784. Sailing to Saint Domingue in 1784 as an agent of the Colonial Machine and head of the official cartographic expedition sent to local waters, Antoine-Hyacinthe-Anne de Chastenet, the count de Puységur (1752-1809), brought with him the full armamentarium of contemporary mesmerism in addition to quadrants and chronometers.[231] Puységur was a convinced partisan, and he had further close connections to mesmerism through his brother and Mesmer disciple, Armand-Marc-Jacques Chastenet, the marquis de Puységur (1751-1825), the discoverer of "mesmeric somnambulism" or hypnotism.[232] The Count de Puységur's official hydrographical expedition was controversial for carrying mesmeric equipment, and it attracted the critical attention of the Académie Royale de Marine and the Société Royale de Médecine even before it left.[233] In this respect, the ambiguities inherent in the mission further delineate the boundaries of the Colonial Machine. Puységur

231. On Puységur in Saint Domingue, see above Part II A, pp. 242-43; Regourd (2008b) and (2000b); Pluchon (1987); McClellan (1992), pp. 175-79; Homer, pp. 316-22.

232. On the Marquis de Puységur, see Peter (1999).

233. The Académie de Marine refused Puységur its imprimatur for his work on the curative power of mesmerism, SRdM, Ms. 10, p. 157 ("Séance du mardi 27 avril 1784"). For debates within Marine Academy at Brest regarding Puységur and the efficacy of mesmeric treatments, see SHMV, Ms. 74 ("Comptes-rendus des séances de l'Académie"), vol. 11, pp. 193-94; 232-35; see also Ms. 77 ("Mémoires des correspondants et étrangers à l'Académie royale de marine."), vol. 2, pp. 313-15.

installed mesmeric apparatus on his ship, the *Vautour*, that sailed from Brest, and kept his crew healthy on the four-month Atlantic crossing. With his beautiful and equally partisan wife in tow, the thirty-two-year-old naval officer came to Saint Domingue as much to bring the benefits of mesmerism as to complete the cartographical duties assigned to him. Upon his arrival in Saint Domingue in June 1784, he set up the colony's first "magnetic tub" in the hospital of the poorhouse in Cap François.[234]

The mesmerist movement spread rapidly throughout the colony, and a network of mesmerist practitioners sanctioned by doctor Mesmer in Paris expanded along with it. By August 1784, mesmerism had penetrated the Caribbean and the southern department of Saint Domingue. In a letter dated 25 August 1784, a certain Mme Millet wrote to her sister about mesmerism:

> A magnetizer has been in the colony for a while now, and, following Mesmer's enlightened ideas, he causes in us effects that one feels without understanding them. We faint, we suffocate, we enter into truly dangerous frenzies that cause onlookers to worry. At the second trial of the tub a young lady, after having torn off nearly all her clothes, amorously attacked a young man on the scene. The two were so deeply intertwined that we despaired of detaching them, and she could be torn from his arms only after another dose of magnetism. You will admit that such are ominous effects to which women should sooner not expose themselves. [Magnetism] produces a conflagration that consumes us, an excess of life that leads us to delirium. We will soon see a maltreated lover using it to his advantage.[235]

Independently, Moreau de Saint-Méry reported piously: "Magnetism had its disciples, its apostles, and consequently its miracles in the southern department [of Saint Domingue]. But it was also ridiculed, and it died. The miraculous was rejected by all faiths, except those that admit the Resurrection."[236]

A report coming out of the central Artibonite Plain in September of 1785 provided another glimpse of the Mesmer affair in Saint Domingue and some more sober reasons why mesmerism could seem attractive and worthwhile. The report in question was from the judicious Artibonite plantation owner, irri-

234. In 1782, the médecin du roi, Duchemin de l'Etang, had already written to the Société Royale de Médecine in Paris about the arguments he used against a local admirer of Mesmer, but at that time there were no tubs nor any kind of magnetic cures available on the island; see Duchemin letter of 1 October 1782, SRdM, Archives, 167 dr. 1.

235. Millet letter transcribed in Debien (1972b), p. 433; quoted in McClellan (1992), pp. 176-77.

236. DPF, p. 275.

gation expert, and member of the Cercle des Philadelphes, Jean Trembley, writing to his Swiss cousin, the naturalist Charles Bonnet:

> The great debates surrounding mesmerism hardly seem to be settled definitively by the very respectable report of the commissioners, academicians, doctors, and *physiciens* [in Paris] that attributed the effects only to the play of a biased imagination. Their decision does not prevent that there are still partisans.... Two mesmeric tubs in this colony were directed by Monseigneur, the Count de Puységur, officer of the royal navy, and by other adepts. Marvelous cures that could hardly be attributed to any play of the imagination have been reported. A cripple brought from the plain to Cap François on a litter walked freely afterward. A female slave paralyzed for fourteen years was entirely cured in a short time without her realizing that she was being treated, etc. A plantation owner on this plain made a big profit in magnetizing a consignment of cast-off slaves he bought at a low price. Restoring them to good health by means of the tub, he was able to lease them at prices paid for the best slaves. The rage for magnetism has taken hold of everyone here. Mesmeric tubs are everywhere. [But] today hardly anyone speaks of them any longer perhaps because of already having spoken too much about them.[237]

The enigmatic suggestion that mesmerism spread covertly among slave owners in order to maximize profits is an astounding one. Mesmerism exemplified just the kind of useful application of science that the Colonial Machine and plantation owners and managers might have been expected to vigorously exploit. Mesmerism was a potential boon for the slave system, and had mesmeric treatments proved truly effective, slave owners certainly would have been allowed to treat slaves. But when mesmerism seemed to pass out of the hands of whites and into the nonwhite population, one can be equally sure that colonial authority would crack down.

Puységur and mesmerism immediately drew the attention and the quick condemnation of the Colonial Machine in Saint Domingue, notably from the médecin du roi and the chirurgien du roi in Cap François, Charles Arthaud and Cosme d'Angerville. A long, skeptical notice about Mesmer and animal magnetism, probably by Arthaud, ran in the Cap François edition of the *Affiches Américaines* on 9 June 1784, the day after Puységur arrived in town. The printer, Mozard, followed for the next few weeks with heated attacks on Puységur from Port-au-Prince. In June and July 1784, Arthaud, Cosme

237. Trembley letter transcribed in Debien (1955), pp. 16-17; quoted in McClellan (1992), p. 178.

d'Angerville, and Alexandre Dubourg, a local botanist and apothecary, formed their own ad hoc investigating committee on Mesmer wholly analogous to the official commissions then meeting in France. And this colonial Mesmer committee independently arrived at the same damning conclusions about mesmerism as the royal commissions in France did simultaneously in their reports – namely, that animal magnetism was a figment of everyone's imagination and without sound scientific basis or hope for medical success. The Colonial Machine in Saint Domingue repudiated Mesmer, in parallel with its other elements in France.

As mesmerism provoked internal and external scientific debates under the watchful eye of royal metropolitan institutions, its spread in the colony proved to be a turning point in the building of a local scientific community. And in fact, on 15 August 1784, an expanded group of anti-Mesmerists centered around Arthaud, d'Angerville, and Dubourg founded the Cercle des Philadelphes at Cap François. That Puységur and mesmerism triggered the formation of the Cercle des Philadelphes in the Mesmer-mad summer of 1784 makes the episode that much more significant for our account.[238] But the focus of attention in the present context extends to a further echo of mesmerism that rippled through the colony subsequent to Puységur's arrival; mesmerism supposedly spilling over from the world of European science and white colonial enclaves into slave communities of blacks from Africa. Was this a case of French science migrating across cultural borders into the world of the "other"?

Voodoo Mesmerism? Judicial authorities, notably the Conseil Supérieur of Cap François, picked up on what they perceived as the penetration of European mesmerism into slave communities. In two rulings of 1786 the Conseil legislated against mesmerism in the colony. In May of 1786 the Conseil denounced "false prodigies due to this would-be magnetism...usurped by Negroes and disguised by them under the name of *Bila*" – a term referring to vodun (voodoo) practices.[239] The edict forbade nocturnal meetings of slaves and proscribed the practice of magnetism for slaves and any person of color. Making the connection to contemporary science, the Conseil declared: "It would be extremely dangerous for this colony to leave in the hands of Negroes an instrument that physical science only handles with great wariness, and which lends itself so easily to excesses and conjurers' tricks common among Negroes and venerable in their eyes."[240] In November of 1786 the Conseil Supérieur at Cap François found four slaves guilty of "having held nocturnal meetings...fraught

238. The foundation of the Cercle des Philadelphes is taken up further below, Part III.

239. "Faux prodiges de ce prétendu magnétisme...usurpé par les Nègres et déguisé par eux sous le nom de Bila," quoted in Pluchon (1987), p. 66-67 with source cited as AN, 27 AP 12 (papers from François de Neufchâteau). See reference to an "arrêt du Conseil du Cap qui déffend aux gens de couleur l'exercise du Magnetisme," SRdM, Ms. 11, fol. 56v ("Séance du Mardi 7 Novembre 1786"). On vodun, besides Pluchon's work, see, Regourd (2008); Weaver; Métraux; Hurbon.

240. Quoted in Pluchon (1987), p. 66.

with superstition and tumult, in several plantations in the La Marmelade district and other nearby places under the pretence of magnetism."[241] One, the mulatto named Jérôme, known as Poteau, was sent to the galleys; his accomplice, Télémaque, was pilloried; the slave Julien was condemned to be hung while the slave Jean was to watch. At the end of the century, Moreau de Saint-Méry evoked this episode in the following gauzy terms: "One certainly does not expect to hear that La Marmelade had been the place chosen to bring to fruition the ideas of magnetism suited to the views of those who propagated them. They appeared in La Marmelade together with the hoaxes of the Illuminated, the disgusting tricks of the Convulsionaries, and excesses of profanation because their aim was to reap the profits of swindle...."[242]

The idea that European mesmerism could cross over into African slave communities alarmed the colonizers and slave owners. What happened at that time in plantations near Cap François, on the north part of the island, was seen by many as much more than a trivial event involving unscrupulous charlatans. However, these events may not have had much to do with European mesmerism directly. They may have had more to do, in fact, with the conceptual categories available to colonists to deal with the frightening rituals of vodun in contemporary Saint Domingue. The Marmelade district was a new place mainly cleared by the work of first-generation slaves, most of them coming from the Congo, and vodun rituals haunted the countryside. There are many reports of vodun activities, including the 1786 Conseil edict itself that observed: "The miraculous operator has the subjects who ask to submit to his power brought to him into the circle. He does not limit himself to magnetizing them in the modern sense of the word. After the magician has caused stupor or convulsions in them using both the sacred and the profane, holy water is brought to him since he pretends it is necessary to break the spell that he had previously cast on the subjects...."[243] Other reports were more vivid.[244] As

241. Quoted in Pluchon (1987), p. 68; see also Debien (1972a), pp. 277-83 for the full indictment.

242. DPF, pp. 275-76. Moreau de Saint-Méry went on to speak about "chimerical mysteries," "superstitions," and "shameless charlatanism" of Mesmerism.

243. Arrêt of 16 May 1786 (Conseil Supérieur du Cap), cited in Pluchon (1987), p. 66.

244. See, for example, AN, 27 AP 12, cited in Debien (1972a), p. 279: "M. Jacquin, économe de M. Estève, dit que dans le courant du mois de juillet dernier, il vit clairement à travers le clissage de la case du nègre Jean Lodot, le nègre Jean au milieu d'une assemblée considérable, ledit nègre à genoux devant une table couverte d'un tapis et éclairée par deux chandelles, élevant à différentes distances un fétage [fétiche], qu'il n'a pas pu bien distinguer les nègres à genoux et en silence pendant cette cérémonie; il ajoute avoir trouvé deux manchettes [machettes] croisées à terre à l'endroit où opérait le nègre Jean. Le nommé Dimanche, nègre esclave de l'habitation de M. Estève, dit qu'il s'est plusieurs fois trouvé aux assemblées que le nègre Jean tenait sur l'habitation de M. Estève, son maître; que ces assemblées se nommaient *mayombe* ou *bila*. Il ajoute le détail des cérémonies qui s'y pratiquaient, telle que de leur mettre dans les mains des feuilles de framboisier, d'avocat et d'oranger, de les faire mettre à genoux et dans cette posture de leur donner à boire du tafia dans lequel il mêlait du poivre, de l'ail, du blanc d'Espagne et que cette boisson les faisant tomber, ledit nègre les relevait à coup de manchette. Il ajoute que le nègre Jean portait sur lui en bandoulière un petit sac dans lequel était un crucifix, du poivre, de l'ail, de la poudre, des cayous, des cloux et un étui."

Gabriel Debien and Pierre Pluchon have shown, the terms of these accounts are unambiguously revealing manifestations of the vodun cult. Yet the judges of the Conseil Supérieur at Cap François saw or feigned to see vodun as a manifestation of the familiar: "animal magnetism" of Puységur and Mesmer. What was at stake was nothing but the place and status of black, hidden religious and magical practices – this in a colony of 350,000 or more black people dominated by a minority of only 25,000 whites.

In the contemporary colonial world, knowledge and practices undoubtedly did travel in both directions across cultural divides, including those separating communities of slaves from slave owners in the plantation system. But whether or how French authorities came to see a contagion of mesmerism in the slave community raises tricky interpretative issues. If true, this part of the story would be of interest as evidence of European knowledge migrating to slave communities. But a persistent doubt lingers concerning the real effects of mesmeric theories and practices in Saint Domingue, and this doubt reinforces the feeling that any appropriation of European knowledge, "usurped" and "disguised" by black people, represented a potential risk for the colony. The major fear for the judges may have been the risk of a charismatic leader taking advantage of such illusions, a repeat, perhaps, of the great slave insurrectionist, Macandal, executed by colonial authorities in 1758. Such a suspicion was a part of the fight conducted over real and symbolic power by both masters and slaves.[245]

Therefore, even if we should not neglect the attractive and appealing possibility of a long journey of mesmeric knowledge from Vienna to Saint Domingue's plantations, passing through Paris and Cap François – always opposed by the Colonial Machine – we must be cautious, as these descriptions of black mesmerism seem to have been a smokescreen set between the rationality of French judges and the Colonial Machine on the one hand, and the frightening manifestation of black nocturnal vodun ceremonies on the other. In that context, the use of words designating a familiar and reassuring form of charlatanism was doubtless a way to publicly disqualify black magic, so to speak, and irreligion. The categories of mesmerism thus gave words to judges to describe and condemn a mysterious phenomenon, characterized by impressive "trances" and hypnotic phenomena. In the same way, branding vodun as mesmerism opportunely precluded further scientific inquiries from the Cercle des Philadelphes, for example, into such elusive, mysterious, and irreducible black knowledge deliberately rejected from European spheres of scientific

245. On the Macandal episode and its lasting impact in Saint Domingue, see Weaver, pp. 89-97; Pluchon (1987), chapts. 7-8; DPF, pp. 629-31. Such a perspective can be compared to what Vincent Brown has brought to light in an article on contemporary Jamaican slave society; studying the question of Obeah and supernatural beliefs underlying the confrontation between white masters and black slaves, Brown shows how much the "political potential" of any supernatural belief, whether its effects were attested or not, was of great concern for white judges; Brown (2003).

interest. European "mesmerism" and its "mesmeric trances" thought of by physicians, judges, and the rest of the Colonial Machine as an occult power, provided colonial authorities with the perfect words and tools they needed to understand and disqualify the strange (and to them incomprehensible) practice of vodun (voodoo).

French Interest in Native Botanicals and Remedies

If rather uncomfortable with the magical practices of vodun, French authorities and the Colonial Machine were more than open to the botanical and medical knowledge of other peoples and cultures. As one commentator put it in 1777, "The Savages have contributed more to the progress of materia medica than all the Colleges of all the centuries."[246] The record overflows with examples of native remedies passing over into French medicine and assimilated by the Colonial Machine, so much so that it is hard to embrace all the details. In a word, the Colonial Machine became a mechanism for absorbing knowledge of plants and medicinal practice among the peoples of the world encountered by the French in their overseas expansion in the Old Regime. This multi-institutional concern with remedies from the world of the "other" occupied – even preoccupied – the Colonial Machine.

A simple list of transmissions across these cultural barriers does not do justice to the scale or the complexity of what became involved. We have seen how important Peruvian bark became to the European pharmacopeia and the treatment of malaria and intermittent fevers. Europeans learned of the effectiveness of ipecacuanha root, sovereign against dysenteries, from the Indians of Brazil. About these appropriations, the Académie des Sciences itself noted in 1703 :

> Modern medicine possesses several remedies unknown to ancient medicine and as infallible as remedies can be. There are some [in ancient medicine], like antimony and mercury, whose usage apparently was foreseen and divined somehow by reason. Others, such as cinchona and ipecacuanha, are pure gifts of Nature. We received them directly from her own hands, or rather those of a savage People who know only her.[247]

246. "les Sauvages ont plus contribué au progrès de la matiere Médicale que les Colleges de tous les siecles," Ellis (1777), p. 39.

247. "La Médecine moderne a plusieurs remèdes inconnus à l'ancienne, & aussi infaillibles que des remèdes peuvent l'être. Il y en a quelques uns, comme l'Antimoine & le Mercure, dont apparemment l'usage a été prévu & deviné par quelque raisonnement, d'autres comme le Quinquina & l'Ipecacuanha sont de pures faveurs de la Nature. Nous les avons reçues immédiatement de ses propres mains, ou plutôt de celles d'un Peuple sauvage qui ne connait qu'elle," HARS 1700 (1703), pp. 46-47.

In Canada the French picked up Canadian ginseng from Native Americans who passed it on to the Jesuits.[248] Sarrazin reported on a citrus tree "whose root is a poison used by the savages to end their days."[249] Jean-François Gaultier, médecin du roi and the Academy's correspondent at Quebec, studied aboriginal remedies, compiling a veritable ethnobotanical inventory of Canadian and Amerindian popular medicine; and he reported on diseases affecting Indians and advised which native practices should be followed.[250]

Promising reports came out of Indian territory in North America. A 1723 manuscript on "the rivers, lands, and savages of Missouri" mentioned that "the Savages have admirable remedies for wounds and for snakebite. For women who have difficulties conceiving, they use herbs, roots, and seeds that cure quickly and without doing any harm. These remedies are very mild."[251] About the Illinois and other tribes to the west and south, another source summarized by saying that "every day the natives produce marvelous cures using medicinal simples that this land produces. There are so many and with such different virtues so as to make a fatter book than Dioscorides."[252] The Iroquois are mentioned, and from Louisiana, too, the Navy minister back in Versailles received Indian medicinals.[253]

A French colonist in Louisiana reported being cured of sciatica by an Indian medicine man, whereas European doctors and surgeons failed.[254] Astonishingly, a French staff surgeon and correspondent of the Académie de Chirurgie, one Couranier Deslandes, became enslaved by local Indians in Florida in the 1760s. He roamed the woods with them for two months, living mostly on honey, and he lived in a village for another month. He observed medical prac-

248. ALS Sarrazin to Bignon, dated "a quebec le 5 9bre 1717," PA-DB (Sarrazin). This episode is treated separately below, Part IIC.

249. "…anapodophyllor canadens. morini Inst. citronier son fruit est acide bon a manger mais fievreux. sa racine est un poison dont les sauvages se servent lorsqu'ils ne veulent pas survivre a leurs chagrins…," "Plantes envoyees de Canada par M.r Sarrazin Conseiller du Conseil supreme et Medecin du Roy en Canada, 1704," BCMNHN, Ms. 944.

250. B. Boivin, ms. "Notice biographique sur J. F. Gaultier (1708-1756), Naturaliste-médecin du Roi au Canada," pp. 8-9 in PA-DB (Gaultier); see also "Journal des observations Meteorologiques &c. de Mr. Gauthier à Kebec depuis le 1 Octobre 1744 jusqu'au 1 Octobre 1745," OdP, Ms. A 7, 6 [64^3], notably entry for April, 1745, where Gaultier reports diseases affecting Indians and the possibility of Indians being poisoned by the English; Gelfand (1987), pp. 90-91, discusses Native-American medicine and its impact in Canada.

251. "…Les Sauvages ont des remèdes très admirables pour les blessures, pour les morsures de serpents. Pour les femmes en mal d'enfant, ils se servent d'herbes, de racines et de grains de fruits qui guérissent en très peu de temps et sans faire de mal. Ces remèdes sont fort doux," unsigned "Mémoire sur les rivières, les terres et les Sauvages du Missouri," AN-COL F^3 24 (1723), fol. 145.

252. "Les sauvages font tous les Jours des Cures Merveilleuses, avec les simples que produit Cette Terre, il y en á de tant de sortes, et de vertus si differentes, qu'on En feroit un Livre plus gros que Celuy de Dioscoride," "Relation de la Louïsiane par Lemaire," BCMNHN, Ms 948.

253. See AN-COL C^{13A} 4 (Louisiana), p. 561 (23 January 1716) and C^{13A} 5, fol. 283 (25 April 1719).

254. De Page de Pratz (1758), vol. 1, pp. 129-36.

tice among this group. He fell sick himself and was treated by a medicine man. In a report to the Académie des Sciences, Couranier Deslandes described native medical treatments for a number of serious ailments (putrid fever, dysentery, spitting blood), and surgical interventions.[255] Couranier Deslandes noted that the aboriginals he was with were brave in battle, but cowards when it came to surgery. His overall take on this other medical tradition that he was forced to experience was attentive and respectful, and as he was a medical professional, his account is that much more relevant to this one.

A 1784 report out of New Orleans announced a Native American cure for leprosy picked up from the Chicachat Indians. Complicating any simple picture of what is cultural transmission, the French learned of the Chicachat treatment from an Englishman who had lived among them for thirty years.[256] The treatment used the herbaceous flowering plant *Solanum americanum*, or American nightshade. The plant enjoyed something of a vogue as news of it reverberated around the colonial medical world. "Waiting the outcome of experiments," this particular report concluded,

> let us admire and bless the effects of Solanum against the ravages of leprosy, and let us strip away this unjust and outrageous contempt our civilization inspires in us against the indigens of America. I have seen these peoples up close, and I dare think and say that they could in many respects be the masters of our botanists and our naturalists in what concerns the properties of plants and the conservation and physical development of the human body.[257]

By the same token, Sarrazin from Canada held less charitable views, believing that "savages observe animal behavior pretty well, the unique part of Philosophy that seems to have been accorded them."[258]

In the Antilles and South America much the same story unfolded in the same way. Father Le Breton recognized medical practice among Caribbean

255. PA-PV 87 (1768), fols. 85r-88v ("Samedy 14 May 1768") and PA-PV-110 Couranier des Landes, 14 May 1768 ("Sur l'Etât de la Médecine chez les Indiens utchyses").

256. Ms. "Mémoire sur la cure de la Lepre Occidentale. À la Nouvelle Orléans 17 Mai 1784," signed "Villay, Ch.er de S.t Louis et Commissaire de S.M.T.C. à la Louisiane," AN-MAR G 102, dossier 3, #118.

257. "En attendant le succès de ces experiences bienfaisantes, admirons et benissons les effets du Solanum contre les ravages de la Lépre et dépouillons ce mépris injust et outrageant que notre civilisation nous inspire contre les indigênes de l'Amérique. J'ai vu ces peuplades de près et j'ose penser et dire qu'ils pourroient être à bien des egards les maitres de nos botanistes et de nos naturalistes dans ce qui concerne la propriété des plantes et la conservation et le dévelopement physique du Corps humain," Villay ms., AN-MAR G 102, dossier 3, #118.

258. "C'est ainsi que le rapportent les Sauvages, qui selon M. Sarrasin, observent assés bien le naturel des Animaux, unique partie de la Philosophie qui leur ait été accordée," HARS 1714 (1717), p. 27.

natives, even if he disapproved.[259] A missionary on Guadeloupe, Valentin Delaporte, concocted a medicinal tonic from plants from a native pharmacopeia in 1787 and sent it to the Société Royale de Médecine for approval.[260] Doctors in Saint Domingue spoke well of the same vetch that was the base of Delaporte's tonic.[261] The Indians in Cayenne were widely acknowledged as masters of poisons.[262] Out of Cayenne, too, through the Jardin du Roi, came simarouba root, a decoction of which was recognized as a specific against dysentery.[263] Authorities systematically distributed simarouba from Guiana to other colonies for tests.[264] As early as 1718, Antoine de Jussieu tried simarouba successfully, and he introduced a syrup made of simarouba root and published a paper on the medical use of the plant in the *Mémoires* of the Academy of Sciences for 1729.[265] The médecin du roi, Jean Prat, reported positively on the root, as well, from New Orleans in 1735.[266] Later in the eighteenth century, Badier on Guadeloupe made his own syrup out of simarouba and sent his observations to the Société Royale de Médecine.[267]

The same flow of information from local and traditional herbals into French medicine is visible in Africa, the Indian Ocean, and South Asia. For example, Adanson reported from Senegal on "true incense" that he had learned about from natives "who have been in the forests."[268] Of a plant shipped from Île de France, all that could be said was "what our slaves told us."[269] A report of 1710 brought a crossover from Indian and Hindu medicine that preoccupied the Academy of Sciences for a while.[270] In another instance, someone took

259. "Relation historique sur l'île caraïbe de Saint Vincent," in Lapierre, pp. 76-78, 97-98.

260. SRdM, Archives, 97 dr. 72.

261. "Duchenin de l'etang Med à St. Domingue dans laquelle il parle des vertus quil a decouvertes pour les maladies de poitrine a une plante de L'amerique à la quelle il donne le nom de vulneraire & Le donne sous La forme de thé on peut le prendre ou a l'eau ou au Lait le matin a dejeuné et en boires quelques tasses dans la journée," SRdM, Ms. 10, p. 40 ("Séance Du Vendredi 19 decembre 1783").

262. Ms. Code de Cayenne, vol. 3 (1740-1760), p. 360, in AN-MAR G 237.

263. See "Cinq mémoires divers sur le simarouba," BCMNHN, Ms. 1152.

264. See AN-COL B[35] (1713), fol. 612 (29 March 1713).

265. A. de Jussieu (1729). See also report by Dehorne, *Journal de médecine, chirurgie et pharmacie militaire,* vol. 5 (1786), pp. 214-15.

266. See Prat letter dated "a la n[lle] orleans le 3e may 1735," in Lamontagne (1963), pp. 124-26.

267. SRdM-H&M, vol. 2 (1778), p. 299.

268. "C'est cette resine qui est regardée ici comme l'Encens et le véritable Encens, non seulement pàr les Blancs, mais encore pàr les Negres: je me suis informé particulièrement du nombre de ces derniers qui ont été dans les forests du Gommier, quel étoit l'arbre qui donne la resine ou l'encens que l'on trouve mêlé dans la gomme que les Maures nous vendent...," ALS, Adanson to [Bernard de Jussieu], dated "Au Senégal le 1er aoust 1750," in PA-DB (Adanson).

269. "...tout ce qu'on a pu savoir de nos Esclaves c'est qu'elle [cette plante] demande la bonne terre des Vallons," undated, unsigned ms., "Annotation sur ce que cette Caisse contient," dossier De Reine, BCMNHN, Ms. 357.

270. See PA-PV 29 (1710), fols. 41v-42 ("Le Mércredi 5. Février 1710"), 44-44v ("Le Samedi 8 Février 1710), 47v ("Le Samedi 15 Février 1710"), 75-78 ("Le Samedi 1. Mars 1710"), 121-121v ("Le Samedi 12 Avril 1710"), and 203v-206v ("Le Samedi 28 Juin 1710"), the latter entry reporting that "M.r Saveur avoit eû ces Drogues de M.r de la Mare natif de Caën Officier des Vaisseaux du Roy qui les avoit apportées du Perou, des Indes, et du Brésil...."

"notes on different drugs unknown to us with their Malabar names commonly found in the bazaars of India."[271]

The case of Nicolas L'Empereur is of note here. L'Empereur was a surgeon major employed by the Compagnie des Indes and posted first to Balasore in India and then to Chandernagor where he was a member of the Council of the colony. He both transmitted (or at least tried to transmit) local botanical knowledge and at the same time acted as a botanical entrepreneur, positioning himself in a revealing way towards the Colonial Machine. In 1719, he received from the abbé Bignon two gold medals, one from the king and one from the regent, for his botanical samples. In 1725, aged sixty-five, he sent to the Académie Royale des Sciences an impressive botanical manuscript he called the "Jardin de Lorixa." This manuscript was a sum of medical and botanical knowledge, patiently gathered in the field, from books, translations, and local informers. Jussieu rejected it. L'Empereur's failure to succeed with his manuscript and with the Colonial Machine was, first, due to the fact that he asked for money to cover his expenses (which were significant and which Jussieu was seemingly reluctant to pay) but also because the work was not of direct use to Jussieu in his botanical program.[272]

Pierre Sonnerat (1748-1814), traveling in India, submitted a nosology of Indian diseases to the Société Royale de Médecine.[273] Sonnerat likewise presented his observations of the medicinal properties of the bark of the tree called *Béla-Aye* from Madagascar, a bark that Cossigny *fils* noted was widely available on Île de France.[274] Based on Indian sources, Cossigny himself experimented with different varieties of ricin oil made from castor beans, and he sent his correspondent in Paris, Le Monnier, a recipe for "the bitter drug of India" for indigestion.[275] Native use of roots seemed to attract a lot of attention. The Société Royale de Médecine endorsed a bitter root from Ceylon for treatment of intermittent fevers.[276] The "potato with two roots" of Île de France presents a revealing example of the transmission of knowledge across cultural frontiers. In 1763, using the appropriate scientific and medical terms and backed up by experi-

271. "Nottes d'où L'on Tirre differents drogues tels que L'on trouve communement dans Les Bazard de L'inde avec Leurs noms En la langue Malabard aussi que cel dont L'usage ne nous est pas Connue," undated 2pp. ms., BCMNHN, Ms. 357.

272. Raj, pp. 27-59, gives full details and an analysis of this case. The manuscript of the "Ellemans botanique des plante du Jardin de Lorixa" [sic] is still in the Library of the Museum National d'Histoire Naturelle in Paris.

273. SRdM, Ms. 8, p. 433 ("Séance du Mardi 17 Juillet 1781").

274. SRdM-H&M, vol. 3 (1779), Mémoires, pp. 689-90; SRdM, Ms. 9, pp. 193-4 ("Séance du Vendredi 14 Juin 1782"). On Sonnerat, see further below, Part IIC.

275. "Recette de la Drogue amere de L'Inde spécifique contre les Indigestions," ALS, Cossigny to Le Monnier, dated "Paris le 3 Mars 1773," BCMNHN, Ms. 1995; see also Cossigny's "Mémoire sur l'huile de Ricin, appellé vulgairement *Palma christi*," dated "Palma Ile de france, le 28 7bre 1785," BCMNHN, Ms. 357.

276. SRdM-H&M, vol. 2 (1778), Histoire, pp. 291-94.

ments, a certain Dr. De la Ruë on Île de France communicated information about native use of this potato with two roots to the Academy of Sciences in Paris. In committee, however, the Academy voted to "suppress [the paper], given the danger of publication." In this case, knowledge of native practices in the East made it to the metropolis only to be extirpated from public knowledge in the West because the potato with two roots could be used as an abortifacient.[277]

From China, Jesuit reports filtered in on the state of Chinese medicine.[278] The botanical traveler and Academy correspondent, Laurent Garcin (1683-1751), reported on Chinese and Indian medicine.[279] D'Incarville sent an herbarium of 400 plants from a Chinese pharmacopeia.[280] The chemist, apothecary, and *pensionnaire* at the science academy, Claude-Joseph Geoffroy (1685-1752) wrote about preparing root compounds in the Chinese manner.[281] Undoubtedly, the world's materia medica flowed into France via the Colonial Machine in other ways as well.

The actors involved seemed cognizant of other medical traditions out in the wider world. Antoine de Jussieu certainly appeared so in writing in 1725 to the abbé Bignon, president of the Académie des Sciences, about the expedition of the Academy correspondent, André Peyssonnel, to North Africa:

> But we owe Mr. Peyssonnel even more for having followed the advice we gave him to go to all the markets of the cities, towns, and villages of this land and to collect roots, leaves, fruits, and seeds used in everyday life because in this way we gain an idea of the difference between their medicine and ours by the different usages of vegetables, edible plants, and specifics that we may lack for illnesses that they use in that country.[282]

277. McClellan (2003B) tells this story, pp. 74-76. For the larger background, see Schiebinger (2004), chapts. 3-4.

278. SRdM, Ms. 11, fol. 115r ("Séance du Mardi 31 Juillet 1787").

279. See Garcin (1730), passim and p. 387, where he writes of species "known to the Indians...," and (1734), passim and p. 240, where he recommends infusions of mangosteen skins to combat "looseness"; see also originals of these papers in French, RS/L Cl.P. X (ii), #12-14 and LBC 21.

280. ALS Incarville to [B. de Jussieu] "à Peking ce 3e 9bre 1751," PA-DB (Incarville).

281. PA-PV 59 (1740), fols. 147-49 ("Le Samedi 9e Juillet 1740").

282. "Mais ce dont on doit savoir meilleur gré a M.r Peyssonel est d'avoir suivi le Conseil que nous lui avons donné de ramasser dans tous les marchés des villes, bourgs, et villages de cette côte, les racines, feüilles, fruits et semences que l'on y porte pour les usages de la vie, parceque nous avons par la une idée de la différence de leur Medecine avec la Notre par les differents usages de plusieurs plantes legumes, des plantes potageres et des spécifiques qui peuvent nous manquer pour des maladies auxquelles ils servent dans ces pays la," ms. "Copie de la lettre écrite a Monsieur l'Abbé bignon par Monr.r de Jussieu De Paris le 29 Juillet 1725," AN-MAR C7 128 (Granger).

This openness to foreign drugs provoked some resistance, however, from official pharmacy in France. For example, from 1753 to 1761 the Compagnie des Indes sent J.-B.-C. Fusée-Aublet to Île de France to develop indigenous drugs from India. The idea was that French drugs were expensive and often ineffective after the long sea voyage to the Indian Ocean. Fusée-Aublet was to find local substitutes. Yet the Compagnie then ordered him to stick to the Paris pharmacopeia. "What an order!," he replied. "It goes directly against the views of the Compagnie, against the interests of the colony, and against the object of my mission. I was expressly sent to this island to use the products of this country and of India in general."[283] In Cayenne, too, local officials turned to native medicines because the cost of imported drugs was so high.[284]

Official French medicine was the standard for Europeans, of course, and throughout the colonial world the practice of medicine was strictly limited to whites.[285] Yet colonists and officials turned to alternative, slave medicine, usually as a last resort, and they displayed the same kind of probing interest in medical traditions out of Africa as they did with those of the rest of the world. They were interested in African treatments against venereal disease and smallpox.[286] They were very fearful of slave knowledge of poisons, yet European doctors looked to slave medicine for treatments *against* poisoning, particularly against accidental poisoning caused by the manchineel apple.[287] An undated manuscript from Guinea-Bissau, probably from the 1760s, compiled a "description of some of the plants blacks of the island of Bissau use to cure different illnesses."[288] The compilation lists 105 plants with botanical descriptions and their uses in native African medical practice. Number 93, for example, was "spiny bran…whose decoction is good against colic and causes women to abort."[289]

The flow of medical botanicals and information about treatments from the world beyond France into French medicine and medical practice in the eighteenth century was constant and relatively open; this stream testifies to the

283. "…*quel ordre*! C'est aller directement contre les vües de la compagnie, contre les interets de la colonie, contre l'objet de ma mission. J'avois eté expressement envoyer pour transporter dans l'isle; et pour faire usage sur les lieux des productions du pays, et de l'inde en general," "Mémoire justicatif" [1763], BCMNHN, Ms. 452.

284. Ms. Code de Cayenne, vol. 5 (1768-1777), pp. 649-50, in AN-MAR G 237.

285. McClellan (1992), pp. 135-36; Weaver (2006).

286. See SRdM, Ms. 7, p. 141 ("Séance du Vendredi 23 avril 1779") and references to venereal disease, AN-MAR, C^7 128 (Granger).

287. On this point in general, see Pluchon (1987), McClellan (1992), pp. 54-55, and Peyssonnel (1758).

288. "Description D'une partie des plantes dont les negres de L'isle du Bisseaux se servient pour La Cure de differentes maladies," BCMNHN, Ms. 1288. Guinea-Bissau was founded as a slave-trading entrepôt on the coast of West Africa in 1765.

289. "L'Epine brane… Sa decoction Est bonne Contre les coliques et elle fait avorter les femmes," BCMNHN, Ms. 1288, #93.

effectiveness of the Colonial Machine in providing conduits for information to pass from frontier contact with natives back to the central councils of French science and government in the motherland. This drumbeat of interest in native botanicals is another defining feature of the Colonial Machine in action. It connects with the interest we have seen in native astronomies, too, particularly those of India and China. We should note en passant that the interests of the French in the expert knowledge of other peoples and cultures was not limited to the medical, as seen in the case of adopting Asian-Indian boatbuilding techniques to preserve wood from the ravages of insects.[290] The point would not need further comment except that two special cases allow us to see more deeply into the functioning of the Colonial Machine as it confronted medical knowledge and medical knowledge systems foreign to its own.

Lizards. The flow of information across cultures was not a straightforward matter, as it might seem at first. Indeed, in assimilating medicines and the medical practices of others, the Colonial Machine functioned as a knowledge-transforming machine. Native knowledge did not map directly onto European knowledge; rather, knowledge that the French came to have of "the other" was fabricated through a process of knowledge-making. This side of the Colonial Machine in action is no better seen than in the case of the lizards from Guatemala.[291]

American Indians in Louisiana, Central America, and South America regularly ingested lizards of the genus *Anolis* to cure a variety of diseases. In 1782 a Spanish physician, Joseph Florès, published a pamphlet on the practice and the medical miracles to be had from eating these lizards. Florès' tract was translated the same year into French by a professor at the Royal College of Surgery at Lyon, Grassot: *Spécifique nouvellement découvert dans le royaume de Guatimala, pour la guérison du cancer & de quelques autres maladies fréquentes* (A Specific Recently Discovered in the Kingdom of Guatemala for Curing Cancer and Some Other Common Illnesses). To this point, it can hardly be said that the Colonial Machine was involved, but Grassot wrote about this discovery to the Minister of Foreign Affairs, Vergennes, and this step triggered a series of events that very much entangled the Colonial Machine. As Grassot reported:

> The Indians...take a small gray lizard, cut off its head and tail, pull out the intestines, strip off the skin, and immediately eat the raw and still warm flesh of this animal. Every day they eat a small lizard in this fash-

290. "Memoire sur l'utilite et L'avantage de la decouverte de M.r frisher Pour préserver La Carêne des vaisseaux de la piqure des Vers," AN-MAR G 235, #8, where one reads: "On sait qu'on fait usage dans toute la partie de L'inde, Et des phillippines, de la Composition appellée galle-galle, et qu'on En tire un grand avantage pour La durée des Vaisseaux...."

291. The details of this story are taken from the thirty-one items preserved in SRdM, Archives, 165 dr. 11. SRdM-H&M, vol. 4 (1780-1781), Histoire, pp. 337-42, also presents information on this episode.

> ion. They say that sometimes one suffices, if not they eat three of them. These Indians swear that by dint of this remedy they have always cured their ulcers and endemic venereal disease.[292]

Calling the practice "very disagreeable," Vergennes nevertheless wrote about it to Joseph-Marie-François de Lassone (1717-1788), First Physician to the King and Queen, member of the Paris medical faculty, veteran *pensionnaire* of the Académie des Sciences, and a leading member of the Société Royale de Médecine.[293] The Société Royale was naturally interested in this information, and it had Joseph-Barthélemy-François Carrère (1740-1802) prepare an official report on medical uses of the *Anolis* lizard, which appeared in the *Histoire et Mémoires* of the Société Royale for 1782. Carrère reported positively, noting that Spanish physicians had used the treatment to good effect, and "if these observations are true, they leave no doubt about the efficacy of this reptile for the treatment of cancerous ailments and perhaps even for many skin diseases." He went on to offer a scientific explanation of the phenomena: "It seems that this reptile acts in purging the blood and in expelling morbific molecules with which the blood is infected."[294] By the same token, he thought the Indian way of swallowing still palpitating lizards "makes it impractical in Europe," and he suggested that alternative preparations be developed.[295]

The Société Royale de Médecine sought to obtain *Anolis* lizards for its own experiments, and using his diplomatic contacts with the Spanish court, Vergennes obtained lizards from America. They were shipped to France in special containers for the long-distance journey.[296] The lizards were handed over to Daubenton and the physician Pierre-Jean-Claude Mauduyt de la Varenne, for experiments, but too few had survived the voyage to make any tests. But all was not lost, seeing that the lizards from America resembled "wall lizards" indigenous to the south of France. For Carrère, differences between French and American lizards notwithstanding, "advantages superior to those remedies that have been

292. "Les indiens...prennent un petit Lezard gris, lui coupent la tête et la queue, lui arachent les intestins, lui enlevent la peau et mangent dans le même moment la chair crue et encor chaude de cet animal. c'est de cette façon qu'ils mangent tous les jours un petit Lézard. ils disent que quelques fois un seul suffit. si non ils en prennent trois. ces indiens assurent qu'au moyen de ce spécifique, ils ont toujours gueri de leurs ulceres et de la maladie venerienne endemique," SRdM, Archives, 165 dr. 11, #1.

293. ALS De Vergennes to M. De Lassone, dated "fontainebleau le 4 9bre 1783," SRdM, Archives, 165 dr. 11, #2.

294. "Ces observations, si elles sont vraies, ne laissent aucun doute sur l'efficacité de ce reptile dans le traitement des affections cancéreuses, & peut-être même de plusieurs maladies de la peau... Il paroît résulter de là que ce reptile agit en dépurant la masse du sang, & en poussant au dehors les molécules morbifiques dont elle est infectée," "Rapport sur les vertus médicales des lézards du royaume du Guatimala," signed and dated, "Au Louvre, le 20 décembre 1782, Carrere," SRdM, Archives. 165 dr. 11. #4; published in SRdM-H&M, vol. 4 (1782), pp. 337-342.

295. "La manière de faire usage de ce remède paroît le rendre impracticable en Europe... Il seroit à desirer qu'on fît avec cet animal quelque préparation particulière," SRdM, Archives. 165 dr. 11, #4.

296. See reference, SRdM, Ms. 9, p. 38 ("Séance Du mardi 11 fevrier 1783").

employed to date can still result."[297] Daubenton and Mauduyt de la Varenne suggested that "there are enough similarities between American lizards and ours for those of our colleagues who think it apropos to test the latter, keeping in mind that beneath a cooler sky, if these animals have the properties attributed to them, they will be weaker, and the dose and duration of use of this remedy will have to be increased."[298]

In its usual fashion the Société Royale de Médecine printed a broadside about the medical use of lizards, which it distributed to its correspondents.[299] Vicq d'Azyr, permanent secretary of the Société Royale, sent this to the Lieutenant General of Police in Paris, Lenoir, who responded confidently that "green lizards are recognized as proper for the cure of different illnesses."[300] Their medical efficacy would seem to have been established at this point, and reports of successful cures began to be returned to the mother society in Paris. Ramel, M.D. and correspondent of the Société Royale, wrote to Vicq d'Azyr from the galley prisons of Marseilles in May of 1785:

> In the Marseilles region one only talks about the happy discovery just made by Spanish medicine of a specific against cancers, ulcers, skin diseases, and certain sores considered incurable. This discovery will in effect be very important, precious. and good for humanity....This discovery was made in the Spanish colonies and sent first to Spain whence it came to France, and it was tested in the crucible of experience around Madrid. Perhaps we have a first step in the long road that can lead us to a very important discovery. How many very necessary remedies do we not owe to a happy accident.[301]

297. "...mais il pourroit toujours en resulter Des avantages superieurs a ceux des remedes qui ont été employés jusqu'ici," SRdM, Archives. 165 dr. 11, #4.

298. "...mais nous croyons qu'il y a assez de rapport entre eux & les nôtres, pour que ceux de nos confrères qui le jugeroient à propos, mettent les derniers en expérience, en se souvenant que sous un ciel moins chaud, si ces animaux ont la propriété qu'on leur attribue, elle sera moins exaltée, & qu'il faudra augmenter la dose & la durée de l'emploi du remède," see published report, signed and dated "Au Louvre, le 23 novembre 1784," SRdM-H&M, vol. 4 (1780-1781), p. 341.

299. See reference to this brochure, ALS Cassagne [to Vicq d'Azyr], dated "à aix le 13 juin 1788," SRdM, Archives, 165 dr. 11, #14.

300. "Lezardes verds [sont] reçonnûe propres à la guérison de differentes maladies," ALS Lenoir to "M. Vicq dazir," dated "À Paris le 19e Xbre 1783," SRdM, Archives, 165 dr. 11, #6.

301. "on ne parle plus à marseille et dans les environs que de l'heureuse Découverte que la médecine espagnole vient de faire d'un specifique assurè contre Les Cancers, Les ulceres [illegible], Les maladies de la peau et certaines playes regardèes comme incurables. cette Dècouverte serait en effet bien importante, bien précieuse et bien interessante pour l'humanité...Cette découverte faite dans les Colonies espagnoles et envoyée D'abord en espagne d'ou elle est parvenue en france, a èté soumise au creuset de l'experience à Madrid et dans ses environs...peut-être avons nous fait un premier pas dans un sentier laborieux qui pourra nous conduire à une dècouverte trés important. Combien de Remedes tres necessaires ne devons nous pas à un hazard heureux," ALS "Ramel le fils MD, cors.pdt. de la societé Royale de med.ne," dated "[Marseilles] au bagne 20 mai 85," SRdM, Archives, 165 dr. 11, #8.

From Besançon came word of a case of smallpox cured by lizards.[302] From the royal hospital at Pontarlier a report arrived of the successful treatment of two cases of venereal disease; one of the patients ate 800 lizards.[303] From Rennes, a physician wrote in about a woman with advanced breast cancer for whom he prescribed lizards; she got only marginally better, but the physician, Degland, still effused that "we must acknowledge that we live in the century of marvels."[304] Other positive reports from elsewhere in France continued to filter in to the Society in 1786, 1787, and 1788.[305] But no one was a bigger supporter of the lizard treatment than one Chantrans, a retired military engineer who had been stationed in Saint Domingue. Chantrans learned of the cure from a diseased colonist in the town of Haut Limbé in Saint Domingue who had cured himself in three months "using this antidote of the savages."[306] A similarly affected slave was able to return to work after nine weeks.

Chantrans and his medical associates at Besançon distilled lizards in the hopes of creating a more palatable treatment. The first distillation involved 50 lizards, the second 114, and a third distilled the skin, feet, tails, and heads of lizards from the first two trials.[307] The resulting liquors worked well in one trial case. Chantrans held out hope that different distillations of gray and green lizards would tell which contains the most "volatile alkali." Tellingly, Chantrans and company operated with scientific assumptions regarding the efficacy of lizards. As he wrote:

> ...the less the body of the small animal remains exposed to evaporation, the more it conserves these volatile salts that probably constitute its principal virtue....Most blacks wolf them down like pills....One abstains from fatty foods that envelope the alkali and destroy its effectiveness....We have not yet conducted chemical analyses of the different kinds of lizards found in this province, but our daily trails have shown us that they all produce more or less the same secretions, and apparently the volatile alkali detaches itself from their flesh and then passes into the circulation....As for this volatile alkali that I had suspected was in the distillate, it may contain the elements, but it is not

302. SRdM, Archives, 165 dr. 11, #16.

303. ALS "Gresset, chirurgien Major de l'hôpital royal de pontarlier," dated "À pontarlier le 1e Xbre 1785," SRdM, Archives, 165 dr. 11, #19.

304. "Il faut convenir que nous sommes dans le siécle des merveilles...," "Extrait d'une Lettre de M. Degland Med. a Rennes 17 7bre 1785," SRdM, Archives, 165 dr. 11, #9.

305. SRdM, Ms. 11, fol. 58r ("Séance du Mardi 14 Novembre 1786"), fol. 59r ("Séance du Vendredi 17 Novembre 1786"), fol. 64v ("Séance du Vendredi 15 Décembre 1786"), and fol 68r ("Séance du Mardi 9 Janvicr 1787"). SRdM, Archives, 165 dr. 11, #22-31, notably #22, ALS, Rognon to "Monsieur et très honoré confrere," dated "Besançon 9 Janvier 1785."

306. "...d'essayer Cet antidote des Sauvages," ALS, "de Chantrans," dated "à Besançon le 17 mars 1784," SRdM, Archives, 165 dr. 11, #17.

307. ALS de Chantrans to SRdM, dated "à besançon le 16 septembre 1786," SRdM, Archives, 165 dr. 11, #21.

completely formed there. I convinced myself of this at Mr. Fumé's, an accomplished chemist from Besançon, in tests that seem decisive."[308]

In contradistinction to the theory espoused by Carrère, Chantrans believed that "...as experience often shows, the flesh of animals and plants becomes more loaded with active principles as the soil and air where they live are less humid. The climate of France being much drier than that of Louisiana, French lizards should be more alkalized than those of Louisiana and consequently better suited to curing venereal diseases.[309] Having written in 1784 that "the use of *Anolis* lizards is spreading here every day," by early 1786 Chantrans confidently asserted that "it seems to me that the lizard remedy against venereal diseases and even scrofula (*dartres*) is pretty well established today that it should be offered to the public which will profit from it."[310]

But then, less optimistic reports began to flow in to the Société Royale. On behalf of the Société, its secretary, Vicq d'Azyr, wrote to the French ambassador in Spain for further information. The report back was that trials in Spain were not successful. Nonetheless, his correspondent, rather than thinking lizards were ineffective, wrote that "one must conclude that the climate and foodstuffs influence the power of the remedy."[311] But the médecin du roi recently returned from Cayenne, Laborde, was unabashedly negative in squashing the lizard cure.

308. "...moins le corps du petit animal demeure Exposé à L'evaporation, plus il conserve de ces sels Volatils qui constituent probablement sa principale vertu...La plupart des négres les avalent comme des pilulles...L'on s'abstient d'alimens gras qui pourroient envelopper L'alcali et détruire son action...nous n'avons pas Encore fait L'analyse Chymique des différentes Sortes de Lézards que L'on trouve dans cette province; mais nos Epreuves journalieres nous ont démontré, que tous produisoient sensiblement les même sécrétions...apparemment que L'alcali Volatile seul, se détache de leurs chaires, et passe ensuite dans La circulation....Quant à L'alcali Volatil que j'y avois Soupconné, elle [the distillate] peut En contenir les elémens, mais on n'e L'y trouve pas tout formé. je m'en suis convaincu chez Mr. *fumé*, habile chymiste de besançon, par des Epreuves qui paroissent décisives." These words are extracted from three different letters of Chantrans to Vicq Azyr, SRdM, Archives, 165 dr. 11, #17, dated "à Besançon le 17 mars 1784"; #20, dated "à besançon le 6 Xbre 1785"; and #21, dated "à besançon le 16 septembre 1786."

309. "...Comme l'expérience le montre souvent, que les chairs des animaux et les plantes sont d'autant plus chargés de principes savoureux, que les sol et l'atmosphère ou il vivent sont moins humides, le climat de france étant beaucoup plus sec que celui de la louisianne, Les lézards de france doivent être plus alcalisés que ceux de La louisiane et par Conséquent plus propres à la guérison des maladies vénériennes...," ALS, "de Chantrans" [to Vicq d'Azyr], dated "à Besançon le 17 mars 1784," SRdM, Archives, 165 dr. 11, #17; see also, SRdM, Ms. 10, p. 138 ("Séance du Vendredi 26 mars 1784").

310. "L'usage des anolis s'etend ici chaque jour...," ALS, "de Chantrans" dated "à Besançon le 17 mars 1784," SRdM, Archives, 165 dr. 11, #17; "...il me paroît que le remède des Lézards, contre les maladies Vénériennes et même les dartes est assez bien constaté des aujourdhuy, pour le doner au public, qui En profitera...," ALS "de Chantrans," dated "à besançon le 5 janvier 1786," SRdM, Archives, 165 dr. 11, #18. For later contact of Chantrans with the Société Royale de Médecine, see ALS de Chantrans, dated "besancon le 10 janvier 1787," SRdM, Archives, 165 dr. 11, # 29; SRdM, Ms. 11, fol 81v ("Séance du Vendredi 9 Mars 1787").

311. "On doit en conclure que le clymat et les alimens influent sur la vertu du remede," ALS "hiviart" to Vicq d'Azyr, dated "St. Sébastien en Espagne le 17 Janv.r 1784," SRdM, Archives, 165 dr. 11, #7.

> Skin diseases such as yaws, scrofula, leprosy, and others of less consequence are extremely common in the colony of Cayenne. The remedy, if one can call it that, would be very advantageous. I am speaking of lizards, of which at least seven or eight species are known. For more than six months colonists and people from everywhere have made great use of them with confidence in their great promise. The ill believed themselves to have experienced the greatest relief. Time showed that this was only an illusion.... This treatment has been entirely abandoned.[312]

From Besançon came new information that a patient originally reported as cured was now in "pitiable state."[313] A surgeon from Aix-en-Provence announced the failure of the treatment in three cases, yet he continued: "For me I wish that we don't lose sight of a means that can be useful when it becomes better known."[314] Similarly, one Calvet from Avignon saw no change in a patient with breast cancer after an elaborate protocol using lizards. Nevertheless, Calvet wrote that "it seems to me that the use of lizards for cancer is not totally fruitless. We might hold even greater hopes that the flesh of snakes or perhaps also that of toads would be of some help."[315] In a later letter to Vicq d'Azyr, Calvet says that "I continue to believe that we must multiply experiments on the treatment of cancer using lizards. I have seen, if not success, at least reasons for hope."[316]

These developments prompted a second official report from the Société Royale de Médecine by Carrère:

312. "Les maladies de la peau telles que le pians, les dartrés, La Lepre et autres de moindre consequence sont extrement communes dans La colonie de caïene: le remede, si on pouvoit lui donner ce titre, Le seroit encore davantage; je veux parler des Lezards, dont au moins, sept à huit especes sont connûes. Les habitans et Les personnes de partou ont fait le plus grand usage pendant plus de six mois avec cete confiance que les plus belles promesses [undecipherable]...Les malades eux mêmes croïaient en eprouver les plus grand soulagement: Le tems a prouvé que ce n'etait quillusion [sic]....Ce remède a été entièrement abandoné," "Memoire sur le peu de succés eprouvés dans la colonie de caïene par L'usage medicinal des Lezards dans plusieurs especes de maladies de la peau," signed "Laborde, paris 16 juin 1786," SRdM, Archives, 165 dr. 11, #10; see also SRdM, Ms. 11, fol. 27v ("Séance du Vendredi 16 Juin 1786").

313. "Elle est dans un etat pitoyable," ALS Rongnon to Vicq d'Azyr, dated "Besançon 6 Juillet 1788," SRdM, Archives, 165 dr. 11, #31; see also SRDM, Ms. 11, fol. 203r ("Séance du Mardi 15 Juillet 1788").

314. "...pour moi je Désire qu'on ne perde pas De Vûe un moyen qui peut etre utile quant il sera mieux connû," ms. "Observations et remarques sur les effets des petits Lézards de muraille dans le traîtement du cancer, par M. Cassagne, maitre en chirurgie à aix en provence," SRdM, Archives, 165 dr. 11, #15; see also SRdM, Ms. 11, fol. 196v ("Séance du Lundi 23 Juin 1788").

315. "...il me semble que, dans le cas que l'usage des lezards dans le cancer ne soit pas tout a fait infructieux, il y auroit encore plus à espérer de la chair des viperes prise de la même maniere. peut être aussi que celle des crapaux seroit de quelque secours...," ALS "Calvet," dated "Avignon ce 9e 9bre 1786," SRdM, Archives, 165 dr. 11, #11; see also SRdM, Ms. 11, fol. 60v ("Séance du Vendredi 24 Novembre 1786").

316. "...Je persiste à croire que nous devons multiplier les expériences sur le traitement du cancer par les lezards, j'ay entrevû, si non des succès, du moins des motifs d'esperance," ALS, Calvet to [Vicq d'Azyr], dated "Avignon ce 9e janvier 1787," SRdM, Archives, 165 dr. 11, #12.

> We cannot ignore that several letters written from Spain have spread doubts on the assertions of physicians who want to extol and doubtless exaggerate the virtues [of lizards]. Mr. de Laborde, one of our most zealous correspondents and who practiced medicine and surgery for a long time in Cayenne, communicated his observations to us, from which it resulted that the lizards used in this colony for the treatment of yaws, venereal disease, cancer, scurvy, etc. seemed at first to bring some relief, but soon after the afflicted received no real advantage from them.[317]

In mid-1787 Mauduyt de la Varenne was in further discussion with Daubenton over the use of lizards, but the affair petered out from there.[318] The case came to naught, but it illustrates well the social construction (and deconstruction) of knowledge that involved the Colonial Machine as it wrestled with medical knowledge and expertise among the world's peoples. A fact in Guatemala ceased to be one in France.

Ant Nests. A particular species of a small ant in Guiana gathers material and makes its nests in the crooks of trees. In the eighteenth century at least, these ant nests were plentiful along the Approuague River in Guiana. The aboriginal peoples of the region seemingly made no use of these nests. Slaves, however, did use nests as tinder (One presumes these were maroon slaves in independent communities as well as plantation slaves.) They would take some of the brownish-red, spongy nest material, put it in a gourd, light it, and then cover the gourd, letting the material smolder until it went out. With this treatment the nest material became an excellent tinder that lit easily with sparks from a flint. The French, however, came to use these ant nests very differently – in a medical capacity to staunch hemorrhages.[319] The instance is a small, but revealing one of the Colonial Machine in action. The beauty of the episode is that it blurs the boundaries between any simple notion of a division between native and Western knowledge, it undermines our sense of the easy transfer of knowledge across cultural barriers, and it complicates our understanding of the actors involved in the process of colonization and the making of knowledge.

The story is quickly told. Laborde, the médecin du roi at the military hospital in Cayenne, experimented with nest material as a substitute for what was

317. "Nous ne devons point leur laisser ignorer que plusieurs Lettres écrites d'espagne, ont repandu des doutes sur les assertions des medecins qui en ont celebré et sans doute exageré les vertus. M. DelaBorde l'un de nos corespondants les plus zèlés; et qui a pratiqué longtemps la médecine et la Chirurgie à Cayenne, nous a communiqué des observations, desquelles il resulte que les Lezards emploiés dans cette Colonie pour le traitement du Pian de la maladie venerienne, du cancer, du scorbut, &c ont paru apporter d'abord quelque soulagement, mais que bientot après les malades n'en ont retiré aucun avantage réel...," report signed "Carrere" and dated "au Louvre le 2 juin 1786," SRdM, Archives, 165 dr. 11, #3.

318. See SRdM, Ms. 11, 113r ("Séance du Mardi 24 Juillet 1787").

319. On this episode see SRdM, Archives, 132 dr. 49, "Cayenne, nids de fourmis provenant de Cayenne." See also relevant entries in the minutes of the meetings of the Société Royale de Médecine, SRdM, Ms. 11.

then used (the fungus agaric) to stop hemorrhaging from amputations, wounds, and sores.[320] The nests proved highly effective as an absorbent and an astringent. Samples were sent to de Castries at the Ministry of the Navy and the Colonies, who in turn forwarded the material to the Société Royale de Médecine with a request for its advice on the properties of this specific.[321] The Société formed a committee consisting of Mauduyt, Vicq d'Azyr, and Fourcroy, who with other physicians in the capital tested the efficacy of the nests to stop blood flow. The material held together well, and its supple and spongy properties made it wonderfully adaptable for treating wounds. The rapporteurs issued a positive report, concluding that "the qualities of the ant nest seemed to us superior to those of agaric and consequently we think it would be useful to provision hospitals with this substance."[322] The minister wrote back thanking the Société for its report, adding: "It seems that this substance can be very useful for use in hospitals and shipboard pharmacies. I have given orders to Cayenne that as much of this material as possible be collected."[323]

So, the French got the local Indians to forage in the forest for ant nests! We know this only because of the bureaucratic exchange over how much to pay the Indians for the ant nests they retrieved. The original positive report of the Société Royale de Médecine on ant nests passed from Minister of the Navy de Castries to Minister of Foreign Affairs Montmorin, who then wrote to the Société about the pricing matter:

> As it is the Indians of Guiana who take care of finding ant nests in the forests, the Administrators would like to know the value of this substance in order to determine the payment due the Indians who bring nests to Cayenne. I am sending you a new sample of the material so that the Société Royale can decide the price to put on this substance. Would you be so kind to let me know the decision.[324]

320. There is a perhaps spurious report from 1775 of ant nests being used in this way; see ms. Code de Cayenne, vol. 5 (1768-1777), p. 832, in AN-MAR G 237.

321. Letter of de Castries, "à Versailles le 29 juin 1786," SRdM, Archives, 132 dr. 49, #1; see also SRdM, Ms. 11, fol. 31r ("Séance du Mardi 4 Juillet 1786").

322. "... les qualités du nid de fourmis nous ont paru Superieures a celles de l'agaric, qu'en conséquence nous pensons qu'il seroit utile d'approvisionner les hopitaux de cette substance," SRdM 132 d 49, #5; see also SRdM, Ms. 11, fol. 60v ("Séance du Vendredi 24 Novembre 1786").

323. "Il paraît que cette substance peut être très utile pour l'usage des hôpitaux et des pharmacies des vaisseaux. Je donne des ordres à Cayenne pour qu'on en rassemble autant qu'il sera possible," ALS de Castries, dated "à Versailles, le 15 décembre 1786," SRdM 132 d 49, #2; see also SRdM, Ms. 11, fol 65r ("Séance du Mardi 19 décembre 1786"), and ms. Code de Cayenne, vol. 7 (1783-1803), p. 363, in AN-MAR G 237.

324. "Comme ce sont les Indiens de la Guiane qui s'occupent de la recherche des nids de fourmis dans les forêts, les administrateurs désireraient connaître la valeur de cette substance afin de déterminer la rétribution à donner aux Indiens qui en apporteroient à Cayenne. Je vous en envoïe un nouvel échantillon pour mettre la Société Roïale en état de prononcer sur le prix à mettre à cette substance. Vous voudrés bien m'en faire connoître la décision," Montmorin letter, dated "à Versailles le 24 septembre 1787," SRdM, Archives, 132 dr. 49, #3; see also, SRdM, Ms. 11, fol. 133v ("Séance du Vendredi 28 Septembre 1787").

The Société said it would try to get the necessary information and wrote to its correspondent, Laborde, previously médecin du roi in Cayenne who by then had retired to the town of Auch in France. The suggestion was made that the Société contact the current surgeon-major of the colony at Cayenne, a M. Noyer. "Mr. Noyer can doubtless give satisfactory information on ant nests. The Company decided to ask him to send us some pristine nests, to have them come to us in this state without any type of preparation or any alteration, with information on their usage, on their properties, and [passing the buck] on the price that one can pay the Indians of Guiana."[325] The point was apparently pressing, for not having heard from the Société Royale for eight months, La Luzerne, as minister of the navy, complained that the Société had not responded to the minister of foreign affairs and that "I wish you to transmit to me as soon as possible the opinion of the Société Royale."[326]

A price was set in July of 1788.[327] One imagines the Native Americans of Guiana, heretofore apparently indifferent to ant nests, now searching them out in the forests and bringing them by canoe or through the jungle for exchange at the post in Cayenne. Whatever and however they were paid, Laborde undoubtedly had it right when he responded to the Société Royale de Médecine that "everything the Indians do for Europeans costs little."[328]

In all of this, from Africa to the forests of French Guiana, from a hospital in Cayenne to government and scientific circles in Paris and Versailles, and then back to the depths of the Amazonian forests, this story – as many others told in these pages – depicts in a striking way a global economy of knowledge coordinated and animated by the Colonial Machine of eighteenth-century France.

325. "...M. Noyer Ch.en Major de la Colonie à Cayenne pourroit sans doute donner des détails satisfaisants sur le nid de fourmis, La Compagnie a arrêté de le prier de nous envoier du nid de fourmis et absolument qu'il se trouve, de nous le faire parvenir dans cet état sans qu'il ait subi aucune espece de préparation et d'une maniere qui ne lui fasse éprouver aucune altération, avec des renseignemens sur son usage sur ses propriétés et sur le prix que l'on peut le paier aux Indiens de la Guiane," SRdM, Ms. 11, fol. 143v ("Séance du Mardi 13 Novembre 1787").

326. "Je vous prie de me transmettre le plus promptement qu'il sera possible l'avis de la Société roïale"; ALS La Luzerne, dated "à Versailles, le 15 mai 1788," SRdM, Archives, 132 dr. 49, #4.

327. "Retribution accordée aux indiens qui s'occupent de leur recherche," ms. Code de Cayenne, vol. 7 (1783-1803), pp. 363, 509, in AN-MAR G 237.

328. "...tout ce que les indiens font pour le service des européens coûte peu de chose," "Réponse de M. de La Borde pour le prix du nid de fourmi," dated "Auch 29 8bre 1787," SRdM, Archives, 132 dr. 49, #6.

Part II

The Colonial Machine in Action
C – Cultivating An Empire

Permettez moï d'avoir l'honneur de vous communiquer quelques reflexions sur les avantages qu'il y auroit de favoriser dans nos Colonies la culture des arbres et des plantes, qui peuvent ou servir a la nourriture des Colons et de leurs negres, ou former des objets de commerce pour la metropole et pour les Etrangers.

– Abbé Tessier (1787)[1]

The success of French overseas colonies in the Old Regime was overwhelmingly based on the commercial production of agricultural commodities, first tobacco, then indigo, then sugar and coffee. Matters botanical formed an essential part of the colonizing process for the French from the outset, and economic botany and applied botanical research constituted a major sphere, probably *the* major sphere of action of the Colonial Machine. From this point of view the botanical activities of the Colonial Machine – notably in creating botanical gardens in the colonies and in promoting economic botany overseas – had a more significant impact in the colonies than did the work of the Colonial Machine in astronomy/navigation or medicine. But the botanical and the navigational are inseparable in the Colonial Machine.

The Power of Sugar and Coffee

The example of coffee is key to understanding the thinking of French officials. The success of French coffee plantations in the Americas and in the Indian Ocean was spectacular, given that a large and lucrative industry arose

1. ALS Tessier to "Monseigneur," dated "a Paris ce 10 fevrier 1787"; AN-MAR G 101, dossier 4, fols. 179-80.

from nothing, indeed, from an innocent scientific exchange that backfired. Coffee was unknown in Europe until the middle of the seventeenth century. Through the early part of the eighteenth century the Dutch monopolized coffee imports into Europe, having transplanted the tree from Mocha on the Arabian peninsula for cultivation in Ceylon (Sri Lanka) and Batavia. The coffee tree seeds reached the Amsterdam Botanic Garden in 1706. Shortly thereafter, a French artillery officer and amateur botanist managed to have a seedling sent to the Jardin du Roi in Paris, and in seeming innocence, in 1714, the burgomaister of Amsterdam sent another one for presentation to the king at Marly, which also made it to the Jardin du Roi.[2] With this stock the French turned around and inaugurated coffee production in the Caribbean, unsuccessfully in 1716, but then successfully introduced the tree into Martinique in 1723 and then to Saint Domingue in 1726. As the eighteenth century unfolded, Saint Domingue became the world's leading producer of coffee, and coffee likewise was largely responsible for the success of the Indian Ocean colony of Île Bourbon.[3] The coffee case provided a powerful example to administrators of the economic value of applied botany.

The same can be said for sugar and indigo, neither of which were native to the Americas and both of which became the major cultivars in the colonial system of production.[4] (Need we point out the connection between these industries and the slave trade?) Similarly, manioc was not native to the Mascarene Islands of the Indian Ocean, but, imported from the Americas, manioc became a staple for slaves there as well as in French colonies in the Caribbean and South America.[5] These examples provided potent models for economic botany

2. Antoine de Jussieu (1713), p. 291-92. The academicians at the Académie des Sciences evidenced an early interest in the chemistry and preparation of coffee; see Boulduc "Observations sur une préparation particuliére du Caffé," PA-PV 32 (1713), fols. 237-39v ("Le Mercredi 12 Juillet 1713"), and further, fols. 339v-40v ("Le Samedi 18 Novemb[re] 1713). On this general background, see also Allorge, pp. 181-204.

3. Haudrère (1992), pp. 66-70; McClellan (1992), p. 66. On the introduction of coffee to Île Bourbon, see ms. Code de l'Île de Bourbon, vol. 1, p. 447, in AN-MAR G 237, whence it seems coffee stock was imported into Île Bourbon from Martinique. Other sources report that coffee on Île Bourbon came directly from Mocha; PA-PV 35 (1716), fols. 133-33v ("Le Samedi 25 Avril 1716"). Coffee likewise passed from the Amsterdam Botanic Garden to Dutch Guiana and thence to French Guiana and Brazil; see Touchet, pp. 142-45; McClellan (1992), p. 316n24.

4. The abbé Tessier makes this point, ALS Tessier to "Monseigneur," dated "a Paris ce 10 fevrier 1787," in AN-MAR G 101, dossier 4, histoire naturelle, fols. 179-80. On sugar, one begins with Mintz; see also Pritchard (2004), pp. 162-86 and pp. 127-30 for indigo.

5. On the importation and establishment of manioc culture in the Mascarenes, see "Extrait du registre général des déliberations de la Compagnie des Indes. Du 26 juin 1742" and "Extrait du registre général des délibérations de la Compagnie des Indes. Du 11 avril 1744" in Lougnon (1940), pp. 102, 146; "Copie de la lettre écrite par la compagnie à M.rs du Conseil supérieur de l'isle de France dattée à Paris le 28 février 1749" in Lougnon (1949), p. 132. See also Pierre Poivre, "Observations sur l'état de l'agriculture chés differents Peuples de l'affrique et de l'asie," BCMNHN, Ms. 575, pp. 25-29; ms. by Malesherbes, "extrait de quelques conversations avec M. poivre en 1758;" BCMNHN, Ms. 1765, fols. 227-50; Delaleu (1777), vol. 2, p. 98: "Ordonnance... concernant la plantation des Maniocs. Du dix Avril 1771."

and colonial development, and they drove the pursuit of utility and application by the Colonial Machine in these areas. From this point of view, the commercial concerns of the Colonial Machine to establish new agro-industries are to be distinguished from scientific botany and natural history or the disinterested pursuit of knowledge of plants and animals still unknown in the vast lands beyond the territory of France. Scientific botany was not neglected, to be sure, but economic interests drove the process.

Officials unrelentingly sought opportunities to expand commodity production and to create new industries and employment based on botanical products. Ginseng from Canada, rhubarb from China, rice, saffron, mulberry trees, and a host of other plants were shipped hither and yon in the mercantile aspiration of bringing new wealth to the kingdom. The royal instructions given to the new intendant posted to Quebec in 1725 noted "the encouragement he is to give to fisheries and to farming, especially to flax, hemp, and in general to everything that France needs for the consumables that it procures from foreign countries."[6] As an English source put it in 1775, "The introduction of the potato, the working of the silkworm, the discovery of cinchona, and the use of the cochineal insect, of lacquer and of indigo clearly indicate the good we can expect from the work of industrious men."[7] Perhaps reflecting an ambiguity to the search for useful products, a project issuing out of the Jardin du Roi in 1788 sought "to charge the Administration of each colony to send seeds and living plants to the Jardin du Roi *not only of the things whose utility is known, but even those that seem to have no property.*"[8] And so it was, to pick a single example, that by 1760 cotton, cinnamon, pepper plants, palm trees, sandalwood, sapan wood, Indian saffron, ginger, rhubarb, ginseng, the cacao tree, poppies (for opium!), and a variety of oil-producing plants and trees had made their way to remote Île de France from various parts of the world.[9]

Our concern in this part is with the official, organized, and institutionalized activities of the Colonial Machine to promote colonial development through economic botany and applied botanical research; but before launching into that inquiry we should say something about the background of private interest in matters botanical. Roger Williams has labeled the explosion of interest in bot-

6. "Encouragements qu'il doit doner à la pêche et aux cultures, surtout celles du lin, du chanvre et généralement à tout ce dont la France a besoin pour sa consommation et qu'elle se procure à l'étranger." AN-COL B^{48} fol. 744, 1725, 8 mai: "Mémoire du roi pour servir d'instruction au Sr de Chazel, intendant de la Nouvelle France." Beavers and trade in beaver fur fit somewhat in this context and are not to be overlooked; see AN-COL C^{11A} 121 ("Registre. Commerce au Canada et spécialement commerce des castors"). Pritchard (2004), pp. 139-62, discusses fish and fur.

7. "L'introduction de la pomme de terre, la conduite du travail des vers-à soie, la découverte du quinquina, l'usage de la cochenille, de la laque & de l'indigo, annoncent visiblement le bien qu'on peut attendre du travail des hommes industrieux...," Lettsom (1775), p. iv.

8. "...de charger l'Administration de chaque Colonie d'envoyer au Jardin du Roi des Semences et des Plantes en Nature, non seulement des choses dont l'Utilité est connue mais même de celles qui ne paraissent avoir aucune propriété," our emphasis; ms. titled "Notte relative a la Correspondance d'Economie rurale, ~~politique~~ [sic!] et de Botanique qui doit etre établie entre les Colonies francaises et le Jardin du Roi," BCMNHN, Ms. 308.

9. "Memoire sur les Isles de France et de Bourbon," AN O^{1} 597, #13.

any from 1750 to 1810 as "botanophilia."[10] Private and unofficial shipments of plants and botanical information circulated uninterruptedly in, around, and through the contemporary French colonial world and the metropolis. The garden of Abbé Rozier in Lyon, for example, contained 3000 foreign plants.[11] A 1785 catalogue of a private garden in Clamart mentions plants from Canada, China, Saint Lucia, Lapland, Japan, the Carolinas, Siberia, India, Ethiopia, the Cape of Good Hope, Peru, Madagascar, and "Africa."[12] An artillery captain in Saint Domingue on his own put together an album of 466 plants from Saint Domingue, for which he painted as many watercolors.[13] Or then, there was the manuscript of Charles-Auguste Leroy de la Potherie, the military man born in Guadeloupe, written in prison in 1764.[14] In the 1730s, a naval ensign in Canada compiled a private collection of plants that attracted the attention of government authorities who ordered it brought to France because it might contain novelties for the Jardin du Roi.[15] The chief naval engineer at Brest, Marc Blondeau, sent Chilean strawberry plants to a friend in 1771 along with the instructions that if he wanted them to fruit they needed to be planted alongside French strawberries.[16] Notably, this kind of private circulation went not only from the world overseas to the colonies, but the other way as well – as when, for example, Adanson in Senegal wrote to his correspondent in France: "It is to be a little free with you to ask of you almost as many plants from your country as I am sending you from mine."[17] These private exchanges formed the backdrop; the official botanical work of the Colonial Machine, to which we turn next, has to be seen as taking place on top of or alongside the private and uncontrolled circulation of plants that took place simultaneously.

Botanistes du Roi and Royal Botanical Voyagers

A key element in the success of French botany in the eighteenth century was the creation of the position of royal botanist or *botaniste du roi*.[18] These

10. Roger Williams, p. 1.

11. Lalande (1775), p. 10n.

12. See Fillassier (1785).

13. Fouchard (1955), p. 40; Ms. 446, BM/Besançon.

14. Institut de France, Bibliothèque Mazarine, Paris (don Marcel Chatillon).

15. AN-COL B^{64} fol. 446 (15 May 1736), "...le Sr. de Muy, enseigne, qui pendant cinq ans a commandé à la rivière St. Joseph, avait fait une collection de plantes. Comme il pourrait s'en trouver d'inconnues que le roi serait bien aise d'avoir pour son jardin, il pourrait le charger de les apporter en France."

16. ALS dated "Brest le 28 Juin 1771," AN-MAR C^7 32 (Blondeau, Marc).

17. "C'est en agir un peu librement avec vous que de vous demander presqu'autant de Plantes de votre Pays que je vous en envoie du mien...," ALS, Adanson to "Messieurs," dated "Au Senegal ce 20 aoust 1751," in PA-DB (Adanson). For an example of 'private' shipment of natural history curiosities through the Compagnie des Indes, see SHML, 1P 284, Liasse 112, #80: Syndics to Buisson, "à Paris le 3 Juin 1769."

18. Bret (1999) unpacks the role of royal botanists in the colonies, pp. 65-89, and especially pp. 72-81; see also remarks by Ly-Tio-Fane (1976), p. 137, regarding a new type of government botanist that began to appear in the 1770s.

were botanical specialists paid by the crown and who staffed royal botanical gardens in the metropolis, traveled on botanical expeditions, and/or occupied stations overseas in the colonies. Many, if not most of the French botanistes du roi trained at the Jardin du Roi and were vetted by the Académie des Sciences. They represent a new, more professional generation of experts concerned with botany and the world overseas. They succeeded royal doctors, who may have dabbled in matters botanical while on colonial station and, before the doctors, independent travelers or individuals with connections to religious orders. The existence and activities of French royal botanists testify to the seriousness with which botany and particularly economic botany came to be pursued by French authorities, and the further institutionalization of the Colonial Machine.

At some points, the dual position of *médecin-botaniste* existed. Louis Prat (or Duprat), for example, was appointed médecin-botaniste du roi for Louisiana in 1725.[19] Another report indicates the presence of a médecin-botaniste on Île Bourbon as early as 1729.[20] But already by the end of the seventeenth century, royal botanists per se had made their appearance, and they underscore the existence of a career path for botanists distinct from that of royal doctors or surgeons, who occupied separate spots in colonial officialdom. The governor-general in Saint Domingue, La Luzerne, made this much plain in 1787 when he sought a replacement for Joubert de la Motte, the deceased royal botanist at the Jardin du Roi in Port-au-Prince. A physician or surgeon would not do for La Luzerne, writing to the Minister, de Castries:

> No physician would give up a private practice that brings sixty thousand livres in Saint Domingue or sometimes more in order to pursue a difficult and hardly lucrative occupation [*métier*]....A botanist and a naturalist (the latter especially knowing enough chemistry to analyze mineral waters and to recognize and test stones and minerals) are two subjects I really want in this colony, but I want them active and always occupied with their craft....In the first place, it seems appropriate to me that this position be full time and not given to a doctor or a surgeon or that they be expressly forbidden to exercise these professions that are very lucrative in Saint Domingue....Someone suitable for such employment does not exist in the colony...but educated individuals are to be found in Paris who have a taste for botany, like the mania a poet has for verses, who long to roam in regions that are new to them or to science.

19. Rousseau (1969), p. 624, reprints the ministerial communication of 31 October 1725 mentioning Prat's appointment; this individual seems not to have been Le Page du Pratz (c. 1695-1775), who traveled independently in the region from 1715 to 1734 and was the author of a *Histoire de la Louisiane* (1758); see Chinard (1957a), pp. 268-69.

20. See Lougnon (1956), p. 78, item #596 (14 February 1729) and reference to "Le sieur Couzier, médecin-botaniste."

At the Jardin du Roi you will even be able to choose among a sizeable number of them.[21]

His fellow minister, Malesherbes, backed up La Luzerne in this view, and, as we saw earlier, the unfortunate Montpellier physician, Bruguière, who threw his hat in the ring for this position, lost out on account of his medical credentials.[22] The person who eventually got the job, Hypolite Nectoux, was a trained botanist, and he himself called for professional botanists to tend royal gardens in the colonies in preference to doctors, apothecaries, or even lawyers.[23]

Louis-Claude Richard. In a backhanded way the case of the royal botanist Louis-Claude Richard (1754-1821) illustrates the existence of a separate professional cadre of botanists employed by the government. Louis-Claude Richard came from a notable family of royal gardeners, being the nephew of Antoine Richard and his uncle Claude Richard who managed the Trianon gardens for the king and then the queen.[24] Louis-Claude studied botany under Bernard de Jussieu at the Jardin du Roi in Paris and was known to the Académie des Sciences, having presented papers to that institution in 1780 and 1781. Richard was posted to Cayenne and the Jardin du Roi there from 1781 to 1785, whence, on the recommendation of the Academy to the government and under its auspices, he moved on to Guadeloupe and Martinique from 1785 through 1789.[25] Louis XVI personally discussed his expedition and wrote instructions for Richard.[26] On the surface, Richard seems perfectly exemplary and wholly integrated into the Colonial Machine. Yet the administrators became increasingly annoyed because he did not send back any seeds or sam-

21. "Aucun médecin ne renoncera à une pratique qui rapporte à St. Domingue annuellement soixante milles livres ou quelquefois plus pour faire un métier pénible et qui n'est point lucratif....Un botaniste et un naturaliste (ce dernier surtout sachant assez de chimie pour analiser les eaux; pour reconnaitre essayer les pierres, les minéraux) sont deux sujets que je désire et beaucoup dans cette colonies, mais je les désire errants et toujours occupés de leur métier....Il convient d'abord, selon moi, qu'on n'accorde cette place ni à un médecin ni à un chirurgien, ou qu'il soit expressement defendu d'exercer ces professions qui sont fort lucratives à St. Domingue et l'occuperaient de nouveau tout entier....Le sujet propre pour un tel emploi n'existe pas dans la colonie...mais il se trouve à Paris des gens instruits ayant le gout de la botanique, comme un poete a la manie des vers, bruland d'errer dans les pays qui, soit pour eux soit pour la science sont nouveaux. Au Jardin du Roi vous aurez même la facillité de choisir sur un assez grand nombre...," ALS La Luzerne, dated "Port-au-Prince, le 3 mars 1787," AN-COL E 54 (Bruguière).

22. Letter from Malesherbes to La Luzerne, dated "à Versailles le 3 juillet 1787," AN-COL E 350 (Richard, Louis), and above, Introduction, pp. 28-29.

23. See undated Nectoux memoir, "arbres à epicerie," BCMNHN, Ms. 308.

24. On Richard, see Touchet, pp. 134-38, 181-88, 301; Lacroix (1938), vol. 3, pp. 91-96; Jandin; see also BCMNHN, Ms. 471, "Notes manuscrites de botanique de L.-Claude Richard (1754-1821)."

25. "L'Auteur étoit, avant son voyage, connu avantageusement de l'Académie, qui, en le recommandant au Gouvernement pour ce même voyage, lui donna un marque de son estime," "Rapport sur les travaux de M. Richard Botaniste du Roy," PA-PS ("Samedi 15 Mai 1790").

26. Lamy, pp. 70-71.

ples and kept everything for himself in order not to be scooped scientifically before his return. Malesherbes objected to the minister of the navy and the colonies, La Luzerne. Personal glory was one thing, but that was not why Richard was sent abroad in the first place nor why he was being maintained at royal expense:

> After the four or five years he has spent in Cayenne at royal expense, he will perhaps bring back some seeds on his return, i.e., seeds to display in camera as objects of curiosity, but he has deliberately not sent nor wished to send a single seed to be sown, and I hardly have to point out to you that seeds collected in Cayenne and brought back to France five or six years later will not be seeds for sowing.... It follows from this that your Richard is someone who for five or six years travels at royal expense for his own glory. On his return he hopes to publish a magnificent work on the plants of America that will be printed if the King helps him further with the costs of engraving, and this work will garner him a great reputation, will give him qualifications to join the Academy, etc....Nevertheless, the effect of this strategy is absolutely contrary to why he was sent.... As for the seeds he should send and that were one of the principle objectives of his voyage, he well knows that this cannot be put right by a handsome book. And thus in this regard, he has been completely unfaithful to his mission.[27]

Plainly, what the Colonial Machine wanted and in so many other instances obtained were not botanical prima donnas, but agents who could and would fulfill its commands. The professionalization of botany and the institutionalization of government botanical agents are already evident at the end of the seventeenth century, as we observed in the case of Charles Plumier, the co-opted Catholic priest who made four state-sponsored trips to the Americas between 1689 and 1704 as botaniste du roi. Between Plumier at one end of the eighteenth century and L.-C. Richard at the other, a whole series of traveling bota-

27. "En sorte que les quatre ou cinq années qu'il a passées à Cayenne aux frais du roi, il rapportera peut etre des graines à son retour, c'est à dire des graines a étaler dans un cabinet comme objet de curiosité, mais il n'a pas envoyé et, de projet déterminé, n'a voulu envoyer aucune graine à semer, car je n'ai que faire de vous observer que les graines recueillies à Cayenne et qu'il rapportera cinq ou six ans après en France ne seront plus des graines à semer... Il en résulte que votre Richard est un homme qui depuis cinq ou six ans voyage aux frais du roi pour sa propre gloire, qu'il compte publier à son retour un magnifique ouvrage sur les plantes d'Amérique, qui sera imprimé si le Roi l'aide encore pour les frais de gravure, que cet ouvrage lui fera un grand nom, lui donnera des titres pour être de l'academie, etc....Cependant l'effet de cette politique est absolument contraire à l'intention dans laquelle il a été envoyé.... Mais quant aux graines qu'il devoit envoyer et qui étoient un des principaux objets de son voyage, il savait bien que cela ne peut pas être reparé par la publication d'un beau livre; ainsi à cet égard il a été absolument infidèle à sa mission...," ALS Malesherbes to La Luzerne, dated "à Versailles le 3 juillet 1787," AN-COL E 350 (Richard, Louis).

nistes du roi fanned out across the globe, exploring the realm of nature under the banner of France and the Colonial Machine.[28]

The Call of the Orient. A goodly number of travelers went to the Middle East. The great French botanist Joseph Pitton de Tournefort (1656-1708) started his career by going to the Levant in 1700-1702 as the emissary of Fagon at the Jardin du Roi, of the Académie des Sciences, and of the government; his *Relation d'un voyage du Levant* appeared posthumously in 1717.[29] Send by Fagon of the Jardin du Roi, Augustin Lippi was the official médecin-botaniste du roi on a diplomatic mission to Africa in 1704-05; Lippi returned three shipments of plants to the Jardin du Roi before his party was massacred.[30] Authorities sent Jean-André Peyssonnel to Egypt in 1714 and then to the Barbary Coast in 1724, before ultimately posting him to Guadeloupe. The Académie des Sciences and the minister of the marine, Maurepas, twice sent Claude Granger on botanical and natural history expeditions to Egypt, once in 1730-1732 and again in 1733-1737.[31] A correspondent of the Academy, Granger traveled with fifty-four volumes of scientific reference books, and after his first voyage he worked closely with the Jussieux and others at the Jardin du Roi. He died in Egypt in 1737; a posthumous and privately published *Relation du voyage fait en Egypte* appeared in 1745.[32] As mentioned earlier, a young Joseph de Jussieu accompanied La Condamine to South America as botanist in the 1730s. Commissioned to gather plants for royal nurseries, Jean-Louis Guérin worked as botaniste du roi in Louisiana from 1737 until his death in Mobile in 1741.[33] Academy correspondent and naval surgeon Laurent Garcin (1683-1751) made three trips to East Indies in the 1720s, and notices of his work continued to appear in the Academy's publications through the early 1740s.[34] A naval officer with connections to the Jardin du Roi, Sonnini de Manoncourt, traveled on royal account to Guiana in 1772-1773 and then to Egypt and Asia Minor from 1777 to 1780.[35] The abbé Galloys, the brother of

28. For these details, see Laissus (1964), p. 299n; Laissus (1981), passim; and also Gasc and Laissus (1981). In this connection botanistes du roi functioned as the botanical equivalents of royal astronomers sent on mission; Louis Feuillée's appointment as "mathématicien et botaniste" straddled these roles.

29. On Tournefort's expedition and the Academy, see PA-PV 19 (1700), fols. 29-30 ("Le mercredi 27 Janvier 1700").

30. "Catalogue Alphabetique des Plantes observées en Egypte, Par Mr. Lippi," BCMNHN, Ms. 944; see also Bonnet, esp. pp. 258, 262 ; Allorge, p. 164.

31. See AN-MAR C^7 128 (Granger); PA-DB (Granger). Granger's real name was Tourtechot.

32. See Alain Riottot, "Granger: un chargé d'histoire naturelle 'retrouvé'," Mémoire de D.E.A., pp. 11-12; PA-DB (Granger).

33. See ministerial correspondence regarding Guérin, AN-COL C^{13A} 26 (1741), fol. 160 ("27 avril 1737" [sic]), and fol. 158 ("25 septembre 1741").

34. See Jacquat; Taillemite (2002), p. 202; PA-DB (Garcin); HARS (1730), pp. 43, 66-67; (1743), pp. 28-32; and PA-PV 49 (1730), fols. 180 ("Le Samedi 15. Juillet 1730"), 183 ("Le Samedi 22 Juillet 1730"), 185 ("Le Samedi 29 Juillet 1730"), and 188 ("Le Mercredi 5.^{e} Aoust 1730").

35. See above, note 28; Allorge, pp. 448-49; Chaïa (1978).

the gardener at Brest, traveled to China in 1764-1769 as Royal Naturalist and Emissary of the Jardin du Roi.[36] All of these voyagers sent seeds and plants back expressly for the Jardin du Roi, and they were generally in contact with one arm or another of the Colonial Machine.

Michel Adanson. The career of the botanist Michel Adanson (1727-1806) is revealing in these regards. Adanson was posted to Senegal from 1749 through 1753 as an employee of the Compagnie des Indes.[37] But the twenty-two-year-old Adanson was very ambitious. Departing from Lorient, he wrote to his mentor, Bernard de Jussieu:

> I have plans other than being employed by the Compagnie des Indes. You know that the illustrious Academy, of which you and your brother are members, has always been appealing to me, and it is with the goal of entering the Academy that I am working on the study of natural history which will occupy a large part of my time, although I will fulfill my duty to the Compagnie with exactitude and honor.[38]

Once he arrived in the African colony Adanson elaborated his ambitions to Jussieu:

> The kindness of Mr. De la Brüe, our Director, is making my research easier. Since I let him know that I only came to this country exclusively to work on natural history and to spend a year or two at most, he is letting me off from working in the office, work which is really demeaning for a man who aims for something other than being a copyist. You never doubted my station to be other than that of Naturalist.... Supposing that on my return from Senegal the Compagnie might want to send me to Île Bourbon or to India, you know that I could very likely not be as free from office work.... That's why if it were possible that I be sent by the King to these regions or others, or if the Compagnie sent me to the Indies exclusively in the role of Naturalist (which I strongly doubt),

36. "...naturaliste du roi et du Jardin royal," Laissus (1973), p. 45. Malleret provides the full details of Galloys' contact with Pierre Poivre and his role in introducing a range of plants and animals from China into Île de France.

37. On Adanson, see Nicolas first and foremost; see also, Broc (1975), pp. 72-73; Allorge, pp. 308-312; Lacroix, (1938), vol. 4, pp. 187-207; Chassagne, pp. 226-30; Roger Williams (2003), pp. 7-8.

38. "j'ai d'autre vües que sur les emplois de la Compagnie des Indes: vous sçavez que l'Illustre Academie dont vous et M.r votre frere êtes les membres, a toujours eüe des attraits pour moi, et que c'est dans la vüe d'y entrer un jour que je travaille à l'étude de l'histoire naturelle, qui occupera la plus grande partie de mon temps, quoique je ne pretende point manquer à mon devoir à l'égard de la Compagnie: je tacherai de m'en aquiter exactement, et de le remplir avec honneur...," ALS Adanson to Bernard de Jussieu, dated "A l'Orient ce 31 mars 1749," in PA-DB (Adanson).

> I could travel for a few more years, unless you had some other view regarding my advancement or if you thought to secure me a post where I could do natural history.[39]

Jussieu apparently secured the desired royal appointment for Adanson, who wrote back:

> The position that you propose for me as naturalist-explorer ["voyageur naturaliste"] sent to the colonies by the King suits me to a tee, and you should have no doubts that I prefer this option to being in the service of the Compagnie, not only because of the advantages and the honor attached to it, but more for the conveniences that I will have in traveling for the King which I cannot have with the Compagnie....With the position of royal naturalist I will be free to arrange my travels when and where I judge necessary....[40]

Bernard de Jussieu actually secured Adanson's transfer to Île Bourbon as a full-fledged naturalist within the Compagnie des Indes at a salary of 2000 livres.[41] But Adanson plainly did not want to get stuck on Île Bourbon, and he pleaded a debilitating seasickness as his excuse for not accepting the offer. As alternatives, he suggested that his position in Senegal be upgraded, that the

39. "Les bontés de M.r De la Brüe notre Directeur me facilitent les recherches que j'ai à faire sur ce sujet; depuis que je lui ai fait connoître que je ne m'étois rendu dans ce pays que pour y travailler à l'histoire naturelle seulement, et pour y passer un an ou deux au plus, il m'a dispensé de travailler au Bureau, ouvrage bien ingrat pour un homme qui tend à autre chose qu'à être copiste; car vous n'avéz jamais douté que mon état ne fût celui de Naturaliste....Vous sçavéz Messieurs que supposé qu'à mon retour du Senégal la Compagnie voulût m'envoyer à Bourbon ou aux Indes, je pourrois fort bien n'y point être aussi libre des occupations du bureau....c'est pourquoi s'il étoit possible que je fusse envoyé par le Roy dans ces paÿs ou dans d'autres, ou que la Compagnie en m'envoyant aux Indes ne m'y regardât qu'en qualité de Naturaliste, (ce dont je doute fort), je voyagerois encore plusieurs années, à moins que vous n'aÿés quelqu'autre vûe sur ce qui regarde mon avancement," ALS, Adanson to [Bernard de Jussieu] dated "Au Senégal le 15 aoust 1749," in PA-DB (Adanson). It is clear that Adanson pressed de Brüe, too; "Copie d'un mémoire présenté à M.r La Brue le 15 Fevrier 1752: Mémoire instructif des facilités que je pense m'être nécessaires pour continuer et augmenter le nombre de mes recherches au sujet de l'histoire naturelle de la concession du Senégal," also PA-DB (Adanson).

40. "La place que vous me proposéz de voyageur naturaliste envoyé par le Roy dans les colonies me convient fort, et vous ne devéz pàs douter que je ne préfere ce parti au service de la Compagnie, non seulement à cause des avantages et de l'honneur qui y sont attachés, mais encore des commodités que j'aurai en voyageant pour le Roy et que je ne puis avoir à la Compagnie....en qualité de naturaliste envoyé par le Roy, je serai libre de disposer de mes voyages quand et comme je les jugerai plùs nécessaires...et dans ce cas je serois charmé si j'avois besoin d'un dessinateur avec moi pour dessiner et tirer les plantes que j'observerai, que mon frere qui est aux Jésuites fût en etat de me suivre pour cet effet....en qualité de naturaliste envoyé par le Roy, je serai libre de disposer de mes voyages quand et comme je les jugerai plùs nécessaires...," ALS, Adanson to [Bernard de Jussieu], dated "Au Senégal le 1er aoust 1750," in PA-DB (Adanson).

41. ALS, Adanson to [Bernard] de Jussieu, dated "Au Senégal, ce 10 fevrier 1751," in PA-DB (Adanson).

Jussieux find him a post in Europe, or at worst that he be transferred to America. In the end, Adanson got nothing from his patrons. He returned to France from Senegal in 1754. He was elected a *correspondant* of the Academy only in 1759; he did not join the institution as a resident member until 1773, and his career in the Academy thereafter was a rocky one.[42] His rank ambition, unseemly even by the standards of the time, does have the virtue of offering us a glimpse into the appeal and possibilities of royal service in botany and natural history.

Philibert Commerson and Pierre Sonnerat. A series of other peripatetic individuals were more successful in securing appointments as royal botanists and naturalists. Philibert Commerson (1727-1773) was one.[43] A Montpellier graduate, disciple of Linneaus, and protégé of Bernard de Jussieu, Le Monnier, and Poissonnier, Commerson sailed with Bougainville as royal physician, botanist, and naturalist (*médecin botaniste et naturaliste du roi*) on the latter's circumnavigation in 1766-1769. Commerson was the first French naturalist to be formally assigned to accompany a royal expedition. On the return trip, at the request of Pierre Poivre, the intendant on Île de France, Commerson and his illustrator, Paul Jossigny, remained on Île de France to work on the botany and natural history of the Indian Ocean at a salary of 3000 livres.[44] Commerson made botanical side trips to Île Bourbon and to Madagascar, and, although never elected a correspondent of the Académie des Sciences in Paris, he did submit work to the Academy through Poissonnier.[45] (According to Poissonnier, one reason for Commerson's stay was to find botanicals that could serve as replacements for pharmaceuticals from France that degenerated on the voyage to the Indian Ocean.[46]) Commerson also sent back botanical and natural history specimens for the Cabinet du Roi, the Jardin du Roi, and the Trianon gardens, including a posthumous shipment of 34 cases containing 5000 species, 3000 of which were new to science.[47] Commerson died on Île de France

42. McClellan (2003B) on Adanson and PA.

43. On Commerson, see Laissus (1971); Lacroix (1938), vol. 4, pp. 1-13; Ly-Tio-Fane (1975) and (1976), pp. 54ff; Cap (1861); Roger Williams (2003), pp. 9-11; Allorge, pp. 344-60; and manuscript sources, AN-COL, E 89 and BCMNHN, Mss. 884-93, #1904.

44. See Cossigny letters to Le Monnier, dated "A l'habitation, Isle de france le 25 juillet 1769," "Aux Raines de Willhem, Isle de france le 19 Juillet 1770," and "A Palma, dans l'Isle de france le 30 Aoust 1771," all, BCMNHN, Ms. 1995.

45. PA-PV 89 (1770), fol. 11v ("Mercredy 24 Janvier 1770") and fol. 81 ("Mercredy 7 Mars 1770"), where one reads that "M. Poissonnier a lu un memoire sur un peuple de Nains qui existe à Madagascar. Ce mémoire a été dirigé [rédigé] par M. Comerson sur les observations de M. le Ch.er de Nodare, lui et M. Le Roy ont été nommés pour l'examiner."

46. See Poissonnier letter to minister dated 7 April 1770, quoted in Ly-Tio-Fane (1976), p. 54.

47. Commerson boasted of his accomplishment in a letter of 25 February 1769 reprinted in Cap (1861), p. 35: "Je rapporte déjà de mon voyage autour du monde, une fois plus de plantes que Tournefort n'en cueillit dans son voyage au Levant. Ma collection seule de fougères et de gramens surpasse celles de Scheuschzer et de Plumier. J'ai enrichi à proportions toutes les autres parties de l'histoire naturelle, sans compter les nouvelles récoltes que je vais faire dans cette île, dans celles de Bourbon et de Madagascar, etc."

in 1773, and Le Monnier and Buffon fought over obtaining his papers and scientific collections.[48]

Pierre Sonnerat (1748-1814) almost literally followed in the footsteps of Commerson, and he certainly associated himself with his botanical predecessor.[49] Sonnerat became a protégé of d'Angiviller at the Bâtiments du Roi and a correspondent of both the Académie des Sciences and Buffon at the Cabinet du Roi, and he always styled himself "Naturaliste Pensionnaire du Roi et Correspondant de son Cabinet" (pensioned royal naturalist and correspondent of the king's cabinet). More significantly, Sonnerat was a distant cousin of Pierre Poivre, the intendant on Île de France, who in 1771 famously sent Sonnerat to the Philippines, the Moluccas, and New Guinea in search of spice plants for cultivation and propagation on Île de France. Sonnerat returned to France in 1773, depositing a substantial collection with Buffon and the Cabinet du Roi, including 300 new bird species (many were birds of paradise), 50 quadrupeds, various fish, reptiles, and insects, and an herbarium of 8000 plants. In 1776 he published his *Voyage à la Nouvelle Guinée*. At the Académie des Sciences he was in contact with Adanson, Jussieu, Daubenton, Lamarck, and Thouin, and he published in the *Savants Étrangers* series.[50] He returned to the Indian Ocean in the period 1774-1781, traveling to Île de France, India, Malaysia, and China, whence he sent back shipments to d'Angiviller and the Jardin du Roi.[51] Back in France in 1782 and with the imprimatur and approval of the Académie des Sciences, Sonnerat published his *Voyage aux Indes orientales et à la Chine, fait par ordre du Roi, depuis 1774 jusqu'en 1781* in two handsome volumes.[52] This work was dedicated to d'Angiviller and well subscribed by the Colonial Machine.

In letters to Michel Adanson, Sonnerat described his research and the pains to which he went to acquire authentic local knowledge. As he put it in one:

48. See material related to this episode, BCMNHN, Ms. 357 Papiers de L. G. Lemonnier; a catalogue of Commerson's collections is to be found, BCMNHN, Ms. 1343.

49. On Sonnerat, see Ly-Tio-Fane (1976); Lacroix, vol. 4, pp. 25-31; Sonnerat (1782); Dorst, p. 596; Chassagne, pp. 248-51; Allorge, pp. 399-400. See also Adanson, "Compte rendu des droits de Sonnerat pour prétendre au titre de correspondant [1776]," HI, AD 294, and Adanson, "Remarques d'histoire naturelle sur un mémoire et un herbier de M. Sonnerat [1777]," HI, AD 296.

50. See Sonnerat (1776a) and materials PA-DB (Sonnerat); PA-PV-110, Sonnerat, 18 December 1773, 19 January 1773, 23 January 1782, 23 February 1782.

51. ALS "Sonnerat" to "M.r le C.te Dangivillers," dated "Duplicata De Canton le 29 Decembre 1776," AN O[1] 2111: Pépinières, 1778-1782; for other Sonnerat shipments to the Jardin, see BCMNHN, Ms. 691, fol. 180.

52. Sonnerat (1782). Sonnerat recaps his career and these developments in two letters to the authorities; ALS "Sonnerat, Commissaire des Colonies a Pondichery" to "Monseigneur le Baron de Breteuil ministre et Secretaire d'Etat," dated "Pondichery le 1er novembre 1788," AN O[1] 1292, #24; and ALS Sonnerat to "M.r le C.te d'angivillers," dated "Pondichery ce 11 mars 1777," AN O[1] 2110: Pépinières, 1774-1777. Sonnerat was in regular contact with the Société Royale de Médecine which also approved publication of his 1782 *Voyages*.

> The scope of my work is immense. Not only am I bringing back a complete collection of all genres of birds, quadruped, insects, butterflies [sic], fish, reptiles, shells, wood, and petrifications, but also curious and useful research and observations in botany and a sizeable herbarium, but also on the mores, arts, and religion of all the peoples of Asia to whom I have traveled: Chinese, Malaysians, Pegorins, people from the Maldives and Madagascar, and especially Indians. In order not to write uselessly like every other traveler, I learned the Tamoul language and translated all their sacred books, having looked at the originals. I stubbornly pursued this work with the help of several learned Brahmen who willingly accompanied me in my travels. They went with me to temples and taught me the history of their gods, and this helped me get to the origin of this people whom we do not know that well.[53]

Sonnerat returned yet again to the Indian Ocean as a colonial administrator, holding down the post of naval commissioner at Pondicherry before being imprisoned in India by the English in 1793 for twenty years. On Île de France, J.-F. Charpentier de Cossigny was a later critic of Sonnerat, yet in 1774 Cossigny could write that "the thirty to forty letters I have from him on Île de France form almost a course on botany for our island."[54]

Joseph Dombey. One of the Colonial Machine's most heroic botanical travelers was Joseph Dombey (1742-1794).[55] The controller-general of finances, Turgot, asked A.-L. de Jussieu to recommend "a botanist to send to Peru to

53. "Mes travaux sont immenses, non seulement je rapporte une Collection Complette en tout genre, oiseaux, quadrupedes, insectes, papilons, poissons, reptiles, Coquillages, bois, et petrifixations, mais encore des Recherches et des observations curieuses et utiles sur la botanique, avec un herbier Considerable, ainsi que sur les moeurs, les arts et la Religion de tous les peuples de l'asie chez lesquels j'ai voyagé, des chinois, de Malais, des pegorins, des maldivois, des malgaches et particulierement des indiens, pour ne point ecrire inutilement Comme tous les voyageurs, j'ai appris la langue tamoule et traduis tous leurs livres sacrés, dont je rapporte les originaux. Ce travail que j'ai suivi avec opiniatreté aidé de plusieurs Brames Savans qui on bien voulu me suivre dans mes voyages et parcourir les temples avec moi pour me faire connoitre l'histoire de leurs dieux m'a aidé a Remonter jusqu'a l'origine de ce peuple sur lequel on a pas porté une attention assez suivie.,"ALS Sonnerat to Adanson, dated "Cadix Ce 25 fevrier 1781," HI, AD 240. In the earlier letter, Sonnerat wrote: "Pour ne point Ecrire inutilement Comme tous les voyageurs et ne point etre trompé, J'ay d'abord traduit lorsque j'ay sçu la Langue tamoule Leurs principaux ouvrages ainsi que l'histoire de leurs dieux, je me suis ensuite attaché á des brames savantes qui *en les payant bien* on parcouru L'inde et les Pagodes avec moi, il m'expliquoient Le sujet de chaque figure," ALS, Sonnerat to Adanson, duplicata dated "Isle de france ce 1er Sep.bre 1779," HI, AD 238; our emphasis.

54. "...les 30 ou 40 Lettres que j'ai de luy à L'Isle de france, forment presque un cours de la Botanique de notre Isle," ALS, Cossigny to Le Monnier, dated "A Paris le 27 fevr. 74," BCMNHN, Ms. 1995.

55. On Dombey, see Reyniers; Laissus (1970); Lacroix (1938), vol. 3, pp. 131-32; Cap (1858); Allorge, pp. 442-48; Bleichmar, p. 231; and extensive manuscript sources, PA-DB (Dombey); AN O^1 2112 and O^1 1292, #39-195; BCMNHN, Ms. 222 ("Correspondance de Joseph Dombey"), and BCMNHN, Ms. 2625 ("Voyage de Dombey"). A comprehensive study of Dombey's voyage is much to be desired.

undertake research on plants that might be naturalized in our climate," and Dombey, a Montpellier M.D., was chosen.[56] Dombey traveled to South America as botaniste du roi from 1776 to 1785.[57] He went with the backing of the Academy of Sciences and the Jardin du Roi; and because he was traveling in Spanish territory the Ministry of Foreign Affairs and French diplomatic personnel in Spain likewise became involved in aspects of his trip, including the delicate job of getting his materials passed through Spanish customs. Dombey received 6000 livres a year plus expenses. He was connected to Turgot and then Necker at the Control General; to Buffon, Jussieu, and Thouin at the Cabinet du Roi and the Jardin du Roi; to Sage, Duhamel de Monceau, and others at the Académie des Sciences; and to d'Angiviller at the Bâtiments du Roi. The Académie des Sciences elected him a correspondent while he was in South America, and it performed chemical analyses of plants he sent back.[58] Dombey was known to the Société Royale de Médecine for a memoir he submitted on scurvy. More than any other expeditionary botanistes du roi of the period, Dombey was an agent of the whole of the Colonial Machine. From Peru he shipped back Inca relics, saltpeter, quinoa, coca, balsa wood, and plant and natural history specimens to the Control General, the Académie des Sciences, the Cabinet du Roi, the Jardin du Roi, and the Bâtiments du Roi. The 1783 shipment from Santiago de Chile to d'Angiviller at Versailles consisted of fifty-three crates.[59] That of 1784, from Lima, was seventy-three crates.[60] Thouin regularly recorded shipments received from Dombey at the Jardin du Roi – 2000 plants and 60 new species – and Dombey's botanical materials were redistributed throughout Europe. Dombey's charge included shipping back the exotic metal platinum. Roughly 200 pounds of platinum seem to have made it to France for distribution to the Academy, the Cabinet du Roi, the Control General, and the Bâtiments du Roi; and the navy astronomer Rochon tried platinum for mirrors in telescopes.[61] Dombey traveled under difficult and dangerous circumstances and was often ailing in one fashion or another. Such was the appeal of the position of botaniste du roi that the premature report of

56. "un botaniste pour l'envoyer au Perou faire la recherche des vegetaux qu'il seroit possible de naturaliser dans nos climats," undated note by A.-L. de Jussieu, BCMNHN, Ms. 2625; Cap (1858), pp. 4-5, adds that Condorcet was involved in this choice.

57. On Dombey's status, see Thouin undated draft letter, "a M.r Dombey Docteur en Medecine et Botaniste du Roy," BCMNHN, Ms. 222. Dombey left France in 1776, but did not reach Peru until 1778.

58. PA-PV-110, Dombet [Dombey], 30 August 1788.

59. ALS "Dombey" to "M. le C.te d'Angiviler De l'academie royale des Sciences," dated "Santiago de chili le 20 aoust 1783," AN O[1] 2112.

60. ALS "Dombey," dated "Lima 8 fevrier 1784" to "Monsieur Le C.te D'Angiviller de l'academie des Sciences," AN O[1] 2112. For the distribution of Dombey materials, see ALS, Thoüin to [G. Fabroni], dated "A Paris ce 31 Mars 1786," and other correspondence, APS-MS Fabroni Papers B/F113 no. 1.

61. See above note 55; on European "discovery" of platinum in the 1740s, see Safier (2008b), pp. 132, 303.

his death provoked an unseemly scramble to take his place.[62] Once back safe and sound, however, Dombey continued to receive his annual stipend of 6000 livres while he completed his herbarium, but the Revolution intervened to bring an end to this project, too.[63]

Palisot de Beauvois, Du Puget, de la Billardière and Michaux. Ambroise-Marie-François-Joseph Palisot, the baron de Beauvois (1752-1820) was another botanical traveler of note. Elected a correspondent of the Académie des Sciences in 1783, Palisot de Beauvois traveled largely on his own account, but with official approval and letters in hand.[64] He botanized avidly in Africa and in the Caribbean from 1786 to 1803. The science academician Fougeroux de Bondaroy provided instructions for de Beauvois, who corresponded with Jussieu and d'Angiviller and who duly sent materials back to the Jardin du Roi and the Bâtiments du Roi.[65] Palisot de Beauvois was forced to leave Africa on account of illness. To restore his health he sailed to Saint Domingue, where he recovered and became part of the local scientific establishment there. In Saint Domingue Palisot de Beauvois botanized with the abbé de la Haye, and he became a colonial associate of the Cercle des Philadelphes.[66]

While not an official botaniste du roi, Edme-Jean-Antoine Du Puget, the count d'Orval (1742-1810) might also be considered in these regards. An army officer, Du Puget served as inspector-general of colonial artillery, spending the years 1784 to 1786 on an inspection trip to the Antilles and Guiana. Lacroix calls Du Puget "a scientist of a wholly different order, the military-naturalist."[67] Du Puget proposed a program of botanical research and a book on the natural history of the Antilles to the Académie des Sciences; he presented memoirs to the Academy on the manchineel tree and its poisonous fruit, on rubber, and on his voyage generally; he came in second to Bougainville in the election of 1789 for the position of free associate in the Academy and was later

62. ALS "Thouïn" to "Monsieur le Comte" dated "A Paris ce 9 Aoust 1784," AN O[1] 2112, where Thouin calls Dombey "un vray Martir de l'histoire Naturelle...."

63. Re his stipend and other 60,000 livres owed to Dombey, see ALS "Dombey" to "Monsieur" dated "Paris le 31 8.bre 1785," AN O[1] 2112. Dombey ended up returning to America in 1794, sent by Revolutionary authorities to bring the new meter and kilogram to the New World; unfortunately, storms, factious colonists on Guadeloupe, and pirates beset his journey, and he died on the Caribbean island of Montserrat in April, 1794; see Adler, p. 238.

64. ALS "Le M.al de Castries" to "M. Le C.te d'angevillers," dated "a Versailles le 9 Avril 1786," AN O[1] 2112. On Palisot de Beauvois generally, see Gillispie (1992) and (2007), pp. 144-53; Lacroix (1938), vol. 4, pp. 209-221; Camus, pp. 109-111; Duprat, pp. 234-35; Allorge, pp. 313-24; and materials in PA-DB (Palisot de Beauvois).

65. For Fougeroux's instructions, see loose sheet dated "1786," APS-MS, B/D87, group 28. On de Beauvois materials received in France, see ALS "L'Abbé Nolin," dated "a Paris Le 7 mars 1789" to "Monsieur Le Comte," O[1] 2113A ("Pépinières 1789"); BCMNHN, Ms. 691, fols. 117-18, 125-26, 129-30, 154-54v; and Ms. 1327, passim.

66. McClellan (1992), pp. 160-61, 266-67.

67. "un savant d'un tout autre ordre,... un militaire naturaliste," Lacroix (1938), vol. 3, pp. 97-106, here, p. 97; see also Jandin, p. 17, and other Du Puget materials, PA-DB (Du Puget); Lesueur focuses on Du Puget.

elected to the Institut de France. Du Puget sent back materials to Thouin at Jardin du Roi, and while in Cayenne, he botanized alongside L.-C. Richard, whom he exhorted to apply himself to more useful botanical researches, particularly concerning trees.[68] Du Puget was a colonial and then (on his return to France) national associate of the Cercle des Philadelphes, and in every way he exemplifies the work and role of the naturalist traveling under the aegis of the Colonial Machine.

Jacques-Julien Houtou de La Billardière (1755-1834) fits a similar bill.[69] La Billardière was sent to England for two years to study the botanical collections of Joseph Banks, and then, sponsored by L.-G. Le Monnier, he undertook an official trip to the Levant in 1786-1788, returning from Cyprus, Lebanon, and Syria with 1000 plants for the Jardin du Roi. The Académie des Sciences commissioned La Billardière to prepare for publication the scientific papers of the Spanish voyager Noroña who had died on Île de France. (Noroña had bequeathed his papers to Cossigny de Palma who in turn sent them to the Academy.[70]) In 1791-1793 La Billardière accompanied d'Entrecasteaux as naturalist on the voyage in search of La Pérouse. The Academy elected him a correspondent in 1790. La Billardière was the first botanist to collect in New Caledonia, but unfortunately he lost his collections, including sixty breadfruit trees, when captured by the Dutch. After being released in 1795, he spent six months botanizing on Île de France before returning to France, where he published his two-volume *Relation du voyage à la recherche de Lapérouse* in 1799 and became a resident member in botany of the Institut de France in 1800.

Botaniste du roi André Michaux (1746-1802) was a protégé of Louis-Guillaume Le Monnier and Bernard de Jussieu of the Académie des Sciences and the Jardin du Roi. Michaux studied with these men and with Claude Richard, chief gardener at the royal Trianon gardens. Early in his career, with a short botanical trip to England under his belt, Michaux apprenticed as a junior member of a botanizing expedition to the Auvergne in 1779 led by the chief gardener of the Jardin du Roi, André Thouin. For this trip Michaux was appointed as *correspondant du Cabinet du Roi* by the great Buffon, the head of the Jardin du Roi.[71] Another botanizing voyage to the Pyrenees and Spain followed. Through Le Monnier, Michaux introduced himself to the Academy of Sciences in 1781 with a well-received presentation of a case of his own design for trans-

68. See remarks to this effect in "Extrait de son Journal de mission a Guyane," PA-DB (Du Puget).

69. On La Billardière, see Chevalier; Allorge, p. 502; Lacroix (1938), vol. 4, p. 63; and materials PA-DB (La Billardière).

70. Allorge, pp. 618-20; PA-PV 109 (1790), p. 28 ("Samedi 6 Février 1790"), pp. 117-19 ("Samedi 15 Mai 1790"); PA-PS (15 May 1790).

71. A copy of Michaux's letters of correspondence as *correspondant du Cabinet du Roi* is located in AN AJ[15] 501-514, #369.

porting delicate plants on long voyages.[72] In 1782-1785 the crown sent Michaux on his own botanical expedition to Persia and the Caspian Sea. Already before this trip, the abbé Nolin and the count d'Angiviller at Rambouillet and the Bâtiments du Roi were aware of Michaux.[73] On his return from the East, officials then posted him to the new United States of America in 1785-1796; although he traveled widely in North America, at this stage in his career Michaux was less a wandering botaniste du roi than he was permanently stationed in America and part of a much larger, formal enterprise of bringing American plants, particularly trees, to France. Michaux's American adventure is of such high importance to our story that it receives separate, extensive treatment further on.

Tipu Sultan. Finally in this connection, the envoys sent to France in 1788 by the Indian prince, Tipu Sultan, spurred the Colonial Machine into unexpected, but for us revealing action.[74] Tipu Sultan inherited power in 1782 at the age of thirty-two from his father, Haïder Ali, the ruler of Mysore. He was trained by the French in India and developed decidedly anti-English politics. He sent emissaries to France in 1788 to secure support in his struggle against the English. Assisted by the French royal secretary and interpreter, Ruffin, the exotic Indian ambassadors made a splash at court and elsewhere. In Paris, on Wednesday, 5 September 1788, they attended a meeting of the Académie Royale des Sciences, where Lavoisier performed various chemical experiments for their entertainment![75] In addition to their political mission, the emissaries came to France to recruit technical personnel for service in India. La Luzerne wrote to Thouin at the Jardin du Roi about choosing the requisite individuals, asking him to be discreet about the matter because otherwise "a bad lot of subjects will present themselves."[76] In the end, a physician (Pierre-Rémy Willemet at a salary of 6000 livres), a surgeon (Barrault, 3600 livres), two gardeners (Mulot and Luhrmann, 2400 and 1600 livres, respectively), two clockmakers, and a family of cloth makers (the Regniers – husband, wife, and three children) signed four-year contracts and constituted the French party that traveled to India for Tipu Sultan and for the Colonial Machine.[77]

72. PA-PV, 100 (1781), fol. 118v ("Samedi 12 Mai 1781"); fols. 203r-204r ("Samedi 1er Septembre 1781"); PA-PS ("Samedi 1er Septembre 1781"); PA-PV-110, André Michaux, 1er Septembre 1781.

73. See AN O[1] 211: Pépinières, 1778-1782, AL draft in d'Angiviller's hand dated "30 janvier 82" to "M.r André [sic] naturaliste du Roi."

74. The details of this story emerge in set of papers and correspondence preserved in the papers of André Thouin, BCMNHN, Ms. 307 and Ms. 308. On Tipu Sultan, see Vincent, chapts. 5-6.

75. Fraser describes impact of this visit on the court of Louis XVI, pp. 312-14. On the visit to the Paris Academy, see PA-PV 107 (1788), fol. 256v ("Mercredy 5. Septembre 1788"): "MM Les Ambassadeurs du Nabal Tipon Sultan Bachadour ont assité [sic] à cette Séance. M. Lavoisier a fait diverses expériences chimiques."

76. "une foule de mauvais sujets se présenteront...," ALS, La Luzerne to Thouin, dated "Versailles, ce huit aoust 1788," BCMNHN, Ms. 307.

77. "Articles de la Convention Générale," dated "le 29 7bre 1788," BCMNHN, Ms. 307. The contract of October 3, 1788 added the two clockmakers.

In preparation for their departure, Thouin assembled seven cases of European seeds and plants that included 250 or so different ornamental trees and plants and stock for kitchen gardens.[78] He prepared an elaborate set of instructions for how to tend live plants during the voyage to India and for the botanizing the gardeners should do on their projected stops at the Cape of Good Hope and on Île de France.[79] In addition, the crown provided one-time payments of 600 livres each to the doctor and surgeon for their expenses prior to departure, and spent another 12,380 livres in equipping the party, 6220 livres of which went for "things necessary for the doctor if we desire him to occupy himself with natural history research in botany, zoology, mineralogy and fisheries for ships' crews and for the Paris Jardin and the Cabinet du Roi."[80]

As this quotation suggests, La Luzerne and Thouin did not go to such lengths for Tipu Sultan or simply to fulfill the political agenda of Louis XVI. They had another plan in mind, namely, instituting a formal chain of botanical exchanges that would link India, Île de France, and Paris. La Luzerne ordered the administrators on Île de France to mobilize the royal garden there to this end, and he indicated that members of the Tipu Sultan party were to take every opportunity to send useful or unusual plants from India to Île de France.[81] Thouin was more explicit, advising the gardeners "on the nurseries they have to set up in India, on how to tend economic plants,...and on how to keep up a correspondence with the royal gardens on Île de France and Paris."[82] He reminded them that "the Minister of the Navy feels strongly that the gardeners should correspond with the botaniste du roi on Île de France and send him the useful productions in India that are lacking at the Jardin du Roi on Île de

78. See various documents contained in BCMNHN, Ms. 307: "Etat des objets contenus dans la Caisse no. I expediée du Jardin du Roi le 3 8bre 1788," "Etât des Arbres fruitiers et Autres contenus dans les Caisses no. 4 et 5. Expediées du Jardin du Roi le 7 8bre 1788 Pour le Nabob Typoo Sultan," "Etât des Plantes vivaces Economiques et Medicinales Contenues dans la Caisse no. 6....[sic] Expediée du Jardin du Roi le 7 8bre 1788," and "Etât des objets contenus dans la Caisse no. 7. Emballée au Jardin du Roi le 10 8bre 1788." Regarding the 730 livre expense for shipping, see Thouin note to La Luzerne dated "a Paris ce 16 fevrier 1789" in the same collection.

79. See ms., "Instruction pour diriger les Jardiniers dans la Culture des Végétaux en Nature destinés au Nabob Typoo Sultan pendant leur Voyage sur Mer," BCMNHN, Ms. 307. Thouin's instructions also included a "Model du Journal d'observations Metheorologiques" and a 4pp. "Etât des Genres de plantes qui croissent au Cap de Bonne esperance et qui manquent au Jardin du Roi."

80. See ALS La Luzerne to Thouin, dated "A Versailles le 10 8bre 1788" and "Etat des objets necessaires au Medecin si l'on desire qu'il s'occupe des recherches d'histoire naturelle pour le Jardin et le Cabinet du Roi de Paris pour des Equipages, de Botanique, de Zoologie, de Mineralogie, de Peche," BCMNHN, Ms. 307.

81. "Je ne doute pas, qu'ils ne s'empressent de profiter des occasions qui se présenteront, pour envoier à l'isle de france les plantes utiles ou curieuses de la Côte de malabar," ALS La Luzerne to Thouin, dated "A Versailles le 10 8bre 1788," BCMNHN, Ms. 307.

82. "je leur ai donne [sic] des Conseils sur l'établissement et la culture des Pepinieres et des plantes économiques qu'il doivent etablir dans L'Inde. Enfin je leur ai indiqués les moyens d'entretenir une correspondance entre les jardins du Roi de l'Isle de france et de paris," AL, Thouin to Pouget, dated "le 14 8bre 1788," BCMNHN, Ms. 307.

France."[83] Thouin wrote more directly to this effect to Jean-Nicolas Céré, the head of the royal garden on Île de France:

> As these gardeners will stay in India for a considerable period, the intention of the Minister is that they maintain a formal correspondence with you, Sir, and that they send to you all the useful productions of India that your cultivations lack.... These same gardeners are also under orders to correspond with our Jardin du Roi and to annually pass on seeds, trees, and herbaria for the Cabinet du Roi.[84]

And writing the same day to Pierre Sonnerat, then naval commissioner at Pondicherry, Thouin repeated:

> ...these gardeners are specially charged by the Minister of the Navy to keep up a regular correspondence with the royal gardens on Île de France and at Paris and to send to these two gardens all the useful vegetable productions that they will encounter in the part of India where they are going to be living, as seeds, living plants or as herberia or wood samples for the Cabinet du Roi.[85]

In the event, the expedition left Brest on 14 November 1788. It did not stop at the Cape of Good Hope as planned, but pushed on directly to Île de France, arriving there on 18 February 1789. For a month or so, the physician and surgeon botanized on the island and mingled with local amateurs and scientific personnel there, while the gardeners tended the plants from Europe kept aboard ship. Before departing, the party sent a botanical shipment back to France.[86] The group finally arrived at Pondicherry in May of 1789. The fact

83. "...le Ministre de la Marine à fort à coeur que les Jardiniers correspondent avec M. Céré et qu'ils fassent passer de L'Inde les Productions utiles qui manquent au Jardin du Roi de L'Isle de France," "Instruction pour diriger les Jardiniers dans la Culture des Végétaux en Nature destinés au Nabob Typoo Sultan pendant leur Voyage sur Mer," BCMNHN, Ms. 307.

84. "Comme ces Jardiniers resteront dans l'Inde un tems asséz considerables [sic] L'Intention du Ministre est qu'ils entretiennent avec vous, Monsieur; une Correspondance reglée et qu'ils vous fassent passer toutes les Productions utiles de l'Inde qui manquent à vos cultures....Ces mêmes Jardiniers ont ordre aussi de correspondre avec nôtre Jardin du Roi et d'y faire passer annuellement des graines des Plantes en nature, des Bois et des herbiers pour le cabinet du Roi...," AL, Thouin to "M. Ceré Intendant du Jardin du Roi a l'Isle de france," dated "de Paris le 14 8bre 1788," BCMNHN, Ms. 307.

85. "Ces jardiniers sont specialement chargé par le ministre de la marine d'entretenir une correspondance reglée avec les jardins du Roi de l'Isle de france et de Paris et d'envoyer à ces deux Jardins toutes les productions végétales utiles qu'ils rencontreront dans la partie de l'Inde qu'ils vont habiter; tant en graines, plantes en nature, qu'en herbiers et bois pour le Cabinet du Roi...," AL, Thouin to "Monsieur Sonnerat Commissaire de la Marine a Pondichery le 14 8bre 1788," BCMNHN, Ms. 307. In this connection, see related ALS Sonnerat to Adanson, dated "Pondichery ce 20 juin 1788," HI, AD 241.

86. ALS, Thouin to "M.M. Mulot et Luhrmann Jardiniers du Sultan Typoo," dated "29 9bre 1789," BCMNHN, Ms. 307.

that most of the plants brought so carefully from France and from Île de France died on this leg of the voyage signaled the hard fate that awaited the travelers there. Apart from the small French entrepôt, the English controlled the region, and the local governor, Macnamara, was actively hostile toward the Indian ambassadors. Terrible things were said about Tipu Sultan and his supposed mistreatment of the French at Mahe. Willemet, the doctor, became disheartened at being told that he was better off starving in France than coming to this country. They left with only a few carts and oxen to make the excruciatingly long trek inland across the desert to Mysore. In a letter back to France dated 15 June 1789 the second in command at Pondicherry, Dufresne, described their departure and painted a depressing portrait of the party as it headed out: "...And lastly, this poor Madame Regnier, wife of the cloth maker [had the husband died during the trip?] had only a pathetic small open cart for herself and her three small children. In it she was exposed to the full intensity of the sun. You will judge how much this poor unfortunate woman had to be discomfited, especially after having undertaken a journey of five hundred miles."[87] This is the last report we have of the French in the service of Tipu Sultan, disappearing as they did into India and out of history.[88]

* * *

As notable as these botanical expeditions were, like scientific expeditions generally, they were transitory and inevitably only partly effective. That may explain the effort from 1779 on to tap the network of diplomatic corps to facilitate botanical exchanges; this effort doubtless failed for want of expertise and commitment, but it signals the need and desire of officials to secure more solid botanical collecting.[89] Just the same, the expeditionary botanists sent out by the Colonial Machine in the eighteenth century constituted a remarkable collection of individuals who made their way (or did not, as the case may be) in an often treacherous world overseas. The botanistes du roi *en mission* form the complement in botany and natural history to the *astronomes du roi*, who also

87. "...enfin cette pauvre Madame Regnier femme du manufacturier En Drap n'a eu pour elle & ses trois petits enfants qu'une mauvaise petite charette non couverte, elle était exposée dedans a toute l'ardeur du soleil, ainsy jugés combien cette pauvre mal'heureuese a du être mal a son aise, surtout pour entreprendre une route de 200 lieues...," "Extrait d'une autre [lettre] du même [Mr. Defrene second Commandant de Pondichery] le 15 juin 1789," BCMNHN, Ms. 307.

88. Allorge, pp. 616-17 and 686-87, has uncovered evidence that Willemet survived for twenty months or so at the court of Tipu Sultan.

89. For this initiative, see draft letter in d'Angiviller's hand to "M. de sartine," dated "Vers.les le 24 aoust 1779," AN O[1] 2111 and printed pamphlet, *État des graines d'arbres, arbrisseaux, plantes, oignons a fleurs, qu'il seroit nécessaire de faire venir du Levant, pour les Jardins Botaniques, & d'agrémens, de Sa Majesté* (Paris: V.e Hérissant, Imprimeur du Cabinet du Roi, Maison & Bâtimens de Sa Majesté, 1779) in AN O[1] 2112.

traveled to distant lands to do their science. As telling as these instances are of the nature and goals of the Colonial Machine, they constituted a lesser part of the botanical apparatus. Unlike the case of astronomy, which did not truly implant itself in the colonies, the botanical arm of the Colonial Machine spanned the oceans, sprouted botanical gardens in the colonies, and then mobilized the whole in an intercontinental system of applied botany and commercial development.

Colonial Gardens – East

Coincident with institutional developments in the first half of the eighteenth century in France, where official botanical gardens arose in Brest, Lorient, Nantes, and Rochefort, so, too, did official gardens with staffs of paid scientific experts begin to emerge in the overseas colonies. In the East, gardens cropped up from time to time alongside French stations in Africa, such as the one maintained by Michel Adanson while in Senegal.[90] But the most significant developments in the East concerning economic botany and colonial botanical gardens occurred on the Indian Ocean outposts of Île de France and Île Bourbon.

The isolation of these islands, their pristine original state, and their transformation into lively centers of horticultural production deserve emphasis. As a report of 1676 from Île Bourbon observed: "This island is a veritable terrestrial paradise. Its soil is so excellent that it produces many things without sowing. The air is so healthy that illness is rare, and there are no poisonous animals on the whole island."[91]

Early Commodity Production. The transformation of these isolated specks of land in the vast body of the Indian Ocean began with the colonization of Île de France by the Compagnie des Indes in 1721. In the early phases of development of the Mascarenes, the policy of the Compagnie was to use Île de France to produce food and staples to replenish its ships traveling to and from China and India, and to use Île Bourbon to cultivate economically useful

90. On this garden, see Nicolas (1963a), pp. 16-30, and reference to this garden, ALS, Adanson to "Messieurs" [de Jussieu], dated "Au Senegal ce 20 aoust 1751," PA-DB (Adanson); see also BCMNHN, Ms. 314, entry for Le Voiturier.

91. "Cette Isle est un veritable paradis terrestre, et la terre en est si excellente qu'elle produit beaucoup de choses sans les semer, L'Air si salubue que les maladies y sont Rares, et quil ny a point d'Animaux Veneneux dans tout LIsle...," in ms. copy of Guillaume Delisle, "Voyage de France aux Indes Orientales en 1676. à L'Isle Mascareigne ou de Bourbon, Mozambique Bombaye &c," AN-MAR 2JJ 53. The common tendency by reporters to exaggerate nature's idylls needs to be discounted in evaluating this quote. Just the same, the great and wildly successful romantic novel by Bernardin de Saint-Pierre, *Paul et Virginie* that appeared over a century later in 1789 was set on the island of Île de France and played off of this same reputation; it is not for nothing that Bernardin de Saint-Pierre (1737-1814) took over as Intendant du Jardin des Plantes and the Cabinet d'Histoire Naturelle in 1791.

plants.[92] The implantation of indigo production represents an early initiative vigorously promoted by the Compagnie des Indes to develop the economic base of the Mascarenes. This effort was encouraged by Jean-Baptiste-François de Lanux *père* (1702-1772). Lanux was a colonist on Île Bourbon, a jurist on the local Conseil Supérieur, and later a correspondent of the Académie des Sciences in Paris. He reported that indigo grew well on Île Bourbon and produced a good quality product, and in 1735 Compagnie offered a 1500-livre credit to any colonist who would clear land for indigo production.[93] The local administrators on Île Bourbon sent word that introducing indigo production was difficult for want of slaves and knowledge of how to produce indigo.[94] The Compagnie responded by importing indigo stock at great expense from Saint Domingue, along with two slaves who were master indigo makers![95] (Other indigo imports came from India, where the plant originated, and local cultivators were to decide which plants did best.[96]) One of the slaves from Saint Domingue was later sent to Île de France to train other slaves of the Compagnie in the production of indigo.[97] The Compagnie put a certain Chandos in overall charge of developing indigo production on Île de France, providing him a house and forty male and female slaves.[98] When not otherwise employed on fortification or shipbuilding, Compagnie workers built vats for indigo production.[99] On their own plantations, two colonists, the royal surveyor Guyomar and a certain Calvert, promoted indigo production, but merchants in Nantes, to whom samples were sent, judged their product not quite up to snuff.[100] Prob-

92. "Extrait du registre général des Déliberations de la Comp.ie des Indes. Du 27 juin 1741," in Lougnon (1940), p. 15. On the points touched on here, see also Lougnon (1956).

93. "M.rs du Conseil Supérieur de l'Isle de Bourbon," dated "A Paris le 11Xbre 1734," in Lougnon (1933), p. 212; letter to "M.rs du Conseil Supérieur à l'Isle de Bourbon," dated "A Paris le 17 fev.er 1738," in Lougnon (1935), p. 122.

94. "A la Compagnie. Du 31e Xbre 1735," in Lougnon (1933), pp. 302-303.

95. See letter from headquarters to "M.rs du Conseil Supérieur à l'Isle de Bourbon," dated "A Paris le 17 fev.er 1738," in Lougnon (1935), p. 118. On the production of indigo dye, see McClellan (1992), p. 67.

96. See letter to "M.rs du Conseil Supérieur à l'Isle de Bourbon," dated "A Paris le 25e Mars 1741" in Lougnon (1935), p. 181.

97. See Compagnie letter to "M.rs du Conseil Supérieur à l'Isle de Bourbon," dated "A Paris le 25e Mars 1741" in Lougnon (1935), p. 182. The slave turned out to be a "mauvais sujet," but "Comme son camarade et lui sçavent certainement bien leur métier, qu'ils ont coûté très cher à la Comp.e, et qu'elle a encoure eu bien de la peine à se les procurer, vous devez sentir que ce sont des noirs qu'il faut ménager et conserver par de bons traittements, autant qu'il sera possible...."

98. "Ordres au Conseil de l'Isle de Bourbon," in Lougnon (1940), pp. 118, 155. See also Letter of La Compagnie to "M.rs du Conseil supérieur à l'isle de Bourbon," dated "A Paris le 9 avril 1745" in Lougnon (1940), p. 200.

99. "Extrait du registre général des délibérations de la Compagnie des Indes. Du 11 avril 1744...Ordres au Conseil de l'Isle de Bourbon," in Lougnon (1940), p. 155.

100. Letter of La Compagnie to "M.rs du Conseil supérieur à l'isle de Bourbon," dated "A Paris le 9 avril 1745" in Lougnon (1940), p. 199; see also letter from Conseil "A Messieurs les directeurs de la Compagnie des Indes à Paris," dated "A l'isle Bourbon ce 12 avril 1747," in Lougnon (1949), p. 47.

lems persisted with establishing productive *indigoteries* in the Mascarenes, and production remained touch and go.[101]

Such was not the case for coffee. Initially, the Compagnie des Indes tried to discourage coffee production in favor of indigo.[102] (For the same reason it at first discouraged sugar and rice production.[103]) Still, coffee production on Île Bourbon reached 1,286,000 pounds in 1727-1731, at least 2,250,000 pounds in 1732-1735, and 3,000,000 pounds by the 1740s.[104] (In 1733, officials on Île Bourbon sought samples of Martinique coffee to compare with theirs.[105]) Not thought of sufficient quality, Île Bourbon coffee was destined for the export market and not for internal consumption in France, but was nonetheless a major colonial product.[106] Cotton production, too, was at first rejected by the Compagnie because the women of Île Bourbon were thought too lazy to work in that industry; but then the Compagnie tried to encourage it by importing seventeen cloth workers from India to begin cloth production on Île Bourbon in 1745.[107]

Officials anxiously sought to diversify production because they feared that reliance on only one or two commodities would leave the colonies vulnerable if a world glut in production drove down prices. They realized that spices offered a grand opportunity to diversify the economy.[108] Early on, the Compagnie des Indes imported pepper, cinnamon, rhubarb, and ravensara from India, China, and Madagascar for distribution to reliable colonists for propagation and production on Île de France and Île Bourbon.[109] With regard to rhu-

101. See Compagnie letter to "M.rs du Conseil Supérieur à l'isle de Bourbon," dated "A Paris le 17 mars 1750," in Lougnon (1949), p. 235: "Quelles que puissent estre les difficultés qui se rencontrent pour l'établissement des indigoteries, il faut faire tout son possible pour tâcher de les surmonter.... "

102. See correspondence to this effect: from La Compagnie to "M.rs du Conseil Supérieur à l'Isle de Bourbon," dated "A Paris le 25e Mars 1741," in Lougnon (1935), p. 157; "Lettre du Conseil à la Compagnie. A St. Paul, Isle de Bourbon le 20 Janvier 1738," in Lougnon (1937), p. 65; and Compagnie letter to "M.rs du Conseil supérieur à l'isle de Bourbon," dated "A Paris le 9 avril 1745," in Lougnon (1940), p. 195.

103. See "Extrait du registre général des délibérations de la Compagnie des Indes. Du 11 avril 1744," in Lougnon (1940), p. 140.

104. Lougnon (1933), p. lxviii; and, letter from Conseil Supérieur to "Messieurs les sindics et Directeurs de la Compagnie des Indes," dated "A l'isle de Bourbon le 10 8bre 1749," in Lougnon (1949), p. 175.

105. "A Mrs. du Conseil Supérieur de l'Isle de Bourbon. A Paris le 17 9bre 1732," in Lougnon (1933), p. 88.

106. At a certain point coffee was thought to be anti-scurvy cure; see "Lettres des administrateurs de l'Ile de France, Ile de France, 13 juin 1744," Lougnon (1956), p. 60.

107. See letter to "M.rs du Conseil Supérieur à l'Isle de Bourbon," dated "A Paris le 17 fev.er 1738," and letter to same, dated "A Paris le 25e Mars 1741," in Lougnon (1935), pp. 118, 160; see also Compagnie letter to "M.rs du Conseil supérieur à l'isle de Bourbon," dated "A Paris le 9 avril 1745" in Lougnon (1940), p. 200.

108. See letter to this effect, "A Mess.rs du Con.seil Provincial à l'Isle de France. A Paris le 24 7bre 1729," in Lougnon (1934), pp. 88-95, here p. 91; Lougnon (1953) documents Compagnie and ministerial correspondence from the late 1720s devoted to plant transfers.

109. Lougnon (1934), p. xxvii, and letters to "M.rs du Conseil Supérieur de l'Isle de Bourbon," dated "A Paris le 31 Xbre 1727," and "A M.rs les directeurs généraux de la Compagnie des Indes," dated "a l'Isle de Bourbon le 20 Xbre 1731," in Lougnon (1934), pp. 26-65, notably p. 50, and p. 154.

barb, for example, already in 1729 the agent of the Compagnie des Indes in Canton sent rhubarb plants and seeds to the Mascarenes. The major source of rhubarb, considered a medicinal plant, was then central Asia, and the Compagnie wished to establish a commercial alternative.[110] Cinnamon was another spice that authorities promoted. By the early 1730s cinnamon was being shipped back to France where it was judged of good quality by Jussieu.[111] The cinnamon from the Mascarenes was not as good quality as cinnamon from Ceylon, but the Compagnie nonetheless paid highly for it and encouraged its production.[112] Already in 1734 the Compagnie instructed its agents on Île Bourbon that "this type of spice product is very sought after, and you should not omit anything to try to make its cultivation widespread."[113] Authorities imported and brought into local cultivation ravensara from Madagascar, "known in France as cloved cinnamon."[114] Cardamom was another early import into these islands.[115] The coconut, thought an antidote for poisons, was likewise successfully introduced.[116] In the 1730s, too, trials took place with tea and the Chinese gum tree from which lacquer was made.[117] The Compagnie des Indes encouraged the export of sandalwood from Madagascar to India and China.[118] By contrast, although it expressed an early interest in the annatto ("rocou") tree for dyeing, the Compagnie abandoned the prospect because colonists did not know how to grow or process the tree. This example raises a

110. See Compagnie letter to "M.rs du Conseil Supérieur de l'Isle de Bourbon," dated "A Paris le 11Xbre 1734," in Lougnon (1933), p. 100. Rhubarb samples were sent to France for analysis. Medical interest in rhubarb was sustained throughout the eighteenth century; the Société Royale de Médecine tested samples from Île de France for trials in 1791; see SRdM, Ms. 11bis, fol. 98v ("Séance du Vendredi 18 mars 1791") and fol. 111v ("Séance du Vendredi 29 Avril 1791").

111. See Lougnon (1933), pp. xxv, 77, and letters "A Mrs. du Conseil Supérieur de l'Isle de Bourbon. A Paris le 17 9bre 1732," "M. Despremenil. A l'Isle Bourbon le 12 mars 1734," and "M.rs du Conseil Supérieur de l'Isle de Bourbon," dated "A Paris le 11Xbre 1734," in Lougnon (1933), 99-100, 183, 204.

112. See previous note and letter to "M.rs du Conseil Supérieur à l'Isle de Bourbon. A Paris le 5 Mars 1735," in Lougnon (1933), p. 289.

113. "Cette espèce d'épicerie étant très recherchée, vous ne devez rien obmettre pour tâcher d'en rendre la culture commune," letter to "M.rs du Conseil Supérieur de l'Isle de Bourbon," dated "A Paris le 11Xbre 1734," in Lougnon (1933), p. 100.

114. "connue en France sous le nom de canelle gérofflée," letter to "M.rs du Conseil Supérieur de l'Isle de Bourbon," dated "A Paris le 11Xbre 1734," in Lougnon (1933), p. 100. Ravensara is a tall tree from Madagascar with an aromatic bark; an oil can be had from it, too.

115. Lougnon (1956), p. 78, item #596 ("Lettres des administrateurs des comptoirs de l'Inde. 1729. au Fort-Louis, 14 février").

116. See Lionnet; Sonnerat (1776a).

117. "Extrait des lettres de l'Isle Bourbon du 20 décembre 1730," in Lougnon (1934), pp. 129-36, here pp. 134-35.

118. "Extrait du registre général des déliberations de la Compagnie des Indes. Du 26 juin 1742," in Lougnon (1940), p. 41.

point that runs through all of the correspondence relative to these developments: the dearth of expertise to expand these productions.[119]

Pamplemousses and Poivre. In 1727 the king appointed a Sieur Couzier as *médecin botaniste du roi* for the East Indies at a salary of 2000 livres paid from the Royal Treasury. He was given free passage on the ships of the Compagnie des Indes and charged to establish a botanical garden on Île Bourbon; for a while Couzier was on Île Bourbon, and at another time he reported botanical observations from Pondicherry.[120] Given the scope of this botanical activity that was quickly reshaping the economic, political, and physical landscape of Île de France and Île Bourbon, it comes as no surprise that the Jardin du Roi in Paris took an interest. In 1735 the Jardin asked the Compagnie des Indes to forward to Paris all the plants and seeds it received from the Mascarenes.[121] In 1736 the Compagnie commissioned an agent, the surgeon Jullia, to venture into the forest to collect samples for the Compagnie to send to the Jardin du Roi; Jullia apparently did this in 1736, but in 1738 his demands were judged too expensive, and the matter was dropped.[122] In any event, the Compagnie agreed to send only seeds to the Jardin du Roi because plants themselves took up too much space aboard ship and were impossible to tend.[123]

It was against this background that in 1735 the French East India Company established the first formal garden, known as *Pamplemousses* (Grapefruits), on Île de France. Mahé de La Bourdonnais, then governor of the Mascarenes, bought and developed the property as an acclimatizing garden that ultimately expanded to over 200 *arpents*.[124] He sold the garden to the Compagnie in

119. On 'rocou', see letter "A la Compagnie. Du 31e Xbre 1735," in Lougnon (1933), p. 303; letter to "M.rs du Conseil Supérieur à l'Isle de Bourbon," dated "A Paris le 12 Janvier 1737," in Lougnon (1935), p. 72. Various animals thought to be useful or economically significant were also brought in from entrepôts in India, including silkworms, rams, geese, ducks, and camels. All but one of the latter died shortly after their arrival. On these importations, see "Extrait du registre général des Déliberations de la Comp.ie des Indes. Du 27 juin 1741" in Lougnon (1940), p. 4; letter to "M.rs les Syndics et Directeurs de la Comp.ie des Indes," dated "A l'isle de France, le 18 9bre 1749," and "Copie de la lettre écrite par la compagnie à M.rs du Conseil supérieur de l'isle de France dattée à Paris le 28 février 1749," in Lougnon (1949), pp. 132-33, 199.

120. See letters from "Le Peletier" to "M. Le Comte de Maurepas," dated "A fontainebleau le 15 7bre 1728" and "A Marly le 26 Juin 1729," AN-MAR B^3 326, fol. 98, and AN-MAR B^3 334, fol. 118, respectively; the latter gives the official title; the former notes that "... S.r Couzier médecin Botaniste qui a esté Envoyé par le Roy dans les Indes orientales aux appointem.^ts^ de 2000# par an payables par Sa M.^te^ au Tresor Royal... [Couzier] estoit veritablement Employé en qualité de Botaniste aux appointements de 2000# sur le Tresor Royal;" see also Lougnon (1953), pp. 4, 15, 25-26, 158.

121. See letter to "M.rs du Conseil Supérieur à l'Isle de Bourbon. A Paris le 5 Mars 1735," in Lougnon (1933), p. 256; and letter to "Les syndics et directeurs de la compagnie des Indes à Pierre Dumas. Paris, 5 mars 1735," in Lougnon (1956), #65, p. 30.

122. Maroon slaves made such collecting dangerous; see letter to "M.re les Directeurs de la Compagnie" dated "A S.t Paul le 20e Mars 1736" in Lougnon (1935), p. 43; "Lettre du Conseil à la Compagnie. A St. Paul, Isle de Bourbon le 20 Janvier 1738" in Lougnon (1937), pp. 73, 136.

123. See letter to "M.rs du Conseil Supérieur à l'Isle de Bourbon" dated "A Paris le 12 Janvier 1737" in Lougnon (1935), p. 94. In 1744 Gaultier in Canada similarly noted the reluctance of the Compagnie des Indes to transport live plants; see Gaultier letter of October 25, 1744 probably to Duhamel, Houghton Library, Harvard University, dossier 57, M-13.

124. Haudrère (1992), pp. 62-66. On the Indian Ocean gardens discussed here, see also Rouillard; Ly-Tio-Fan (1967), (1970b), (1996).

1738.[125] After 1750, colonial administrators upgraded and enlarged the Pamplemousses garden, which then functioned as an experimental horticultural station.

The story behind this story and the further development of the French Indian Ocean islands as centers for economic botany concerns the eponymous Pierre Poivre (1719-1786) and his dramatic raids on the Dutch East Indies and Spanish possessions in Southeast Asia to obtain pepper, cinnamon, nutmeg, and other valuable spices from France's colonial competitors who jealously guarded their botanical riches.[126] Poivre undertook the first two of these expeditions himself in the 1750s.[127] After the French government formally assumed control of the Indian Ocean territories from the Compagnie des Indes, Poivre then returned as administrator of the Mascarenes, whence he sent out two more successful expeditions to South Asia. The one in 1769 returned with 400 nutmeg plants and 70 clove trees; the second in 1771-1772 returned with 28 nutmeg plants and 500 clove trees. As the commissioners in the Paris Academy reported shortly thereafter, "The number of nutmegs from this latter importation, either having germinated or in a condition to germinate, surpassed 40,000," and they concluded that the clove and nutmeg samples they received from the Indian Ocean were of commercial quality.[128] Local officials distributed these and other spices to private and public gardens on Île de France and Île Bourbon for propagation, and to this end in 1772 they printed a nine-page *Instruction for Colonists on the Îles de France and Bourbon on How to Plant and Successfully Cultivate Clove and Nutmeg Plants and Seeds.*[129] At this time, too, it became illegal and formally treasonous to export any spice plant from these islands; violators were subject to fines, imprisonment, and confiscation of their property.[130]

125. "Extrait de la lettre de la compagnie des Indes au Conseil supérieur de l'Isle de France, en datte du 17 Février 1738," in Lougnon (1935), p. 141.

126. This story has been fully told by the historian Madeleine Ly-Tio-Fane; among her other works listed in the bibliography, Ly-Tio-Fane (1967) sets the details presented here of the larger story of the French presence in the Indian Ocean and Pacific in the eighteenth century. On Poivre, see also Spary (2005), Grove (1995), chapt. 5; Lacroix (1938), vol. 3, pp. 191-209; Doublet; Cordier, ed.; Malleret; Allain, chapt. 6; Allorge, chapts. 19-20; Tessier (1789). See also ms. source, "Relation abrégee des Voyages faits par le Sr. Poivre pour le Service de la Compagnie des Indes depuis 1748 jusqu'a 1757," BCMNHN, Ms. 575. Le Gouic updates the Poivre story with full and well-documented details.

127. Le Gouic, p. 110, documents four more expeditions sent out later by Poivre as Intendant on Île de France; see also "Procès-Verbal de vérification des plants et noix de muscade," Toussaint, p. 360, #69.

128. "...le nombre des muscades de cette derniere importation, soit germées soit propre à la germination, passent le nombre de 40 mille," "Rapport des commissaires [sur] des gérofliers et des muscadiers," ms. dated "17 fev.r 1773," ms. copy in HI, AD 343.

129. *Instruction sur la maniere de planter et cultiver avec succès les plants et graines de gerofliers et muscadiers. A l'Usage de M. M. Les Habitans des Isles de France & de Bourbon.* A L'Isle de France, de l'Imprimerie Royale. 1772; AN-MAR G 101, dossier 4, fols. 171-75.

130. Delaleu (1777), vol. 1, pp. 331-32 ("Ordonnance, portant défense de transporter hors de la Colonie aucuns plants ou fruits propres à germer soit de Muscadiers, soit de Gerofliers. Du seize Juillet 1770"); and, ms. Code de l'Île de France, vol. 2, under "Muscadiers et Gerofliers" in AN-MAR G 237.

Beyond his own efforts, Poivre received plants from his correspondents in India and China, notably Abbé Galloys traveling in China, who sent specimens of Chinese trees, tea, star anise, and the lotus plant to Poivre on Île de France.[131] As a government official, Poivre naturally maintained close contact with the botanical establishment back in France, notably with Le Monnier and Bernard de Jussieu. Records indicate that he shipped clove and nutmeg plants to Le Monnier for the Trianon gardens and likewise for the Jardin du Roi on the banks of the Seine.[132] Poivre was in touch with Malesherbes, and he was elected a correspondent of the Academy of Sciences in 1754, having already been in contact with Réaumur from 1749. In 1772 Poivre gave an account of his activities to the Academy in Paris, which then lauded him in its report:

> The event of which we are giving an account under this head is a victory for both botany and the commerce of the kingdom.... The Academy felt it necessary to give a detailed summary of this expedition and to immortalize, so to speak, the name of our French Argonauts. Those that made the famous conquest of the Golden Fleece doubtless did not intend such a useful goal, nor perhaps anticipate such fearsome perils.[133]

Poivre's Successor: Jean-Nicolas Céré. In 1772 Poivre retired to France with a pension of 12,000 livres.[134] He sold the Pamplemousses garden to the crown, at which point it officially became the *Jardin du Roi de l'Île de France.*[135] In 1775, backed by Poivre and appointed by Turgot, control of the Jardin du Roi fell to the talented Creole, Jean-Nicolas Céré, who greatly expanded the activities of the Pamplemousses garden, making it a major, indeed *the* major, horticultural center in Asia.[136] Already in 1777 glowing reports flowed back to Paris: "The Jardin du Roi is superb. It is the nursery that

131. Galloys was a protégé of Le Monnier and was promised a bonus of 10,000 livres for securing the tea plant; Malleret, p. 118. Malleret lists the plants sent to Poivre by Galloys.

132. See letter, Poivre to Le Monnier, dated "A l'Isle de France, le 19 juillet 1772," in Laissus (1973), pp. 49-50.

133. HARS (1772), pp. 56-61, regarding Poivre report to Academy, "sur le transport des plants...à l'Ile de France."

134. See certificate dated 3 March 1777, AN-COL E 337 and PA-DB (Poivre).

135. "Copie de l'acte de vente du jardin "Mon Plaisir" à l'île de France, cédé au roi par Poivre en 1772," dated "12 8bre 1772," BCMNHN, Ms. 303; original: AN-COL, C^4 32.

136. On the strong support for Céré and the royal garden on Île de France by metropolitan officials, see exchanges with d'Angiviller and La Luzerne, AN O^1 2112 and O^1 1292, #20. At first, the elder Cossigny, tending his garden at nearby Palma, was less enthusiastic about Céré, but later warmed to him; see Cossigny letters to Le Monnier dated "Au Port Louis Isle de france le 28 Xbre 1775," "Au Cap de Bonne Espérance le 25-8bre 1775," and "Au Port Louis Isle de france le 12 février 1776," all BCMNHN, Ms. 1995. In his letter to Le Monnier dated "Au Port Louis Isle de france le 25 févr. 1776" Cossigny wrote: "Ce M. Céré est un trés-honnête garçon, et très obligeant, créole de notre Isle."

is providing trees for everyone in the colony. In the care he gives to it, M. Céré brings singular zeal and a lot of intelligence. A few native clove are fruits from last year, the ravensara is in fruit, as well as the rima or breadfruit."[137] Sonnerat described the Jardin du Roi on Île de France to Adanson in 1779: "I had the pleasure of finding the rarest trees here in the Jardin du Roy, most of which were in flower or with fruit. I made very accurate drawings of the false mangosteen, litchi, uampi [?], clove, nutmeg, trevi or the Cythera tree (the fruit of which is not as good as that Mr. Commerson made of the traveler's tree, revenala), ravensara, breadfruit, different cardamoms, eagle wood, different soapberry trees, caper bushes, large aloes, candlenut trees, etc."[138]

Céré received plants from everywhere. A 1785 inventory of the Jardin du Roi on Île de France lists approximately 650 different species and their geographical origins: Europe (France, Spain, Italy, Holland, Moldavia), the Near East (Crete, Lebanon, Arabia, Mocha), the Americas (Virginia and elsewhere in North America; Saint Domingue and elsewhere in the Antilles; Mexico, Brazil, Peru), India, (Bengal, Ceylon, Malabar, Goa), Africa (Guinea, Madagascar, St. Thomé, Senegal, Mozambique, the Cape of Good Hope), China, Japan, and Southeast Asia (Batavia, Indochina, the Moluccas, Surate, Tahiti, the South Seas, the Philippines, Île Bourbon, the Seychelles, Île Rodrigues).[139] Similarly, Céré sent out boxes of specimens to his correspondents all over, not just to the Jardin du Roi and the Academy of Sciences in Paris.[140] He was in regular contact with Thouin at the Jardin du Roi in Paris (Thouin sending him lists of plants and trees lacking in Paris), with the count d'Angiviller of the Bâtiments du Roi, and with de Castries, and then with La Luzerne at the Ministry of the Navy and the Colonies.[141]

137. "Le jardin du Roi est superbe. C'est la pépiniere qui fournit des arbres à toute la Colonie. M. Céré apporte dans les soins qu'il lui donne, un Zele unique, et beaucoup d'inteligence. Il y a quelque Girofliers créoles des fruits de l'année d.re Le Ravinesara est en fruit, ainsi que le Rima ou Arbre à pain...," Cossigny to Le Monnier, dated "Palma le 2-8bre 1777," BCMNHN, Ms. 1995.

138. "J'ay eu le plaisir de trouver ici dans le jardin du Roy, les arbres les plus Rares, la plus part etoit en fleurs ou en fruits et je rapporte des dessins tres Correct du faux Mangoustan, du litchi, du Uampi [?], des gerofliers et muscadiers, du trevi [?] ou arbre de Cythere dont le fruit n'est pas aussi bon que nous l'avoit fait M.r de Commerçon du Ravenala, Du Ravensara, du Rima, des differens cardamones, du bois d'aigle, des differens savoniers, des Capriers, des grans aloës, de la noix de bancoul, &c. Je vous les aurois envoyé par cette occasion, si le tems m'est permis d'en lever des doubles et de faire un choix dans mes herbiers...," ALS, Sonnerat to Adanson, duplicata dated "Isle de france ce 1er Sep.bre 1779," HI, AD 238.

139. Céré, "Recensement de tout ce que renferme Le Jardin du roi," BCMNHN, Ms. 303.

140. On the latter, see PA-PV, vol 89 (1780), fol. 140v ("Du Mercredi 31 Mai 1780, M. l'abbé Texier est entré et a lû une notte sur l'état des Gérofliers et des muscadiers, aux Îles de france et de Bourbon par M. de Ceré"). See also PA-PV-110, Céré, 5 September 1787 ("Plusieurs mémoires relatifs aux différentes espèces d'arbres, arbustes et plantes renfermées dans le Jardin du Roi et l'île de france, &c)."

141. Re Thouin, note dated "fait le 14 fev. 1788"; ALS "Thouïn" to "Monsieur" dated "A Paris ce 10 Aoust 1784," see BCMNHN, Ms. 47, and 4pp. ms., "Liste d'Arbres Cultivés a l'Isle de france qui sont desirés au Jardin du Roy," AN O[1] 2112. Re La Luzerne, see letter from Céré to Thouin, dated "à Belle Eau, isle de france, le 16 Mars 1789," BCMNHN, Ms. 47. For other Céré shipments to the Jardin du Roi in Paris, see mss. entitled "État de ce que contient la Caisse marquée No. 1. *Graines pour le roi*" and "Copie de ce que contient la Caisse marquée No. 2 *graines pour le roi*," BCMNHN, Ms. 1995.

More than anything else, Céré was intent on distributing plants from the Pamplemousses nursery to colonists so that they might begin commercial production of spices. In 1783, 1785, and again in 1787 he printed up announcements notifying colonists on Île de France and Île Bourbon that they could pick up plants gratis at the king's garden.[142] The 1785 list noted 107 species, including over 10,000 clove plants ("for 38 of which, two strong slaves are needed to carry one plant; for 894, a slave for each tree; for 484, two trees for each slave; and for 9000 four-to-six-month-old trees one slave can carry four at a time in a basket.")[143] Clove production had succeeded so well that Céré alerted the public that he was cutting back in order to encourage other productions, such as the 500 Cuban tobacco plants and the campeche wood he mentioned. The 1787 list enumerates 87 species in toto, including "10,000 young Ceylanese cinnamon trees, 10,000 rose-apples, 3,000 young clove plants, 1,200 coconut trees from Pondicherry, 1,000 Chinese Litchi, 1,000 young mango trees, 1,000 Madagascar sago plants, 1,000 Moluccan sago plants, 600 Chinese dwarf latanias, 1,000 Madagascan gum shrubs, sweet marjoram, and 73 Rima or breadfruit." Céré likewise distributed thousands of cacao and clove plants to colonists on Île Bourbon.[144] The scale of this enterprise underscores that what Céré et al. had first and foremost in mind was commercial development, not scientific botany, and in this they succeeded as never before. For his remarkable efforts, in 1788 the Société Royale d'Agriculture in Paris awarded Céré a commemorative medal, an award that came to him on Île de France with a letter of praise from the minister, La Luzerne.[145] Colonial administrators held a special ceremony on the island to mark the event.

Cossigny and the Garden at Palma. At least three other botanical gardens arose on Île de France. The most important was the garden founded by the Cossigny family on Île de France at Palma in 1764. A royal engineer and chief engineer for the Compagnie des Indes on Île de France, Jean-François Charpentier de Cossigny (Cossigny père, 1690-1780), initiated this garden on his property.[146] It eventually had litchi, jasmine, other plants from China and India, and even ferns from Canada. Cossigny was elected a formal correspondent of the Académie des Sciences in 1733, attached first to Réaumur and then to Portal. The elder Cossigny maintained regular contact with Réaumur, send-

142. See Le Gouic, p. 122; for list of 1783, see BCMNHN, Ms. 308; for 1785 and 1787, see documents preserved in BCMNHN, Ms. 303; and materials in AN O[1] 2112. See also remarks about this distribution by Sonnerat (1782), vol. 2, p. 86.

143. "10,416 Girofliers, dont 38, il faudra 2 forts noirs pour en porter un; 894, un noir pour un arbre; 484, deux pour un noir; & 9000 âgés de 4 à 6 mois, un noir pourra porter 4 de ces derniers dans un moyen panier...," BCMNHN, Ms. 303.

144. "Recensement de tout ce que renferme Le Jardin du roi, le Montplaisir; remit par Mr. De Ceré," BCMNHN, Ms. 303.

145. See printed notice, BCMNHN, Ms. 47. On the significant colonial interests of Royal Society of Agriculture, see above Part I and Regourd (1998).

146. On Cossigny père, see Lacroix (1938), vol. 3, pp. 151-59; Crépin, passim and p. 104.

ing botanical samples and meteorological and declination data back to France.[147] His Creole son, Joseph-François Cossigny de Palma (1736-1809), traveled widely in China, India, and Southeast Asia, settled on the family property, and inaugurated and maintained a close correspondence with Le Monnier at the Jardin du Roi in Paris.[148] In 1774 the Académie des Sciences elected the younger Cossigny its correspondent, attached to Le Monnier, and every year, often with Poivre's help, Cossigny sent a substantial botanical shipment to France aboard the vessels of the Compagnie des Indes.[149]

Sonnerat called the younger Cossigny "one of the most zealous cultivators in this region....Mr. Cossigny owns the most beautiful garden in the colony, and even more he hastens to multiply and to share with other colonists the rare and precious plants that he has had shipped at great cost from Europe, the Cape, Batavia, China, and India."[150] In 1775 Cossigny *fils* published a brochure on coffee cultivation that was distributed in the East and West Indies through the offices of the ministry; later he produced substantial pieces on indigo production and on brandy.[151] In these works he outlined the successful transplantation to Île de France of clove, nutmeg, coconut, mango, *le porcher* [?], tamarind, grapefruit, *le margofier* [?], *le teque* [?], star fruit, agarics, jackfruit, *le monongue* [?], *le bilimbi* [?], *le chérimbelle* [?], *le nourouk* [?], Moluccan and Ceylonese cinnamon, three kinds of rose apples, the false mangosteen, durian, and coffee. Through Cossigny's offices, too, Batavian sugarcane came to Île de France, whence it made its way to the Antilles, becoming the main variety of sugar produced there. Even the mangosteen seemed to be flourishing in the garden at Palma, given that in 1771 there were only two plants in the whole colony, their male/female varieties not yet discovered.[152] In 1780 Cos-

147. See Cossigny correspondence with Réaumur in PA-DB (Cossigny); see also "Treize lettres de Cossigny à Réaumur," and Cossigny, "Mémoire sur l'Ile de France [1764]," reprinted in Crépin, pp. 51-91. On Cossigny and Réaumur, see also MARS 35 (1733/1735), pp. 417-37; MARS 36 (1734/1736), pp. 553-63; MARS 36 (1735/1738), pp. 545-76; and MARS 40 (1738/1741), pp. 387-403.

148. On Cossigny *fils*, see Lacroix (1938), vol. 4, pp. 33-37; Allorge, pp. 397-98; and materials PA-DB (Cossigny de Palma), including unsigned chronology of "Famille Cossigny." See also retrospective account of his life and works, in 12-page Cossigny letter to "Citoyen La Cépède," dated "Paris le 21 Nivôse, an 7," AN-COL F[3] 162. Wanquet details Cossigny's career after 1789.

149. See the 124 Cossigny letters to Le Monnier preserved in BCMNHN, Ms. 1995. One shipment included two orangutans preserved in alcohol; ALS, Cossigny to Le Monnier dated "A L'Ile de France le 12 Xbre 1781."

150. "M. *de Cossigni*, l'un des plus zèlés cultivateurs de cette contrée ...M. De Cossigni qui posséde le plus beau jardin de la colonie, s'est encore empressé de multiplier & de partager avec les habitans les plantes rares & précieuses qu'il a fait venir à grands frais d'Europe, du Cap, de Batavia, de la Chine & de l'Inde...," Sonnerat (1782), vol. 2, pp. 81, 86.

151. See Cossigny de Palma (1775) and the presentation copy to Adanson, HI, AD 40. See also Cossigny de Palma (1779 and translation, 1789) and (1781/1782). See also Cossigny letters to Le Monnier, dated "A Paris le 18 Xbre 73," "A Paris le 19 Xbre 73," and "A Paris le 22 Xbre 73," BCMNHN, Ms. 1995. See also ms., "Récapitulation Générale de L'Essai sur la fabrique de l'Indigo;" BCMNHN, Ms. 357.

152. See Cossigny (1775), passim and pp. 17-18. On the latter point, see Sonnerat (1782), vol. 2, p. 81.

signy could rightly claim that "Palma is today very rich in exotic trees."[153] At one point, one of the children of the Richard family of gardeners at the Trianon gardens performed what amounted to an internship at Cossigny's garden.[154]

Cossigny and the garden at Palma maintained close contacts with the official royal garden at Pamplemousses as the latter developed. Although they apparently ended up feuding, Cossigny worked closely and enthusiastically with Céré, and to a real extent the garden at Palma functioned as another royal nursery and garden. At one point in 1781, for example, Cossigny and Céré teamed up on a contract with the local administrators to plant 120,000 acacia trees, pending the minister's approval to raise that number to 500,000.[155] (Special plantations of acacia were maintained for use in gunpowder manufacture.)[156] Cossigny was intent on following Céré's example and making his nurseries a center of production and distribution; at one point he boasted that his gardens contained 200,000 coffee trees and 4000 cinnamon plants.[157] Cossigny also collaborated with Commerson during the latter's years in the region.

Le Réduit, Mon Goust, and the Jardin du Roi on Île Bourbon. Likewise on Île de France the Compagnie des Indes established a garden in 1748 called *Le Réduit*.[158] Le Réduit came to possess the cinnamon plant, cacao, the Malabar pepper, mangosteen, and over 300 other species of plants from India, Batavia, Europe, the Cape Verde Islands, America, Brazil, the Near East, Senegal, the Cape of Good Hope, Madagascar, Sumatra, the Seychelles, the Malay peninsula, and elsewhere. A visitor in 1756 reported the splendors of the Réduit gar-

153. "Palma est aujourd'hui très-riche en arbres exotiques," ALS, Cossigny to Le Monnier, dated "A L'Isle de France le 16 Avril 1780," BCMNHN, Ms. 1995.

154. Cossigny *fils* was not happy with this Richard, writing to Le Monnier at one point, "Le fils de Richard, entend très bien les parterres, et la distribution d'un Jardin, et sait éxécuté les plans à merveille. Il sait planter; Mais il ne connoit pas la taille des arbres, ni tout ce qui a rapport à notre climat. Il est jeune et se formera. Il m'étoit plus inutile qu'a tout autre habitant, et parce que mon jardin est fait, et par ce que je m'occupe de sa culture et que depuis 15 ans je forme un jeune noir qui a de la bonne-volonté. Je gardois donc chez moy le jeune Richard, par consideration pour son Pere, et par bonne-volonté pour le jeune homme, jusqu'à ce que je trouvâlles à le placer...," ALS, Cossigny to Le Monnier, dated "Au Port Louis, Isle de france le 3 Avril 1776," BCMNHN, Ms. 1995.

155. See ALS, Cossigny to Le Monnier, dated "Palma le 25 Janvier 1781," BCMNHN, Ms. 1995. The administrators agreed to a subvention of 20,000 livres for this project; see "Propositions faites à M.M. les Administrateurs de l'Ile de France par M.M. de Cossigny et Céré relativement à l'entreprise d'une plantation de Bois Noir," dated "Au Port Louis Ile de France Ce 16 9bre 1781," BCMNHN, Ms. 357., mentions this matter.

156. Ly-Tio-Fane (1996), p. 11; Delaleu, Supplément 1 (1783), pp. 278-281: "Ordonance...qui défend d'envoyer paître des bestiaux dans les terreins aux environs de la Ville du Port-Louis, qui doivent être ensemencés pour le compte du Roi, de graines de bois noir. Du 9 Janvier 1782."

157. ALS, Cossigny to Le Monnier, dated "Palma le 29 Mars 1777," BCMNHN, Ms. 1995. Cossigny defended the climate of Île de France as suitable for the production of spices, against critics who said it was too mild; see his *Lettre sur les arbres à épiceries* (1775), pp. 17-18.

158. Laissus (1973), p. 38; Ly-Tio-Fane (1976), pp. 49-54; Fusée-Aublet dossier, "Séjour à l'Ile de France, 1753-1761. Jardin botanique de Réduit," in BCMNHN, Ms. 452; and, retrospective account of the gardens at Réduit by Charpentier Cossigny de Palma, letter to "Citoyen La Cépède," dated "Paris le 21 Nivôse, an 7," AN-COL F^3 162.

den and that it might someday rival the Dutch garden at the Cape of Good Hope.[159] From 1752 through 1761, the Réduit garden was staffed by the royal botanist, J.-B.-C. Fusée-Aublet (1723-1778), who had studied with the Jussieux in Paris and who was in contact with the Academy in Paris.[160] Cossigny *fils* helped supply the garden at Le Réduit. Notably, one of Fusée-Aublet's functions was to supply the hospital of the Compagnie des Indes with pharmaceutical supplies for which he was reimbursed over 128,000 livres.[161] Fusée-Aublet was active in developing local botanical resources, particularly cinnamon cultivation, but he feuded with Poivre and eventually transferred to Cayenne as part of the Kourou adventure.[162]

Fusée-Aublet was replaced as director of the garden by a certain Desportes-Milon.[163] The gardens at Le Réduit continued to be staffed and maintained into the 1770s, but they eventually declined in the face of the ascension of the Pamplemousses garden as the official Jardin du Roi on the island. According to a report of 1776, "This garden, entirely neglected for a long time, now has nothing. Everything there is in the worst possible state."[164]

A third, private garden on Île de France, *Mon Goust*, was the handiwork of a local judge, the well-named Le Juge.[165] Le Juge launched a substantial garden in 1750, and it soon harbored a great variety of plants and trees from India, China, Africa, and America.[166] Le Juge likewise sent sizeable botanical and natural history samples back to France, at various times through the Duc d'Orléans and the abbé de Lacaille.[167] Le Juge was in contact with d'Après de

159. "Extrait d'une Rélation d'un voyage aux Indes orientales Par M. de Maudave," pp. 41-42, BCMNHN, Ms. 1765.

160. See Laissus (1973), p. 36; Le Gouic, p. 114; Allorge, pp. 384-86; and references in Lougnon (1953), pp. 74ff.

161. "Mémoire justicatif" (1763) in BCMNHN, Ms. 452.

162. Spary (2005) details this dispute and the deeper factors at play. On Fusée-Aublet's later career in Guiana, see Touchet, passim and pp. 131-34, 300; Chaïa (1966), and Froidevaux (1897). Joseph Banks purchased Fusée-Aublet's herbarium on the latter's death in 1778. See also Fusée-Aublet ms. "Notes de Botanique en particulier sur l'Ile de France," BCMNHN, Ms. 453, which was apparently intended for presentation to the Paris Academy of Sciences.

163. See Cossigny letter to "Citoyen La Cépède," dated "Paris le 21 Nivôse, an 7," AN-COL F³ 162.

164. "Ce jardin négligé entièremt. depuis longtems n'a plus rien. Tous y est dans le plus mauvois état," ALS, Cossigny to Le Monnier, dated "Au Port Louis Isle de france le 12 février 1776," BCMNHN, Ms. 1995. See also Pierre Poivre's report of his inspection of the Réduit garden, "Conduite Du sieur aublet anciennement Chargé de cette Pharmacie. Mardi 4 aoust 1767," BCMNHN, Ms. 452.

165. On these points, see Malleret, p. 124, and Le Juge material, BCMNHN, Ms. 293.

166. See in particular, "Etat des arbres fruitiers et autres des différentes parties du Monde dans le Jardin, et les habitations du Sr. Le Juge Conseiller à l'Isle de france," BCMNHN, Ms. 293 ; see also references in Lougnon (1953).

167. See undated "Memoire des curiosités naturelles rassemblée Par le Sr. Le Juge pour etre Envoyée En France," BCMNHN, Ms. 293; ms. "Memoire des Curiosités naturelles Envoyées a Monseigneur Le Duc D'Orleans," dated Au Port Louis isle de france le 30 Mars 1749," BCMNHN, Ms. 2619; and PA-PV, vol. 69 (1750), pp. 15 ("Samedi 24ᵉ Janvier 1750"), and 68-90 ("Mercredi 11ᵉ Mars 1750").

Mannevillette, and mention of his natural history researches appeared in the *Histoire* section of the Academy's publications.[168] Le Juge's widow kept up the garden until her death in 1778, at which point it seems to have declined.

Finally, a garden arose at Saint-Denis on nearby Île Bourbon in the Indian Ocean.[169] The Compagnie des Indes established a garden there in 1761, but the site was poor, and a new facility was installed beginning in 1767. With the transfer of power to the crown in 1769, this facility became the Jardin du Roi on Île Bourbon. Other botanical stations and transplant gardens may have existed on the island as well. Poivre transferred exotic products from Île de France to the king's garden on Île Bourbon for acclimatization, propagation, and distribution to locals.[170] The relative weakness of botanical infrastructures on Île Bourbon was made up for in part by the activities of Jean-Baptiste-François de Lanux (sometimes spelled La Nux or Nux, 1702-1772) who was *conseiller* at the Conseil Supérieur on l'Île Bourbon and an official correspondent of the Académie des Sciences attached to Réaumur and then to Jussieu. From the 1730s, Lanux *père* sent botanical shipments and passed on information about apiculture, silkworms, and other natural history matters on Île Bourbon to Réaumur and the Paris Academy.[171]

These gardens and the transformation of the French Indian Ocean possessions into industrial-scale centers of commodity production was impressive enough. The next step was to link them to their sister gardens on the other side of the world.

Colonial Gardens – West

A corresponding set of gardens grew up in the New World.

Quebec and Louisiana. Nothing much of note appears in North America, although a small botanical garden was maintained at the intendance at Quebec

168. See ALS, Le Juge to "Monsieur," "A lisle de france le 26 decembre 1752," BCMNHN, Ms. 293.

169. On the royal garden on Île Bourbon, see Ribes.

170. Laissus (1970), p. 39n. Regarding Joseph Hubert (1747-1825), Laissus writes, "Originaire de l'Ile Bourbon, agronome et botaniste. C'est à lui que Poivre confia, en juillet 1772, le premier giroflier qu'il voulait voir acclimater à l'Ile Bourbon. Hubert le planta dans sa propriété du Bras Mussard, à Saint-Benoît, où il avoit réalisé un véritable jardin d'essai."

171. On Lanux père, see material in PA-DB (Lanux), particularly his twelve letters to Réaumur in 1750s. Officials of the Compagnie des Indes wrote praising Lanux père: "C'est un fort honnette homme, qui la sert depuis longtems et est au fait de cette colonie...," see letter "A M.rs les directeurs généraux de la Compagnie des Indes," dated "20e Xbre 1731," in Lougnon (1934), p. 169. See also letter regarding Lanux, "A Mrs. du Conseil Supérieur de l'Isle de Bourbon. A Paris le 17 9bre 1732," in Lougnon (1933), pp. 99-100. On the Lanux contact with the Academy of Sciences, see PA-PV 69 (1750), p. 368 ("Samedi 8eme Août 1750"); PA-PV 90 (1771), fols. 83v-84 ("Samedi, 13 Avril 1771"), 178 ("Samedy 27. Juillet 1771"), 182 ("Samedy 3 Aout 1771"), and 213v-218v ("Mercredi 4. Septembre 1771"); PA-PV-110, Delanux, 4 Septembre 1771, where the rapporteurs Daubenton and Bailly noted Lanux was "Invité à continuer son travail."

in Canada through the 1750s. The royal physicians there, Sarrazin and then Gaultier, worked in this garden. Sarrazin was charged to send back useful plants from the Quebec garden to the Jardin du Roi in Paris.[172] Governor de La Galissonnière and Duhamel du Monceau especially encouraged Gaultier in his botanical researches at the Quebec garden.[173] The administrators maintained another small garden and menagerie at the general hospital at Quebec.[174]

[Jeton: Colonies françaises de l'Amérique, 1751]

From the mid-1720s in Louisiana there was a small formal garden associated with the hospital in New Orleans, whence officials shipped plants to the acclimatization gardens in the Atlantic port cities of France and to the Jardin du Roi.[175] A private surgeon in the colony at the time, Bernard Alexandre Vielle (known as Alexandre), was elected a correspondent of Dortous de Mairan by the Académie des Sciences in 1722, and Alexandre seems to have been the active intermediary through the late 1730s between the garden in Louisiana and the Paris institution.[176] Various references indicate the continuation of this garden under the aegis of the chief medical officer in the colony through the 1750s and until the French bequeathed Louisiana to the Spanish

172. AN-COL B^{27} (1705), fol. 108v (15 July 1705) and B^{29} (1729), fol. 28v (29 June 1707), the latter regarding a royal order concerning the transport of plants from Canada; Rousseau (1957) and (1969), and above, Part II B. Banks, p. 159, notes shipments from France to Canada for the purposes of experimental botany. Mathieu, pp. 195-96, 201, 206, points to earlier, private gardens in New France.

173. Allard, pp. 77-78; Taillemite (2008), p. 121; ms. by B. Boivin, "Notice biographique sur J. F. Gaultier 1708-1756), Naturaliste-médecin du Roi au Canada," p. 27 in PA-DB (Gaultier).

174. AN-COL B^{23} (1703) fol. 240v (June 20, 1703).

175. Langlois (2003), pp. 344-47; AN-COL C^{13} passim; and inventory by Menier et al. Le Page des Pratz, vol. 2, chapts. 1-13, provides a notable botanical and natural history survey of Louisiana at this time.

176. See PA-DB (Alexandre); see also Huard et al., p. 110n; Kernéis et al., p. 65; Langlois (2003), pp. 345-46.

after the Seven Years War.[177] In 1738 the new médecin du roi, Jean Prat, started a second botanical garden and nursery in New Orleans, the government providing the land and two slaves.[178] From 1735 through 1746, with the backing of Count Maurepas and the local administrators, Prat faithfully corresponded with Bernard de Jussieu and acted as a botanical agent for the Jardin du Roi, regularly sending cases of botanical and natural history materials to Paris.[179] Botanical samples continued to be received at the Jardin du Roi from Louisiana through the 1780s, indicating the possibility that these gardens continued to function after the colony's transfer to Spain.[180]

Guiana's Gardens. The French invested more significantly in South America.[181] From its beginnings the colony in Cayenne was oriented toward commodity production, with cacao and coffee among its early productions and with sugar, indigo, and annatto (*rocou*) not far behind.[182] Introduced from Surinam in 1716, coffee culture in Cayenne antedated production in the Caribbean.[183] Early on, Cayenne was thought of as an ideal locale for spice production, and already in 1694 colonists were ordered to grow vanilla.[184]

One Pierre Barrère (1690-1755) served as royal botanist (médecin-botaniste du roi) in Cayenne in South America from 1722 to 1725, an early date.[185] In 1720 the minister solicited a talented botanist to collect and describe medicinal plants in French Guiana, and Antoine de Jussieu recommended Barrère, a former student. Barrère departed with instructions from the Academy of Sci-

177. See ministerial correspondence, AN-COL C^{13A} 8 (1724), fol. 136 (12 October 1724); C^{13A} 22 (1737), fol. 155 (20 May 1737); C^{13A} 23 (1738), fol. 119 (15 March 1738); C^{13A} 27 (1742), fol. 95 (24 March 1742); C^{13A} 36 (1752), fol. 133 (28 September 1752); C^{13A} 38 (1754), fol. 61 (20 June 1754); Langlois (2003), p. 350.

178. Lamontagne (1963), p. 138.

179. Langlois (2003), pp. 347-52; Lamontagne (1963) reprints selected letters from Prat to Jussieu written from New Orleans. Samples from the Jardin du Roi may also have been sent to Louisiana; see Prat correspondence just mentioned and letter to Bernard de Jussieu of 29 June 1740 transcribed in Lamontagne (1962b), pp. 222-24.

180. Langlois (2003), p. 350.

181. On French botany in Guyana, see McClellan and Regourd (2000), pp. 40-44; Regourd (1999), pp. 39-63; Regourd (2000), pp. 469-89; Lescure, p. 59.

182. Ms. Code de Cayenne, vol. 1, p. 189; vol. 2, pp. 47-49, 523, in AN-MAR G 237; ALS La Barre to minister, dated 29 January 1689, AN-COL C^{14} 2, fols. 60-61; Pritchard (2004), pp. 125-27 provides background.

183. Broc (1975), p. 122; Lacroix (1938), vol. 3, p. 29; Froidevaux (1892), p. 223, and administrative correspondence, AN-COL 11, fols. 81-82, indicate that coffee production in Cayenne did not begin until 1719-1720.

184. Ms. Code de Cayenne, vol. 1, pp. 194, 253, in AN-MAR G 237.

185. On Barrère, see Touchet, passim and p. 299; Chaïa (1964), pp 17-26; Froidevaux (1895); Lacroix (1938), vol. 3, pp. 31-34; Broc (1975), pp. 123-24; Lescure, p. 63; Regourd (2000), pp. 317-18; Allorge, pp. 454-55. See also Barrère materials, BCMNHN, Mss. 683-87; PA-DB (Barrère), including Barrère letter, dated "À Rochefort le 21e mars 1722." See also PA-PV-110, Barrère, 14 Juin 1741 ("Rapport sur l'histoire naturelle de la nouvelle France") and "Diverses observations de phisique et d'histoire naturelle. Histoire Naturelle de la France Équinoctiale," HARS 1741, p. 24.

ences, which elected him a correspondent after his return.[186] While in South America he earned a salary of 2000 livres a year. Barrère shipped materials to the Académie, and he went on to publish his natural history of French Guiana, *Essai sur histoire naturelle de la France equinoxiale*, in 1741, followed by his *Nouvelle Relation de la France equinoxiale* in 1743 under the privilege of the Academy. Jacques-François Artur (1708-1779), his ultimate successor, criticized Barrère thus:

> You are much in the right of it, Sir, when you say that Mr. Barrere has rather excited than satisfyed our curiosity. He was very well able to give us something better than what he has publish'd in his Essay, having nothing else to do; and natural history being the only object of his commission, when sent to Cayenna, but 1o he did not reside long Enough in the country. 2o his pension was not Equal to the charges, which more strict Enquirys might have occasioned. 3o he found no assistance in this place, Either on the part of those, who were capable of promoting his researches, or on that of the principal officers, who alone could furnish him with the necessary Accommodations. 4o he was deficien in point of diligence, taking very little trouble, and spending most of his time in the city of Cayenna, in drawing many things that presented themselves or were brought him to his house. 5o he frequently took things upon trust from illiterate and ignorant people, and often depended on reports. And lastly, and this is the greatest defect, he bethought himself too late of writing his memoirs, when out of the country, at a time, when his memory could not furnish him with many particulars, which he had neglected to.[187]

A certain Duhaut replaced Barrère, but the next man of note was Jacques-François Artur who was trained at the Jardin du Roi in Paris and who spent thirty-five years in Cayenne from 1735 to 1770 as médecin du roi, ultimately with a salary of 2400 livres.[188] Artur studied at the Jardin du Roi and was the first of Buffon's correspondents of the Cabinet du Roi. From 1753 he was also a correspondent of the Academy of Sciences, attached first to Réaumur and

186. Barrère was also elected a correspondent of the Bordeaux academy; see previous note and Barrière [sic], pp. 66, 71.

187. "A Letter from Dr. Arthur, a French Physician at Cayenne to Dr. Maty, F.R.S.," RS/L, L&P II, #368, contemporary translation.

188. On Artur, see now the critical apparatus with its ample biographical documentation supplied by Marie Polderman in Artur (2002), pp. 17-101, esp. pp. 28-56; see also Le Seigneur; Touchet, passim and pp. 127-31, 299; also, Chaïa (1963), (1968), (1977); Lacroix (1938), vol. 3, pp. 49-51; Froidevaux (1892), p. 226; see also Artur ms. materials, BnF, n.a.f., 2571-2583, that include his unpublished manuscript opus, *Histoire des colonies françoises de la Guianne*, published by Polderman in 2002.

then to Antoine de Jussieu. He regularly sent materials to Réaumur, to Dufay at the Jardin du Roi, and to Minister Maurepas, whom he asked to instruct ships' captains to exert particular care in handling shipments from Artur.[189] As médecin du roi, Artur complained that his responsibilities as a physician and a member of the local Conseil had limited his ability to do science:

> I have therefor contented myself hitherto with answering those friends, who do me the honour to correspond with me, and satisfying in the best manner I can to their querys; to Send to the King's Garden, and to my friends those things relating to natural history which I think may be agreable to them, with what observations they desire, and I am able to give.[190]

J.-B.-C. Fusée-Aublet was first posted to Île de France in 1753, as we saw earlier in this Part, and he later transferred to the Americas to become royal botanist in Cayenne from 1762 to 1764, part of the effort to colonize Kourou.[191] In 1775 he published his great *Histoire des plantes de la Guyane française* in four volumes, with 400 plates describing 800 new plants. Jean-François Patris served as médecin-botaniste in Guiana from 1764 to 1786.[192] We have similarly mentioned Bertrand Bajon who was a royal surgeon in Cayenne from 1764 to 1776 and correspondent of the Parisian academies of science and of surgery. As part of Bajon's duties he was charged with naturalizing plants from China and India and with sending plants back to Europe from the South American colony.[193] Bajon was in contact with Adanson, Daubenton, and Buffon of the science academy over matters botanical and natural historical. In 1775 the Academy of Sciences specifically charged Bajon to be its field agent for research on rubber, a request formally backed up by the ministry and local administrators.[194] On his return to France Bajon published his *Mémoires pour servir à l'histoire de Cayenne et de la Guiane française* (2 vols., Paris, 1777-78) with the privilege of the Academy of Sciences.[195] Bajon was a very

189. See Artur letter, dated "A Cayenne, le 2 8bre 1736," in PA-DB (Artur); see also Artur's "Description du cacaoyer" (1737), BCMNHN, Ms. 1865.

190. "A Letter from Dr. Arthur, a French Physician at Cayenne to Dr. Maty, F.R.S.," RS/L, L&P II, #368, contemporary translation.

191. Re Fusée-Aublet, see above at note 160 and Lescure, p. 64; ms. Code de Cayenne, vol. 4, pp. 31, 201, in AN-MAR G237.

192. On Patris, see Chaïa (1970); Touchet, passim and pp. 97-99, 301; Ms. Code de Cayenne, vol. 4, p. 609, and vol. 7, pp. 215, 551, 777, in AR-MAR G 237.

193. On Bajon, see above Part I (Académie de Chirurgie); Dordain, pp. 117-18; Touchet, passim and p. 300; Chaïa (1958); see also ms. sources, AN-COL E 15 (Bajon); PA-DB (Bajon); and Adanson's "Catalogue des Plantes demandée le 19 9bre, à M. à Caienne, par les moien de Mr. Bajon, chir. Correspdt l'acad," HI, AD 131.

194. ALS from de Montigny and Daubenton to the minister of the navy, dated "Paris 22 may 1775," and ALS reply "Versailles, 27 juin 1775," PA-DB (Bajon).

195. Bajon (1777-1778); PA-PV 97 (1778), fols. 59-62 ("Mercredi, 25 Février 1778").

active, if unofficial, correspondent of the Société Royale de Médecine, reporting on cures for snakebite, on the electric eel, and on elephantiasis.[196] In addition, he published several of these memoirs in the *Journal de Physique* and the *Journal de Médecine*. Beyond that, he reported to the navy ministry about conditions in the military hospital in Cayenne, and he advised the minister on the further colonization of South America.

Following Bajon, another correspondent of the Academy of Sciences, Jean-Baptiste Leblond traveled as "physician-naturalist" in Cayenne in the 1780s, and his reports (on indigo manufacture, quinine, the mineralogy of Guiana, and the biogeography of Peru) duly filtered back to the science academy in Paris and ultimately earned him a pension of 3000 livres.[197] Leblond's nephew, Hypolite Nectoux, took the post of royal botanist in 1787, first in Cayenne and then moving on to Saint Domingue.[198] Other royal botanists and physicians served in the equatorial jungles of French Guiana and Cayenne in the eighteenth century, one of their charges being to seek out the cinchona plant.[199] By 1773 at least three gardens existed in Guiana: at the hospital, at the residence of the governor, and at the intendance.[200] In this context the royal horticultural station on the La Gabrielle plantation needs to be highlighted. This was Rambouillet in Guiana. The facility was established in 1778 specifically as a nursery to receive and multiply spice plants from the Indian Ocean. It became a key station in the intercontinental exchanges that developed in the 1780s.[201]

196. See SRdM, Ms. 7, p. 29 ("Du mardy 29 Xbre 1778...M Mauduyt nous a dit qu'il avoit appris de M. Bajon qu'un medecin nommé de la Sorte [Porte] partant pour Cayenne se prétendait envoyé par la Société; M Poisonnier nous a assuré que c'étoit lui qui a donné a M. De la Sorte [Porte] une place de Med en service pour cette Isle et qu'il lui avoit mandé de rendre a Bordeaux pour s'y embarquer"), Ms. 7, p. 300 ("Séance du vendredi 17 7bre 1779"); Ms. 9, p. 114 ("Séance du Mardi 5 Mars 1782"); Ms. 10, p. 264 ("Séance du Vendredi 24 Septembre 1784"); Ms. 11, fol. 45v ("Séance du Mardi 5 Septembre 1786"), fol. 61v ("Séance du Vendredi 1 Décembre 1786"), fol. 65v ("Séance du Mardi 19 décembre 1786"), fol. 280r ("Séance du Mardi 19 Mai 1789").

197. On Leblond, see Touchet, pp. 155-57, 302; Lacroix (1938), vol. 3, pp. 75-82; Allorge, pp. 455-59; Smeaton, p. 113; see also PA-PV 107 (1788), fols. 264v, 265v ("Samedy 15. Novembre 1788"), 268v ("Mercredy 26. Novembre. 1788."); PA-PV-110, Le Blond, 7 April 1786, 23 August 1786, for Leblond contact with the Academy; also Leblond (1813).

198. On Nectoux, see Bret (1995), (1999a), (2000); Touchet, passim and pp. 155-57; McClellan (1992), pp. 157-62, Nectoux later accompanied Napolcon to Egypt.

199. Raymond Laborde, who feuded violently with Bajon, was premier médecin du roi from 1770 through the late 1780s. In 1745, a certain doctor Gaultier was commissioned for natural history research in the colony; see ms. Code de Cayenne, vol. 3, p. 367, in AN-MAR G 237.

200. Touchet, p. 180.

201. On La Gabrielle, see Touchet, pp. 151-58ff; Chaïa (1979b), pp. 52-53; Lescure, p. 65; Froidevaux (1892), p. 224; ms. Code de Cayenne, vol 7, p. 57, in AN-MAR G 237; see also ms., "Mémoire présenté à l'institut national, par le C.en Joseph Martin, botaniste, chargé de la direction des jardins et pépinières coloniales, dans la guyane française," BCMNHN, Ms. 48; and inspection report by one Provost: ALS, Provost to "Monseigneur," dated "Cayenne le 5 aout 1781," and "Compte des travaux et autres operations relative à la partie agraire et hidraulique pour Monseigneur de Sartine Ministre de la Marine," both BCMNHN, Ms. 357.

Gardens in Martinique and Guadeloupe. Even earlier than in South America or the Indian Ocean, at the turn of the eighteenth century the effort began to establish botanical gardens in the Caribbean on the islands of Guadeloupe and Martinique. We have already encountered many of these details. On Guadeloupe, the Lignon brothers served as royal botanists sometime prior to 1707, making theirs the first instance of French government botanists per se *stationed* in the colonies. The younger and more active sibling, Jean-Baptiste Lignon (1667-1729), was a correspondent of Tournefort already in 1699. He sent Caribbean specimens back to the Jardin du Roi, but his primary role seems to have been to naturalize French honeybees, and mulberry and olive trees, for which, as mentioned, he was granted 1000 livres annually in addition to receiving a medal and a gold chain from the regent.[202] In 1729 another correspondent of the Academy, Jean-André Peyssonnel, was appointed royal physician and botanist in Guadeloupe.[203] Peyssonnel had already compiled an extensive record as royal botanist prior to his posting to the Caribbean, having previously traveled to Louisiana and the Antilles and then to Egypt and the Barbary Coast. He was charged to take over the Guadeloupe garden after Lignon's death, but the facility seems to have fallen into mortal decline by the end of the 1720s. Nevertheless, Peyssonnel continued to send botanical and natural historical samples back to France during his long tenure as a government botanist on Guadeloupe. Officials upgraded the royal gardens in Guadeloupe only in 1775, with Jacques-Alexandre Barbotteau becoming superintending botanist for the whole of the Windward Islands. In 1776 Barbotteau was elected correspondent of Duhamel du Monceau at the Academy of Sciences, and the former transmitted botanical, natural historical, and mineralogical specimens and meteorological reports to the Academy of Sciences and to A.-L. de Jussieu in Paris.[204]

Similar efforts on nearby Martinique came to less. In Part I we saw how the Paris Academy of Sciences became involved in creating a botanical establishment there, sending a specially-named correspondent, Michel Isambert, to Martinique in 1716 precisely for this purpose. Isambert brought with him bees, silkworms, and coffee trees. Unfortunately, he and his coffee plants died

202. See above Part I, p. 83 at n. 149 and n. 150. On the Lignon brothers, see materials PA-DB (Lignon) and Lignon letter dated 27 April 1717 reproduced in Lacroix (1938), vol. 3, plate IV, p. 25. See also PA, Archives, DG 31 ("Histoire de l'Académie") and letters from the Duc d'Orléans to "M. du Lignon à la Guadeloupe, 22 décembre 1716," which confirms the salary, and from Abbé Bignon of 11 January 1717 in same folder. Bignon estimated the value of the medal and chain at 860 livres.

203. On Peyssonnel, see manuscript letters and other materials in PA-DB (Peyssonnel); Lacroix (1938), vol. 3, pp. 23-24, 39-47; Gasc and Laissus, p. 26. We take up Peyssonnel again in Part III, esp. pp. 439-42.

204. See materials relating to Barbotteau AN-COL E 17 (Barbotteau); PA-DB (Barbotteau); also, APS-MS, B/D87, Group 14 (ms., "Mouche masonne de la Guadeloupe"), PA-PS (9 February 1774), (22 March 1775), (15 November 1777); HARS (1776/1779), pp. 19-20.

shortly after his arrival, and little resulted from this initial effort.[205] The state prosecutor before the Conseil Souverain of Martinique, Bernard-Laurent d'Hauterive, did much to encourage cinnamon production on the island and, as we saw, he facilitated botanical shipments from Martinique back to France. But his were purely personal efforts. Much later in the century the Intendant, first on Guadeloupe and then on Martinique, François-Joseph Foulquier de Bastide (1744-1789), was instrumental in reviving botanical and scientific activity on those islands. Before leaving for the islands, Foulquier proposed himself to the Academy of Sciences as its agent, a proposal gratefully accepted, and he was elected a correspondent of the Paris Academy of Sciences in 1781.[206] He annually sent cases of botanical specimens to Thouin at the Jardin du Roi and to the garden at Toulouse where he had family connections.[207] In his "Observations on Growing Manioc and Banana Trees" of 1787, he noted that "we should introduce into Martinique breadfruit and mangosteen from Île de France."[208] Although there was no official botanical garden on Martinique, Foulquier saw to it that the royal botanist, who had previously served in Cayenne, Louis-Claude Richard, was posted there; and Richard established his operations on a private plantation and was in independent contact with the Académie des Sciences.[209] Richard finally returned to France in 1789, bringing back a rich scientific trove of nearly 2500 plants (about half from Guiana and half from the Antilles), 1500 mineral samples, not to mention birds, quadrupeds, and insects. The manuscript catalog for this collection is 487 pages long.[210] Minister Malesherbes, as we saw, had grave misgivings about the value of Richard's botanical harvest, but Thouin praised it highly.

205. Letter from a member of the Conseil Supérieur de la Martinique, [Charles] Mesnier, dated 14 July 1716, AN-COL, C^{8A} 21, fol. 156: "Isambert, envoyé en mission par l'Académie des Sciences, est mort le 11."

206. According to one set of minutes, PA-PS (5 August 1780), Foulquier "envoyé par le roi à St. Domingue a proposé à l'Académie faire à St. Domingue les différentes observations dont l'Académie voudroit le charger. L'Académie l'a remercié de son attention." PA-PV 99 (1780), fol. 199 ("Du Samedi 5. Août 1780") for the same meeting reads: "M. de foulquier membre du Parlement de Toulouse qui doit incéssamment partir pour St. Domingue est entré et a offert à l'Académie de se charger des commissions qu'elle voudroit lui donner; l'Académie a accepté son offre et l'en a sur le champ remercié." On Foulquier, see further PA-DB (Foulquier), Lacroix (1938), vol. 3, pp. 83-90.

207. On Foulquier shipments to the Jardin du Roi, see BCMNHN, Ms. 1327.

208. Foulquier letter to administrators of 21 July 1787, AN-COL C^{8A} 87, fol. 169.

209. PA-PV-110, Richard, 1 February 1783 ("Description de quatre arbres ou arbrisseaux nouveaux").

210. On these materials, see "Catalogus plantarum in Guyana et Antillis collectarum a Ludovico-Claudio Richard, a fine anni 1781 ad medium annum 1789, Parisiis, ineute anno 1790," BCMNHN, Ms. 1320 and BCMNHN, Ms. 1608 ("Carnets de notes de voyage aux Antilles, par L.-Cl. Richard"). See also Jandin, pp. 32-33; Lamy, p. 71, puts the number plants brought back by Richard at 3000. For Richard's contact with the Paris Academy at this time, see "Rapport sur les travaux de M. Richard Botaniste du Roy," PA-PS ("Samedi 15 Mai 1790"); also, PA-PV 108 (1789), fol. 162 ("Du Samedi 13 Juin 1789"); PA-PV-110, Richard, 1 May 1790, 12 May 1790, 15 May 1790.

Saint Domingue's Botanical Outposts. With the appearance of royal gardens in Saint Domingue in the later 1770s, the infrastructure of French applied botany overseas saw its completion. By the end of the colonial period, a number of formal botanical gardens could be found in Saint Domingue. Louis XVI lent his name to the Jardin Royal, established in Port-au-Prince in 1777. In the later 1780s this garden gave way to a better situated Jardin du Roi in Port-au-Prince, and a third garden existed there at the headquarters of the colonial administrators. In the north, on the outskirts of Cap François the Brothers of Charity maintained a substantial botanical garden on the grounds of the Hôpital de la Charité. Likewise in the Cap François area, the Cercle des Philadelphes established the first of its two small botanical gardens with government support in 1785. A.-J. Brulley, a colonist, maintained a sizable private garden and nursery on his property in the Marmelade district in the northern mountains, as did another colonist, Paul Belin de Villeneuve in Limbé on the north coast.

In large measure, the botanical gardens in Saint Domingue grew out of efforts to establish cochineal dye production in the colony. The cochineal insect (*Dactylopius coccus*) is a parasite of cactus plants, and its dried residue makes an intense red dye. Through the 1770s, the Spanish guarded their monopoly on the production of cochineal in Mexico, for which the French paid dearly to supply the Gobelins dye works in Paris. One J.-N. Thiery de Menonville had studied with Bernard de Jussieu, and in early 1777, with permission from the highest levels of government and with a substantial government purse, he traveled from Saint Domingue to Mexico to steal samples of cochineal for husbandry in Saint Domingue.[211] In an adventure fraught with dangers, Thiery de Menonville succeeded in his mission, returning to Saint Domingue with his precious cargo of so-called *fina* cochineal in September of 1777. The Jardin du Roi in Port-au-Prince was established especially to receive Thiery de Menonville's treasure, and experimental cochineal husbandry proceeded satisfactorily under his guidance as a very well-paid botaniste du roi.[212]

The colonist Brulley seeded his own nopalry (cactus plantation) with 4000 cacti and succeeded with several harvests. In 1787, he sent samples of his cochineal product to France, where the Paris Academy of Sciences conducted tests that showed the samples to be very nearly equal to that of Mexican *fina* cochineal.[213] Brulley also supplied the starter stock for the garden of the Cercle des Philadelphes in Cap François, and by November of 1785 the Cercle had produced three cochineal harvests of its own. The Cercle sent samples of the

211. On Thiery de Menonville and this episode, see McClellan (1992), pp. 154-55, 220-21; materials in BCMNHN, Ms. 40 ("Voyage économique à Guaxa... par Nicolas-Joseph Thiéry de Menonville") and BCMNHN, Ms. 1081-1082. Buti sets the story in a larger frame.

212. Thiery de Menonville received 2-½ times the salary of the chief government doctor in the colony; see McClellan (1992), pp. 152-156; DPF, p. 1020.

213. PA-PV-110, Brulley, 3 February 1790; PS (1787), 'plumitifs' for 13 June 1787, 4 August 1787.

dyestuff back to Paris for testing. Swatches from these tests survive, still brilliantly scarlet after more than 200 years. As part of this effort, the Cercle des Philadelphes also undertook publication of Thiery de Menonville's opus, *Traité de la Culture du Nopal et ... de la Cochenille*, which appeared in 1787 in a handsome two-volume octavo set of 436 pages with four colored plates. The second volume, especially, was a practical manual of cochineal husbandry destined to instruct potential growers. In his review of this book, Charles Mozard, the editor of the colonial newspaper, the *Affiches Américaines*, suggested that cochineal production might someday rival that of coffee in economic importance to Saint Domingue. While far from challenging coffee, the nascent cochineal industry followed the established model of applied botany provided by coffee cultivation, and it held great promise for the long-term economic development of the colony. Colonial authorities had taken the necessary first steps, and in the process the institutions and infrastructures of applied botany arrived in Saint Domingue.

Botanical activity continued to mount in Saint Domingue as the 1780s progressed. In 1785 the Cercle des Philadelphes gave a public course on botany in its garden on Tuesdays and Saturdays from 2:00 to 4:00 p.m. One could find mango, the Senegal date, Chinese mulberry, the Cape of Good Hope palm, and "the precious breadfruit," all in the gardens of the Charity Hospital in Cap François.[214] A 1788 inventory of the Jardin du Roi in Port-au-Prince depicts an even more vibrant establishment. Its nopalry numbered 400 to 500 cacti covered with cochineal provided by Brulley. Seventy-two different types of exotic trees and plants graced the botanical garden, including clove, cinnamon, ipecac, aloe, fig trees from various parts of the world, various types of palm tree, jasmine from the Cape of Good Hope, Madagascar indigo, Surinam mustard, green tea plants, litchi trees, Chinese rose bushes, chestnut trees from Virginia, wax bushes from Louisiana, Cuban cedar trees, genuine sago, and other trees and plants from Egypt and India. Bamboo grew well in the nearby government garden, and it probably grew in the Jardin du Roi as well. Pecan trees from Mississippi were likely to be found there, too, as the government had distributed seedlings for propagation trials in 1787.[215]

In the 1780s, then, with substantial botanical gardens in Saint Domingue, in Cayenne, and on the Indian Ocean islands, not to mention lesser outposts in the colonies and the numerous royal gardens and nurseries dotting the motherland and its ports, the French had established an unprecedented set of botanical infrastructures in support of economic botany and colonial development. The stage was set to mobilize these gardens to function as a collective whole.

214. DPF, p. 575, where Moreau de Saint-Méry speaks of "le précieux arbre à pain." The mulberry tree was brought by a colonist who got it from Buffon at the Jardin du Roi in Paris.

215. DPF, pp. 1019-22; McClellan (1992) p. 162, citing an inventory published in the colonial newspaper, the *Affiches Américaines*, March 8, 1788.

The Interhemispheric System at Work

By the 1780s, then, a *system* of state botanical gardens in France, in the Indian Ocean, and in the Americas had fallen into place.[216] As is already apparent, many exchanges took place between and among these and other outposts. But more than that, the creation of this system culminated in a formal, state-sponsored program of transshipping economically valuable spices and other plants from the Indian Ocean to South America and the Caribbean. This new, intercontinental system operated to secure, nurture, and bring valuable plant commodities into large-scale, commercial production. Here again, coffee provided the model and the central government the stimulus. At least by 1698, officials broached the possibility of transferring spice plants to the Americas for cultivation.[217] The early transfers to Martinique and Guadeloupe from France in the 1710s are part of this story, as are the Academy's sustained connections to correspondents there and the efforts to send plants from Martinique to the royal garden at Nantes through the colonist Hauterive. In the 1770s, readers will recall, a plan was in place and operating to transship plants from Asia to America through the port and royal garden at Lorient.[218] As the infrastructures developed, the idea of shipping plants directly from the Indian Ocean to the Americas became more and more a practical and desirable possibility. Already by the mid-1770s French successes in transplanting spices and inaugurating commercial spice production in the Indian Ocean and in the Americas provided an example to the British for emulation and imitation.[219]

The Beginnings of Formal Plant Transfers. Cayenne was the intended center for development in the Americas, and the initial functioning of the intercontinental system of botanical gardens may be said to have begun in 1763 with pepper plants being transferred from the Indian Ocean to Cayenne.[220] In 1770 Maillart Dumesle, the naval commissioner on Île de France and himself

216. In a slightly different context Ly-Tio-Fane speaks of the "system" of French colonial botany; see her (1976), p. 5. Ly-Tio-Fane (1996) sketches the larger background to the story presented here. These points are also taken up in McClellan (1992), chapt. 9; McClellan and Regourd (2000), pp. 40-44; Regourd (2000), pp. 469-89; Regourd (1998), pp. 162-69; and Regourd (1999). The article by Bourguet and Bonneuil, "De l'inventaire du monde à la mise en valeur du globe. Botanique et colonisation (fin XVIIe siècle-début XXe siècle)," is a larger and landmark presentation of botany and colonialism that needs to be consulted in this connection; Bleichmar provides a useful contrast with related endeavors by the Spanish at the end of the eighteenth century.

217. See ministerial directive of 22 January 1698 to the ship's captain Desaugiers, AN-COL B^{21} (1698), fol. 18v.

218. See above, Part I.

219. Ellis (1777), pp. 35-36.

220. Touchet, Part 3, provides the most detailed account of plant transfers to Guiana; see also "Lettre du Ministre à M. de Chanvalon en lui faisant passer de la graine de poivre des indes. Instructions sur cet objet," ms. Code de Cayenne, vol. 4, pp. 179-180 (17 December 1763), in AN-MAR G 237. The subsequent importation in 1767 of "buffles" from Italy into Cayenne may be seen in this light; see idem., p. 723. Le Gouic, pp. 119-122, also documents the series of plant exchanges discussed here.

a former administrator in Cayenne, sent clove and nutmeg plants to Cayenne from Île de France.[221] In 1771 the Minister of the Navy, de Boynes, ordered further shipments from the Indian Ocean to Cayenne. A few clove, cinnamon, and nutmeg plants made it from one hemisphere to another.[222] Cossigny then sent mangoes to Martinique via France in 1772.[223]

A consignment from the Indian Ocean arrived in Saint Domingue in 1773 with plants brought aboard the *Artibonite*. Most died, but some apparently survived to be distributed to colonists, doubtless producing some local impact. In 1774, once again Maillart Dumesle on Île de France, this time as director of the Pamplemousses gardens, saw to another modest shipment to South America.[224] In 1776 the minister allocated funds for the creation of special propagation nurseries in Cayenne, and the model established itself for distributing plants from royal nurseries for local production.[225] The burgeoning production of royal and private gardens on Île de France under Céré and Cossigny contributed to these developments, with the first harvest of nutmeg there dating from late 1778.[226] Other shipments from the Indian Ocean may have followed, given reports of 1780 and 1781 relating the growing success of clove and cinnamon production in Cayenne.[227] Cossigny's shipment of gum arabic seeds as well as other seeds and plants directly to Governor-General Bellecombe in Saint Domingue in 1782 was also part of the program.[228] A heightened sense

221. Tessier (1779), p. 50; see also SRdM-H&M, vol. 2 (1778), Histoire, pp. 76-77.

222. Jandin, p. 16; ms. Code de Cayenne, vol. 5, pp. 553, 631, in AN-MAR G 237; this may or may not have been the same shipment that arrived in Guyana in 1773 aboard the *Prince de Condé*, per Touchet, p. 151.

223. ALS, Cossigny to Le Monnier, dated "A Paris le 30 Juillet 1772," BCMNHN, Ms. 1995.

224. Jandin, p. 16; ms. Code de Cayenne, vol. 6, pp. 658, 666; vol. 7, p. 147, in AN-MAR G 237.

225. Ms. Code de Cayenne, vol. 6, p. 47 (21 August 1776: "Lettre du Ministre aux administrateurs portant avis d'un envoi de fonds pour être employé en semis et pepinières") in AN-MAR G 237; "Compte des travaux et autres operations relative à la partie agraire et hidraulique pour Monseigneur de Sartine Ministre de la Marine" (1780), BCMNHN, Ms. 357.

226. See reports by Cossigny to Le Monnier, ALS, Cossigny to Le Monnier, dated "Isle de france le 6 janvier 1779," and ALS, Cossigny to Le Monnier, dated "Palma le 16 Aoust 1777," BCMNHN, Ms. 1995, the latter reading: "...procurer des noix de Ravinesara de Madagascar; Mais j'en ai un arbre à Palma, jeune encore. Il y en a un au jardin du Roi qui rapporte fruit cette année. J'espere que M. Céré m'en donnera des noix pour vous et pour moi: il est porté de la meilleur volonté pour nous deux. Il a une 30.ne de jeunes créoles, provenant des graines de Girofliers de l'année d.re. Ces arbres sont maintenant en fleurs. Si nous n'avons pas de coup de vent de bonne heure, nous aurons de quoi les multiplier. Les Muscadiers sont aussi en fleur...."

227. See Tessier (1779); PA-PV-110, Tessier, 8 May 1779 ("M.re sur le Géroflier, dont la culture est établie à Cayenne"); previously cited ms. of 1780, "Compte des travaux et autres operations relative à la partie agraire et hidraulique pour Monseigneur de Sartine Ministre de la Marine," and similar document of 1781, "Compte des travaux et autres opération relatives a la Partie Agraire et hidraulique pendant les mois d'Avril, May et Juin [1781] Pour Monseigneur le Marquis de Castries," the latter two BCMNHN, Ms. 357.

228. ALS, Cossigny to Le Monnier, dated "Palma le 19 Avril 1782," BCMNHN, Ms. 1995, and as reported in "Recensement de tout ce que renferme Le Jardin du roi, le Montplaisir; remit par Mr. De Ceré," BCMNHN, Ms. 303.

of activity is likewise felt in a series of regulations for Cayenne from 1779 to 1782 mandating the declaration and deposit of all spices at the "dépôt des épiceries," the guarding of spice nurseries, the prohibition of private export of spices, the setting aside of land by the royal engineer for cultivating imported spice plants and trees, and fines for illegal harvesting of spices.[229]

The American War interrupted further efforts until 1783-84 when the *Aimable Indienne* brought pepper plants expedited by Céré from Île de France in the Indian Ocean to Cayenne on the northeast coast of South America.[230] All but two of the nutmeg plants died, and the staff on Île de France was encouraged to improve propagation. A new series of encouragements and regulations followed in Cayenne in the mid-1780s. The Minister explicitly encouraged cinnamon production; the government distributed spice plants to colonists for propagation and offered special encouragements for growing nutmeg, pepper, and tea. The regulations repeated the prohibition against private export of spices, and the local "ingénieur agraire" produced an instructional pamphlet on growing spices.[231] Rocou production, encouraged from the mid-1750s seems to have flourished.[232] The first shipments of clove from Cayenne made it back to France where Lavoisier judged them as good as ones from the Moluccas.[233]

Later in 1787 and again in 1788, on orders from La Luzerne in Paris, the peripatetic Louis-Claude Richard made trips from Cayenne to Saint Domingue, Martinique and Guadeloupe, transferring clove, cinnamon, nutmeg, and other spice plants that had been sent from Île de France to South America for further distribution within the Caribbean.[234] (Only 15 of 400 plants survived the arduous transatlantic journey.[235]) In Saint Domingue, plants were

229. "Compte des travaux et autres operations relatives à la partie agraire et hidraulique pour Monseigneur de Sartine Ministre de la Marine," and letter from Provost to "Monseigneur," dated "Cayenne le 5 aout 1781," BCMNHN, Ms. 357; see also legislation, ms. Code de Cayenne, vol. 6, pp. 435, 517, 658, 659, 666, 725 in AN-MAR G 237.

230. See reference (p. 15) in the ms. by Céré, "Instructions Pour le Sr. Martin Eleve Jardinier," BCMNHN, Ms. 47; Jandin, passim and ms. sources listed pp. 69-70, reporting that L.-C. Richard made an inventory of the shipment and planted the surviving plants; and "Recensement de tout ce que renferme Le Jardin du roi, le Montplaisir; remit par Mr. De Ceré," BCMNHN, Ms. 303, where Céré speaks of shipments to Cayenne and Guiana "de Girofliers; de Muscadiers mâles et de Muscadiers femelles; de caneliers, Espece de Ceylan; de 24 Plants de Véritable Poivriers Blanc de Mahé; de The vert; Cardamorne, Cardamum minus; de Roting; de Sayou, Voakoat; Litchis; Rotissiat, &c." On this shipment, see also Desrivierre-Gers ms. "Observations sur les causes qui ont dû contribuer aux maladies et à la mort de différentes plantes envoyées au commencement de cette année de l'isle de France à Cayenne par ordre du Roy. 1784," BM/Bordeaux, Ms. 828, CII, doc. 4.

231. On these points and for other ministerial encouragements of spice production in Cayenne, see ms. Code de Cayenne, vol. 7, pp. 59, 131, 241, 483, 613, 621, in AN-MAR G 237.

232. On rocou, see Froidevaux (1892), p. 222; ms. Code de Cayenne, vol. 7, pp. 59, 305-307, 347-49, in AN-MAR G 237; Cayenne, Annales du Conseil 1713-1780, p. 78, in AN-MAR G 237.

233. Ms. Code de Cayenne, vol. 7, pp. 571, 655, in AN-MAR G 237.

234. At least two trips from Cayenne to the Antilles and possibly four are involved here; for details, see Jandin, pp. 20-21; Bret (1999a), pp. 81-82.

235. See also reference of the government botanist in Cayenne, Leblond, who indicates that a new variety of indigo from India arrived from Île Bourbon at this time; Leblond ms. "Essais sur L'Art de Lindigotier," AN-MAR G 101, dossier 4, fols. 183-213, here fol. 185.

distributed to the Cercle des Philadelphes and to private individuals for propagation. In July of 1788, another botanical cargo, this one directly from the Indian Ocean, arrived in Saint Domingue aboard the merchant ship *Alexandre*. Sent by Céré from the Île de France gardens and accompanied by another government botanist, Darras, the lot included pepper plants, cinnamon trees, mango trees, mangosteen fruit, and a few breadfruit trees from Tahiti.[236] Authorities allocated 52,000 thousand francs/livres for the Darras/*Alexandre* shipment.[237] This was an enormous sum, the equivalent of the salary for a top military officer or government minister. (The navy's inspector-general, for example, received only 12,000 livres a year.) Such an expense powerfully underscores the importance government authorities placed on establishing spice cultivations in the New World. Unfortunately, kept in the ship's hold, nearly all the plants aboard the *Alexandre* died before arriving in Saint Domingue. But seventeen different types of plants and sixteen different types of seed did survive, including the "precious" breadfruit. The scientific establishment back in France was kept informed of these developments.[238]

Curiously, at the same time that Darras and the *Alexandre* wended their ways to Saint Domingue in 1787 and 1788, Captain William Bligh was transporting the breadfruit tree to the Caribbean for the English aboard HMS *Bounty*.[239] The mutiny aboard the *Bounty* incidentally secured a victory for the French in the race to be the first to get breadfruit to their respective West Indian colonies. Colonial authorities recognized the value of the breadfruit as especially suitable for the climate of their Caribbean possessions and for the hundreds of thousands of captive mouths that had to be fed there.[240] The royal botanist in Saint Domingue, Hipolyte Nectoux, distributed breadfruit to twelve private and public stations around Saint Domingue, including the garden in the government compound, the second garden of the Cercle des Philadelphes, and the gardens of the Charity Hospital in Cap François. The full effect of the breadfruit project remains unknown, but the instance illustrates again the mounting official efforts to promote economic botany in the French imperium in the 1780s.

236. McClellan (1992), pp. 158-59; this may or may not have been the same 1788 shipment reported in ms. Code de Cayenne, vol. 7, pp. 9, 555-57 in AN-MAR G 237. Apparently animals ("bestiaux") were part of this shipment, also; Céré kept the Paris Academy informed of this shipment; PA-PV 107 (1788), fol. 236 ("Mercredy 27. Aout 1788").

237. See figure reported in ALS Martin to Thouin, dated "De false baye de 2 Juillet 1788," BCMNHN, Ms. 47. Before the franc became the official currency of France in 1795, the term, franc, was used colloquially for the *livre tournois*.

238. See report before the Société Royale d'Agriculture, APS-MS B/D37, group 35 ("Séance de la Société d'agriculture, du jeudi 11 decembre [1788]") ; also, Touchet, pp. 188-89.

239. See Sylvie Lacroix; surprisingly, her account only concerns the British experience; Allain, chapt. 5, likewise documents the story of Bligh and the breadfruit tree, but he overlooks the success of the French to import the breadfruit into the Americas before the Revolution.

240. See Ellis (1775) and French translation (Ellis, 1779), where Ellis calls the breadfruit "le plus utile de tous les fruits des Indes Orientales."

We should pause to say a word about the technology for transporting trees and plants around the world aboard the sailing ships of the day. That technology needed attending to, and it improved over the decades.[241] On the French side, and indicating this concern, Duhamel du Monceau and the marquis de la Galissionnière published their *Avis pour le transport par mer des arbres, des plantes vivaces, des semences et de diverses autres curiosités d'histoire naturelle* in 1752 and an expanded edition in 1753.[242] The chevalier de Turgot, brother of the famous physiocrat, published his *Avis pour le Transport par mer, des Arbres, des Plantes vivaces, des Semences & de diverses autres Curiosités d'Histoire naturelle* in 1758. André Thouin designed containers, and they sailed with the La Pérouse expedition.[243] Michaux presented a case of his own to the Academy, as we saw.[244] D'Angiviller at the Bâtiments du Roi and Céré at the Jardin du Roi on Île de France wrote back and forth about building transport containers and paying for them.[245] D'Angiviller arranged with the royal printer, Dupron, to print an instruction pamphlet on transporting foreign trees.[246] Le Juge from Île de France contributed a manuscript.[247] And, from an early date, some attention was paid to transporting insects.[248] The French also took what they could from the English. They translated John Coakley Lettsom's *The Naturalist's and Traveller's Companion, Containing Instructions for Collecting and Preserving Objects of Natural History* in 1775, the year after it appeared in London.[249] John Ellis' *Description of the Mangostan and the Bread-Fruit* of 1775 had a section on transporting these plants, and Ellis appeared in French in 1779.[250] André Michaux reported constructing his bird

241. Allain provides a substantial work on this subject; see esp. his chapt. 3; Rousseau (1957), p. 154, gives a vivid account of a failed effort to ship plants from Sarrazin in Canada to France.

242. La Galissonnière's name does not appear on this work, but he is generally recognized as the coauthor; see Bonnnault, p. 175; Roussel et al., p. 19.

243. Turgot (1758); "Projets, instructions, mémoires et autres pièces, relatifs au voyage de découvertes, ordonné par le Roi, sous la conduite de M. de La Pérouse," planches, pp. 247-53, Bibliothèque Mazarine, Ms. 1546; see also Regourd (2000), p. 744, and (2005a), pp. 36-37; Musée National de la Marine (2008), where Thouin's cases were displayed; see also Allain, pp. 16-18.

244. See above p. 319 at n. 72.

245. AL draft d'Angiviller, dated "Vers.es le 16 9.bre 1784" to "M. de Cerné, directeur du jardin R.l des plantes a l'isle de France," AN O[1] 2112. Interesting cooperation is shown here between Bâtiments du Roi and La Marine regarding paying for these containers.

246. See exchange, AN O[1] 2112: Pépinières, 1783-1787.

247. "Maniere de transporter les jeunes plants de toutes sortes d'arbres, sans embarras, et sans depense d'eau pour les arroser dans les Vaisseaux" (1763), BCMNHN, Ms. 293.

248. Draft letter in d'Angiviller's hand to "M. de Montigny" in China, dated "Vers.lles. 5 mars 1786" ["Si vous pouvés enrichir notre cabinet du roy de quelques insectes rares, M. de Buffon et moi vous en auront la plus grand obligation."], AN O[1] 2112; for an earlier report on transporting insects, see PA-PV 29bis (1710), fols. 409v-411 ("Le Mecredi 19 Novemb[re] 1710").

249. Lettson (1775). This was a translation of the second, enlarged edition; Lettson (1774). Lettson (1772) was the first edition. Lettson was also known for his expertise concerning tea and its transport.

250. See Ellis (1775), (1779), pp. 28-34.

cages based on English models.[251] The physical transportation of plants was an obstacle, but not an insuperable one.

La Luzerne's Project. After the American War, rather than becoming discouraged by the high mortality associated with intercolonial plant transfers, the French government redoubled its efforts. The documentation surrounding the next, largest, and last botanical expedition of the Old Regime is especially rich, and it allows a close-up view of workings of the contemporary French inter-hemispheric botanical system.

The initiative began at the highest levels of the government. After having served as ambassador to the United States and then as governor-general of Saint Domingue, César-Henri, the count de La Luzerne (1737-1799), took over as minister of the navy and the colonies in late December of 1787. La Luzerne had a strong interest in the economic development of the colonies through plant transfers and agronomic projects; already coincident with his posting as governor-general of Saint Domingue in 1785, he worked on plans to import useful plants and animals to Saint Domingue from South America, Senegal, and the Cape of Good Hope.[252] La Luzerne brought with him to ministerial office an ambitious plan for stimulating botanical activities in France and in the colonies. His plan called for the Jardin du Roi to ship useful European plants to the colonies; for the provincial "entrepôt" gardens in Brest, Lorient, La Rochelle, and elsewhere to be reinvigorated; for a massive transportation of plants of commercial significance from the Indies directly to America; and for further botanical exchanges between and among the colonies in the Americas.[253] La Luzerne made overtures to Thouin to this effect on 5 January 1788, and he followed up two days later with a memoir wherein he underscored that "our great possessions are in the West. The colonies in the East must provide useful plants and animals that those of the West will exchange among themselves."[254] Five days after that he produced an eight-point proposal, "Project for Initiating Correspondence between the French Colonies and the Jardin du

251. ALS "A. Michaux," dated "A Charleston le 27 X.bre 1787," to "Monsieur Le Comte," AN O[1] 2113A (in folder "Pépinieres. Commission du S. Michaux Pour l'Amerique"), where Michaux writes, "La construction de cette Cage a ete faite sur un Modele envoyé de Londres et peut servir a transporter les oiseaux les plus sauvages, sans danger qu ils perissent en se heurtant eux même de frayeur."

252. McClellan (1992), p. 157. On departing for Saint Domingue and claiming "je suis fort ignorant en histoire naturèlle," La Luzerne approached Adanson about which plants and animals to promote; ALS La Luzerne to Adanson, dated "Paris, ce vendredi 7 octobre 1785" and Adanson's three-page reply, HI, AD 207.

253. See dossier, "Pieces relatives au Projet d'une Correspondance Agriculto-Botanique entre les différéntes Colonies francoises et le Jardin du Roi. Ebauchée En Janvier 1788 d'après le Projet de M. le C.te de La Luzerne," BCMNHN, Ms. 308.

254. "Nos grandes possessions sont dans L'Ouest. Il faudroit que les Colonies de L'est y fournissent les plantes & animaux utiles, que Celles de L'Ouest se les communiquerassent entrelles," letter from La Luzerne to Thouin, dated "Versailles le 5 Janvier 1788," and related "Notte remise par M. le C.te de la luzerne le 7 Janvier 1788," BCMNHN, Ms. 308.

Roi."[255] As minister in charge of colonial affairs, La Luzerne energized the botanical establishment and the colonial administration in this project. He directed local administrators to catalog their botanical needs and resources so that the whole might make up for the deficiencies of the parts; and much activity ensued as subordinates around the globe scurried to fulfill his orders.[256]

Coincidentally, in early February of 1788, an opportunity arose for La Luzerne to realize his plan concretely. A certain captain Fournier was slated to sail the navy packet boat *Stanislas* from Le Havre to the Indian Ocean without a cargo. Fournier volunteered his services to science and to Thouin, who then wrote to La Luzerne: "I do not know if similar occasions will often present themselves, but I believe that this one cannot be more favorable, my lord, to begin to put into play your great projects to enrich our colonies with the most useful items from Europe and to bring back on the return trip those that can be of some use to France."[257]

A wily bureaucrat, La Luzerne knew he had a problem in that the aged Buffon was still nominally in charge of the Jardin du Roi. Conspiring with Thouin, La Luzerne maneuvered to make it seem as if the ensuing voyage was actually Buffon's idea. Thouin got Buffon to suggest that he, Buffon, would pay for the expenses of importing rare plants from the Indian Ocean to the Jardin du Roi, while the Navy should underwrite sending European plants outward to Île de France. Appearing as the benevolent "protector" of Buffon's initiative, La Luzerne graciously approved the endeavor when it came across his desk at the end of February of 1788.[258]

Two days before he gave his formal approval, La Luzerne wrote to Céré on Île de France to prepare for the unstated component of the expedition: transshipping from the Indian Ocean to the Americas "plants and seeds of the male and female nutmeg, gray pepper, cardamom, rattan, mangosteen, litchi, and breadfruit if possible."[259] Thouin and La Luzerne agreed that, rather than

255. "Projet d'Etablissement de correspondance entre les Colonies francaises et le Jardin du Roi," dated "12 Janvier 1788," BCMNHN, Ms. 308. It is revealing perhaps that the draft of this proposal is entitled, "Notte relative a la Correspondance d'Economie rurale, ~~politique~~ [sic!] et de Botanique qui doit etre établie entre les Colonies francaises et le Jardin du Roi."

256. See, again, BCMNHN, Ms. 308.

257. "Je ne sais s'il se presente souvent de pareilles occasions, mais je crois que celle-ci est on ne peut pas plus favorable, monseigneur, pour commencer l'exécution de vos grands projets d'enrichir nos Colonies des plus utile productions de l'Europe et d'en rapporter au retour celles qui peuvent être de quelque usage a la France," ms. memoir [from Thouin], dated "Paris ce 14 fevrier 1788," BCMNHN, Ms. 47. See also [Thouin:] "Precis historique qui a donné lieu au Voyage du Sr. Joseph Martin Garçon Jardinier Botaniste du Jardin du Roi au Cap de Bonne Esperance et aux Isles de france et de Bourbon pendant les Années 1788 et 1789," also BCMNHN, Ms. 47.

258. See AL unsigned [from Thouin] to "Monsieur Pouget Intendant des Classes de la Marine le 25 fev. 1788," Ms. [Thouin:] "Precis historique," and letter from La Luzerne to Thouin, dated "A Versailles le 29 fevrier 1788," all BCMNHN, Ms. 47.

259. "Plants et graines de Muscadier male et femelle, Poivre gris, cardamome aloës, Rotaing, Mangoustan, Litchi, et de l'arbre a pain s'il est possible," letter from Pouget [to Thouin], dated "Paris 27 Fevrier 1788," BCMNHN, Ms. 47.

sending along an expensive and perhaps uppity royal botanist, an apprentice gardener would suffice to superintend the strictly botanical arm of the expedition, and the young Joseph Martin received the appointment.[260] Buffon finally died as Martin waited at Le Havre to depart on his adventure. Thouin unveiled the rest of the plan to his young protégé:

> It seems to me that the Minister strongly wishes that you return via Cayenne and that you take charge of bringing to this colony all the spice trees that are ready in the royal garden on Île de France. This will extend your voyage and will interfere a little in our project to bring back on your return trip via the Cape of Good Hope those items gathered for you in your absence. But you must accommodate yourself in this and zealously fulfill the Minister's orders.[261]

On the outward journey from France, the *Stanislas* brought hundreds of European fruit and nut trees (450 chestnut; 400 almond) and plants for kitchen gardens, for medical and pharmaceutical use, for animal forage, and for use in spinning and dyeing.[262] After an unusually quick sail of only three months, Martin arrived at Île de France on 26 July 1788, apparently without losing a single plant on the way. Céré wrote to Thouin that this was the largest and most successful botanical shipment ever received in Île de France – a communication that did not fail to be presented to the Academy of Sciences.[263] Martin took a quick botanizing trip to Madagascar, while Céré prepared the major consignment for Cayenne: thirty cases and 105 barrels of plants and seeds,

260. Letter from La Luzerne to Thouin, dated "A Versailles le 29 fevrier 1788," BCMNHN, Ms. 47.

261. "Il me paroit que le Ministre desire fort que vous revenir par Cayenne et que vous vous chargiez de Porter a cette Colonie tous les Arbres à Epices qui se trouvent assez multipliés dans le Jardin du Roi a l'Isle de france. cela alongera vôtre voyage et contrariera un peu nos Projets de prendre à votre retour par le Cap les choses qu'on auroit pris y rassembler pendant votre absence. Mais il faut vous arranger la dessus et remplir avec Zel les ordres du Ministre…," undated letter from Thouin to Martin, BCMNHN, Ms. 47. The date of mid-April 1788 is inferred by the order of the letters in the file. See also 30pp. of instructions prepared by Thouin detailing what Martin was to do at every step of the way, "Instruction pour Le Sr. Joseph Martin realtivement à son voyage a L'Isle de france et à la Conservation des Plantes qu'il transporte d'Europe et à celles qu'il doit y rapporter," BCMNHN, Ms. 47. In this connection, see also, BCMNHN, Ms. 56: "Pieces relatives au Voyage de M. Joseph Martin a l'Isle de france en 1788 et 1789."

262. "Etât des Végétaux utiles qui se trouvent en france et qui manquent à nos Colonies des Iles de france et de Bourbon en 1788," BCMNHN, Ms. 47.

263. ALS Ceré to Thouin, dated "à Belle Eau, Isle de france, le 6 août 1788," BCMNHN, Ms. 47. In same collection see related letters from Martin [to Thouin], dated "De Lisle de france ce 7 Août 1788," copy of letter from Ceré to La Luzerne, dated "isle de france, le 5 août 1788," and Le Fournier to Thouin, dated "Isle de france 25 Aoust 1788." For evidence of the Paris Academy of Sciences being kept informed of Martin's voyage, see APS-MS, B F8245, "Séances de l'Académie Royale des Sciences, 1786-1789" ("le mercredi des Cendres 25 fevrier [1789]) ("...lettre de M. Ceré de lisle de Bourbon sur la demande quon lui a fait darbres et Epiceries transportés a Cayenne et St-Domingue –"), and APS-MS, 580, D881, Box 58, #20.

including the clove, nutmeg, cacao, litchi, and various palm trees for a total of eighty-six different species.[264] As luck would have it, the true pepper plant arrived from India for the first time in Île de France just days before Martin sailed, so that plant also was included in the lot.[265] As Thouin had done, Céré prepared an extensive set of instructions for Martin.[266] The previous expedition, led by Darras, had failed because he had kept his plants in the ship's hold. Martin was to bring the plants up top as soon as he reached warmer latitudes after passing the Cape. Céré advised caution in other regards, too. Martin was to say that the ship was returning to France with a shipment for the Jardin du Roi, not that it contained spices for transplantation to the Americas. If war broke out and it looked like the *Stanislas* would be taken, Martin was to "set himself to pulling up, breaking, and throwing into the sea all the clove, nutmeg, cinnamon, pepper, Ravensara, and Cardamom plants in the various cases and barrels."[267]

Martin sailed from Île de France on 6 March 1789. He wrote back to Céré from the Cape of Good Hope that cold, wet, and rats were big problems. Captain Fournier reported to Thouin that the pepper plants were especially affected.[268] Yet the cargo arrived in Cayenne on May 15th in "pretty good shape," according to Martin's report to Thouin.[269] There, Martin coordinated with the royal botanist, Leblond, in off-loading the spice plants intended for Cayenne and in boarding additional productions, notably clove, from Cayenne for further transport to Martinique, Saint Domingue, and the Jardin du Roi in Paris. In June of 1789, the *Stanislas* made quick calls at Martinique and Saint Domingue, again exchanging South American and Indian Ocean specimens for Caribbean ones, before regaining Le Havre by late July 1789. Martin's barrels and cases were off-loaded, transported to Rouen, and put on a barge to Paris where, via the Seine, they pulled right up to the Jardin du Roi.[270]

264. "Instructions pour servir aux 30 caisses et 105 bariques d'arbres, plantes et graines remises au Sr. Martin pour le jardin royal des Plantes de Paris," BCMNHN, Ms. 47.

265. Letter from Céré to Thouin, dated "A Belle Eau isle de france le 8 9bre 1788," and letter from Martin to Thouin, dated "a Lisle de france ce 30 Janvier 1789," BCMNHN, Ms. 47.

266. "Instructions Pour le Sr. Martin Eleve Jardinier du Jardin royal des plantes de Paris chargé de la collection d'arbres & de graines et d'Animaux embarquée sur le Paquebot le Stanislas,... au jardin du Roy le Monplaisir, Isle de France, le 5 février 1789," "Supplément aux instructions données au S.r Martin," and "Instructions pour servir aux 30 caisses et 105 bariques d'arbres, plantes et graines remises au Sr. Martin pour le jardin royal des Plantes de Paris," BCMNHN, Ms. 47.

267. "... se diposer à arracher, briser, et jetter à la mer tout ce qui s'appelle Girofliers, Muscadiers, Canelliers, Poivriers, Ravenesara et Cardamome planté dans les différentes caisses et bariques." "Instructions Pour le Sr. Martin Eleve Jardinier du Jardin royal," BCMNHN, Ms. 47.

268. ALS Martin to Céré dated "Au Cap Bonne Esperance le 4 Avril 1789," and ALS Le Fournier, presumably to Thouin, dated "Cayenne le 19 may 1789," BCMNHN, Ms. 47.

269. "...en assès bonne état...," ALS from Martin to Thouin, dated "A Cayenne ce 19 May 1789," BCMNHN, Ms. 47.

270. ALS Le Fournier, dated "Du Havre 8 Aoust 1789," BCMNHN, Ms. 47. Some of Martin's shipment likewise headed for the government gardens in Brest; see ALS Céré to Thouin "à Belle Eau, isle de france, le 16 Mars 1789" in same collection.

Joseph Martin arrived in Paris with 136 barrels containing more than 1200 individual plants.[271] He brought 107 different species from Saint Domingue, 21 from Martinique, 32 from Cayenne, 42 from Île de France, 15 from Madagascar, and one each from the Cape of Good Hope and Jamaica.[272] These included clove, cinnamon, sago, palm trees, litchi, hevea, and the cucumber tree. Martin also brought back a "handsome species of monkey" from Île de France. La Luzerne must have been pleased. Thouin was exuberant: "This shipment is the largest and most precious of all those that have been made to the Jardin du Roi since its foundation."[273] At its public meeting on the 28th of December 1789, the Royal Society of Agriculture awarded Martin a gold medal:

> ...for having transported from Europe to the colonial garden on Île de France a collection of fruit trees, garden plants, and a large quantity of useful seeds; for having transported spice plants and the breadfruit tree from the Île de France to our colonies in the Antilles; for having returned from these different countries the largest collection of plants, living trees and seeds ever brought to France; and finally by dint of great effort, considerable care and new procedures, for having enriched the National Garden [the Jardin du Roi] and the whole world with precious plants the multiplication of which can contribute to the wealth of the State and the happiness of Humanity.[274]

For a while it looked as if the momentum of the expedition of 1788-89 would continue. The number of shipments to the Jardin du Roi from Île de France and the East continued to grow.[275] As a reward for his efforts, Martin was sent back to Cayenne in the spring of 1790 as a full-fledged royal botanist with the handsome salary of 3000 pounds. He held the post for the next twenty years.[276] Martin tended the very plants he had brought from the Indian Ocean

271. For lists of plants Martin collected at Île de France, Madagascar, the Cape of Good Hope, Cayenne, Martinique, and Saint Domingue, see BCMNHN, Ms. 691 ("Catalogues des graines semées au Muséum pendant les années 1790-1801"), fols. 2-8.

272. See previous note and "Catalogue des plantes raportées par Mr. Martin des iles de St. Dom, de l'amartinique de L'ile De France &c a paris en 1789," BCMNHN, Ms. 953.

273. "Cet envoi est le plus considerable et le plus precieux de tous ceux qui ont été faits au Jardin du Roi depuis sa fondation," draft memo by Thouin, "de M. Joseph Martin reçu en 7bre 1789 un Envoi d'Arbres et de Plantes," BCMNHN, Ms. 47. See also in same collection, ALS Thouin to Céré, dated "A Paris ce 30 9bre 1789."

274. *Mémoires d'agriculture, d'économie rurale et domestique, publiés par la Société Royale d'Agriculture de Paris*, 1789 (3rd trimester), p. xiii.

275. See BCMNHN, Ms 691.

276. See BCMNHN, Ms. 48: "Correspondance de Joseph Martin (avec André Thouin) depuis son installation à Cayenne, comme directeur des cultures d'arbres à épiceries (1790-1809); Touchet, pp. 158-65, 189-201, 303; also Letouzey.

the previous year. Pepper, breadfruit, clove and cinnamon all did well under his ministrations. Taking a cue from Céré, Martin distributed plants from the royal nurseries to local residents. In the period 1791-1793, he gave away some 60,000 plants to colonists in Cayenne, not counting almost 7 million clove berries distributed locally and sent to Martinique and Marie Galante. Four thousand clove trees at the royal nursery at La Gabrielle in Cayenne were producing over 20,000 pounds of product in the early 1790s, with 18,000 new plants under cultivation since 1791 and due to bear fruit soon. In 1791 and 1792 Martin sent botanical shipments to Martinique, Saint Domingue, the Navy garden at Brest, and the Jardin des Plantes in Paris.[277] In short, the French and La Luzerne in particular had succeeded in establishing a new branch of industry and rural economy in their American possessions.

The success of the intercontinental system of botanical exchanges elaborated by French scientists and government administrators at the end of the eighteenth century can be seen in the list of 600 types of seeds sown by Thouin at the Jardin du Roi in Paris in the spring of 1790. Plants came from Saint Domingue, Martinique, Cayenne, Île de France, Île Bourbon, Madagascar, Senegal, the Cape of Good Hope, Florida, Louisiana, China, India, Manila, the Dutch East Indes, and Botany Bay (Australia).[278]

No further opportunity for exchanges between the Americas and the Indian Ocean occurred after 1789. On his return to France, Captain Fournier proposed to La Luzerne that a copper-sheathed vessel be sent to the Moluccas to get more spice plants and thence to sail directly to Cayenne. No doubt frustrated, La Luzerne replied that he would be glad to help, but no ships of the royal navy were presently available.[279] It was 1789, after all, and the French Revolution put an end to the ambitious plans of Old Regime functionaries and this remarkable episode of applied botany.

Still, in France, a plan pushed by André Thouin, Dutrône La Couture, and the Royal Society of Agriculture before the National Assembly in 1791 called for the creation of a set of commercial gardens in Saint Domingue to raise spices and exotic plants.[280] In Saint Domingue itself, in the midst of a civil

277. On these points, see ALS Martin to Thouin, dated "A Cayenne ce 28 8bre 1791," and "Mémoire présenté à l'institut national, par le C.en Joseph Martin, botaniste, chargé de la direction des jardins et pépinières coloniales, dans la guyane française; Le 29 frimaire, an 5e de la République française, une et indivisible" ["9bre 1796"], BCMNHN, Ms. 48. See also Laurent letters to Thouin, dated "Brest Le 15 frimaire, Lan 2.e de La Republique une et indivisible" [interpolated date: 5-12-1793] and "Brest Le 10 Germinal La 2 De la Rep.que françoise, une et indivisible" [interpolated date: 30-3-1794], BCMNHN, Ms. 848. See also ms. Code de Cayenne, vol. 7, pp. 351, 571, 573, 655 in AN-MAR G 237.

278. See sixteen-page manuscript list, BCMNHN, Ms. 47; see similar list, BCMNHN, Ms. 691 ("Catalogues des graines semées au Muséum pendant les années 1790-1801"), fols. 2-18 for plantings for 1790.

279. ALS La Luzerne to Thouin, dated "A Versailles le 26 7bre 1789," BCMNHN, Ms. 47.

280. McClellan (1992), pp. 278-79; ALS Thouin to Martin, dated "A Paris ce 15 8bre 1791," BCMNHN, Ms. 48; Regourd (1998), pp. 180-83.

war and a raging slave revolt, as late as January of 1792, a colonial Provincial Assembly was calling for a new botanical garden for Port-au-Prince. These stillborn efforts testify to just how deeply economic botany was felt to be necessary for colonial development and the economic well-being of the motherland.

Joseph Martin's voyage of 1788-89 provides a wonderful concluding vignette that captures something of the state of colonial botany among the great powers at the end of the eighteenth century. On the way outward from France to the Indian Ocean in 1788, Martin and the *Stanislas* put in for provisioning at the Cape of Good Hope. There he crossed with the Darras expedition outward from Île de France on its way to Cayenne and the Caribbean. In a letter dated 2 July 1788, Martin wrote back to Thouin in Paris from False Bay at the southern tip of Africa:

> A merchant vessel is here bound for Cayenne that took on a cargo of spice trees at the Île de France at a cost of 52,000 francs. [This is the Darras cargo aboard the *Alexandre*.] It put in at the Cape of Good Hope. There was also an English packet boat in port at False Bay that was going to the Tahitian islands to get breadfruit. Their ship was sent for this express purpose. The ship is carrying greenhouses, a botanist, two astronomers, and a geographer.[281]

This antipodean intersection of two French botanical expeditions to and from the Indian Ocean and the HMS *Bounty* on its way to Tahiti in 1788 epitomizes the forces and factors at play in the history of contemporary European colonialism and the story of the Colonial Machine and French economic botany.

The Problem of Naval Timber

> *Vous ne sauriez croire, Monsieur, dans quelle misère la Marine est aujourd'hui au sujet des bois. On en apporte ici de tous les côtez du royaume, et ils se trouvent tous mauvais, ou dans le cas de le devenir bientôt. Les navires bâtis avec toutes les précautions possibles, ont besoin en moins de trois ans d'un radoub considérable, et qui égale presque la première construction. Celà a fait chercher divers expé-*

281. "Il y a Monsieur un vaisseau marchand qui a frête pour 52 mille francs a lsle de france d'arbres a Epices pour Cayenne. Il a relaché au cap de Bon Esperence. Il avoit aussi un paquebot anglais en relache a falsebaie qui alloit aux Isles des taitis pour chercher l'arbre a pain leur vaisseau a été expedié pour cet seul chose il y dans ce vaisseau des serres chaudes un botaniste et deux observateurs astronomes et un geographe," ALS Martin to Thouin, dated "De false baye de 2 Juillet 1788," BCMNHN, Ms. 47.

> *dients, et pour conserver les bons bois et pour tâcher de corriger les défectueux, ou les suspects. On a même tenté des expériences à ce sujet, qui ont coûté de fort grosses sommes: je ne sçai si ce sera utilement pour la suite....*
>
> – Boureau-Deslandes, Navy Controller at Brest, to Abbé Bignon, "A Brest, ce 26e d'avril 1731."[282]

France and the Colonial Machine faced a timber famine.[283] Ships were the literal sine qua non of French colonial expansion in the Old Regime, but shipbuilding ate whole forests, and as the French Navy and merchant marine expanded from the 1660s onward, the dearth of naval timber, particularly of oak, became an increasingly serious problem keenly felt by contemporaries. The felling of trees in France was strictly controlled by law and policed by government forestry officials; elaborate legislation existed concerning naval timber in particular. Already in 1667 Colbert expressed his hope that timber from Canada would spare France having to purchase wood from Scandinavia, and he ordered the administrators in Canada to set aside vast stretches of Canadian forest for the supply of naval timber.[284] The Colonial Machine naturally came to concern itself with this matter, and no more central activity occupied the Colonial Machine than grappling with the problem of securing reliable, quality supplies of wood for French shipbuilding.

Researching Wood. The problem only got worse as the eighteenth century began to unfold, and a crisis point seems to have been reached in the later 1720s. The perception mounted that the quality as well as quantity of wood available for naval construction had seriously declined over time and that no systematic, scientific work had been undertaken on preserving wood for construction.[285] As one report had it:

> It is well proven that ships built in the last few years last only a short time compared to those built in the past. After a period, substantial repairs are required. Some vessels have not been able to withstand a voyage on the high seas without considerable refitting and had to be condemned on their return. Still others have not been able to put to sea

282. PA-DB (Deslandes); reprinted in Laurent, pp. 103-104.

283. On these points, see Buridant; Boudriot (1991), (2000); Corvol-Dessert, Pritchard (1987a), pp. 163-67; and materials in AN-MAR G 61, vol. 1, fols. 73-93 ("Bois"); Llinares (2005), pp. 173-74, also points to the problem of wood for naval timber, as does Allard, p. 32.

284. Letter from Colbert to Talon, AN C^{11A} (1667), vol. 2, fol. 290 (April 5, 1667).

285. On the latter point, see "Discussion sur les differens moiens emploiés pour la conservation des Bois, 1729," AN-MAR G 61, vol. 1, fol. 83.

> without refitting beforehand because essential components were entirely rotten.[286]

The scarcity was so bad on Île de France that, in order to conserve wood and prevent fires, no wooden buildings were permitted in Port-Louis.[287] In 1730 the Minister of the Marine, the count de Maurepas, formally directed the Académie Royale des Sciences to investigate: "For this purpose, the Count de Maurepas wrote a letter to the Academy in which he explained to the Company the uncertainty rampant in ports where royal vessels are built. He noted several abuses that continue there only out of habit and supported only by routine and prejudice. He invited the Academy to look on this work as most worthy of its care and attention.[288]

The next year, Maurepas, who seemed very knowledgeable about various techniques for preserving wood, again pressed the Academy through its president, the marquis d'Argenson; and the year after that he provided an annual budget of 2000 livres for the Academy's project and explicitly commissioned Duhamel du Monceau to work on the problem.[289] (This commission preceded Duhamel's appointment as inspector-general of the navy in 1739.) The Academy was not lax in its response to Maurepas, appointing Réaumur, de Mairan, and Duhamel du Monceau as its commissioners. In 1735 Buffon joined the team. Duhamel solicited information from ministry and port officials, from navy engineers, from royal forestry officers, and from his correspondents abroad, including Gaultier in Canada and Cossigny on Île de France.[290] Buffon planted an experimental forest of thirty *arpents* on his property at Montbard in Burgundy, and because his trials might violate existing forestry regulations, the

286. "C'est une chose bien prouvée que les Vaisseaux que l'on construit depuis quelques-années ne durent que peu en comparaison de ceux qu'on construisoit autrefois, au bout de quelques tems il leur faut faire des reparations de consequence, il y en a meme qui n'ont pu suporter un voyage de long cours sans des radoubs considerables et qu'il a falu condamner a leur retour, bien plus quelques-uns n'ont pu être mis en Mer sans avoir êté auparavant radoubés, des membres essentiels estant entierement pouris," "Projet d'un travail entrepris par ordre de M. le Comte de Maurepas sur les Bois de Construction," APS-MS, B/D87, group 27.

287. Delaleu, vol. 1, p. 371: "Ordonnance...portant défense de bâtir en bois dans tout l'étendue de la Ville du Port-Louis. Du trois Novembre 1774."

288. "Pour cela M. Le Comte de Maurepas ecrivit une Lettre a L'academie dans laquelle il exposoit a la Compagnie l'incertitude qui regnoit dans les Ports ou l'on construit des vaisseaux pour le Roy et plusieurs abus qu'il avoit remarqué qui n'y subsistoient que parce qu'ils y êtoient en usage n'etant soutenus que par la routine et par le prejugé, invitant L'academie a regarder ce travail comme des plus digne de ses soins et de son attention," "Projet d'un travail entrepris par ordre de M. le Comte de Maurepas sur les Bois de Construction," APS-MS, B/D87, group 27; see also related materials in this collection.

289. "Lettre à M. D'Argenson Président de l'Academie des Sciences pour exiter l'Academie a s'occuper des moïens de conserver les bois" (June 20, 1731), AN-MAR G 61, vol. 1, fols. 88-89; see also ms. in same collection, "M. Duhamel chargé de faire des recherches sur les Bois," fol. 93 (27 August 1732). Allard, pp. 80-91, discusses the Canadian chapter of Duhamel's research into naval timber.

290. See previous notes and Gaultier correspondence with Duhamel, Houghton Library, Harvard University, dossier 57, M-13.

king's forestry agents were instructed not to interfere with Buffon's experiments.[291]

Duhamel incorporated the results of these studies in his later published work on trees and their preservation, notably his *Avis pour le transport par mer des arbres* (1752) and his *Traité des Arbres et Arbustes* (1755). A projected second edition of the latter was to include related work by Richard *père* of the Trianon gardens, academicians and professors at the Jardin du Roi (Le Monnier, B. de Jussieu, A.-L. de Jussieu), Gaultier in Canada, someone named Sonette in Louisiana, and the abbé Nolin.[292]

Other papers and reports on how best to secure and preserve naval timber followed in the course of the century. Some came before the Académie de Marine.[293] These studies were not strictly empirical, but sought theoretical and chemical explanations for the phenomena behind the degradation of wood and means of preventing it. Sap was the key element in these scientific accounts. The observations of one Coulomb from Toulon in 1775 give a flavor of the scientific underpinnings this research:

> Before entering into the subject, we must acknowledge a principle of *physique* concerning the nature and composition of oaks and their sap....Several experiments on woods made by a great *physicien* and academician [doubtless Duhamel du Monceau] seem to prove that wood is formed by a fixed and light earth whose parts are united and consolidated by an adhesive and resinous substance with different salts....[After trees are cut down,] the adhesive substances lose their moisture and contribute to the hardness of wood. One can even see them as a preservative balm that acts against the corruption of wood....With this principle of the nature of woods and their sap established, every solvent must act adversely on the conservation of woods, since it strips them of the adhesive substance and ultimately can effect their utter decomposition.[294]

291. See entry ("Bois"), 5 January 1735, AN-MAR G 61, vol. 2, 434v.

292. See Fougeroux ms., "Préface inédit pour une seconde édition jamais parue du Traité des Arbres et Arbustes," APS-MS, 580 D881 #24.

293. SHMB, Série A, Sous-série 3A, entry for 10 September 1775.

294. "Avant que d'entrer en matiere, il faut convenir d'un principe de phisique sur La Nature et La Composition des chesnes et de leur seve...Plusieurs experiences faites sur les bois par un grand phisicien et academicien doubtless Duhamel, semblent prouver que le bois est formé par une terre fixe et Légere, dont Les parties sont Réunies et Consolidées par une substançe gommeuses, Raisineuse remfermant Les differentes Sels....[After trees are cut down...] Les substances gommeuse en perdant Leur humidité contribuent à la dureté des bois. on peut encore les regarder comme un beaume conservateur, qui s'oppose a Leur corruption....d'aprés ce principe Etabli de la nature et de la composition des bois et de leur Seve, tout dissolvant doit etre contraire à la conservation des bois, puisqu'il les depoüille de La substance gommeuse et qu'on parvient jusqu'à en faire une decomposition absolüe...," "Reponse aux observations sur le resultat du Conseil de Marine tenu Le 12 May 1773," dated "À Toulon Le 10 Juillet 1775" and signed, "Coulomb," AN-MAR G 235 #120. See also ms., "Réponse au Mémoire de M.r de Ruis présenté à l'Accadémie Le 16 Janvier 1772" ("Le premier objet du Mémoire où on considère La cause de la pouriture des Bois employé à la Construction des Vaisseaux..."), AN-MAR G 141, and remarks, AN-MAR G 61, vol. 2, fol. 89.

Hope from Overseas. These reflections were all well and good, but the practical problem remained of securing alternative sources of naval timber; from overseas and Canada, in particular, was the obvious answer. Colbert's remark of 1667, quoted at the beginning of this section, was in response to a letter the previous fall from the local administrator, Talon, who said he was himself going to explore the forests richest in oak, maple, and ash trees. Colbert sent carpenters from France to report on the quality of wood available in Canada for naval construction; Talon had built a six-ton port sloop out of Canadian wood, and he brought master founders from France to head shipbuilding activities.[295] By 1700, masts from Acadia in guaiac, elm, oak, maple, and ash regularly made their way to ports in Toulon and Rochefort. In 1703, authorities imported a forestry master to Canada to supervise this activity, and in 1705, naval personnel were forbidden to import masts or other Canadian wood on their private account. At other moments in the 1720s, however, with the opening up of the Baie de Saint Paul, procuring naval timber was privatized, and colonists engaged in this trade were to be encouraged.[296] Agents were on the lookout for curved and twisted pieces as well, but the ministry only wanted large masts, saying that it could procure small ones and yardarms at much better prices in France.[297] As late as July 1755 the ministry in France was writing to the governor-general and intendant in Canada, saying that "they should always be involved in pursuing research into wood of every species, particularly those suitable for masts."[298]

Much the same story repeated itself in Louisiana.[299] In 1729 the local *ingénieur du roi*, Baron, and other local authorities proposed to seek out easily exploitable construction timber in the vicinity of New Orleans.[300] In 1740, plans went ahead to log oak forests at Barataria and to build a 74-cannon ship on the spot in Louisiana.[301] A master shipbuilder, a certain Olivier Millous, proceeded from Brest to Louisiana in 1742 to undertake this project, and a second carpenter followed in 1743.[302] The outcome of this undertaking remains

295. See correspondence to this effect, AN C[11A] (vol. 2), fols. 199, 216 (letters dated April 5, 1666, November 13, 1666).

296. Concerns over masts and naval timber crop up frequently in ministerial correspondence between Versailles and officials in Canada through this period; on these points see AN-COL B[21], B[22], B[23], B[27], B[42] passim.

297. See ministerial letter of May 29, 1725, AN-COL B[48] (1725), fol. 850.

298. Letter of July 15, 1755; AN-COL B[101], fol. 16. For more on bringing wood from Canada, see AN-MAR G 62.

299. See, for example, AN-MAR G 61, vol. 2, fol. 442v, entry "Bois," dated 1740.

300. See letter of November 25, 1729, AN-COL C[13A] 12 (1729-1730), fol. 30.

301. See letter dated June 21 [1740], AN-COL C[13A] 25 (1740), fol. 121.

302. See correspondence related to this episode, AN-COL C[13A] 26 (1741), fols. 232, 243; AN-COL C[13A] 27 (1742), fol. 3 (23 March 1742); fol. 179 (12 September 1742); AN-COL C[13A] 28 (1743-1744), fol. 171 (20 July 1743).

unclear, but it illustrates how much the French and the Colonial Machine looked across the sea to relieve their timber famine.[303]

These were all sporadic and not very effective efforts, however. By the 1780s, another alternative emerged: not to send masts and construction timber or to exploit standing forests thousands of miles away, but to send seeds and seedlings from America for propagation in nurseries in France. This was the heart of André Michaux's mission to America.

André Michaux's Mission to America

> *Si le voyage Etoit moins celebre que celuy du Capt.e Cook, il seroit non moins Interessant a bien des Egards, & Jose meme le repeter, ajouterait beaucoup aux Liens qui unissent aujourdhuy L'americains et Le Francois – et Tendroit a Etablir quelques nouvelles branches de commerce....*
>
> – St. Jean de Crevecœur, "New York, 15 Nov.bre 1784."[304]

On the evening of 27 September 1785 the botaniste du roi André Michaux (1746-1802) set sail from Lorient and headed west to North America on an official expedition to survey American forests and to send back specimens of trees for cultivation and propagation in France.[305] Commissioned by Louis XVI, the expedition had been planned by the director-general of the Bâtiments du Roi (the count d'Angiviller), by André Thouin of the Jardin du Roi, and by Abbé Nolin, director of royal nurseries under d'Angiviller, and it built on Thouin's efforts from 1781 to formalize botanical exchanges with agents in America.[306] These had not succeeded, and Michaux sailed for America as an agent of the Colonial Machine in response to the shortage of naval timber and

303. In the Mascarenes a similar difficulty was felt of having to import logs from France to use as masts, and the attempt was made to find Indian sources; see ms. Code de l'Ile de France, vol. 2, in AN-MAR G 237.

304. Crevecœur letter of this date to unnamed recipient, probably d'Angiviller at the Bâtiments du Roi, AN O[1] 597. Crevecœur is referring to an expedition of his own proposing, but knowing that Michaux was already on his way to America for the same purposes.

305. For the modern literature on Michaux, see Chinard (1957a); Savage and Savage; McClellan (2004); Taylor and Norman (2002); Roger Williams (2001), pp. 168-69; Allorge, pp. 597-600; Lacroix (1938), vol. 3, pp. 101-107 and vol. 4, pp. 73-80; Duprat presents important documentation. See also Deleuze "Notice historique sur André Michaux," (1804) and translation of Deleuze in a modern edition by Charles Williams et al. PA-DB (Michaux) contains off-prints of articles by Houth (1931) and M. Caron, "Notice historique sur André Michaux" (Société libre d'Agriculture de Seine et Oise, meeting of 28 Prairial an 12). The present account is also based on ms. materials in AN O[1] 2113A: Pépinières d'Amérique. The papers by Ewans, Robbins, Camus, and others from the conference sponsored in 1956 by the CNRS are also notable sources.

306. Duprat, p. 249.

with the aim of invigorating French silviculture and enhancing resources for French shipbuilding.[307]

A Botaniste du Roi for North America. By the time he left for America, Michaux was a seasoned government naturalist. Born in 1746, he was the son of a farming family from Satory near Versailles. Depressed after the death of his wife in 1770 soon after childbirth, he developed an interest in botany encouraged by Louis-Guillaume Le Monnier (1717-1799). Michaux studied with Claude Richard, chief gardener at the nearby Trianon gardens, and through Richard and Antoine Laurent de Jussieu, he met and studied with Bernard de Jussieu, the academician and naturalist attached to the Jardin du Roi in Paris. We previously noted the training Michaux received as a junior member of botanical forays in France and Spain, his appointment as correspondent of the Cabinet du Roi, his contact with the Académie des Sciences, and the expedition he led to Persia in 1782-1785. In 1785, with enhanced court and scientific connections, André Michaux, aged 39, was plainly prepared and the person most qualified to undertake the American adventure promoted by Thouin, Nolin, and the count d'Angiviller.[308]

Copies survive of the royal commission appointing Michaux "to the title and position of Botanist attached to the Nurseries cultivated under the orders of the Director and Superintendent General of the King's Buildings."[309]

> His Majesty, being at Versailles and having in mind the desire that has occupied him for several years to introduce into his realm and to acclimatize there through an intelligent and studied cultivation…all the trees and forest plants that Nature has given up until now only to foreign regions, but which are of the greatest interest for the arts as well as for construction,…sought a subject who joined intelligence nurtured through experience with the faculties and strength necessary for travel in any and all countries, for the careful study and collection for His Majesty of plants, seeds, and fruits of all trees and shrubs and even grasses which cannot be multiplied enough for use as forage for animals. To take on in addition every kind of research pertaining to botany and to establish correspondences by dint of which the Administrator of

307. On this point, see also Spary (2000), p. 133, and Chinard (1957b), p. 508, where he writes about "the systematic policy pursued by the royal government to obtain from the United States timber for the use of the French Navy and to acclimatize American trees in France. It is well known that these efforts culminated in the mission of André Michaud...."

308. On Michaux, see above, pp. 318-19. Thomas Jefferson apparently was also behind the idea; see Broc (1975), pp. 375-76; Ewan, p. 22; on Jefferson's extensive botanical interests and contact with Thouin, see Chinard (1957a), pp. 282-83.

309. "au titre & état de Botaniste, attaché aux Pépinieres cultivées sous les ordres du Directeur & Ordonnateur-Général des Bâtimens du Roi...," AN-MAR C^7 207. The printed *brevet* is dated July 18, 1785. Other copies survive in AN O^1 2112 and AN O^1 2113A.

> the King's Buildings can extend research and the advantages they should produce. Being informed of the deep knowledge possessed by Sieur André Michaux who was born at Versailles of a family engaged in agriculture and who to the present, led by his love of the sciences, has not ceased to devote himself to the most difficult and useful voyages in different climates, notably in Persia,...His Majesty has resolved to attach him to his service, and as a result His Majesty has taken and retained the aforesaid Sieur André Michaux to the title and position of Royal Botanist...to undertake useful voyages at His Majesty's expense...under the protection and immediate safeguard of His Majesty, in consequence of which he is entitled to all the help, backing and support, even pecuniary, that may be necessary for him from all public or private agents of France everywhere....[310]

Michaux received this official commission as botaniste du roi on 18 July 1785. He was given specific instructions at Versailles on 9 August 1785, when he presumably met the king, and after a long delay in port at Lorient, he finally departed for America that evening in September. He was accompanied by his son, François-André Michaux; by a gardener, the royal nurseryman (*pépiniériste du roi*) Pierre-Paul Saunier (aged thirty-four); and by a servant, Jacques Renaud. The elder Michaux received 2000 livres a year for his services, Saunier 800. Saunier was supplied with several manuals, including Duhamel's *Traité des Arbres et Arbustes*.[311] Michaux carried with him official

310. See previous note. "Sa Majesté étant à Versailles, & ayant fixé ses idées sur le desir qui l'occupe déjà depuis plusieurs années, d'introduire dans son Royaume & d'y acclimater, par une culture intelligente & suivie...tous les Arbres & Plans forestiers que la Nature n'a donné jusqu'à présent qu'à des régions étrangeres, mais qui sont du plus grand intérêt pour les travaux des Arts, ainsi que pour les constructions en charpente,...[cherchait] un Sujet qui joigne à des lumieres mûries par l'expérience, les facultés & les forces nécessaires pour voyager en quelque pays que ce soit, en étudier les productions & rassemblent avec soin pour Sa Majesté, des plans, graines, & fruits de tous arbres & arbustes, même des plantes herbacées, propres à multiplier les espèces de fourrages qu'on ne peut trop accroître pour les bestiaux; embrasser, au surplus, toutes les recherches qui se rapportent à la Botanique, & établir des correspondances, à la faveur desquelles l'Administrateur des Bâtimens puisse perpétuer les recherches & les avantages qu'elles doivent produire; & Sa Majesté étant informée des connoissances profondes, qu'a acquises le sieur André Michau, né à Versailles, d'une famille livrée à l'Agriculture, & qui entraîné par son amour pour les Sciences, n'a jusqu'à présent cessé de se livrer aux voyages les plus pénibles & les plus utiles dans différens climats, notamment en Perse... Sa Majesté a résolu de l'attacher spécialement à son service: en conséquence, Sa Majesté a pris & retenu ledit sieur André Michau, au titre & état de Botaniste... faire aux frais de Sa Majesté les voyages qui seront jugés utiles,...sous la protection & sauve-garde immédiate de Sa Majesté, & qu'en conséquence, il éprouve de la part desdits Agents [publics ou privés pour la France] de tout ordre, tout appui, aide & secours, même pécuniaires, qui pourront lui être nécessaires...."

311. "Reglement qu'on pourroit faire pour Le Jardinier Pépineriste D'Amerique," O[1] 2113A (folder: "Pépinières d'Amérique," marked in pencil "1791"). Originally the "garçon-jardinier," Jean-Louis Thuillier, whom Thouin recommended for his docility, was supposed to accompany Michaux; see ALS "Thouïn" to d'Angiviller, dated "A Paris ce 5 Aoust 1785" AN O[1] 2113A ("Correspondance Générale pour la Mission du S. Michaux, Botaniste du Roi, Année 1785"). For Saunier's title, see reference, BCMNHN, Ms. 691, fols. 143v-144v.

letters of recommendation to U.S. and French officials, including one from Lafayette, and a letter of credit for 15,000 livres.[312] After a turbulent crossing of forty-seven days that included hitting a sandbank on the way in, Michaux and company safely dropped anchor in New York Harbor on 13 November 1785.

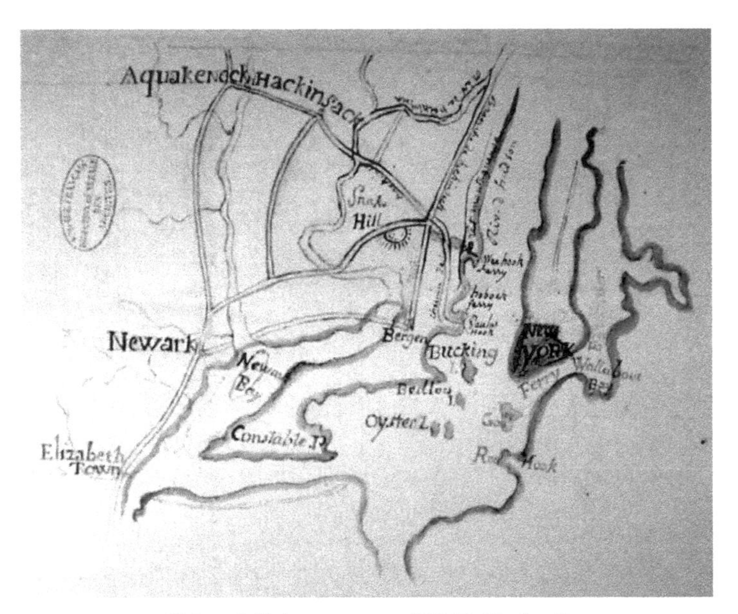

[Map: Michaux map of N.Y. Harbor]

Two Royal Gardens in the United States of America. Michaux remained in North America just short of eleven years. He established two formal gardens and plant nurseries there as outposts of the Colonial Machine – the first in 1786 in New Jersey, just a few miles west of New York (now the old Hoboken cemetery in Union City, New Jersey), the second in 1787 near Charleston, South Carolina near the present Charleston airport. For the New Jersey garden Michaux paid $750 (3862 livres) to Nicolas Fish of New York for thirty acres in Bergen's Wood, and he spent another $800 to build a house there and to improve the property. At the time, foreigners who bought property in New Jersey had to pledge allegiance to the State, but, as an agent of the king of France, Michaux refused to make such a pledge. It took a special act of the New Jersey state legislature to accomplish this favor, which was partly predicated on French

312. See ALS "A. Michaux," dated "A L'Orient ce 18 7bre 1785," AN O^1 2113A ("Correspondance Générale pour la Mission du S. Michaux, Botaniste du Roi, Année 1785"), and letter to "M.r Rob.t Morriss A Philadelphie," dated "Paris le 21 Aoust 1785," AN O^1 2113A ("Pépinières d'Amérique, 1786")

trees and plants being shipped from France to the United States.[313] (French plants were indeed sent to North America.[314]) The next year, Saunier stayed behind to tend the New Jersey garden, while Michaux and his son ventured south to better botanical climes and to found the royal garden at Ten-Mile-Station, north of Charleston. There, in 1787 Michaux finalized the purchase of a property of 111 acres for 100 louis/100 guineas (roughly 2000 livres).[315]

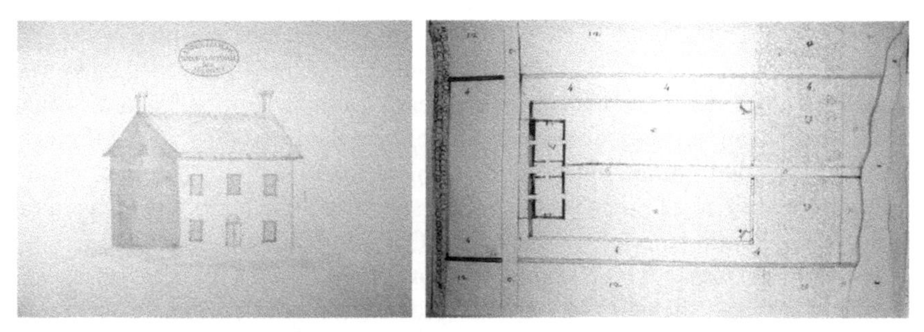

[Figure: Michaux's house and garden in New Jersey]

How Michaux conceived of the gardens he established in New Jersey and in South Carolina is telling of his ties to the Colonial Machine and the place of these gardens in the larger structure of the Colonial Machine. In a word, Michaux thought of himself as part of the royal, scientific administration, viz., the Colonial Machine, and he thought of the gardens in America as extensions of royal gardens in France. For example, he referred to the New Jersey garden as "the Royal Establishment."[316] He refused to let his son cultivate the Caro-

313. On these points see ALS "A. Michaux," dated "A New York Ce Ce [sic] 13 May 1786," to "Monsieur Le Comte," AN O[1] 2113A ("Pépinières d'Amérique. Correspondance Générale pour la Mission du S. Michaux, Botaniste du Roi, Année 1786"); Michaux letter of 26 January 1786 in Duprat, p. 245. See also related material with this letter, including "Traduction d'un acte de l'Etat de New jersey qui permêt au S.r André Michaux d'acheter des terres dans le dit Etat sous de certaines conditions, en date du 2 Mars 1786," and "Boundaries of Col.l Nicolas Fish Lot of Land. N.o 20, in the County of Bergen, N. Jersey…29 ¾ Acres." In the same folder, see also "Copie de la lettre de M. Otto à M. le C.te de Vergennes. À New york le 14 Mars 1786," which also reports these details; Robbins, pp. 44-46.

314. See draft letter [from d'Angiviller's hand] on this subject, dated "27 7.bre 1786," to "M. Limosin negociant au havre," AN O[1] 2113A ("Pépinières d'Amérique. Correspondance Générale pour la Mission du S. Michaux, Botaniste du Roi, Année 1786"); see related material in same folder regarding botanical exchanges with Richard Cary, Oster, and d'Angiviller; see also ALS "Rich.d Cary" to "C.te Dangivillers," dated "Virginia Warwick County March 20th 1787," AN O[1] 2113A ("Pépinières d'Amérique. Correspondance relative à la Mission du S. Michaux, Botaniste du Roi. Année 1787").

315. See ALS "A. Michaux," dated "A Charleston le 2 X.bre 1786," to "Monsieur Le Comte," AN O[1] 2113A ("Pépinières d'Amérique. Correspondance Générale pour la Mission du S. Michaux, Botaniste du Roi, Année 1786"); and, ALS "A. Michaux," dated "Charleston le 2 Avril 1787," to "Monsieur," AN O[1] 2113A ("Pépinières d'Amérique. Correspondance relative à la Mission du S. Michaux, Botaniste du Roi. Année 1787"); Ewan, p. 22.

316. "…l'Etablissement du Roy," ALS "A. Michaux," dated "A New York le 2 Aoust 1787," to "Monsieur Le Comte," AN O[1] 2113A ("Pépinières d'Amérique. Correspondance relative à la Mission du S. Michaux, Botaniste du Roi. Année 1787").

lina garden for his own profit, being contrary to both "the commission conferred upon me" and "to service in the department."[317] At several points Michaux spoke of being sent by and of working in "the service of the King," where his usage suggests not personal service of an earlier age but of "le service" in the modern sense that anyone familiar with French bureaucracy today will immediately recognize.[318] At another point Michaux spoke of threatening to have Saunier "leave the service."[319] When François-André Michaux referred to the "nurseries at New York and at Charleston," he plainly thought of them as outposts attached to Rambouillet.[320] The Carolina garden was known by local Charlestonians as "the French Botanic Garden," and they may well have meant something more by the phrase than its being run by Frenchmen.[321] It is thus not too much of an exaggeration to suggest that these French gardens in the United States had the same status as French royal gardens in the Caribbean, South America, or the Indian Ocean colonies.

New World Trees for the King of France. With bases in New Jersey and South Carolina, Michaux was an intrepid traveler and peripatetic botanizer over the course of his eleven years in North America. He was "the first trained botanical explorer to travel extensively in North America."[322] Over the years his travels took him variously along the eastern seaboard in the New York-New Jersey-Connecticut region, from coasts to mountains in and around Georgia and the Carolinas, along the frontier areas of Tennessee, Kentucky and Illinois, down to Spanish Florida and the Bahamas, up to British Canada almost to Hudson Bay, and twice west to the Mississippi River. In all, he botanized in

317. "...étrangeres a la commission qui m'a été confiée,...au Service du Départment," ALS "A. Michaux," dated "A New York le 2 Aoust 1787," to "Monsieur Le Comte," AN O[1] 2113A ("Pépinières d'Amérique. Correspondance relative à la Mission du S. Michaux, Botaniste du Roi. Année 1787").

318. "...le Service du Roi," ALS "A. Michaux," dated "Charleston le 12 Novembre 1786," to "Monsieur Le Comte," AN O[1] 2113A ("Pépinières d'Amérique. Correspondance relative à la Mission du S. Michaux, Botaniste du Roi. Année 1787"). For other Michaux references to "le Service du Roy," see ALS "A. Michaux," dated "Abord du navire Adriana pres Charleston le 10 Mars 1787," addressed to "Monsieur Cuvilliers, Secretaire des Batimens du Roy A Versailles," AN O[1] 2113A ("Pépinières d'Amérique. Correspondance relative à la Mission du S. Michaux, Botaniste du Roi. Année 1787"), and ALS "A. Michaux," dated "Charleston le 5 Janvier 1789," to "Monsieur Le Comte," AN O[1] 2113A ("Pépinières d'Amérique. Année 1789").

319. "...quitter le service," ALS "A Michaux," dated "Charleston le 15 Avril 1791," to "Monsieur Le Comte," AN O[1] 2113A ("Pépinières d'Amérique" and in pencil: "1791").

320. "...les Pépinieres de New-York et de Charleston," ALS "Michaux, Rue de la Harpe N.o 133," to "Monsieur le Comte," dated "A Paris ce 25 fevrier 1791," AN O[1] 2113A ("Pépinières d'Amérique," marked in pencil "1791").

321. Savage and Savage, p. 150.

322. Savage and Savage, p. 4; the point is in part belied by the example of Pehr Kalm, who was trained by Linnaeus and who traveled to America in 1748-1751; the Bartram family of Philadelphia needs to be noted in this connection, especially William Bartram (1739-1823) who, although not formally trained, also did extensive traveling and botanical work in North America in 1773-1777 prior to Michaux that resulted in the publication of his influential *Travels through North and South Carolina, Georgia, East and West Florida, the Cherokee Country* (Philadelphia, 1791).

thirteen of the sixteen contemporary states of the United States. André Michaux was at home in rude circumstances. He traveled thousands of miles on foot, horseback, by ship and canoe, often accompanied by Indian guides, and he regularly stayed in villages and in the company of Native American hosts. He reported that at one point he spent sixty consecutive nights out under the stars.[323] With like facility he rubbed shoulders with the scientific and political elites of the new United States, including Benjamin Franklin, George Washington, Thomas Jefferson, William Bartram and his brother, John Bartram, Jr., Benjamin Rush, and other associates of the American Philosophical Society in Philadelphia. In 1793 Jefferson drew up a contract for Michaux to explore the lands west of the Mississippi that is clearly a precursor to the instructions Jefferson gave to Lewis and Clark ten years later. As a foreigner on an official mission, Michaux developed an intimate knowledge of the highways and byways of the new United States, and he left trenchant observations of the lives of contemporary Americans. As a botanist, Michaux was in his element:

> I confess that I could not desire a more agreable situation than to have an immense country to visit and immense harvests to make. For these, I would employ my whole life and all my time and fortune, even if otherwise provided with all possible assistance. I endure the resulting weariness with the greatest satisfaction, and the most minor outings bring me gleanings for which other botanists would have to go from one end of France to the other.[324]

But the goal of the Michaux mission was not scientific botany so much as it was the utilitarian project of shipping trees back to France for propagation and dissemination as sources of naval and building construction timber. The success of the enterprise depended on sending large numbers of seeds and seedlings and on having nurseries and beds ready in France to acclimatize and propagate specimens as they arrived from the New World. For the latter purpose the royal establishment at Rambouillet, founded in 1784, a year before Michaux's departure, was the natural terminus. For their part, beginning in 1786, Michaux and

323. See ms. originals, Michaux, "Journal de mon Voyage," APS-MS. C. S. Sargent (Michaux 1889) published Michaux's journal in its original French in 1889; Thwaites (Michaux, 1906) and others have translated portions of Michaux's travel journals; Dr. Eliane Norman is preparing a complete, modern translation of these journals.

324. "Je confesse que je ne puis désirer de situation plus heureuse que celle davoir un immense Pays a visiter, des recoltes immenses a faire et pour lesquelles j'employerois toute ma vie, mon temps, ma fortune et non obstant cela d'être fourni de tous les secours possibles. J'éprouve avec la plus grande satisfaction les fatigues qui en resultent et les moindres herborisations me procurent des Recoltes pour lesquelles les Botanistes iroient d'une extremité de la france a l'autre," ALS "A. Michaux," dated "New jork Ce 18 Aoust 1786," to "Monsieur Le Comte," AN O[1] 2113A ("Pépinières d'Amérique. Correspondance Générale pour la Mission du S. Michaux, Botaniste du Roi, Année 1786").

Saunier regularly expedited large-scale shipments to France from the American gardens. Of Michaux, J. Ewan writes that "the breadth and depth of his botanical collecting was without parallel in all the history of French exploration."[325] In the winter of 1785-86 Michaux sent 12 packages of 80 species of seeds and more than 4800 plants and trees; he sent a total of 60,000 plants and 90 cases of seed in the six years from 1785 to 1790, an average of about 10,000 plants and 15 cases a year.[326] He must have sent more in the totality of his stay in America. Michaux and Saunier sent sizable numbers of plants and trees with each consignment, usually in the hundreds. An early shipment, sent just weeks after Michaux's arrival, consisted of twelve cases including one of junipers "for the king's pleasure."[327] A Saunier shipment of 1788 consisted of over 2000 plants, including 1100 Northern Catalpa trees (*Acadia mignonia Catalpa*) and 600 sorrel trees (*Andromeda Calimtata*).[328] Another shipment from Saunier to d'Angiviller three months later contained 800 white cedar trees (*Cupressus thyoïdés*), 400 chestnut trees (*Castanea chinquapin*), 200 holm oaks ("Chêne verd"), 180 spruce trees ("sapinette a feuille d'if"), and 150 chestnut oaks and willow oaks ("Chêne-saule, chêne-chataignier").[329] Clearly, these were not shipments intended for "scientific" gardens or herbaria; instead, they approached the industrial scale needed to address the problem of naval timber. Still, Michaux's strictly botanical impulses were not to be denied, and along with selecting trees, he botanized at every opportunity and sent every new plant he found – to the extent that Nolin at one point tellingly complained, "We have trees to plant, not botanical gardens to create."[330]

Most of the tree specimens sent back by André Michaux ended up in Rambouillet via Abbé Nolin and under the immediate care of Abbé Tessier. Thirteen cases arrived at Rambouillet in 1786.[331] Nolin received twenty-one crates from Michaux in 1787.[332] Nolin was likewise in contact with Saunier.[333] As

325. Ewan, p. 22.

326. Michaux quoted in Boott, pp. 127-29; Houth (1931), p. 71.

327. Michaux letter of 26 January 1786, reprinted in Duprat, pp. 244-46, here p. 245.

328. ALS Saunier to "Monsieur," dated "New York le 18 fevrier 1788," AN O[1] 2113A ("Pépinières d'Amérique, Année 1788"). The species denominations are Saunier's.

329. ALS Saunier to "Monsieur Le Comte D'Angiviller" dated "New York le 15 may 1788," AN O[1] 2113A ("Pépinières d'Amérique, Année 1788").

330. "…ce n'est pas des jardins botaniques que nous avons à former, mais des arbres à planter," ALS "L'abbé Nolin," dated "Paris le 21 janvier 1787," to "Monsieur le Comte," AN O[1] 2113A ("Pépinières d'Amérique. Correspondance relative à la Mission du S. Michaux, Botaniste du Roi. Année 1787").

331. ALS "Brumard," dated "Lorient le 24 fevrier 1786," to "M. Le C.t Dangivilier," AN O[1] 2113A ("Pépinières d'Amérique. Correspondance Générale pour la Mission du S. Michaux, Botaniste du Roi, Année 1786").

332. ALS "A. Michaux," dated "A Charleston le 10 Mars 1787," to "Monsieur Le Comte;" AN O[1] 2113A ("Pépinières d'Amérique. Correspondance relative à la Mission du S. Michaux, Botaniste du Roi. Année 1787"); see also ALS copy, "Signé l'abbé Nolin," to "M. Michaux à Charlestown," dated "Paris 7 Juin 1787," AN O[1] 2112.

333. ALS copy "Signé l'Abbé Nolin," dated "Paris 10 Juin 1787," to "M. Saulnier," AN O[1] 2112.

one source later, somewhat imaginatively, wrote of Rambouillet: "In this park one can admire forests of larch, Lord Weymouth's pines, American red oak, these magnificently massive cypress trees from Louisiana, wax trees, Virginia cedars, American sugar maples, tulip trees, the liquidambar with leaves with balsam and other exotic specimens sown or planted by André Michaux, the voyager sent on the orders of Louis XVI."[334]

Institutional Contacts and Supports. As a royal botanical agent, one of Michaux's principal points of institutional contact was also with the Jardin du Roi, and every year from 1786 on Michaux and Saunier sent parcels to Thouin and Le Monnier there. Shipments from America also went to Monsieur, the king's brother, and Daubenton at the Jardin.[335] In 1792 André Thouin prepared a catalog of his correspondence which lists 32 letters received from Michaux through 1791; another compilation of Thouin's correspondence lists 69 missives from Michaux father and son through 1806.[336] The former catalog also lists 60 letters from Saunier in the period 1781 through 1789, including some from "Newierk" (Newark, New Jersey). In another notebook, Thouin carefully logged all of the consignments he received at the Jardin du Roi, including eighteen from Michaux in the period 1786 to 1797.[337] (There is a notable hiatus because Michaux apparently ceased shipments during the Terror in 1793-94.[338]) The first of Thouin's entries for Michaux was in 1786: "From M. André Michaux received thirteen species of seed from trees collected from vicinity of New-yorck [sic] in America."[339] For one cargo in 1789, Thouin recorded: "Received from M. André Michaux forty species of seed collected in Florida last year of which three quarters are new plants or trees."[340] The added parenthetical remark recording a 1792 lot was not exceptional: "Seeds collected in the year 1791 by Michaux botanical traveler in New jorck [sic]. 186 rare species. (A very interesting shipment.)"[341] Other entries recorded specimens col-

334. "Si la ferme expérimentale n'existe plus, on peut du moins admirer dans le parc ces hautes futaies de mélèzes, de pins du lord Weymouth, de chênes rouges d'Amérique; ces magnifiques massifs de cyprès de la Louisiane, de ciriers, de cèdres de Virginie, d'érables à sucre, de tulipiers, de liquidambar aux feuilles balsamiques, essences exotiques semées ou plantées par le voyageur Michaux sur l'ordre de Louis XVI," Heuzé, p. 24.

335. ALS "Saunier" to "Monsieur," dated "New York le 24 Octobre 1787," AN O[1] 2113A ("Pépinières d'Amérique. Correspondance relative à la Mission du S. Michaux, Botaniste du Roi. Année 1787").

336. BCMNHN, Ms. 314, "Etat de la Correspondance de M. A. Thouin [1792]," and BCMNHN, Ms. 2310, "Liste des Correspondans du Museum dont les lettres se trouvent dans les cartons de la correspond.ce" ; Duprat, Appendix I, reproduces some of this correspondence.

337. BCMNHN, Ms. 1327, "Journal des Envois de Plantes, Arbres et Graines faits au Jardin du Roy par ses Correspondants, avec ceux faits par le Jardinier dudit Jardin du Roy aux Correspondants d'icelui."

338. Thouin does record one shipment from Michaux in 1795; BCMNHN, Ms. 691, fol. 227v-28.

339. "De M. André Michaux recu 13 Especes de Graines d'arbres recoltées aux environs de New-yorck en Amerique…," BCMNHN, Ms. 1327.

340. "Reçu de M. André Michaux 40 Especes de Graines recoltées dans la floride pend.t l'année derniere dont les trois quarts sont de Plantes nouvelles ou d'arbres," BCMNHN, Ms. 1327.

341. "Michaux Voyageur Botaniste a New jorck. en Graines recoltées pend.t l'année 1791. 186 Especes rares. (Envoi très interessant)," BCMNHN, Ms. 1327.

lected in New York, the Carolinas, Canada, and elsewhere in temperate North America. Entries often noted the state of the lot on arrival, such as that in 1791 labeled "fresh and well conserved."[342] Yet another of Thouin's notebooks, this one recording the seasonal plantings that took place at the Jardin du Roi, also shows the broad impact that Michaux had on activities at the king's gardens in Paris. An entry for 1790, for example, recorded the planting of thirty-four species of seed, including *Magnolia grandiflora*, collected in Carolina by Michaux and delivered to the Jardin du Roi by his son, François-André.[343] The *Feuille du cultivateur*, a broadside for amateur botanists, stated matters succinctly in 1792 when it labeled Michaux "the French botanist sent to North America to study plants and to correspond with the Jardin du Roi in Paris."[344]

Important intermediaries also linked Michaux to the Jardin du Roi and other French scientific institutions. On at least one occasion, the count d'Angiviller himself transmitted one of Michaux's packages to Thouin, and another time, so, too, did the minister, the count de la Luzerne.[345] D'Angiviller and La Luzerne were influential academicians at the Académie Royale des Sciences, as were Buffon, Le Monnier, Thouin, and Tessier. In other words, although his primary institutional connections were with Rambouillet and the Jardin du Roi, Michaux also had ties that ramified through other institutions of the Colonial Machine. For example, Michaux was elected a member of the Société Royale d'Agriculture; in 1786, specimens from "M. André Michault de Newyorck" were presented to the Société, and in 1787 a memoir of his was read at a public meeting of the Société on "the nurseries maintained in America."[346] Michaux's place in the institutional firmament of French science was confirmed by his election in 1796 to the first class of the Institut de France in the section for rural economy and veterinary arts.

Michaux's activities in America reverberated throughout the Colonial Machine. In 1786 Thouin, for example, presented a paper before a public meeting of the Société Royale d'Agriculture "on the cultivation of foreign trees to replace the scarcity of wood."[347] Fourgeroux de Bondaroy of the

342. "de M. André Michaux reçu le 30 [janvier 1791] une petite Caisse remplie de 40 Esp. de Gr. d'Arbres, d'Arbustes et de Plantes presque toutes nouvelles récoltées dans l'Amerique tempe-rée. Elles etoient bien fraiches et bien conservées…," BCMNHN, Ms. 1327.

343. BCMNHN, Ms. 691 "Catalogues des graines semées au Muséum pendant les années 1790-1801," passim and here, fols. 126v-127v. For other plantings of Michaux consignments, including plants from Florida and the Bahamas and Carolina rice, see also fols. 133-37, 141-42v, 146v, 151-52, 160v-63v, 192-97, and 206v. For plants from Saunier, see fols 143v-144 in this same collection.

344. "renseignements envoyés par M. *Michaux*, botaniste français envoyé dans l'Amérique septentrionale, pour y étudier l'histoire des végétaux, & pour y correspondre avec le Jardin des Plantes de Paris…," *Feuille du cultivateur* #2 (1792), p. 138. This remark was apropos of a short article, *Sur l'Orobanche de Virginie*.

345. See BCMNHN, Ms. 1327, entries for 29 March 1781 and 28 March 1791.

346. "des pépinieres entretenues en amerique...," APS-MS B/D87, Group 26: Loose ms. sheets headed: "societé dagriculture – du jeudy – 9 mars 1786" and "Séance Publique de la Société d'agriculture du 19 juin 1787."

347. "memoire sur la culture des arbres Etrangers pour remplacer la rareté des bois," APS-MS B/D87, Group 26.

Académie des Sciences and the Société d'Agriculture wrote papers on "the use to be had of certain woods lately cultivated in France."[348] Saint Jean de Crevecœur's sending of wood samples from America to the Académie des Sciences is doubtless part of the same effort.[349] So successful was Michaux that in 1787 d'Angiviller projected a new Jardin du Roi in Perpignan to acclimatize plants from warmer places.[350] More significantly – as is evident in the document "Notes on shipments of seeds coming from China, Île de France, New York, and Charleston" – in his own mind d'Angiviller set Michaux and the American connection in the larger context of French botanical exchanges that extended to the Indian Ocean and to China.[351] D'Angiviller noted that Michaux in America and the French consul in Canton, C.-L.-J. de Guignes, "have the same purpose."[352] He made the connection even more explicitly in a 1785 letter to the director of the Compagnie des Indes:

> You probably know, Sir, that having the royal nurseries under my department and, more, being the next Intendant of the Jardin du Roi, I take a lively interest in the transplantation of trees and plants that can contribute to expanding our vegetable kingdom. I have therefore asked the garden on Île de France, as well as in India and especially a few of the principal agents of the Compagnie des Indes in Canton to send me various plants and seeds I have indicated for the King....His Majesty is particularly interested in these matters, and in effect the king takes such an interest in them that to this end he has just sent a botanist to North America to use every occasion to send new things here.[353]

That Michaux's outposts in America formed part of the larger system of French botanical gardens is confirmed by an unprepossessing 1794 list of

348. "Sur l'employ que l'on peut faire de certains Bois cultivés depuis peu en France," APS-MS, B/D87, group 24, and 580/D881, no. 32.

349. PA-PV-110, de Crevecœur, 22 June 1785; Robbins, p. 44.

350. ALS "Le M.al de Noailles," dated "Paris 25 février 1787," to "M. le C.te d'Angiviller" and ALS "Raymond de St. Sauveur," dated "A Perpignan le 20 avril 1787," to "M. le C.te Dangiviller;" AN O[1] 2113A ("Pépinières d'Amérique. Correspondance relative à la Mission du S. Michaux, Botaniste du Roi. Année 1787").

351. "Nottes sur Les Envois De Graines Venues de La Chine, De L'isle de france, de newyorck et de Charlestown," AN O[1] 2113A ("Pépinières d'Amérique. Correspondance Générale pour la Mission du S. Michaux, Botaniste du Roi, Année 1786"); also above, Part I.

352. ...comme ayant le meme objet," see ministerial note added to de Guines letter, "Quanton le 30 janvier 1786," AN O[1] 2113A ("Pépinières d'Amérique. Correspondance Générale pour la Mission du S. Michaux, Botaniste du Roi, Année 1786").

353. "Vous scavès, M.r probablement qu'ayant les pep. Du Roi dans mon departement et de plus etant intendant en survivance du jardin du roy je prens un interet vif a la transplantation des arbres et plantes qui peuvent contribuer a l'augmentation de notre regne vegetal. J'ai en consequence demandé tant au jardin de l'isle de france que dans l'inde et surtout à quelques employés principaux de la Comp.e des indes à Canton de me faire pour le Roi des envoys de diverses plantes et graines que je leur ai designées... comme des objets qui interessent specialemen S.M. et en effet le roy y prend un tel interet qu'il vient d'envoyer a cet effet un botaniste dans l'amerique septentrionale pour y profiter de toutes les occasions d'envoyer ici des choses nouvelles...," draft letter in d'Angiviller's hand to "M. de mars Directeur de la Comp.e des indes," dated "Vers.es le 25 X.bre 1785," AN O[1] 2112.

"National Gardens" found in the papers of André Thouin. Among other establishments and their staffs it lists:

> At Cayenne. In America. Spice nursery. Joseph Martin director....At Port au Prince on the Island of Saint Domingue, Nectoux gardener....At Lorient. Depot of the National Jardin [formerly the Jardin du Roi.]....At Brest. Garden of the Naval Hospital. Laurent, Gardener and Professor....At Île de France in Africa. Céré director....Trianon. Richard gardener....New York/Charlestown. Michaux director, Saunier gardener.[354]

André Michaux undoubtedly became involved in the establishment of the *Académie des Sciences et Beaux-Arts des Etats-Unis* in 1786. The chevalier de Quesnay proposed this French university (not an academy) for Richmond, Virginia, and the idea received wide support and subscriptions from various levels of French society and officialdom.[355] Given the oncoming French Revolution, nothing much developed from the proposal, but apparently a botanical garden was set up in Virginia associated with this Académie; Thouin from the Jardin du Roi sent a substantial shipment of seeds in 1789, noting in his logbook: "For the Chevalier Quesnay and for the botanical garden at Richmond in America, gave seeds of 180 species of trees and shrubs, medicinal and economic plants, flowers and ornamentals."[356]

In America, the French consuls posted in New York, Philadelphia, Richmond, and Charleston provided indispensable bureaucratic support for Michaux's endeavor. The Maréchal de Castries, minister of the navy in 1785, alerted the diplomatic corps in the United States to Michaux's coming, and once he arrived Michaux regularly interacted with local French administrators, if only to process his letters of exchange on the accounts of the Bâtiments du Roi.[357] Louis-Guillaume Otto, the count de Mosloy (1754-1817), variously the

354. "a Cayenne, En Amerique. Pepinière d'arbres a epiceries Joseph Martin Direct....au Port au Prince, Isle S.t Domingue, Nectoux Jard....a L'orient, Dépot du Jardin National....a Brest. Jardin de l'hopital de la Marine. Laurent Jardinier et Professeur....a L'Isle de france, *En Afrique*. Ceres direct....Trianon. Richard Jardinier....New York/Charlestown. Michaux, directeur, sanier Jardinier...," "Projets pour la conservation et L'Etablissement des Jardins...," dated "du mois Germinal an 2," BCMNHN, Ms. 308.

355. McClellan (1985), pp. 144, 289; and printed *Prospectus*, AN O[1] 1292, #29.

356. "a M. le C.lier Quesnay pour le jardin de botanique de Richemont en Amerique donné 180 Especes de Graines d'Arbres et d'Arbustes, de Plantes medicinales et Economiques de fleurs et de Plantes pittoresques," BCMNHN, Ms. 1327, shipments for 1789.

357. See ALS "Le M.al de Castries" to "M. Le.C.te d'Angivillers," dated "A V.lles le 7 Juillet 1785," AN O[1] 2113A ("Pépinières d'Amérique. Correspondance Générale pour la Mission du S. Michaux, Botaniste du Roi, Année 1785"). See also a series of related letters from St. Jean de Crevecœur from New York, 1784-1785, AN O[1] 597. Various of Michaux's bills of exchange are preserved in AN O[1] 2113A: see "Notte des Lettres d'échange tirées par M. Michaux" ("Pépinières d'Amérique. Correspondance relative à la Mission du S. Michaux, Botaniste du Roi. Année 1787"), "Traites Michaux en 1788 sur M. Dutartre" ("Pépinières d'Amérique, Année 1788"), "Trois traittes faites par M. Michaux botaniste deputé par le Roy" ("Pépinières d'Amérique," marked in pencil "1791").

"Royal Chargé d'Affaires to the United States of North America" in New York and the "Royal Chargé d'Affaires to the Continental Congress of the United States," was a particularly important intermediary.[358] Prior to Michaux's arrival and in response from requests from the abbé Nolin, Otto himself had previously sent trees and seeds back to Rambouillet.[359] Michaux also had contact with a M. de la Forest, French consul general in New York; one Oster, variously vice consul and consul in Virginia; Barbé-Marbois, after his posting in Saint Domingue, and an unnamed "Vice-Consul de Philadelphie"; one de Chateaufort, and then one M. Petry who succeeded one another at the French consulate in Charleston. In addition to providing general institutional support for Michaux, these diplomats were ordered to see that all botanical material from North America went through Michaux.[360]

In Saint Domingue the local administrators became aware of Michaux's existence and his enterprise, having been so informed by their superiors, a commission they acknowledged in a letter dated January 14, 1786 from Port-au-Prince:

> We received the dispatch with which you honored us, that the king has judged it appropriate to attach to his service as a botanist Mister Michaut [sic] and in this capacity to charge him to search out in foreign countries all the trees and other productions that can be naturalized in France. If Mr. Michaut were to travel to this colony, we would not fail, my Lord, to procure for him, as you instruct, every resource at our disposal that he might successfully fulfill His Majesty's goals, including financial help for his mission.[361]

358. "Chargé des affaires du Roi près des Etats Unis de l'Amérique septentrionale" in New York; "Chargé d'affaires du Roi près le Congrès des Etats-Unis de ce Continent…." For this styling see "Copie de la lettre de M. Otto à M. le C.te de Vergennes. À New york le 14 Mars 1786" and ALS "De Vergennes" to "M. le C.te d'Angiviller" dated "Versailles le 8 Août 1785, " both in AN O[1] 2113A ("Pépinières d'Amérique. Correspondance Générale pour la Mission du S. Michaux, Botaniste du Roi, Année 1786").

359. See ALS Nolin, "a paris le 20 juin 1785," and Otto letter, dated "A New York le 27 8bre 1785," AN O[1] 2113A ("Pépinières d'Amérique. Correspondance Générale pour la Mission du S. Michaux, Botaniste du Roi, Année 1785").

360. Draft letter in Angiviller's hand, dated "Vers.es le 29 7bre 1786," to "Mr. Oster," AN O[1] 2113A ("Pépinières d'Amérique. Correspondance Générale pour la Mission du S. Michaux, Botaniste du Roi, Année 1786").

361. "Nous avons recu la depêche done vous nous avés honnoré...que le roy a jugé a propos d'attacher à son service le Sr. Michaut comme botaniste et la chargé en cette qualité d'aller dans les pais etrangers chercher tout les arbustes et autres productions qu'il sera possible de naturaliser en France. Si le S.r Michaut par une suite du plan de ses voyages, vient dans cette colonie, nous ne manquerons pas, Monseigneur, de lui procurer comme vous nous le recommandés toutes les facilités qui dependront de nous pour qu'il puisse remplir avec succès les vues de Sa Majesté et dans le cas où aurait besoin de secours pécuniaires relativement à sa mission, M. de Marbois les lui fera fournir ainsi que vous luy autorisé et aura attention, Monseigneur de vous rendre compte," letter dated "Au Port-au-Prince 14 janvier 1786," signed by Coustard and de Marbois, AN-COL E 311.

In terms of the material conditions underlying his efforts – for getting his shipments to France until they were suppressed in 1788 – Michaux depended on vessels of the state-chartered shipping and mail company, Régie des Paquebots, which plied regularly scheduled routes to and from the Americas. This arrangement naturally entailed complicated bureaucratic procedures whereby the Bâtiments du Roi reimbursed the Régie des Paquebots for this service.[362] After 1788, Michaux had to pay private captains to transport his shipments to France. Initially, Michaux's shipments went through the Atlantic port of Lorient, the headquarters of the French Compagnie des Indes. This organization vigorously policed the port and everything passing through it. Significant high-level negotiations were required before botanical materials were allowed to pass freely through Lorient.[363] Even then, the trip overland from Lorient to Paris and Rambouillet proved disastrous for the transport of botanical specimens, and soon the route shifted to the Channel port of Le Havre, with transport to Paris through Rouen via the Seine. For Nolin, Le Havre was "decidedly to be preferred, given its proximity to Paris and the ease of having shipments brought there."[364] Still, there were problems with carters in Le Havre who would only take cargo as far as Rouen.[365]

Partridges for Rambouillet. As much as the primary goal of the Michaux mission was to provide seeds and seedlings for naval and construction timber, or even, secondarily, to extend botanical science, other aspects of Michaux's activities tell a distinctively Old Regime story, notably the desire to import birds for the royal aviaries and stock at Rambouillet.[366] Michaux seems fixated on getting "Summer docks" (his English for "summer ducks"), turkeys, and "dwarf deer" to Rambouillet.[367] Most all of the ducks died on the voyages to France, but some ultimately made it. Michaux's biggest challenge, however, involved partridges which were delicacies at royal and gourmand tables. Despite strict rules restricting the hunting or trapping of partridges, they had

362. See ALS "Le M.al de Castries" to "M. Le C.te d'Angiviller," dated "Versailles Le 10 Aoust 1785," AN O[1] 2113A ("Pépinières d'Amérique. Correspondance Générale pour la Mission du S. Michaux, Botaniste du Roi, Année 1786"); McClellan (1992), p. 81.

363. See correspondence between and among de Castries, d'Angiviller, Thévenard (Commandant de la Marine à Lorient), and Controlleur Général Calonne in contemporary folder marked "Pieces relatives au Premier envoy de M. Michaux," AN O[1] 2113A ("Pépinières d'Amérique, 1786").

364. "…le havre, ce dernier doit être préféré à tous égards, attendu sa proximité de Paris et la facilité d'en faire venir les envois," annotation to list of seeds on document, "Copie du 3 Juillet 1789," AN O[1] 2113A ("Pépinières d'Amérique. Année 1789").

365. ALS "Limozin" to "Monsieur Le Comte d'Angiville," dated "Havre le 16 février 1788," AN O[1] 2113A ("Pépinieres. Commission du S. Michaux Pour l'Amerique").

366. See reference to "la faisanderie du Roy a Rambouillet" in ALS "Labbe Nolin" to "Monsieur Le Comte," dated "A paris Le 23 janvier [1787]," AN O[1] 2113A ("Pépinières d'Amérique. Correspondance relative à la Mission du S. Michaux, Botaniste du Roi. Année 1787").

367. ALS "A. Michaux," dated "Charleston le 23 fevrier 1790," to "Monsieur Le Comte," AN O[1] 2113A ("Pépinières d'Amérique. Mission Michaux. 1790").

become scarce.[368] In Grenoble, for example, there was a parlementary proscription against taking partridges out of the province.[369] The idea surfaced to stock Rambouillet with American partridges, and in 1786 Michaux sent eighteen birds ("excellent for eating") through Nantes.[370] "Destined for the King's pleasure," these were well received at Rambouillet.[371] Another shipment of sixty birds followed in 1787 through Le Havre.[372] These birds were carried from Le Havre and Rouen to Paris by a 'birdman' in cages.[373] That Michaux's mission involved its own Papagano making his way to Rambouillet with prizes for the king's table speaks volumes about the circumstances of the day.

The French Revolution and Michaux's Return to France. Although Michaux was a confirmed patriot who enthusiastically endorsed the Republic when it came, the French Revolution complicated his life, given his dependence on the increasingly fragile and unreliable infrastructures of the Colonial Machine after 1789. Michaux had ever greater difficulty in cashing his letters of credit, and d'Angiviller exhorted him to reduce expenses.[374] Already by October 1790 Michaux was writing to d'Angiviller from Charleston: "To respond to your intentions, Monsieur le Comte, I have cut back on horticultural and shipping expenses for the setup in New York. I have also made cutbacks here, and I have stopped traveling."[375] Michaux survived this period in the States largely

368. See "Ord.ce defendant de tuer des perdrix du 15 mars au 15 juillet," 28 January 1721, AN-COL F[3] 10, fol. 124; and "Ordonnance qui defend de chasser des perdrix depuis le 15 may jusqu'au 15 juillet," 23 March 1727, AN-COL F[3] 11, fol. 31.

369. ALS "De Berulle," dated "Grenoble ce 6 9bre 1786," to "M. le Comte d'Angiviller," AN O[1] 2113A ("Pépinières d'Amérique. Correspondance Générale pour la Mission du S. Michaux, Botaniste du Roi, Année 1786").

370. "...tres bonnes a manger," ALS "A. Michaux," dated "A New-york le...[sic] X.bre 1785," to "Monsieur le Comte," AN O[1] 2113A ("Pépinières d'Amérique, 1786"). On partridges and shipments through Nantes, in same folder see also ALS "Berger" to "Monsieur," dated "nantes 4 fevrier 1786," and ALS "Ruellan, au havre le 23 May 1786." See also letter copy "Signé l'Abbé Nolin," dated "Paris 10 Juin 1787," to "M. Saulnier," AN O[1] 2112.

371. "...destinées aux Plaisirs du Roy," letter, dated "Paris le 17 Mars 1786," from "Les fermiers Généraux de Messagerie" to "M. Dangevilliers," AN O[1] 2113A ("Pépinières d'Amérique. Correspondance Générale pour la Mission du S. Michaux, Botaniste du Roi, Année 1786").

372. ALS "[Limosin?]," dated "Du havre le 4 f.er 1787," to "Monsieur Le Comte," AN O[1] 2113A ("Pépinières d'Amérique. Correspondance relative à la Mission du S. Michaux, Botaniste du Roi. Année 1787").

373. On this point, see letters from one Limosin, the Le Havre merchant acting as d'Angiviller's agent; ALS "Limosin," dated "Havre le 29 Janvier 1787," to "Monsieur Le Comte d'Angivillers"; ALS "Limozin," dated "Havre le 30 Janvier 1787," to "Monsieur Le Comte d'Angivillers"; and ALS "Limozin," dated in Minister's hand "au Havre 2. f.er 1787," all in AN O[1] 2113A ("Pépinières d'Amérique. Correspondance relative à la Mission du S. Michaux, Botaniste du Roi. Année 1787").

374. These are recurring themes in AN O[1] 2113A ("Pépinières d'Amérique. Mission Michaux. 1790") and ("Pépinières d'Amérique," marked in pencil "1791").

375. "Pour repondre a vos intentions Monsieur Le Comte j ay retranché les dépenses dans l'Etablissement de New York relativement a la culture et relativement aux Envoys. Jay fait aussi des retranchemens ici, et jay cessé les Voyages," ALS "A. Michaux," dated "Charleston le 4 Octobre 1790," to "Monsieur Le Comte," AN O[1] 2113A ("Pépinières d'Amérique," marked in pencil "1791").

though help from American friends. Finally, with the harshest moments of the Revolution passed, he returned to France in 1796, but not without the further trial of being shipwrecked off the coast of Holland on his return. He himself nearly drowned, and only with difficulty was he able to save his botanical cargo and all he brought back from his decade-long stay in the United States.

While in America, Michaux sent Thouin and the Jardin du Roi a twenty-page manuscript, "Trees of North America," listing 128 different species of American trees and mentioning the size to which each grows, the proper soil, best uses of each, and so on.[376] Other Michaux manuscripts likewise denominate American trees of potential utility for France. One enduring result of Michaux's American adventure was the publication of his study of – what else? – American oaks. His *Histoire des chènes de l'Amérique septentrionale* (1801) was a work treating twenty different species and five varieties that was published in an edition with thirty-six copper plates. Illustrated by P.-J. and H.-J. Redouté, this volume was the most substantial and important treatise on American silviculture up to that date. Michaux's orientation toward matters arboreal can likewise be seen in the work later published by his son, François-André, the full title of which we can translate as *History of Forest Trees of North America, Considered Principally with Regard to Their Commercial Application as well as for the Advantages that They Can Offer to European Governments and to Persons Wishing to Form Large Plantations* (1810-1813). The central element of this work ended a family mission concerned with timber and the timber famine in contemporary France.[377]

Back in France, André Michaux became a member of the first class of the Institut de France (the body that in 1795 replaced the former Académie Royale des Sciences that had been closed in 1793). He was a respected colleague in the scientific establishment of the day, but he was unable to find a permanent scientific or academic appointment, and so he signed on for another botanical adventure, this time to Madagascar, where he died unexpectedly in 1802 after a brief illness. His death was a loss to science, of course; but by that time the eighteenth century's Colonial Machine and the first French colonial empire were receding into the past.

Agronomy and Other Applied Projects

The Colonial Machine was a giant research engine tackling the practical problems of sailing the ocean, establishing and maintaining colonies overseas, and stimulating commerce and the economy. This dimension of the Colonial

376. "Arbres de l'Amérique Septentrionale"; see this and various other Michaux manuscripts, BCMNHN, Ms. 357.

377. François-André Michaux (1810), esp. vol. 1, p. 7, where the younger Michaux references his father's work.

Machine has already shown itself in the wide-ranging set of special research projects previously encountered, such as the navy's efforts to preserve fresh water at sea or initiatives centering around the Société Royale de Médecine to combat tetanus. Similar R&D-oriented projects also arose concerning plant husbandry and agrarian development. Like their counterparts, none of these agronomy projects was large, nor were many very successful. But they were all of a piece and rounded out larger programmatic initiatives of the Colonial Machine – in the present, case the great programs of sugar and coffee, colonial gardens and the spice trade, and addressing the problem of naval timber. The projects enumerated in this section reinforced the structures and functioning of the Colonial Machine, and they show the lengths to which the Colonial Machine could go in pursuit of its aims.

In agronomy, private planters played a larger part in the study of cultivars and in improving agricultural techniques, particularly of coffee, sugar, and cotton, than we have encountered elsewhere so far.[378] This finding is not unexpected, but our concern here is with official, institutionalized efforts of the Colonial Machine to promote cultivars and their products.

Sugar sweet. Not surprisingly, as a center of science and for approving new technologies, the Académie des Sciences at the Louvre in Paris heard many reports and considered many projects linked to sugar and its production. Sugar mills, sugar refining, running a sugar plantation, making rum and schnapps, and the chemistry of sugar were subjects of a stream of reports and projects. One example would be the improved sugar mill suggested by one Lemonnier from Port Margot on Saint Domingue.[379] The mill proposed by a certain Cazault in 1780 would be another. The Academy not only approved this mill, but it blessed Cazault's manual on cultivating sugarcane and running a sugar plantation.[380] (Incidentally, Cazault's manuscript contains twenty-three "rules" for how to organize slave labor in the operation of a sugar manufactory.)

The chemistry of sugar was the subject of considerable scientific attention.[381] Jacques-François Dutrône La Couture (1749-1814) became the recog-

378. Such studies, printed and manuscript, abound. See, for example, "Mémoire sur la cueillette et la manière de préparer le cacao de Cayenne" [1784] of Desrivierre-Gers; BM/Bordeaux, Ms. 828, CII, doc. 3. They show up in the *Mémoires d'agriculture, d'économie rurale et domestique, publiés par la Société royale d'agriculture de Paris*, published in the later 1780s, as well as in the archives of Chambres d'Agriculture founded in the Antilles in the second half of the century, as can be found in AN-COL F^3 124-126.

379. PA-PV-110, "Lemonnier (au Port Margot Isle S.t Domingue)," 7 July 1779 ("Projet d'un nouveau Moulin à presser les Cannes à Sucre").

380. PA-PV-110, Cazauls, 15 March 1780 ("Sur la Culture des cannes a Sucre et la manutention d'une Sucrerie"); PA-PV 99 (1780) fols. 46v-49 ("Du mercredi 15. Mars 1780"). See also in this connection, APS-MS, Fougeroux de Bondaroy Papers, B F8245, "Séances de l'Académie Royale des Sciences, 1786-1789," "du mercredi 11 février 1789...maniere annoncée par M. le C.e du puget pour Empêcher la mutilation des Negres par les moulins a Sucre – examiner l'art de raffiner le sucre de Mr. duhamel –."

381. McClellan (1992), pp. 67-68 explores the details.

nized expert on sugar. His *Précis sur la canne et sur les moyens d'en extraire le sel essentiel* appeared in 1790 with kudos from the Academy of Sciences and the Société Royale de Médecine.[382] Dutrône's 400 pages in octovo saw subsequent editions in 1791 and 1801. He provides a detailed chemical analysis as well as a practical manual for sugar and indigo production. In 1788 the minister asked the Academy about Dutrône's work. The Academy welcomed the opportunity "to pronounce upon and enlighten the Minister on this important production of the colonies."[383] It urged the government not only to publish and disseminate Dutrône, but to send Dutrône himself to the colonies to promote improvements in sugar manufacture.

Indigo Blue and Cochineal Red. We in the twenty-first century tend to forget how drab the world was in the eighteenth century. In that period dyes and dyeing agents attracted great interest and were drivers of development for the Colonial Machine. Colorfast indigo, the dye that makes our bluejeans blue, was a principal commercial crop and a focus of research concern, more so than for sugar because sugar production was already well understood and established.

We have touched on efforts to introduce indigo production onto Île de France in the 1730s and 1740s. In the 1750s while in Senegal, Michel Adanson undertook a not insubstantial series of tests of African and American indigo dyes and processing techniques for the Compagnie des Indes.[384] The Compagnie was very interested in and supported this work.[385] Bernard de Jussieu sent Jean Hellot's 1750 book on dyeing to Adanson, and Adanson conducted experiments and sent back sample swatches for further evaluation in France.[386] In his manuscript "Essais de Teintures; ou Expériences faites au Senegal, en 1751 et 1752 sur différentes espèces d'Indigo naturelles au Sénégal; sur celle de l'Amerique, et sur plusieurs autres Plantes qui peuvent servir dans les tein-

382. See PA-PV-110, Dutrône, 31 May 1788 ("Ouvrage Sur la Canne et sur les Moyens d'en extraire le Sucre") and related entries; SRdM, Ms. 11bis, fol. 24r ("Séance du Vendredi 11 7bre 1789") and SRdM, Archives, 170 dr. 24, # 1.

383. "de prononcer et d'éclairer le Ministre sur cette importante production des Colonies…," PA-PV 107 (1788), fols. 136-140v ("Samedy 31 Mai. 1788"), here, fol. 136.

384. ". je souhaiterois avoir réüssi et être utile en quelque chose à la compagnie, câr l'on prétend ici que si cette teinture se trouvoit aussi belle que celle de l'Amerique, elle en auroit fait un objet de commerce. Je vous en envoye un petit échantillons; vous jugerez de sa beauté en la comparant à celle des Isles," ALS, Adanson to [Bernard] de Jussieu, dated "Au Senégal, ce 10 fevrier 1751," PA-DB (Adanson); in this connection in the same file see also ALS, Adanson to "Messieurs," dated "Au Senégal ce 24 Juin 1751" and *Collection Académique, Partie Française* vol. 12 (1786), p. 218.

385. See ALS "Les Syndics et Directeurs de la Comp. des Indes" to "M. Adanson Au Senegal," dated "A Paris le 3 Avril 1751," HI, AD 181, and ALS "Les Syndics et Directeurs de la Comp. des Indes" to "M. Adanson Au Sénegal," dated "A Paris le 3 may 1752," HI, AD 182.

386. ALS, Adanson to "Messieurs," dated "au Senégal le 20 Fevrier 1752," PA-DB (Adanson); see also Hellot (1750).

tures," Adanson was attentive to "the method used by blacks from Guinea for extracting indigo."[387]

As it was for sugar, the Académie des Sciences was the clearinghouse for information and innovation in indigo processing. In 1760, for example, one M. Fage, a colonist from Saint Domingue and a member of the Academy at Mayence, came to the Academy in Paris and read a paper on indigo manufacture. The rapporteurs for the occasion, Duhamel and Jussieu, noting that one of the goals of the Academy had always been "the perfection of manufactures and the progress of the arts," approved Fage's memoir, and he was duly elected a correspondent of the Academy.[388] Cossigny *fils* continued the tradition of attention paid to indigo on Île de France, submitting to the Academy in 1781 his "Essay on the Manufacture of Indigo," which was likewise approved for publication by the institution.[389] Sonnerat commented positively on Cossigny's indigo work: "M. Cossigny, one of the most zealous cultivators in this region, has made very interesting discoveries in this matter that one can read in his treatise on indigo manufacture that the government has printed at Île de France."[390]

A revealing little dustup erupted between Cossigny and another colonist on Île de France, a certain Champeau, over their views on indigo production, and the Academy was called upon to adjudicate in 1788. Chemists Fourcroy and Berthollet performed independent analyses for the Academy. They dissolved samples in acid and then measured coloration via another acid. In explaining their procedures, Fourcroy and Berthollet used the brand-new language of Lavoisierian chemistry, though they translated some phlogiston terms. As a result of these tests and relying on the power of number, the authors confidently concluded that the dyeing power of M. Champeau's indigo excelled over indigo from Saint Domingue in a ratio of 16 to 23.[391] The Academy in

387. "façon dont les Nègres de Guinée tirent la teinture de l'Indigo," loose sheet, HI, AD 254. See further remarks on this set of tests by Adanson, ALS Adanson, "Copie de la lettre écrite à la Comp.e Du Senégal le 20 Février 1752," in PA-DB (Adanson), where he speaks of having "...fait l'épreuve de plusieurs espèces d'Indigo, qui croissent naturellement et qui sont for communs aux environs du Senégal....il ne s'en trouvé qu'une qui m'ait donné le véritable Indigo, et c'est précisément celle que les naturels du pays emploient pour la teinture de leurs pagnes....J'en envoie par cette même occasion une pareille montre à M.rs de Jussieu....La comparaison que vous feréz de cet Indigo avec celui des Isles vous décidera mieux de la qualité que toutes les notions que je pourrois vous en donner."

388. "la perfection des manufactures et les progrès des arts," PA-PS (29 March 1760); on this episode, see also PA- PV 79 (1760), fols. 201v ("Mercredy 26. Mars 1760), 208-209v ("Samedi 29. Mars 1760"), 257 ("Mercredi 30 Avril 1760"). Fage was later stricken from the list of correspondents.

389. PA-PV-110, Cossigny, 28 February 1781 ("Ouvrage intitulé: *Essay sur la fabrique de L'indigo*").

390. "M. *de Cossigni*, l'un des plus zèlés cultivateurs de cette contrée, a fait des découvertes très intéressantes sur cette matière: elle sont consignées dans son traité de l'Indigoterie, que le Gouvernement a fait imprimer à l'île de France," Sonnerat (1782), vol. 2, p. 81.

391. "les proprietés colorantes de ces deux indigos sont dans le rapport de ces nombres," PA-PV 107 (1788), fols. 111-14, here fol. 112 ("Vendredy 9. Mai 1788"); in same volume see related entries, fols. 75 ("Samedy 12 Avril 1788") and 132-32v ("Mercredy 28. Mai. 1788.").

Paris sent this information back to Île de France for publication and dissemination by local authorities.

Also in 1788 the Academy received a paper on indigo from Dutrône La Couture. Fourcroy and Berthollet again did the honors, encouraging Dutrône to continue his observations.[392] And, from Cayenne in 1789, Jean-Baptiste Leblond submitted his substantial "Essay on the Art of Indigo Dyeing," which was given a favorable Academy report.[393] Leblond styled himself in this manuscript as "physician-naturalist pensioned by the king, correspondent of the Royal Academy of Sciences, the Royal Society of Agriculture and the Royal Society of Medicine of Paris, etc., commissioned by his Majesty to find cinchona in French Guiana."[394] For its part in the Colonial Machine, the royal Gobelins dye works had already pronounced Cayenne indigo "excellent" in 1786.[395]

Trees and tree bark were other sought-out sources of tincture. The benzoin tree was traded for dyeing in the Indian Ocean region, and it was protected on Île Bourbon under penalty of heavy fine.[396] The mustard "tree" from Guiana attracted brief attention as a promising dyeing agent around 1787. That year, the ministry brought samples in from Guiana for tests by the Academy, but the tests proved negative, and that was that.[397] Out of North America in the 1730s came a report of a species of ash that dyed a solid blue.[398] And in the 1720s in Louisiana there were systematic collections of herbs and tinctorial barks.[399]

The story of dyestuffs and the Colonial Machine culminates with the cochineal bug and the successful implementation of cochineal production in Saint Domingue in the 1770s. To recall, Thiery de Menonville's surreptitious trip to

392. PA-PV-110, Dutrône La Couture, 14 June 1788 ("Observation sur deux sortes de fécules que donne la plante nommée *Indigo fera Tinetoria*"), the rapporteurs wishing the Academy to "Engager L'auteur à continuer ses intéressantes Observations," and PA-PV 107 (1788), fols. 168v-169 ("Samedy 14 Juin 1788").

393. PA-PV-110, Le Blond, 23 December 1789 ("Mém.re Intitulé: Essai sur l'art de L'Indigotier"). For further Leblond contact with the Academy, see APS Manuscripts, Fougeroux de Bondaroy Papers, B F8245, "Séances de l'Académie Royale des Sciences, 1786-1789," "Séance du Samedi 15 Novembre [1788]...."

394. "médécin naturaliste pensionnaire du roy, correspondant de L'accademie royale des sçiences, de la société royale D'agriculture et de celle de médécine de paris &c. Commissionné par sa Majesté pour faire la recherche Du Quinquina dans la Guyane française," ms. "Essais sur L'Art de Lindigotier pour servir à un ouvrage plus Etendû," signed and dated, fol. 213, "A Cayenne ce 1er Janvier 1789 Le Blond," AN-MAR G 101, dossier 4, histoire naturelle, fols. 183-213.

395. Ms. Code de Cayenne, vol. 7, p. 297 in AN-MAR G 237.

396. "Ordonnance...portant défense aux Cordonniers & Tanneurs de dépouiller les arbres de Benjoin de leurs écorces. Du quinze Janvier 1770," Delaleu, vol. 2, p. 91.

397. The Ministry under de Castries and the Academy worked closely on this matter for a period; see PA-PV-110, Castries, 4 July 1787, PA 'plumitifs' 1787 (4 July, 10 March, 23 June, and 8 August), and PA 'plumitifs' 1790 (3 February); see also ms. Code de Cayenne vol. 7, p. 437, in AN-MAR G 237, for the Academy's report sent to Cayenne.

398. AN-COL B^{64} (1736), fol. 446 (15 May 1736).

399. AN-COL C^{13A} 6 (1720-1722), fol. 398 (24 January 1723).

Mexico to procure Spanish cochineal in 1777 not only introduced cultivation in Saint Domingue, it was the trigger for the creation of the position of royal botanist and the foundation of the Jardin Royal in Port-au-Prince. Taken up by the Cercle des Philadelphes and enlightened colonists, the practice of cochineal husbandry spread in Saint Domingue, and in this connection the Cercle des Philadelphes gets significant credit for also publishing Thiery de Menonville's *Traité de la Culture du Nopal et ... de la Cochenille* in 1787.[400] That Thiery's cochineal work also passed through the Academy of Sciences made it that much more a product of the Colonial Machine.[401] But Thiery de Menonville and French cochineal did not appear out of nowhere. Earlier, in 1710, Feuillée had reported from South America on cochineal and the desirability of alternatives, and Réaumur suggested to the régent that the French start their own cochineal production.[402] In Louisiana, in the 1720s, searches were conducted to find *fina* cochineal.[403] And, to a limited extent at least, some theoretical questions were raised regarding the action of dyes.[404] As part of the solidifying links between the regions, by 1784 cochineal husbandry had spread eastward from the Caribbean to Île de France where Cossigny *fils* received a sample and began to grow nopal in his garden at Palma on the island "for the useful arts and to contribute to the advantage of the state and our colonies."[405]

Silkworms and Mulberry Trees. A national effort was made by the government to spread mulberry trees and grow silkworms in France.[406] Up to the mid-eighteenth century, France still depended on imports of raw silk, silk cloth, and thread for further processing and manufacture by its own silk industry around Lyon. Mulberry trees grew in France, and the government had incentives to promote the practice. Controller General Philibert Orry led one of these initiatives in the 1740s, recognizing "the importance of establishing silk manufacture." He thought that "several million" would be saved for the nation, adding that raising silkworms did not otherwise interfere with agriculture and was especially suitable work for the peasant women and young children of the provinces.[407] Extending himself through the system of intendants,

400. See above, this part (Colonial Gardens – West).

401. PA-PV-110, Thiery, 19 December 1787.

402. So says Charles Arthaud in his éloge of Thiery de Menonville, SRdM, Archives, 136 dr. 1, #15; on the Feuillée report, see PA-PV 29bis (1710), fols. 466-70 ("Le Mecredi 17 Décemb. 1710").

403. AN-COL C^{13A} 11 (1728-1729) fol. 51 (31 July 1728).

404. Hellot (1750); PA-PV-110, Mazeas, 25 Janvier 1758 ("Deux Mémoires sur la cause physique de l'adhérence de la couleur rouge, tant aux toilles peintes qu'aux Echevaux de coton qu'on apporte des côtes de Malabar").

405. "de leur être utile [les arts] et de contribuër à l'avantage de l'état et de nos colonies," ALS Cossigny to "M. le C.te d'Angivillers," dated "A L'Ile de France le 31 Mars 1784," AN O^1 1292. Cossigny also makes some theoretical suggestions about dyeing.

406. "Memoire concernant le Commerce de la Soye, La plantation et la culture des Meuriers, dans les Provinces de France," AN O^1 1293; see also Pritchard (2004), pp. 126-27 provides larger background.

407. "l'importance de l'etablissement de la fabrique des soyes," ALS Orry to Barentin, dated "A Versailles le 20 fevrier 1741," AD La Rochelle, Série C 195.

Orry felt that the first need was for more trees. He ordered that "the King wishes that mulberry nurseries be established in all the provinces of the Kingdom," and he mobilized intendants and existing royal gardens and nurseries in the provinces to plant more mulberry trees.[408] He saw that seeds were distributed and mulberry trees given away to anyone interested. Orry's annoyance at one Intendant who failed in his duties in this regard ("It is regrettable to see how little attention has been paid to this matter. The implantation of mulberry trees is very worthwhile for your province, and you should have had this attended to...") suggests how important extending mulberry horticulture was to him.[409]

There was a colonial connection, of course, one part of which played out over many years in Louisiana. Already in the 1680s and 1690s, efforts unfolded to establish silk production in Louisiana. Mulberry trees and silkworms were sent there, and a Mme de la Calle, either in Louisiana or back in France, was involved in the project with the minister in 1687.[410] Silk manufactured in Louisiana was freed from import duties, but this was doubtless a theoretical incentive. The ministry intervened again in 1713, discovering this time that mulberry trees were plentiful, but silkworms lacking.[411] Further efforts to engage colonists and train children in the orphanage to raise silkworms ran up against this shortage of cocoons.[412]

Things took another turn with Orry's initiative of 1742. Instead of silkworms coming from France, mulberry trees were sent from Louisiana to La Rochelle. Here the Colonial Machine clearly came into play. On the orders of Maurepas, the local administrator in New Orleans duly and carefully bundled up three cases of mulberry trees and seed and saw them aboard the royal vessel *Le Maurepas* (sic). Six more cases stood by for further shipment.[413] The plants arrived safely from Louisiana, but to avoid the heat of summer and mortality from further transportation, Orry ordered that they be planted and watered tem-

408. "le Roy qui veut qu'il soit fait dans toutes les Provinces du Royaume des establissements de pepinieres de muriers," ALS Orry to Barentin, dated "A Bercy le 24 Juillet 1741," AD, La Rochelle, Série C 195.

409. "Il est facheux de voir le peu d'attention que l'on a eu a cette partie. L'etablissement des muriers est trés interessant pour votre Province et vous auriés dû y faire veiller...," ALS Orry to Barentin, dated "A Versailles le 29 Janvier 1742,"AD, La Rochelle, Série C 195.

410. See ministerial correspondence relating to this episode, AN-COL B^{13} (1687), fol. 7 (25 February 1687), fol. 11v 19 March 1687), fol. 14v (30 March 1687), fol. 56 (22 August 1687); B^{14} (1688), fol. 48 (22 September 1688), fol. 486v (13 May 1693).

411. "Mémoire pour répondre aux instructions envoyées par M. le Comte de Pontchartrain au Sr. Duclos, Commissaire de la Marine, Ordonnateur à la Louisiane…,"AN-COL F^3 24, fol. 61 (15 July 1713): "…Les muriers sont en grande abondance. Il serait très avantageux d'y introduire les vers à soie…."

412. See ministerial correspondence related to this initiative, AN-COL C^{13A} 1 (1678-1706) [1718] p. 47; C^{13A} 11 (1728-1729), fols. 111, 305 (30 January 1729); C^{13A} 13 (1732) fol. 43 (19 January 1732); C^{13A} 14 (1732) fol. 160 (19 February 1732).

413. ALS "Salmon," dated "A la N.lle Orleans. Le 5 Avril 1742," to "Monseigneur,"AN-COL C^{13A} 27 (1742) fol. 113.

porarily in a shady spot at the *pépinière* at La Rochelle.[414] He followed up in the fall, inquiring about "the present state of the seedbeds and nurseries established in your Department and whether you have transplanted the plants from Louisiana."[415] What happened from there is unknown to us, but there may be mulberry trees in southwest France today descended from this American stock.

In the 1740s, not coincidentally, silkworms and mulberry trees began to be imported from China and India to the Mascarenes by the Compagnie des Indes.[416] But the Compagnie des Indes lacked the expertise to make silk from silkworms, and initially had to ask informants how the Chinese did it![417] In 1750 Orry, again, and the Compagnie des Indes installed silk workers on Île de France.[418] Dame Arnald and her two daughters from Romans in France "engaged to serve the Compagnie des Indes for six years on the Île de France and Île Bourbon and to give all their care to raising silkworms, to unwinding cocoons and related operations."[419] The Compagnie also charged Lanux with studying and supervising silkworm husbandry on Île Bourbon, and Lanux was in contact with Sauvages in Paris over the matter.[420] The Compagnie encouraged private colonists to raise silkworms, and at one point had special metal boxes built to import silkworms and cocoons from China.[421] A shortage of laborers was a principal problem, but so were pests and predators. Silkworms had to be raised indoors, but even so "the quantity of insects, like mice [sic], ants, and others abounding on this island devour these worms."[422] This difficulty points to other, ecological issues to be treated separately; but to conclude with silk, we should indicate the involvement of the Academy of Sciences,

414. ALS Orry to Barentin, dated "A Versailles Le 20 May 1743," AD La Rochelle, Série C 195. One imagines he meant the royal naval garden at Rochefort, not La Rochelle.

415. "Je vous prie de m'instruire de l'etat actuel des semis et des Pepinieres etablies dans votre Departement et si vous avés fait transplanter les plantes de la Loüisiane," ALS Orry to Barentin, dated "A fontainebleau le 7 Octobre 1743," AD, La Rochelle, Série C 195.

416. "Extrait du registre général des déliberations de la Compagnie des Indes. Du 26 juin 1742," in Lougnon (1940), p. 37.

417. "Extrait du registre général des déliberations de la Compagnie des Indes. Du 26 juin 1742... Ordres pour le Conseil de direction de Canton," in Lougnon (1940), p. 67.

418. Compagnie letter to "M.rs du Conseil Supérieur à l'isle de Bourbon," dated "A Paris le 17 mars 1750," in Lougnon (1949), pp. 235-36.

419. "... dame Arnal et de ses deux filles, de Romans, diocèse de Valence, lesquelles s'engagent de servir la Compagnie des Indes aux îles de France et de Bourbon pendant six années et de donner tous leurs soins à l'éducation des vers à soie, au tirage des cocons et à toutes les opérations qui seront relatives, suivant les instructions," SHML, 1P 74-76, (February, 1749); also Legrand et al., p. v.

420. See particularly Lanux letter to Sauvages, dated "a l'Isle de Bourbon Le 29 Xbre 1754," and "Réponses aux questions de Monsr. L'abbé Sauvage," PA-DB (Lanux).

421. See letter, Syndics to Buissière, dated "A Paris le 6 7bre 1766," SHML, IP 284 Liasse 107, #30; and ALS "Les Sindics et Directeurs de la Compagnie des Indes," dated "A Paris, le 26 Jan.er 1765," to "M. De la Vigne Buisson. A Lorient," SHML, 1P 284, Liasse 104.

422. "la quantité d'insectes, comme souris, fourmis, et autres dont cette isle abonde, qui détruisent ces vers," Bourbon Conseil letter to "Messieurs les sindics et Directeurs de la Compagnie des Indes," dated "A l'isle de Bourbon le 10 8bre 1749," in Lougnon (1949), p. 177.

which treated the craft in an article for the *Description des Arts* and which approved seven papers on various aspects of silk manufacture by Jean-François Paulet.[423]

Rubber. Mayans and other New World indigenes knew of the rubber plant and tapped it as a resource. Through contact, Europeans came to know of caoutchouc/rubber, but not its plant source. Jussieu's "Histoire du cachou" in the *Mémoires* of the Académie des Sciences for 1720 brought caoutchouc to the attention of the French.[424] The substance was thought to have many usages, such as making syringes or possibly in dyeing or as a substitute for ambergris, and Jussieu made the suggestion that rubber trees, when found, could be transplanted to the colonies. The tree in question was identified by the French at the time of the passage of La Condamine through Guiana in the early 1740s on his way back from the meridian expedition to Peru. François Fresneau (1701-1770), the chief royal engineer in Cayenne since 1732, was the actual discoverer for France, knowing of the rubber tree from Indians.[425] Fresneau worked on the chemistry and properties of *résine élastique* from Cayenne.[426] La Condamine brought the rubber tree back to France and reported on it.[427]

The state and the instrumentalities of the Colonial Machine noticed rubber. In Cayenne, as Fresneau and La Condamine began to tap rubber, they sent out slaves and runners into the jungle with clay models of the fruit and plant to aid their search for the plant. Later, in Cayenne, the royal botanists L.-C. Richard and then Leblond were sent into the jungles to find more rubber trees.[428] Latex tapped in Guiana had to be preserved in liquid form to be useful in manufactures, and in the 1770s state minister Bertin asked Macquer and Darcet of the Academy of Sciences to research how best to transport the raw material; Vaucanson of the Academy likewise was involved.[429] Further testing was proposed on sap from plants that Poivre had transplanted from Madagascar to the

423. Note in PA-DB (Jean-François Paulet) lists Paulet papers dating from 1772 to 1777; Jacques de Vaucanson (1709-1782), *pensionnaire* of the Academy of Sciences, devoted significant attention to improving machines involved in silk manufacture; see his several papers from the 1740s through the 1770s appearing the MARS, listed in Halleux et al., vol. 2, p. 59.

424. Jussieu (1720); see also manuscript, BCMNHN, Ms. 1150.

425. Llinares (2005), pp. 174-80, provides an overview and details of the rubber story; on these particular points see also Regourd (2000), pp. 479-80; Chaïa (1977), p. 25; Allorge, p. 293; Touchet, pp. 82-83; La Condamine (1751), p. 323; HARS (1751), pp. 17-22.

426. See previous note and "Mémoire du Sieur Fresneau… ingénieur en chef à Cayenne, sur divers sucs laiteux d'arbres qu'il a découvert en cherchant la raisine élastique," BCMNHN, Ms. 2299.

427. See La Condamine (1751).

428. Letter of Minister de Castries, dated 26 September 1783, AN C[14] 56, fol. 88, reprinted in Jandin, p. 69.

429. PA-PV 69 (1750), fol. 3 (" Samedi 13 Janvier 1770"); Vaucanson (1770); 1779 ms. by the chemist Macquer, "Mémoire sur les moyens a essayer pour transporter en france de la Résine Elastique de Cayenne," AN O[1] 1292, #5.

Jardin du Roi on Île de France.[430] At one point, Turgot issued instructions for testing sap on Île de France. At another point, state minister de Castries ordered d'Angiviller at the Bâtiments du Roi and Cossigny on Île de France to evaluate resinous plants and to send samples to France and possibly to grow plants there.[431] This work took place from 1779 though 1786. Cossigny experimented with "gomme élastique" and sent a description and sample of a promising plant from Madagascar for d'Angiviller to present to the Académie des Sciences.[432] Another Cossigny shipment from Île de France in 1786 brought sap in good shape, and Lavoisier, Berthollet, Darcet, and Fourcroy of the Academy were urged to examine it.[433] Du Puget reported to the Paris Academy on the "gomme elastique" in 1790.[434]

D'Angiviller at the Bâtiments du Roi promoted the project heavily. A note from Intendant d'Angiviller tells the tale of the bureaucracy enlisted to promote understanding and use of rubber sap: "Write to Mr. Darcet [of the Academy] and ask him on behalf of the Minister for a paper on the experiments to be made of the elastic gum to be sent by the Minister to the Governor or Intendant at Cayenne where surely the tree in question exists."[435] In 1783 he asked de Castries to communicate with administrators in Cayenne to get Richard to perform more experiments, writing: "I am even of the opinion that this branch of commerce will result in a particular advantage for the Cayenne colony that might perhaps become considerable, given the great number of usages to which this material is subject."[436] D'Angiviller solicited de Guignes in China for

430. See previous note and related documents in AN O[1] 1292, notably 1781 ms. ("Mémoire sur les moyens a essayer pour transporter en france de la resine elastique dans un état de liquidité ou de molesse assez grande pour qu'on puisse l'employer à divers ouvrages tres importants pour les Arts") and identically titled ms. of 1785; in same collection see also correspondence related to this research: AL Cossigny to d'Angiviller, dated "A L'Ile de France le 16 Avril 1782"; d'Angiviller draft to "M. de cossigny," dated "Versailles le 9 7bre 1783"; ALS "D'arcet" to "Monsieur," dated "Paris le 26 May 1783"; ALS Cossigny to "M. le C.te D'Angivillers," dated "A L'Ile de France le 15 X.bre 1783"; and, Cossigny letter to "M. le C.te d'Angiviller," dated "A L'Ile de France le 26 Février 1784"; also, ALS "Cossigny, Ingénieur du Roi" to "Monsieur," dated "A L'Ile de France le 6 9.bre 1782," AN O[1] 2110.

431. Reported in ALS, Cossigny to Le Monnier, dated "Palma le 25 Janvier 1781," BCMNHN, Ms. 1995.

432. Cossigny letter to "M. le C.te d'Angiviller," dated "A L'Ile de France le 26 Février 1784," AN O[1] 1292.

433. ALS "Thouïn" to "Monsieur le Comte," dated "A Paris ce 18 7.bre 1786," AN O[1] 2113A ("Pépinières d'Amérique. Correspondance Générale pour la Mission du S. Michaux, Botaniste du Roi, Année 1786").

434. PA-PV-110, Du Puget, 7 August 1790; PA-PS 1790, 'plumitifs' for 13 March 1790.

435. "Ecrire à m. d'arcet et lui demander de la par de M. le comte un memoire sur les experiences a faire sur la gomme elastique pour etre envoyé par le ministre de la marine au gouverneur ou intendant de Cayenne où il y a surement l'arbre en question," administrative note penned to Cossigny letter to d'Angiviller, dated "A L'Ile de France le 16 Avril 1782," AN O[1] 1292, #7.

436. "je pense meme qu'il resulteroit un avantage particulier pour la colonie de cayenne de cette branche de commerce, qui peut etre deviendroit considerable, vu le grand nombre d'usages auxquels cette matiere est applicable," d'Angiviller draft letter to "M. le marechal de Castries," dated "Vers.es le 5 7bre 1783,"and 11pp. ms., "Memoire sur La Gomme ou plutot la resine Elastique," attached to cover letter, AN O[1] 1292, #14, #25.

information regarding "gomme" and "résine élastique" in China and India, noting that they would be of great utility for European arts.[437]

The Louisiana Wax Plant. The case of the Louisiana wax plant is a minor one, but like so many others it shows the Colonial Machine in action, and it adds its weight to the overall record of the Colonial Machine. The case displays once again the inexorable pressure to extract utility from every source – this time wax from a holly bush native to Louisiana.[438] In the early 1720s Alexandre, the surgeon-botanist in Louisiana and correspondent of Dortous de Mairan at the Academy of Sciences in Paris, sent information to the Academy about the Louisiana wax plant, whose berries yielded a waxy residue when boiled. Alexandre said he learned of the plant from the local Indians and the technique from the slaves of Carolina, and mentions appear in the *Histoire* section of the Academy's volumes for 1722 and 1725.[439] Authorities then made a push to commercialize the wax plant with Jean Prat, brother of the original médecin du roi in Louisiana, installed as the same alongside the surgeon Alexandre.[440] Prat promoted holly-berry plantations as "an object of commerce advantageous to the colony" that would complement the production of indigo, tobacco, and cotton.[441] Prat sent samples to the ministry and to Jussieu at the Jardin du Roi, and he received a favorable reception.[442] The project envisioned planting of 9000 plants to grow 54 million berries to produce upwards of 5000 livres of wax.[443] A labor shortage was the main problem, but the idea to force colonists to grow the plant was rejected in favor of encouraging them by example.[444] Apparently, some production emerged out of New Orleans, as in 1747

437. D'Angiviller letter, dated "Vers.les 7 X.bre 1785," to de Guines posted to Macao as "agent du roi," AN O[1] 1292, #18.

438. On this episode, see archival materials in AN-COL C[13A] 27 (1742), fols. 8, 10, 12, 13, 16, and C[13A] 28 (1743-1744), fols. 22, 184; see also PA-DB (Alexandre).

439. PA-PV 41 (1722), fol. 247-48 ("Le Vendredi 14 Aoust 1722") and PA-PV 42 (1723), fol. 1 ("Le Samedi 9 Janvier 1723"); HARS 1722, p. 11; 1725, pp. 39-41; see also Langlois (2003), p. 346, and Lacroix letter of 10 March 1934 to J. Usher, Librarian of Howard Memorial Library, New Orleans, regarding Alexandre's contact with the Paris Academy of Sciences, PA-DB (Alexandre).

440. See letter from local administrators to minister, dated "A la N.lle Orleans, Le 24 Mars 1742," AN-COL C[13A] 27 (1742), fol. 8, and ALS Prat to "Monseigneur," dated "A la N.lle Orleans le 5 avril 1742," fol. 12; also, Langlois (2003), p. 347; Lamontagne (1962b), p. 219.

441. "Les plantations de ces arbrisseaux deviendroient dans la suite un objet de Commerce avantageux a La colonie," ALS Prat to "Monseigneur," dated "A la N.lle Orleans le 5 avril 1742," AN-COL C[13A] 27 (1742), fol. 12; see also 1743 ms., "Memoire du S.r Prat medecin concernant Larbrisseau de la Loüisianne qui porte des Graines propres a faire de la Cire," AN-COL C[13A] 28 (1743-1744), fol. 184; see also indications provided by Lamontagne (1963), p. 141.

442. See internal ministry document of September 1742, AN-COL C[13A] 27 (1742), fol. 10, and Prat ms., "Reponse au Memoire d'informations sur l'arbrisseau de la Loüisiane qui port la Cire," fol. 13, indicating involvement of Jussieu; in same collection see also Alexandre ms. of April, 1742, "Reponses aux questions du Memoire concernant les Arbrisseaux a Cire de La Louisianne," fol. 16.

443. See previous notes and ALS "Vaudreüil" and "Salmon" to minister, dated "A la N.lle Orleans, Le 21 Juillet 1743," AN-COL C[13A] 28 (1743-1744), fol. 22.

444. "Memoire du S.r Prat medecin concernant Larbrisseau de la Loüisianne qui porte des Graines propres a faire de la Cire," AN-COL C[13A] 28 (1743-1744), fol. 184.

Governor-General Vaudreuil wrote to the minister reporting a good harvest of indigo and wax.[445] Jean Prat's successor as médecin du roi in New Orleans, one Bénigne de Fontenette, sent seven cases of the bush aboard the *Parham* to Buffon at the Jardin du Roi, along with a report of his researches.[446] In 1749 another local administrator, Michel, promised to increase production and in 1752 he reported "great production of wax."[447] But Louisiana and the wax plant soon passed out of French hands.

Hemp and Fodder. Rope was another of those unobtrusive products on which the Royal Navy (much less the rest of the sailing world) and French colonial and overseas expansion depended. As inspector-general of the navy, Duhamel du Monceau concerned himself with ropes and cordage, and in 1739 the navy performed tests and experiments under his direction.[448] The manufacturing facility (the *Corderie Royale*) at Rochefort turned out rope for the navy, and the navy used immense amounts of it. High-grade hemp was imported from Ukraine at great expense, and alternatives were sought after for the commodity. A very early report alluded to a colonial connection and to experiments in making rope from palm, coconut, and latania trees.[449] The Académie des Sciences became tangentially involved with hemp late in the century.[450] By 1700, authorities introduced the cultivation of hemp and linen into New France, sending hemp and linen seeds to Canada.[451] This effort was such a success that by the 1730s the same authorities reduced the price for hemp hoping to moderate the enthusiasm of colonists.[452] Two decades later they regretted their decision, raised the price, and proposed sending several families knowledgeable about hemp production to the colony.[453]

Attention to sources of fodder and forage similarly occupied the Colonial Machine. The Société Royale d'Agriculture became interested in a grass grown in New England for fodder, and on its behalf d'Angiviller instructed André Michaux to send thirty to forty pounds of seed for growing at Ram-

445. Vaudreuil letter to minister of 16 May 1747, AN-COL C[13A] 31 (1747) fol. 93.

446. See undated letter from Fontenette to the minister, AN-COL C[13A] 33 (1749), fol. 137, wherein Fontenette solicited the formal position of botaniste du roi and a position on the local council; on the latter point, see also letter of 27 April 1753 from local administrators in New Orleans to minister, AN-COL C[13A] 39 (1755-1757), fol. 245; see also Langlois (2003), p. 350, who notes that Fontenette send a herbarium from Louisiana to Buffon in 1750.

447. See Michel correspondence, AN-COL C[13A] 34 (1749-1750), fol. 96 (28 July 1749) and C[13A] 36 (1752), fol. 269 (23 September 1752).

448. See report of 13 August 1739 from aboard "La Vénus," SHMB, Série A, Sous-série 1A, 1-A2; Pritchard (1987a), pp. 175-78, provides the larger background.

449. See Seignelay letter of 13 March 1684, AN-COL B[11] (1684), fol. 59.

450. PA-PV-110, "Isle de France, Assemb.ée Provinciale de L'," 5 September 1788 ("Questions relatives au ruissage du Chanvre").

451. See ministerial correspondence, AN-COL B[22] (1700), fol. 92v (28 April 1700); B[23] (1702), fol. 49 (5 April 1702).

452. See ministerial letter of 21 March 1730, AN-COL B[54] (1730), fol. 396.

453. See ministerial letter of 15 April 1750, AN-COL B[91] (1750), fol. 19.

bouillet.[454] On another occasion d'Angiviller was in contact with Broussonet and the Société d'Agriculture over a forage plant in the Carolinas reported by the English.[455] A grass from Saint Domingue called "La Luzerne" also found success.[456] The Société d'Agriculture investigated, and "Luzerne du Cap" became naturalized on Île de France.[457]

Ginseng. The ginseng plant has already popped up in this account. What needs emphasis here are the ways world trade in ginseng arose and how, through the agency of the Colonial Machine, the French managed the resource from Canada to China. By 1711, reports of this almost magical plant reached Europe from China.[458] Demand from China spurred a lively trade in Canadian ginseng, which the Iroquois knew and used as a stimulant. The Jesuit Father Lafiteau "discovered" the plant for Western science and published a description and illustration in the *Lettres édifiantes* of the Order.[459] Either through Lafiteau or on his own, Sarrazin, the royal doctor at Quebec, sent samples to the Jardin du Roi in Paris.[460] In 1718 Antoine de Jussieu investigated the plant.[461] Jussieu used the Canadians and other contacts to compare Canadian and Chinese ginseng, noting the potency of both.[462] In China, Father d'Incarville, who had seen the plant while on station in Canada and knew of Lafiteau's description, reported from Beijing that ginseng grew in Tartary, but it was more difficult for him than for Europeans to say whether the Chinese and Canadian plants were the same or different. One thing he did know, however, was that because of "the vogue for ginseng in China ...it sometimes sells for more than the price of gold."[463] French government authorities and the Compagnie des Indes jumped on this opportunity and created a pipeline for moving

454. See draft in d'Angiviller's hand, dated "Vers.es le 13 juin 1788" and at top left, "M. Michau," AN O[1] 1293, #349.

455. Ministerial draft letter to "M. Michaux," dated "3 9bre 1788," and ministerial draft letter to "M. Broussonet," dated "Vers.es le 5 9bre," AN O[1] 1292, #21 and #22; see also materials relative to "Gramen de Walter," AN O[1] 1293 #391.

456. McClellan (1992), p. 246.

457. APS-MS, B/D87, Group 26, loose sheet headed, "du 29 Nov. 1786"; see also undated, unsigned 1p. ms in Cossigny's hand, "Etat des Graines envoyées à Monsieur Le Monnier, par la Corvette La Dauphine," that included "*Luzerne du Cap*, naturalizée à Palma," BCMNHN, Ms. 1995.

458. See letter from Father Jartoux, dated Beijing, 12 April 1711, reprinted in Vissière and Vissière (1979), pp. 176-77.

459. Rousseau (1957), p. 152; this "discovery," notes Harris, p. 78, "was merely a matter of asking a Mohawk medicine woman to find the plant for him."

460. Lortie, p. 11; see also Vaillant letters, dated "A Paris le 25 Janvier 1718" and "A Paris le 28 Fevrier 1718," RS/L, Sh, #530 and #531.

461. Antoine de Jussieu letters, "de Paris ce 17 Mars. 1718" and "de paris ce 26 Mars. 1718," RS/L, Sh, #281, #286.

462. ALS, Jussieu to Sherard, "de Paris ce 14 Fevr. 1718.," RS/L, Sh #280.

463. "Je crois ce que a donné tant de vogue au *gin seng* en Chine, c'est L'Effet subit de ce remede...[ginseng] se vendant quelques-fois audela du poids de L'or...," "Catalogue Alphabetique des plantes...," marked "Par P. d'Incarville de Peking Mars 9, 1748," RS/L, L&P II, #153.

Canadian ginseng to China. In 1752 a ministerial report to the administrators in Canada noted that "the success of Canadian ginseng in China has considerably augmented the price in Canada, selling last year up to 20 livres a pound, even 30 livres at La Rochelle. There's a great advantage here for Canada, but it is likely that this plant is also found in the English provinces, whence it will be brought as contraband"![464]

Rice and Potatoes. In the 1760s, on the urging of state minister Bertin, the newly formed royal societies of agriculture in Paris and the provinces focused their attention on rice.[465] The Paris institution that ultimately became the Société Royale d'Agriculture concerned itself with rice and spreading and improving rice production through to the end of the Old Regime.[466] Africans had brought rice cultivation to the Carolinas, and Michaux shipped "Carolina dry rice" to Thouin for planting.[467] From Île de France, Cossigny sent in another rice from Cochinchina.[468] The ministerial instructions to the administrators in Canada are the most revealing: that they should not cease to increase emergency food supplies and that rice provides the possibility of feeding lots of mouths inexpensively. The ministry sent an instructional booklet to be distributed to colonists on preparing rice.[469]

For the same reasons French authorities recognized the value of cultivating potatoes, but this practice was new in France, being more common in Great Britain and Germany, and a certain wariness prevailed. Just as it recommended rice, so did the ministry recommend potatoes to its agents in Canada. The potato nourishes man and beast, they said, staves off hunger and famine, and it offers an abundant harvest for minimal labor. Potatoes grew in New England, where the English had introduced the plant, and its cultivation in Canada was to be encouraged. By the same token administrators were afraid that colonists and Indians would become less dependent on central authority if they gave up cultivating wheat in favor of potatoes.[470] The French agronomist, player in the Société Royale d'Agriculture, and one of the editors of the *Feuille du Cultiva-*

464. "Le succès du ginseng du Canada en Chine en a augmenté considérablement le prix en Canada, s'étant vendu l'année dernière jusqu'à 20 livres la livre, et même 30 livres à La Rochelle. Il y a là un grand avantage pour le Canada, mais il est probable que cette plante se trouve également dans les provinces anglaises, ou qu'on l'y portera en contrebande," ALS Le Président du Conseil de marine à M. de Montaran, AN-COL B[96] (1752), fol. 125 (3 August 1752); see also ALS Syndics to Roth, dated "A Paris le 15 may 1758," SHML, 1P 278, Liasse 14, #48bis.

465. Letter "à Monsieur de Boisdeveuil du Bureau d'Angoulesmes, subdelegué de M.r L'intendant de Limoges," dated "16 May 1764," and "Response de la Societé d'agriculture aux eclaircissements demandés par Monsieur Bertin, ministre et sécretaire d'etat," dated in margin, "25 juillet 1764," AC/La Rochelle, Série E 1186 "Société d'agriculture de la Rochelle, 1762-1767."

466. APS-MS, B/D87, Group 26, folded ms. sheet headed, "Séance du jeudy – 16 Mars 1786...."

467. "ris sec du Caroline," BCMNHN, Ms. 691, 133-34.

468. BCMNHN, Ms. 691, 178v-179.

469. See ministerial correspondence, AN-COL B[107] (1758), fol. 29 (20 March 1758).

470. See ministerial correspondence, AN-COL B[107] (1758), fol. 25 (24 February 1758).

teur, Antoine-Augustin Parmentier (1737-1813) famously did much to promote the potato.[471] Moreau de Saint-Méry wrote on the potato, and by 1782 the plant was introduced into cultivation on Île de France.[472] Compared to its concern with other, more commercial products, however, the Colonial Machine displayed somewhat less interest in the potato.

Miscellaneous. The range of matters botanical and agronomic undertaken by the Colonial Machine was wide and inclusive of every conceivable option. Interest in certain vegetables and Asian trees as possible sources for oil is to be noted.[473] Indians in the Illinois region grew grapes, and the Jesuits used these to make wine for their services.[474] Gum arabic attracted attention, too. The Senegalese monopolized commerce in gum arabic, and in 1780 over 1 million pounds of the stuff were traded out of Africa and consumed in Europe.[475] Adanson estimated the production at over 3 million pounds per annum, trade that was worth more than the slave trade and more reliable, to boot![476] Efforts were made to encourage French trade in Senegal, given that "commerce finds both gum arabic itself, so necessary for manufactures, and a fairly large number of blacks of which the French islands in America are deprived."[477] Chinese lacquer – from the lac insect and symbiotic trees – and the development of French alternatives for lacquer and various varnishes and shellacs likewise drew sustained attention. The Paris Academy, for example, heard early reports and displayed an interest in lacquer in India.[478] D'Incarville in Beijing sent a memoir on Chinese varnish and lacquer techniques to Paris so that the Chinese manner might be practiced in Europe, especially

471. See Parmentier (1789); Anonymous (1931) gives details on Parmentier and the colonies. On Parmentier's active role in the Société Royale d'Agriculture, see minutes of the Société, APS-MS, B/D87, passim; on his contact with the Jardin du Roi, see BCMNHN, Ms. 1327, passim and "Distribution de Graines de Plantes Potageres, Economiques, officinales, et de fourrages"; on his contact with the Société Royale de Médecine, see SRdM, Ms. 33, "Recueil de lettres," 135; Ms. 8, p. 356 ("Séance du Vendredi 23 Mars 1781...M. Thouret [?] a fait voir à la societe, un Echantillon qui lui a été remis par M. Girard Méd de S.t Domingue, d'un Biscuit de Mer fait avec des Pommes-de-Terre, suivant le Procedé de M. Parmentier").

472. Moreau de Saint-Méry (1789); Sonnerat (1782), vol. 2, p. 85.

473. PA-PV-110, Guignes, 24 March Mars 1789 ("Lettre sur une nouvelle manière d'Employer l'huile du *Mingeon* ou L'huile de Bois").

474. "Relation de la Louïsiane par Lemaire," BCMNHN, Ms 948.

475. See 1780 document, "Mémoire du Roi pour servir d'instructions au Sieur _____ Commandant du Sénégal," BCMNHN, Ms. 293.

476. See "Sur les Gommiers du Sénégal," *Collection Académique, Partie Française*, vol. 16 (1787), p. 169.

477. "le commerce y trouve à la fois la gomme Arabique si nécessaire aux manufactures et un asses grand nombre de noirs dont les isles francaises de l'amérique sont dépouvûes...," "Mémoire sur le Sénégal," SHMV, Ms. 361.

478. PA-PV 29 (1710), fols. 5 ("Le Samedi 11 Janvier 1710"), 7-9 ("Le Mecredi 15 Janvier 1710"), and 39-41v ("Le Mecredi 5. Février").

given that a lacquer tree was found in Mississippi.[479] In 1750 Jean Hellot presented a manuscript to the Paris Academy on Chinese varnish.[480]

The Colonial Machine was interested not just in plants and their economically useful products. Animals, too, attracted some attention, but not to the same extent, doubtless because of the difficulties of transporting large animals over transoceanic distances. The Compagnie des Indes shipping exotic animals for the royal menagerie was one thing, as was the importing of camels onto Île de France.[481] The problem in the colonies was the constant dearth of work animals – asses, donkeys, and horses. These were always needed but not easy to transport in substantial numbers.[482] Authorities managed to send four asses to Canada in 1750, and that seemed like an accomplishment.[483]

Mines and Mineralogy. The vigorous pursuit of mineral resources by the French state in its colonies, particularly in Canada, does not fit in automatically with botany or applied agronomy, but the subject demands consideration, and the spirit and investments made to identify and exploit mineral resources were of the same type and order that motivated the applied botany projects surveyed here. They were a further facet of the Colonial Machine.

On assuming power with Louis XIV in the 1660s, Colbert displayed an active interest in finding and exploiting mineral resources in Canada, and he directed operations through an ongoing correspondence with his agents on site. He sent two engineers to Canada in 1664, and reports came back of vast mineral resources: iron, lead, copper, coal, even silver and gold.[484] In 1687 a director of the forges in Brittany was sent to "study means of exploiting mines discovered in New France."[485]

The lead mines at Gaspé were the first to be developed. A local man knew the location, and in 1665 a party of twenty-five departed to begin working the vein.[486] Colbert and the Compagnie des Indes *twice* sent a German smelter to Canada to investigate the Gaspé lead mine site.[487] The Compagnie des Indes

479. ALS d'Incarville to B. de Jussieu, dated "à Peking le 22e 8bre 1747," in PA-DB (Incarville).

480. PA-PV 69 (1750), p. 173 ("Vendredi 15eme May…L'on scait maintenant en Europe que le Vernix de Chine n'est point une composition, mais une Gomme de Résine qui coule d'un arbre que les Chinois appellent *Tsi chou* ou Arbre du Vernix.").

481. "Nottes d'animaux utiles a naturaliser en France," BCMNHN, Ms. 308, and above Part I at note 208.

482. On the shortage of work animals in Saint Domingue, see McClellan (1992), p. 33; see also Roche (2008).

483. See ministerial correspondence related to this matter, AN-COL B^{91} (1750), fol. 2 (3 February 1750), B^{92} (1750), fol. 23, (3 February 1750), and fol. 56 (6 March 1750).

484. Colbert letter to Tracy, AN C^{11A} 2, fol. 99 (1664); Colbert letter to Talon, dated 11 February 1671, AN-COL F^{3} 4, p. 52.

485. "étudier les moyens d'exploiter les mines découvertes en Nouvelle France," ministerial letter of 11 May 1687, AN-COL B^{13} (1687), fol. 197v.

486. Talon letter to Colbert, dated La Rochelle, 22 April 1665, AN C^{11A} 2, fol. 124.

487. See Talon letter to Colbert, dated Quebec, 4 October 1665, and Colbert letter to Talon, dated Versailles, 5 April 1666, AN C^{11A} 2, fols. 143, 199.

provided specific instructions, and mineral samples were sent to France.[488] Later, royal instructions directed the intendant and the governor-general to begin exploratory mining for lead at the Baie St. Paul, and at another point, as Governor-General, La Galissonnière visited the lead mines in the company of Gaultier, the royal doctor in Canada.[489]

Colbert was no less interested in sources of iron ore in Canada.[490] In 1665 the Compagnie des Indes organized tests of sands that seemed to be iron ore. Sand samples were taken from three spots, but the results of the smelting undertaken in France proved negative and Colbert was disappointed.[491] In 1708 a seam of iron ore was discovered near Trois-Rivières, and a mining and smelting operation became established there.[492] These were the Forges de Saint-Maurice, and in 1736 running them cost the state 36,000 livres.[493] For at least two decades the Forges de Saint-Maurice turned out cast-iron bars and forged-iron implements for use in French Canada and across the Atlantic.[494] A master founder ran the forges, and his was a nontrivial expertise that deserves to be noted.[495] The Ministry of the Navy had tests made of Canadian iron, and these tests were performed at the royal mint, La Monnaie.[496] With the quality of Canadian ore found satisfactory, the minister authorized the intendant and governor-general in Canada to ship to Rochefort annually whatever quantity of iron the navy requested for its purposes.[497] The Canadian forges turned out forged iron, steel, and some kind of bomb. The Navy tested these at its iron-works at Rochefort, finding the iron satisfactory, but not the steel. In reference to the bombs, the report said only that they "were not made according to standard specifications."[498]

488. See "Mémoire de la Compagnie des Indes occidentales" (1665) and "Etat des affaires du Canada en l'année 1665 qui sont à régler par la Compagnie," AN C^{11A} 2, fols. 172, 174.

489. "Mémoire du roi à MM. de Beauharnois et Hocquart," AN-COL B^{71} (1740), fol. 50 (13 May 1740); Lamontagne (1960), p. 520; Taillemite (2008), pp. 120-21.

490. Colbert letter to Talon, dated 11 February 1671, AN-COL F^3 4, p. 52.

491. "Etat des affaires du Canada en l'année 1665 qui sont à régler par la Compagnie" and Colbert letter to Talon, dated Versailles, 5 April 1666, AN-COL C^{11A} 2, fols. 174, 199.

492. See ministerial correspondence of 6 June 1708 to the marquis de Crisafy, AN-COL B^{29} (1708), fol. 371.

493. "Mémoire du roi à MM. Beauharnois et Hocquart," dated 15 May 1736, AN-COL B^{64} (1736), fol. 438v.

494. See Abénon and Dickinson, p. 84, and register of accounts for the Forges de Saint-Maurice for 1729 to 1741, AN-COL C^{11A} 110. The Forges de Saint-Maurice deserve further study.

495. "Mémoire du roi à M. de Sérilly," AN-COL B^{82} (1745), fol. 30 (16 March 1745).

496. See ministerial letter to the administrators in Canada, dated 6 May 1738, AN-COL B^{66} (1738), fol. 25v.

497. See ministerial correspondence, dated 7 May 1740, AN-COL B^{70} (1740), fol. 33v.

498. "Les bombes n'ont pas été fabriquées suivant les règles ordinaires," ministerial letter to de la Galissionnière et Hocquart, dated 31 May 1748, AN-COL F^3 13, fol. 308. For another Canadian iron mine see "Brevet qui autorise le S. r Poulin de Francheville à ouvrir et exploiter pendant 20 ans des mines de fer en Canada," AN-COL F^3 11, fol. 275 (25 March 1730); Pritchard (1987a), chapt. 9, provides larger background.

Word of Canadian sources of copper ore came early and trickled back to France.[499] But the discovery, through Indian contacts, of raw copper around Lake Superior in the 1730s provoked a spirited response from French officialdom. The Monnaie tested a sample sent to France through government channels, finding the copper "very good, very pure, and very malleable."[500] The director general of the mint himself reported on the results, emphasizing their "great promise" and the hope that silver mines might be found in the neighborhood.[501] The next year, Sieur de la Ronde set out from the port of Saint Louis on a boat of his own making for Lake Superior "to visit the island where the savages assure him that the copper found there is very pure." "It is not to be doubted," continued the minister in the king's voice, "that the mineral from this mine is rich and that its exploitation could be something considerable."[502] He hoped that the report by natives who claimed to know an island covered with copper would be confirmed, not least because the the mine situated on an island would provide some protection against the troubles caused by these selfsame natives![503] The party found raw copper on the island that could be cut away with scissors and a hammer, and the minister ordered its exploitation to proceed. "We are sending plans for a furnace that should only use charcoal. The copper should be let to flow into molds and ingots for easier transport. This enterprise is not to be neglected."[504] As a result, the government engaged two, likely German, miners, Jean Adam Forster and Christophe Henry Forster, father and son, "for the service and work of the copper mines in Canada."[505] They left on a royal vessel out of Rochefort for the long reach to Lake Superior.

499. Colbert letter to Talon, dated 11 February 1671, AN-COL F[3] 4, fol. 54; see also "Ordre pour permettre l'ouverture des mines de cuivre trouvées dans l'établissement du S.r Le Sueur, habitant du Canada," dated 21 May 1698, AN-COL F[3] 8, fol. 84.

500. "L'échantillon dont on a fait l'essai est très bon, très pur et très malléable," "Mémoire du roi à MM. de Beauharnois et Hocquart," AN-COL B[63] (1735), fol. 462v (11 April 1735).

501. "ce qui est propre à donner de grandes espérances," ministerial letter to colonial administrators, dated 15 May 1735, AN-COL B[63] (1735), fol. 506v.

502. "A vue que le Sr. de la Ronde était parti du havre St. Louis, dans le lac Supérieur, pour aller, sur le bâtiment qu'il a fait construire, visiter l'île où les sauvages lui assure que le cuivre qu'on y trouve est pur. Il n'y a pas à douter que le minerai de sa mine est riche et que son exploitation peut être d'un objet considérable," "Mémoire du roi à MM. Beauharnois et Hocquart," AN-COL B[64] (1736), fol. 438v (15 May 1736).

503. "Mémoire du roi à MM. de Beauharnois et Hocquart," AN-COL B[65] (1737), fol. 423v (10 May 1737). "Espère que le rapport des sauvages qui assurent connaître une île remplie de cuivre, se trouve confirmé, il y aurait de la sorte pluys de sécurité contre des troubles que pourraient causer les sauvages."

504. "On envoie un plan du fourneau dans lequel il n'y aura que du charbon de bois à employer. Le cuivre sera coulé dans des moules en lingots convenables pour le transport. Cette entreprise ne doit pas être négligée," ministerial letter to local administrators, Beauharnois and Hocquart, AN-COL B[64] (1736), fol. 445 (15 May 1736).

505. Ministerial letter to M. Saur, AN-COL B[65] (1737), fol. 47v (16 April 1737).

An abundance of coal in Canada provoked the attention of authorities, too. Colbert thought that Canadian coal could be used as ballast "for every ship returning from New to Old France," thus saving on coal imported from England.[506] Coal mined at Cap-Breton was especially prized for a perhaps unexpected reason, that it could be shipped to the Antilles with relative ease for use in processing sugar.[507] Coal thus provided the basis for a sort of northern "triangular trade," with Canadian coal going to the Caribbean, sugar to France, and goods and people back to Canada across the North Atlantic.

These ventures bespeak a high level of official interest and species of technical expertise not to be overlooked, but the information gathered about the mineral resources of Canada was not lost for the higher reaches of science. In particular, Gaultier in Canada sent mineral samples to the mineralogist and academician Jean-Etienne Guettard (1715-1786), who incorporated this information into a glorious colored map of Canadian mineral resources published in the *Mémoires* of the Paris Academy.[508]

Louisiana, too, and mineral resources in the Indian lands up the Mississippi and Ohio rivers likewise drew the attention of authorities. Reports popped up of silver, lead, and antimony mines in Illinois territory.[509] While he was stationed in Louisiana in the 1720s, the engineer Baron undertook a trip sponsored by the Compagnie des Indes to seek out potential mining sites.[510] For some years one Renaud had the concession to mine in Illinois for the French. Some lead was shipped down to New Orleans, and the local administrators there liked the idea of using lead rather than bricks as ballast.[511] After Renaud's return to France, the local Indians in Illinois apparently continued mining for their own profit, prompting the administrators to ask for convicts condemned to French mines for evasion of the salt tax in France to work in the lead mines of Illinois and Missouri.[512]

In the Caribbean, the army sent fifty miners to Martinique in 1692.[513] Another early project sent miners to Saint Domingue.[514] Later, the Cercle des

506. "on en pourra lester tous les vaisseaux qui reviendront de la nouvelle en l'ancienne France," Colbert letter to Talon, dated Saint-Germain-en-Laye, 5 April 1667, AN C^{11A} 2, fol. 290,

507. Talon letter to Colbert, dated Quebec, 11 November 1671, AN C^{11A} 3, fol. 184.

508. See Guettard (1752a) and (1752b); ALS, Gaultier to Guettard, dated "québec 21 8bre 1752," and related material, BCMNHN, Ms. 293; Lamontagne (1964b) and (1965); McClellan (2003c), pp. 44-45.

509. La Mothe-Cadillac letter to the minister, dated 2 January 1716, AN-COL C^{13A} 4 (1716), p. 509.

510. Baron memoir to Compagnie, AN-COL C^{13A} 10 (1726-1727), fol. 162.

511. See ALS Salmon, dated "A la N.lle Orleans. Le 20 Juillet 1723," AN-COL C^{13A} 28 (1743-1744), fol. 105, and previous communication from local administrators, dated 25 April 1741, C^{13A} 26 (1741), fol. 11.

512. ALS, "Vaudreüil," dated "De la N.lle Orleans ce 6 X.bre 1744," AN-COL C^{13A} 28 (1743-1744) fol. 245.

513. Service historique de l'Armée de Terre, Vincennes, Série A, 2470, #3.

514. Service historique de l'Armée de Terre, Vincennes, Série A, 1599.

Philadelphes published a chemical analysis of mineral samples from Saint Domingue.[515] In 1711-12, Indians brought in samples of gold and silver to the station in Cayenne, but the agent sent into the jungle to discover the source returned empty-handed.[516] Efforts continued in the following decades to identify mineral resources in French Guiana.[517] Later, in the 1790s, Du Puget reported to the Paris Academy on the mineralogy of Cayenne and Antilles.[518] Mining took place on Île de France, and a forge was set up there. The government botanist Fusée-Aublet investigated the mineralogy of the island while in the region.[519]

Ecological Cause and Effect

Colonization inevitably left its mark on local ecologies, and local ecologies (including the sea) likewise affected the course of French colonization in the Old Regime.

The Colonial Footprint. The building of towns, hamlets, ports, docks, highways, way stations, and military fortifications in the colonies necessarily transformed local surroundings.[520] Large human population shifts from Africa and from Europe were involved, and colonists, free and slave, brought with them their customary plants and animals that transformed the native flora and fauna.[521] Commodity production of tobacco, sugar, coffee, cotton, and indigo also reshaped local landscapes. The colonial footprint was always large locally and often devastating, as on the turtle population on Île Rodrigues in the Indian Ocean, where 30,000 were taken by Europeans in 1760-61.[522] By the same token, all kinds of plant exchanges took place, private ones as well as the large-scale ones sponsored by the government that we have seen. And so, back and forth between Europe, America, and Asia, new routes and reservoirs of pest and pathogen opened up. European colonialism and imperialism ultimately won (or at least seems to have won) a long battle against nature that continued into the twentieth century (think yellow fever), but in the period we are concerned with the ecology bit back with force, and ecological effects reg-

515. DPF, 149.

516. See ministerial correspondence regarding this episode, AN-COL C[14] 6, fols 31-52; C[14] 7, fols. 7-8.

517. See correspondence to this effect, AN-COL C[14] 12-15; HARS 1741, pp. 23-25.

518. PA-PS, 'plumitifs' for 1790, 26 February 1791, 6 April 1791, 18 May 1791.

519. See "Extrait d'une Rélation d'un voyage aux Indes orientales Par M. de Maudave," pp. 41-42, BCMNHN, Ms. 1765; ms. Code de l'Ile de France, vol. 1, p. 179, in AN-MAR G 237; see also Fusée-Aublet ms., "Notes diverses de minéralogie," BCMNHN, Ms. 452.

520. McClellan (2006), pp. 69-70, 73-75; McClellan (1992), p. 32; Roger Williams, chapt. 11, takes up "Botanofiles Confront Deforestation."

521. Crosby (1972) and (1986) famously provides the larger background here.

522. Gasc and Laissus.

ularly proved a problem for colonial development and the colonial system. Call it the beginning of Nature's revenge. Actors at the time recognized and responded to these ecological effects.

Given its characteristic rigidities, the French government installed planned towns in its colonies, each replete with a church, square/parade grounds, government and military quarters, markets, and housing.[523] Beyond that, colonial authorities sought to control development and the natural world around them through legislation. On Guadeloupe, for example, one tenth of all properties were to remain forested or planted with trees. From Grands Fonds to Basse Terre on Guadeloupe, on Marie Galante and also on the Indian Ocean islands, one quarter of the land was to be left as woods. Inspection was by the commander of each parish; fines were 150 livres.[524] Deforestation and a resulting dearth of timber and firewood became problems wherever the French (and other Europeans) settled[525]. The "environmental policy" of Pierre Poivre on Île de France has been already fully detailed.[526] Everywhere ordinances proscribed or tightly controlled the cutting of trees.[527] Woods were to be preserved along streams and beaches. Trees were ordered to be planted along highways and around houses. Trees over six inches in diameter were reserved for the navy's use.[528] Officials planted cactus along the seashore at Martinique as a defensive measure![529] Regulations dictated the planting of bamboo and quitch (couch grass) on Île de France and Île Bourbon.[530] Already by the 1720s, colonists on Île Bourbon had pursued turtles almost to extinction, and authorities now forbade catching or killing turtles and tortoises, reserving the resource for provisioning ships.[531] Likewise in Martinique, at the beginning of the eighteenth century, local authorities fought against pollution of rivers near sugar plantations and techniques used by slaves for fishing – a method that

523. Pérotin-Dumon; McClellan (1992), chapt. 5.

524. Durand-Molard, vol. 2, pp. 538-42.

525. See Grove (1995), chapt. 6: "Climate, conservation and Carib resistance: The British and the forests of the Eastern Caribbean, 1760-1800", and pp. 474-486.

526. Grove (1995), chapt. 5, pp. 168-263.

527. McClellan (1992), p. 32; ms. Code de l'Ile de France, vol. 2, in AN-MAR G 237.

528. "Réglement économique. De M.M. les Général & Intendant sur le défrichement des terres & la conservation des bois de l'Isle de France. Du quinze Novembre 1769," Delaleu, vol. 1, pp. 322-27.

529. "Ordres au sujet de la plantation des Raquettes le long des Bords de la mer," AN-MAR G 62, Martinique (1731), fol. 86v.

530. "Lettre circulaire,...concernant les exploitations de bois dans les différens Quartiers de l'Isle. Du 4 Janvier 1777," Delaleu, Supplément 1, pp. 232-33; "Chiendent Planté aux Quartiers St. paul par ordonnance des administrateurs," ms. Code de l'Ile de Bourbon, vol. 2, pp. 469, 653, in AN-MAR G 237. For further regulations concerning planting of trees on Île de France and Île Bourbon, see 26 articles of the "Ordonnance...pour la plantation d'arbres dans la rue du Chef-Lieu de Saint-Denis & sur les chemins publics. Du quinze Janvier 1770," Delaleu, vol. 2, pp. 92-96; and entries under "Arbres" and "Bois," ms. Code de l'Ile de France, vol. 2, in AN-MAR G 237.

531. See Gaubil ms. of 1721, "Voiage a la chine, de france, dans la fregatte La Danaé," PA-DB (Gaubil); original BCMNHN, Ms. 581II.

consisted in poisoning water with toxic wood, or other substance, in order to kill fish.[532]

Nature Out of Control? But efforts by local authorities to control the local environment pale before the ways in which the natural world at the edges of the colonial world made life difficult. Plant and animal transfers often backfired, and introduced species exploded out of control. Fusée-Aublet reported that one plant on Île Bourbon "multiplied, sowed itself [and was] found on all parts of the island; indestructible; in the sand by the sea." Of another he wrote: "we cannot destroy it."[533] Cactus grew out of control on Île Bourbon, and the Compagnie outlawed planting it, regulations renewed by the king in 1774.[534] In what for us is a paradigmatic example, the Séguineau brothers, presumably for delectation, imported escargots from their native La Rochelle to their plantation in Saint Domingue. The snails succeeded all too well in the local struggle for existence and threatened coffee trees. The Séguineau brothers not only set their slaves to collect the snails, but paid them and gave them salt with which to cook and eat snails. The escargot example is wholly minor, but it illustrates well the principles involved in considering the ecological impact of colonialism or, more particularly in this case, French colonialism.[535]

Introduced birds, waves of insect pests, and that great human congener, the rat, voraciously attacked colonial agriculture and the roots of French colonialism itself. On Martinique rats infected the cane fields and were hunted and trapped.[536] The rats on Île Bourbon apparently devoured fowl, sheep, and even pigs![537] Rats and monkeys were animal pests endemic to Île de France.[538] Wild dogs were a problem, and they were killed on Île de France.[539] Insects were a problem everywhere, especially pesky mosquitoes, and there were worms that ate whole libraries and made it difficult to maintain public records.[540] Birds from Madagascar, India, Java, and China were resident nui-

532. AD/Martinique B2, fols. 152-53 (30 April 1718: " Ordonnance du lieutenant général François de Pas de Feuquières au sujet de l'énivrement des rivières "); AD/Martinique B6, fol. 165 (10 May 1737: "Ordonnance du lieutenant général Champigny et de l'Intendant Pannier interdisant la pollution des rivières, Fort Royal, du 8 novembre 1736"). Regourd (2000), pp. 490-91.

533. "…multiplié; leve de luy mesme; venant partout; se trouve en toutes les parties de l'isle; indestructible; au bord de la mer dans le sable…nous ne pouvons la detruire," ms., "Plantes cultivées au Reduit en 1759," BCMNHN, Ms. 452.

534. "Ordonnance...pour la destruction des Raquettes dans tous les Quartiers de l'Isle de Bourbon. Du vingt Septembre 1774," Delaleu, vol. 2, pp. 134-36.

535. Per McClellan (1992), p. 33; McClellan (2006), p. 75.

536. See remarks to this effect, ms. "Plantes de Martinique/Guadeloupe," HI, AD 371.

537. Letter from La Compagnie to "M.rs du Conseil Supérieur à l'Isle de Bourbon," dated "A Paris le 25e Mars 1741," in Lougnon (1935), pp. 180-81.

538. "Extrait des lettres de l'Isle Bourbon du 20 décembre 1730," in Lougnon (1934), p. 135. See also ordonnances listed in ms. Code de l'Ile de France, vol. 1 and vol. 2 and ms. Code de l'Ile de Bourbon, vol. 2, pp. 543, 759, in AN-MAR G 237. See also Fusée-Aublet comments on rat and grasshopper problems, "Mémoire justicatif" (1763), BCMNHN, Ms. 452.

539. "Mémoire justicatif" (1763), BCMNHN, Ms. 452.

540. McClellan (1992), pp. 30-32; "Plantes de Martinique/Guadeloupe," HI, AD 371.

sances in the Indian Ocean colonies. Birds were a problem on spice plantations in Guiana, too.[541] In 1782 Pierre Sonnerat gave a vivid picture of what colonists were up against on Île de France:

> The siskin and the grosbeak from Java, first imported as curiosities and carefully kept in cages, have so multiplied today that they devour almost all harvests. To keep them off seeded fields, one is obliged to set several slaves as sentinels who constantly cry and clap their hands. The rats are of such a great quantity that often they devour a field of grain in a single night. They also eat fruits and destroy young trees by their roots. This is the reason, it is said, that the Dutch abandoned this island.
>
> These pernicious animals have attracted the attention of the Government. Each colonist is required to destroy a certain number according to how many slaves he possesses and to send the bureau of the Police the birds' heads and rats' tails he has had killed. But all these precautions proved pointless....Grasshoppers are of such prodigious numbers that when a cloud of these insects lands on a field of rice, wheat, or corn, nary a trace remains. Martins, a kind of blackbird imported from India, feed off of this insect, and the Government has stopped destroying them, but...despite prohibitions many martins are killed every day.[542]

As Sonnerat indicates, the colonial government in the Mascarenes mobilized campaigns to fight these pests. The 1774 ordinance on Île Bourbon, for example, required each colonist to deposit four rat tails each month for each slave on their property.[543] The 1770 regulations for Île de France demanded ten bird heads and twenty rat tails per slave per year per plantation. So significant was this effort that the administration promised to award a slave to the colonist who provided the most bird heads and rat tails.[544] Already by 1745, on Île Bourbon colonists were ordered to kill 100 pest birds a year and deposit their heads.[545] It was illegal, as Sonnerat noted, to kill good birds (martins and shrikes), but an especially bad bird called the "Calfat" counted for two

541. See remarks to this effect in "Mémoire sur la Guyane, par M. Victor Hugues," AC/Bordeaux, Ms. 141 (Fonds Delpit, Guyane).

542. Sonnerat (1782), vol. 2, p. 83. Sonnerat gets his ornithology wrong here; martins are not blackbirds.

543. "Ordonnance de M. M. les Commandant & Ordonnateur de Bourbon, pour la destruction des Oiseaux & des Rats dans toute la Colonie. Du 12 Septembre 1774," Delaleu, vol. 2, pp. 132-34. There was a 500 livre fine for killing martins.

544. "Réglement...pour la destruction des Oiseaux & des Rats. Du sept Mai 1770," Delaleu, vol. 1, pp. 327-29, esp. article 4.

545. Ms. Code de l'Ile de Bourbon, vol. 1, p. 692, in AN-MAR G 237.

heads.[546] These procedures fell into disuse, if they were ever successful, and the regulations were renewed in 1775 and again in 1778. The lack of lead shot for shooting birds seems to have been a stumbling block, and the government promised to provide additional lead.[547]

As Sonnerat likewise indicated, on Île de France and Île Bourbon grasshoppers and locusts came as plagues and devastated crops.[548] A report of 1731 out of Île Bourbon presaged Sonnerat's observations a half a century later:

> On this island grasshoppers continue to wreak great havoc on wheat, rice, corn, and vegetables, and for this reason colonists are at their last resorts for feeding themselves and their slaves. They can't raise poultry or pigs and are consequently deprived of every means of existence. This island is in dire need of emergency food supplies, until it pleases Providence to deliver it from this terrible scourge...The colony has nonetheless subsisted for five years without any help from abroad, despite the grasshoppers, hurricanes, and the rats that have not left the 20th part of the harvest that might have been expected.[549]

The government gave orders for colonists to have their slaves destroy grasshoppers. In 1730 all other work on plantations was ordered to stop, and colonists and slaves were to form brigades to burn grasshoppers. Other episodes occurred in 1729 and 1768.[550] In 1766, colonists themselves solicited a government directive "concerning the destruction of rats, monkeys, birds, and grasshoppers that are devastating crops."[551] The regulations of 1768 for Île de France created a grasshopper corvée of sorts, with one out of every ten slaves to form work gangs in January to combat grasshoppers wherever they cropped up – something like an entomological fire brigade.[552]

546. So, too, did Java and Chinese sparrows ("moineau de Java" and "moineau de la Chine"); see "Ordonnance...sur la destruction des Oiseaux. Du quatorze Octobre 1775," esp. article 5, Delaleu, Supplément 2, pp. 152-55.

547. "Ordonnance...sur la destruction des Oiseaux. Du premier Janvier 1778," Delaleu, Supplément 1, pp. 256-58; see also "Ordonnance...concernant le débit de la poudre à canon...," Delaleu, vol. 1, pp. 386-87, concerning providing colonists powder with which to shoot birds.

548. See above and Malleret, p. 122; and reports: ms. Code de l'Ile de France, 1556-1806, vol. 1; ms. Code de l'Ile de Bourbon, 1673 à 1789, vol. 1, pp. 375, 479, in AN-MAR G 237.

549. Letter from Bourbon Conseil Supérieur to "M.rs les directeurs généraux de la Compagnie des Indes," dated "20e Xbre 1731," in Lougnon (1934), pp. 161-73, here p. 164.

550. Ms. Code de l'Ile de Bourbon, vol. 1, p. 397; see related regulations, ms. Code de l'Ile de France, vol. 1, p. 501; ms. Code de l'Ile de Bourbon, vol. 1, p. 375, vol. 2, p. 479, in AN-MAR G 237.

551. "12 juillet 1766. Pétition présentée par Messieurs les syndics et députés de la colonie de l'île de France demandant qu'il soit publié un règlement concernant la destrucion des rats, singes, oiseaux et sauterelles qui dévastent les plantations," Toussaint, p. 364, #87.

552. "Ordonnance...pour la destruction des Sauterelles. Du dix-huit Janvier 1768," Delaleu, vol. 1, pp. 306-309.

These kinds of ecological problems bring us to consider more generally the obstacles thwarting French colonial expansion and the actions of the Colonial Machine. (These we address in the next Part of this book.) With a notable exception, the episodes surrounding rats, birds, grasshoppers, and the like did not directly involve the Colonial Machine, except insofar as colonial government came to be involved. The exception concerns the ravages caused by ants in the cane fields and the prize of 1 million livres sponsored by the Conseil Souverain of Martinique to find an effective ant repellant.[553]

A report appearing in the *Philosophical Transactions* of the Royal Society of London for 1790 described the threat that ants posed for the sugarcane fields. Grenada suffered an invasion of ants that had arrived in contraband from Martinique, and for several years ants had marched across the island destroying sugarcane plantations as they went. From Grenada, John Castles, Esq. wrote: "Their numbers were incredible. I have seen the roads coloured by them for miles together; and so crowded were they in many places, that the print of the horses feet would appear for a moment or two, til filled up by the surrounding multitude. This is no exaggeration."[554]

The ant problem was a serious one that engaged the Colonial Machine through the Conseil Souverain on Martinique and the instrumentalities of colonial administration. In 1775, meeting at Fort Royal, Martinique deputies from the different districts on the island proposed that the colony fund a prize of 1 million colonial livres for "a sure means of destroying ants."[555] The Conseil Souverain registered the resolution, published it, and with the blessing of the local governor and intendant, sent it on to Versailles, where it was duly homologized as a royal edict by the king in 1776. The king approved the 1 million livre prize to be paid by assessments on sugar plantation owners over three years, based on the number of their slaves. A panel of seventeen representatives from the different quarters of Martinique were to adjudicate the prize after tests were made in all the districts.[556] Although initiated and nominally controlled by local planters, the involvement of the Conseil Souverain, the administrators, the minister, and the king made this more than a local initiative. The contest – with its extraordinary prize – was advertised widely in Europe, and submissions poured in, many from German states.[557] The royal surveyor

553. On this episode, see AD/Martinique, Série B, Registre 13, fol. 152; AN-MAR G 146 ("Procédés pour détruire les fourmis, 1775-1778"); Durand-Molard, vol. 3, pp. 256-58.

554. "Observations on the Sugar Ant. In a letter from John Castles, Esq. To Lieut. Gen. Melvill, F.R.S.," *Phil. Trans.*, 80 (1790), pp. 346-58, here p. 347.

555. "Arrêt du Conseil d'Etat du Roi, portant homologation de la Déliberation des Députés de la Colonie, qui assure *un million* [sic] argent des iles à celui qui indiquera un moyen sûr de détruire les Fourmis, Du 8 juin 1776," Durand-Molard, vol. 3, pp. 256-58. (The colonial livre was worth 2/3rds of a *livre tournois*).

556. AD/Martinique, Série B, Registre 13, fol. 152.

557. AN-MAR G 146 contains several of these submissions.

from Saint Lucia sent one; the Académie des Sciences in Paris received one, and rejected it.[558] The prize was never awarded. The ants won.

* * *

The record of the Colonial Machine in the conquest of maritime and colonial space; in attempting to manage the physical health of sailors, colonists, and slaves; and in creating and activating the infrastructure and personnel for applied botany and economic development was impressive and historically unprecedented. The Colonial Machine achieved its successes in action through an effective chain of command and by mobilizing its constituent parts towards specific ends. The British colonial administrator Edward Long understood this when, in 1774, he wrote the words we chose to head this study. They bear repeating here: "There is a spirit in the French monarchy, which pervades every part of their empire; it has select objects perpetually in view, which are steadily and consistently pursued; in their system the state is at once the sentient and the executive principle. It is in short, *all soul*; motion corresponds with will; action treads on the heels of contrivance; and sovereign power usefully handled and directed, hurries on, in full career to attain its end."

By the same token, many features of life in the eighteenth century operated to limit the success, if not the ambition, of the French and their Colonial Machine. In order to properly delineate the nature and record of the Colonial Machine, we need to investigate those limits where failure and alternatives defined identity and success.

558. Ms. "traduction de la lettre adressé a lacademie par Mr. jean ludolphe fager licensier en medecine a Lipsic du 2 fevrier 1778" and ms. report with marginal note, "lu a lacademie le 7 mars 1778. Commissaires Mrs. fougeroux et adanson," APS-MS, B D87, Group 13.

Part III

Limits and Alternatives

So far, we have traced the coming-into-being of the Colonial Machine, the details of its organization and functioning, and the range of its activities across broad areas of scientific work and research in cartography and navigation, colonial medicine, and botany and agronomy. Through the actions of the men and institutions that constituted it, the Colonial Machine reveals itself to have been an historical entity of importance for the story of French science and overseas expansion in the seventeenth and eighteenth centuries.

As much as the Colonial Machine draws our attention in these regards, we need to be careful not to exaggerate what it was or what it achieved. Therefore, in this part we examine factors that constrained the Colonial Machine, that limited its success, and that challenged the Colonial Machine as a model for the organization and pursuit of science. In this way we can define the Colonial Machine negatively, so to speak, by focusing on its limitations and what it failed to achieved. The disastrous attempt to establish a new French colony in Kourou in French Guiana in the 1760s provides an excellent entrée into this line of inquiry.

Kourou as a Test Case

In 1763 and 1764 roughly 10,000 white settlers, mostly recruited from Alsace and Germany, sailed from French ports to found a new settlement at Kourou about thirty miles north of Cayenne in Guiana. Such a massive undertaking did not come about by accident. With little lag time, it had been quickly decided upon, but thoroughly planned. It was a huge colonizing project organized and carried out by the French state. The Colonial Machine was centrally involved in the whole affair, a fact that naturally draws our attention to the Kourou episode. But in addition, the project planned for French Guiana by the minister of the navy in 1761 exposes fault lines between scientific expertise and political authority in colonial expansion, and it reveals the limits of cooperation between science and polity inherent in the Colonial Machine. Indeed, despite the full mobilization of the Colonial Machine and teams of botanists, cartographers, and scientific academicians, the enterprise to implant a new colony of several thousand white settlers in faraway Guiana led to about 7000 deaths, partly because of cupidity and an excess of ambition on the minister's part, partly because of climatic conditions, but also because of a too theoretical and abstract conception of knowledge and development put forward by the scientiific experts and politicians involved. The fiasco at Kourou is a test case gauging the ambitions of the Colonial Machine and the interactions of expert knowledge and political power.

Experts on Site. The story of the aborted French colony at Kourou is well documented.[1] Suffice it to say here that never, in effect, had so much information been brought together about a territory with the aim of informed, rational colonization. Never had so many learned experts been mobilized for a colonial endeavor – to the point of these experts feeling empowered to guide the steps of administrators and financiers who were calling on them. And yet, Kourou proved an unprecedented disaster in European colonial history, which entailed the deaths of several thousand colonists in appalling conditions.

In 1761 Étienne-François, the duc de Choiseul (1719-1785), came to head both the ministry of the navy and the Ministry of the Army and was the principal government authority of the day. Choiseul seriously thought to make Guiana a French tropical riposte to the English colonies in North America. Given the inevitable loss of Canada, in his mind the goal was to implant a settlement colony, a "white colony" in Guiana. The strategic aim of this colony was to serve as a logistical base and resource for the Antillean colonies and as a rapid-response military outpost of French soldiers. The project was not wholly unrealistic, and if the numerous reports that drifted into the *Bureau des Colonies* in the Ministry of the Navy and the Colonies in those years are to be believed, French Guiana offered real hopes for economic and strategic development.

Wishing to better understand the potential of Guiana, but without a clear project in view, Choiseul and the head of the Bureau des Colonies in the Ministry, Jean-Augustin Accaron, pressed the administrators of Guiana for information. In 1762 Choiseul and Accaron sent a retired army lieutenant colonel, Béhague de Septfontaine, to Cayenne with instructions (in the form of a veritable questionnaire) to study the possibilities for the development and expansion of the colony.[2] He was asked to determine the plants best suited to grow

1. On the expedition to Kourou, see Regourd (2005), whence this account is drawn; see also the somewhat romanticized Thibaudault; Michel, esp. pp. 175-76, for an overview of manuscript sources; Chaïa (1958), (1979a), (1979b); Petto, pp. 122-23; Froidevaux (1892), (1893) and (1899); and the volume produced by the Ministère de la Marine et des Colonies in 1842, *Précis historique de l'expédition du Kourou (Guyane française). 1763-1765*; also Godfroy-Tayart de Borms (2009a, 2009b).

2. Foidevaux (1892), p. 225, makes this point.

there in order to augment the colony's commerce, and to study the possibilities of introducing cinnamon, clove, and nutmeg. (These developments preceded the implantation of spice production in Guiana in the 1770s and 1780s.) In the following months Béhague sent seventy papers to Choiseul, in which he expressed numerous reservations about a sudden and enormous colonizing project.[3] These reservations fell on deaf ears.

At the same time, the Bureau des Colonies sent two additional expert agents to Cayenne and assigned them important tasks. The first of these agents was the naturalist Fusée-Aublet, who was transferred from Île de France in the Indian Ocean. Fusée-Aublet went to Guiana as royal apothecary-botanist. The royal order of 2 May 1762 instructed him as follows: "The scope of your mission must focus on seeing and examining everything relative to your knowledge about the production of this new land, to prepare good reports, and to give an account of everything that can be done for a country that merits more attention than it has received to the present."[4] Fusée-Aublet was forty-two-years-old. He had trained at the Jardin du Roi, where he studied under Rouelle and Bernard de Jussieu, and he had spent eight years spent on Île de France in the Indian Ocean. He was a field botanist and an expert in exotic botany. He volunteered for a round-the-world expedition to collect useful plants for introduction to Kourou, but this was refused.[5] During his two years in Guiana, he did substantial work and sent many samples to the Jardin du Roi and the government, but also to his protector, the collector Bombarde, a financier close to Choiseul. Fusée-Aublet's shipments do not seem to have been directly useful to the establishment of a new colony, but a decade or so later, in 1775, using the results of his conscientious botanizing at that time, he published the four volumes of his *Histoire des plantes de la Guyane française*, the first flora of Guiana worthy of the name.

The second expert agent was a cartographer, who gained a name for himself in Canada and in mapping the right bank of the Rhine: Joseph-Charles Dessaingy. He was appointed as a surveyor to make "exact maps" of Guiana. Dessaingy became one of the ingénieur-géographes du roi in 1768 and served in Guiana until 1786, making topographical maps of Cayenne and various areas along the coasts of Guiana.[6]

It was in this context that in June 1762, a colonist from Guiana with the redolent name Jean-Antoine Brûletout de Préfontaine arrived back in France. A twenty-year veteran of Guiana and a former fugitive slave hunter, Brûletout de Préfontaine owned two properties in Guiana, 120 head of cattle, and a sugar operation with sixty-two slaves. Wishing to advance the development of the

3. Froidevaux (1892), pp. 229-30.

4. Fusée-Aublet (1775), pp. xii-xiii; Regourd (2005), pp. 236-37.

5. Froidevaux (1893), p. 9, citing AN COL, C^{14} 26, pp. 349-52.

6. Berthaut, vol. 1, p. 31; Chaïa (1979b).

sleepy colony, Préfontaine presented Bernard de Jussieu, he of the Academy of Sciences and the Jardin du Roi, with a thought-out project to settle 200 or 300 colonists and 600 hundred slaves on the banks of the Maroni in the north of the colony. Word of this project rapidly came to the ears of Accaron at the Bureau des Colonies, who informed Choiseul. The minister was naturally receptive, given the context, but whereas Préfontaine envisioned a modest establishment of a few hundred colonists aided by a few hundred slaves, Choiseul used the moment to launch the project he had been brewing for several months; a huge and well-thought-out colonial undertaking. Pushed by the desire to counter the English in America, but also by his ambition and appetite for profit that the new colony seemed to offer, the minister threw himself into a hasty and excessive endeavor that brings to mind, at least on paper, the pipe dream of the Mississippi less than a half a century earlier.[7]

While teams of financiers and politicians hatched this gigantic project in Choiseul's offices (the talk was of several thousand colonists), other meetings around the minister brought together figures close both to power and to the scientific circles of the capital. These included the naturalist Thibault de Chanvalon, the financier Bombarde, the two Turgot brothers, and the colonist Brûletout de Préfontaine, the only one with practical experience of the geography and character of Guiana.

Etienne-François Turgot (1721-1788), a Knight of Malta, was the brother of the intendant of the Limousin and later finance minister, the famous Anne-Robert-Jacques Turgot. Close to the learned world, the chevalier Turgot had a military background, but was equally enamored of botany and natural history, having published in Lyon in 1758 a *Mémoire instructif sur la manière de rassembler, de préparer, de conserver et d'envoyer les diverses curiosités d'histoire naturelle*. One of the founders of the Paris Société d'Agriculture in 1761, he was named governor-general.

The financier Bombarde is more mysterious. He was a great German collector of "curiosities" and the protector of the royal apothecary-botanist Fusée-Aublet, who, as we saw, was sent to Guiana a few months earlier. Bombarde was also a relative of the duc de Choiseul.

Choiseul named Jean-Baptiste-Mathieu Thibault de Chanvalon (1725-1788) as intendant. Born in Martinique, Thibault de Chanvalon became a jurist in Bordeaux, but was an amateur, passionate about *physique* and natural history. He was a member of the Academy at Bordeaux, and he proved a diligent contact of the Académie des Sciences in Paris, which elected him a correspondent in 1754, first of Réaumur and then of Jussieu. He gained a name for himself

7. Choiseul and his cousin and fellow minister, Choiseul-Praslin, tried to have the immense territory deeded to them between the left bank of the Kourou River and the right bank of the Maroni, the place where the future colony was mostly likely to be situated; Ministère de la Marine et des Colonies (1842), pp. 14-15.

in writing his *Voyage à la Martinique*, which for the most part concerned natural history and the social development of his native island where he sojourned again from 1751 to 1756. In his own mind, this text was only an introduction to future, more detailed studies that he never had the opportunity to compose. His *Voyage* appeared in 1763, having received the official approbation of the Académie Royale des Sciences in 1761.[8]

Toward the end of 1762, Choiseul mobilized the cartographical establishment to work on the Guiana project. Under the supervision of Philippe Buache, first royal geographer, Louis Charles Buache and Brûletout de Préfontaine prepared a new "Carte géographique de l'Isle de Cayenne" dedicated to Choiseul.[9] This new map contained information on the existing colony at Cayenne, as well as the new colony envisioned on the banks of the Maroni. In addition, Philippe Buache assembled as much information as he could about Guiana and, according to his *éloge*, drew "forty-five manuscript maps for the use of the minister and for M. Turgot, member of the Academy of Sciences, named as governor of this colony."[10] For his own part and as part of this effort, in 1763 Jacques-Nicolas Bellin, chief hydrographer at the naval Dépôt des Cartes et Plans, published a substantial, 283-page volume, the full title of which gives a sense of its genesis and amplitude.[11]

> Geographical Description of Guiana, Encompassing the Possessions and Settlements of the French, Spanish, Portuguese, and the Dutch in this Vast Country. The Climate, Agricultural Products, Animals, Inhabitants and their Mores and Customs. And, the Commerce that Can be Had. With Remarks on Navigation, and Maps, Plans, and Diagrams. Drawn at the Navy's Depot of Maps and Plans by Order of M. the Duke of Choiseul, Colonel-General of the Swiss Guards, Minister of War and the Navy. By Mr. Bellin, Engineer of the Navy and the Depot of Maps, Royal Censor, Member of the Academy of Naval Sciences and the Royal Society of London.

This compilation, inspired by Barrère's *Nouvelle relation de la France Equinoxiale* published twenty years earlier, came with twenty or so maps, half of which covered the territory of French Guiana.

8. See Thibault de Chanvalon (1763) and (2004); Lacroix (1938), vol. 3, pp. 63-69; Lescure, p. 62; materials in PA-DB (Thibault de Chanvalon); Thibault mss. at BCMNHN, Mss. 690, 1337, 1338. The catalogue of the BnF has his name as Chanvalon.

9. BnF, Département des Cartes et Plans, Ge C 5003.

10. "Éloge de M. Buache," HARS 1772 (1776), p. 146. The documents and maps produced by Buache for the Kourou expeditions are preserved at SHMV, SH Ms. 350, and AN 2JJ 18 et 19. NB: Turgot became a Free Associate of the Académie des Sciences only 1765.

11. See Bellin (1763), and SHMV, SH Ms. 346 3818A; see also Broc (1975), p. 124; Petto, pp. 121-23.

Beyond all that, a new ingénieur-géographe was sent to Guiana to assist the cartographer Dessaingy, already there. His name was Simon Mentelle, and he was charged to draw up a precise map of Guiana using advanced triangulation techniques.[12] Arriving in Guiana with Préfontaine during the summer of 1763, Mentelle had the task of laying out the new colony, as much to aid its development as to keep the government informed of the project's progress. A student of Buache and the astronomer Lalande, and trained in the field on the Cassini map of France, Mentelle had all the requisite qualities. He was joined by another cartographer, François-Etienne Haumont, aged around thirty, charged to map Guiana between the left bank of the Macouria River and the right bank of the Kourou River. (Haumont completed this map in 1766; he died in 1767.) In 1764, the mathematician and future historian of science, Jean-Etienne Montucla (1725-1799), known for various mathematical works, was named the astronomer for the new colony, on top of his position as secretary to the governor.

The full armamentarium of the Colonial Machine became engaged in the Kourou project. We should not overlook other experts already in place or on their way to Guiana, who augmented the scientific weight of the endeavor planned from Versailles. We have encountered most of them. A student of Jussieu, the médecin du roi for Cayenne, Jacques-François Artur, had been in place since 1735 and knew the colony well. (Artur, one recalls, was Academy correspondent of Réaumur and then of Duhamel du Monceau and was attached to Buffon and the Cabinet du Roi.) Surgeon-major Bertrand Bajon was posted to Guiana at the beginning of 1764 and stayed through 1776, becoming an established scientific figure. Likewise in 1764, Jean-Baptiste Patris, a graduate of the medical faculty at Reims, was named as médecin du roi at Cayenne on the recommendation of Bernard de Jussieu, with whom he had studied at the Jardin du Roi.[13] On their arrival in the colony all these correspondents sent off papers and samples of which some have come down to us. Official draftsmen-naturalists were also attached to the colony. On the same boat that carried Préfontaine back to the Americas, Louis-Pierre Desmon arrived at Kourou; he produced some panoramas that give an idea of the distortion that existed between the project as it was conceived and the concrete conditions of its realization.[14] Here is where Charlotte Dugée enters the story, the mulatto woman from Saint Domingue, who became licensed as a royal artist for Cayenne ("dessinatrice du roi pour Cayenne") in 1764 and who botanized with Patris for the next few years before vanishing, driven insane, in the forest.

12. These would be trigonometrical and mensural techniques developed out of the Cassini map project and its successors, not the chronometric and longitudinal techniques available two decades later with the perfection of the chronometer, as we saw in Part IIA.

13. See Patris (1777) and Chaïa (1975).

14. Thibaudault, p. 205.

[View of the Kourou Colony in 1764]

Choiseul likewise consulted Michel Adanson in the planning of the Kourou project. Adanson, that botanical iconoclast with experience in Senegal a decade before, was then a recognized authority on extra-European botany and had been appointed a royal censor for works relating to colonization. Adanson was invited to go to Kourou to prepare for the arrival of colonists and supervise the beginnings of cultivation; but having refused such postings in the past, as we have seen, and comfortably ensconced in Paris, Adanson declined the job, perhaps frightened by the prospect of having to assume responsibilities on the spot in the colony.[15] But Adanson did work closely with Choiseul and the Chevalier Turgot in the planning stages of Kourou.[16] Adanson's idea was to create a sister colony at Île Gorée in West Africa that would be linked to Kourou. He suggested indigo and gum arabic as commercial and trade items the two colonies could develop.[17] He went on to propose an intercontinental botanical exchange of the sort that fully developed with the Indian Ocean islands later in the 1770s and 1780s, envisioning a system embracing Africa, Cayenne, Guadeloupe, Martinique, Saint Domingue, and Louisiana.[18] He

15. Froidevaux (1893) and (1899).

16. Froidevaux (1899), p. 86.

17. Froideveaux (1893).

18. See "Pièces instructives concernant l'île Goré, voisine du Cap Verd en Afrike avec un projet et des vues utiles relativement au nouvelle etablissement de Kaiene" and "Avantages qu'on peut tirer de l'isle Goré," presented in Froidevaux (1889); also Froidevaux (1893). Note Adanson's idiosyncratic, phonetic spelling.

thought, too, that African elephants might be introduced into the new colony at Kourou to good effect.[19]

The intellectual quality of the administrators appointed to bring into being this new French South American colony and their connections to the rest of the Colonial Machine augured well for the adventure. The impressive deployment of scientific and technical expertise allowed the administration to assemble in record time all the information necessary to make rational decisions. Information was collected in the domains of cartography, agronomy, botany, and medicine and made available to the responsible authorities.[20] Undeniably, such a conjuncture of knowledge and power (and the human talent underlying both) could have provided a model for colonial development. Yet the political authorities seem to have acted as if they had no information at all at their disposal.

Unmitigated Disaster. Colonists were recruited through a vigorous propaganda campaign in the hospitals of the kingdom, but especially in Alsace, in Germany, as well as in Holland and Ireland. The first of a dozen or so massive convoys left in May 1763 and debarked in Guiana at the height of the rainy season to find that sufficient shelter had yet to be built. This, despite all the works and memoirs on Guiana that indicated precisely the periods when heavy rains precluded any productive work. Administrative disorganization showed itself not only in the lack of means to accommodate the immigrants, but also in the delay of ships bringing food for colonists. Beyond that, colonists were forbidden to use slaves, even for the hard work of clearing land, despite the constant appeals of the local administrators, who underscored the necessity of employing slaves. Medical and natural history works touching on tropical fevers, such as those by Barrère (1743) and Poissonnier-Desperrières (1763), underlined the disease conditions faced by Europeans in confronting the climate of the tropics. The royal physician Artur, in Cayenne, addressed a memoir to the financier Bombarde to argue against the creation of a colony of white immigrants.[21] But Choiseul, for the strategic reasons touched on above, maintained the interdiction against slave labor in the new colony. And then, when the illnesses that typically affected half of new European immigrants broke out, the sick were confined in squalid hospitals and left without resources. Needless to say, the means put in place for treatment of the ill often contradicted the advice found in the medical literature.[22] On top of that, convoys of

19. "Catalogue des plantes utiles qu'on peut tirer du Sénégal pour Kaiene," in Froidevaux (1899).

20. See Turgot's "Registre contenant les copies de différens Mémoires sur les objets utiles à l'établissement de la colonie," BnF, f. fr., Ms. 6235; see related documents, BnF, n.a.f., Ms. 3605.

21. BnF, n.a.f., Ms. 2572, fol. 779, cited by Chaïa (1958), p. vii; see also Polderman in Artur (2002), pp. 44-47.

22. On these epidemics and the medical personnel attempting to contain them, see Chaïa (1958).

immigrants, many themselves sick and malnourished, continued to wash up in Guiana during all of 1764, putting all of the colony at Cayenne in peril.

In Cayenne, Préfontaine and Chanvalon ran up against the hostility of the local administrators, who were jealous of them as outsiders and newcomers to colonial administration; but hostility also came from planters already there, who were threatened by the unreasonable influx of new colonists. In the meantime, Turgot, the governor-general, took his time in coming to Guiana, preparing agricultural projects without troubling, or so it seems, to deal with the surge of people piling up by the thousands in French ports and awaiting departure for Kourou. Only in November of 1764 did Turgot leave for Guiana, while thousands of colonists who had preceded him (run out of French ports, where they represented a menace to public order and safety) were already there and in the grip of terrible epidemics. Turgot did not stay, but quit the colony at the beginning of April 1765, hoping to lay blame on Chanvalon for the catastrophe – something Turgot succeeded in doing perfectly. Chanvalon was arrested and thrown into the Bastille, where he languished for over a decade.[23]

Of the 15,000 colonists who planners hoped to bring to Kourou, a little more than 10,000 seem to have arrived on the shores of Guiana in 1763 and 1764. Today, we can estimate at 6000 to 9000 the number of deaths of victims, weakened by hunger and forced labor and struck down by typhoid, malaria, scurvy, dysentery, and pulmonary diseases. Some hardy souls survived and remained at Kourou and other points in the colony. One notable group of Kourou survivors washed up in Saint Domingue, where authorities organized them in a new settlement known as Bombardopolis![24] Fusée-Aublet was charged to create this settlement. Only 2000 to 3000 thousand colonists succeeded in regaining Europe. The dramatic failure of the endeavor in human terms was compounded by a financial disaster, with estimated losses at some 25 million livres.

The causes of this debacle were numerous. It can be seen as the result of the cynicism, blindness, and cupidity of Choiseul and his entourage, to which should be added the administrative inexperience of the men chosen, at least officially, for their scientific and agronomic competence, men such as the Chevalier Turgot, the colonist Brûletout de Préfontaine, or the naturalist Thibault de Chanvalon. It is equally possible to blame the virulence of tropical pathologies that were particularly ferocious against populations more attuned

23. Lescure, p. 62. Thibault de Chanvalon was, in effect, the scapegoat and clearly paid for others, notably Choiseul and the Chevalier Turgot, whose responsibilities seem greater, insofar as we can judge today. Chanvalon was imprisoned for many years, his worldly goods seized, and his family sorely treated. He was not rehabilitated until 1781, a few years before his death in 1788. For his part, the Chevalier Turgot suffered only a short period of exile that was soon overlooked, no doubt thanks to his brother's position. In 1765 he became a Free Associate of the Académie Royale des Sciences, and on his death in 1789 Condorcet pronounced his academic éloge.

24. Regourd (2000), pp. 499-501; McClellan (1992), pp 61-62.

to the rigors of cold than the heat and humidity of the tropics, and who were also weakened by long voyages at sea.

The Kourou affair marks the paradoxical failure of a project that more than any other to that point sought to put science at the forefront of overseas expansion and to place knowledge and learned experts in the service of the ambitions of colonial administration. Undeniably, the information received by the administrators, be that at Versailles or in Guiana, was precise, abundant, and of quality. The scientific wheels and gears of the Colonial Machine showed themselves particularly effective in assuring the intellectual logistics for such a government-sponsored project: finding specialists, collecting information, and producing maps and printed material destined for larger distribution. All of this of knowledge was duly organized and approved by unchallengeable learned authority. The Kourou disaster cannot be attributed to a failure of the central units of expertise and scientific validation such as the Jardin du Roi, the Dépôt des Cartes et Plans, the Observatoire Royal, or the Académie Royale des Sciences, which strove to respond to the needs of the state and colonial administration. On the other hand, the Colonial Machine as a whole seized up and failed in the Kourou endeavor because of a failure at the level of decision making. More than once, the minister and his office rejected warnings from specialists and local informants. Béhague, Artur, Préfontaine, and Thibault de Chanvalon vainly protested in the face of the outlandish project of Choiseul.[25] In this case, the political wheels and gears the Colonial Machine disconnected from the scientific and expert components furnishing the information, and spun off into disaster. Yet in the end, Choiseul blamed the experts. As he wrote in his exculpatory memoir to Louis XV in 1765: "My errors stemmed from what I was told, and I could not have been worse informed by the colonial office I found in place. To this misinformation I added my own ideas, which proved false because I was ill informed."[26] That notwithstanding, Accaron, head of the colonial bureau in 1763 was promoted to intendant-general of the colonies in 1765.

This episode thus poses the question of the status of expert knowledge in the making of political decisions and decisions pertaining to colonial policies in France in the second half of the eighteenth century. The Kourou affair reminds us that learned expertise, as valued as it was, remained subordinate in the minds of the administrators of the time, subordinate to military, political, and commercial imperatives. Knowledge and expertise came "in support" of a project; they "served" it. It was a question of providing maps for the military, practical advice to colonists for agronomy and commerce, and effective remedies to treat troops and colonists. In their way, too, the scientists and experts of the Colonial Machine offered the elements of an intellectual legitimation of

25. Froidevaux (1892), pp. 230-31.
26. Quoted in Chaïa (1958), p. lxxv.

the colonial enterprise for the prestige and glory of the king. In no case could the savant or expert (even if brought to the forefront of political action like Thibault de Chanvalon or Brûletout de Préfontaine) hope to constrain political power, which remained master of decision making.

Obstacles to Success

More than one obstacle inherent in the wider material world of the seventeenth and eighteenth centuries gummed up the works and limited the success and historical impact of the Colonial Machine. Collectively, these impediments – amply evident in the Kourou affair – define the practical limits of the Machine and were so serious that by present-day standards it seems hardly credible that the Colonial Machine functioned at all.

Disease and Shipwrecks. Disease conditions and the very real prospect of illness faced by colonists abroad and especially in the tropics constituted major hurdles obstructing the growth of the French colonial enterprise overseas. What this barrier represented is plain in the Kourou affair, and we have examined what it meant to the Colonial Machine in the previous Part of this book that explored living and dying in the colonies. Diseases attacked not only humans, but their animals and their cultivars. The animal disease of glanders became a serious problem in Saint Domingue, for example; and in collaboration with the royal veterinarian, the royal physician Charles Arthaud conducted experiments to try to inoculate donkeys and asses against the disease.[27] An otherwise unidentified "*poux*" or louse attacked coffee trees on Île de France in the 1740s, threatening the entire coffee production.[28] In 1749, the previous 3 million pounds of the coffee harvest was reduced to 120 thousand pounds.[29] The problem resurfaced in 1751.[30] And then, in 1777, most of the coffee trees on Île Bourbon were wiped out by an epidemic.[31]

Finally, death and mental illness stalked the agents of the Colonial Machine. Credible reports of Dombey's death arrived in France, setting off a shameful scramble for his post![32] We have already encountered Charlotte Dugée, the

27. McClellan (1992), pp. 233-34. On disease as a factor affecting colonization, see also Pritchard (2004), p. 77; Boucher, pp. 133-34, 263.

28. See correspondence related to this matter: "Copie de la lettre écrite par la compagnie à M.rs du Conseil supérieur de l'isle de France dattée à Paris le 28 février 1749"; Compagnie letter to "M.rs du Conseil supérieur de l'isle de Bourbon," dated "A Paris le 8 Juillet 1749"; and Compagnie letter to "M.rs du Conseil Supérieur à l'isle de Bourbon," dated "A Paris le 17 mars 1750," in Lougnon (1949), pp. 139, 158, 235.

29. Bourbon Conseil Supérieur letter to "Messieurs les sindics et Directeurs de la Compagnie des Indes," dated "A l'isle de Bourbon le 10 8bre 1749," in Lougnon (1949), pp. 174-75.

30. See entries in Lougnon (1956), #124 and #133, pp. 37-38.

31. ALS Cossigny to Le Monnier, dated "Palma le 2-8bre 1777," BCMNHN, Ms. 1995.

32. Thouin called Dombey "un vray Martir de l'histoire Naturelle...," ALS "Thouïn" to "Monsieur le Comte," dated "A Paris ce 9 Aoust 1784,"AN O[1] 2112.

botanical artist, who supposedly lost her mind and disappeared into the Guyanese jungle, and we have seen the murder of the brutal commander of the gold mines of the Compagnie in Senegal.[33] Berry, a clockmaker for Compagnie des Indes, committed suicide; Guyomar, a surveyor on Île de France, went insane.[34] Homesickness was reputed to be a major killer of sailors.[35]

The Colonial Machine faced other, perhaps less obvious obstacles as well. The danger of being shipwrecked, for example, beset every vessel that ventured out upon the waters, and every year ships disappeared at sea. The record is replete with instances that impacted the Colonial Machine. In 1700, for example, Pierre Couplet was shipwrecked off the coast of Picardy on his return from Brazil. Couplet was an *élève* of the Paris Academy, sent to collect astronomical and natural history observations; he managed to save himself, but he "lost all his papers and everything he was bringing back of interest to science."[36] The naturalist-priest Plumier lost his herbarium in a shipwreck.[37] In 1760 one of the voyages planned to test preserving fresh water at sea had to be cancelled because the designated ship, *Le Soleil royal*, sank before reaching port.[38] In 1772 the *Jason*, a ship of the Compagnie des Indes, sank in a typhoon, but the botanical shipment it carried managed to be offloaded onto another vessel.[39] In 1789 the naturalist Bajon was shipwrecked off the coast of Spain on his way back to France from Cayenne.[40] Likewise in 1789, the *Nancy* on its way from Charleston with a cargo of seeds from André Michaux went down off the Île d'Oleron; and we have already seen that André Michaux himself almost lost his life and all his botanical samples on his return voyage in 1795.[41] The naturalist and correspondent of the Academy of Sciences Palisot de Beauvois lost all of his natural history collections from Africa, the Carib-

33. See above Part I, Compagnie des Indes; "Mémoire du Roi pour servir d'instructions au Sieur [blank] Commandant du Sénégal" (1780), BCMNHN, Ms. 293.

34. Lougnon (1956), #924, p. 110; Bourbon Conseil letter to "Messieurs les sindics et Directeurs de la Compagnie des Indes," dated "A l'isle de Bourbon le 10 8bre 1749," in Lougnon (1949), pp. 175-76.

35. According to Poissonnier, referring to patients in naval hospitals, "La Nostalgie en a détruit un grand nombre surtout parmi les provençaux," SRdM, Ms. 7, p. 353 ("Séance du mardi 9 9bre 1779"), and similar reference SRdM-H&M, vol. 10 (an VI, 1798), Mémoires, p. 464.

36. "M.r Couplet le fils qui a été dans le Bresil ou il avoit fait diverses observations Astronomiques, et fait plusieurs remarques par rapport a l'histoire naturelle, a Lû le relation du naufrage quil a fait a son retour sur la coste de Picardie, dans lequel il a perdu tous ses papiers et tout cc quil rapportoit de Curieux," PA-PV 19 (1700), fol. 133 ("Le Mércredi 31 Mars 1700"); see also fols. 235-39 ("Le Mércredi 23 Juin 1700").

37. Jacquet, p. 85.

38. See ministerial document, dated "Versailles le 22 Mars 1760," AN-MAR D^3 42, fol. 74.

39. ALS, Cossigny to Le Monnier, dated "A L'Isle de france le 1er Avril 1772," BCMNHN, Ms. 1995.

40. See report of this incident, SRdM, Ms. 11, fol. 280r ("Séance du Mardi 19 Mai 1789") and ("Séance du Mardi 26 Mai 1789").

41. ALS "Pevrey," dated "Paris ce 30 X.re. 1789," to "M. le Comte d'angivillers," AN O^1 2113A ("Pépinières d'Amérique. Année 1789").

bean, and the United States – twelve years worth of effort – when the ship carrying them sank off Halifax.[42]

Working in the Field. Living, traveling, and conducting research in the field presented other hardships. For example, writing to Delisle in 1755 about his travels in Canada, Father Bonnecamp agreed that good longitude data was a desideratum: "But when you understand the way one travels in this country you will readily admit that such a thing is almost impossible. Our conveyance is a bark canoe that can barely hold the essentials. And, we leave at one or two o'clock in the morning and we camp well after sunset....For any exact observations the geographer must be in charge and not obliged, as I was, to follow a military detachment."[43] The mere recitation of the transits of Venus expeditions obscures the real conditions under which the men involved traveled and worked. Here, one thinks of poor Chappe d'Auteroche racing across Siberia in 1761 before the spring thaw made travel impossible, or the same Chappe again in 1769 in Baja California, ill and taking longitude observations before dying there at the age of forty-one. The epic trip of Jean-Baptiste-Barthélemy de Lesseps across Asia in 1787-1788 fits this category; de Lesseps left the La Pérouse expedition in Kamchatka on 7 October 1787, charged to transport expedition documents and samples to Paris, and he arrived in Paris a year later on 17 October 1788 after unbelievable trials.[44] In Peru in the 1770s and 1780s, the voyager Joseph Dombey was attacked by Indians and by bandits and actually killed a man.[45] In the course of his adventures Dombey also lost his hearing and most of his eyesight, got scurvy, and developed badly bleeding gums.[46] In Cayenne in the 1730s, the naturalist Artur faced starvation along with other colonists, asking his correspondent, "What can you expect from someone in my state? Do you believe that he can maintain enough equilibrium and enough tranquility of mind amid such awful misery to busy himself with collecting plants, drawing fish, and dissecting animals?...One must live and one's first consideration will always be to stave off death, especially the horrible death by

42. Gillispie (2007), p. 151; Banks, p. 72, mentions "at least thirty three major vessels" that sank in the Saint Lawrence between 1650 and 1760.

43. "...Je convien avec vous que rien ne seroit plus propre [than longitude data] a rectifier l'estime des distances qui malgré les plus grandes attentions n'est Jamais sans erreur. mais quand vous sçaurez la façon dont on voyage dans ce païs, vous n'aurez pas de peine a avouer que la Chose est presque impossible. On as pour voiture un canot decorce qui peut a peine contenir les choses les plus necessaires a la vie. d'ailleurs lon part a une ou deux heurs du matin et lon ne campe que longtems aprés le coucher du Soleil...pour avoir quelque chose de bien exact il faudroit que le geographe fut maitre de diriger sa route, et non pas obligé, comme j'ai eté, de suivre un detachement de troupes...," ALS Bonnecamp to Delisle, dated "a Quebec ce 23 octobre 1755," AN-MAR 2JJ 68. Banks, pp. 88-89, writes about native technologies and their employ by European travelers.

44. See above, Part IIA, and Dunmore (2007), pp. 235-36.

45. ALS, Dombey to Thouin, dated "Lima Le 11 Xbre 1778," BCMNHN, Ms. 222; Cap (1858), p. 7.

46. ALS Dombey to Jussieu, dated "Santiago De chili le 20 aoust 1783," BCMNHN, Ms. 222.

starvation."[47] Pace *Manon Lescaut*, the threat of starvation likewise regularly threatened colonists in Louisiana.[48]

Even under less horrific circumstances, fieldwork was taxing. So much is plain in Pierre Sonnerat's report to Adanson from Canton in 1776:

> Since I left you I have traveled all over India. In each locale I amassed new things. I worked a lot because one must at the same time study, learn, and collect. Things follow so quickly that the night is not long enough to catch up on the day's observations. I haven't had a moment's rest since I left Europe. For me ports of call are very tiring and very expensive....I regret neither the hardships nor the considerable expenses that I have to make for the Cabinet du Roi without the hope of someday being reimbursed. I am happy if I might one day make myself useful to my country.[49]

In the same letter Sonnerat goes on to make plain that the incredulity of local populations in the face of European collecting was something with which he and presumably other field naturalists had to contend. Even in civilized China, he reported,

> As nothing I do accords with their ordinary forms of thought, they can't conceive why I stuff birds that I hang in my room, why I collect plants that I put between sheets of paper with a few words of writing, and why I offer substantial sums to wander across mountains that are not inhabited except for tigers. All that is beyond them, and I seem to them so unusual that with all the intelligence we attribute to them they can't

47. "...Que peut-on attendre d'un homme en l'etat ou je suis. Croyez vous qu'il puisse conserver assés de sang froid, assés de tranquillité d'esprit dans une misere si affreuse, pour s'occuper a considerer des plantes, a dessiner des poissons, a dissequer des Animaux...il faut vivre, et son premier objet sera toujours d'eloigner la mort, surtout l'affreuse mort de faim," ALS Artur, dated "A Cayenne le 28e d'avril 1738," PA-DB (Artur).

48. Memoir to "Messieurs," dated "Au Biloxy le 25e avril 1721," signed [Bienvilly?] and Delorme, SHML, 1P 274 Liasse 2, Pièces diverses intéressant le comptoir de Biloxi, 1719-1721, #20.

49. "Depuis que je vous ay quitté, j'ay parcouru toute l'inde, dans chaque pays j'ay receuilli de nouvelles choses, j'ay beaucoup travaillé parce qu'il falloit en même temps etudier apprendre et Recueillir, les objets se succedent si Rapidement que la nuit n'est point assez longue pour ebaucher les observations de la journée je n'ay pas eu un Moment de Repos depuis mon depart d'europe, les Relaches sont pour moy bien fatigantes et bien dispendieuses...je ne Regrette ni les peines ni les depenses Considerables que je suis obligé de faire pour le Cabinet du Roy sans Espoir d'etre un jour defrayé, je suis Content si je puis un jour me rendre util à ma patrie," ALS Sonnerat to Adanson, dated "de Canton le 29 decembre 1776," HI, AD 237.

imagine why I don't do what others customarily do, which is to come to China to purchase porcelain or tea.[50]

Time and Distance. We should not overlook the isolation of the French colonies of the Old Regime and the distance that separated them from France and from each other. In 1733 there were no clocks on Île de France, and the colony had received no newspapers for a year.[51] In notable ways the time it took for letters and packages to reach their destinations syncopated the pace of activities of the Colonial Machine. Except where they occurred, it would be anachronistic to speak of delays in the mail, but instantaneous communications that we take for granted today were entirely lacking, and because of the inherent time lag between sending and receiving communications in the eighteenth century, any exchange of information was inevitably disjointed between sender and receiver. This disjuncture holds true for all contemporary communications, but it is especially pertinent when considering the great distances involved to reach the colonies on the one hand *and* the efforts of central authority in the metropolis to govern it all on the other.

J.-N. Delisle's correspondence with missionaries in China and India provides the paradigmatic example, with letters and packages taking eighteen months or more to arrive at their respective destinations. Delisle did not receive Father Gaubil's letters of May and June, 1732 from China, for example, until November of 1733.[52] The Jesuit Michel Benoist's letter from Beijing of October 1759 did not get to Paris until January of 1761.[53] Father Cœurdoux's letter from Pondicherry of February 1757 was not received in France until March 1758; his earlier letter of October 1755 did not show up until April 1758.[54] And so on. In this connection we should note the Russian caravan that wended its way between Moscow and Beijing and that provided a notable means of transport, particularly for shipping books and parcels between

50. "...comme toutes mes autres actions ne s'accordoient point avec leur façon de penser ordinaires, il ne pouvoient Concevoir pourquoi j'écorchois des oiseaux que je pendois dans Ma chambre pour quoy je ramassois des plantes que je mettois entre des feuilles de papier avec quelque mots d'ecriture et pour quoy j'offrois des sommes considerables pour parcourir les Montagnes qui ne soient habitées que des tigres, tout Cela etant audessus de leur portée et je leur parai si extraordinaire que ne pouvant imaginer avec tout leur Esprit (qu'on leur a supposé) pourquoi je ne suivois pas la Coutume des autres, qui est d'aller chez eux pour acheter de la porcelaine ou du thé...,"ALS Sonnerat to Adanson, dated "de Canton le 29 decembre 1776," HI, AD 237.

51. See letter from the Bourbon Conseil Supérieur, "A la Compagnie. A l'Isle Bourbon le 4 janvier 1733," in Lougnon (1933), pp. 47-48. For more complaints about not receiving any gazettes or mercures, see letter "A la Compagnie. Du 31e Xbre 1735," in Lougnon (1933), p. 323. Banks, pp. 52-56, discusses the mechanisms for what was generally an effective communication system for the bureaucracy.

52. See Gaubil letters to Delisle, dated "à Pekin Ce 28 May 1732" and "a Pekin Ce 13 Juin 1732," AN-MAR 2JJ 62; Safier (2008a), p. 213, points to these and related problems discussed here.

53. ALS Benoist to Delisle, dated "Pekin 27 Octobre 1759," AN-MAR 2JJ 66.

54. Both AN-MAR 2JJ 66.

Europe and East Asia; in 1732 Gaubil wrote to Delisle of his anticipation of the arrival of the Russian caravan in 1735![55] Contemporaries were more than aware of these temporal dislocations; Delisle, for example, expressed his annoyance to his correspondent that his instructions regarding the 1753 passage of Mercury arrived in Macao three months after the fact.[56] Yet the slow pace of communications seems to have been an accepted fact of life; Father Gaubil, for example, wrote to Delisle in December 1736: "I have many things to write to you about, and I will do this at the end of next year."[57]

To a greater or lesser degree, every communication that rippled throughout the Colonial Machine evidenced these same delays. Charles Pierre Thunberg's letter of February 1775 from the Cape of Good Hope, for example, did not reach André Thouin at the Jardin du Roi in Paris until August of 1776.[58] The letters patent issued in May of 1789 transforming the Cercle des Philadelphes into the officially chartered Société Royale des Sciences et Arts du Cap François in Saint Domingue did not arrive in that colony until August of that year, and it is a revelation to read of the unfolding of the French Revolution at a remove of weeks and months in the pages of the Caribbean colony's newspaper, the *Affiches Américaines*. The moral is plain: each locality was and remained isolated from the rest of the Colonial Machine, and a peculiar kind of relativity applies to the functioning of its parts.

Climate and Insects. The climate of France's tropical possessions posed its own problems. These we have already seen in terms of disease conditions and mortality, but tropical conditions imposed particular constraints when it came to books and papers. Insect attacks on paper threatened the very basis of colonialism, for in tropical conditions anything written or printed on paper soon became a meal for various bugs, larvae, and worms. Not only could colonists hardly keep libraries – an obstacle to the pursuit of science and learning recognized by all concerned – but, as serious, public, legal, and commercial documents similarly succumbed within a short period. Moreau de Saint-Méry provides a vivid description of this problem:

> One can imagine the ravages caused by insects that devour paper in the Colony [of Saint Domingue]. Old registers and acts are rendered absolutely illegible. Insects convert the paper they eat into a type of earthy

55. "une Caravane Russienne sera icy en 1735. on pourra se servir de Cette Caravane pour me remettre un Cataloguc detaillé des livres chinois et tartares qu'on souhaittera...," ALS Gaubil to Delisle, dated "A Pekin Ce 15 May 1732," AN-MAR 2JJ 62.

56. ALS Delisle "au P. de Neuville missionaire Jesuite à Macao... à Paris le 27 Octobre 1754," AN-MAR 2JJ 68.

57. "J'ay beaucoup de choses à vous Ecrire, et je le feray à la fin de l'an prochain," ALS Gaubil to Delisle, dated "à Peking Ce 31 Xbre 1736," AN-MAR 2JJ 64.

58. ALS, Charles Pierre Thunberg to Thouin, dated "Cap de bonne Esperance ce 14 fevrier 1775," marked by Thouin, "reçu en Aoust 1776," BCMNHN, Ms. 1989.

> gluten, and this gluten hardens, causing the sheets of paper to stick together and to tear if separated....The age of the pieces is not always a guide to the damage, for the nature of the paper and the humidity of the place contribute to accelerate the destruction. In 1783 at the registry of the Sénéchaussée in Cap François I saw parish records dating from 1774 that were already partly illegible.[59]

Other courts and bureaux were affected. The Conseil Supérieur at Port-au-Prince, for example, at one point had its entire records copied on account of the problem with insects. The Conseil Souverain on Martinique proposed spending 6600 livres for a law library, but because "the climate does not allow libraries with many books, which worms and humidity devour or decompose, the Government decided to limit this request to the most standard law books."[60] In part because of the susceptibility of colonial records to insect attacks, in 1776 the national government created a separate colonial archive in France and ordered officials to keep duplicate records.

Naturally, various remedies appeared, including the potion sold for 33 livres by the printer, Mozard, in Port-au-Prince and good for a hundred books: "A halfway intelligent slave can apply it," advertised Mozard, and, in fact, colonists tried to combat the problem by having slaves manually pick out bugs and worms from among books and papers.[61] In 1787 the Cercle des Philadelphes proposed a well-funded prize for the best means of preserving paper from the ravages of insects. Unfortunately, even though the Cercle and its correspondents worked diligently on the problem over the next several years, no satisfactory solution was forthcoming. The Cercle received numerous samples, and all but one succumbed to insects. The sole sample not to be eaten had been treated with arsenic and was deemed too dangerous for actual use. The Cercle wrote up and published its disappointing results in 1788 as its *Dissertation sur le papier.*

Fire and Rain. Fire and rain also constrained the Colonial Machine.[62] The seminary at Quebec, for example, burned to the ground in 1709; the government hospital there did the same in 1752, with the natural history and minera-

59. DPF, p. 337; see also McClellan (1992), pp. 217-18 and notes, whence this material is drawn.

60. "...le climat ne permet pas de former des bibliothèques volumineuses en livres, que les vers et l'humidité dévorent ou décomposent, le Gouvernement a jugé devoir borner l'objet de cette demande aux livres de Droit les plus usuels...," AD/Martinique, Série B, Register 15, fol. 153-153v, and follow-up, Registrer 16, fol. 3; see also Durand-Molard , vol. 3, pp. 658-59. An irony here is that the record of the Martinique courts is itself a nineteenth-century copy, the original having perished; it, too, is in bad shape.

61. *Affiches Américaines*, Port-au-Prince edition, 22 May 1781, quoted in McClellan (1992), p. 218.

62. On fire as a shaping material agent, see also McClellan (1992), p. 28.

logical collections of the royal doctor and Academy correspondent, Gaultier, going up in smoke.[63] The naval hospital at Brest caught fire in 1776 with the loss of 45 galley-prisoners, burned to death because they were chained up.[64] Five powder magazines blew up on Île de France in 1780.[65] Officials on Île Bourbon worried about fire because storerooms were made of wood with palm frond roofs.[66] Pierre-Jean-Nicolas Turpin (1775-1840), the pharmacist, botanist, and botanical artist attached to the Napoleonic invasion force seeking to recapture Saint Domingue in 1802-03, lost his herbarium and years' worth of botanical collections when insurgent Haitians burned down Cap François.[67]

As for rain, on Île de France during Lacaille's three months in the bush in 1753, it poured solidly every day ("We can't leave the tent"), and he nearly died from dysentery.[68] Consecutive typhoons in 1760 and 1761 hit Île de France hard, and the typhoon of 1776 virtually destroyed Cossigny's garden: "My litchis, my mangosteens, my teas all perished in the storms....hurricanes also felled the two pecan trees that I had in my garden."[69] Over in the Americas, wind damage in 1796 caused a total loss of clove production in the government gardens tended by Joseph Martin.[70] The hurricane that hit Martinique in 1756 destroyed all the papers and collections of the above-mentioned Academy correspondent, Thibault de Chanvalon. That poor Thibault, later the scapegoat for the Kourou affair, was subsequently captured on his way back to France that same year by English corsairs illustrates another material factor that shaped the history of the Colonial Machine – war – the one that destroyed Turpin's collections.[71]

Wartime. Gavin de Beer's study, *The Sciences Were Never at War* (1960), remains a classic account of how in many instances international and supranational allegiances transcended the bellicose instincts of nations at war in the

63. ALS, Gaultier to Guettard, dated "a quebec 2 9bre 1752," BCMNHN, Ms. 293.

64. ALS, Laurent to Thouin, dated "a Brest ce 24 Novbre 1776," BCMNHN, Ms. 848; Roussel et al., p. 65; on related fire dangers, see Pritchard (1987a), p. 120.

65. ALS, Cossigny to Le Monnier, dated "A L'Isle de France le 16 Avril 1780," BCMNHN, Ms. 1995.

66. See letter from Bourbon Conseil Supérieur to "M.rs les directeurs généraux de la Compagnie des Indes," dated "20e Xbre 1731," in Lougnon (1934), p. 168.

67. See ALS Turpin to "Monsieur jussieu Botaniste, et president de L'institut national," dated "philadelphie 1e fevrier 1804," PA-DB (Turpin); Hocquette.

68. "La pluye est continuelle tous le jour: nous ne pouvons sortir de la tente...," "Journal historique de mon voyage au Cap de bonne Esperance," OdP, Ms. C 3, 26. Compare Lacaille's sedate paper in MARS 1754, p. 109, where none of his hardships are apparent. On the weather as a factor affecting colonization, see also Pritchard (2004), pp. 76-77.

69. "Mes Letchis, Mes Mangoustans, Mes Thés, sont péris dans les Ouragans... Les 2 Pacaniers que j'avois laissées dans mon jardin, ont aussi péri par les Ouragans," ALS, Cossigny to Le Monnier, dated "Au Port Louis Isle de france le 25 févr. 1776," BCMNHN, Ms. 1995; see also Gunny, p. 300.

70. "Mémoire présenté à l'Institut national, par le C.en Joseph Martin, botaniste, chargé de la direction des jardins et pépinières coloniales, dans la guyane française," BCMNHN, Ms. 48.

71. Lacroix (1938), vol. 3, pp. 61-69; Boucher, pp. 18-20 broaches these issues.

Age of Reason. Although we do not know the outcome of his request, that Delisle could pen the following to a British royal agent in North America, William Mildmay, following the seizure of the French vessel the *Pondichéri* at the start of the Seven Years War, is evidence of this spirit of international scientific cooperation:

> I take the liberty of addressing you…to obtain from those who captured the French vessel of the Compagnie des Indes named the Pondicheri…that they would be so kind as to hand over to me the letters and other things the Jesuit missionaries addressed to me through their French Jesuit correspondents. Because these letters and parcels concern only matters of science and the state of the missions and perhaps some curiosities relative to them, I hope that no difficulties will be had in returning these things to us. As they are of no consequence for you in the present war, only the curious like yourself, Sir, or a few members of the Royal Society might use them for the progress of scientific knowledge for which I am in contact with these missionaries.[72]

For all his internationalist sympathies, Delisle at the same time cautioned Cœurdoux against "imprudently sending the new map of the Compagnie des Indes in this time of war as well as any other observations on geography or our efforts to perfect the cartography of India."[73] And Cœurdoux, a staunch French patriot, replied that "the fear of working for the English or the sea is slowing me down considerably."[74] Toward the end of that war Cœurdoux likewise complained, "We are trying in vain to obtain from the English minister the communication of his observation of the passage of Venus. It would have helped determine the longitude of Tranquebar. Such is the way of the English

72. "...Je prends la liberté de m'addresser à vous...pour obtenir de ceux qui ont fait la prise du vaissau François de la Compagnie des Indes nommé le Pondicheri...qu'ils veuillent bien me rendre les lettres et autres effets des missionaires Jesuites, qui me sont addressés, aux Jesuites de France leurs correspondans; comme ces lettres et pacquets ne concernent que des affaires de sciences et l'etat des missions; et peutêtre quelques curiositez relatives; j'espere que l'on ne fera pas de difficultés de nous les rendre; comme n'etant de nul consequence pour vous dans la guerre present; il ne peut avoir que quelques curieux comme vous Monsieur, ou quelques membres de la société Royale qui pourraient s'en occuper pour le progres des connoissances scientifiques sur les quelles je suis en Correspondance avec ces missionaires," Delisle to William Mildmay, dated "à Paris 12 Avril 1757," OdP, B 1-8 (Supplement, #53).

73. "…mais qu'il y auroit de l'imprudence de l'envoier [la carte de la Compagnie des Indes] dans ces tems de guerre aussi bien que quelques autres remarques sur la Geographie et les tentatives que l'on faites pour perfectionner celle des Indes," Delisle to Coeurdoux, dated "à Paris le 16 Fevrier 1759," AN-MAR, 2JJ 66.

74. "…la crainte de travailler pour les Anglois ou pour la mer me ralentissent beaucoup…," ALS Coeurdoux to Delisle, dated "a Pondicheri ce 18 aout 1758," marked "reçue le 8 mars 1759," AN-MAR 2JJ 66.

everywhere to communicate little and to be anxious to profit from the work of others."[75]

Even when the values of the Republic of Letters prevailed, periods of warfare severely interrupted the smooth functioning of the Colonial Machine.[76] At an earlier period, the elder Lignon, royal botanist on Guadeloupe, had shipments seized by privateers.[77] A substantial botanical shipment to the Jardin du Roi by the royal doctor Prat in 1745 was lost when the English seized the French ship, the *Elephant*.[78] The Académie de Marine at Brest ceased operations during the Seven Years War and the American War, and the Marine Academy suffered a significant loss when one of its officers, Jean-Jacques de Marguerie, was killed off Grenada in 1778.[79] Many other, smaller examples attest to the debilitating effect of war on the Colonial Machine. In the Seven Years War, for example, Pierre Poivre and his botanical samples were interned in Ireland on his return from the Indian Ocean in 1757.[80] In the same conflict, Father Pingré was held and roughly treated by the English for three and a half months on Île Rodrigues, where he was "reduced to the ignoble drinking of water."[81] In the American War Cossigny held off sending his memoir on gunpowder to France until after hostilities.[82] One of Cossigny's tests in the Indian Ocean to preserve fresh water aboard ship failed because the ship was attacked, and the shock from cannon fire accidentally broke the barrels with the samples.[83] In 1778 an insect collection from Cayenne, commissioned by officialdom for the Cabinet d'Histoire Naturelle, was taken by the English.[84] Sonnerat's herbarium from the Indian Ocean was captured by English in 1781.[85] Later, during in the Revolution, La Billardière, sailing with d'Entreca-

75. "On a travaillé en vain a obtenir du ministre Anglois la communication de son observation du passage du Venus. Elle auroit contribué a determiner la longitude de cette ville. C'est le gout des anglois en tout pays d'etre peu communicatifs, et curieux de profiter du travail d'autres," ALS Coeurdoux to Delisle, dated "a Trinquebar ce 28 Aoust 1762," marked by Delisle, "reçu le 27 septemb. 1764," AN-MAR 2JJ 67.

76. Pritchard (2004), Part 2, details the devastating effects of war in and on the American colonies from 1672 through 1713.

77. See undated, unsigned ministerial document, PA-DB (Lignon).

78. See Prat letter dated "a la n^{lle} orleans le 10e 9bre 1746," transcribed in Lamontagne (1963), pp. 147-48.

79. DDP V:25.

80. ALS Poivre to Réaumur, dated "à Cork ce 19 mars 1757," PA-DB (Poivre).

81. "...J'ai été prisonnier des Anglois à Rodrigue, et je suis resté trois mois et demi sur cette Isle, réduit entre autres à l'ignoble breuvage de l'eau," ALS, Pingré to "Monsieur le Président," dated "Au Port-Louis de l'Isle de France ce 19 7bre 1761," PA-DB (Pingré).

82. ALS Cossigny to Le Monnier, dated "A L'Isle de France le 20 9bre 1779," BCMNHN, Ms. 1995.

83. "Essai sur la maniere de conserver à la mer l'eau potable dans les voyages de long cours," test #10, 7 October 1759, AN-MAR D^3 42 fols. 27-37.

84. ALS "Castries," dated "À Versailles le 28 X.bre 1781," to "M. le C.te d'Angivillers," AN O^1 2111.

85. ALS Sonnerat to Adanson, dated "Cadix Ce 25 fevrier 1781," HI AD 240.

steaux in search of La Pérouse, was "captured" by the Dutch when stopping in Java; the British got the scientific collections, which were eventually returned to France.[86] Not, however, that the French were always victims, as when the Rouen botanical garden came to possess seeds from North America taken from a British ship by French privateers in the War of Jenkins' Ear.[87] In that same conflict, the Jesuit Incarville's botanical specimens from China, destined to go to Bernard de Jussieu and the Jardin du Roi, were taken by the English, who subsequently (to return to the theme that began this discussion) were shipwrecked off the French coast![88]

It was not only European wars that proved disruptive. Periodic conflicts with Indian tribes in North America likewise impeded the work of the Colonial Machine there. Early in the eighteenth century Michel Sarrazin in Canada complained to Antoine de Jussieu that war with the "Renards" held up his plan to send botanical samples back to Paris and the Jardin du Roi.[89] Decades later, André Michaux's collecting was set back because "colonists in Georgia and Carolina declared war on a confederation of Creeks, Cherokees, and Chickasaws."[90]

In 1780 an English corsair seized Dombey's natural history collections, which included 238 botanical drawings and thirty-eight pounds of platinum.[91] This booty, on its way to Paris from Peru, ended up in Lisbon, where the Académie des Sciences and Jardin du Roi, aided by the Ministère des Affaires Étrangères, had to ransom the collections.[92] This example illustrates another obstacle faced by the Colonial Machine: bureaucratic hurdles. Spanish customs authorities, for example, ruined Sonnerat's collections as they passed through that country.[93] The French Compagnie des Indes, although an integral part of the Colonial Machine, itself proved a hindrance to its smooth functioning. We saw previously the difficulties encountered by André Michaux and others, as the Compagnie des Indes impounded his specimens for their inspections before the procedure was changed to have shipments pass through le Havre.[94] The same Compagnie des Indes was reluctant to board private botanical shipments

86. See Chevalier, pp. 194-99.

87. See minutes of Rouen academy for "Le Mardi 8 fevrier, 1746," BM/Rouen, B4/1, p. 12.

88. ALS Incarville [to B. de Jussieu] "à Peking le 11e 9.bre 1748," PA-DB (Incarville).

89. Sarrazin letter to A. de Jussieu, 11 October 1728, reprinted in Rousseau (1969), p. 621; see also Sarrazin remarks quoted p. 622.

90. "...la guerre est declarée de la part des Géorgiens et des Caroliniens avec les Creeks, les Cheroikies et les Chicasaws confédérés," ALS "Michaux," dated "A Charleston le 25 Septembre 1787," to "Monsieur Le Comte," AN O[1] 2113A ("Pépinières d'Amérique. Correspondance relative à la Mission du S. Michaux, Botaniste du Roi. Année 1787").

91. ALS "Thoüin" to "Monsieur le Comte," dated "A Paris ce 22 Mars 1780," AN O[1] 2111; ALS Dombey to Thouin, dated "huanuco Le 20 juin 1780," BCMNHN, Ms. 222; Cap (1858), p. 9.

92. AL draft in d'Angiviller's hand "Pour le S. Thouin 24 mars 1780," AN O[1] 2111; see also in the same collection, draft letter with corrections in d'Angiviller's hand to "M. Le C.te de Montmorin," dated "Versailles le 4. Septembre 1780."

93. See correspondence regarding this episode, AN O[1] 2112; they did the same regarding Dombey's collections, per Cap (1858), p. 12.

94. See above, Part IIC.

without express order from the king.[95] Cossigny's shipments for the Jardin du Roi faced a double blow from the Compagnie and from French customs. He wrote to Le Monnier that "M. Essaim…gave strict orders in Lorient about this. He said that the cases must be placed in the storehouses of the Compagnie on their arrival. They must then be forwarded to customs authorities in Paris, where he will inspect them, that is to say, declare them inspected, but for form's sake they must go through Customs."[96]

The Soul of the Machine. In many respects the Colonial Machine was a wonderful organization. It was complex and socially significant. It tapped the best scientific and administrative talent of the Old Regime. It had access to resources. It mobilized action and knowledge. It got things done. By the same token, like other collections of people and institutions, the Colonial Machine had a temperament and a personality. A public ethos governed its operations. It embodied a set of values, and woe be unto those who contravened the Machine or the authorities running it.[97]

On the positive side, the Colonial Machine was open to talent and provided career pathways for many. In the late 1720s the regent, Philippe d'Orléans, launched a special initiative to recruit talent for science and the Colonial Machine. Solicited to do so, he personally engaged naval officers to pursue subjects that "could be very useful in the advancement of science."[98] Through administrative channels and in the name of the king, he circulated a memoir about promoting the sciences and the arts. In Canada at least, local administrators were ordered to seek out and encourage talent, an order by which: "His Majesty invites those of its subjects who can to contribute to the design it has formed to have the sciences and the arts flourish more and more."[99] Undoubtedly it was because of this initiative that authorities in New Orleans in 1728 accorded subsidies to "young Saussier" to encourage him to "attach himself to the sciences."[100]

95. See complaints by Cossigny about his packages to Le Monnier, BCMNHN, Ms. 1995, e.g., ALS, Cossigny to Le Monnier dated "A L'Isle de france le 27 Mars 1770."

96. "M. d'Essaim…a donné des ordres severes à L'Orient sur cet objet. Il dit qu'il faut faire mettre dans les magazins des fermes les caisses qui arrivent, en débarquant, et les faire expédier pour la Douanne de Paris, où elles seront visitées par luy, c'est-à-dire censées visitées, mais il faut pour la forme qu'elles aillent à la Doüanne," ALS Cossigny to Le Monnier, dated "A Paris le 12 Mars 1775," and follow-up letter, ALS Cossigny to Le Monnier, dated "A Paris le 22 Mars 75," BCMNHN, Ms. 1995.

97. Douglas's classical study develops the point of "institution thinking" in a more general context.

98. "Mémoire au Régent, réclamant des instructions aux officiers de marine, qui peuvent être très utiles à l'avancement des sciences," BCMNHN, Ms. 1140, #5. On d'Orléans as a patron, see Pedley (2005), pp. 83-84.

99. See ministerial correspondence to local administrators, referring to "…un mémoire dressé par ordre du roi par lequel S. Mte. Invite ceux de ses sujet qui le peuvent à concourir au dessein qu'elle a formé de faire fleurir de plus en plus les sciences et les arts," AN-COL B^{52} (1728), fol. 570v (27 April 1728). See also letter the following year, repeating the desire "de chercher les moyens de faire fleurir les sciences et les arts," AN-COL B^{53} (1729), fol. 494v (12 April 1729).

100. See reference of local administers to "...subsides accordés au petit Saussier pour l'inciter à s'attacher aux sciences,"AN-COL C^{13A} 11 [Louisiana] (1728-1729) fol. 66 (30 March 1728).

But the Colonial Machine was not one big happy family. Bureaucratic infighting and conflict disrupted its smooth operations. We have already seen this feature of the Colonial Machine, too. Disagreements between the Société Royale de Médecine and the Marine Royale over sea rations were institutional in nature. The feud between Abbé Rochon and Ensign Grenier was between individuals, conducted through the channels of the Colonial Machine. A further example of the latter saw Cossigny spar with the intendant, who succeeded Pierre Poivre on Île de France and who opposed developing a spice industry on the island.[101] In addition, the Colonial Machine was not an intrinsically "nice" institution. It policed itself carefully. For example, the ministry apparently feared that Gallois, the royal gardener at Lorient, was selling plants for his personal profit. Cossigny's response to Le Monnier when he heard about the matter reveals just how true the accusation was and how carefully people were looking: "I can't say, however, that the suspicions you have against [Gallois'] fidelity are without foundation. But when such is true to a certain degree, I fail to see that you can find here [at Lorient] a more worthy correspondent because you will be less exposed to a loss from a small infidelity that does not make you any less rich."[102]

The Colonial Machine has its nasty side. We saw in Part IIB hospital inmates used as experimental subjects in trials of the Poissonnier-Desperrières diet. In 1779 the Société Royale de Médecine conducted forced medical experiments on prisoners to test antisyphilis treatments, the Colonial Machine's own Tuskegee.[103] In 1752 the navy ministry stopped the distribution of the supplement to the *Dictionnaire de Trévoux* because the geography articles it contained "are absolutely contrary to the truth and favorable to the pretensions of the English regarding the boundaries of Acadia." Sales were suspended. The author was to be brought in and instructed to correct his errors.[104]

101. "Notre Intendant actuel qui a déclaré la guerre à M. Poivre est ennemi déclaré de toutes les épiceries; et c'est bien malgré luy qu'elles s'avisent de fleurir dans notre Isle," ALS, Cossigny to Le Monnier, dated "Au Port Louis Isle de france le 28 Xbre 1775," BCMNHN, Ms. 1995. In the same collection see also ALS Cossigny to Le Monnier, dated "Au Port Louis Isle de france le 12 février 1776," where Cossigny writes, "...L'Intendant qui est Anti-poivre, est par conséquence anti-Muscadier, Anti-Giroflier, a été scandalisé de ma petite brochure sur les Epiceries. Il m'en a parlé, comme ignorant que j'en fûsse l'auteur...."

102. "Je ne puis pas cependant vous affirmer que les soupçons que vous avés contre la fidélité, soient sans fondement. Mais quand cela seroit vrai à un certain point, je vois pas que vous puissiés trouver ici de correspondant qui vaillent mieux, parce que vous serés moins exposé à perdre par une petite infidélité qui dans le fond ne vous rend pas moins riche," ALS, Cossigny to Le Monnier, dated "L'Orient le 15 May 75," BCMNHN, Ms. 1995.

103. SRdM, Ms. 7, p. 311 ("Séance du 28 7bre. 1779"), and other, related entries, pp. 328, 332, 337, 339.

104. "Le supplément du dictionaire de trévoux, qui vient de paraître, contien des articles sur la géographie qui sont tout-à-fait contraires à la verité et favorables aux prétentions anglaises touchant les limites de l'Acadie," ministerial correspondence of 6 April 1752 to Malesherbes, Directeur de la Librairie, AN-COL B^{96} (1752), fol. 43.

Another vignette is particularly revealing in this regard. It concerns one Laureau, a professor of mathematics and hydrography at Rochefort around 1780.[105] In a manuscript he circulated, Laureau strongly criticized the textbook by Bézout used in the cadet training schools throughout France. Laureau complained that Bézout's approach was too theoretical and abstract and was not useful in practice.[106] Laureau's attack was pointed, but not overly personal in nature. But Bézout's textbook was officially mandated and in use from 1763 (to 1864!), and he was powerfully positioned at the Academy of Sciences and in the Colonial Machine. Bézout apparently maneuvered against Laureau in Rochefort. The minister asked Borda for a report on Laureau's manuscript. Borda concluded that Laureau "had neither the talent or the probity necessary for his position…I will say only that it seems to me appropriate to dismiss this professor, who merits no special treatment after the rage, insubordination, and violence of which he has made himself guilty, beyond the fact that his presence in the department can only injure the spirit of docility and instruction necessary to be maintained in the schools."[107] That was on the 21st of September 1780. On the 30th, the minister obtained a full-fledged royal order from Versailles stripping Laureau of his post and ordering him out of Rochefort within three days.[108] On the 3rd of October 1780 an official informed the minister from Rochefort that Laureau had died there the day before. Three days later, on October 6th, Bézout wrote to the minister that he would find a replacement. It is not far-fetched to imagine that Laureau's death was a suicide or that the Colonial Machine was responsible.

Obedience, loyalty, a sense of one's place in a hierarchy, and commitments to royalty and presumably to Catholicism formed part of the unstated requirements of the Colonial Machine and the set of values that lay behind it. The state monitored and enforced these values strictly. Laureau's death is one example. Another concerns a local surgeon in Brest, a certain Flame or Flamme, who opposed Poissonnier's desalinization machine and who circu-

105. See AN-MAR C[7] 170 (Laureau) for these details; Neuville (1882) makes passing mention of Laureau.

106. See 16-page Laureau memoir, "Recherches sur l'utilité et la perfection de la Marine," AN-MAR C[7] 170 (Laureau).

107. "…l'examinateur n'a ni les talens ni la probité necessaires pour sa place…vous voyés d'après cela, Monsieur, que l'examen de ce memoire ne me regardant pas, je vous dirai seulement qu'il me paroit Convenable de renvoyer promptement ce professeur qui ne merite d'être menagé en aucune maniere, après l'emportement l'insubordination et la violence dont il s'est rendu coupable d'autant mieux que sa presence dans le departement ne peut manquer de nuire à l'esprit de docilité et d'instruction qu'il est necessaire de maintenir dans les écoles...," Borda note, dated "à Paris le 21 7bre 1780," AN-MAR C[7] 170 (Laureau).

108. "De Par Le Roy. Sa Magesté n'étant point satisfaite de la conduite du s.r Laureau Professeur de Mathematiques des Gardes de la Marine à Rochefort elle lui a ordonné de quitter le service et de sortir du département de rochefort dans trois jours de la signification qui lui sera faite du present ordre, sous peine d'être arrêté et mis en prison," one-page royal order, dated "à Versailles le 30 septembre 1780," AN-MAR C[7] 170 (Laureau).

lated a pamphlet saying that Poissonnier's distilled water had insalubrious effects. A local thespian also made fun, suggesting a sleight-of-hand substitution of fresh water for seawater. Informed by his agent in Brest, the minister summarily penned the order: "Put Sieur Flame in prison until he admits his error. Warn the actor."[109] In other words, it was dangerous to go up against Poissonnier or Bézout, and you risked threats, imprisonment, and death if you challenged the Colonial Machine.

Norms and Incommensurability. The Colonial Machine was an assemblage of institutions in action that produced and promoted a certain order of knowledge based on sophisticated procedures for the evaluation, validation, and diffusion of knowledge. European contemporaries seemed to have thought of this institutional structure and order of knowledge as universal and unrivaled, a hubris that naturally led to political and cultural shocks and mutual incomprehensibility, particularly in a colonial context.

Devoted as it was to French overseas expansion, the Colonial Machine was a cultural and intellectual weapon deployed to reduce non-Europeans and "otherness" by words, numbers, and categories, as well as by the sword and religion. In that engagement, the Colonial Machine showed another limiting feature in acting as a filter through which only calibrated knowledge could pass. As the eighteenth century advanced, intellectual norms became more precise, more constraining, and more strongly prescriptive. Plants and minerals not accompanied with a precise account giving details of their place of origin were of no use; accounts presented too informally were not even read by men of science comfortably sheltered in official institutions; and the increasingly required use of proper nomenclatures to describe and classify plants, shells, or animals implied an evermore rarefied selection of contributors. As a direct consequence, science in the field became more and more the exclusive domain of experts.[110]

At least partly as a result, geographical, cartographical, or medical knowledge gathered up from others, as we have seen the Colonial Machine doing in India, China, Africa, and the Americas, was generally not directly or immediately assimilated. The necessity of the transformation, organization, and reduction of local knowledge into written accounts, drawings, or lists suitable for Europeans sometimes involved complex local negotiations, including local informants and local assistants.[111] Such phenomena of assimilation and reduction to standardized forms of communication and knowledge led to inevitable losses of knowledge.

109. "Mettre en prison le sr. flame jusqu'a ce qu'il ait prouvé & a faire avertir le comédien...," ministerial note added atop ALS Rigaud to "Monseigneur," dated "à Brest le 30 janvier 1764," AN-MAR D^3 41, fol. 46.

110. See Regourd (2005a).

111. Raj (2006), pp. 43-52; Safier (2008b); Schaffer et al.

The naïve comments of the Dominican missionary Raymond Breton in 1665 gives an insightful example of what "otherness" could represent for an early modern European confronted with different cultures: "Savages have names for only four colors: yellow, red, white, and black...the Caribes do not practice the liberal or the mechanical arts at all; they do not even know the words."[112]

Such remarks underline a fundamental failure to communicate and lead to the idea of the "incommensurability" of knowledge as the concept has been used by historians, philosophers, sociologists, and ethnohistorians for many years. This notion appears in a striking way in the pages written by another missionary, the Jesuit Le Breton, who sojourned among Amerindians in Saint Vincent around 1700, when he evoked a "kind of knowledge about the stars" of the Amerindians: "Everyone but the Caribs knows that arithmetic is almost the first step one takes toward knowledge of the stars, but these are two very different sciences for them, or, rather, they do not know this second one and they are completely ignorant of it." [113] What he explains about their charts for sailing is even more revealing for us: "They content themselves in their navigation to make certain marks on wooden planks they rely on to travel. You have to be Carib to understand this science."[114]

A nuanced story of the concrete contacts among experts, amateurs, and local informants and of their inability to communicate is too large a subject to be treated here. It suffices for us to say that instances of cultural incommensurability illustrate the inefficiency of the Colonial Machine in assimilating and processing information. Here lies another limit to the possibilities of the Colonial Machine. The case we encountered earlier in these pages of the botanical manuscript compiled by Nicolas L'Empereur, his "Jardin de Lorixa," that never succeeded in getting beyond the Jardin du Roi's library despite the richness and qualities of the work, may confirm this analysis.[115] Along these lines, the way Europeans perceived the rites and the medical practices of their black slaves through a veil of mystery – regarding their ceremonies for making rain as well as their medical knowledge (not to mention their vodun practices) as magic secrets – made the assimilation of this native knowledge via the wheels and gears of the Colonial Machine almost impossible.[116] The royal physician

112. Breton (1999), pp. 97, 121: "Les Caraïbes ne font point profession des arts libéraux ni mécaniques, ils n'en savent pas même les noms"; "les Sauvages n'ont des noms que pour quatre couleurs, jaune et rouge, blanc et noir."

113. Lapierre ed., p. 99: "Ils ont acquis une espece de connaissance des astres... Tout le monde, excepté les *Karaybes*, sçait que l'arithmétique est presque le premier pas ou l'on monte à la connoissance des astres, mais cela forme chez eux deux sciences bien différentes, ou plutôt ils ne connoissent pas cette seconde, ils y sont parfaitement ignorants."

114. Lapierre ed., p. 100: "Ils se contentent dans leurs voyages de faire certaines marques sur des planches à l'aide desquelles ils font route: il faut être Karaybe pour connoitre cette science"; more in Regourd (2000), pp. 155-159.

115. Raj (2006), pp. 27-59.

116. Regourd (2000), pp. 162-64, (2008b).

Arthaud gave a striking judgment on that point, saying with a great arrogance: "Blacks only use plants in treating the ill, but they do so with the mystery of ignorance."[117] For the Colonial Machine, there was no place for mystery or unalloyed otherness, or any way to penetrate them.

The Republic of Letters

The eighteenth-century Republic of Letters was a glorious thing – that great, supranational and transnational intellectual and scientific community, distinctive to the Old Regime, that united scientists and intellectuals across Europe and the world in a crusade for progress and Enlightenment. Voltaire and his universalist perspectives can stand as paradigms of the values and practice of the contemporary Republic of Letters. The Republic of Letters had its limits, of course; it was naïve and romantic in some ways, and its ideology was stronger, perhaps, than the reality.[118] But the eighteenth-century Republic of Letters was, nevertheless, a genuine and remarkable cultural phenomenon. More to the point here, the Republic of Letters offered another way of organizing the learned world and an alternative to the structures and ethos of the Colonial Machine. Delineating that alternative and the borderlands between the Colonial Machine and the rest of the Republic of Letters, of which it itself was undoubtedly a part, helps us further clarify the nature and character of the Machine.

Religious Networks. Before turning to other elements of the Republic of Letters, we should not lose sight of "republics of faith" and the organization of religious communities that continued in the eighteenth century as they had in the seventeenth, complementary to the Colonial Machine, but at the same time separate from it and following their own agenda.

Catholic missionaries owed their primary allegiance to their orders and then to Rome and not to the French state or the Colonial Machine per se. Recent literature tells the story of how missionaries dealt with scientific discoveries and knowledge of nature, how they organized their networks in order to collect and transmit findings about the natural world, and how they integrated knowledge of nature into their theological perspectives.[119]

Circles of the religious represented another way of organizing people and purposes distinct from the Colonial Machine itself. The international organization and networks of Jesuits stand out in this regard, especially given the commitments of the Jesuits to science and learning. We have seen how important

117. SRdM, Archives, 136 dr. 1, #27, fol. 2: "Les nègres n'emploient que des plantes dans le traitement des maladies; mais ils le font avec le mystère de l'ignorance."

118. McClellan (1993b); Lilti; Roche (1988, 1993).

119. Romano (2000, 2002).

the Jesuits were as conduits of information from India and from China. The Colonial Machine would have been diminished without these informants. But as intellectuals and teachers, the Jesuits developed their own knowledge-making apparatus that they shared with outsiders, of course, but that served primarily the purposes of their mission. Private, intramural correspondence formed part of this apparatus, as, for example, when Father Mongin sent cometary observations from Saint Christopher in the Caribbean to his former astronomy teacher back in Paris, Father Fontenay, professor of mathematics at the Collège de Clermont.[120] Mongin was not alone in this type of activity.[121] It is telling in these regards that Father Nicolson, a Dominican and author of a natural history of Saint Domingue, bequeathed his papers to his order and not to the Jardin du Roi. On their own, the Jesuits published their *Lettres édifiantes et curieuses*; these were missionary reports that often included medical, botanical, and astronomical information. Father Souciet's compilation of 1729, *Observations mathématiques, astronomiques, géographiques, chronologiques et physiques, tirées des ancien livres chinois, ou faites nouvellement aux Indes et à la Chine par les Pères de la Compagnie de Jésus*, indicates just how much the Jesuits created their own knowledge-making processes as well as being part of those of the Colonial Machine.[122]

One bit of evidence suggests a certain tension between the knowledge-making instrumentalities of the Colonial Machine and those of the Jesuits. The Spanish Jesuit Juan Magnin (1701-1753) had written a *Noticias auténticas del famoso Río Marañón y misión apostólica de la Compañía de Jesús de la Provincia de Quito*. As a result of this trip to South America, La Condamine obtained the work and, after some kind of negotiation with the Jesuits, he took on translating the work into French. The royal doctor in Cayenne, Artur, asked La Condamine in 1746 about the state of the translation. He replied: "Our manuscript of Father Magnin remains in the same state. If I had left it for the Jesuits, they would have made another work of it, and we would not have been the masters of it or gotten anything from it." Likewise indicative of this tension and the more purely scientific spirit of the Colonial Machine, La Condamine went on to say: "In any event, it needs a new map, that is to say that I will have to make one keeping only a few details from the author's original and subjecting them to points determined astronomically or geometrically."[123]

120. Letter dated 3 March 1681, BM/ Carcassonne, Ms. 82, fols. 74-77.

121. Debien (1976), p. 43.

122. On the Jesuits, Souciet, and this work, see also above, Part I, pp. 155ff, Overseas Missions and Missionaries, The Jesuits in Asia.

123. "Notre manuscrit du P. Magnin est resté au même etat. Si je l'avois abandonné aux Jesuites, ils en auroient fait un ouvrage nouveau et nous n'en aurions plus eté maitres ny rien tiré. D'ailleurs il faut l'accompagner dune Carte, c'est a dire quil m'en faut faire une en ne conservant de celle de l'autheur que les details et les assujetissant aux points determinés Astronomiquement ou Geometriquement...," ALS Condamine [to Artur], dated "A Montmorency le 22 Juin 1746," with p.s. dated "a Paris le 25," PA, Fonds La Condamine; Safier (2008b), pp. 71-76, 290n, notes this other instances where La Condamine exploited Magnin as a source.

The Permeability of the Colonial Machine. The Colonial Machine was not walled off from the rest of the world, but was porous. It existed contemporaneously with the Republic of Letters, and the men and institutions that composed the Colonial Machine had their own official and unofficial contacts with the wider scientific and learned communities of contemporary Europe. This point might be made by alluding to the international contacts of the Cassinis and others at the Observatoire Royal.[124] It might also be seen in the formal and institutionalized "correspondences" of the Académie Royale des Sciences with other academies and learned societies across Europe and America, including the Royal Society of London, the Imperial Academy at Saint Petersburg, the *Académie Royale des Sciences et Belles-Lettres* in Berlin, the *Kungl. Vetenskapsakademie* in Stockholm, and the American Philosophical Society in Philadelphia.[125] But nowhere is this permeability more clear than in considering the connections and contacts of the Jardin du Roi with non-French botanists and other agents in Europe and around the world.[126] Partly, the advance of botany and the enriching of botanical gardens depended on international botanical exchanges; to that extent the Jardin du Roi was one of several major European scientific gardens operating somewhat together. Like the Jardin du Roi, the botanical garden at Uppsala was another, and Carl Peter Thunberg (1743-1828), Linnaeus' successor as professor of botany at the university, was a decades-long correspondent of Thouin at the Jardin du Roi; Thouin and Thunberg regularly exchanged seeds and plants for their respective gardens.[127] Thouin and the Jardin du Roi were also in regular and sustained contact with William Aiton, Director of the English royal gardens at Kew.[128] The Jardin du Roi likewise received materials from Sir Joseph Banks, president of the Royal Society of London.[129] Pallas sent northern plants from Russia, and so on.[130]

The Jardin du Roi had some contact with the Asiatic Society of Bengal, and the presence in Jamaica of Dr. Thomas Clarke, a correspondent of Thouin at the Jardin du Roi, reveals a particularly colonial twist to these international exchanges of the Jardin.[131] In addition to the direct Jamaica-Paris connection,

124. To our knowledge, this study remains to be done; the Cassini papers at the Observatoire de Paris provide ample resources.

125. McClellan (1985), pp. 164-69, 174-75.

126. This internationalism is plain in the archival records of the botanists at the Jardin du Roi; see, for example, Tournefort's correspondents, BCMNHN, MS 253.

127. See Thouin's "Journal des Envois de Plantes, Arbres et Graines faits au Jardin du Roy par ses Correspondants, avec ceux faits par le Jardinier dudit Jardin du Roy aux Correspondants d'icelui," BCMNHN, Ms. 1327. This catalogue itself gives ample evidence of the European and worldwide scope of Thouin's contacts; on this point see, Spary (2000). Other Thunberg contact with the Jardin du Roi/Muséum is evident in BCMNHN, Ms. 691, fol. 158v: "Graines recoltées au Upsale, envoyées par M. Thunberg. Semées sous chassis le 7 Avril 1792."

128. See entry, BCMNHN, Ms. 2310, "Liste des Correspondans du Museum...."

129. See entry, BCMNHN, Ms. 691, fol. 181-81v.

130. On a Pallas shipment of seventy-five kinds of seed, see BCMNHN, Ms. 691, fols. 148v-149v.

131. See entries, BCMNHN, Ms. 2310 "Liste des Correspondans du Museum...."

in a spirit of fraternity following the American War, botanical exchanges began between and among British and French gardens in the Caribbean. Nectoux, the botaniste du roi at the Jardin du Roi in Port-au-Prince in Saint Domingue, twice traveled to Jamaica where he and his English counterpart, Clarke, exchanged interesting specimens from their respective gardens, including tea.[132] The main English royal garden in the Caribbean was on Saint Vincent (1765), and the Jardin du Roi in Paris duly received seeds from Saint Vincent, forwarded through England from a Mr. Forsyth.[133] Céré at Jardin du Roi on Île de France sent "a sizeable and very interesting shipment" to Joseph II of Austria in 1782.[134]

The spirit of the Republic of Letters did sometimes pervade the operations of the Colonial Machine. For example, the Swedish botanist Pehr Kalm undertook a notable botanizing trip through Canada and North America as an emissary of Linnaeus from 1747 to 1751.[135] When word of Kalm's travels came to the ministry, it ordered the governor-general and the intendant in Canada (de La Galissonnière and Bigot) to extend to "Sieur Kalm, professor of botany in Sweden, who is traveling to Canada, every aid and resource at their disposal in his researches in this science."[136] The royal physician, Gaultier, escorted Kalm for the two months while he was in the Quebec region.[137] Two years later, when the request came through the Swedish ambassador in Paris for Kalm, who was then in English Canada, to pass back through French Canada to examine the flora of the lake regions, the governor-general and the intendant were instructed to have the highest regard for him and that no further orders were required for Kalm to pursue his botanical studies in Canada.[138] This response from French authorities is less dramatic, but similar to the ecumenical reception given to Captain Cook a decade or so later.

132. Bret (2000); McClellan (1992), p. 160.

133. BCMNHN, Ms. 691, fols. 135v-136, where the name is "Forsyst"; see also Ly-Tio-Fane (1996), pp. 8, 11; Anderson and Hiebert, Appendix, which reprints the Reverand Lansdown Guilding's *An Account of the Botanic Garden in the Island of St. Vincent* (1825).

134. "un Envoi considerable et très interessant," "Recensement de tout ce que renferme Le Jardin du roi, le Montplaisir; remit par Mr. De Ceré," BCMNHN, Ms. 303; Ly-Tio-Fane (2006), p. 60.

135. On Kalm and his travels, see Müller-Wille, pp. 39-48; Chartrans and Duchesne, p. 60; Rousseau (1966), pp. 669-73; Bonnault, pp. 171-73; Skottsberg; Granit.

136. "...donneront au Sr. Kalm, professeur de botanique en Suède qui passera au Canada, tous les secours et facilités qui dépendront d'eux dans les recherches qu'il doit faire se rapportant à cette science," ministerial correspondence, AN-COL B^{87} (1748), fol. 57 (28 September 1748). Granit, p. 210, notes that when Kalm was in Canada "French officials received him in princely fashion and paid his traveling expenses within the colony."

137. B. Boivin, "Notice biographique sur J. F. Gaultier 1708-1756), Naturaliste-médecin du Roi au Canada," in PA-DB (Gaultier), pp. 4, 27.

138. See ministerial correspondence of 30 March 1750, AN-COL B^{91} (1750), fol. 5, and related letter of the same date, AN-COL B^{92} (1750), fol. 75.

In what has been called "a remarkable instance of Anglo-French collaboration" incidentally involving the Colonial Machine, in the 1760s and 1770s the French cartographer of the Compagnie des Indes, d'Après de Mannevillette, and the Englishman, Alexander Dalrymple of the East India Company, maintained a significant contact and exchange of cartographical information.[139] The 1775 edition of d'Après de Mannevillette's *Neptune Oriental*, published as we have seen through the offices of the Colonial Machine, contained four East India Company maps supplied by Dalrymple from East India House. But after that, England and France went back to war.

Other institutions and individuals associated with the Colonial Machine had other, outward connections that forged a web of national and international contacts, contacts that generally benefited the Colonial Machine. From 1726 through 1747, for example, Joseph-Nicolas Delisle served as imperial astronomer at the Imperial Academy of Sciences at Saint Petersburg.[140] The Jesuits in China maintained official connections with this institution, performing corresponding observations for the Russian expedition to Siberia in 1732.[141] The Marine Academy at Brest was in contact with the Société Royale des Sciences de Montpellier, and it received the scientific publications of the Saint Petersburg Academy, the royal provincial academy in Dijon, and the *Società Italiana delle Scienze* in Verona, among others.[142] The American Philosophical Society in Philadelphia became a haven for André Michaux, Moreau de Saint-Méry, Palisot de Beauvois, and other French refugees during the Revolution.[143] Among the larger set of institutions complementing the components of the Colonial Machine, one stands out: the Royal Society of London.

The Royal Society of London. In the eighteenth century, academies and learned societies were the "capitals of the Republic of Letters," and, as such, the Royal Society of London occupied a high place in that Republic.[144] Although they were different kinds of institutions in many respects, to contemporaries the Royal Society and the Académie Royale des Sciences in Paris were the leading scientific institutions of the day and complementary centers in the Republic of Letters. But more than that, the Royal Society of London provided a complementary pole for the organization of contemporary science,

139. Andrew Cook, p. 182.

140. As noted in Gaubil's letter to Delisle, dated "A Pekin Ce 20 May 1732," AN-MAR 2JJ 62; see also Part IIA.

141. This link between Saint Petersburg and Beijing is confirmed in Delisle's 21-page letter to Gaubil, marked "envoiée le 15 janvier 1735," AN-MAR 2JJ 63.

142. DDP VI:28.

143. Gillispie (2007), p. 148; BCMNHN, Ms. 691, fol. 221-221v. Moreau de Saint-Méry published his *Description de la partie française de l'Isle de Saint-Domingue* in Philadelphia in 1797-1798.

144. On the trope of academies as capitals or colonies of the Republic of Letters, see Hahn (1971), pp. 35, 43-45; McClellan (1985), pp. 5-10; McClellan (1993b), pp. 153-54; McClellan (1999).

especially on the international level, and a key alternative to the Colonial Machine within the Republic of Letters.

Because of its prestige and *their* prestige, the Royal Society elected not a few of the men of the Colonial Machine as Fellows. Adanson, Artur in Cayenne, Buffon, Bellin, Bougainville, Broussonet, Bouguer, all of the Cassinis, Chabert, Daubenton, Delisle, Duhamel du Monceau, Father Gaubil in Beijing, Godin, Antoine and Bernard de Jussieu, Lacaille, La Condamine, the brothers Le Monnier, Maupertuis, Peyssonnel in Guadeloupe, Poissonnier, the abbé Raynal, and others with whom we are familiar were F.R.S.[145] Like many, Bellin presented material to the Royal Society of London in the hope of getting elected, and he was.[146] In sending his book on African shellfish to the Royal Society in 1758, Adanson made clear the stature of the institution in the Republic of Letters and why he and others of the Colonial Machine would want such an association: "A society as illustrious as yours and that communicates with all the learned of the Universe [!] has a legitimate right to all works relative to the sciences regardless of their country of origin."[147]

As might be expected, a handful of individuals in our story published in the *Philosophical Transactions* of the Royal Society of London, which we know was among the leading scientific journals of the day. The Cassinis published there regularly; Jussieu, Delisle, and Maupertuis, less so. A report from Father Feuillée on the longitude of Buenos Aires appeared in Volume 32 (1722-23); La Condamine's account of an Indian poison is in Volume 44 (1746-47).[148] The astronomer Pingré had a number of pieces and astronomical reports in the *Phil. Trans.* over a ten-year period.[149] In 1787 the marquis de Chabert, former inspector-general of the Dépôt des Cartes et Plans, presented to the Royal

145. See membership list in Royal Society of London. Re Artur, see "A Letter from Dr. Arthur, a French Physician at Cayenne to Dr. Maty, F.R.S., concerning the Birds of Passage in that country," RS/L, L&P II, #368.

146. Ms. translation labeled "M. Bellin's Letter to the Royal Society," dated "Paris 20 April 1752," marked as "Read at R.S. May 14, 1752," RS/L, L&P II, #299.

147. "Une société aussi Illustre que la vôtre, et qui communique avec tous les savans de l'Univer, a un droit legitime sur tous les ouvrages qui sont relatifs aux sciences, de quelque païs qu'ils sortent. C'est ce qui m'autorise, Messieurs, à vous présenter ce premier volume d'un grand ouvrage que j'ai commencé sur l'histoire Naturelle du Sénégal," ALS "Michel Adanson, Correspondant de l'acad.e Roy.e des sciences de Paris," to "Messieurs," dated "à Paris ce 1 Janvier 1758," RS/L, Misc. Manuscripts.

148. "The Longitude of Buenos Aires, Determin'd from an Observation Made There by Pere Feuillee," *Phil. Trans.* 32 (1722-1723), pp. 2-4; "A Letter from Richard Brocklesby M.D. and F.R.S. to the President, concerning the Indian Poison, Sent Over from M. de la Condamine, Member of the Royal Academy of Sciences at Paris," *Phil. Trans.* 44 (1746-1747), pp. 408-12.

149. "Observation of the Transit of Venus over the Sun, June 6, 1761, at the Island of Rodrigues; By Mr. Pingre, of the Royal Academy of Sciences at Paris," *Phil. Trans.* 52 (1761-1762), pp. 371-77; "A Supplement to Mons. Pingre's Memoir on the Parallax of the Sun," *Phil. Trans.* 54 (1764), pp. 152-60; "Observations of the Transit of Venus on June 3, 1769," *Phil. Trans.* 59 (1769), pp. 374-77, with Pingré's observations reported on p. 376; "A Letter from M. Pingre, of the Royal Academy of Sciences at Paris, to the Rev. Mr. Maskelyne," *Phil. Trans.* 60 (1770), p. 497-501.

Society a general map of the Atlantic Ocean prepared at the Dépôt, which duly appeared in the *Phil. Trans.* for 1788.[150] This kind of publication seems perfectly normal and exemplary of the connections of the Colonial Machine to the wider world and to the Royal Society of London in particular.

The Royal Society itself effected connections of its own with French Jesuits in China, a connection that modifies somewhat our sense of the place of these religious figures in the Colonial Machine. The Royal Society made overtures to Father Pierre Incarville in Beijing in 1748.[151] Incarville received materials from the Chelsea gardens in London and requested seeds from Carolina.[152] In return, he sent a broad range of botanical, natural historical, and technical information from China to the Royal Society.[153] Incarville also used the connection with the Royal Society to ship materials to Bernard de Jussieu in Paris via London.[154] But when again solicited by Cromwell Mortimer, P.R.S., to send more materials to England and the Royal Society, Incarville got nervous and insisted that the minister in France be informed first and be sent duplicates of anything destined for England.[155] The Royal Society also engaged Incarville's older Jesuit contemporary in China, Father Antoine Gaubil. Gaubil forwarded a raft of materials to the Royal Society, including paper money, manuscripts on chronology, astronomical data (Jupiter, eclipse, and longitude information), maps, and Chinese books.[156] In return, the Royal Society elected Gaubil F.R.S., and sent its *Phil. Trans.* to Gaubil in China.

The connections of Incarville and Gaubil with the Royal Society suggest that individuals on the outer edge of the Colonial Machine could have dual allegiances or that they operated in the middle between the Colonial Machine

150. *Phil. Trans.* 77 (1788), p. 435, under "Presents."

151. This overture is clear in d'Incarville's response, "A Letter from Father d'Incarville to Dr. Mortimer Secr. R.S. Dated at Peking, Nov. 11: 1748," RS/L, L&P II, #153.

152. See Incarville letters to Mortimer of 11 December 1752 and 25 October 1753, reprinted in Bernard-Maître, p. 709 (off-print, p. 46).

153. See "Catalogue Alphabetique des plantes...," marked "Par P. d'Incarville de Peking Mars 9, 1748," RS/L, L&P II, #153; ALS "d'Incarville, J." dated "à Peking ce 15e 9bre 1751," L&P II, #359; "A Letter from Father D'Incarville, of the Society of Jesus, at Peking in China, to the Late Cromwell Mortimer, M. D. R. S. Secr.," *Phil. Trans.* 48 (1753-1754), pp. 253-60. On the request of the Royal Society for information about fossils from Incarville, see "An Answer to the Questions upon the Natural History of Fossils," RS/L, L&P II, #359.

154. Incarville reports this information in letters to Bernard de Jussieu, dated "à Peking le 11e 9.bre 1748" and "à Peking le 2d. 9.bre 1750," PA-DB (Incarville).

155. See Incarville letter to Mortimer, dated 21 December 1752, reprinted in Bernard-Maître, pp. 694-95 (off-print, pp. 31-32).

156. See Gaubil letter to Mortimer, dated "Peking Nov. 9 1748," RS/L, L&P II, #33, and associated paper, "A Letter from Father Anthony Gaubil Jesuit, to Dr. Mortimer, Secr. R.S. Containing Some Account of the Knowledge of Geography among the Chinese, and of Paper-Money Current There," *Phil. Trans.* 46 (1749-1750), pp. 327-30; ALS Gaubil to "Dr. Cromwell Mortimer, M.D. Secretary to the Royal Society," dated "Peking, Nov. 2, 1752," L&P II, #385, and associated paper, "Extracts of Two Letters from Father Gaubil, of the Society of Jesus, at Peking in China," *Phil. Trans.* 48 (1753-1754), pp. 309-17; and, ms. trans. of Gaubil letter to Cromwell Mortimer, dated "Peking 30 Octob. 1751," RS/L, L&P II, #422.

and other modes for the organization of science as represented by the Royal Society and the Republic of Letters. The Swiss botanical traveler and former employee of the Dutch East India Company (the *VOC*), Laurent Garcin (1683-1751), provides a good example of someone who straddled multiple worlds. From 1730 on Garcin was a correspondent of Antoine de Jussieu and the Academy of Sciences in Paris, and his travels to Batavia, Java, the Moluccas, and Bengal and his reports to the Academy made him a valuable contributor to the work of the Colonial Machine. But Garcin was also elected F.R.S. in 1730 by the English institution; he published several articles in the *Phil. Trans.* and sent botanical samples to the Royal Society.[157] In 1732 Garcin wrote to Sir Hans Sloane about mangosteens, still a not well-known fruit: "If my observations are agreeable to you, I will continue to keep communicating them to you as much as I am able. I will count myself happy if I can satisfy your curiosity and your tastes as well as those of your illustrious Society."[158]

The great French botanists Antoine and Bernard de Jussieu at the Academy and the Jardin du Roi were in contact with the Royal Society and circles around the Society in the early decades of the eighteenth century.[159] But a letter written by Antoine de Jussieu in 1727 to the English botanist, collector, and F.R.S. William Sherard (1659-1728) shows the limits of good will of the Colonial Machine to engage outsiders, or at least the reciprocity expected in these relations, be it the reciprocity of collectors or, in the case of the Royal Society, the bestowing of recognition and prestige in exchange for scientific gifts.

> You say you wish that I should now send you what I have had collected in Barbary by Mr. Peyssonnel and in Cayenne by Mr. Barrere. More than willingly! But is it fair that I let you, in preference to all others, enjoy the pleasure of new things when I do not profit from your collections except after they are known to the whole world and made known

157. "Mémoires envoyés par Monsieur Garcin a [Paul de] Saint Hyacinthe," RS/L, Cl.P. X (ii), #12-14, and "Memoirs Communicated by Mons. Garcin to Mons. St. Hyacinthe, F.R.S. Containing a Description of a New Family of Plants Called Oxyoides; Some Remarks on the Family of Plants Called Musa; And a Description of the Hirudinella Marina, or Sea Leach," *Phil. Trans.* 36 (1729-1730), pp. 377-94 and plate; "A Letter from Monsr. Garcin to Sir Hans Sloane," dated "Neuchatel en suisse ce 23:8bre:1732," LBC 21, pp. 69-70, and "The Settling of a New Genus of Plants, Called after the Malayans, Mangostans; By Laurentius Garcin, M.D. and F.R.S.," *Phil. Trans.* 38 (1733-1734), pp. 232-42; "A Letter from Dr. Laurence Garcin, of Neuchatel, F. R. S. to Sir Hans Sloane Bart. Late P.R.S. concerning the Cyprus of the Ancients," *Phil. Trans.* 45 (1748), pp. 564-78; "The Establishment of a New Genus of Plants, Called Salvadora, with Its Description; By Laurence Garcin M.D. F.R.S.," *Phil. Trans.* 46 (1749-1750) pp. 47-53.

158. "Si mes observations Vous sont agréables, je continuerai de Vous en communiquer autant, et du mieux qu'il me sera possible. Je m'estimerai heureux, si je puis satisfaire vôtre curiosité et vôtre goût de même que celui de vôtre illustre Societé...," "A Letter from Monsr. Garcin to Sir Hans Sloane," dated "Neuchatel en suisse ce 23:8bre:1732," and marked as "Read May: 23: 1734," RS/L, LBC 21, p. 70.

159. Antoine de Jussieu correspondence is RS/L, Sh. 276-288; Bernard de Jussieu correspondence is RS/L, Sh. 289-292.

> to me by botanists other than yourself. Start, then, by keeping your word to me and send me all the plants from Ray's *Synopsis* and your doubles from the Levant and Africa, and I will share with you the new collections that come to me.[160]

Renegades. The strictly national logic encouraged by the Colonial Machine (and by the ministers of the navy and the colonies, in particular) was not always upheld by the actors in the field. The rigidities of the Colonial Machine meant that it was also an organization that excluded and alienated people. And so, the Royal Society of London was not simply a key element of the international organization of science and the Republic of Letters at the time, but it provided an effective alternative to the Colonial Machine. Disgruntled scientific specialists, who became disaffected with the Colonial Machine, turned to the Royal Society in opposition to the Colonial Machine.

Joyeuse the elder may just have been a crank. Frustrated that Poissonnier's method of distillation had won over everyone in France to the disadvantage of his own method of preserving fresh water at sea, in desperation he thought to turn to the English court and the Royal Society of London, writing to Adanson in 1767:

> People at our Court are so persuaded that Mr. Poissonnier's distillation process can supplement in this regard any other supplies for ships headed to sea that they consider the notion of conserving fresh water useless and superfluous, and they will not stop there. It must first be proven that, outside of calm weather, there will be a shortage of distilled water, and for that one needs someone at Court who is in a position to be heard when he explains all that I have to say on this subject. Secondly, one must decide to clash almost head on with Mr. Poissonnier, who has such strong grounds for being pleased with his discovery. These two reasons have caused me to renounce developing my sweet water in France....This is what has led me to think of the English. My initial idea was to address myself, as to you, first to the Minister of this Nation who could then have my proposal examined by the Society in London. But on reflection, given that I know no one at the Court of

160. "Vous voudriez a present, dites vous, que je vous fasse part de ce que j'ay fait ramasser en Barbarie par Mr. Peyssonel et en Cayenne par Mr. Barrere, tres volontiers! [sic] mais est il juste que je vous fasse gouter le plaisir de la nouveauté preferablement a tout autre, lorsque je ne profite de vos collections que lorsqu'elle sont connues de tout le monde, et que ce son d'autres Botanistes que vous qui me les font sçavoir. Commencer donc par tenir la parole que vous m'aviez donnée de m'en faire par de meme que de toutes les plantes [illegible] *Synopsis de Rai* et de vos doubles [illegible] du levant et d'afrique, que je partageray avec vous les nouvelles collections que je reçois...," letter dated "de Paris ce 31 Janvier 1727," RS/L, Sh. 288. The reference is either to John Ray's *Synopsis methodica Animalium Quadrupedum et Serpentini Generis* (1693) or to his *Synopsis methodica Avium et Piscium* (1713).

> England, I thought that if you would have the kindness to alert people you know at the Society of London of my request, especially the President or Secretary of this society, they could take it upon themselves to speak with one of the English Ministers....Here is what I would wish you write to the President of the Society of London or to the Secretary: 1) that someone of your acquaintance has found the means of preserving fresh water and whose discovery is backed up by ten years of experimentation, and 2) that, given the bias of the French Court today in favor of Mr. Poissonnier's distilled water, despite being able to show the inadequacy of his method in all but calm weather, and the conservation of fresh water being viewed perhaps as superfluous, this person thought to propose it to the English.[161]

The astronomer from Marseilles, Saint-Jacques de Silvabelle (1722-1810), was not a crank, however. Silvabelle went on to direct the royal naval observatory at Marseilles in 1763 and to publish with the Academy of Sciences. Earlier, however, in 1750 he submitted his Newtonian paper, *Les variations célestes*, to d'Alembert, who promised to present the work to the Paris Academy, but never did. After pleading with d'Alembert for years, in frustration Silvabelle finally sent his work to the Royal Society of London; it appeared in the volume of the *Phil. Trans.* for 1754 and was dedicated to the members of the London institution.[162]

As much or as little as these examples show that some individuals saw and used the Royal Society of London as an alternative to the Colonial Machine,

161. "Tant qu'on sera persuadé, dans nôtre Cour, que l'eau dessalée par M. Poisonnier, peut suppliér a tout approvisionnement quelconque, en cette partie, sur les Vaisseaux qui vont a la mer, on regarde comme inutile et superflüs la proposition de conserver l'eau douce, et on ne s'y arrêtera pas. Il faudroit commencer par prouver l'insuffisance de l'eau dessalée, hors les temps calmes; et, pour cela, il faudroit avoir quelqu'un a la Cour, qui fût en état de se faire écouter, lorsqu'il exposeroit tout ce que j'ai à dire a ce sujet; en second lieu, il faudroit se déterminer presque a heurter de front, M. Poissonnier, qui a si fort lieu d'être content de sa découverte. Ces deux raisons sont cause que j'ay renoncé à faire valoir mon eau douce en France...C'est ce qui m'a fait penser aux Anglois. Ma premiere idée avoit êté, comme a vous, de m'adresser d'abord au Ministere de cette Nation, qui ensuite ferait examiner ce que je proposais, par la Societé de Londres; mais faisant reflexion que je ne connoissois personne dans la Cour d'Angleterre, j'ay crû que si vous vouliés avoir la bonté de prévenir les persones que vous connoissés dans la Societé de Londres, surtout le Président ou le Secretaire de cette société, sur ce que je demande, ils pourroient se charger eux même d'en parler avec quelqu'un des Ministres Anglais,...Voici donc ce que je voudrais que vous ecrivissiés au Président de la Societé de Londres, ou au Secrétaire: 1o. qu'il y a une personne de vôtre connoissance, qui a trouvé le moyen de conserver l'eau douce, et qui s'est assuré de sa découverte par des experiences dc 10. ans. 2o. que cette personne considerant que, par la prévention où l'on est actuellement a la Cour de France, en faveur de l'eau desallée de M. Poissonnier, quoy qu'elle soit en état de démontre l'insuffisance, hors les temps calmes, on regarderoit peut être la conservation de l'eau douce, comme superflüe, elle a pensé de la proposer aux Anglais...," ALS "Joyeuxe l'aîné" to Adanson, dated "a Marseille le 5 Juillet 1767," HI, AD 199.

162. Stephan; "A Treatise on the Precession of the Equinoxes, and in General on the Motion of the Nodes, and the Alteration of the Inclination of the Orbit of a Planet to the Ecliptic. Inscribed to the Gentlemen of the Royal Society, by M. De St. Jaques Silvabelle," *Phil. Trans.* 48 (1753-1754), pp. 385-441.

the case of Jean-André Peyssonnel (1694-1759) is unambiguous.[163] For the most part, Peyssonnel was a creature of the Colonial Machine, and his career path was built around opportunities afforded by the Machine. Born in Marseilles, by the time he was eighteen he had sailed to the Antilles and the mouth of the Mississippi. He was part of a state-sponsored voyage to Egypt in 1714, and in 1724, after receiving an M.D. from the Paris faculty, Peyssonnel headed a botanical and natural history expedition to Tunisia and the Barbary Coast, sent by Abbé Bignon and the Académie des Sciences. The Academy elected Peyssonnel a correspondent prior to his departure, attached to Geoffroy and Antoine de Jussieu. Peyssonnel's North African trip proved a scientific success.[164] He returned seeds and plants for the Jardin du Roi. He provided Delisle with geographical and cartographical information about the region. He made the notable discovery of the animal nature of coral, and in general the Academy was pleased with Peyssonnel.[165] He narrowly missed election to a position as adjunct academician in the Academy of Sciences, so in 1726 the minister posted him to Guadeloupe as médecin-botaniste du roi at a salary of 1000 livres.[166] There, as we have seen, he did research in botany and natural history and fulfilled the various duties of a colonial médecin du roi for three decades. Peyssonnel would seem to be the very model of an expert functionary in the Colonial Machine.

And yet Peyssonnel ran afoul of the scientific establishment, and he found himself on the scientific as well as the colonial periphery. His paper on the animal nature of coral was read at the Académie des Sciences on 28 June 1726. Réaumur and Bernard de Jussieu took strong exception to Peyssonnel's views that contradicted the accepted wisdom that coral were plants.[167] The Academy

163. See mss. materials PA-DB (Peyssonnel), AN-MAR C^7 128 (under Granger); and AN-COL E 335 (Peyssonnel); see also Rampal; Lacroix (1938), vol. 3, pp. 23-24, 39-47; Gasc and Laissus, p. 26; Lescure, p. 62; Lucette Valensi, in her Introduction to Peyssonnel (2001), pp. 7-38, provides further biographical background, esp. regarding his trip to North Africa.

164. "Memoires sur le royaume de Thunis en Afrique par Mr. Peyssonel, Docteur en Medecine envoié par le Roy pour les recherches d'histoire naturelle en 1724," AN-MAR 2JJ 54. Peyssonnel (2001) reprints Peyssonnel's letters to the abbé Bignon and to Delisle reporting on his trip.

165. At least according to Antoine de Jussieu; see "Copie de la lettre écrite a Monsieur l'Abbé bignon par Monr.r de Jussieu De Paris le 29 Juillet 1725," C^7 128 (Granger), that praises Peyssonnel while speaking of "plantes marines de la coste de Tripoli, sur-tout de ces plantes pierreuses et Eponges, sur lesquelles il a plusieurs observations singuliéres a communiquer a l'Academie...."

166. See ministerial memorandum, dated "19 décembre 1726," AN-COL E 335 (Peyssonnel), and administrative correspondance, dated "du 1re fevrier 1729, La Guadeloupe," AN-COL E 286 (Lignon), copy in PA-DB (Peyssonnel).

167. See above and Chaïa (1968), p. 56; PA-PV 45 (1726), fols. 187 ("Le Samedi 8 Juin 1726...On a commencé a lire un Memoire de M.r Peyssonnel sur les Plantes pierreuses de la Mer, qu'il croit être des matiéres animales"), 207 ("Le Vendredi 28 Juin 1726"), and 209 ("Le Mercredi 3 Juillet 1726...On a Lû un Ecrit de M.r Peyssonnell par lequel il pretend prouver que l'écorce des Correaux est formée par de petits Animaux, et M.r de Jussieu a fait voir des petrifications qui pouvoient avoir rapport a ce sujet"); see also PA-PS for these dates; Réaumur (1727); see also Peyssonnel mss., "Traité du corail...," BCMNHN, Mss. 1035-1036; "Dissertation sur le corail," BCMNHN, Ms. 1260; and Bernard de Jussieu ms., "Copie d'un mémoire de Peyssonnel sur les plantes marines," BCMNHN, Ms. 677.

did not look favorably on his paper on tides either the following year.[168] Rejected by the Academy, yet still a trained and committed man of science, Peyssonnel turned – was forced to turn – to other outlets.

The provincial academies of France provided one such outlet. Possibly as a native son, Peyssonnel maintained regular contact with the *Académie des Belles-Lettres, Sciences et Arts* of Marseilles for over two decades from 1726, sending many papers, including ones on coral and on leprosy, the latter stemming from his work as médecin du roi on Guadeloupe. He entered a prize competition in 1749, and the Marseilles Academy published his article on the Souffrière volcano on Guadeloupe. The Academy elected Peyssonnel an official correspondent in 1756 and made him an associate academician the following year.[169] Peyssonnel became a foreign associate of the *Académie Royale des Sciences, Belles-Lettres et Arts* of Rouen late in his career, in 1755, and he sent papers to Rouen.[170] He was similarly associated with the *Académie Royale des Sciences et Belles-Lettres* in Bordeaux, to which he also sent papers, and he was a correspondent of the prestigious *Société Royale des Sciences* of Montpellier.[171] We examine the provincial academies and their place in, and relationship to, the Colonial Machine next, but for Peyssonnel they provided a substantial institutional opportunity for him and his science beyond the central core of the Colonial Machine and the Academy of Sciences headquartered in Paris.[172]

But it was toward the Royal Society of London that Peyssonnel turned most emphatically as an alternative to the Colonial Machine. In 1751 he wrote to the Royal Society:

> The Republic of Letters has for its limits the whole Earth, and all the learned, amateur, or curious who can enrich it or at least enjoy the study of nature and cultivating the arts and sciences are received into it, regardless of their nationality or religion....My qualities as a French gentleman, a doctor of medicine, and an associate of French academies does not exclude me from a correspondence with the civilized nations of the world. It seems that some amateurs in France are giving themselves credit and taking credit for my work and my discoveries, and

168. As reported in ALS "L'abbé Bignon" to "M. le Comte de Maurepas," dated "A Lislebelle Le 13 Septembre 1727," AN-MAR G 234, #79.

169. Barrière, p. 66.

170. BM/Rouen, B4/1, p. 84 ("Du Mercredi 7 janvier 1756"), and p. 91 ("Le Mercredi 10 mars 1756"); BM/Rouen, C23 Correspondance (Peyssonnel).

171. Barrière, pp. 44, 67; see also traces of Peyssonnel correspondence with the Bordeaux academy, BM/Bordeaux, Ms. 1696 (XXVIII), 18; Ms. 828, XX, 63, fol. 3. Peyssonnel styled himself a correspondent of the Montpellier Society in his communications with the Royal Society of London; Rampal, p. 327, suggests that Peyssonnel was elected to the Académie Royale des Belles-Lettres of La Rochelle.

172. See also Peyssonnel publications in the independent scientific press, notably report on the manchineel apple in the *Recueil périodique d'observations de Médecine, Chirurgie, Pharmacie, &c.* 7 (1757), pp. 411-12.

> without friends, I am daring to take the liberty of addressing myself to you, Mylords and Messieurs, begging you to be so kind as to insert in your literary treasury the present work if you judge it worthy.[173]

The work in question was Peyssonnel's *Traité du Corail*, a 400-page manuscript in quarto on corals. William Watson, F.R.S., prepared a report on Peyssonnel and his work for the Royal Society.

> In France some lovers of natural history do attribute and even appropriate to themselves his labours and his discoveries, of which they have had the communication; and that himself, retired [sic] to the West Indies, and not having the means of giving to his work the perfection he desired, for want of books, and yet more for want of judicious persons, with whom he might not only consult, but who might also enable him to give a more full explanation to such passages of his work, as might be thought obscure, and even correct the faults thereof; for which reason he takes the liberty to request this good office of the Royal Society....[174]

Watson praised Peyssonnel's originality ("M. de Peyssonnel, if his system is admitted, has made a great alteration in that part of natural history") and went on: "I cannot conclude this account, without observing, that, in my opinion, the Royal Society is greatly obliged to M. de Peyssonnel, for his transmitting this manuscript, which I consider as a very valuable literary present." The Royal Society duly elected Peyssonnel F.R.S. and printed a twenty-four-page abstract and summary of his treatise on corals prepared by Watson in the *Phil. Trans.* for 1750-51.[175]

173. "La republique des Lettres a pour ses Limites toute la terre, tous les amateurs, curieux ou scavants qui puissent l'enrichir ou au moins qui aiment a etudier la nature, a cultiver les arts et les sciences y sont recus, de quelque etat nation ou religion qu'ils puissent estre....Ma qualité de gentilhomme francois, de docteur en medecine, associé aux accademies francoises ne m'exclut pas d'une Correspondance avec les nations civilisées, comme je m'apercois qu'en france quelques amateurs de l'histoire naturelle s'attribuent et s'aproprient meme mon travail et mes decouvertes, plus faute d'amis j'ose prendre la liberté de m'adresser a vous, Mylords et Messieurs, vous priant de vouloir inserer dans vostre trésor litteraire cet ouvrage si vous le juges digne," Peyssonnel letter dated "a la guadeloupe ce 1e may 1751," to "Milords et Messieurs," RS/L, L&P II, #279; see copy PA-DB (Peyssonnel).

174. "An Account of a manuscript treatise presented to the Royal Society, entitled *Traité du corail*," RS/L. L&P II, #294. According to Watson, Bernard de Jussieu had apparently changed his mind about the animal nature of coral after Abraham Trembley's discovery of the fresh-water polyp in 1740 and did not rush to credit Peyssonnel whose manuscript was read at the Academy in Paris fifteen years before.

175. "An Account of a Manuscript Treatise, Presented to the Royal Society, Entitled, Traite du Corail...by the Sieur de Peyssonnel, M.D. Correspondent of the Royal Acad. of Sciences of Paris, of That of Montpelier, and of That of Belles Lettres at Marseilles; Physician-Botanist, Appointed by His Most Christian Majesty in the Island of Guadalupe, and Heretofore Sent by the King to the Coasts of Barbary for Discoveries in Natural History. Extracted and Translated from the French by Mr. William Watson, F.R.S.," *Phil. Trans.* 47 (1750-1751), pp. 445-69. Also, RS/L, L&P II, #294.

More than that, the *Philosophical Transactions* of the Royal Society of London became Peyssonnel's main publication outlet, and his scientific corpus is to be found there and not in the records or publications of any branch of the Colonial Machine. One paper followed another until his death in 1759, translated in the pages of the premier English scientific journal: on coral, on the Soufrière volcano, on sea currents, on leprosy in Guadeloupe, on the *Limax* snail, on sponges, on sea algae, on earthquakes and their causes, on the manchineel apple, on the "American sea-sun-crown," and on sea millipedes.[176] Peyssonnel's shift of allegiances to the Royal Society and to the *Philosophical Transactions* casts a revealing shadow on the Colonial Machine and makes plain that scientific authority as well as political authority controlled or sometimes failed to control the Colonial Machine.

The Société d'Histoire Naturelle. The *Société d'Histoire Naturelle* was founded in Paris in August 1790 in the wake of the private and ephemeral *Société Linéenne de Paris* (1787-1789). Both organizations emerged in response to the great popularity of contemporary botany and to fluid social conditions on either side of 1789.[177] The main idea behind the Société d'Histoire Naturelle was to create a freer and more open institution than royal ones, as well as promoting natural history and the work of Linnaeus (1707-1778). The interesting point for us is that this new society rapidly offered another valuable platform for naturalists from overseas. Louis-Claude Richard, former director of the botanical garden in Cayenne, was one of the 63 members on the membership list in 1792, but above all, more than 15 percent of the 88 *associés* recruited by the Society between September 1790 and March 1792 were living in French colonial space. Among them, we may notice enlightened colonists like Genton in Saint Domingue, as well as specialists already familiar to us, such as the physician Cassan in Saint Lucia, the agronomist Dutrône La Couture in Saint Domingue, or the naturalists Leblond and Martin in Guiana.[178] This Society's publications provided an outlet for overseas contributions, publishing, for example, catalogs of mammals, shells, and insects specimens gathered by Leblond in Cayenne.[179]

176. For Peyssonnel's originals and the manuscript translations of these papers, see RS/L, L&P II, #294; for the papers themselves, see Bibliography.

177. Duris makes a complete presentation of these two societies.

178. "Liste des membres et associés de la Société d'histoire naturelle de Paris, par ordre de réception," in Société d'Histoire Naturelle de Paris, pp. 131-32.

179. Société d'Histoire Naturelle de Paris (1792): "Catalogue des mammiferes envoyés de Cayenne par M. Le Blond" by Alexandre Brongniart (p. 115); "Catalogue des coquilles envoyées de Cayenne, à la Société d'Histoire naturelle de Paris, par M. Le Blond" by Jean-Guillaume Bruguière (p. 126); "Catalogue des Insectes envoyés de Cayenne, à la Société d'Histoire naturelle de Paris" by Guillaume-Antoine Olivier D.M. (pp. 120-25); "Catalogus plantarum, ad societatem, incunte anno 1792, e cayenna missarum a domino Le Blond. Conscriptus a L. Richard, hujusdem Societatis Membro" by Louis-Claude Richard (pp. 105-14), and a "Catalogue des oiseaux envoyés de Cayenne, à la Société, par M. Le Blond, associé, publié par MM. Richard et Bernard," by Louis-Claude Richard and Jean-Philippe Bernard (pp. 116-19).

Even more strikingly for our story was the fact that this Society was called upon by the National Assembly to work alongside the Académie Royale des Sciences to provide to the minister detailed instructions, lists of books, instruments, and personnel recommendations of gifted naturalists ready to go overseas for the d'Entrecasteaux expedition. Eminent members of the Society like Richard and Dolomieu wrote eighteen accounts to the minister. Dutrône La Couture, national associate of the Cercle des Philadelphes, provided information on sugarcane. Among other reasons, the case is of note for the history of the Colonial Machine in showing that with the French Revolution underway, the Royal Academy of Sciences was losing influence and power and faced vibrant alternatives.[180]

Provincial Academies. More than two dozen literary and scientific societies dotted the French provinces in 1789.[181] Although graced with royal recognition, French provincial academies and societies were not national institutions – the Académie Royale de Marine at Brest excepted – and cannot be considered as linked so much to the state or the Colonial Machine. Rather, provincial academies expressed regional identities and those of their sponsoring urban elites. In fact, provincial academies vigorously resisted efforts by the state to organize them on a national level or to subsume them under the authority of the kingdom's central learned institutions in Paris.[182] Interestingly, a proposal to the minister of the navy and the colonies in 1780 sought to have all French provincial societies orient themselves toward navigation.

> It would be easy, my lord, to have the nation focus on nautical matters in a specific way by engaging the academies spread out in the provinces to offer prize programs exclusively on topics related to the perfection of this highly important art. Professors of mathematics, be that of the Collège Royal, the University of Paris, or provincial universities, in offering works on Navigation instead of Optics, Dioptrics, and Catoptrics, which are only matters of pure curiosity, would awaken the attention of young people and stimulate their creativity without it costing anything but a word from the Monarch. The French today are sufficiently enlightened to work towards the good. The institutions are there. They only await orders or at least a simple, tacit invitation. Purely literary societies could even contribute towards this same end by looking to

180. For the d'Entrecasteaux expedition, see Richard (1986). Richard (1984) provides the detailed story of this involvement of the Société d'Histoire Naturelle for this expedition's preparation work; see also Duris, p. 95.

181. Roche's magisterial work remains the keystone of a large literature; see also McClellan (1985), pp. 89-99, 133-40; McClellan (2002).

182. McClellan (1985) tells this story, pp. 182-87.

> antiquity to discover procedures that have fallen into disuse, but that could be of the greatest utility.[183]

This proposal by a private individual came to naught, but it reveals both the felt possibilities of provincial academies to address matters of importance to French colonial expansion and the extent to which they were not part of the royal and central organization of the Colonial Machine. Provincial academies were more open to flattering the vanity of their correspondents, but they, too, provided an alternative to the Colonial Machine and an independent outlet for the voices of scientific amateurs – colonial and otherwise.

As much as they were rooted locally, French provincial academies were rife with the colonial and acted as separate centers receiving information from overseas. The Académie Royale des Sciences, Belles-Lettres et Arts of Rouen (1744) provides a case in point. Colonial and overseas reports and materials regularly filtered in to the institution.[184] In the 1760s the Rouen Academy received letters and botanical shipments from merchants, colonists, and doctors in Saint Domingue, Guadeloupe, Martinique, and Dominica.[185] Plants arrived in Normandy from America, coral from China, bugs from Martinique, monstrosities from Saint Vincent, "tea" from Paraguay, minerals from Île Bourbon, and nutmegs from Île de France. Correspondents presented papers on the anatomy of blacks, diseases of Madagascar, indigo production, and the Chinese mulberry. Lots of papers concerned naval matters, navigation, compasses, ship design and construction, and the health and diet of sailors.[186] The Rouen Academy superintended the town's botanical garden and took pride in cultivating bananas and pineapples, the latter ritually offered annually to local author-

183. "...il seroit facile, Monseigneur, de tourner les vûes de la nation du côté de la Marine d'une maniere toute particuliere – en engageant les academies repandues dans les provinces à ne donner pour programes de leurs prix que des sujets analogues à la perfection de cet art si important. Les professeurs de mathematique soit du Colege royale, soit de l'université de Paris ou des universités des provinces en donnant des traités de Navigation au lieu des traités d'Optique, de Dioptrique et de Catoptrique qui ne sont que de pure Curiosité, reveilleroient l'attention des jeunes gens exciteroient leur genie sans qu'il en coutât rien qu'un mot du Monarque. Les Français sont aujourdhui asses éclairés pour operer le bien. Les etablissemens sont faits, ils n'attendent que les ordres ou du moins une simple invitation tacite. Les societés purement litteraires peuvent encore concourrir au même but en faisant des recherches dans l'antiquité pour y découvrir des procedés tombés en désuétude qui pourroient être de la plus grande utilité...," ALS Pingeron to "Monseigneur," dated "à Paris ce 22e avril 1780," AN-MAR G 102, dossier 2, #63.

184. See BM/Rouen, B4/1, Registre no. 1 (1744-1763), "Registre journal des Assemblées et Deliberations de l'Academie des Sciences, Belles-Lettres, et Arts, De Rouen etablie en 1744," and B4/2, Registre no. 2 (1764-An XII), "Registre des Assemblées et Deliberations de l'Academie des Sçiences, Belles lettres et arts De Roüen Etablie en 1744. Deuxiéme Volume commencé le 11 Janvier 1764."

185. See "Lettres des négociants de Saint-Domingue et de la Martinique à Le Cat (1766-1767)," BM/Rouen, C6.

186. See minutes of the meetings, and "Possibilité de rendre l'eau de mer potable (1763)" and "Le 'Sauer-Kraut'," BM/Rouen, C10.

ities and to visiting dignitaries.[187] The institution performed independent tests of desalinization, timber for naval use, dyeing agents from Central and South America, cinchona, and *jalap*.[188]

Whether because they were rejected by the Colonial Machine or were otherwise not qualified, or simply to add to the attributes displayed on so many title pages, men sought admission to the Rouen Academy as correspondents. Peyssonnel, as we saw, sought out the Rouen Academy and sent his papers to it.[189] Blondeau from the Marine Academy sent material and solicited membership, as did Father Pingré and Parmentier. Lefebvre-Deshayes forwarded his tract on coffee liqueur from Saint Domingue.[190] And Moreau de Saint-Méry sent all of his works to Rouen, including the volumes of his *Loix et Constitutions des Colonies Françoises de l'Amérique Sous-le-Vent*, and he was duly elected a correspondent.[191]

The Rouen Academy effected its own contacts with other learned institutions. It maintained independent contact with the Bordeaux Academy from 1772.[192] And most notably for our purposes, from 1785 on the Rouen group forged direct and substantial contacts with the Cercle des Philadelphes in Saint Domingue. The two institutions enjoyed reciprocal exchanges of plants and seeds and performed in common medical and dyeing experiments using Caribbean plants. These contacts culminated in an official correspondence between the two organizations.[193] In all these respects, the Rouen Academy functioned in ways analogous to the Académie Royale des Sciences in Paris, except with its own regional orientation and identity and a difference in scale and intensity. Essentially the Rouen Academy enjoyed no contact with Paris officialdom, and in this way was independent of the Colonial Machine.

Much the same points can be made about the *Académie Royale des Sciences et Belles-Lettres* of Bordeaux (1712). The Bordeaux academy was connected

187. See, for example, academy meeting "Du Mercredi 10 decembre 1766... Mr. L'intendant du jardin et Mr Le Professeur de Botanique ayant represente quil y avoit deux beaux ananas dans la serre du jardin des plantes de la Compagnie et dont elle pouvoit disposer, il a ete arreste que le premier seroit donné au Corps de ville le second a Madame de Miromesnil et que lun et lautre seroit presenté par Mr Le Professeur de botanique," BM/Rouen, B4/2, p. 63.

188. "Le cochenille et le bois de teinture. Essai de teinture sur les végétaux de Saint-Domingue (1790)," BM/Rouen, C8; BM/Rouen, B4/2, unpaginated ("Du Mercredi 24 Mars 1790"); "Examen analytique du Jalap," BM/Rouen, C6.

189. BM/Rouen, B4/1, p. 84 ("Du Jeudi 18 Decembre 1755") and p. 91 ("Le Mercredi 10 mars 1756"); also, Peyssonnel correspondence (two letters), BM/Rouen, C23.

190. BM/Rouen, B4/2, pp. 400v ("du Mercredi 21 juillet 1784"), 402 ("Du Mercredi XI aoust 1784"), and 409 ("du Mercredi 20 Juillet 1785").

191. See McClellan and Regourd (2006); Moreau was in contact with the Rouen Academy from January 1786 through May 1790; see BM/Rouen, B4/2, fols. 413v, 414, 414v, 425, 426, and unpaginated, meeting for "Du Mercredi 5 May 1790." See also Moreau correspondence, BM/Rouen, C25 and C27. In addition, Moreau sent the Rouen Academy a "Mémoire sur un nouvel Equipage de chaudière à sucre pour les colonies avec le plan dud. Equipage" and a "fragment sur les Mœurs de S.t domingue."

192. BM/Rouen, B4/2, p. 159 ("Du Mercredi 26 fevrier 1772").

193. On the contact between the Rouen Academy and the Cercle des Philadelphes, see McClellan (1992), pp. 161, 267; and further below.

to the overseas and the colonies through such correspondents as Moreau de Saint-Méry, Peyssonnel, Cassan, Barrère, and others, and it received regular reports from Martinique, Guadeloupe, Saint Lucia, and Guiana.[194] There were colonial correspondents in other academies in Dijon, Orleans, Amiens, Angers, Grenoble, Metz, La Rochelle, and, of course, Marseilles.[195] Through his extensive correspondence network, Dubois de Fosseux, permanent secretary of the *Académie Royale des Belles-Lettres* of Arras, disseminated news of the colonial world, not least as a national associate of the Cercle des Philadelphes.[196] Further research into the archives of other provincial academies will show the penetration of the colonial and the overseas into the deepest recesses of French cultural life. By the same token – the Brest Marine academy and the Cercle des Philadelphes of Saint Domingue notably excepted – provincial academies were distinct from the Colonial Machine, and lesser in status.

Colonial Autonomy and the Local Circulation of Knowledge

All kinds of tensions existed between the colonies and the motherland. Royal government and royal power were not always respected. Poorer colonists resented their social superiors and the strictures imposed by the central government and the Compagnies des Indes. Planters and merchants chafed against trade restrictions dictated by the mercantilist policies of the day, and they often engaged in contraband trade with the Dutch or the English. Residents resented transients from France. At various points, too, the Church seems not always to have been respected. Race and slavery introduced other, fundamental complications. Richer colonists and wealthy urban residents, such as they were and even if co-opted into ruling elites, began to feel the pull of a distinct colonial identity as the colonies, and especially Saint Domingue, developed over the course of the eighteenth century. The contemporary French colonial world was politically and sociologically complicated, governed by a dialectical tension that played itself out between the central and royal forces directing colonization and other tendencies toward colonial autonomy.[197] There were no formal movements toward independence, however, in any of the

194. See BM/Bordeaux, Fonds Ancien, ms. 828 (I-CVI) and ms. 1696 "Archives de l'Académie de Bordeaux. Correspondance."

195. Regourd (2000) pp. 533-35; Roche (1978), vol. 2, pp. 325ff, provides precise maps showing the distribution of the correspondents for each academy: see esp. map #36/3 (Amiens); #36/4 (Angers), #36/14 (Bordeaux), #36/23 (Grenoble), #36/26 (Marseilles), #36/27 (Metz : two correspondents in Guadeloupe, and three in Saint Domingue!), #36/32 (Orléans), #36/34 (La Rochelle), #36/35 (Rouen), and #36/37 (Toulouse).

196. On Dubois de Fosseux, see Berthe; McClellan (1985), pp. 185-86; on de Fosseux's membership in the Cercle des Philadelphes, see Cercle des Philadelphes (1787).

197. This theme of centrifugal and autonomous forces in the colonies structures James Pritchard's account of French colonization in the Americas through the 1730s; see Pritchard (2004), passim and pp. 254-60, 420-22; see also Banks, chapt. 5.

French colonies prior to the French Revolution, and the centrifugal tensions in the colonies to which we allude were vague and implicit, felt perhaps, but not necessarily articulated.

The implications of this polarity for the Colonial Machine were twofold. First, as an arm of central government, the Colonial Machine itself stirred up animosities, and this circumstance may help explain the general lack of participation of planters and traders in the activities of the Colonial Machine. Secondly, as the colonies grew, we can trace the beginnings of local communities of educated individuals there interested in the natural world and scientific questions. At no point did these communities ever become independent or crystallize in opposition to central authority and the Colonial Machine, but we can see their beginnings nevertheless. These communities represented alternative possibilities for organizing communities and for communicating information distinct from and sometimes in opposition to royal and central authority emanating from Versailles. These nascent tendencies toward political and scientific independence constitute a part of the story that cannot be omitted, and they throw the Colonial Machine into further high relief.

To the extent that distinct identities did arise in the colonies of Old Regime France, they were not without institutional bases on which to build a future scientific and political independence. The Cercle des Philadelphes/Société Royale at Cap François was the foremost of these, but the various Conseils Supérieur and Chambres d'Agriculture in the colonies also embodied autonomous forces that existed at some remove from central authority and the Colonial Machine. We have seen how the provincial academies in France offered institutional options for colonials. The press in France and in the colonies also helped support incipient colonial communities and helped provide a basis for the circulation of knowledge outside and apart from the Colonial Machine. These structures were not, for all that, orthogonal to the Colonial Machine. The story is not in black and white, a pitting of absolute opposition in the colonies to Versailles and the Colonial Machine. We are dealing, rather, with a nuanced set of tensions and accommodations with royal, central authority and with degrees of distance from France and the Colonial Machine.

Local Contacts. With the cadres of royal physicians, surgeons, botanists, hydrographers, and engineers posted overseas and the infrastructures in place to support them, the Colonial Machine became established in the colonies. This presence defined an interface with the rest of colonial society, especially as the positions mentioned were associated with colonial administration and central authority from France. At other points, the passage back and forth of scientific emissaries of the Colonial Machine could not leave cultivated elites in the colonies indifferent, especially where such distractions were infrequent. In these cases, illustrious individuals blessed with the prestige of official recognition of their status as experts, not to speak of the spectacular deployment of instruments that they sometimes used, made their presence felt. Yet, the reality of the colonial reactions to encounters with the Colonial Machine is hard to gauge, so rare and underdeveloped are the references found in the archives.

But contacts did arise between emissaries of the Colonial Machine and interested amateurs on the ground in the colonies. That was the case in Martinique when, in the 1690s, botaniste du roi Charles Plumier stayed with the missionary Jean-Baptiste Labat.[198] The same was true in 1744 when the academician La Condamine found himself at Cayenne in the company of the engineer Fresneau, who had been in the colony for a number of years.[199] Earlier, in 1735, when the academicians from France passed through Saint Domingue on their way to Peru, enlightened colonists received them with interest. One local resident, for example, with a "extreme passion for everything concerning natural history," offered his services to the expedition.[200] On this trip, Joseph de Jussieu stayed with the Montpellier-trained and converso physician De Pas, and botanized with him in the southern district of Saint Domingue.[201] At Petit Goave, the academician Bouguer had local workers forge a pendulum in steel under the direction of the royal clockmaker associated with the expedition.[202] Thirty years later, the royal ingénieur-géographes in Saint Domingue enlisted the aid of local people and taught them advanced surveying techniques.[203] The correspondent of the Paris Academy, Palisot de Beauvois made friends with the abbé de la Haye and botanized with him while in Saint Domingue in the 1780s. And, in 1784, the instructions of Chastenet de Puységur, the naval officer and hydrographer charged to map the coasts of Saint Domingue, ordered him to welcome aboard officers of the merchant marine interested in learning about hydrographical techniques and the use of the latest instruments, including the chronometer and Borda's circle.[204]

Either in place or passing through, agents of the Colonial Machine interacted with local amateurs and brought advanced European science to them, and in so doing they catalyzed the development of local communities. But it is also plain that in some instances representatives of the Colonial Machine provoked hostility from local residents. One anonymous letter writer in Saint Domingue, for example, commented on the airs of the academicians on the Peru expedition:

198. Labat (1979), vol. 2, p. 342.

199. La Condamine (1751), p. 328.

200. "...passion extrême pour tout ce qui fait l'objet de l'histoire naturelle. Il y a une plus de huit ans que j'en fais mon principal amusement dans cette isle," ALS "de Noyëlles," dated "de l'île à Vache, le 13 août 1735," BCMNHN, Ms. 179, no. 10. On the visit of the Paris academicians in Saint Domingue, see McClellan (1992), pp. 120-21.

201. BCMNHN, Ms. 179, no. 11.

202. Bouguer (1735), p. 526.

203. Glénisson, pp. 146-47.

204. ALS "Le Cte Mac-Nemara" ["directeur général de la marine au Cap"] to Puységur, dated "à bord de l'Amphion, rade des Gonaîves, le 14 juin 1784," AN-MAR 2 JJ 104, div. 1, doc. 3.

> One speaks of nothing else but the important expedition of our academicians. I learned from some trustworthy people, whom the babble of these *savants* could not deceive, that they are taking a lot of money with them to South America. They spent a fair bit in Cap François before leaving. To my face a merchant familiar with the destination predicted that if they do not leave off with their haughty airs and this French petulance that so displeases foreigners, they run the risk of finding themselves in some disagreeable situations, despite the protection given them by the Spanish Court. Beyond that, everyone rushes to receive them well.[205]

In the preface to his *Nouveau voyage aux isles de l'Amérique* (1722/1742), Labat, who is otherwise invisible in the archives of the Colonial Machine, raised his voice against what he took to be an arrogant scientific establishment, which he thought arbitrary and ill-adapted to his description of the Antilles:

> Some estimable persons have wished that I maintained a more methodical order to my account and that I organized things with each species placed in its genus. They had their reasons for wishing this, and I had my own for not satisfying them. Beyond the fact that this dogmatic approach is not at all to my taste, I constantly would have had to interrupt the flow of my diary...and I preferred to follow its course.[206]

Further on in his account, he confirmed his reluctance to follow the techniques of standard botanical nomenclature, classification, and description: "I don't know what the Indians call this tree or whether Father Plumier or some other Botanist has baptized it and enlisted it in some regiment of trees supposedly of the same species. For the rest of us, who are not so fancy and without getting tied up in knots over the name of this tree, we are content to call its fruit serpent's nut."[207] In the face of procedures for astronomical observation perfected by Cassini for determining longitudes, Father Labat displayed the same skepticism, reducing the astronomers of the Academy, blinded by their technical knowledge, to the rank of vulgar astrologers:

> As for longitude, I won't say a thing. I haven't measured it. The prime meridian is too far away, and there are so many differences and so many errors in the measurements of our astronomers, that the surest thing is to have good eyes and to use them well in approaching the islands, in

205. "Voyage du Comte de **** à Saint-Domingue," quoted in McClellan (1992), p. 121.
206. Labat (1979), vol. 1, preface, p. 17; Regourd, 2005a.
207. Labat (1979), vol. 2, p. 111.

> order not to break your neck in following the opinions of these gentlemen, the surveyors of the planets, who are usually as sure of what they put forward as the makers of almanacs and horoscopes.[208]

And finally, Labat seemed to enjoy humiliating Plumier, the botanist sent by the Academy of Sciences, in showing that slaves on his plantation knew the dyeing properties of shellfish, which Plumier thought he alone had discovered.

> Here, Father Plumier told me, is the treasure. I have discovered in this country the purple of Tyre. It will bring more wealth to this country than all the mines of Peru and Mexico. I examined the handkerchief and immediately discovered the constituent principle of this color....And to show him that his purple was not a new discovery, in the presence of several of our slaves I asked how this piece of cloth had been dyed, and they all responded that it was with the "dyeing snail" that one finds every day on the seashore.[209]

True or not – Plumier had died almost twenty years before the appearance of Labat's work – the anecdote is edifying. The recourse to the testimony of slaves suggests a cultural clash between the model of a bookish, theoretical, and all-conquering science embodied by Plumier and the ordinary, empirical knowledge of colonists and even slaves, presented as more to the point than all the confabulations of the academicians. Over and over, Labat condemned the acculturating and arrogant reach of the Colonial Machine, whose institutional power alone justified its claims in the eyes of the missionary. The critique offered by Father Labat of Pierre Plumier and his science is notable in and of itself, and it reveals the existence of another kind of knowledge at play in the colonies: practical, common knowledge little interested in commensurability or universality.

The Fragility of Local Communities. The only Old Regime French colony that turned out to be capable of sustaining its own community of scientific inquirers was Saint Domingue. Île de France was not without possibilities, as we will see, and Cayenne, too, but Cayenne was staffed by royal agents only and lacked a colonial base large enough and wealthy enough for a local community to arise distinct from the Colonial Machine. There were never enough people in Canada or Louisiana for that outcome either. Martinique and Guadeloupe may have had the population and infrastructures, but they lacked a large enough body of established (royal) experts and interested amateurs together in an urban context. In Saint Domingue, by contrast, the white population in the 1780s was a substantial 30,000; and it was the most urbanized French colony in 1789, with the largest colonial town of Cap François having a population of

208. Labat (1979), vol. 3, p. 270.
209. Labat (1979), vol. 2, p. 342.

18,500 and Port-au-Prince having a population of almost 10,000. The Colonial Machine was well entrenched in Saint Domingue, as we have seen, with royal doctors, royal surgeons, and royal botanists stationed there, and these men brought expertise and training with them. Saint Domingue and Cap François in particular developed independently and in fairly complex ways as colonial outposts – notably with the rise of a medical community rooted in the colony, apart from the royal medical establishment and the naval medical corps. Saint Domingue in the 1780s had come a long way since its primitive beginnings a century or so earlier, and the colony was in a position to support a new haven of science, albeit a tiny one, in the Americas.

To whatever extent they became involved in scientific questions and research, colonials remained largely peripheral and dependent on the structures and resources of organized science in the Colonial Machine. The entomological research of Palisot de Beauvois illustrates the difficulties of doing science locally in the colonies. While in Saint Domingue he created his own system for classifying insects, and he sought to have colleagues in France send him samples to work on in Saint Domingue.[210] As he wrote to Jussieu at the Jardin du Roi in Paris,

> As for insects, my research has produced, not a system, but a method that seems to me more simple and easier than those of Linnaeus, Geoffroy and Fabricius. As it is nothing less than complicated and long, I include herewith the complete method that I ask you to present to the Academy....
>
> You can see, dear Sir and friend, that to perfect this work I need books as much as insects. If some zealous insectologists wish to send me insects from Paris,...I pledge my honor to quickly return them. I have doubles of the many insects of this country, and with those that I can procure through my correspondents in Martinique and New England and ships surgeons, who sail to Africa, I can make a fairly complete collection so that my work will become close to perfection.[211]

210. See undated ms. memoir of 1789 or 1790, "Nouvelle Methode Entomologique Par M. Le Baron de Beauvois de la Société Royale du Cap et Correspondant de l'Academie Royale des Sciences de Paris"; PA-DB (Palisot de Beauvois).

211. "...Quant aux insectes mes recherches m'ont donné lieu à la rédaction, non pas d'un système, mais d'une méthode qui me paroit plus simple et plus facile que celles de Linné, Geoffroy et Fabricius. Comme elle n'est rien moins que compliquée et longue je joins ici l'ensemble de cette méthode que je vous prie de présenter à l'Académie.../... Vous voyés, Monsieur et cher ami, que pour ce travail j'ai besoin pour le perfectionner d'avoir des secours tant en livres qu'en insectes. Si quelqu'insectologistes zélés veut m'envoyer les insectes de Paris...je m'engage d'honneur à les lui renvoyer et de suite, car j'ai des doubles, pleines des insectes de ce païs, ceux que je me procurerai par mes correspondances à la Martinique, à la Nouvelle Angleterre et par quelques chirurgiens de navire qui vont à la côte je pus me faire une collection assées compète pour que mon ouvrage acquiere à peu près la perfection." This quotation is drawn from two letters from Palisot to Jussieu, dated "Du Cap le 9 8bre 1789" and "Commencée au Cap le 10 May finie le 28 aoust 1790," in PA-DB (Palisot de Beauvois).

Palisot raised the prospect of science based in the colonies – Martinique, New England, Africa, but he lacked resources, especially books, and remained isolated in Saint Domingue.[212] After his return to France he published his *Insectes recueillis en Afrique et en Amérique* in fascicles between 1805 and 1821, but could do so only in the metropolis. In his *éloge* of Palisot, Cuvier made plain the difficulties he faced: "It is true that the author had left France, and ideas that are not presented and defended by the person who conceived them are more likely than others to fall into oblivion. Truth itself needs patrons to succeed in the world, however evident it may be, and even more so for positions for which proof is still incomplete."[213]

The dependence of colonials on the resources at the center is likewise evident in the case of the physician Dumonville, a colleague of Peyssonnel on Guadeloupe. Dumonville sent his 122-page manuscript concerning a disease that appeared in the Windward Islands in 1748 to the ministry with the request that it be distributed to the medical schools in Paris and Montpellier. He added, "Thus, the counsel and remedies that these two faculties might give will subsequently be of such great utility that one must employ all sorts of ways and means to procure them for the inhabitants of these colonies."[214]

Their local and peripheral circumstances vis-à-vis the Colonial Machine limited more than one colonist's potential career in science. The colonist and merchant Jean-Baptiste Auvray provides one example. Auvray was a founding member and later president of the Cercle des Philadelphes, a syndic of the Chambre d'Agriculture of the Cap, and a correspondent of the Société Royale d'Agriculture in Paris. Auvray worked on the mineralogy of Saint Domingue, epizootics, quinine, and local fodder, and he published in the memoirs of the Cercle des Philadelphes.[215] Despite this activity, he was stuck on the edges of the Colonial Machine with no possibility for scientific advancement. The eclipse prediction by Adam List, the royal surveyor at the Môle Saint-Nicolas outpost in Saint Domingue, who was praised as "one of the best mathematicians in the colony," is touching in its testimony to the limitations and obstacles faced by those who wished to pursue science on the colonial periphery.[216]

212. See Palisot repeated requests for books and specimens, ALS Palisot to Jussieu, "Cap le 7 avril 1791," in PA-DB (Palisot de Beauvois).

213. "Il est vrai que l'auteur avait quitté la France, et que des idées, qui ne sont pas présentées et défendues par celui qui les a conçues, sont plus sujettes que d'autres à tomber dans l'oubli. La vérité elle-même a besoin de patrons pour se produire avec succès dans le monde, quelque évidente qu'elle puisse être à plus forte raison des vues dont la preuve est encore aussi incomplète," Cuvier, "Éloge historique de M de Beauvois, lu à la séance publique de l'Académie des sciences, du 27 mars 1820," in PA-DB (Palisot de Beauvois).

214. "Ainsi les avis et les rémedes, que pourroient donner ces deux facultés, seront, dans la suite, d'une utilité si grande, qu'on doit employer toutes sortes de voïes et de moïens pour les procurer aux habitans de ces Colonies…," "Essay de Dissertation sur une Maladie observée dans quelques Îles du Vent de l'Amerique en 1748," AN-MAR G 102, dossier 3, #112.

215. DPF, p. 1446; McClellan (1992), p. 201.

216. McClellan (1992), p. 127, citing the *Affiches Américaines*, Port-au-Prince edition, for 13 September 1787.

But the most poignant example is that of Étienne Lefebvre-Deshayes, the successful plantation owner, active member of the Cercle des Philadelphes, and official correspondent of Buffon and the Cabinet du Roi, who was in contact with Lalande, Father Cotte, and the scientific community in France. Lefebvre-Deshayes would have done more in science if he could, and he was neurotically sensitive to the limitations imposed on him by his colonial remove. As he wrote, with perhaps some exaggeration, in an article published in Rozier's *Journal de Physique* in 1785:

> That one cannot know everything or examine everything thoroughly is a true means of encouragement for those engaged in the immense undertaking ["carrière"] of natural history. But so as not to take wasted steps, or at least not to put out old knowledge as new, one must read all that has been written on the subject one is pursuing. Otherwise, how can one be assured of not simply repeating what has been said before? How can one know if one is adding anything to what our predecessors have established?
>
> There are nevertheless cases where it is not possible to follow this rule that reason prescribes. Such is the case, for example, of an inhabitant of a distant country, especially when he finds himself destitute of all scholarly resources. Should this man refuse to give in to his penchant for observation?...
>
> It is our misfortune not to have any book that can instruct us in this very interesting part of natural history, and we don't even know if someone has already spoken about the strange animal that is the subject of this notice....Unfortunately, there are no libraries in our area or even from far about. What a multitude of impediments and obstacles one encounters here at each step in the pursuit of science ["dans la carrière des sciences"].[217]

At another point, the poor colonial opined: "I will be very satisfied if I can take a few steps along the route that the learned alone can travel, especially if I succeed in drawing their attention to the landscape I have in view. May they decide to dig deeper into a subject that I have only skimmed."[218] More than

217. "Notices sur l'Anéomone de mer à plumes ou animal-fleur," *Journal de Physique* 27 (1785), pp. 373-81, here pp. 374, 377; quoted in McClellan (1992), p. 271.

218. "...je serai très satisfait si je puis faire quelque pas dans une carrière que les savans seuls peuvent parcourir, et surtout si je réussis à leur faire jetter les yeux sur les eaux que j'ai en vue; puissent-ils se déterminer à approfondir une matière que je n'ai fait qu'effleurer," "Essai analytique sur eaux thermales minérales dites de la Grande-anse ou du bras-gauche de la grande rivière de Jérémie," SRdM, Archives, 89 dr. 30.

anyone else, the royal physician in Cap François, Charles Arthaud, was responsible for forging a local community of experts and amateurs interested in the sciences, yet even, or most especially, he was sensitive to the hurdles faced by any effort to implant science in the colonies: "Science is an exotic plant that has hardly yet taken root here. We don't yet know how to cultivate it, and I think it be naturalized only with difficulty."[219] More bluntly, he spoke of Saint Domingue as "a colony where one does not yet know the word science, much less the word eloquence."[220]

The Metropolitan and Colonial Press. The press and the technology of printing formed essential elements of the Colonial Machine and of contemporary intellectual life in general. The Enlightenment would have been impossible without the press, and in a more limited fashion the publications of the men and institutions of the Colonial Machine – their books, broadsides, and volumes of learned memoirs – constituted sine qua non of the Colonial Machine and colonial administration in general. By the same token, both at home in France and abroad in the colonies, the press also provided outlets and a voice for colonials interested in science and the natural world on the margins of the Colonial Machine.

Metropolitan journals and newspapers were willing to open their columns to contributions from the colonies. The *Journal de Physique*, the *Journal des savants*, and even more specialized sheets like the *Feuille du cultivateur* or the *Gazette de santé* published numerous articles written in the tropics. If many of these were, in fact, transmitted by distinguished members of the various royal academies, in many other instances authors addressed editors directly, bypassing official channels. The *Journal de Physique*, or Rozier's *Journal* provides the premier case in point. Abbé François Rozier and successive editors published the *Observations sur la physique, sur l'histoire naturelle et sur les arts*, to give it its full and official title, monthly in Paris from 1772 to 1826.[221] The speed of publication and the fact that Rozier's was exclusively a scientific journal publishing original material made it a landmark in the history of the scientific press. The Académie Royale des Sciences elected Rozier a correspondent, and his *Journal* quickly became an unofficial publication of the Académie, which supplied material to him either to speed publication or to see into print papers that would not otherwise appear in its *Histoire et Mémoires*

219. "La science est une plante exotique qui n'a point encore pris racine icy; on ne sait pas encore l'y cultiver, et je crois qu'elle s'y naturalisera difficilement," ALS Arthaud to Vicq-d'Azyr, dated "au Cap, le 10 septembre 1785," SRdM, Archives, 136 dr. 1, #26.

220. "...une colonie où l'on ne connoit pas encore le mot de science, et moins encore celui d'éloquence,"Arthaud letter probably to Vicq d'Azyr of 24 September 1785, SRdM, Archives, 136, dr. 1, #15bis; in this same letter Arthaud speaks of himself as "un médecin colonial qui est isolé et qui cherche à se faire un atmosphère dans lequel on puisse respirer les élémens qui lui conviennent."

221. McClellan (1979); McClellan (2003b), pp. 87-88.

series. Rozier's *Journal* thus contained plenty of colonial and overseas material forwarded through the Academy, such as Pingré's 1773 "Rapport des Observations Faites sur mer pour la détermination des longitudes" or a 1775 report on wasps from Guadeloupe, sent to Rozier by the Academy.[222] As much as his *Journal* exercised this semiofficial role within the Colonial Machine, Rozier also provided a publication outlet for colonials, who otherwise would not have had one. The volume for 1773, for example, contained reports on beehives from Lanux in the Indian Ocean, on plants sent from Île de France and Île Bourbon to the Jardin du Roi, on the interior of Guiana by Laborde, on inoculation against yaws in Jamaica, on tetanus in Antigua, and on the natural history of California. Other volumes kept up this drumbeat of information flowing in to Rozier from the colonies overseas. From Saint Domingue articles appeared by Simeon Worlock, Lefebvre-Deshayes, Charles Artaud, Amic, and Genton, whose paper on the mineralogy of Saint Domingue appeared in the *Journal de Physique* in 1787.[223] In sum and unsurprisingly, Rozier's *Journal* published scientific and medical reports and provided yet another window for encountering and adjudicating science from the colonies and the world overseas.

Primarily as an adjunct of colonial administration, the government licensed royal presses in the colonies. Arriving in Saint Domingue from Dijon in 1723, Joseph Payen was the first printer-bookseller authorized in that colony, and his *Code Noir* was the first work printed there. But Payen soon feuded with the local administrators, who accused him of trafficking in 'dirty books'. It was not until 1762 that, backed by the Chambre d'Agriculture, presses came to be established in Cap François and Port-au-Prince; a third press arose in Port-au-Prince in 1788. In 1728 in Martinique authorities granted a license for a press to a certain Vaux. For Guadeloupe it was not until 1764 that Jean Bénard became a printer there[224]. No press appeared in Canada before the fall of the colony, despite an initiative by La Galissonière in 1749. A French press began operations in Louisiana in 1764, only to announce the Spanish takeover.[225] The press arrived in Guiana in 1770.[226] By 1789 there were only four or five printers and a dozen bookstores in the Antillean colonies and one on Île de

222. Pingré (1773); McClellan (2003c), p. 88.

223. Genton (1787); McClellan (1992), pp. 270, 356n. Rozier proposed to the Bordeaux Academy to publish material sent to him; see Rozier letter, dated Paris, 14 September 1776, BM/Bordeaux, Ms. 828, XXI, 57.

224. McClellan (1992), pp. 99-102; Regourd (2000), pp. 512-15, and (2001), pp. 186-87. See also ministerial correspondence to the administrators in Martinique, dated 10 October 1728, AN-COL C^{8A} 39, fols. 165-168, and 4 April 1730, AN-COL C^{8A} 41, fols. 34-36.

225. Banks, p. 180; the press did not arrive in Canada until the 1760s after the English takeover of that colony; see de Lagrave; Mélançon (2007), pp. 31-34; Fleming et al.

226. Touchet, pp. 119-20.

France.[227] Pierre Poivre established the first press on Île de France in 1767; the first press on Île Bourbon dates only to 1792.[228]

The primary function of the colonial press was to print official documents for the administration (notices, posters, announcement, broadsides, and so on.), and the press was tightly controlled and censored by local colonial administrators. Various almanacs and reprints of important laws such as the *Code Noir* made for good sales, but the publication of newspapers marked a new stage in the cultural life of the colonies. The *Gazette de Saint-Domingue* (1764-65), followed by the *Affiches Américaines* (1765-1793), the *Gazette de la Martinique* (1766-1793), the *Gazette de Sainte-Lucie* (1787-1793), and the *Affiches de la Guadeloupe* (1789-90) were the principal publications of a fairly dynamic colonial press. Irregular imprints include the *Gazette des Petites Antilles* (1774-1776, 1788, 1784-85), *Gazette de Médecine et d'Hyppiatrique* (1778-79), *Follicules Caraïbes* (1785), *Gazette de la Grenade* (1779-1783), and the *Mercure des Antilles* (1783). The *Affiches des Îles de France et de Bourbon* appeared from 1773 into the 1790s. The royal press on Île de France apparently did not amount to much; Lalande called it inactive, and Cossigny complained that "it is the only one that exists on this island, and it is not significant. It lacks type, workers, and paper."[229] Yet in 1783 that press did print a prospectus for a journal of medicine, physics and politics (*Journal de Médecine, de Physique et de Politique*) for Île de France, but doubtless the "politics" involved precluded its launch.[230]

Newspapers published in the colonies served primarily as vehicles for circulating shipping information, commodity prices, news from Europe, official and private announcements (regarding maroon slaves, for example), and other reports.[231] Beyond these strictly utilitarian notices, one can also read agronomic memoirs, mathematical demonstrations, technical tracts, medical and meteorological reports, and even poetry. If some of the inserts in colonial newspapers are attributable to local administrators, most of the submissions of colonists were published on the initiative of editors, who like Charles Mozard, publisher of Saint Domingue's *Affiches Américaines*, played a significant role in endowing local scientific statements with a degree of legitimacy. We have seen how Mozard organized the collection of weather data in Saint Domingue through the pages of the *Affiches Américaines*; and such was the independent role of the colonial press that through the *Affiches Américaines* the Academy

227. See Mellot and Queval, passim.

228. On these publications, see entries by Alain Nabarra in Sgard, ed.; see also Ménier and Debien.

229. "c'est la seule qui existe dans l'Isle, et elle n'est pas considérable; on manque de caractéres, d'ouvriers et de papier," ALS Cossigny to Le Monnier, dated "A L'Isle de France le 20 9bre 1779," BCMNHN, Ms. 1995; Lalande (1775), p. 49.

230. See mention, SRdM, Ms. 9, p. 418 ("Séance Du mardi 1er avril 1783").

231. McClellan (1992), pp. 97-101; for the larger social context, see Fouchard.

in Rouen became aware of yet another process for the desalination of seawater and undertook experiments of its own.[232]

Pornography formed a part of the book trade in the colonies, to be sure, but sale catalogues and the advertisements of booksellers in colonial newspapers make clear that the entire range of contemporary literary and scientific works were available to colonists, just as they were to their counterparts in the provinces of France. Titles ranged from ecclesiastical and secular history, to law and jurisprudence, to belles letters and poetry, including works by Rousseau, Corneille, and Molière, and the dictionary of the *Académie Française*. Scientific subjects concerned anatomy, architecture, chemistry, and medicine; and at various points in time colonists could purchase several editions of the *Encyclopédie*, including the original, a set of thirty-three volumes of the *Histoire et Mémoires* of the Académie des Sciences, Buffon's *Histoire naturelle*, Voltaire's and Mme de Châtelet's *Élements de Newton*, Lalande's astronomy, Bélidor's hydraulics, and the standard texts by Bézout and Camus.[233] The availability of these works attests to the existence of a clientele hungry for books more challenging than smut, a clientele that alone could justify the costs of shipping books to the colonies. In addition, a significant portion of the books making their way to colonists were purchased directly in France. Frequently, commercial correspondents, relatives, or friends sent books through the mail or slipped them into the luggage of a trusted traveler headed to the colonies. This private circulation of books left little trace in the archives, but did not cease to exist all through the eighteenth century.

The interest in books on the part of colonists did not pass unnoticed by the ever-observant Father Labat in Martinique. Already in the 1690s he could write scathingly:

> Although our creoles and other inhabitants have hardly degenerated from the bravery of their ancestors, they have given into the tastes of the rest of the world. They want to appear learned. They read everything or want to seem to have read everything....Women are also wrapped up in this, and instead of sticking to their spindles and spinning wheels, they read fat books and think themselves scholars. I know one who explains Nostradamus at least as poorly as minister Jurieu explained the Apocalypse....There are doctors and apothecaries, and we have plenty of surveyors, engineers, botanists and astronomers....books are needed for all that. Although most readers don't understand any-

232. See "Mémoire pour servir de moyen de dessaler l'eau de mer et de la rendre potable, par Le. Sr. Chervain habitan de Mok a quartier de Limonade, dépendance du Cap et Inserré dans l'affiche américaine de St. domingue du mercredy 22 8bre. 1766," AN-MAR D^3 40, fol. 117; BM/Rouen, B4/2, p. 68 ("Du Mercredy 17 mars 1767,...Du Mercredi 29 Mars 1767").

233. McClellan (1992), pp. 101-102; Regourd (2000), pp. 516-21, and (2001), pp. 190-91.

> thing, they wish to appear learned, and for that bookcases are needed that in time can turn into libraries, which leads me to say that a well-stocked bookstore could do a good business here.[234]

Labat was right, and, despite the problem of insects, across the century private libraries became larger and more numerous. Moreau de Saint-Méry's library contained roughly 3000 volumes at the beginning of the 1780s.[235] Moreau's collection was exceptional, but notarial records and newspaper announcements reveal the existence of a number of not insubstantial libraries in the colonies, sometimes with cabinets of curiosities attached. That was the case of the library left by the planter Barthélemy Badier on Guadeloupe. In addition to a cabinet of curiosities, the inventory of Badier's collection cites 150 titles in 300 volumes, in addition to 113 miscellaneous or incomplete volumes and 40 or so volumes of the *Journal de Physique*. The collection contained 35 volumes of the *Encyclopédie* of Diderot and d'Alembert, Buffon's complete works, the *Histoire des plantes de la Guyane française* of Fusée-Aublet, the *Telliamed* of de Maillet, a number of travel accounts, a pocket atlas, and books on mathematics, botany, mineralogy, and astronomy. Typical of such colonial libraries, few works of literature were to be found, but a few titles by Voltaire, Rousseau, and Racine complemented Badier's essentially scientific library. The inventory of the notary on Guadeloupe, Antoine Mercier de La Ramée, presents a collection of 500 volumes, wherein one also finds the complete works of Buffon, the *Encyclopédie*, Voltaire's tract on the trial of Callas, and the *Histoire philosophique des deux Indes* by Abbé Raynal.[236] On his death, the library of the surgeon Pierre Teytaud contained almost 100 volumes, including 9 volumes of the *Histoire et Mémoires* of the Académie Royale des Sciences, while the inventory of the library of the colonist Tanturier des Essarts, dating from 1768, contained more than 1000 titles.[237] Other studies have shown the existence of dozens of personal libraries in Saint Domingue, some with several hundred volumes.[238] Finally in this connection, in Saint Domingue and doubtless other French colonies of the day, the private reading rooms (*cabinets de lecture*) that cropped up in the 1770s and 1780s testify to changes in the practice of reading in the colonies.[239]

Toward the end of the eighteenth century, then, in Saint Domingue and elsewhere social and institutional structures were in place to begin to support inde-

234. Labat (1979), vol. 2, pp. 329-30, in Regourd (2001), p. 189; the reference is to the French Protestant theologian, Pierre Jurieu (1637-1713), and his *Accomplissement des propheties* (1686).

235. Regourd (2001), p. 190; Moreau de Saint-Méry papers, AN-COL F^3 74, fol. 47.

236. Bégot.

237. Regourd (2001), p. 191.

238. Fouchard, p.72-73.

239. McClellan (1992), p. 97. For Canada, see Mélançon (2005) and (2007).

pendent local communities of colonists with serious interests in studying the natural world around them. Connections to Europe and to the academies in Europe put colonists in touch with up-to-date science; resident experts of the Colonial Machine provided a core of working professionals to seed local communities. In addition, with colonial Conseils Supérieurs, Chambres d'Agriculture, Masonic lodges, bookstores, colonial newspapers, private libraries, and reading rooms, a critical mass of social institutions and educated individuals fell into place, and the stage was set for the emergence of colonial scientific societies.

Commerson's Colonial Fantasy. The Cercle des Philadelphes, founded at Cap François in Saint Domingue in 1784, was the first (and only) scientific society established in the colonies of Old Regime France, but the first plans for a colonial scientific society had been made fourteen years earlier, in 1770, for the colony on Île de France in the Indian Ocean. There, the intendant, Pierre Poivre, may have thought of organizing a colonial academy, but the botanist Philibert Commerson became the motive force behind the idea on his arrival on Île de France in 1769, after having accompanied Bougainville on the round-the-world expedition led by that navigator.[240] Commerson wrote to Lalande in France about his proposal:

> In the First Class, that of the sciences, will be mathematics, natural history, *physique*, medicine, and subordinate disciplines. This Academy will deal only with exotic subjects, i.e., observations and research in astronomy, geography, extra-European hydrography, productions of the three kingdoms of nature coming from outside of Europe, tropical diseases, examination of the terrain and plants native to this country, changes to European plants grown or transplanted here, comparisons of their products, etc.
>
> I am in contact with experts appropriate to start out each class: the Abbé Rochon, Mr. Veron, and a royal naval officer for mathematics; Mr. Poivre, colonel Puguet, Mr. Munier, and myself for natural history; Mr. Bourdier and the royal physician on Île Bourbon for medicine; and a goodly number of excellent and well-intentioned agriculturalists, the group to which we look to spark emulation because it will be from them that the colony will soonest reap the rewards. There is an idle, but well-equipped press here, and one will learn that there is an Academy in this

240. On Commerson and his plans for a colonial scientific society for Île de France, see his "Academia politica sive universalis" and related documents, BCMNHN, Ms. 1904, Commerson: "Mélange de manuscrits botaniques et autres," folder III; see also Lacroix (1938), vol. 4, p. 11; Cap, p. 21; Doublet (1920), pp. 9-11; Allorge, pp. 395-97.

> part of the world through the appearance of a volume of memoirs for which I will furnish three quarters of the text, if necessary.
>
> I communicated a summary of my project to Mr. Poivre, that excellent man of goodwill and right thinking. It pleased him immeasurably, and he eagerly awaits the details. I flatter myself to think that this project will no less please Mr. Poissonnier, to whom I ask you to send the *Prospectus* for his consideration with the request that, if he approves, he will seek the support of the minister.[241]

Commerson makes plain his understanding of the chain of command within the Colonial Machine (Poivre, Poissonnier, and the minister) and the fact that a nascent community of experts (Rochon, Véron, Poivre, himself, and others) was in place to support the possibility of a colonial scientific society for Île de France. Yet Commerson was a fantasist. He sketched elaborate plans for an entire "Academic City" ("Ville Académique") devoted to the study of all the arts and sciences. That he imagined his colonial academy could be staffed by 100 academicians at 3000 livres apiece indicates just how out of touch he was. Most importantly, Commerson seems not to have understood that French provincial or colonial learned societies were not top-down operations imposed by central authority, but had to begin modestly with locals organizing themselves and then petitioning government for recognition and support. Had Commerson and friends begun that way, the first French colonial learned society might well have arisen on Île de France instead of in Saint Domingue.

In 1770 Commerson was in contact about his proposal with the younger Cossigny on Île de France.[242] In the early 1780s Cossigny himself went on to develop contacts with the Dutch colonial society in Batavia, the *Bataviaasch Genootschap van Kunsten en Wetenschappen*, founded in 1778. Cossigny's work on indigo was well received by the Dutch group; it published his paper in volume three of its memoirs and elected him a correspondent in 1782.[243] Cossigny was also a *correspondant* of the Académie Royale des Sciences in Paris, and he proposed to his contact there, Le Monnier, that he, Cossigny, form a center for organized correspondence throughout the Indian Ocean for Le Monnier and the Académie in Paris.[244] Cossigny's activities in this connec-

241. Commerson, quoted by Lalande (1775), p. 46; Doublet (1920), p. 11, repeats this quotation.

242. ALS Cossigny to Le Monnier, dated "Au Port Loüis Isle de france 30 mars 1770," BCMNHN, Ms. 1995.

243. On these points, see Cossigny letters to Le Monnier, dated "Palma le 27 Août 1780," "A L'Ile de France le 12 Xbre 1781," and "A L'Ile de France le 30-8bre 1782," all BCMNHN, Ms. 1995.

244. "Je me propose de former à l'Isle de france le centre de toutes vos correspondances...," ALS Cossigny to Le Monnier, dated "A Besançon le [?] Xbre 72," BCMNHN, Ms. 1995.

tion are notable in their own right, and they, too, reveal that a threshold had been reached in the 1770s and 1780s so that the formal organization of science in the Indian Ocean was a real possibility. But that possibility was not to be, and, as we know, the Cercle des Philadelphes of Cap François in Saint Domingue became the first and only colonial scientific society of the first French colonial empire.

The Beginnings of a Colonial Learned Company: The Cercle des Philadelphes. The Cercle des Philadelphes originated in August of 1784, and it received formal letters patent elevating it to the status of the *Société Royale des Sciences et des Arts du Cap François* in May of 1789. The Cercle des Philadelphes/Société Royale represents the chief manifestation of an indigenous community of scientists/intellectuals in the colonies of Old Regime France. The organization rounded out the Colonial Machine and brought it to its apogee on the eve of the French Revolution. And, for present purposes, the Cercle des Philadelphes provides a test case for gauging the balance of centripetal versus centrifugal forces defining the Colonial Machine and French colonial history generally in the Old Regime. The Cercle des Philadelphes of old Saint Domingue has been well studied, and the historiography of the institution itself has been the object of study.[245] Here, we can content ourselves with the main features of its history and historiography.

Similar in a way to what happened with Commerson and his proposal for an academy on Île de France, the idea of a colonial academy of some sort popped up in Saint Domingue in 1769 and again in 1776, but nothing eventuated from these proposals.[246] Instead, the hydrographical expedition to Saint Domingue led by A.-H.-A. de Chastenet, count Puységur triggered the formation of the Cercle des Philadelphes. As we saw, Puységur brought with him not just advanced chronometric techniques for cartography, but also the full armamentarium of contemporary mesmerism, and Puységur and mesmerism provoked a vigorous, negative response among the scientific elite of the colony.[247] Charles Mozard attacked Puységur in the *Affiches Américaines*, and Charles Arthaud and a few like-minded individuals formed themselves into an ad hoc anti-Mesmer committee that in June and July of 1784 worked to stamp out what they saw as the charlatanism of Mesmer and Puységur. That group included besides Arthaud, the médecin du roi in Cap François; Alexandre Dubourg, a local apothecary; and probably J. Cosme d'Angerville, the chief royal surgeon in the northern department of the colony. After a break they met again in early August with another purpose, and on 15 August 1784 the nine charter members founded the Cercle des Philadelphes.

245. Much of this discussion of the Cercle des Philadelphes is drawn from McClellan (1992), Part III, and McClellan (2000c).

246. McClellan (1992), pp. 188-91.

247. See above, Part IIA and Part IIB; and Regourd (2008b).

Unlike Commerson on Île de France, Arthaud and company understood the steps they needed to take to secure their new organization, and a week after their charter meeting, on 22 August 1784 the group of now twelve Philadelphes petitioned the Saint Domingue governor-general and intendant for permission to hold meetings. This granted, they followed up with a *Prospectus* announcing the Cercle and its goals to the public at large:

> We aspire to having a general description of the Colony. We are asking for specific descriptions of different districts. We would like observations on the soil, on minerals found there, on trees and plants that grow there [and] on agriculture and manufactures undertaken there. The history of insects will be interesting and quite valuable to us, as will histories of birds and shellfish.
>
> We seek astronomical and meteorological observations. We will be obliged by research on the constitution of the air, on temperature, on the winds, and on the quality of the water, sweet or mineral. [We seek information] on reigning illnesses and on diseases particular to each district. We would like philosophical observations on the constitution and habits of people born in the colony [and] on the changes in temperament and in physical and moral constitution experienced by Europeans [in coming to the colony. We desire information] on the character, talents, and mores of slaves, and on ways of improving their lot, without harming the interests of colonists [!]. Finally, we wish to have observations concerning diseases affecting livestock, ways of treating these diseases, and especially of preventing them.[248]

The prospectus went on to note the Cercle's interest in promoting cochineal husbandry and in forming a public library, a scientific *cabinet*, and a botanical garden. Thus, from its beginning the Cercle des Philadelphes articulated a multifaceted but unified program of scientific inquiry and useful applications, involving economic and natural historical surveys, studies of colonial medicine and diseases, investigations into colonial agriculture and manufactures, and a pragmatic concern for slaves.

The Cercle drafted a set of formal statutes that were approved by the administrators in November 1784 and published in January 1785. The statutes themselves are unexceptional in setting forth a typical eighteenth-century learned society. They called for twice-monthly meetings and provided for three classes of members: resident associates living in or near Cap François, colonial associates elsewhere in the colony and the West Indies, and national/foreign asso-

248. Quoted in McClellan (1992), p. 209.

ciates abroad. The Cercle soon added a class of honorary associates. The effective officers were a president and a permanent secretary (Arthaud). The Cercle did meet twice a month and funded itself through dues imposed on its resident members, which at the beginning amounted to 1000 colonial livres a year, a substantial sum.[249] Over the years the Cercle des Philadelphes elected a total of 163 associates, of whom about 110 were active at any one time. The Cercle des Philadelphes came to be composed of roughly 20 resident members, 40 colonial members, 30 national members, and 15 or so honorary members. At roughly one third of the membership, physicians and those in the medical profession formed the largest segment of members. Colonial planters (but not merchants) constituted about a fifth of the membership, followed by members with military, legal-judicial, governmental, or engineering backgrounds. Thirty-four members, or 21 percent, held a royal appointment of one sort or another; they earned their livings as paid employees of the state and were thus tied directly to the Colonial Machine. The Cercle otherwise carefully crafted ties to institutionalized power in electing intendants, governors-general, and colonial jurists.

Several times in the spring of 1785 the Cercle des Philadelphes approached the minister of the navy, de Castries, seeking "letters patent and the title of an academy." The local administrators generally supported the Cercle, but knowing that it would be asking for money, they opposed letters patent until the institution had "displayed somewhat greater continuity." Over the next eighteen months the Cercle did what it needed to do to secure the approval of the authorities. It held regular meetings, including annual public meetings. It expanded its membership, including electing of Benjamin Franklin as an honorary member. It began its series of prize contests with questions concerning tetanus, fertilizers, and "the best means of constructing buildings required for colonial agriculture and manufacture, from a simple slave's hut to the most complicated mill." The Cercle launched its botanical garden on a plot granted by the government and began its public course on botany. And it published two scientific volumes: the late Thiery de Menonville's *Traité du Nopal et de la Cochenille* (appearing in France in 1786), and its collection of papers, *Dissertation et observations sur le tétanos* (1786). By December 1786, with La Luzerne as the new governor-general and with J.-B.-G. de Vaivre, a former intendant and head of the colonial department within the Ministry of the Navy, backing the plan, the Cercle received royal approbation and provisional royal recognition. The administrators attended a special meeting of the Cercle at its rooms on the Rue Vaudreuil in Cap François. Unveiling a bust of Louis XVI on the occasion symbolized the new relationship. From Versailles, de Castries wrote to his subordinates in Saint Domingue:

249. A colonial livre was worth two-thirds of a *livre tournois*.

> The society formed at Cap François under the name of the Cercle des Philadelphes has undertaken useful work that can encourage people to improve production in Saint Domingue and that can even speed progress to other colonies. I proposed to His Majesty to authorize this society provisionally in its present form.... You will convey to the Cercle des Philadelphes this mark of the King's goodwill, which it should consider as a very glorious recompense for its first efforts and as a powerful spur in endeavoring it to realize the hopes that it has raised. The Cercle des Philadelphes can hope to acquire an even higher regard by publishing interesting memoirs and by devoting itself to objectives of recognized utility.[250]

Over the next two years, the Cercle des Philadelphes did just that. It continued holding its private and public meetings and building its membership. A new round of prize questions asked about preserving libraries and papers from the ravages of insects, on preserving flour in government stores, on sugar production, and, reflecting anxieties over slave resistance, on different types of poison. At the request of the government, the Cercle carried out work on epizootic diseases, publishing its *Recherches sur les maladies épizootiques de Saint-Domingue* in 1788. In 1788, too, the Cercle published its first (and only) volume of formal *Mémoires du Cercle des Philadelphes*, a collection primarily devoted to investigations of local mineral waters. In the fall of 1787 the Cercle launched a general survey of colonial agriculture and rural economy that included a separate section devoted to gathering information on the colony's slaves. (Some results of this survey appeared in the *Affiches Américaines*, but 1789 interrupted the planned volume.) The Cercle began publication of Abbé de la Haye's botanical work, *Florindie, ou histoire physico-économique des végétaux de la Torride*, a work de la Haye dedicated to the Saint Domingue learned society.[251] Along these lines, "wishing to fulfill the desires of the administration," in 1788 the Cercle distributed clove bushes it received from the royal botanists in Saint Domingue and Cayenne, Nectoux and Richard. At this time, too, the Cercle gave away grass seed intended for eroded soils and as additional forage for animals.

From 1786 onward the government provided an annual subvention of 2000 livres drawn on the general fund of the colony, and in 1788 the Cercle used some of this money to have its own *jeton* struck in France. On the front of the coin, facing right in positively grotesque relief (he was then thirty-four-years-old) is Louis XVI and the ominous yet quintessential date of 1788. On the back were the insignia and legends of the Cercle des Philadelphes: a beehive and

250. Letter to "Mrs. De la Luzerne et du Marbois, Le 29 Xbre 1786," AN C^{9B} 36 (2), quoted in McClellan (1992), p. 232.

251. This work never appeared.

hovering bees beneath the rays of a meridian sun encompassed with the motto, *Exercet sub sole labor* (We do our work beneath the sun), and in the exergue the inscription *Cercle des Philadelphes, Etablie au Cap, 1784*. Each associate received a silver *jeton*, and the Cercle also sent these to the colony's administrators and their wives. Silver coins were valuable, but the payoff to members of the Cercle for their financial sacrifices was literally and figuratively token. The Cercle overflowed with enthusiasm and gratitude in writing to La Luzerne, now transferred to Versailles as the minister of the navy and the colonies. They begged La Luzerne to assume the title of protector of the Cercle and to send his portrait for "the first temple erected to the Sciences in the colonies of France."[252] La Luzerne refused out of modesty.

[Jeton of the Cercle des Philadelphes]

A Colonial Center and its Networks. Befitting its status as an emerging "provincial" academy, the Cercle des Philadelphes forged contacts with other academies and organizations in France. We have already mentioned the contact of the Cercle des Philadelphes and the Academy in Rouen.[253] The first word of the existence of the Cercle arrived in the Rouen meeting rooms at the Hôtel de Ville on 10 August 1785, a year after the foundation of the Saint Domingue institution, when one of Rouen's stalwart members presented "a letter, program, prospectus, and membership list of the Cercle des Philadelphes, a society formed at Cap François. They hope that the academy would take an interest in a society that meets only for the pleasure of doing work and being useful."[254] Jean-Baptiste Auvray, one of the Cercle's founders and later its president, and a certain Levavasseur had family ties to Le Havre and Rouen, and

252. See Cercle letter to "Monseigneur," dated "17 Avril 1788," AN C^{9B} 38, quoted in McClellan (1992), p. 242.

253. See above, this Part, p. 445 at note 193.

254. "...M dambourney a prêsenté une lettre, un programe, un prospectus et un Tableau Du Cercle des philadelphes société qui sest formê au Cap francois. ils espérent que Lacad... [sic] voudra bien prendre intéresse à la naissance dune société qui ne se Réunit que p. le plaisir de Travailler et d'Être utile," BM/Rouen, B4/2, fol. 411v. ("Du Mercredi 10 aoust 1785").

these colonials were instrumental in establishing this Norman connection. Auvray sent several scientific shipments from Saint Domingue, one of which contained over 100 different plant species for the Rouen Academy's botanical garden.[255] Levavasseur, too, sent seeds and a catalog of the Cercle's botanical garden; and he sought the approbation of the Rouen Academy in order to publish in Rozier's *Journal*, styling himself, "Le Vavassseur, Directeur du Jardin de la Sociëté Royale des Sciences et des arts du Cap francais."[256] The Cercle elected Dom Gourdin, librarian of the Rouen Academy, as a member, and the Rouen Academy itself sent seeds to Saint Domingue.[257] Initiated by Auvray in 1789, this growing contact resulted in a formal "correspondence" and affiliation between the two institutions.[258]

The Cercle des Philadelphes developed similar contacts with the Académie des Sciences et Belles-Lettres of Bordeaux.[259] From May 1785 on, the Cercle sent all of its materials to the Bordeaux Academy, including its prospectus, prize programs, Thiery de Menonville's book on cochineal, and its own publication on tetanus. The Bordeaux Academy in turn sent its memoirs and prize programs to Saint Domingue, and at one point it enlisted the Cercle in tests of Parmentier's sea biscuits.[260] At one point the Bordeaux Academy pushed to upgrade its own botanical garden because "the garden that the Society of Philadelphes proposes to establish in Saint Domingue will make Bordeaux's absolutely necessary, if only as a site for the deposition and gradual acclimatizing of plants destined for the Jardin du Roi."[261] After it received its formal letters

255. See BM/Rouen, B4/2, unpaginated, ("Du Mercredi 18 juin 1788") and ("Du Mercredi 18 9.bre 1789").

256. "Mémoire de Levavasseur sur le quinquina caraïbe approuvé par l'Académie de Rouen," *Journal de Physique* 36 (1790), p. 241-255; see also BM/Rouen, B4/2, unpaginated, ("Du Mercredi xi mars 1789"), ("Mercredi 14 Avril 1790"), and ("Du Mercredi 28 Avril 1790"); BM/Rouen, C6, "Catalogue Raisonné des graines envoiées a M. d'ambourney," and other Levavasseur material, BM/Rouen, C25.

257. BM/Rouen, B4/2, unpaginated meeting ("Du Mercredi 9 fevrier 1791"), and references to shipments from Rouen in Le Vavasseur letters, dated "Cap Le 1er 7bre 1789," and "Cap Le 1.r mai 1790," BM/Rouen, C25.

258. BM/Rouen, B4/2, unpaginated, meeting ("Du Mercredi 18 9.bre 1789"), where one reads: "En sa qualité De président *Du Cercle Des philadelphes*: M. Auvray nous a Adresse Copie de la Délibération et une lettre de sa Compagnie par lesquelles Elle Exprime son Voeu de lier Correspondance avec LAcadémie... [sic] La Compagnie a chargé ses secrétaires de Remercier M. Auvray de Lenvoy des graînes et de La faveur quelle Etoit sensible a Loffre de la société Royale des sciences Et Arts du Cap francois et que La Correspondance offerte Lui serait infiniment agréable"; see also in this connection ALS "Auvray, President," dated "...30 Juin 1789," with marginal notation, "18 9bre 1789. Commissaires MM. Dambournay, Roudeaux, pinard," and ALS Le Vavasseur, dated "Cap Le l.er août [1789]," wherein he writes, "lacademie de rouen n'aura pas a rougir d'accepter... correspondance," BM/Rouen, C25.

259. See BM/Bordeaux, Ms. 1696 (XXIX), "Archives de l'Académie de Bordeaux. Correspondance," vol. II.

260. McClellan (1992), p. 219, and ALS François de Neufchâteau to Lafon de Ladebat, dated "à La Charité, près le Cap, le 25 novembre 1785," BM/Bordeaux, Ms. 828, CV, 63.

261. "Le jardin que la Société des Philadelphes se propose d'établir à Saint-Domingue rendrait celui de Bordeaux absolument nécessaire ne fut ce que pour y déposer & y aclimater peu à peu les plantes destinées pour le Jardin du Roi," undated, unsigned "Mémoire," BM/Bordeaux, Ms. 1696 (XXVI), dossier 17, #12.

patent, Charles Arthaud on behalf of the new Société Royale in Saint Domingue requested a formal affiliation with the Bordeaux Academy, writing, "We would be delighted if we could interest you by our works, and if, through a worthy and for us useful correspondence, we could profit from the enlightenment you are in a position to spread."[262]

Through Moreau de Saint-Méry and another Saint Domingue jurist and Cercle associate, François de Neufchateau, the Cercle des Philadelphes forged contacts with the *Musée*, or free school, in Bordeaux. As had become his custom, the Cercle's permanent secretary, Arthaud, regularly sent the Cercle's materials to the Bordeaux Musée on behalf of the Cercle.[263] And in 1787 the two groups formed another official correspondence, with the Bordeaux Musée sending an official "diploma of association" to its Cap counterpart.[264] In addition, the Cercle des Philadelphes developed ties to the *Académie Royale des Sciences, Inscriptions et Belles-Lettres* and the *Musée* in Toulouse, and had dealings with provincial academies in Dijon, Metz, Arras, Caen, La Rochelle, Montpellier, Orléans, and Poitiers.[265] To this extent, the Cercle des Philadelphes was in many ways a typical French provincial academy that happened to be located in the Antilles.

But the Cercle moved beyond these strictly "provincial" contacts to integrate itself fully into the Colonial Machine. The Cercle's *jeton* and its recognition by the authorities in Saint Domingue and by the Ministry of the Navy and the Colonies in France indicate this assimilation. The institutional correspondence established with the Société Royale d'Agriculture is another indication of the elevated position of the institution.[266] The close contact it forged with the Société Royale de Médecine in Paris is yet another. As chief médecin du roi in Cap François, Charles Arthaud was already in contact with and a formal correspondent of the Paris royal medical society, and he used his position to establish a connection between the Cercle des Philadelphes and the Société Royale de Médecine. In early 1785 he sent the Cercle's initial prospectus to the Paris institution along with the request that "the society help back its establishment."[267] From that point, Arthaud forwarded everything the Cercle produced

262. "Nous serions très enchantés si nous pouvions vous intéresser par nos travaux et si par une correspondance honorable et utile pour nous, nous pouvions profiter des lumières que vous êtes dans le cas de répandre," ALS Arthaud to the Académie de Bordeaux, written from Cap François and dated 5 October 1789,BM,/Bordeaux, Ms. 1696 (XXIX), vol. 2, pp. 143-44.

263. See Arthaud letters, dated "Au Cap, le 20 juillet 1786," "Au Cap, le 17 avril 1787," and "au Cap, le 1er juin 1787," BM/Bordeaux, Ms. 829, 1786 (17), 1787 (32), 1787 (37).

264. McClellan (1992), p. 267; "Extrait des registres du Cercle des Philadelphes. Séance du 24 juillet 1786," and letter from one Prévost, dated "Au Cap, le 25 août 1786," BM/Bordeaux, Ms. 829, III-IV, p. 220, and 1786 (26), respectively.

265. McClellan (1992), pp. 267-68.

266. DPF, p. 348.

267. "... une Lettre de M. Arthaud med. à Cap [sic] avec un Prospectus du Cercle des Philadelphes qui y est établie. il desire que la société emploie son Credit pour cet établissement," SRdM, Ms. 10, p. 341 ("Séance du Mardi 8 fevrier 1785").

to the Société Royale in Paris. The Cercle participated in the Société's projects to investigate elephantiasis, tetanus, and mineral waters, to the point where the Cercle might be seen as an extension of the Société in the tropics.[268] Its status in the Colonial Machine is seen in a small way in the case of Jean-Baptiste-Théodore Baumes (1756-1828), a physician from Nîmes, who in 1788 won a prize from the Cercle for his paper on infantile convulsions; the Société Royale de Médecine in turn examined Baumes' paper and approved its publication "with the approbation and under the imprimatur of the Society."[269] At one point in 1789, Arthaud in Saint Domingue even proposed to fund a prize on poisons under the auspices of the Société Royale de Médecine.[270]

The Société Royale des Sciences et des Arts du Cap François. In the meantime, in early 1789 the Cercle des Philadelphes renewed its quest for letters patent by sending a delegation to meet with La Luzerne at Versailles. The delegation consisted of Jean Barré de Saint-Venant, a rich colonist and the current president of the Cercle, Moreau de Saint-Méry, the colonial jurist sponsored by the government to work on a compilation of colonial ordinances, and the count Du Puget d'Orval, the nobleman, military officer, and colonial associate of the Cercle, elected in 1786 during his inspection trip to the West Indies. At the time of their meeting in late 1788, Du Puget was well connected at court as the tutor to the dauphin. La Luzerne approved the request presented to him, and on documents circulating through the ministry relative to the Cercle des Philadelphes he noted laconically: "Give them letters patent."[271] The twenty-two articles of the formal letters patent that emerged in the spring of 1789 changed the name of the Cercle des Philadelphes to Société Royale des Sciences et des Arts du Cap François, increased the size of the membership somewhat, and raised the funding to 10,000 livres. The new Société Royale was charged "to make its principal occupation everything pertaining to the physical and natural history of the colonies and everything that might perfect farming, running plantations, the sciences and arts relative to manufactures, and the extension of commerce."[272] But otherwise the letters patent left the institution as it was.

268. See above, Part IIB; McClellan (1992), pp. 267 and 355n, and SRdM, Mss. 10-11, passim; SRdM-H&M, vol. 5 (1782-1783), p. v (noting receipt of Arthaud's 1785 *Discours prononcé à l'ouverture de la première séance publique du Cercle des Philadelphes, avec une description de la ville du Cap*); vol. 8 (1786), p. 74 (his 1788 *Dissertation sur le papier, dans laquelle on a rassemblé tous les essais qui ont été examinés par le Cercle des Philadelphes, sur les moyens de préserver le papier de la piqûre des insectes, par M. Arthaud, Médecin du Roi, au Cap-François, Secrétaire perpétuel du Cercle*).

269. "...imprimer avec l'approbation et sous le privilége de la société," SRdM, Ms. 11, fol. 228v ("Séance du Vendredi 17 Octobre 1788") and fol. 205r ("Séance du Mardi 22 Juillet 1788"). Baumes's *Des Convulsions dans l'enfance, de leurs causes et de leur traitement* duly appeared in 1789, published in Nîmes by C. Belle.

270. "M. Artaud desire que la Société veuille proposer pour sujet d'un prix pour lequel il donnera une Somme de 300#, une question sur les poisons," SRdM, Ms. 11, fol. 249v ("Séance du Vendredi 30 Janvier 1789").

271. Memorandum dated "11 Xbre 1788," AN C^{9A} 162, cited in McClellan (1992), p. 248.

272. McClellan (1992), p. 352.n67 lists the manuscript original and several published versions of the letters patent of the Société Royale at Cap François.

Louis XVI himself signed the document on 17 May 1789. At that very moment elsewhere on the grounds at Versailles, the Estates General of France – called for the first time since 1614 – had been meeting for close to two weeks and was deadlocked over the verification of deputies. The declaration of a National Assembly and the storming of the Bastille were just weeks away. The signing of letters patent for the Cercle des Philadelphes took place on the very threshold of revolution in France, and the moment epitomizes both the Old Regime and the history of the Colonial Machine of Old Regime France.

The granting of letters patent was a capital achievement for the group in Saint Domingue, but success crowned their fortunes in another way in early 1789: a formal union with the great Académie Royale des Sciences.[273] On Wednesday, 25 February 1789 Barré de Saint-Venant joined again by Moreau de Saint-Méry and, this time, by Barré de Saint-Leu, a naval officer and French national associate of the Cercle, addressed the royal science academy in its large meeting room in the Louvre palace.[274] Barré probably thought his audience largely unfamiliar with Saint Domingue, so, after an opening in praise of the Academy of Sciences, he described the lush Caribbean paradise whence he came. But Barré did not lose sight of his audience, and as much as possible he kept his focus on points of scientific interest. For example, he remarked on an unusual diurnal cycle of barometric variations evidenced in the colony; and, relative to geology, he spoke about measuring the erosion of mountains through the silt content of rivers in order to establish a baseline for geologic time. But most of all Barré wanted to present the Cercle des Philadelphes to the Parisian academicians. He mentioned the Mesmer controversy and the origination of the Cercle in 1784. He noted the Cercle's achievements and the recognition it had received from the government to date. "Although just beginning and the only one of its type established at the end of the world, the Society seems to have been in existence for a long time." Barré let on that letters patent and 10,000 livres had just been approved for the Cercle, but he continued, "All these happy preliminaries will not satisfy us; we see here only the dawn of our existence and of our prosperity."

Barré elaborated the now familiar program of the institution he represented, emphasizing the useful at every turn. The practical benefits forthcoming from

273. McClellan 1992, pp. 251-56, provides these details; see also PA-PV-100, Barré de St Verrant [sic], 28 March 1789 ("Discours sur l'objet, les projets et les Voeux de la Société des philadelphes /Les membres de cette Société auront la Liberté d'assister aux séances particul.re de L'Academie/ Les Officiers de L'acad.e, Jussieu, Le Roy"); APS Manuscripts, Fougeroux de Bondaroy Papers, B F8245, "Séances de l'Académie Royale des Sciences, 1786-1789" ("le mercredi des Cendres 25 fevrier [1789] ... sur la culture de Nopal et la cochenille qui sen nourit – du cercle des philadelphes – du cape françois – etabli en 1784"); PA-PS, 'plumitifs' for 25 February 1789 and 28 March 1789.

274. See the fourteen-page "Discours prononcé à l'Académie royalle des Sciences, le mercredy 25 Fev. 1789 par Mr. Barré de Ste Venant Président du Cercle des Philadelphes du Cap François," PA-PS (25 February 25 1789); copied into PA-PV 108 (1789) ("Du Mercredi 25. Fevrier 1789").

medical research received considerable attention, especially regarding tetanus and venereal diseases. Botanical research promised to yield medically useful plants and possibly a substitute for cinchona. The cultivation of spices in the colony was potentially a fruitful area of further institutional activity. Research regarding colonial mineral waters would surely prove useful in medical practice, said Barré, and veterinary medicine had already seen considerable advance through the work of the Cercle des Philadelphes on epizootic diseases. Barré envisioned a large field for the practical application of scientific principles in agriculture, manufacture, commerce, and the useful arts, and he emphasized the theme that science and enlightenment could overturn "dull routine and thoughtless prejudice."

> Such are the objects of the Cercle's work and energy. The tasks the Cercle has imposed on itself would exceed its forces if it could not hope that the learned of Europe and especially the most illustrious men of science of this realm would favor its efforts and encourage it through their counsel....Our society is worthwhile, therefore, and if it did not exist, the Academy would doubtless want to create one like it.[275]

Only at the very end of his speech did Barré make clear that, in coming before the Academy of Sciences, the Cercle des Philadelphes had something more in mind than simple institutional goodwill or receiving the blessings of a senior society, or even establishing an ordinary institutional "correspondence." "In seeking support within the maternal embrace of the Academy of Sciences," Barré de Saint-Venant and colleagues proposed something more formal and specific: an actual union of the institutions, akin to (and even beyond) the special affiliations that linked the Paris institution with the Société Royale des Sciences of Montpellier and the Académie Royale de Marine at Brest. Barré pleaded his case for a similarly special relationship with his society in Cap François:

> We await the moment when the Royal Academy of Sciences will give light to our eyes in granting us the favor it did not refuse the Society in Montpellier. Doubtless the academy [in Montpellier] possessed glorious titles that we cannot have yet, but given our distance, we need your help even more. Without it we cannot prosper. Without it, interesting discoveries will remain in the shadows. The Academy knows all too well how important it is to enlighten this part of the world, and it will not refuse to adopt the Cercle des Philadelphes and invigorate it with its counsel...[for which the Paris Academy will earn] the gratitude, affection,

275. PA-PS (25 February 1789), cited in McClellan (1992), p. 253.

love, and respect that a weak and poor child owes the kind mother who adopts it.[276]

Barré concluded his presentation by asking the Academy, regardless of its decision on affiliation, to accept the Cercle's jeton, and he ventured to request one of the Academy's in return. "May this exchange and mutual gift put a seal on the union we desire."

The assembled academicians doubtless applauded this spirited representation from Saint Domingue, but the request for an institutional union required separate consideration, and the Academy named a seven-person committee to consider the matter. La Luzerne had been elected an honorary member of the Academy in 1788, and he served as vice president of the organization in 1789. So, La Luzerne, the former governor-general in Saint Domingue and current minister of the navy and the colonies, was on the committee along with the Academy's permanent secretary, the marquis de Condorcet; the other annual officers; and *pensionnaires* A.-L. de Jussieu, and J.-B. Le Roy. Le Roy wrote the committee report. For the most part, it merely summarized Barré's speech and information about the Cercle des Philadelphes gleaned from the statutes left at the Academy by Moreau de Saint-Méry. The laudatory report noted candidly that "it is not hard to form a clear idea of the Society of Philadelphes. In effect they all want to apply knowledge they acquire to perfect agriculture and augment the productions of the colony."[277] The report did not reject the Cercle on that account, but, rather, it underscored the scientific character of the institution, for "*physique*, medicine, chemistry, botany, agriculture, meteorology, and mechanics form the bases of its activities."[278]

About the proposed institutional union of the two institutions, the committee pretended not to have an opinion, "it being up to the Academy to decide." But the rapporteurs went on to observe that the Academy in Paris had long recognized the advantages of having representatives on the spot rather than relying on travelers or emissaries. More than that, the Cercle des Philadelphes would become an actual agent of the Paris Academy, ready and, with its own institutional resources, able to undertake scientific commissions in Saint Domingue as assigned by the Academy in Paris.

276. PA-PS (25 February 1789), quoted in McClellan (1992), p. 254.

277. "...Il n'est pas difficile de se former une juste idée de ses project. En Effet ils doivent tous tendre à mettre en œuvre les Connoissances qu'elle pourra acquérir pour perfectionner la Culture et augmenter toutes les productions de La Colonie," PA-PV, 108 (1789), fol. 88v ("Du Samedy 28. Mars 1789), quoted in McClellan (1992), p. 255.

278. "...ainsi La Phisique, la Médecine, La Chimie, la Botanique, Lagriculture, la Météorologie, La Mécanique doivent faire les bases de ses occupations," PA-PV, 108 (1789), fol. 87v ("Du Samedy 28. Mars 1789), quoted in McClellan (1992), p. 255; see also fol. 59 ("Du Samedi 28. Fevrier 1789").

> By favoring this nascent society, the Academy will align itself with a number of well-informed people who will make the most of the opportunity. The Academy will accelerate the progress of useful knowledge on the island of Saint Domingue in a marked manner, and thereby render an essential service to the colony. Finally, it is important to remark that the Society of Philadelphes established at Cap François on the island of Saint Domingue in the Torrid Zone is such a special case that the Academy should not fear that any other society might impose on the Company for similar treatment.[279]

In the end the committee recommended that the Academy treat the resident associates of the Cercle in the same manner as its individual *correspondants*. In other words, resident associates of the Cercle des Philadelphes would become *correspondants* en masse of the Academy of Sciences. This action to confer the status of *correspondant* on a body collectively was unprecedented in the annals of the Academy. The Academy approved a formal association with the Société Royale du Cap François at its meeting on 1 April 1789. The oldest and the newest of France's science academies thus forged a remarkable new association. It is telling, not only for this account of the Colonial Machine, but for the history of science and European colonial expansion in the period generally, that the Royal Society of Sciences and Arts in colonial Saint Domingue became a formal overseas extension of the venerable Royal Academy of Sciences in Paris.

The new letters patent and news of this affiliation with the Paris Academy of Sciences arrived in Saint Domingue in August 1789 along with word that La Luzerne had at last agreed to become the protector of the Society. Moreau de Saint-Méry offered a prayer for the Cercle des Philadelphes at this, its climatic moment: "Happy association, may you last as long as the New World, and may my feeble pen preserve for your courageous and generous founders the debt of gratitude that is their due."[280]

Given the spiral of events from mid-1789, however, the great promise of affiliating the scientific societies in Paris and in Saint Domingue proved barren. The Paris Academy did receive descriptions and illustrations of colonial plants from de la Haye, a paper from Gauché on cinchona in Saint Domingue, and an article from Dutrône La Couture on the chemistry of sugarcane. After their affiliation, the Paris Academy contacted the Royal Society in Cap François through Abbé Tessier about reports of freezing in the mountains of Saint Domingue, and in early 1790 the Academy in Paris heard a favorable report on a chain-link suspension bridge designed by Barré de Saint-Venant.

279. PA-PV 108 (1789), p. 91, cited in McClellan (1992), p. 255.

280. DPF, p. 349.

The new West Indian royal society elected academicians Tessier and Lalande in 1789, and in September 1789 the Paris Academy received the volume of the Cercle's *Mémoires*; but after December 1789 the record of the contact between the two institutions is silent.[281]

The Cercle des Philadelphes/Société Royale itself soldiered on in Saint Domingue through 1792. In July 1790 it awarded a prize, its last, to Moreau de Saint-Méry for his *éloges* of the founders of the poorhouses in Cap François. Its annual public meeting scheduled for 17 August 1790 was disrupted by unknown political disturbances, but the Société Royale did manage to hold a public meeting the following year, less than week before the outbreak of the great slave revolt in Saint Domingue on 22 August 1791. Throughout this period, the Cap royal society negotiated with various powers on the island for patronage and support, but for naught, as the winds of war and revolution swept all before them.

The Cercle des Philadelphes and the Dialectic of Colonial Identity. The story of the Cercle des Philadelphes/Société Royale of old Saint Domingue caps the history of the Colonial Machine and French colonial science in the Old Regime. Except for two disputed points, one minor, one major, that story seems a straightforward tale of local amateurs and experts forming themselves into a provincial academy set in the tropics that government authorities fast-tracked to royal recognition and support in order to further colonial development.

The first disputed point found in the literature suggests that the Cercle des Philadelphes was not the scientific society it pretended to be, but rather was founded as a Masonic organization with a secret agenda of political reform and colonial independence. This interpretation connects with notions of a Masonic plot sparking the French revolution.[282] True, Saint Domingue was a hotbed of Masonic activity in the period leading up to the French Revolution. At the height of the movement, a total of twenty Masonic lodges and twice that many chapters saturated Saint Domingue with a thousand or so Masons. The spirit of Freemasonry permeated the colony at the time, and most of the colonial members of the Cercle des Philadelphes were indeed Masons. But ample evidence shows that the Cercle des Philadelphes was exactly the kind of institution it claimed to be: a regional organization devoted to sciences and the practical arts, with the aim of improving the local economy and the quality of life in Saint Domingue generally. The significant question regarding Freemasonry and the Cercle des Philadelphes is not whether the members of the Cercle were

281. McClellan (1992), pp. 256-57; PA-PS, 'plumitifs' for 2 September 1789, 28 November 1789, and 19 December 1789; PA-PV 108 (1789), fols. 217v ("Mercredy 2. Septembre. 1789"), 228v ("Samedy 28. Novembre, 1789"), and 242v ("Samedy 19. Décembre, 1789).

282. Pluchon (1985b).

Masons, but why any of Saint Domingue's many Masons were also Philadelphes.[283]

The other, and for us more analytically significant point, concerns the extent to which we need to see the Cercle des Philadelphes principally as an instrument of royal power and the absolutist state or whether it is more appropriate to envision the Cercle primarily as a regional group animated by colonial patriotism and a desire to promote colonial autonomy through regional development. This is a more serious historiographical issue, one that goes to the heart of how we are to interpret the role of the Cercle des Philadelphes within the Colonial Machine.[284]

The argument in favor of what might be labeled the monarchical thesis and seeing the Cercle des Philadelphes as a clear-cut cog in the Colonial Machine is uncomplicated, and we have encountered all of the evidence for it already: early government recognition and approbation, ministerial subventions, the extraordinarily quick path to *royal* letters patent, members in the scientific and political establishment (including La Luzerne as its protector), links between its botanical garden and the official Jardin du Roi in Port-au-Prince, the Cercle's *jeton*, and its affiliation with the Académie Royale des Sciences. In addition to this last connection, one might also signal the unofficial ties effected by Arthaud and the Cercle des Philadelphes with the Société Royale de Médecine and the Société Royale d'Agriculture. In sum, the Cercle seems straightforwardly a royal institution, solidly linked to the metropolis, central power, colonial administration, the monarchy, and colonial policy.

But the matter is not clear-cut, and there is an argument for what might be labeled the colonial thesis. The founders of the Cercle des Philadelphes were all inhabitants of Saint Domingue, and they seem to have been proud of their American home and wished it success in its economic, social, and scientific development. Charles Arthaud was explicit in aligning the Cercle des Philadelphes with the interests of colonists and the colony: "Our organization can be seen in a good light only if we attach ourselves fundamentally to the interests of colonists."[285] Regarding its 1785 prize on infantile convulsions given to Baumes, the doctor from Nîmes, the Cercle observed that "in crowning this work, the Cercle has only one regret, not to have had the satisfaction of awarding its prize to a physician in the colonies.[286] Furthermore, from the outset of its history, Saint Domingue was known as a rude and crude frontier colony where people went to earn their bundle and escape back to civilization as

283. McClellan (1992), chapt. 11 documents this point.

284. McClellan (2000) explores these points.

285. AN-COL F^3 152, fol. 232, quoted in McClellan (2000), p. 85.

286. "...Le Cercle en couronant cet ouvrage, ne doit avoir qu'un regret, de ne pas avoir eu la satisfaction de décerner son prix à un médecin des colonies," "Extrait du registre du Cercle des Philadelphes du Cap Français," BM/Bordeaux, Ms. 828 XLII 6.

quickly as possible. The Philadelphes were sensitive to this reputation and struggled against it.[287] Then, when one considers that 62 percent of the members of the Cercle des Philadelphes were resident or colonial members, we can appreciate to what point the weight of the institution centered in the Antilles and to what point it needs to be considered from purely colonial points of view. The patriotic and decidedly colonial orientation of the Cercle des Philadelphes/ Société Royale that emerged briefly in the period from 1789 to 1792 reinforces this view of the institution.[288]

Today, a better interpretation of the Cercle des Philadelphes and its role in the Colonial Machine embraces both the monarchical and colonial theses and acknowledges that the Cercle des Philadelphes balanced the contradictory pulls of colonial identity and colonial autonomy on the one hand and the pull of central authority and the Colonial Machine on the other. Different and inconsistent ideologies, aims, and policies played themselves out within the Cercle des Philadelphes in the period of the 1780s as Saint Domingue and the world headed toward revolution. This mixed conclusion better defines the Cercle des Philadelphes and its place in the Colonial Machine.

* * *

The Colonial Machine was not a steamroller, and it did not roll triumphant. So much operated to limit its effectiveness and success, from the forces of nature with which the actors had to contend, to centrifugal political currents in tension with central authority – in France and in the colonies – to alternatives to the organization and pursuit of science such as the Republic of Letters that commanded allegiances distinct from the Colonial Machine. These historically countervailing tendencies delimit and so further define the historical reality and specificity of the Colonial Machine as it existed as an instrument of state and colonial expansion in Old Regime France.

287. McClellan (1992), pp. 57-58, 210.

288. McClellan (1992), chapt. 15 discusses these events.

CONCLUSIONS

Two sorts of conclusions bring this account to an end. The first is chronological and marks the abrupt end of the Colonial Machine. Here, once again, the historian can only be grateful for 1789 and the radical break with the Old Regime brought about by the French Revolution and the Napoleonic era. The 14th of July 1789 signaled the end of one era and the start of another in France, the colonies, and the world. Emblematic for us, in later July and early August of 1789 the letters patent of the Société Royale du Cap François bobbed their way across the Atlantic at the same time that a following ship brought news of the fall of the Bastille. Marking this transition in the greatest of France's Old Regime colonies, the one ship dropped anchor in Saint Domingue just days before the other.

True, the Colonial Machine – like the Old Regime itself – did not collapse overnight. The Revolution was a process that took time to unfold and for its consequences to be felt, and for a while the Colonial Machine continued as normal. Correspondence flowed back and forth. André Michaux sent plants from America. Joseph Martin tended clove in Guiana. La Pérouse was still missing. For some of the men of the Colonial Machine, the revolutionary moment was propitious and greeted with enthusiasm. Two reasons explain this dawn of hope. One, the Colonial Machine was, in its way, a progressive element of the late Old Regime, and the Revolution offered the possibility of continuing colonial development without the encumbrance of tradition that so weighed on the French polity. Secondly, not all, but most of the men of the Colonial Machine were of bourgeois backgrounds and rooted in the Third Estate. (This would not be true of the officer class of the navy.)[1] Thouin at the Jardin du Roi wrote a letter in late November 1789 that captures something of this optimism. He addressed his letter to the French government gardeners on their way to serve Tipu Sultan in India.

> Since your departure everything has changed in France. The People have thrown off the Yoke of Ministerial Despotism, the Nobility has

1. On the decidedly bourgeois and Third-Estate character of the membership of the Académie Royale des Sciences, see Roche, vol. 2, pp. 286-87; McClellan (1981), pp. 556-58, and (2001), pp. 17-18. The Société Royale de Médecine was at least of the same order, the Société Royale d'Agriculture probably more so.

> lost its privileges, the Nation has confiscated the Property of the Clergy to pay the Debts of the State. The Parlements are limited to their sole function as Judges. The Venality of offices is abolished. With such wise and salutary laws France is going to take on a new life, and when you return from your voyages you will find her happy and flourishing. But all these great changes could not happen without occasioning some troubles and even calamities. Nevertheless, I think the most critical moment has passed. The King and the Royal Family, as well as the National Assembly, are in Paris. Peace begins to be reborn there, and I think it will be durable.[2]

But soon the Revolution and political events began to take their toll. The year 1789 gave way to 1790, and on to the collapse of the monarchy and the declaration of a republic in 1792, the Jacobins and the Terror that followed through 1794, and the establishment of the Directory in 1795. Already by September 1789 the minister of the navy and the colonies, La Luzerne, as we saw, was unable to keep up botanical shipments from Île de France to South America and the Caribbean. André Michaux was forced to fend for himself in America. Rambouillet declined precipitously after 1792, with local people stealing wood from the park and garden, depredations that effectively destroyed Rambouillet and the grand project to import trees from abroad.[3] Questions concerning the government and the monarchy stood at the center of the debates and revolutionary turmoil in this period; and let us not forget that the Colonial Machine was through and through a royal institution and part of the governing bureaucracy, and so was deeply affected by events as they unfolded.

In the end, the French Revolution destroyed the Colonial Machine, and France's first colonial empire collapsed in its wake. The biggest blow to the Colonial Machine itself was the decree of the National Convention of 8 August 1793 that disbanded all royal academies and societies.[4] With the stroke of a pen, the major wheels and gears and a large part of the scientific core of the Colonial Machine – the Académie Royale des Sciences, the Société Royale de

2. "Depuis votre départ tout a changé de face en france, Le Peuple a secoué le Joug du Despotisme Ministeriel, la Noblesse a Perdu ses privileges, les Biens du Clergé sont confisqués par la Nation pour payer les Dettes de l'Etât, les Parlemens sont restreint a la seul fonction de Juges. la Vénalité des charges est abolie. la france sous des lois aussi sages que bienfaisantes va prendre une nouvelle vie et l'orsque vous reveindrez de vos voyages vous la Trouverez heureuse et florissante. Mais tous ces Grands changemens ne peuvent s'opperer sans occasionner des Troubles et même des malheurs cependant je crois que le moment le plus critique est Passé. Le Roi et La famille Royale, ainsi que L'Assemblée Nationale habitent Paris, la Paix commence a y renaitre et je crois qu'elle sera durable," ALS Thouin to "M.M. Mulot et Luhrmann Jardiniers du Sultan Typoo," dated "29 9bre 1789," BCMNHN Ms. 307.

3. See materials in AN O[1] 2113A, "Pépinières, 1791-1792."

4. Hahn (1971), chapt. 8; Gillispie (2004), chapt. 3. Gillispie's 2004 volume devoted to science and polity in revolutionary and Napoleonic France frames the remarks sketched here.

Médecine, the Académie Royale de Marine, the Société Royale d'Agriculture, and the Société Royale du Cap François – ceased to exist. Other units suffered variously from a lack of resources and as more pressing matters increasingly preoccupied people and purses. Although not formally shuttered, the Observatoire was inactive and essentially closed for two years, from 1793 through 1795.[5] The navy suffered its own revolutionary turmoil and was severely weakened as an institution. The execution of Lavoisier in 1793, not to mention the same fate of Louis XVI at the guillotine the year before, tells the tale of those times and of the dismantling of the Colonial Machine.

France's colonial system obviously became caught up in the Revolution.[6] It suffered a debilitating blow, beginning in 1791, as political events in Saint Domingue led to the great slave revolt that broke out in August of that year. The uprising of hundreds of thousands of slaves in Saint Domingue was the largest and most successful slave revolt in history. Fighting continued on and off on Hispaniola for more than a decade. The Royal Society of Sciences and Arts at Cap François limped along for a few years after 1789, but this distinctive element of the Colonial Machine was totally destroyed in the course of the Haitian revolution. The town of Cap François burned to the ground in June 1793, and remaining members of the Cercle des Philadelphes/Société Royale died or hurriedly sailed into exile.[7] In 1801 Napoleon sent an impressive expeditionary force to Saint Domingue to recapture the colony for France, an army famously defeated by the insurgents and by the ravages of tropical disease against which the Colonial Machine had long fought. The Republic of Haiti declared its independence on 1 January 1804.

The ruin and loss of Saint Domingue, the richest colony of France, sounded the death knell for the first French colonial empire, now known in French as its "old colonies," the "vieilles colonies." France had already lost Canada, Louisiana, and India in the Seven Years War; the crown-jewel, Saint Domingue, went in the Revolution, and the gem, Île de France, passed to the British in the Napoleonic wars to become Mauritius. There was little left. France maintained some of its colonies, notably in Guadeloupe, Martinique, Guiana, and Île Bourbon – the latter rechristened Île de la Réunion – and ultimately these extensions became integrated into France as its overseas departments.[8] These possessions never again recaptured the role at the forefront of French colonial policy or as scientific or medical frontiers that they had in the Old Regime, except perhaps today in Guiana where the French launch rockets

5. Gillispie (2004), pp. 298-306; Pelletier (2002), p. 72.

6. Pluchon (1991), chapt. 10, and Meyer et al., Part II, chapt. 3 ("L'effondrement du domaine colonial") recount the course of the French Revolution and the colonies.

7. For the Revolution in Haiti, see previous note; Dubois (2004a, 2004b); Geggus; Ott; Regourd (2006); McClellan (1992), pp. 273-74. One of the more notable exiles was Moreau de Saint-Méry in Philadelphia, where he ran a print and bookshop.

8. Regourd (2006); see also Bouche; Meyer et al., Part III.

from Kourou, the site of the greatest failure of the Colonial Machine. By dint of circumstances or limited personal vision, Napoleon did not have a strong colonial policy, the invasions of Saint Domingue and Egypt notwithstanding; by 1815, the remnants of the first French empire incorporated barely 7000 square kilometers with fewer than 1 million inhabitants.

Not everything fell apart or stayed crushed. Under Thouin and other leaders, the Jardin du Roi transformed itself seamlessly into the new National Museum of Natural History (the Muséum National d'Histoire Naturelle). Unlike the haughty academies of science or medicine, summarily eliminated, the Jardin/Muséum harmonized with attitudes toward nature on the part of revolutionaries and so found support and continuity.[9] In 1795, following Thermidor and the most radical phase of the Revolution, the new national government founded the Institut de France, a scientific and intellectual umbrella organization that more or less regrouped the old academies.[10] The first class of the Institut, in particular, directly succeeded the proscribed Académie Royale des Sciences. These changes brought back many of the scientists and intellectuals of the former academies, and together they formed the institutionalized basis for colonial outreach in which the new regime, like the old, tapped the agency of science – anew, but differently from the ways the Colonial Machine had been organized and functioned. The famous expedition to Egypt led by Napoleon and his armies from 1798 to 1801 is often regarded as part of the story of French colonialism.[11] We saw that the French were interested in Egypt from early in the eighteenth century and sent several scientific emissaries there. That Napoleon went to Egypt with a cadre of scientific and technical experts or that the Institut d'Egypte took the Institut de France as its model and parent reinforce the idea that the expedition to Egypt fits in the tradition of the Colonial Machine. That Hipolyte Nectoux, the former botaniste du roi at the Jardin Royal in Port-au-Prince, having abandoned Saint Domingue, accompanied Napoleon to Egypt as a botanist shows some links with the Colonial Machine that did not absolutely disappear with the Revolution.[12] André Michaux's ultimate posting to Madagascar in 1800 reflects these same continuities, as did the formal scientific expedition of Nicolas Baudin (1754-1803) to Australia and the Pacific in 1800-1804.[13]

9. Spary (2000) elaborates this theme, passim and pp. 212-27; Duris, pp. 79ff; Laissus (1964) and (2003); Blanckaert et al.; see also Gillispie (2004), pp. 167-183.

10. Hahn (1971), chapt. 10; Gillispie (2004), pp. 445-58; see also Crosland. Notably in this connection, per Levot, p. 65, the Académie de Marine was not reconstituted until 1815, and per Ganière, today's Académie Nationale de Médecine was not founded until 1820.

11. Gillispie, "Introduction," in Gillispie and Dewachter, pp. 1-22; Bourguet et al.; Gillispie (2004), pp. 557-600; Bret, ed. (1999c); Allorge, p. 579-86.

12. Bret (1995) and (1999a).

13. Baudin had previously undertaken a botanical collecting voyage to the Caribbean and South America and was connected to Antoine de Jussieu and the Muséum d'Histoire Naturelle; Ly-Tio-Fan (2006), pp. 61-62; Ly-Tio-Fane (1996), p. 11; Allorge, pp. 586-611; Fornasiero et al.; see also BCMNHN, Ms 2310 ("Liste des Correspondants du Museum").

The Pacific was indeed the new New World in the early decades of the nineteenth century, and the Muséum National d'Histoire Naturelle and the Institut's Académie des Sciences continued the extension of science and the French overseas presence in that still new arena in much the same pioneering spirit as before.[14] But, as France and its scientific and colonial establishments rebuilt themselves following the Revolution, the Old Regime and its Colonial Machine were increasingly and decidedly things of the past.

The nineteenth century presents a much different reality than what came before. After a few decades of a modest overseas ambition, the successive governments of France led the country toward a new colonial destiny, and from 1830 and the invasion of Algeria to the eve of the Third Republic in the 1870s, France progressively extended its domination over a growing part of the planet, challenging the English, especially in Africa and Asia. In the last third of the nineteenth century, a further phase of expansion united this colonial renewal into an impressive conglomeration of about 15 million square kilometers of colonies, protectorates, and the like, incorporating more than twenty-five times the area of France and almost 50 million overseas inhabitants. But French imperialism and its "civilizing mission" in the nineteenth and twentieth centuries is another story.[15]

* * *

The second sort of conclusions to be drawn are the intellectually more interesting ones of what we are to make of the story of the Colonial Machine. We started with a few simple assumptions to frame this inquiry. The first was that modern world history has been significantly shaped by two great historical developments: European colonial expansion from the fifteenth century, and the advent of modern science since the scientific revolution of the sixteenth and seventeenth centuries. Another was that the nation of France in the eighteenth century was a colonial power that strongly rivaled England. And finally, there was the assumption that contemporary France was the leading scientific nation of Europe of the age. We suggested that, balanced as it was as both a scientific and a colonial power, France in the Old Regime represents the perfect case for investigating the historical interactions of the two great world-historical forces of modern science and European colonial expansion in the seventeenth and eighteenth centuries.

14. Daugeron (2007) and (2009) details these developments and reveals further, deep-structure changes occurring through the 1830s and 1840s that made the Old Regime and the Colonial Machine more and more things of the past; see also Blais.

15. For a synthesis and bibliographical orientation and for the scientific side of French colonialism in this latter period, see Regourd (2006); Pyenson (1993); Osborne (1994); Obsorne (2005).

The history of the Colonial Machine presented here would seem to bear out the validity and utility of these framing perspectives. In the case of the Colonial Machine, state-supported institutions incorporating experts and expert knowledge functioned collectively to support French overseas expansion and colonial development. At the outset of this book we offered a preliminary conclusion that, not only were contemporary French science and French colonial and expansionist efforts overseas deeply intertwined, but that, to a real extent, they depended on one another for their mutual success. The present account would seem to support this conclusion, too.

At the end of this study, the historical reality and specificity of the Colonial Machine of Old Regime France – the assemblage of men and institutions we have surveyed in fine detail – would seem established. The Colonial Machine arose in the 1660s with the reign of Louis XIV and with Colbert's firm hand on the tiller of state. Parts of the Colonial Machine were core parts of the French state, governments, and courts of the time, starting with the monarchy but including the Ministère de la Marine et des Colonies, colonial government through resident intendants and governors-general, the Marine Royale, and those key court agencies, the Bâtiments du Roi and the Maison du Roi. Although they ended up sheltering scientific and technical expertise of their own, these were secular institutions, so to speak, in that they were not themselves expert or specialized organizations devoted to science, medicine, or other technical specialties. But these organizations were great bureaucratic entities, the source of unprecedented and comparatively munificent state support that gave rise to the rest of the Colonial Machine. Thus arose that distinctive set of specialized and expert royal institutions and royally funded positions with which we are now familiar: the Académie Royale des Sciences and paid academicians, the Observatoire Royal and its official astronomers, the Jardin du Roi and its elaborate staff of professors and gardeners, the Société Royale de Médecine, the Académie Royale de Marine, the Société Royale d'Agriculture, the Dépôt des Cartes et Plans de la Marine, the royal gardens in France and in the colonies, the Société Royale des Sciences et Arts du Cap François, expert scientific and medical positions within the navy, médecins du roi and botanistes du roi, court positions, engineering cadres, teaching posts in medical and hydrography schools, and miscellaneous appointments here and there. All told, the Colonial Machine represents an impressive body of state-sponsored science bent toward the colonial effort.

Of course, to royal authority and central government we have to add the Compagnies des Indes and various French Catholic missionary orders, both as supporting authorities themselves and as harbors of expertise. Yet, another concluding observation to be made is how much more important royal and central authority became in the functioning and direction of the Colonial Machine as the eighteenth century progressed. Two developments account for the evermore royal and centralized character of the Colonial Machine. One, the government takeover of the Compagnie des Indes in the period 1764-1767 strengthened the state hand immeasurably. Secondly, from early in the eighteenth century the creation of its own cadres of scientific and technical experts essentially eliminated the need to rely on the personnel or networks of missionary orders.

Thus, in going about their business, the specific sets of men and institutions we have examined fused to become what we are identifying as the Colonial Machine. There is, in other words, an essential unity to our tale and to the story of the Colonial Machine. The coalescing of its various parts to create the Colonial Machine occurred through a variety of means. Bureaucratic connections and subordination to central authority was one (think of the role of the Ministry of the Navy and the Colonies). Actual institutional links were another (think of those that joined the Académie des Sciences in Paris with the Marine Academy in Brest and with the Société des Sciences et Arts in Cap François). Common members and overlapping memberships played a big role. Certain key personnel directed the action of the Colonial Machine (think of d'Angiviller at the Bâtiments du Roi, Duhamel du Monceau and Poissonnier at the navy, or de Castries or La Luzerne as ministers). Many common projects united its elements (think of the tests of the chronometer or the La Pérouse expedition). The whole functioned under central direction from Versailles and related centers of power governed by the Compagnies des Indes and French religious authorities.

Over the span of a century and more, from Louis XIV to his unfortunate great, great, great grandson, Louis XVI, the Colonial Machine expanded and engaged in one issue after another where scientific, medical, and technical expertise were seen to play a role on the forefront of colonial development for France. The Colonial Machine thus achieved a real, if not unmixed record of accomplishment that ranged from its cartographical triumphs, to its war against disease and the travails of sailing, to reshuffling the world's plants and instituting new branches of the economy.

At the same time, reflecting the dual nature of the Colonial Machine as an instrument of the state and as an instrument of science, the bounty for science brought by the Colonial Machine was significant, and the world of science was considerably enriched by work that took place under the aegis of French science and the Colonial Machine. Cartographers and geographers knew the world as never before and had the means to perfect their science. Astronomers used the new world platform for new observations and for pursuing their work and researches on a global scale. Doctors learned of a whole new world of diseases and developed whole new pharmacopeias to enrich their understanding and practice. Physical scientists expanded their knowledge of the world and its constitution. Botanists collected, classified, and compared a cascade of new plants from around the world. The scientific legacy bequeathed by the Colonial Machine to nineteenth-century French and European science was substantial.

These findings concerning the scientific impact of the French colonial experience in the Old Regime carry some notable historiographical implications. Suffice it to say that the study of French science in the Old Regime is highly developed and an historiographical touchstone in even larger literatures. That has been true for decades now, and in recent years the world overseas has received increased attention in the literature of Old Regime science. Just the same, what emerges from the present study is an enlarged sense of the reality and the importance of the colonial and the world overseas in the history of French science in the period. Placing colonial and overseas concerns at the center of our analysis reveals new elements for consideration, opens new perspectives, and enriches established understandings of French science in the Old Regime. More than simply peripheral, the world overseas formed a constituent element of early modern French science. Actors were more attuned to the world outside of France as a field of action and a scientific resource than has ordinarily been taken to be the case. The world overseas enlarged their horizons, as it should ours historiographically. In Old Regime France knowledge was still forged largely at the center, but the Colonial Machine gave the illusion of unrestricted access to the world as a whole and, even more, that the colonial and overseas worlds were immediately at hand and no further away, in effect, than nearby Passy or Rambouillet. Founded on that illusion, the claim of the universality of science found a double truth in the physical as well as intellectual mastery of nature.

The prime beneficiaries of the Colonial Machine were the French state and the Bourbon monarchy. It existed first and foremost as an agent of the French state, and its story can be usefully set in the context of the making of the modern state. The history of the state is a complicated topic with a large research literature of its own, but the connections between the burgeoning state and the simultaneous beginnings of European colonial expansion are close, as the early Spanish and Portuguese states and their empires in the Americas exemplify. Beyond that, the state could not exist without cadres of specialists performing essential work for its maintenance, and scientific expertise was quickly tapped in European overseas expansion after 1492. In the period of concern here – the long eighteenth century – organized science in France and the Colonial Machine represent historical high-water marks in the support of science by the early modern state...in general and for colonial development in particular.

This understanding prompts further historiographical reflections. Men of science, as well as soldiers, sailors, merchants, and priests, functioned as agents of French power and imperialist expansion, and therefore, the actors and institutions of science should receive greater emphasis in the history of contemporary French colonialism. However, more is involved than just seeing the union of science and state power as a straightforward step in the ongoing march of European civilization overseas. Rather, the particular approach we have taken (starting with institutions and the apparatus of contemporary

French bureaucracy) proves essential for understanding the historical specifics of the *French* entrée onto the world stage. Other approaches notwithstanding, the French case and its role in the context of early globalization cannot be properly evaluated without taking into account the agencies of the contemporary state and the institutions it supported, and particularly in the case of the Colonial Machine, scientific and technical institutions and the cadres of experts underwritten by the state. In these ways, increased awareness of the scientific dimension of the French colonial experience can and should extend our grasp of contemporary French colonial and overseas history generally.

As much as political power and scientific expertise found a satisfying mutual embrace in the Colonial Machine, the failure of the effort to implant a colony in Kourou in Guiana in the 1760s points to an important conclusion to be drawn from this study: namely, that experts, for all their utility, remained subservient to political power. They were, to use that old expression, on tap, but not on top.

Even so, the level of organized science and state-sponsored expertise oriented toward the colonies overseas left the French state and the Colonial Machine without a rival among the nations of Europe in the eighteenth century. At the outset we reviewed contemporary English and Dutch institutions that paralleled those in France. Britain and Holland were not without similar structures and potentialities for Colonial Machines of their own, but no contemporary European power matched France or its Colonial Machine for engaging organized science and expert knowledge in the service of colonial and overseas expansion. John Gascoigne points out that something comparable to the French Colonial Machine began to coalesce in Britain, but only after Joseph Banks assumed the presidency of the Royal Society of London in 1778.[16] But Edward Long already knew of the preeminence of the French Colonial Machine in 1774 when he compared the situation in France to England's "own torpid machine" in the epigram that heads this work.

One might reasonably ask, *quo vadis*? Scholars have long argued that what is needed in pursuing historical research into science and European colonialism and imperialism is a turn to comparative perspectives and comparative studies.[17] Some of the more recent literature does this. The 2005 volume edited by Charlotte de Castelnau-L'Estoile and François Regourd, *Connais-*

16. See particularly Gascoigne, pp. 185-98 and Epilogue; Gascoigne notes of Britain, p. 185, "In the late eighteenth century, however, imperial affairs had no clear bureaucratic home and were often untidily spread around a number of different departments." Telling in this connection, the British Ordanance Survey, the state cartographical bureau, dates only form 1791; furthermore, only in 1801 did colonial affairs in Britain shift from the Home Office to the newly renamed Secretary for War and the Colonies. The equivalent French ministry had been in place for over a century and a quarter. Once again, we might note that Pritchard (2004), p. 72 and passim, is critical of the notion that "the French state [w]as the primary agent in the development of the French colonies."

17. McClellan (1993).

sances et pouvoirs. Les espaces impériaux, XVIe-XVIIIe s. France, Espagne, Portugal, is a good example of this comparative thrust. So, too, are the *ISIS* "Forum" on colonial science edited by Londa Schiebinger, likewise in 2005; the 2007 volume edited by Nicholas Dew and James Delbourgo, *Science and Empire in the Atlantic World*; and the 2009 volume, *The Brokered World*, edited by Simon Schaffer et al. focusing on go-betweens and global intelligence in the later part of our period. More remains to be done, particularly of a synthetic sort. There are many exciting ways the literature is moving forward. Our work invites further studies of national styles of colonial development between and among France, England, Holland, and the Iberian monarchies. An example would be to compare the nature and operations of the French Compagnies des Indes, the British East India Company, and the Dutch *Vereenigde Oost Indische Compagnie*. Focusing on tensions between various European centers and colonial peripheries opens the door to larger reflections on the concrete sites where knowledge was made on the colonial periphery – such as science aboard ships, or in and around colonial hospitals, religious missions, botanical gardens, plantations, or on different frontier settings. A comparison of the three colonial scientific and learned societies of France, England, and Holland of that era – the Cercle des Philadelphes (1784), the Asiatic Society of Bengal (1784), and the *Bataviaasch Genootschap van Kunsten en Wetenschappen* (1778) – gives an idea of what might be done in this connection.

The Colonial Machine can be used as a map to guide further explorations of the French case. Up until now the history of French colonial science in the Old Regime has remained largely a terra incognita and has lacked specificity. We have now charted the real place of institutions in the field, traced biographical trajectories, and sketched cohorts of amateurs and experts with their multiple connections. This map highlights the main roads and byways used by travelers, explorers, experts, and officials, but also those taken by plants, seeds, samples, scientific accounts, drawings, geographical reports, and even ideas all around the world. As with every map, this one has its qualities and its limitations. But we hope that it will be useful to other explorers, who can continue to perfect it. Historical work remains to be done concerning, among other matters, colonial engineers, East- and West Indian cartography, colonial medical personnel and practices, colonial agronomy, European exchanges with indigenous knowledge systems, and like subjects.

A last thought strikes us as we take leave of this story. It is too easy to say that the Colonial Machine existed in a preindustrial era, when France and Europe had yet to be transformed by the Industrial Revolution, as they would be in the nineteenth century. In some respects, of course, activities close to the heart of the Colonial Machine occurred at the forefront of contemporary technology and economic development. Here, one thinks of the stimulus of shipbuilding and port activity and the wealth generated by commodity production in the colonies. On balance, however, particularly in France, the Colonial Machine and the Old Regime were part of a preindustrial society and a preindustrial world strikingly different from the one we know today. It was not a world or a society without machines, and the great ships that sailed the oceans were, to quote Moreau de Saint-Méry once again, "the most astonishing machines created by the genius of Man." But these ships were made of oak, and they were handmade. The machines of the day were useful and essential for life in Old Regime France and for the functioning of the Colonial Machine. But for motive power, the Colonial Machine relied in essential ways on wind and tide, rope and human muscle, and animal power, when possible. Daily life was more sinewy and more difficult then, perhaps especially for sailors and colonists struggling as they did at sea and on various colonial frontiers. Set in this other, harder world of preindustrial, Old Regime France, a world not so long ago, yet long ago nonetheless, the nature and record of the Colonial Machine seem even more remarkable and worthy of a history.

Biographical Appendix

[Portrait of Moreau de Saint-Méry, 1789]

The Colonial Machine did not exist independently of the people who populated it. This appendix contains the names of the men (and one woman) associated with the institutions of the Colonial Machine in the period of roughly 1660-1790. Here, the Colonial Machine reveals itself in the people connected to it.

The names have been culled from learned society membership lists, standard biographical sources (in particular Maurel in DPF, Taton, Chapuis, Vergé-Franceschi, Glénisson, Homer, Dictionary of Canadian Biography, and Institut de France's *Index Biographique*), and our own archival researches.

This biographical appendix began as a list of *correspondants* and others of the Académie Royale des Sciences and the Société Royale de Médecine who either lived in the colonies or who had significant contact with the world overseas. Along the way we kept an informal account of people and the positions they occupied in the Colonial Machine. But it was only in the late stages of this project that we realized how essential it is for grasping what was the Colonial Machine to have a sense of the universe of individuals who occupied positions within and animated the Colonial Machine. Unfortunately, although we have made every effort under the constraints we faced to be inclusive and accurate, we cannot vouch for every last date and detail for each and every one of the individuals surveyed here. We invite other scholars to correct and augment the work on view here and to pursue further prosopographical research. Such future corrections and augmentations notwithstanding, we are confident that our list is a solid one that provides more than a first approximation of the personnel and demographics of the Colonial Machine.

Three criteria governed inclusion in this list. First, involvement in the contemporary colonial project in one way or another. Secondly, a formal connection to one or another of the institutions making up the Colonial Machine. And thirdly, the possession of scientific or technical expertise of one sort or another, or, in the case of non-expert administrators, active interest in and support of scientific and technical people and projects.

This list cannot pretend to be exhaustive. Given the imbalance in our sources, for example, the list is probably weak on government medical and engineering staffs, particularly for lesser positions, in Canada and the Indian Ocean colonies. Then, judgment calls had to be made. For example, Alexis Clairaut (1713-1765) is not included, although an outstanding member of the French scientific establishment and an official voyager (to Lapland, 1736-37) because he seems to have had only a loose connection to the colonial project. Conversely, his fellow traveler to the North, Pierre-Louis Moreau de Maupertuis (1698-1759), is included because of his connection to the Dépôt des Cartes et Plans de la Marine. Similarly, most government administrators and officers of the Marine Royale, even though essential to the functioning of the Colonial Machine, are not included because their involvement with or promotion of expert knowledge seems weak or non-existent. Thus, for example, the obscure caretaker of the royal botanical garden in Port-au-Prince in Saint Domingue in 1783, one Lamotte, figures below, but the well-known State Secretary for the Navy from 1723 to 1749, Jean-Frédéric Phélypeau, the Count de Maurepas (1701-1781), does not. These cautions notwithstanding, the list identifies the major figures associated with French science and overseas expansion in the Old Regime, and it can serve as a reliable first approximation of the prosopography of the Colonial Machine.

Corresponding members (*correspondants*) of the Académie Royale des Sciences, the Société Royale de Médecine, the Académie Royale de Marine, and the Société Royale d'Agriculture are included by dint of being a *correspondant* if they a) lived or worked in the colonies, b) had significant maritime or overseas experience, or c) had significant port or naval connections. Resident or ordinary members of these institutions were chosen if they had significant colonial and naval experience or other connections to the Colonial Machine.

Membership in provincial academies or the Cercle des Philadelphes/Société Royale des Sciences et Arts in Saint Domingue did not *ipso facto* lead to inclusion on the list. (N.B.: Dates of membership in the Cercle des Philadelphes for the most part reflect the dates of published membership lists, not necessarily the dates of election.) Ordinary associates of the contemporary Chambres d'Agriculture or the Chambres de Commerce without other ties are NOT included, as these institutions were ancillary to the Colonial Machine.

Entries marked with a double asterisk (**) indicate that the individual named published a volume or volumes listed in the Bibliography (Contemporary Sources).

Accaron, Jean-Augustin.

Head of the Bureau des Colonies in the Ministry of the Navy and the Colonies.

Intendant-General of the Colonies, 1765.

Adanson, Michel (1727-1806).**

Botanist.

Clerk, Compagnie des Indes, Senegal, 1748-1753.

Académie Royale des Sciences: Correspondant (Réaumur, then Jussieu), 1750; Adjunct (botany), 1759; Associate (botany), 1773; Pensionnaire (botany), 1782.

Royal censor.

F.R.S.

Albert.

Navy Surgeon-Major (Chirurgien-Major), Pondicherry, India; active circa 1710-1720.

Albert de Luynes, Charles-Hercule d'.

Navy officer, rising to Commodore.

Director of the Dépôt des Cartes et Plans de la Marine, 1720-1722.

Albert du Chesne, Antoine d' (1686-1751).

Director of the Dépôt des Cartes et Plans de la Marine, 1734-1750.

Académie Royale des Sciences: Free Associate, 1736.

Alliot, Jean-Baptiste-Fauste, alias Alliot de Mussay ([~1695]-†1730).

M.D. (Paris).
Médecin du Roi in Saint-Domingue, 1721.
Académie Royale des Sciences: Correspondant of Antoine de Jussieu, 1721-1730.
Died in Martinique.

Amelot.
Chief royal engineer in Saint Domingue, 1761.
Previously in Louisiana.

Amic.
Médecin du Roi, Guadeloupe.
Active circa 1789.
In contact with the Société Royale de Médecine.

Ancteville, Louis Floixel Cantel, chevalier d' (1738-1785).
Royal engineer, 1762. (Previously, a military engineer).
Ingénieur du Roi, Cap François, Saint Domingue, 1773-1785.
Ingénieur [du roi] en Chef, Môle Saint-Nicolas, Saint Domingue, 1773.
Directeur Général des Fortifications des Îles sous le Vent, 1784.

Angiviller, Charles-Claude de Flahaut de La Billarderie, comte d' (1730-1809).
Directeur-Général of the Bâtiments du Roi from 1774 to 1790.
Académie Royale des Sciences: Supernumerary Associate (chemistry), 1772; Pensionnaire Emeritus, 1777.

Antoine d'Etannion, Jean-François.
Ingénieur-Géographe, Saint Domingue, 1763-1765.

d'Anville, Jean-Baptiste Bourguignon (1697-1782).
Académie Royale des Sciences: Adjunct (geography), 1773, replacing Buache.
First Royal Geographer (Premier Géographe du Roi), 1773.
Royal Geographer (Géographe du Roi), 1719.
Pensionnaire of the Académie des Inscriptions et Belles-Lettres, 1773.

d'Après de Mannevillette, Jean-Baptiste-Nicolas-Denis (1707-1780).**
Officer with the Compagnie des Indes from 1724, rising to Captain and Inspector general.

Compagnie hydrographer/cartographer; "Garde du Dépôt des Cartes et Journaux" of the Compagnie des Indes (1762) and then the Marine Royale, Lorient (1770).

Académie Royale des Sciences: Correspondant of Pierre-Charles Le Monnier, 1743-1780.

Free Associate of the Académie de Marine.

Ennobled by Louis XV.

Archange.

Franciscan.

Académie Royale des Sciences: Elected a Correspondant in 1765 as he departed for Louisiana, but appointment differed because institutional limit of 100 correspondents had been exceeded.

Posted to Louisiana.

Arnaud.

Government veterinarian on Guadeloupe to 1787.

Government veterinarian in Saint Domingue, 1787.

Arthaud, Charles (1748-1793[?]).**

M.D. (Nancy).

Médecin du Roi, 1772, Cap François, Saint Domingue.

Founder, Resident Associate, and Permanant Secretary of the Cercle des Philadelphes/Société Royale des Sciences et Arts du Cap François.

Correspondant of the Société Royale de Médecine (1777).

Correspondant of the Académie Royale de Chirurgie.

Artur, Jacques-François (1708-1779).**

M.D. (Caen).

Médecin du Roi, Cayenne, 1735-1770.

Counselor of the Conseil Supérieur of Cayenne.

Académie Royale des Sciences: Correspondant of Réaumur, 1753; then Antoine de Jussieu, 1757; struck from the ranks of Correspondants, 1766.

Correspondant of the Cabinet du Roi.

Trained at the Jardin du Roi, Paris.

F.R.S.

Auvray, Jean-Baptiste.

Merchant; Colonist, Saint Domingue.

Founder and officer of the Cercle des Philadelphes/Société Royale du Cap François.

Correspondant of the Société d'Agriculture.

Badier, Barthélemy.**

Voyer/Surveyor on Guadeloupe. Amateur botanist.

Correspondant of the Société d'Agriculture.

Informal correspondent of the Société Royale de Médecine; active circa 1770s-1780s.

Bajon, Bertrand (fl. 1751-1778, † after 1790).**

Royal surgeon in Cayenne; Surgeon-Major of the Hôpital militaire in Cayenne (1764-1776).

Académie Royale des Sciences: Correspondant of Daubenton, 1774.

Correspondant of the Académie Royale de Chirurgie; winner of its gold medal.

Unofficial correspondent of the Société Royale de Médecine.

Baradat.

Médecin du Roi, Cap François, Saint Domingue, 1763-1784 and 1786.

Barbotteau, Jacques-Alexandre.

Académie Royale des Sciences: Correspondant of Duhamel du Monceau, 1776.

Counselor, Conseil Supérieur de la Guadeloupe.

Superintending Botanist, Windward Islands, 1775-1784.

Baron.

Arpenteur du Roi (Royal Surveyor), Cap François, Saint Domingue, circa 1780s.

Colonial Associate of the Cercle des Philadelphes, Saint Domingue, 1787-1791.

Baron, Pierre.

Ingénieur du Roi, sent to Louisiana, 1729-1731.

In contact with the Académie Royale des Sciences, Paris.

Barrault.

Royal surgeon sent to India, 1788.

Barrère, Pierre (1690-1755).**
M.D. (Perpignan).
Médecin-Botaniste du Roi in Cayenne, 1721-1725.
Académie Royale des Sciences: Correspondant of Fantet de Lagny, then of Bernard de Jussieu, 1725.
Later, professor of medicine, the University of Perpignan.

Bayet.
M.D. (Montpellier).
Médecin du Roi, Les Cayes, Saint Domingue, 1767-1769.

Beautemps-Baupré, Charles-François.(1766-1854).
Ingénieur-Hydrographe and draftsman at the Dépôt des Cartes et Plans de la Marine, 1785.
Ingénieur-Géographe accompanying d'Entrecasteaux, 1791.

Beauvernet, Louis de.
Ingénieur du Roi in Saint Domingue, circa 1775.

Beauvollier, Antoine de (Father Barnabé) (1657-1798).
Jesuit. Mathématicien du Roi.
Missionary to China.

Bégon, Michel (1638-1710).
Trésorier de la Marine du Levant, Toulon, 1677.
Commissaire Général de la Marine in Brest (1680) and in Le Havre (1681).
Intendant aux Îles Françaises d'Amérique, 1682-1684.
Intendant in Marseilles (1685), in Rochefort (1688-1710) and in La Rochelle (1694-1710).
Charles Plumier gave his name to the *Begonia.*

Bellin, Jacques-Nicolas (1703-1772).**
Cartographer-hydrographer.
Attached to the Dépôt des Cartes et Plans de la Marine from 1721.
Ingénieur-Hydrographe de la Marine attached to Dépôt des Cartes et Plans de la Marine, 1741-1772.
Navy commissioner (Commissaire de la Marine), 1762.
Royal censor.
Free Associate of the Académie de Marine.
F.R.S.

Belloc, Jean-Daniel? (†1765?).
M.D. (Reims).
Médecin du Roi, Petit-Goave, Saint Domingue, 1745-1746.
Médecin du Roi, Port-au-Prince, 1746-1765.

Belval.
Ingénieur du Roi on Ile de France, circa 1773.

Bénigne de Fontenette.
Médecin du Roi, New Orleans, circa 1746.
In contact with the Jardin du Roi, Paris.

Beraud, Pierre (†1790).
Surgeon.
Director of the hospital at Port-au-Prince, Saint Domingue, circa 1780s.

Bernard, Pons-Joseph (1748-1816).
Mathematician/astronomer.
Oratian brother.
Adjunct director at Royal Navy observatory at Marseilles, 1781.
Académie Royale des Sciences: Correspondant of Méchain, 1786.

Bernizet, Gérault-Sébastien (†1788?).
Cartographer.
Royal Ingénieur-Géographe accompanying the La Pérouse expedition, 1785.

Berthoud, Ferdinand (1727-1807).
Horologer Mécanicien du Roi et de la Marine.
F.R.S.

Berthoud, Pierre-Louis (1754-1813).
Horloger de la Marine, 1784.

Bertin, Antoine.**
Médecin du Roi, Guadeloupe.
Correspondant of the Société Royale de Médecine (1783); medal winner (1785); active circa 1782-1785.
Non-resident member of the Académie Royale de Chirurgie.

Bertin, Henri-Léonard-Jean-Baptiste (1720-1792).
Minister and Secretary of State.
Controller General from 1759-1763.
Académie Royale des Sciences: Honorary, 1761; Vice-President, 1763, 1769; President, 1764, 1770.
Leading member of Société d'Agriculture.
Académie Royale des Inscriptions et Belles-Lettres, Honorary.

Bertrand de Saint-Ouen, Jean-Joseph (†1787).
Hydrological engineer.
Royal commissioner for irrigation of the Artibonite plain, Saint Domingue, 1784.
Colonial Associate of the Cercle des Philadelphes, Saint Domingue, 1785.

Bézombes.
M.D. (Montpellier).
Médecin du Roi, Cayes Saint-Louis, Saint Domingue, from 1766.

Bézout (also Bezout) Étienne (1730-1783).**
Examiner of the Gardes de la Marine, 1764-1783.
Académie Royale des Sciences: Adjunct (mechanics), 1758; Associate (mechanics), 1768; Pensionnaire Emeritus, 1779.
Member of the Académie Royale de Marine, 1769.

Bignon, Jean-Paul (1662-1743).
Oratian priest and abbot.
Royal Librarian.
Académie Royale des Sciences: Honorary, 1699; multiple times vice-president and president.
Academician, Académie Française, 1693.
Académie Royale des Inscriptions et Belles-Lettres, Honorary, 1701.

Binoist (or Binois) (†1737).
Government Ingénieur, then Ingénieur en Chef.
Grenada mostly, Martinique, Guadeloupe, and various other islands in the Antilles.
Active circa 1700 to 1737.

Blondeau, Etienne-Nicolas (†1783).
Professeur of mathematics and hydrography for the Gardes Marine, Brest.

Member of Académie Royale de Marine.
Director of the Navy's Atelier de Boussoles, Brest (1778-1783).
Editor, *Journal de Marine* (1778-1783).
Informal correspondent of the Société Royale de Médecine.

Blondeau, Marc (1742-93).
Ingénieur en Chef des Ports et Arsenaux de la Marine, Brest.

Blondeau.
Royal engineer sent to the Antilles, 1666.

Blondel, Nicolas-François (1618-1686).
Appointed to the West Indies on a mission to inspect fortifications and to assess the defensive power of the colonies (1666-1668).
Made maps of the French West Indies.
Académie Royale des Sciences: Académicien géometre (1669).
First Director of the Académie Royale d'Architecture (1671).

Boisforest, Nicolas-Honoré-Marie Taverne de (1731-1788).
Government engineer in Saint Domingue, 1771-1788, rising to Ingénieur en Chef and Directeur Général des Fortifications de Saint-Domingue, 1786.

Boispinel, Nicolas Pinel de (1685?-1723).
"Ingénieur ordinaire," Louisiane, 1720-1723.

Bompar, Maximin de (1698-1773).
Navy officer, rising to Commodore, 1757.
Governor-General of Martinique, 1750-1757.`
Director of the Dépôt des Cartes et Plans de la Marine, 1757-1762.

Bonamy, François (1710-1786).
M.D.
Professor of Botany and Doctor-Regent of University of Nantes.
Director of Botanical Garden, Nantes, Head of royal gardens, Nantes, 1730-1786.
Correspondant of Société Royale de Médecine.

Bonne, Rigobert (1727-1792).
Ingénieur-Géographe of the Dépôt des Cartes et Plans de la Marine, 1775.
Premier hydrographe du Dépôt des Cartes et Plans de la Marine, 1776-1789.

Bonnecamps (also Bonnécamps), Joseph-Pierre de (1707-1790).
Jesuit.
Professor of Hydrography and Mathematics, Quebec, 1741-1759.

Bony de la Vergne, Charles de.
Ingénieur-Géographe, Saint Domingue, 1763-1765.

Borda, Jean Charles de (1733-1799).**
Navy officer, Captain in 1779.
Voyaging with Verdun de La Crenne expedition, 1771-1772.
Inspector of Navy constructions and head of the École des Élèves Ingénieurs Constructeurs de la Marine, 1784.
Académie Royale des Sciences: Adjunct (geometry), 1756; Asociate (geometry), 1768; Pensionnaire (geometry), 1772; Associate Director, 1776; Director, 1777.
Honorary of the Académie Royale de Marine.

Bory, le Chevalier Gabriel de (1720-1801).**
Navy officer, 1767, rising to Captain and Commodore (Chef d'Escadre).
Governor-General of Saint Domingue, 1761-1763.
Académie Royale des Sciences: Free Associate, 1765.
Ordinary Member of the Académie Royale de Marine.

Bouchage, le chevalier du.
Astronomer with Bougainville expedition, 1766-1769.

Boucher, Pierre-Jérôme (1688?-1753).
Draftsman (Dessinateur), and then Engineer, Louisbourg, Canada, 1717-1745 and 1749-1753.

Bougainville, Louis-Antoine de (1729-1811).**
Navy officer, rising to Vice-Admiral.
First French circumnavigation, Pacific explorer, 1766-1769.
Académie Royale des Sciences: Free Associate, 1789.
F.R.S.

Bouguer, Pierre (1698-1758).**
Royal hydrographer, Le Croisic, 1714-1730, and Le Havre, 1730-1745.

Académie Royale des Sciences: Associate (geometry), 1731; Pensionnaire (astronomy), 1735; Associate Director, 1747, 1754; Director, 1748, 1755.

Voyager to South America with La Condamine, circa 1735.

Honorary of the Académie Royale de Marine (1752).

F.R.S.

Bourdier.

Médecin du Roi, Île Bourbon and in India, circa 1770.

Informal correspondent of the Société Royale de Médecine.

Boureau-Deslandes, André-François (1690-1757).**

Born in Pondicherry.

Commissioner-General of the Navy (Commissaire Général de la Marine) from 1736.

Académie Royale des Sciences: Élève (geometry), 1712; Supernumerary Adjunct (geometry), 1716.

Boutin (1672-1742).

Jesuit.

Missionary-astronomer in Saint Domingue, 1705-1742.

In regular contact with Académie Royale des Sciences and Observatoire Royal, Paris.

Bouvet, Joachim (1656-1730).

Jesuit.

Mathématicien du Roi.

Correspondant of Father Thomas Gouye, 1699.

Jesuit astronomer, Beijing.

Brethon.

See Le Breton, Paul.

Brossard, François.

Chief surveyor for the Southern and Western departments of Saint Domingue, circa 1722.

Broutin, Ignace-François (1685?-1751).

Military officer for Le Blanc and Belle-Île, Louisiana, 1719-1731.

Acted as a chief engineer in Louisiana, after Baron's departure, without an official title, 1731-1751.

Bruguière, Jean-Guillaume (1749-1798).
M.D. (Montpellier).
Médecin-Naturaliste sailing with Kerguelen, 1773.
Undertook a state-sponsored botanical mission to the Ottoman Empire, Persia, and Egypt (1792-1798).
Editor, *Journal d'histoire naturelle*, (1791-).

Bruix, Étienne-Eustache (1759-1805).
Creole from Saint Domingue.
Navy officer, rising to Lieutenant de Vaisseau, 1786.
Ordinary Member of the Académie Royale de Marine, 1789.
Assisted in Puységur cartographical expedition to Saint Domingue, 1785.

Brûletout de Préfontaine, Jean-Antoine.
Colonist in Guiana.
Consultant for Kourou expedition, and then Commandant de la Partie nord de la Guyane (with Cross of Saint-Louis) 1762-1764.

Brulley, Augustin-Jean.
Planter/Colonist, Saint Domingue.
In contact with the Académie Royale des Sciences, 1780s.
In contact with the Cercle des Philadelphes, Saint Domingue.
Received 3000 livre government gratification for his work with cochineal and nopal.

Brun.
Navy Chirurgien-Major, Cap François, Saint Domingue, circa 1765.

Buache, Philippe (1700-1773).**
Premier Géographe du Roi.
Académie Royale des Sciences: Adjunct (geography), 1730, first holder of position.
Attached to the Dépôt des Cartes et Plans de la Marine, 1721-1737.

Buache de la Neuville, Jean-Nicolas (1741-1825).
Premier Géographe du Roi, 1782.
In Dépôt des Cartes et Plans de la Marine from 1762.

Chief Royal hydrographer in the Dépôt des Cartes et Plans de la Marine, secretly in 1779; officially in 1789.
Académie Royale des Sciences: Adjunct (geography), 1782; Associate (geography), 1785.

Buffon, Georges-Louis-Marie Leclerc, Count de (1707-1788).**
Intendant, Jardin du Roi, Paris, 1739-1788.
Académie Royale des Sciences: Adjunct (mechanics), 1734; Associate (botany) 1739; Treasurer, 1744.
Of the Société d'Agriculture.
One of the immortals of the Académie Française, 1753.
F.R.S.

Calon, Etienne-Nicolas (1726-1807).
Ingénieur-Géographe, Saint Domingue, 1763-1765.

Calon de Felcourt, Jean-Pierre (1729-?).
Brother of Etienne-Nicolas Calon.
Worked with Cassini on the map of France, 1752-1760.
Ingénieur-Géographe, Saint Domingue, 1763-1765.
Worked with La Cardonnie to map the islands north of Saint Domingue, 1768.
Government engineer, Saint Domingue, from 1776.

Campet, Pierre (1723-1801).**
Chief Surgeon, Hôpital de Cayenne, 1754-1772.
Informal correspondent of Académie Royale des Sciences.
Correspondant de l'Académie Royale de Chirurgie, 1774.

Camus, Charles-Étienne-Louis (1699-1768).
Abbé.
Académie Royale des Sciences: Adjunct (mechanics), 1727; Associate (mechanics), 1733; Pensionnaire (geometry) 1741; Associate Director, 1749, 1760; Director, 1750, 1761.
On the expedition to Lapland.
Honorary of Académie Royale de Marine.
Secretary of the Académie Royale d'Architecture.

Carrel.
Médecin du Roi, Martinique, circa 1726-1729.

Cassan, Jean-Baptiste (1768-1803?).**

M.D. (Toulouse).

Graduate of the École de Médecine Pratique, Brest.

Médecin du Roi, Saint Lucia ("Médecin des Hôpitaux Militaires des Colonies Françaises").

Correspondant of Société Royale de Médecine (1790).

Correspondant (1786), then (1790) Associate of the Académie Royale de Marine.

Informal correspondent of the Société d'Agriculture.

Occasional contact with the Académie Royale des Sciences.

Colonial Associate of the Cercle des Philadelphes, Saint Domingue, 1791.

Cassini, César-François (Cassini de Thury [III], 1714-1784).

Royal astronomer, Observatoire Royal, Paris.

Académie Royale des Sciences: Supernumerary Adjunct (astronomy), 1735; Adjunct (astronomy), 1741; Associate (mechanics), 1741; Pensionnaire (astronomy), 1745; Associate Director, 1757, 1766, 1770; Director, 1758, 1767, 1771.

F.R.S.

Cassini, Jacques [Cassini II] (1677-1756).**

Royal astronomer, Observatoire Royal, Paris.

Académie Royale des Sciences: Associate (astronomy), 1699; Pensionnaire, 1712; Pensionnaire Emeritus, 1746; Associate Director, 1714, 1731, 1738; Director, 1715, 1726, 1729, 1732, 1739.

F.R.S.

Cassini, Jean-Dominique [Cassini I] (1625-1712).**

Royal astronomer; head of the Observatoire Royal, Paris, 1669-1712.

Académie Royale des Sciences: Academician/Pensionnaire (astronomy), 1669.

F.R.S.

Cassini, Jean-Dominique [Cassini IV] (1748-1845).**

Royal astronomer, Observatoire Royal, Paris.

Académie Royale des Sciences: Adjunct (astronomy), 1770; Associate (astronomy), 1785; Assistant Permanent Secretary, 1792.

F.R.S.

Castries, Charles-Eugène-Gabriel de La Croix, Marquis de (1727-1801).
Minister of the Navy and the Colonies, 1780-1787.
Académie Royale des Sciences: Honorary, 1788; Vice President, 1790; President, 1791.

Catalogne (also de Catalougne) Gédéon de (1662-1729).
Arrived in Canada circa 1683, as surveyor in the Navy.
Second Ingénieur en Résidence in Montreal, 1712-1720.
In contact with the Académie Royale des Sciences.

Caylus (also Cailus), Jean-Baptiste de Giou de (†1722).
Ingénieur du Roi, Martinique, 1691-1706. Retired after 1706.

Céré, Jean-Nicolas.
Director of the Jardin du Roi, Île de France, 1775-1806.
Correspondant of the Société Royale d'Agriculture (1787); gold medal winner (1788).
Informal correspondent of the Jardin du Roi, Paris.

Chabert.
Directeur and Inspecteur Général des Écoles Royales Vétérinaires of Paris, circa 1789-1791.
National Associate of the Cercle des Philadelphes, Saint Domingue.

Chabert, Joseph-Bertrand de Cogolin, Marquis de (1724-1805).**
Navy officer, rising to Chef d'Escadre (Commodore).
Assistant Inspector of the Dépôt des Cartes et Plans de la Marine, 1758-1773.
Adjunct inspector of the Dépôt des Cartes et Plans de la Marine, 1773-1776.
Inspector General of the Dépôt des Cartes et Plans de la Marine, 1776-1792.
Académie Royale des Sciences: Free Associate, 1758.
Ordinary Academician of the Académie Royale de Marine (1752).
F.R.S.

Chapel.
Graduate of the École des Mines, Paris.
Government engineer sent to Guiana in the mid-1780s.

Chappe d'Auteroche, Jean-Baptiste (1722-1769).
Astronomer. Voyager.

Académie Royale des Sciences: Adjunct (astronomy), 1759; observer of the Transits of Venus for the Academy.

Chardon, Daniel-Marc-Antoine.**
Intendant, Saint Lucia in 1763-1764.
Member of Académie Royale de Marine, 1787.

Charles de la Blandinière, Louis.
Director of the Dépôt des Cartes et Plans de la Marine, 1722-1734.

Charlevoix, Pierre-François-Xavier de (1682-1761).**
Jesuit.
Professor of hydrography, Quebec; explorer.
Government-sponsored voyage to Louisiana, 1720-1722.

Charlevoix de Villiers.
Government engineer in Saint Domingue, 1746-1767.

Charnières, Charles-François-Philippe de (1740-1780).**
Navy officer, rising to Captain, 1780.
Member of the Académie Royale de Marine, 1769.
In contact with the Académie Royale des Sciences.

Chatard, Pierre François.
Royal apothecary (Apothicaire du Roi), Cap François, Saint Domingue, circa 1772.

Chatelard, J.-B. du.
Jesuit.
Professor of hydrography, Toulon, 1725-1757.
Informal correspondent of the Académie Royale des Sciences.

Chaussegros de Léry, Gaspard (1682-1756).
Chief engineer, Canada, 1716-1756.

Chazelles, Jean-Mathieu de (1657-1710).
Professor of Hydrography with galleys, Marseilles, 1685; cartographer.
Académie Royale des Sciences: Associate (astronomy), 1695; Associate (mechanics), 1699.

Chomel, Antoine (1668-1702).
Jesuit astronomer in China.
Correspondant of Father Thomas Gouye, 1699.

Clairin-Deslaurières, François.
Surgeon-Major, Cap François, Saint Domingue, circa 1772.

Clesmeur (Clefmeur).
Botanical envoy to Guiana, circa 1780s.

Cochon Dupuys (also Dupuis), Jean (1674-1757).
M.D. (Paris).
Médecin du Roi, Médecin de la Marine; Director, École de Médecine, Rochefort.
Académie Royale des Sciences: Correspondant of Winslow, 1726.

Cochon Dupuys (also Dupuis), Gaspard (1710-1788).
Chief Navy Physician, Rochefort, from 1757.

Collet (†1763).
Médecin du Roi, Port-au-Prince, 1763.

Collignon, Nicolas (1761-1788?).
Royal botanist sailing with La Pérouse expedition, 1785.

Commerson, Philibert (1727-1773).
Royal physician, botanist and naturalist ("Médecin Botaniste et Naturaliste du Roi") sailing with Bougainville, 1766-1768, and then on Ile de France.
Informal correspondent of the Académie Royale des Sciences.
Active contact with the Cabinet du Roi, the Jardin du Roi, Paris, and the Trianon gardens.

Cosme d'Angerville, Jean (†1787).
Chirurgien du Roi, Cap François, Saint Domingue, circa 1780.
One of founders and Resident Associate of the Cercle des Philadelphes, Saint Domingue.
Informal correspondent of the Société Royale de Médecine.

Cossigny, Jean-François Charpentier de (1690-1780).**
Ingénieur du Roi and Chief Engineer for Compagnie des Indes on Île de France and Île Bourbon.
Founded botanical garden, Palma, Île de France, 1764.

Académie Royale des Sciences: Correspondant of Réaumur, 1733; then of Morand, 1757, then of Portal, 1774.

Cossigny de Palma, Joseph-François Charpentier de (1736-1809).**

Creole colonist on Île de France.

Ingénieur du Roi, Île de France.

Académie Royale des Sciences: Correspondant of L.-G. Le Monnier, 1774.

Maintained and expanded botanical garden, Palma, Île de France.

Informal correspondent of the Société Royale de Médecine (1780s).

Correspondant of the *Bataviaasch Genootschap van Kunsten en Wetenschappen* (1782).

Cotte, Louis (1740-1815).**

Oratian priest; Abbé.

Curate and professor, Montmorency, France.

Meteorologist.

Académie Royale des Sciences: Correspondant of Tillet 1769.

Member of the Société Royale de Médecine from 1778.

Member of the Société d'Agriculture.

Couranier Deslandes.

Military staff surgeon.

Enslaved by local Indians in Florida in the 1760s.

Correspondant of the Académie Royale de Chirurgie.

Contacts with the Académie Royale des Sciences.

Courcelles, Étienne Chardon de (1705-1775).**

M.D. (Reims).

First Royal Physician (Premier Médecin de la Marine), Navy Hospital, Brest.

Professor and Director of the École d'Anatomie, Brest (1757-1775).

Académie Royale des Sciences: Correspondant of Duhamel du Monceau, 1742.

Member of the Académie Royale de Marine.

Couré, Marc.

Chirurgien du Roi, Fort-Dauphin, Saint Domingue, circa 1772.

One of the founders of the Cercle des Philadelphes, Saint Domingue, 1784.

Courrejolles, Auguste François Gabriel chevalier de.
Infantry officer.
Royal engineer (Ingénieur du Roi) in Saint Domingue.
Resident, then Colonial Associate of the Cercle des Philadelphes, Saint Domingue, 1785-1791.

Courtanvaux, François-César Le Tellier de (1718-1781).**
Scientific amateur.
Académie Royale des Sciences: Honorary, 1765; Vice President, 1768, 1774; President, 1769, 1775.
Promoter of voyage to test chronometers, 1767-1768.

Couzier.
Médecin-Botaniste, Île Bourbon, 1729.

Crespy.
Chirurgien-Major, Senegal.
Informal correspondent of the Société Royale de Médecine, circa 1786.

Crosnier (also Crosiner Jeune, Cronier, Cromier).
Colonist, Cap François, Saint Domingue.
Correspondant of the Cabinet du Roi.

Crossat.
Jesuit missionary and astronomer in Guiana, circa 1724.
In contact with Académie Royale des Sciences.

Crozet, Julien-Marie (1728-1780).
With Compagnie des Indes, then Naval officer.
Sailed with d'Après de Mannevillette in Indian Ocean, 1750.
At the Dépôt des Cartes et Plans de la Marine, circa 1773.

Cury, Étienne de (†1763).
Académie Royale des Sciences: Correspondant of Dortous de Mairan, 1744.
Chief Surveyor (Arpenteur-Général) in Saint Domingue.

Daché.
Royal engineer in Saint Domingue; Chief Engineer, 1755-1777.

Dagelet (also d'Agelet), Joseph Le Paute (1751-1788?).
Astronomer.
Professor of Mathematics, École Militaire, 1778-1785.

Académie Royale des Sciences: Adjunct (astronomy), 1785; Associate (astronomy), 1785.
Chief royal astronomer sailing with La Pérouse, 1785.
Astronomer sailing with Kerguelen, 1773-1774.

Damien-Chevalier, Jean.
M.D. (Paris).
Médecin du Roi, Petit-Goave, Saint Domingue, 1741-1745.
Author of anonymous *Lettre à M. de Jean sur les maladies de Saint-Domingue*, 1752.

Darras.
Government botanist, circa 1780s.
Botanist aboard the *Alexandre* from Île de France to Saint Domingue, 1788.

Dausse.
Royal Engineer/Ingénieur of the Ponts et Chaussées; hydrological engineer.
On mission in Saint Domingue, 1777.

Daussy, Magloire-Thomas (1758-1826).
Draftsman (Dessinateur) in the Dépôt des Colonies, 1786.

David, Joseph.
Médecin du Roi adjoint, Cap François, 1780.

Dazille, Jean-Barthélemy (1732-1812).**
M.D. (Douai).
Surgeon in Quebec, circa 1761.
Surgeon-Major in Cayenne, from 1763.
Surgeon-Major in Île de France, from 1766-1768, and again circa 1775.
Médecin du Roi (honorary) in Saint Domingue, 1776-1784.
Inspector of Colonial hospitals.
Back in France in 1784.
Correspondant of Société Royale de Médecine (1777); National Associate, 1781.
National Associate of the Cercle des Philadelphes, Saint Domingue, 1787-1791.

Decombes (also Descombes).
Ingénieur du Roi, in Martinique, Tobago and Saint Christopher, circa 1680.

Degaulle, Jean-Baptiste (1732-1810).
Ingénieur de la Marine, 1777.

Navy Hydrographer and Professor of Hydrography, Le Havre and Honfleur.
Académie Royale des Sciences: Correspondant of Méchain, 1782.

Deherme, Étienne (1745~1812).
Draftsman for military and ministry of foreign affairs, 1766-1779.
In Dépôt des Cartes et Plans de la Marine, 1781.
Ingénieur in the Dépôt des Colonies, 1784.

de la Haye.
Abbé. Curé du Dondon, Saint Domingue.
Colonial Associate of the Cercle des Philadelphes, Saint Domingue, 1784-1791.
Botanical illustrator and author, botanical manuscript, *Florindie, ou Histoire phisico-économique des végétaux de la Torride.*
Informal correspondent of the Académie Royale des Sciences and the Jardin du Roi, Paris.

de la Martinière, Joseph (†1788?).
Royal naturalist sailing with the La Pérouse expedition, 1785.

Delaporte, Valentin.
Missionary of the order of Frères Precheurs, Guadeloupe, active circa 1775-1788.
Informal correspondent of the Société Royale de Médecine (1787-1788).

Delaporte fils.
Médecin du Roi, Cayenne.
Correspondant of the Société Royale de Médecine (1780).

de La Roulais.
Government Ingénieur in Martinique and the Lesser Antilles, 1716-1723.

De la Sorte (also De la Porte).
"Médecin en service," sent to Cayenne by Poissonnier in 1778.

De la Vergne (also Vergnies, Vergnier, Verynie, de la Vergne).
Médecin du Roi, Guadeloupe, circa 1770-1790.
Correspondant of the Société Royale de Médecine (1777).
Colonial Associate of the Cercle des Philadelphes, Saint Domingue, 1787-1791.

de Lesseps, Jean-Baptiste-Barthélemy (1766-1834).
French diplomat.
Travels with La Pérouse as Russian interpreter; returns overland in 1787-1788.

Delisle, Guillaume (1675-1726).**
Premier Géographe du Roi, 1718-1726.
Académie Royale des Sciences: Élève (astronomy), 1702; Adjunct (astronomy), 1716; Associate (astronomy), 1718.

Delisle, Joseph-Nicolas (1686-1768).**
Astronome-Géographe de la Marine, attached to the Dépôt des Cartes et Plans de la Marine, 1754-1768.
Académie Royale des Sciences: Élève, 1714; Adjunct, 1716; Associate Emeritus, 1741; Pensionnaire Emeritus, 1761.
Imperial astronomer and academician, Imperial Academy of Sciences, Saint Petersburg (1725-1747).
Professor of Mathematics, Collège Royal, 1718-1768.
Member of the Société d'Agriculture.
F.R.S.

Demeure, Jacques-Marie.
Head of the Atelier des Boussoles, Brest.

Denesle, Jacques Amable Nicolas (1735-?).
Director of the Jardin Botanique, Poitiers, circa 1787.
Secrétaire de la Société d'Agriculture, Poitiers.
National Associate of the Cercle des Philadelphes, Saint Domingue, 1788.

Denis, Sylvestre-Florentin. (†1764?).
Ingénieur-Géographe, Saint Domingue, 1763-1764.

De Reine.
Habitant of Isle de France.
Informal correspondent of the Académie Royale des Sciences and the Jardin du Roi, circa 1755.
Mauscript *Mémoire sur les plantes cultivées dans la zone torride.*

De Saint-Georges.
M.D.
Army surgeon and Royal Surgeon (Chirurgien du Roi), Port-au-Prince, Saint Domingue.

Colonial Associate of the Cercle des Philadelphes, Saint Domingue, 1789-1791.

De Saint-Michel.
Médecin du Roi, Île de France.
Correspondant of the Société Royale de Médecine (1777).

Desaulnes.
Ingénieur-Géographe, Guadeloupe, from 1763. Map in 1769.

Deschamps (†1784).
Médecin du Roi, Île Bourbon.
Correspondant de la Société Royale de Médecine, 1777.

Deschisseaux.
Médecin-Botaniste du Roi, Martinique, circa 1731-1735.

Deslauches.
Ingenieur Militaire, Senegal and Gorée, circa 1780s.

Despallets (†1767).
M.D. (Montpellier).
Médecin du Roi, Port-au-Prince, 1766-1767.

Desportes-Milon.
Director of the garden at Le Réduit, Île de France, circa 1770.

Desrivierre-Gers, Henri-Louis-Jérôme.
Army major stationed in Guiana, 1779-1785.
Amateur botanist and naturalist.

Dessaingy, Joseph-Charles. (Also Dessingy or Dessingi)
Cartographer in Canada.
Ingénieur-Géographe du Roi, Guiana, 1762 to 1785.

Devergès, Bernard (†1766).
Draftsman (Dessinateur), Louisiana, 1720-1735.
Military Engineer, Louisiana, 1735-1751.
Chief Engineer, Louisiana, 1751-1766.

Dizard de Kerguette, Jean.
Professor of Hydrography, Le Croisic, 1755-1764.
Professor of Mathematics, navy cadet and hydrography schools, Rochefort, 1765.

Dombey, Joseph (1742-1794).
M.D. (Montpellier).
Botaniste du Roi. Traveler to Peru, 1776-1785.
Académie Royale des Sciences: Correspondant of Antoine-Laurent de Jussieu, 1783.
In contact with the Société Royale de Médecine and the Jardin du Roi, Paris.

Dubois Berthelot de Beaucours (also Boisberthelot de Beaucourt), Josué (1662?-1750).
Infantry officer. Worked on Québec's ramparts, circa 1690.
Chief engineer (with Cross of Saint-Louis), Canada, 1712-1715.
Chief engineer at Louisbourg, Isle Royale, Canada, from 1715.
Governor of Montreal, 1733.

Duboscq, Jean.
Ingénieur du Roi and Grand Voyer (Superintendent of Roads), Saint Domingue, circa 1763.

Dubourg, Alexandre (1747-1787).
Actor, apothecary, amateur botanist, Cap François, Saint Domingue.
Founding Member of the Cercle des Philadelphes, Saint Domingue; Director of Cercle's botanical garden.

Dubrusquet, Bernard.
Ingénieur du Roi and Grand Voyer (Superintendent of Roads), Saint Domingue, circa 1786.

Dubry.
Military Surgeon-Major in Saint Domingue.
Inspector of the Eaux de Boynes spa, Saint Domingue, 1784.

Dubuc-Duferret, Jean-Baptiste (1717-1795).
Head of Bureau des Colonies, 1764-1770.
Heads mission to Cayenne, 1768.

Ducaille.
Ingénieur-Géographe, Guadeloupe, from 1763. Map in 1769.

Duché de Vancy (†1788?).
Landscape and portrait artist accompanying the La Pérouse expedition, 1785.

Duchemin de L'Etang, Julien-François.
M.D. (Montpellier).
Médecin du Roi, Les Cayes, Saint Domingue, circa 1775-1790 (with many interruptions).
Médecin du Roi adjoint, Cap François, Saint Domingue, from 1780.
Editor/publisher of the *Gazette de Médecine et d'Hyppiatrique* (Saint Domingue), 1778-1779.
Correspondant of the Société Royale de Médecine (1780).

Duchoisel.
Jesuit.
Apothecary for mission, Pondicherry, India.
Informal correspondent of the Société Royale de Médecine.

Du Coudreau (†1748).
Government engineer, Saint Domingue, 1740-1748.

Du Dresnay des Roches.
Ship's captain and Governor-General of Île de France and Île Bourbon for the Compagnie des Indes.
Member of Académie Royale de Marine.

Duez de Fontenay.
Ingénieur-Géographe in India, circa 1754-1756. Maps of Coromandel and Malabar coast.

Dufour.
Chirurgien-Major Auxiliaire de la Marine.
In the Antilles circa 1783-1784.

Dugée, Charlotte (†1767?).
Royal botanical artist (Dessinatrice du Roi pour Cayenne), 1764-1767.

Duhamel.
Médecin du Roi, Léogane, Saint Domingue, circa 1730s.
Académie Royale des Sciences: Correspondant of Dufay, 1737.

Duhamel.

Graduate of the *École des Mines*, Paris.

Government engineer sent to Guiana in the mid-1780s.

Duhamel du Monceau, Henri-Louis (1700-1782).**

Inspector General of the Navy (Inspecteur Général de la Marine), 1739-1782.

Académie Royale des Sciences: Adjunct (chemist), 1728; Associate (botany), 1730; Pensionnaire (botany), 1738; Assistant Director: 1742, 1755, 1767; Director: 1743, 1756, 1768.

Honorary of the Académie Royale de Marine.

Founder of the *École du Génie Maritime* (Paris).

Editor of the *Description des Arts et Métiers*.

F.R.S. (1735).

Duhaut.

Botaniste du Roi, Cayenne, circa 1724-1725.

Du Lignon, Alexandre.

Royal Gardener on Guadeloupe, circa 1699.

(Elder brother of Jean-Baptiste Lignon, known as Lignon le jeune, likewise a royal botanist on Guadeloupe).

Dumenil, Louis.

Arpenteur du Roi, Plaisance, Saint Domingue, 1787-1789, 1791.

Colonial Associate of the Cercle des Philadelphes, Saint Domingue, 1787-1791.

Dumoulceau, Mathias-Henri.

Military engineer.

Director-General of Fortifications, Saint Domingue, 1772-1778.

Dunezat.

Army Lieutenant and cartographer, Cayenne, circa 1726-1732.

Dupetit-Thouars, Aubert (1758-1831).**

Botanist.

Traveled to Île de France, Île Bourbon, and Madagascar in 1792-1802.

Dupin.
Apothicaire of the Admiralty, circa 1787-1788.
National Associate of the Cercle des Philadelphes, Saint Domingue.

Dupont, Joseph (1758/1759-1809).
Navy surgeon.
Chief surgeon, military hospital, Basse Terre, Guadeloupe, 1777-1792.
Duportal (also Du Portal), Antoine-Jean-Jacques (1701-1773).
Military officer.
Director-General of Fortifications, Saint Domingue, 1764-1769.

Du Puget, Edme-Jean-Antoine, Count d'Orval (1742-1801).
Army officer.
Commander of artillery school, Strasbourg, circa 1765.
Inspector-General of Colonial Artillery (Inspecteur Général de l'Artillerie Coloniale), 1784-1786; in Antilles 1784-1786.
Académie Royale des Sciences: Active, if informal correspondent. Lost to Bougainville in 1789 as Free Associate.
Informal correspondent of the Société Royale de Médecine.
Colonial, then National Associate of the Cercle des Philadelphes, Saint Domingue, 1786-1791.
Studied at the École de Génie de Mézières.

Durand.
Surgeon-Major of the Admiralty, Cap François, Saint Domingue, 1783.

Dutrône La Couture, Jacques-François (1749-1814).
M.D., agronomist.
Informal correspondent of the Académie Royale des Sciences.
National Associate of the Cercle des Philadelphes, Saint Domingue.

Duvalain.
Médecin du Roi, Cap François, Saint Domingue, 1719.

Eloy de Beauvais, François (1743-1815).
Veterinarian; sent by Ministry to Île de France in 1771.

Ennery, Victor-Thérèse Charpentier, Count d' (1731-1776).
Military officer.
Director-General of Colonial Fortifications.

Governor-General of Saint Domingue and the Antilles, 1776.

d'Entrecasteaux, Joseph-Antoine-Raymond Bruny (1737-1793).
Navy officer, rising to Contre-Amiral, 1785.
Associate Director of Ports and Arsenals, 1783.
Governor General of Île de France and Île Bourbon, 1787-1790.
Led the search for La Pérouse, 1792-1793.

Fage († around 1767).
Colonist/planter in Saint-Domingue.
Naturalist of the Elector of Mayence; member of Electoral Academy at Mayence.
Académie Royale des Sciences: Correspondant of Duhamel du Monceau, 1760; struck from the list of Correspondants, 1767.

Fagon, Guy-Crescent (1638-1718).
Intendant/Superintendant of the Jardin du Roi, Paris, 1693-1718.
Demonstrator of Plants and Pharmacy, the Jardin du Roi, Paris.
First Royal Physician.
Académie Royale des Sciences: Honorary, 1699.
Sponsor of scientific missions.

Fanthome.
Ingénieur-Géographe in India, circa 1755-1756. Maps of Pondicherry, Siringham and Trichinopoli.

Ferrier (also Ferrié).
Médecin du Roi, Saint Domingue, 1789.
Colonial Associate of the Cercle des Philadelphes, Saint Domingue.

Feuillée, Louis (1660-1732).**
Minim monk.
Mathématicien du Roi and Astronomer, headed Navy observatory with Minims, Marseilles from 1714.
In contact with the Observatoire Royal, Paris.
Government naturalist/traveler: the Levant (1700-1701), Antilles (1703-1706), South America (1707-1711).
Académie Royale des Sciences: Correspondant of J. D. Cassini, 1699.

Fleurieu, Charles-Pierre Claret, count de (1738-1810).**
Navy officer, rising to Captain, 1776.
Directeur des Ports et Arsenaux, 1777.

Minister of the Navy, 1790-1791.
Assistant Director of the Dépôt des Cartes et Plans de la Marine, 1776.
Involved in chronometer tests, on the royal frigate *Isis*, 1768-1769.
Member of the Académie Royale de Marine.

Fleuriot de Langle, Paul-Antoine (1744-1787).
Navy officer, Captain in 1782, Lieutenant de Frégate in 1785.
Second in command of the La Pérouse expedition.
Member of the Académie Royale de Marine, 1774.

Fleury.
Royal engineer, Saint Domingue, 1699-1700.

Fontaine.
M.D. (Montpellier, 1714).
Médecin du Roi, Cap François, Saint Domingue.

Fontaney, Jean de (1643-1710).
Royal Professor of Hydrography, Nantes, 1675-1677.
Jesuit astronomer in China.
Académie Royale des Sciences: Correspondant, 1684 ("avec trois Pères de sa Compagnie"); Correspondant of Thomas Gouye, 1699.

Fortin.
Ingénieur-Géographe, Miquelon, from 1764.

Foulquier de la Bastide, François-Joseph (1744-1789).
Intendant, Guadeloupe, 1781-1785, then Martinique, 1786-1789.
Académie Royale des Sciences: Correspondant of J. D. Cassini IV, 1781.
Honorary associate of the Cercle des Philadelphes, Saint Domingue.
Informal contact with the Jardin du Roi, Paris, and the Observatoire Royal, Paris.

Fournerie de Juville (†1786).
Royal Surveyor (Arpenteur du Roi) and Superintendent of Roads (Voyer), Saint-Marc in Saint Domingue, circa early 1780s.

François de Neufchâteau, Nicolas-Louis (1750-1828).
Public prosecutor, Cap François, Saint Domingue.
In contact with the Académie Royale des Sciences.
Informal correspondent of the Société Royale d'Agriculture.
Honorary, then National Associate of the Cercle des Philadelphes, Saint Domingue.

Franquelin, Jean-Baptiste (1650-17..?).
In Canada, made various maps, some of them sent to France, from 1672.
Géographe du Roi, 1686.
Royal Hydrographer, Quebec, Canada, 1687.
Active as a cartographer in Canada from ~1677 to ~1694.

Franquet, Louis-Joseph (1697-1768).
Brother of Charles-Joseph Franquet de Chaville.
Chief civil engineer, and Director and Inspector of Fortifications for Canada, circa 1750.
Chief Engineer in Louisbourg, Isle Royale, Canada, 1750-1757.

Franquet de Chaville, Charles-Joseph (1696-1775).
Brother of Louis-Joseph Franquet.
"Ingénieur ordinaire," Louisiana, 1720-1723?

Fresneau, François, sieur de La Gataudière (1703-1770).
Chief royal engineer (Ingénieur en Chef), Cayenne, 1732-1749.
Performed botanical research and chemical experiments regarding rubber.

Frézier (or Frésier), Amédée-François (1682-1773).**
Ingénieur du Roi, Saint Domingue, 1719-1725.
Director-General of Fortifications, Brest, 1740.

Fusée-Aublet, Jean-Baptiste-Christophe (1720-1778).**
Royal Botanist, Réduit garden, Île de France, 1753-1761.
Royal Botanist (Apothicaire-Botaniste du Roi), Cayenne, 1762-1764.
Director-General of the Môle Saint-Nicolas outpost, Saint Domingue, 1765.
In contact with the Académie Royale des Sciences.
In contact with the Jardin du Roi, Paris.

Gabaret de l'Hérondière.
Army captain, functioning as a government engineer and explorer in Guiana, 1711-1716.

Gallois (or Galloys), François (1719-1779).
Director of the Royal Botanical Garden, Lorient, 1768-1779.
In contact with the Jardin du Roi, Paris.

Galloys, abbé.
Traveled to China in 1764-1769 as Royal Naturalist and Emissary of the Jardin du Roi, Paris.
Brother of the gardener at Brest.

Garcin.
Chirurgien du Roi, Saint Lucia, circa 1785.

Garcin, Laurent (1683-1751).**
M.D. (Reims).
Navy surgeon, traveler and botanist.
Swiss; one-time VOC employee.
Académie Royale des Sciences: Correspondant of Antoine de Jussieu, 1730.
F.R.S.

Garreau, Claude-Jean-Emmanuel (1732-1791).
Ingénieur of the Ponts et Chaussées.
Ingénieur-Géographe, Saint Domingue, 1763-1765.

Garreau de Boispreau, François-Blaise (circa 1739-1775).
Ingénieur of the Ponts et Chaussées.
Ingénieur-Géographe, Saint Domingue, 1763.
Sent by the King to Madagascar, 1773-1775.

Gaubil, Father Antoine (1689-1759).**
Jesuit missionary to China from 1722. Died in Beijing.
Académie Royale des Sciences: Correspondant of Delisle, 1750.
Correspondant of the Imperial Academy of Sciences, Saint Petersburg.
F.R.S.

Gauché, Joseph.
Administrateur des Eaux de Boynes, Port à Piment, Saint Domingue, from 1786.
Colonial Associate of the Cercle des Philadelphes, Saint Domingue, 1784-1791.
Informal correspondent of the Société Royale de Médecine circa 1787-1789.

Gaultier, Jean-François (1708-1756).
Médecin du Roi, Quebec, 1742-1756.
Counselor, Conseil Supérieur of Quebec.
Académie Royale des Sciences: Correspondant of Duhamel du Monceau, 1745; Published in the *Savants Étrangers*.
Naturalist, meteorologist.

Gautier.
Navy Officier, Chemist.

Gélin, Jean.
Vétérinaire du Roi, Saint Domingue, 1782; subsequently stationed on Guadeloupe.
Graduate of royal veterninary school, Alfort.
Correspondant of the Société Royale de Médecine (1790).
Resident Associate of the Cercle des Philadelphes, Saint Domingue, 1784-1790.

Gemosat.
Government engineer in the Antilles, from 1672.

Gense, Louis (†1768?).
Ingénieur-Géographe, Martinique, 1763-1767.

Genty.
Lawyer and royal notary, Cap François, Saint Domingue.
Resident Associate and officer of the Cercle des Philadelphes, Saint Domingue, 1784-1791.

Gerard (also Gérard and Girard).
"Docteur en médecine," Cap François, Saint Domingue, circa 1760-1780.
Correspondant of the Société Royale de Médecine (1778).

Gerbillon, Jean-François (1654-1707, died in Beijing).
Jesuit astronomer and missionary.
Académie Royale des Sciences: Correspondant of Thomas Gouye, 1699.

Godeheu de Riville.
Director of the Compagnie des Indes.
Académie Royale des Sciences: Correspondant of Réaumur, 1748; then Duhamel du Monceau, 1757.

Godin, Louis (1704-1760).
Académie Royale des Sciences: Adjunct (geometry), 1725; Adjunct (astronomy), 1727; Associate (astronomy), 1730-1745; Pensionnaire (astronomy), 1756.

Head of Académie-sponsored voyage to South America, 1730s; Professor of Mathematics, University of Lima until 1751.
F.R.S.

Gollet, Jean-Alexis de (1664-1741, died in Macao).
Jesuit astronomer and missionary.
Académie Royale des Sciences: Correspondant of Thomas Gouye, 1699.

Gouye, Thomas (1650-1725).
Jesuit.
Académie Royale des Sciences: Honorary, 1699; Vice-President, 1707, 1709, 1710, 1712, 1713, 1715; President, 1711.

Granger, Claude, also known as Tourtechot (died in Persia, 1737).**
Navy surgeon.
Académie Royale des Sciences: Correspondant of Réaumur prior to 1735.
Royal Naturalist sent to Egypt, 1730-1732 and 1733-1737.
Informally connected to the Jardin du Roi, Paris.

Grenier, Jacques-Raymond.
Chevalier, Navy officer: Captain, circa 1760-80s.
Captained official, scientific voyage to Indian Ocean, 1767-1770.

Grivelé.
Royal engineer in Saint Domingue, circa 1698-1699.

Grognard du Justin, Benoît (1742- † after 1809).
Military engineer.
In the Dépôt des Cartes et Plans de la Marine, 1773-1793.
Detached to Lorient, 1775, to work with d'Après de Mannevillette on the *Neptune oriental*.

Groignard, Antoine (1727-1799).
Chief Navy Engineer (Ingénieur Général de la Marine) in Toulon.
Graduate of the *École du Génie Maritime.*
Académie Royale des Sciences: Correspondant of de Montigny, 1779; Bory, 1784; winner of Academy prizes.
Ordinary Member of the Académie Royale de Marine.

Guéry, Pierre (†1788?).
Watchmaker and armorer on La Pérouse expedition, 1785.

Guidy, Honoré de (†1784).
Navy officer and astronomer.
On cartographical mission with Borda to the Canaries, 1776-1777.
On Puységur cartographical expedition to Saint Domingue, 1784.
Adjunct, then (1784) Ordinary Member of the Académie Royale de Marine.

Guignes, Chrétien-Louis-Joseph de (1759-1845).
French Consul in Canton.
Académie Royale des Sciences: Correspondant of [L.-G.] Le Monnier, 1783.
Correspondant of the Académie Royale des Inscriptions et Belles-Lettres, 1783.

Guyomar.
Royal surveyor, Île de France, 1740s.

Hapel Lachênaie, Thomas-Luc-Augustin (1760-1808).
Royal veterinarian and apothecary.
Pharmacien en Chef des Hôpitaux de la Marine, Guadeloupe.
Apothicaire du Roi, Pointe-à-Pitre, Guadeloupe.
Previously at Alfort veterinary school.
Informal correspondent from 1785, then Correspondant of the Société Royale de Médecine, 1789; medal winner of the Société Royale de Médecine.
Colonial Associate of the Cercle des Philadelphes, Saint Domingue, 1789-1791.

Haumont, François-Etienne (†1767).
Ingénieur-Géographe, Guiana, 1763-1767.

Hauterive, Bernard-Laurent d'.
Chief prosecutor (Procureur général) of the Conseil Supérieur of Martinique from 1713-1721.
Académie Royale des Sciences: Correspondant of Dortous de Mairan, 1724.

Henault.
Appointed Second Chief Engineer, Saint Domingue, in 1700.

Herlin, Jean-Baptiste (†1777).
Premier Médecin du Roi (de la Marine), Brest.
Correspondant of the Société Royale de Médecine (1777).
Correspondant of the Académie Royale de Marine, 1770; Ordinary Member, 1776.

Hesse.
Ingénieur-Géographe in Martinique and Saint Domingue, 1770-1789.

Hormepierre, Jean-Baptiste (†1765).
M.D. (Salernes).
Premier Médecin Inspecteur Général des Hôpitaux et Pharmacies de Saint Domingue, 1763-1765.

Houël.
Government engineer in the French Antilles, 1727-1745.

Hugot.
Clockmaker accompanying La Condamine to Peru, 1735.

Incarville, Pierre d' (1706-1757).
Jesuit missionary in Beijing.
Académie Royale des Sciences: Correspondant of Claude-Joseph Geoffroy, 1750.
Astronomer and botanist.

Isambert, Michel (†1716 in Martinique).
Académie Royale des Sciences: Correspondant of the Academy as a whole in 1716.
Botanical envoy to Martinique, 1716.
M.D. (Montpellier).
Royal naturalist.

Isle-Beauchaine, Jean-Jacques (1747- † after 1792).
Navy officer, Majeur de Vaisseau.
Adjunct Inspector of the Dépôt des Cartes et Plans de la Marine, 1790-1792.

Jauvin.
Commissaire Général de la Marine and Ordonnateur, Cap François, Saint Domingue, 1788.
Honorary of the Cercle des Philadelphes, Saint Domingue, 1788-1791.

Jay [also known as Le Jay].
Médecin de la Marine pour les Colonies, Guadeloupe.
Correspondant of the Société Royale de Médecine (1781).

Jolliet, Louis (1645-1700).
Royal Hydrographer in Canada, 1697.
The famous explorer and discoverer of the Mississippi, 1673.

Joubert de la Motte, Rene-Nicolas (†1787).
M.D. (Angers).
In Saint Domingue from 1768.
Médecin du Roi ("sans appointements"), from 1772.
Director of the Jardin Royal, Port-au-Prince, Saint Domingue, 1781.
Botaniste du Roi and Médecin-Naturaliste, 1784.
Correspondant of the Société Royale de Médecine, 1781; then National Associate, 1784.
Colonial Associate of the Cercle des Philadelphes, Saint Domingue, 1785.

Jullia.
Compagnie des Indes botanical agent, Mascarenes, 1736-1738.

Jussieu, Antoine de (1686-1758).**
M.D.; Doctor Regent of the Paris faculty.
Professor of Botany, Jardin du Roi, Paris, 1710.
Académie Royale des Sciences: Élève (botany), 1712; Pensionnaire (botany), 1715.

Jussieu, Antoine-Laurent de (1748-1836).
Nephew of Antoine, Bernard and Joseph de Jussieu.
M.D.
Assistant Demonstrator of Plants, later Professor, Jardin du Roi, Paris, 1778.
Académie Royale des Sciences: Adjunct (botany), 1773; Associate (botany), 1782; Pensionnaire (botany), 1786; Associate Director, 1790; Director, 1791.

Jussieu, Bernard de (1699-1777).
Younger brother of Antoine de Jussieu.
M.D. (Paris).
Demonstrator of Plants, Jardin du Roi, Paris; Professor of Botany, 1722-1777.

Garde du Cabinet des Drogues, Jardin du Roi, Paris, 1732.
Académie Royale des Sciences: Adjunct (botany), 1725; Associate (botany), 1739; Pensionnaire (botany), 1739; Associate Director, 1753; Director, 1754.

Jussieu, Joseph de (1704-1779).**
Youngest brother of Antoine and Bernard.
M.D. (Paris).
Académie Royale des Sciences: Adjunct (botany), 1742; Associate (botany), 1743; Associate Emeritus, 1758.
Botanist on La Condamine expedition; in South America until 1771.

Kerbiquet-Lunven.
Captain for the Compagnie des Indes.
Did navigational work on longitudes and latitudes.
In contact with the Académie Royale des Sciences.

Kerguelen (also Kerguelen-Trémarec), Yves-Joseph de (1734-1797).
Navy officer and Captain.
Hydrographer.
Explorer of Antartic, 1771-1772, 1773-1774.
Adjunct member of Académie Royale de Marine, 1769.

Kervéguen, Gaultier de.
Military engineer.
In Saint Domingue, 1763, accompanying the Governor-General d'Estaing.
In Saint Domingue in the American War; made maps of the colony.

La Billardière, Jacques-Julien Houtou de (1755-1834).**
M.D. (Reims) [Another source says Paris].
Académie Royale des Sciences: Correspondant of the Académie, 1792.
Royal naturalist, botanist, botanical traveler.

Laborde, Raymond?
Médecin du Roi, Cayenne, 1770-1785.
Informal correspondent of the Société Royale de Médecine, circa 1784-1786.

La Broue, Pierre (†1705).
Government engineer in Saint Domingue, circa 1700-1705.

Lacaille, Nicolas-Louis de (1713-1762).**
Abbé, astronomer.
Académie Royale des Sciences: Adjunct (astronomy), 1741; Associate (astronomy), 1745.
Professor of Mathematics, Collège Mazarin, Paris.
Scientific voyager for the Académie Royale des Sciences, Cape of Good Hope and the Mascarenes, 1750-1754.
Trained at the Observatoire Royal, Paris.
F.R.S.

La Cardonnie, Jacques Boutier, le chevalier de la (1727-1791).
Navy officer: Garde Marine, 1747, rising to the rank of Commodore (Chef d'Escadre).
Adjunct (1752), then Ordinary Member (1753) of the Académie Royale de Marine.
Cartographer, official scientific voyager to Saint Domingue, 1752-1753.
Captain and head of hydrographical expedition to map the islands north of Saint Domingue, 1768.

La Condamine, Charles-Marie de (1701-1772).**
Astronomer, cartographer, scientific voyager.
Académie Royale des Sciences: Adjunct (chemistry), 1730; Associate (geometry), 1735; Pensionnaire (chemistry), 1739; Pensionnaire Emeritus, 1772; Associate Director, 1748; Director, 1749.
Official voyager to Peru, 1735-1744.
Member of the Académie Française, 1760.
F.R.S.

La Coudraye, François-Célestin de Loynes, chevalier de (1750-1815).
Navy officer, ensign.
Sailed to Saint Domingue, 1770-1771, on test of navy diet and reported on same.
Author, *Théorie des vents et des ondes* (1786).

Lacq.
Médecin du Roi, Cap François, 1750-1758.

Lafitte de Courteil, Denis-Louis-Henry, count (1744-1787).
Military engineer, 1763.
Studied at the École de Génie de Mézières.
Chief Engineer, Cap François, Saint Domingue, and Director-General of Fortifications, 1786-1787.

Lafosse de Laborde, Jean (†1788).**
M.D. (Montpellier).
Docteur en Médecine, Miragoane, Saint Domingue, 1777-1780.
Médecin du Roi, Port-au-Prince, 1780-1785.
Correspondant of the Société Royale de Médecine (1787).

La Galissonnière, Roland-Michel Barrin, Marquis de (1693-1756).
Grandson of the Intendant Bégon.
Navy officer; Captain, 1738; Commodore, 1750.
Lieutenant General of the Navy.
Governor-General of Canada, 1747-1749.
Director, Dépôt des Cartes et Plans de la Marine, 1750-1756.
Académie Royale des Sciences: Free Associate, 1752.
Honorary of the Académie Royale de Marine, 1752.

La Lance, Louis-Joseph de (†1739).
Chief Engineer in Saint Domingue, circa 1730s.

Lalande, Joseph-Jérôme Le François de (1732-1807).**
Chief astronomer of the Navy (*Astronome de la Marine*).
Attached to the Dépôt des Cartes et Plans de la Marine.
Editor, *Connaissances des Temps*.
Académie Royale des Sciences: Adjunct (astronomy), 1753; Associate (astronomy), 1758; Pensionnaire (astronomy), 1772; Associate Director, 1781; Director, 1782.
Professor of Astronomy, Collège Royal, 1761.
Member of the Académie Royale de Marine.
National Associate of the Cercle des Philadelphes, Saint Domingue, 1789.

La Luzerne, César-Henri, comte de (1737-1799).
Nobleman, diplomat, bureaucrat.
Minister of the Navy and the Colonies 1787-1790.
Ambassador to the United States, 1779-1784.
Governor-General of Saint Domingue, 1785-1787.
Académie Royale des Sciences: Honorary, 1788; Vice-President, 1789; President, 1790.
Protector of the Société Royale des Sciences et Arts du Cap François (the Cercle des Philadelphes), 1789-1793.

Lamanon, Jean-Paul de (†1788?).
Mineralogist and meteorologist with the La Pérouse expedition, 1785.

Lamarque, Bernard.
Surgeon-Major and Director of the Eaux de Boynes spa, Saint Domingue, circa 1770.

Lamarque, Michel.
Chirurgien du Roi, Port-au-Prince, Saint Domingue, circa 1786.

La Martinière, Marc de Vaux de (†1716 in Martinique).
M.D.
Médecin du Roi, Martinique, 1691-1716.
Académie Royale des Sciences: Correspondant of Nicolas Lémery, 1699.

La Merveillière, Pierre-Antoine-Jérome Frémond de (1737-1805).
Military engineer and cartographer.
Studied at the École de Génie de Mézières.
Director-General of Fortifications in Saint Domingue, 1788-1792.

Lamotte.
Attendant/custodian, Jardin du Roi, Port-au-Prince, Saint Domingue, 1783.

La Motte-Aigron (†1728).
Navy officer, Lieutenant de Vaisseau and Lieutenant du Roi.
Undertook explorations in Guiana, circa 1688.

Lanux (also La Nux or Nux), Jean-Baptiste-François de (1702-1772).
Colonist/planter, Île Bourbon.
Counselor, Conseil Supérieur of l'Île Bourbon.
Académie Royale des Sciences: Correspondant of Réaumur, 1754; then of de Jussieu, 1757.

Lanux, fils.
Colonist/planter, Île Bourbon.
Amateur astronomer.
Informal correspondent of Pingré and Académie Royale des Sciences.

La Pérouse, Jean-François de Galaup, count de (1741-1788?).
Navy officer, Captain.
Pacific explorer, 1785-1787/1788?

La Peyre, Jean Arnaud.
Surgeon; Army Surgeon-Major for Île de France and Île Bourbon.
Correspondant of Société Royale de Médecine, active circa 1777-1788.

Laporte, Dominique (†1779).
Apothicaire du Roi, Cap François, Saint Domingue, 1774.

La Poterie, Élie de.
Médecin de la Marine; Premier Médecin du Roi, Brest, 1780-1784.
Informal correspondent of the Société Royale de Médecine.

La Roche.
Civil engineer.
In the Dépôt des Cartes et Plans de la Marine, 1772-1807.

La Rozière, Louis-François Carlet, Marquis de.
Military officer.
Member of the Lacaille expedition to the Cape of Good Hope, 1750-1752.

Laurent, Antoine.
Director and Royal Gardener, Navy Botanical Garden, Brest, 1771-1815.
In regular contact with the Jardin du Roi, Paris.

Laval, Antoine (1664-1728).**
Jesuit astronomer.
Founder of the Navy observatory at Marseilles, 1702; Director, 1702-1728.
Professor of Hydrography at Navy training school in Toulon, 1697-1717.
Close connections with the Observatoire Royal, Paris.
Mathématicien du Roi voyaging to Louisiana, 1720.

Lebeau.
Médecin du Roi, Quebec, 1756.

Leblond (also Le Blond), Jean-Baptiste (1747-1815).
Naturalist, traveler in Caribbean and South America, 1766-1785.
Médecin Naturaliste du Roi, Cayenne, 1786-1790, 1792-1803.
Académie Royale des Sciences: Correspondant of Abbé Rochon, 1786.
Correspondant of the Cabinet du Roi.
Correspondant of the Société Royale de Médecine (1786).
Correspondant of the Société Royale d'Agriculture.

Le Blond de la Tour, Louis-Pierre (1673-1723).
Chief engineer, Louisiana, 1719-1723.

Lebrasseur, Joseph-Alexandre (1741-?).
Commissaire de la Marine et des Colonies, 1773.
Administrateur-Général, Gorée, Africa, 1774.
Ordonnateur and Intendant, Cap François, Saint Domingue, 1779-1780.
Ordonnateur and Intendant, Île de France and Île Bourbon, 1785-1787.
Intendant-Général des Fonds de la Marine, 1788-1791.
National Associate of the Cercle des Philadelphes, Saint Domingue, 1787-1790.

Le Breton, Adrien (Jesuit) (1642-??).
Missionary in Martinique and Saint Vincent, 1693....
In contact with the Académie Royale des Sciences.

Le Breton, Paul.
Médecin du Roi, Port-au-Prince, Saint Domingue, 1752-1773.
Inspecteur Général des Hôpitaux et Pharmacies de Saint Domingue, 1769-1773.

Le Crom.
Médecin Privilégié du Roi et Botaniste sent to the Antilles.
Académie Royale des Sciences: Correspondant of Antoine de Jussieu, 1719.

Le Dran, François-Antoine (1690-1724).
Médecin du Roi, Martinique, 1716-1721.
Académie Royale des Sciences: Correspondant of Louis Lémery, 1717.

Lefebvre-Deshayes, Étienne, chevalier (1732-1789).
Colonist/habitant, Tivoli, Saint Domingue; amateur botanist.
Correspondant of Buffon and the Cabinet du Roi, 1778.
Correspondant of the Société Royale de Médecine (1785); medal winner (1785).
Colonial Associate of the Cercle des Philadelphes, Saint Domingue.

Lefort de Latour.
Surveyor General, Saint Lucia, circa 1780.

Le Gentil de la Galaisière, Guillaume-J.-H.-J.-B. (1725-1792).**

Astronomer, voyager.

Académie Royale des Sciences: Adjunct (astronomy), 1753; Associate (astronomy), 1763; Associate Emeritus, 1770; Associate (astronomy), 1772; Supernumerary Pensionnaire (astronomy), 1782; Pensionnaire (astronomy), 1785 ; Assistant Director, 1789; Director, 1790.

Royal astronomer and voyager, Indian Ocean and Pacific, 1760-1771.

Le Grand, Jean Louis (†1730).

Mining assayer employed by the Compagnie de Sénégal.

Le Juge.

Jurist on Île de France.

Founder of private botanical garden, 1750 (Île de France).

In contact with scientific establishment in France.

Lémery, Louis (1677-1743).

M.D. (Paris).

Professor of Chemistry and Pharmacy, Jardin du Roi, Paris, 1730.

Académie Royale des Sciences: Élève (1700); Associate (chemistry), 1715; Associate Director, 1716, 1717.

Lémery, Nicolas (1645-1715).

M.D. (Caen).

Royal Physician; Apothicaire du Roi.

Professor of Chemistry, Jardin du Roi, Paris.

Académie Royale des Sciences: Associate (chemistry), 1699; Pensionnaire (chemistry), 1699; Pensionnaire Emeritus, 1715 ; Director, 1712.

Le Moine.

Intendant in Guiana, circa 1749.

Promotor of medical and botanical researches.

Le Monnier, Louis-Guillaume (Le Monnier the physician) (1717-1799).

M.D.

First Ordinary Physician to the King.

Professor of Botany, Jardin du Roi, Paris 1759-1786.

Head of Trianon Gardens.

Académie Royale des Sciences: Adjunct (botany), 1743; Associate (botany), 1744; Supernumerary Pensionnaire (botany), 1758; Pensionnaire (botany), 1779; Pensionnaire Emeritus, 1779.

Societaire of the Société Royale de Médecine.

F.R.S.

Le Monnier, Pierre-Charles (Le Monnier the astronomer) (1715-1799).**

Navy Astronomer.

Académie Royale des Sciences: Adjunct (geometry), 1736; Associate (geometry), 1741; Pensionnaire (astronomy), 1746; Associate Director, 1751, 1764; Director, 1752, 1765.

Veteran of Maupertuis expedition to Lapland, 1736-1737.

Professor of *Physique Universelle*, Collège Royal, Paris.

Member of the Académie Royale de Marine.

F.R.S.

Le Moyne, François-Pierre (1713-1795).

Cartographer in the Dépôt des Cartes et Plans de la Marine, 1737.

Chief Engineer (Premier Ingénieur et Garde du Dépôt) in the Dépôt des Cartes et Plans de la Marine, 1775-1792.

Le Moyne, Joseph-Antoine (1749-1811).

Engineer and draftsman (Ingénieur Dessinateur), Dépôt des Cartes et Plans de la Marine, 1763.

Engineer, Dépôt des Colonies, 1782-1792.

L'Empereur, Nicolas (1660-1742).

Senior surgeon, Compagnie des Indes, Chandernagor, India.

Recipient of two gold medals, one from king, one from the Regent (1719).

In contact with Académie Royale des Sciences.

In contact with the Jardin du Roi.

Author, ms. *Ellemans botaniques des plante du Jardin de Lorixa* (sic).

Leperchais, Jean Marie (1745- about 1800).

Draftsman in the Dépôt des Colonies, 1784-1800.

Le Roy, Jean-Baptiste (1719-1800).

Académie Royale des Sciences: Adjunct (geometry), 1751; Associate (mechanics), 1766; Pensionnaire (mechanics), 1770; Pensionnaire (general physics), 1785; Associate Director, 1772, 1777; Director, 1773, 1778.

Member of the Académie Royale de Marine.

Le Roy, Pierre-Nicolas (1737-1815).
Ingénieur-Géographe du Roi.
Draftsman (Dessinateur), Dépôt des Cartes et Plans de la Marine, 1772.

Leroy de Bosroger (also Boisroger or Boscroger), Louis-Adam (1737-?).
Ingénieur-Géographe, Saint Domingue, 1763.

Leroy du Fay, Baptiste-Augustin (1744-?).
Brother of Leroy de Bosroger.
Ingénieur-Géographe, Saint Domingue, 1763-1765.

Le Royer.
Surveyor and Superintendent of Roads (Grand Voyer) in Gonaïves, Saint Domingue, circa 1775.

Lestrade.
Médecin du Roi, Martinique, 1791.
Colonial Associate of the Cercle des Philadelphes, Saint Domingue, 1791.

Levasseur de Néré, Jacques (1662?-1723?).
Ingénieur du Roi, Chief engineer, Canada, 1694-1712.

Levavasseur.
Army artillery captain, Cap François, Saint Domingue.
Resident Associate and Officer of the Cercle des Philadelphes, Saint Domingue, 1789-1791.
Director of the Cercle's botanical garden.

Leysses, Jean.
M.D. (Bordeaux), 1779.
Médecin du Roi adjoint, Cap François, 1780.

Lhermite (also Lhermitte), Jacques.
Miltary engineer, Chief Engineer at Louisbourg, Isle Royale, Canada, circa 1703-1715.

L'Hermitte, Raymond-Mathieu.
Hospital administrator, Léogane, Saint Domingue, circa 1750.

L'Huillier de la Serre, Anne-François-Victor (†1790).
Military engineer. Then, Ingénieur-Géographe attached to the Navy, 1764.
Accompanied Bougainville to the Falklands as Ingénieur-Géographe, in 1764.

Garde du Dépôt des Cartes et Plans de la Marine, 1772-1775.
Head of the Corps des Ingénieurs Géographes des Colonies, 1773.
Chief Engineer and Garde du Dépôt des Cartes et Plans des Colonies, 1778.

Lignon, Jean-Baptiste, the younger (1667-1729).
Botaniste du Roi on Guadeloupe, 1699-1729.
Académie Royale des Sciences: Correspondant of Tournefort, 1699.
In contact with the Jardin du Roi, Paris.

Lislet, Jean-Baptiste (Lislet-Geoffroy from 1794) (1755-1836).
Engineer-Geographer, cartographer on Île de France.
Académie Royale des Sciences: Correspondant of the Duke de la Rochefoucauld, 1786.

List, Adam.
Arpenteur du Roi, Môle Saint-Nicolas, Saint Domingue, circa late 1780s.
Colonial Associate of the Cercle des Philadelphes, Saint Domingue, 1784-1791.

Lopez de Pas, Michel.
M.D. (Montpellier).
Médecin du Roi, Petit-Goave, Saint Domingue, 1714-1732.

Loubers (1755-1786).
M.D. (Toulouse).
Médecin du Roi, Basse-Terre, Guadeloupe. Appointed 1786; dies 1786.
Graduate of the l'École de Médecine Pratique de Brest.
Correspondant of the Académie Royale de Marine, Brest.

Loupia Fontenailles, Claude (1721-1788).
Worked with Cassini on the map of France, 1751-1753.
Ingénieur-Géographe, Martinique, 1763-1767.

Luhrmann.
Royal gardener sent to India, 1788.

Macquer, Pierre-Joseph (1718-1784).
M.D.; Doctor-Regent at Paris.
Professor of Chemistry, Jardin du Roi, Paris, 1777-1784.
Chemist, associated with the royal Gobelin dye works.
Ordinary Member of the Société Royale de Médecine.

Académie Royale des Sciences: Adjunct (chemistry), 1745; Associate (chemistry), 1766; Pensionnaire (chemistry), 1772; Associate Director, 1773.
Editor of the *Journal des Savants*.
Royal Censor.

Maillart Dumesle.
Navy Commissioner on Île de France, 1770.
Director of Pamplemousses gardens, Île de France, circa 1774.
Previously, Ordonnateur in Cayenne.

Mallet de la Brossière.
Surgeon-Major (Chirurgien-Major), Juda, Guinea.
Médecin du Roi, Saint-Marc, Saint Domingue.
Correspondant of the Société Royale de Médecine (1777); National Associate, 1781; Medal winner, 1788.
Colonial Associate of the Cercle des Philadelphes, Saint Domingue, 1787-1788.

Mansuy, Sigisbert (1742-?).
Ingénieur-Géographe, Saint Domingue, 1763-1769.
Government engineer, Saint Domingue, 1769-1772.
Chief Engineer, Cap François, Saint Domingue, 1772-1773 (interim).

Mansuy, Charles-Victor.
Son of Sigisbert Mansuy.
Government engineer, Saint Domingue, from 1787.

Maraldi, Jacques-Philippe (1665-1729).**
Royal astronomer, Observatoire Royal, Paris.
Premier Géographe du Roi, 1726-1729,
Académie Royale des Sciences: Élève (astronomy), 1699; Associate (astronomy), 1699; Pensionnaire (astronomy), 1702; Associate Director, 1711.

Maraldi, Jean-Dominique (1709-1788).**
Royal astronomer, Observatoire Royal, Paris.
Académie Royale des Sciences: Adjunct (astronomy), 1731; Associate (astronomy), 1733; Pensionnaire (astronomy), 1758; Pensionnaire Emeritus, 1772.

Marange.
Ingénieur-Géographe, to Madagascar, 1773.

Marchand, Nicolas-Jean-Baptiste (1728?-1764).
Ingénieur of the Ponts et Chaussées.
Ingénieur-Géographe, Saint Domingue, 1763-1764.

Marguerie, Jean-Jacques de (1742-1779).
Navy officer, mathematician.
Secretary of the Académie Royale de Marine.
Director of the Atelier des Bousoles, 1777-1779.

Martin, Joseph.
Royal, Government Botanist, Cayenne, 1790-1810.
Botanical envoy to Indian Ocean and Cayenne, 1788-1789.
Trained at the Jardin du Roi, Paris.
Recepient of gold medal from the Société Royale d'Agriculture.

Maudet, Marie-Rémy.
Navy surgeon in Guadeloupe, circa 1780.

Maupertuis, Pierre-Louis Moreau de (1698-1759).**
Astronomer-voyager.
Académie Royale des Sciences: Adjunct (geometry), 1723; Associate (geometry), 1725; Pensionnaire (geometry), 1731-1746; Pensionnaire Emeritus, 1756; Associate Director, 1735, 1741; Director, 1736, 1742.
Led Academy's expedition to Lapland, 1736-1737.
Navy astronomer attached to the Dépôt des Cartes et Plans de la Marine, 1740-1744.
Member of the Académie Française, 1743.
F.R.S.

Méchain, Pierre-François-André (1744-1804).
Astronomer.
Ingénieur-Hydrographe in the Dépôt des Cartes et Plans de la Marine, 1772.
Navy Astronome-Hydrographe in the Dépôt des Cartes et Plans de la Marine, 1789.
Académie Royale des Sciences: Adjunct (astronomy), 1782; Associate (general physics), 1785.

Mentelle, François-Simon (1731-1799).
Ingénieur-Géographe in Guiana/Kourou, from 1763.

Merdier.
Monk (Father Séraphin), Brothers of Charity.
Superior, Hôpital de la Charité, Cap François, Saint Domingue from 1770.

Developed the botanical garden, Charité hospital of Cap François, 1770s-1780s.

Messier, Charles-Joseph (1730-1817).
Astronomer.
Attached to the Dépôt des Cartes et Plans de la Marine from 1765.
Astronome de la Marine, 1771.
Académie Royale des Sciences: Adjunct (astronomy), 1770; Associate (astronomy), 1782; Pensionnaire (astronomy), 1792.

Meynier (†1740).
Master hydrographer.
Ingénieur de la Marine.
Chief Engineer in Saint Domingue, circa 1739.

Michaux, André (1746-1802).**
Botaniste du Roi, botanical explorer.
In America, 1785-1795.
Founder of botanical gardens in New Jersey (1786) and South Carolina (1787).
Correspondant of Buffon and the Cabinet du Roi, 1779.
Led expedition to Persia and the Caspian Sea, 1782-1785.
With Baudin expedition, dead at Madagascar.
Trained at Trianon gardens and the Jardin du Roi, Paris.
Correspondant of the Société d'Agriculture, 1786.

Milhas.
Médecin du Roi, Saint-Marc (and sometimes in Les Cayes), Saint Domingue, from 1769.

Modley.
Government engineer on Martinique, circa 1742.

Mollin (†1787).
Docteur en médecine, Saint Domingue.
Correspondant of the Société Royale de Médecine (1786).

Monge, Gaspard (1746-1818).
Examiner of Navy cadets, 1783.
Académie Royale des Sciences: Correspondant of Bossut, 1772; Adjunct (geometry), 1780; Associate (general physics), 1785.
Minister of the Navy, 1792-1793.

Mongez, Jean-André (1750-1788?).
Abbé.
Editor of the *Journal de physique*, 1780-1785.
Naturalist accompanying the La Pérouse expedition, 1785.
Correspondant of the Cabinet du Roi.
Ordinary member of the Société d'Agriculture, 1784.
In contact with the Académie Royale des Sciences.

Monier (also Monnier).
Médecin du Roi, Saint-Marc, Saint Domingue, 1784?/1787-1791.
Colonial Associate of the Cercle des Philadelphes, Saint Domingue, 1787-1788.

Monneron, Paul Mérault de (1748-1788).
Military engineer trained at the École de Génie de Mézières.
Posted to Guadeloupe, 1778.
Transferred to Navy engineering department, 1782.
Chief Engineer and Cartographer sailing with the La Pérouse expedition, 1785.

Montaudouin, Daniel-René (1715-1754).
Commercial and slave merchant, Nantes; city counselor.
Académie Royale des Sciences: Correspondant of Bouguer, 1749.
F.R.S.

Montaudouin, Jean-Gabriel, alias La Touche-Montaudouin (1722-1780).
Commercial and slave merchant, Nantes; property owner in Saint Domingue.
Académie Royale des Sciences: Correspondant of Bouguer, February, 1758; then Duhamel du Monceau, August, 1758.

Monteil, François-Aymar, baron de (1725-1787).
Navy officer, rising to Lieutenant Général des Armées Navales, 1783.
Adjunct of the Académie Royale de Marine, 1752; Director in 1778.

Montucla, Jean-Etienne (1725-1799).
Secretary to the Governor-General; astronomer to colony at Kourou. 1764-1765.
Historian of mathematics.

Morainville, de.
Draftsman (Dessinateur) accompanying La Condamine expedition to Peru, 1735.

Moran [Moras?].
Médecin de la Marine.
Correspondant of the Société Royale de Médecine (1791).

Morancy.
Ingénieur-Géographe, Saint Lucia, circa 1764.

Moreau.
Surgeon-Major in Saint-Louis and Port-au-Prince, Saint Domingue, 1759-1774.

Moreau, Jean-Baptiste (1730?-1790).
Ingénieur of the Ponts et Chaussées.
Ingénieur-Géographe, Saint Domingue, 1763-1768.
Ingénieur with irregular gratifications, Les Cayes, Saint Domingue, 1768-1790.

Moreau de Saint-Méry, Méderic-Louis-Élie (1750-1819).**
Lawyer, judge in Saint Domingue; on Conseil Supérieur de Saint Domingue.
Government colonial expert, Paris.
Cercle des Philadelphes/Société Royale du Cap François, Resident Associate, 1784-1791; prize winner, 1790.
Assistant Secretary of the Chambre d'Agriculture of Cap François, 1783-1788.
Correspondant of the Société d'Agriculture, 1787; Ordinary Resident Member, 1789.
Informal correspondent of the Société Royale de Médecine, circa 1785-1791.

Moreau du Temple, René, (1736- ?).
Ingénieur of the Ponts et Chaussées.
Ingénieur-Géographe, Martinique, from 1763. Map in 1770. Still in the colony circa 1783.

Morogues, Sébastien-François Bigot, viscount de (1703-1781).
Navy officer, rising to Captain, Commodore, and Inspector-General of Navy Artillery; Lieutenant General of the Navy.
Académie Royale des Sciences: Correspondant of Dortous de Mairan, 1735; then Duhamel du Monceau, 1763.
Founder and Director of Académie Royale de Marine.

Mouchet.
Ingénieur-Géographe, in India, circa 1750-1754. Maps of Chandernagor.

Mulot.
Royal gardener sent to India, 1788.

Narbonne-Pelet, François-Bernard de (†1775).
Head of the Dépôt des Cartes et Plans de la Marine, 1762-1763.

Nassau, Claude-François.
Royally licensed surveyor (Arpenteur Breveté du Roy) in Guadeloupe, 1770s-1780s.

Naudet.
Surveyor-General of Northern Department, Saint Domingue, circa 1780.

Nectoux, Hypolite (1759-1836).
Royal Botanist in Guiana, 1787.
Botaniste du Roi in Saint Domingue, 1788.
Director, Jardin Royal, then National, Port-au-Prince, Saint Domingue, 1788-1796.
Correspondant of the Société Royale d'Agriculture, 1791.

Noguez.
M.D. (Reims).
Démonstrateur d'Histoire Naturelle and Garde du Cabinet des Drogues, Jardin du Roi, Paris, circa 1720.
Médecin du Roi, Petit Goave, Saint Domingue, 1725-1741, and again 1747-1749.
Médecin du Roi, Cul-de-Sac, Saint Domingue, circa 1750. Retired in 1751.

Nolin, Pierre-Charles.
Abbé.
Controller General of French Nurseries, 1773-1792.

Nouet, Nicolas-Antoine (1740-1811).
Cistersian priest.
Astronomer-chaplain, Observatoire Royal, Paris.
Accompanied Puységur expedition to Saint Domingue, 1784-1785.

Noyer.
Surgeon-major at Cayenne, circa 1787.

d'Oisy, Gabriel-Joseph, chevalier (†1776).
Navy officer, 1744, rising to Captain, 1767.
Head of the Dépôt des Cartes et Plans de la Marine, 1773-1776.
Ordinary Member of Académie Royale de Marine, Brest, 1769.

Ozanne, Nicolas-Marie (1728-1811).
Attached to the Dépôt des Cartes et Plans de la Marine.
Involved in voyage to test chronometers, with Courtanvaux, 1767-1768.
Member of the Académie Royale de Marine.
Dessinateur de la Marine, Brest, 1750.

Ozanne, Pierre (1737-1813).
Navy artist and construction engineer, Brest.
Voyaging with Verdun de La Crenne expedition, 1771-1772.
Voyaging with Borda hydrographical expedition to the Canaries, 1776.
Voyaging with the Puységur expedition, 1784-85.
(Brother of Nicolas Ozanne.)

Pagès, (†1763).
M.D. in Saint Domingue circa 1756.
Médecin du Roi, Cap François, 1759-1763.

Pagès, Pierre-Marie-François, viscount de (1740-1792).**
Navy officer, rising to Captain.
Sailed with Kerguelen to Antarctic waters in 1773.
Académie Royale des Sciences: Correspondant of Duhamel du Monceau, 1781; then of Jeaurat, 1784.
Ordinary Member of the Académie Royale de Marine.
Retired to Saint Domingue. † in Saint Domingue in the course of the slave revolt.

Palisot de Beauvois, Ambroise-Marie-François-Joseph, baron de Beauvois (1752-1820).**
Nobleman, traveler, naturalist.
Académie Royale des Sciences: Correspondant of Cadet AND Fougeroux de Bondaroy, 1783.

In contact with the Jardin du Roi, Paris, and the Bâtiments du Roi.
Colonial/National Associate of the Cercle des Philadelphes, Saint Domingue.

Paquine.
Royal engineer, Cayenne, circa 1690.

Pasteur.
Catholic priest.
Physician and surgeon.
Administrator of Léogane hospital, Saint Domingue, circa 1780.

Patris, Jean-Baptiste.**
M.D. (Reims).
Médecin-Botaniste, Guiana, 1764-1786.
Studied at the Jardin du Roi, Paris.

Pauger, Adrien de (1685?-1726).
"Ingénieur ordinaire," Louisiane, 1720-1726.

Payen.
Government engineer on Martinique, 1680-1692.

Payen, Marc.
Royal engineer in Saint Domingue, 1687-1697 (royal commission in 1694).

Pélays (†1734).
Chemist, assayer.
In Senegal for the Compagnie des Indes, 1730-1734.

Pellerin de la Bruxière (also Pélerin de la Bussière), Jean.
Médecin du Roi sans appointements, Les Cayes, Saint Domingue, 1765-1790?

Périer, Antoine-Alexis (1691-1757).
Director of the Dépôt des Cartes et Plans de la Marine, 1756-1757.

Petit, François-Ignace-Nicolas.
Surveyor General (Arpenteur Général), Western Department, Saint Domingue, 1786.

Petit, Thimothée (†1723).
Royal surveyor, Martinique, 1690-1723.

Peyré, Joseph Bénoît.
M.D. (Montpellier).
Médecin du Roi, Môle Saint-Nicolas, Saint Domingue, 1777.
Médecin du Roi, Port-au-Prince, Saint Domingue, circa 1787-1789.
Informal correspondent of the Société Royale de Médecine, circa 1787-1788.
Founder and Resident Associate of the Cercle des Philadelphes, Saint Domingue, 1784-1789.

Peyssonnel, Jean-André (1694-1759).**
M.D. (Paris).
Botanical envoy to Egypt, 1714-1715.
Botanical envoy to North Africa, 1724.
Médecin-Botaniste du Roi in Guadeloupe (1729-1759).
Académie Royale des Sciences: Correspondant of Étienne-François Geoffroy, 1723; then of Antoine de Jussieu, 1731.
F.R.S.

Pézenas (1692-1776).**
Jesuit (Father Esprit).
Hydrographe du Roi and Royal Professor of Hydrography, Marseilles, 1728-1749.
Académie Royale des Sciences: Correspondant of J.-N. Delisle, 1750; then Lalande, 1767; published in SE.
Director of the Observatoire Royale de la Marine, Marseilles, 1749-1762.
Ordinary Member of the Académie Royale de Marine.

Phelipeaux.
Royal Ingénieur-Géographe, Saint-Domingue, 1763.

Picaudeau Derivières, César.
Ingénieur du Roi in the Antilles, 1737.
Ingénieur en Chef, Martinique, 1745-1763.

Pinard de la Rozière.
Surveyor-General (Arpenteur Général), Saint-Marc, Saint Domingue, 1778.
Inspector of dikes along Artibonite River, Saint Domingue, 1781.

Pingré, Alexandre-Guy (1711-1796).**

Augustinian priest.

Navy astronomer.

Académie Royale des Sciences: Correspondant of P.-C. Le Monnier, 1753; Supernumerary Free Associate, 1756; Free Associate, 1785.

Official voyager to Indian Ocean, 1761.

Involved in voyage to test chronometers with Courtanvaux, 1767-1768.

Official voyager to Saint Domingue, 1769.

Official voyager to Saint Domingue, 1771-1772.

Correspondant of Académie Royale de Marine.

Plumier, Charles (1646-1704).

Priest in Minim order.

Botaniste du Roi.

Made three government-sponsored voyages as naturalist to the Antilles and South America, in 1687, 1689 and 1694.

Poireau de Lance, Jacques-Sébastien-Nicolas (1735?- ?)

Ingénieur of the Ponts et Chaussées.

Ingénieur-Géographe, Saint Domingue, 1763-?.

Poissonnier, Pierre-Isaac (1720-1798).**

M.D. (Paris, 1744); Doctor Regent of Paris faculty.

Counselor of State (Conseiller d'État).

Inspecteur et Directeur Général de la Médecine des Ports et des Colonies, 1764-1791.

Académie Royale des Sciences: Free Associate, 1765.

Professor of Chemistry and Dean, Collège Royal.

Member of the Société Royale d'Agriculture.

Member of the Société Royale de Médecine.

Honorary of the Académie Royale de Marine, Brest.

Royal censor.

In contact with the Jardin du Roi, Paris.

F.R.S.

Poissonnier-Desperrières, Antoine (1723- around 1793).**

M.D. (Paris).

Inspector of Navy and Colonial Hospitals ("Inspecteur des Hôpitaux de la Marine et des Colonies"), 1775.

Royal Physician-Botanist (Médecin-Botaniste) in Saint Domingue from 1748 through 1751.
Ordinary Member of the Société Royale de Médecine.
Royal censor.
Member of the Académie Royale de Marine.

Poivre, Pierre (1719-1786).
Botanical agent for the Compagnie des Indes, 1750s; traveled to Indochina (1749), Manilla (1751), the Moluccas (1753 or 1755).
Intendant on Île de France (1766-1772).
Académie Royale des Sciences: Correspondant of Réaumur, 1754 ; then of Antoine de Jussieu, 1757.

Polchet, Jean-François Hyacinthe (1726-1782).
Military engineer.
Government engineer and Director-General, Môle Saint-Nicolas, Saint Domingue, 1764-1768. Later, "Ingénieur en Chef"

Polony.
M.D. (Montpellier).
M.D. in Saint Domingue, 1766-1772.
Médecin du Roi adjoint, Cap François, 1780.

Pondavy, Julien.
Royal Surveyor (Arpenteur du Roi), Jacmel, Saint Domingue, circa 1737.

Pontis d'Hurtis, Count de (†1745 in Martinique).
Navy Officer.
Académie Royale des Sciences: Correspondant of Duhamel du Monceau, 1739.

Pouppé-Desportes, Jean-Baptiste-René (1704-1748).**
M.D. (Reims). Also studied botany at the Jardin du Roi, Paris.
Médecin du Roi, Saint-Domingue 1732-1748.
Académie Royale des Sciences: Correspondant of Dufay, 1738; then of Bernard de Jussieu, 1745.
In contact with the Jardin du Roi, Paris.

Prat.
Médecin du Roi, Jérémie, Saint Domingue, 1791.

Resident, then Colonial Associate of the Cercle des Philadelphes, Saint Domingue, 1787-1791.

Prat, Jean.
M.D., Montpellier, 1731.
Médecin-Botaniste du Roi, Louisiana, 1735-1746.
In contact with the Jardin du Roi, Paris.

Prat, Louis.
M.D., Montpellier, 1719.
Médecin-Botaniste du Roi, Louisiana, 1725-1735.

Prévost, Guillaume (†1788?).
Botanical artist accompanying the La Pérouse expedition, 1785.

Prévost, Jean-Louis Robert (†1788?).
Natural history artist accompanying the La Pérouse expedition, 1785.
Nephew of G. Prévost.

Pruvost (†1786).
Engineer/draftsman (Ingénieur Dessinateur), Dépôt des Cartes et Plans de la Marine, 1764-1786.

Pruvost, Charles-François (1765-?).
Engineer/draftsman (Ingénieur Dessinateur), Dépôt des Cartes et Plans de la Marine, 1786-1829.

Putod (also Putod De Thievant).
Navy Chirurgien-Major.
Correspondant of the Société Royale de Médecine (1778).

Puységur, Antoine-Hyacinthe-Anne de Chastenet, Count de (1752-1809).
Navy officer, rising to Captain and Major de Vaisseau. Hydrographer.
Captain and head of hydrographical expedition to Saint Domingue, 1784-1785.
Captain of second ship in Borda hydrographical expedition to the Canaries, 1776.
Adjunct academician, Académie Royale de Marine, 1785; Ordinary Member, 1787.

Rallier, Louis-Anne-Esprit (1749-?).
Military engineer, 1768.

Studied at the École de Génie de Mézières.
Posted to Guadeloupe, 1775; then, Saint Domingue, 1781-1784 and 1786-1790.

Raussin.
Government surveyor in the Antilles, 1722.
Arpenteur en Chef, Arpenteur Général des Îles du Vent, 1726 to at least 1739.

Réaumur, René-Antoine Ferchault de (1683-1757).
Académie Royale des Sciences: Élève (geometry), 1708 ; Pensionnaire (mechanics), 1711; Associate Director, 1713, 1718, 1722, 1723, 1726, 1730, 1734, 1739, 1746, and 1752; Directeur 1714, 1716, 1717, 1720, 1724, 1727, 1731, 1735, 1740, 1747, 1753.
Involved in colonial botanical and technological matters.

Regnaudot.
Docteur en Médecine, Guadeloupe.
Correspondant of the Société Royale de Médecine (1778).
Colonial Associate of the Cercle des Philadelphes, Saint Domingue, 1787-1790.

Rémy.
Hospital Surgeon, Cayenne, from at least 1779.
Correspondant of the Société Royale de Médecine (elected 1790).

Renau, Bernard, alias Renau d'Eliçagaray (1652-1719).**
Ingénieur Général de la Marine et Capitaine de Vaisseau, 1691.
Académie Royale des Sciences: Honorary, 1699; Vice-President, 1714, 1716, 1717; President, 1715.
Inspection trip to the Antilles, 1699-1700, as Ingénieur Général de la Marine.

Ribart, Charles-François (1732-?).
Ingénieur of the Ponts et Chaussées.
Ingénieur-Géographe, Saint Domingue, 1763-1766.

Richard, Antoine (1734-1807).
Gardener-Botanist, Trianon Gardens, 1762-1782.
Chief Gardener, Trianon Gardens, from 1782.

Richard, Claude (1705-1784).
Chief Gardener, Trianon Gardens, 1750-1782.

Richard, François (1736-1780).
Médecin du Roi, Port-au-Prince, Saint Domingue, 1774-1780.

Richard, Louis-Claude-Marie (1754-1821).
Naturaliste du Roi, Cayenne, 1781-1785.
Royal naturalist, Martinique, 1785-1789.
Informal correspondent of the Académie Royale des Sciences, 1781-1789.
Studied at Jardin du Roi, Paris.

Richaud, Pierre-Paul.
Surgeon in Guadeloupe, 1753-1755.
Surgeon in Cap François, Saint Domingue, 1755-1763.
In contact with the Académie Royale de Chirurgie.
Chirurgien-Major, Fort-Dauphin, Saint Domingue, from 1765. Still in Saint Domingue circa 1782.

Richer, Jean (1630-1696).**
Astronomer, voyager.
Académie Royale des Sciences: Elève (astronomy), 1666; full academician,1679.
Académie-sponsored voyage to Canada, 1670.
Mission to Cayenne, 1672-1673.
In contact with Observatoire Royal, Paris.

Rigaut (also known as Rigaud), Louis-François (1732-1785).
Navy physician attached to the Royal Navy, Calais, then Saint-Quentin.
Navy Naturalist-Chemist ("Médecin Naturaliste Physicien et Chimiste de la Marine") (1764-1777).
Académie Royale des Sciences: Correspondant of Courtanvaux, 1775; then Macquer, 1781; then Desmarest, 1784.
Correspondant of the Académie Royale de Marine.

Rizzi Zannoni, Giovanni Antonio (1736-1814).
Ingénieur-Géographe et Hydrographe de la Marine attached to the Dépôt des Cartes et Plans de la Marine, 1772-1776.
Premier hydrographe du Dépôt des Cartes et Plans de la Marine in 1772.

Robert, François (†1759).
Inspector-General and Superintendant of Roads (Grand Voyer), Arcahaye, Saint Domingue, circa 1750s.

Robert de Vaugondy, Didier (1723-1786).
Géographe du Roi.

Robert de Vaugondy, Gilles (1688-1766).
Géographe du Roi, 1760.

Rochemore, Henri de (1718-1768).
Directeur Général des Fortifications des Îles du Vent, circa 1764.

Rochon, Alexis-Marie de (1741-1817).**
Navy astronomer and voyager.
Member of the Académie Royale de Marine; Librarian and Keeper of Maps and Instruments ("Garde de cette Bibliothèque et des Instruments astronomiques"), 1765.
Astronome de la Marine, 1766.
Astronomer on official voyage (as Astronome de la Marine attached to the Dépôt des Cartes et Plans de la Marine) to Indian Ocean, 1768-1770.
Astronomer on Kerguelen voyage of 1771.
Directeur d'Optique and Astronome Opticien de la Marine for the Navy from 1787.
Co-Director of the Cabinet de Physique et d'Optique from 1783.
Académie Royale des Sciences: Correspondant of Cassini de Thury, 1767; Adjunct (mechanics), 1771; Associate (mechanics), 1780; Pensionnaire (mechanics), 1783.

Roland.
Médecin Breveté des Armées du Roi, 1784.
Médecin du Roi, Léogane, Saint Domingue, 1787-1789, 1791.
Colonial Associate of the Cercle des Philadelphes, Saint Domingue, 1784-1791.

Rolland.
Ingénieur Pensionné du Roi, Les Cayes, Saint Domingue, 1789-1791.
Colonial Associate of the Cercle des Philadelphes, Saint Domingue, 1787-1791.

Rolland de la Feubraye, Jean-Jacques.
Surveyor, Saint Domingue, from 1754 to 1764.
Ingénieur-Géographe, 1764 (title given by the Governor-General d'Estaing for his cartographical work in the colony).
Arpenteur Général, Southern Department, 1764.

Romain.
Royal Engineer in the Antilles, and Ingénieur en Chef, in Grenada island, circa 1741-1743.

Romainville, Charles Routier de (1742-?).
Royal engineer on Falkland islands, 1764-1766.
Royal engineer and cartographer on Bougainville voyage, 1766-1768.
Chief government engineer on Île Bourbon, circa 1770s.

Romme, Nicolas-Charles (1745-1805).
Royal Professor of Mathematics for Navy Cadets, Rochefort, 1782-1805.
Académie Royale des Sciences: Correspondant of Bézout, 1778.

Rosily, François-Étienne de (1748-1832).
Navy officer, rising to Vice-Admiral.
Adjunct of the Académie Royale de Marine, 1776; Ordinary member, 1784.
Ensign and hydrographer with Kerguelen expedition, 1771-1773.
Director and Inspector-General of the Dépôt des Cartes et Plans de la Marine, 1795.

Rossel, Élisabeth-Paud-Édouard de (1765-1829).
Navy officer and navigator, rising to Captain (1793).
Sailed with d'Entrecasteaux in Indian Ocean, 1785-1789.
Sailed with d'Entrecasteaux expedition to find La Pérouse, 1793-1795.

Rouillé, Antoine Louis, count de Jouy (1689-1761).
Minister of the Navy and the Colonies, 1749-1754.
Académie Royale des Sciences: Honorary, 1751; Vice President, 1752; President, 1753.

Roulin, Barthélémy.
Chirurgien-Major du Roi, Cap François, Saint Domingue, 1784-1791.
A Founder and Resident Associate of the Cercle des Philadelphes, Saint Domingue, 1784-1791.

Sabatier, Antoine-Chaumont (1740-1798).
M.D.
Navy physician, Brest.
Professor of Medicine, Navy training school, Brest.
Correspondant of the Société Royale de Médecine (1778).
Ordinary Member of Académie Royale de Marine.

Sabatier, Rapheaël-Bienvenu (1732-1811).
Académie Royale des Sciences: Adjunct (anatomy), 1773; Associate (anatomy), 1784.
Member of the Société Royale de Médecine.
Member of the Académie Royale de Chirurgie.
Correspondant of the Académie Royale de Marine.
Royal censor.

Saint-Hillier.
Ingénieur Ordinaire, Guadeloupe, 1762-1767.

Saint-Leu. (1668-1700).
Jesuit, missionary to China; died in Mozambique.
Académie Royale des Sciences: Correspondant of Father Thomas Gouye, 1699.

Saint-Mihiel, R.-M. de.
Médecin du Roi, Île de France, circa 1780.
On the Conseil Supérieur of Île de France.

Saint Romes, Charles Durand de.
Military engineer, 1753.
Chief Engineer, Western Department, Saint Domingue, 1758.

Sarrazin (also Sarrasin), Michel (1659-1734).**
M.D. (Reims).
Navy surgeon, Surgeon-Major.
Médecin du Roi, Quebec, 1697-1734.
Counselor, Conseil Souverain of Quebec.
Académie Royale des Sciences: Correspondant of Tournefort, 1699; then Réaumur, 1717.
In contact with the Jardin du Roi, Paris.

Sarrebource de Pontleroy, also (Sarrebrousse de Pontleroy), Nicolas (1717-1802).
Chief engineer, Canada, from 1756.

Saulon, Louis-Auguste (†1764).
Ingénieur-Géographe, Saint Domingue, 1763-1764.

Saunier (also known as Saulnier), Pierre-Paul (1751?-1817).
Royal nurseryman (Pépiniériste du Roi).
Michaux's assistant in New Jersey, 1785-[1800?].

Sauveur.
Government engineer in Saint Domingue, circa 1717.

Sauveur, Joseph (1653-1716).
Royal Professor of Mathematics and Royal Tutor.
Examiner of engineers.
Académie Royale des Sciences: Academician (geometry), 1696; Associate (mechanics), 1699; Associate Emeritus, 1699.
Co-author of the *Neptune français*.

Seniergues, Jean.
Surgeon accompanying La Condamine expedition to Peru, 1735.

Sigalloux, Charles-Emmanuel (1686-1744).
Minim monk (Father Esprit), Marseilles.
Co-director of second Navy observatory in Marseilles with his uncle, Louis Feuillée, 1714; Director, 1732.
Académie Royale des Sciences: Correspondant of Maraldi, 1729.

Silvabelle, Guillaume de Saint-Jacques de (1722-1810).
Director of the Navy Observatory (Directeur de l'Observatoire de la Marine), Marseilles, 1763.
Published in SE.

Sonnerat, Pierre (1748-1814).**
Cousin (removed) of Pierre Poivre.
Voyager, royal emissary in Indian Ocean and Southeast Asia, 1771, 1774-1781.
Navy Commissioner, Pondicherry, India, 1784(?) to 1793.
Académie Royale des Sciences: Correspondant of Adanson, 1774; published in SE.
Correspondant of the Cabinet du Roi.
Informal correspondent of the Société Royale de Médecine, circa 1781-1782.

Sonnini de Manoncourt, Charles-Nicolas-Sigisbert (1751-1812).
Naval officer, naturalist.
Government-sponsored voyager to Guiana, Egypt, and Asia Minor.
Correspondant du Cabinet du Roi.

Sorrel des Rivières, Antoine-François (1737-1816?).
Ingénieur-Géographe, Saint Domingue, 1763-1769.
Government engineer in Saint Domingue, 1769.
Ingénieur du Roi, Port-au-Prince, Saint Domingue, circa 1782.
Supervising engineer for canals and fountains, 1787.

Surian.
Physician and former Mimim monk.
Sent by the abbé Bignon to accompany Plumier expedition to the Americas, 1689-1690.
In contact with the Jardin du Roi, Paris.

Surville, Jean-François-Marie de (1717-1770).
Navy officer, rising to Captain.
Led expedition from India to Peru, 1769-1770.

Tausia-Bcurnos.
Principal surveyor, Port-au-Paix, Saint Domingue, circa 1779.

Tessier, Henri-Alexandre (1741-1837).**
Abbé.
Director, royal gardens, Rambouillet, France, 1784-1792.
Académie Royale des Sciences: Adjunct (botany), 1783; Associate (natural history), 1785.
Ordinary member of the Société Royale de Médecine.
Member of the Société Royale d'Agriculture.
National Associate of the Cercle des Philadelphes, Saint Domingue, 1789.
In contact with the Jardin du Roi, Paris.

Thévenard, Antoine-Jean-Marie, count (1733-1815).
Captain with the Compagnie des Indes, 1758-1769.
Hydrographer in Labrador and Newfoundland, 1751-1755.
Commandant at Lorient for the Compagnie des Indes, 1764-1769.

Navy officer (1770), rising to Vice-Admiral and Commodore, 1784.
Minister of the Navy, 1791.
Ordinary Member of the Académie Royale de Marine, 1771.
Académie Royale des Sciences: Correspondant of Borda, 1778.

Thevenet, François (1731-1780).
Ingénieur-Géographe, Guadeloupe, from 1763. Map of Guadeloupe in 1769.
Government engineer, Port-au-Prince, Saint Domingue, 1776.

Thibaud.
Médecin du Roi, Dunkerque, 1780s.
National Associate of the Cercle des Philadelphes, Saint Domingue, 1787-1791.

Thibault de Chanvalon, Jean-Baptiste-Mathieu (1725-1788).**
Born in Martinique.
Jurist, naturalist, botanical amateur, traveler.
Intendant at Kourou, 1763-1765.
Académie Royale des Sciences: Correspondant of Réaumur, 1754; then of Antoine de Jussieu, 1757.

Thiery de Menonville, Nicolas-Joseph (1739-1780).**
Lawyer, colonist in Saint Domingue, 1774.
Botaniste du Roi, Port-au-Prince, Saint Domingue, 1777.
Government-sponsored covert voyage to Mexico, 1777.
Founder of the Jardin Royal, Port-au-Prince, 1777-1780.
Studied at the Jardin du Roi, Paris.

Thouin, André (1747-1824).
Chief Gardener, Jardin du Roi, Paris 1764-1824.
Académie Royale des Sciences: Associate (botany), 1786.
Member of the Société d'Agriculture, 1783.

Tilleul (or Dutilleul) (†1787).
Royal Veterinarian, Saint Domingue, 1782-1787.
Graduate of the royal veterinary school, the École d'Alfort.

Tinglek.
Vétérinaire Breveté du Roi, Guadeloupe, 1785-1787.
Colonial Associate of the Cercle des Philadelphes, Saint Domingue, 1787-1788.
Graduate of the royal veterinary school, the École d'Alfort.

Tondu.
Astronomical assistant to the Intendant, Guadeloupe, circa 1781-1784.
In contact with the Observatoire Royal, Paris.

Tosgobbi, François-Antoine (†1772).
Médecin du Roi, Martinique, 1755-1767.
Médecin du Roi, Les Cayes, Saint Domingue, 1770-1772.

Tournefort, Joseph Pitton de (1656-1708).**
M.D. (Paris).
Professor of Botany, Jardin du Roi, Paris.
Official botanical traveler to the Levant, 1700-1702.
Académie Royale des Sciences: Academician (botany), 1691; Pensionnaire (botany), 1699.
Professor, Collège Royal, 1702.

Turgot, Etienne-François, Chevalier de (1721-1788).**
Governer-General in Cayenne (Kourou), 1763-1765.
Académie Royale des Sciences: Free Associate, 1765.
Founding member of the Société d'Agriculture, Paris.

Vacherot.
Chirurgien-Major, Jacmel, Saint Domingue.
Correspondant of the Société Royale de Médecine.

Vaivre, Claude-Marie-Henri (1741-1775).
Engineer, 1763.
Government engineer in Martinique, 1766-1771.
Government engineer in Saint Domingue, 1773-1775.

Vaivre, Jean-Baptiste Guillimin de (†1818).
Head of the Colonial Department within the Ministry of the Navy, 1782.
Intendant in Saint Domingue, 1774-1782.

Valincour, Jean-Baptiste-Henry du Trousset de (1653-1730).
Secrétaire Général de la Marine.
Académie Royale des Sciences: Honorary, 1721.
Member of the Académie Française, 1699.
Academician at the Académie Royale des Inscriptions et Belles-Lettres.

Vallière, Jean-Florent de (1667-1759).
Lieutenant-Général d'Artillerie.

Académie Royale des Sciences: Free Associate, 1731.
Member of the Académie Royale de Marine.

Vallière, Louis-Florent, Marquis de (1719-1775).
Military officer.
Directeur de l'Artillerie et du Génie, 1747.
Governor-General of Saint Domingue, 1772-1775.

Varaigne.
Royal Engineer/Ingénieur of the Ponts et Chaussées; hydrological engineer.
On mission in Saint Domingue, 1777.

Verdun de la Crenne, Jean-René-Antoine, Marquis de (1741-1805).**
Navy officer, rising to Commodore.
Associate Member of the Académie Royale de Marine, 1771; Ordinary Member, 1777.
Captained two voyages (1768, 1771) to test marine chronometers.
Correspondant of the Académie des Sciences Morales et Politiques (Section de Géographie), 1796.

Vergnies.
Médecin du Roi, Guadeloupe, circa.1784.

Verguin, Jean-Joseph (1701-1777).
Port Surgeon and Navy Engineer, Toulon; Director of Navy medical school, Toulon, 1766-1773.
Académie Royale des Sciences: Correspondant of Bouguer, 1746, then La Condamine, 1758.
Accompanied Bouguer, Godin, and La Condamine to Peru as « ingénieur de la marine ».
Member of the Académie Royale de Marine.

Véron, Pierre-Antoine (1736-1770).
Astronome de la Marine.
Astronomer with the Bougainville expedition, 1766-1769; dies on Ile de France.

Verret, Jean-Joseph (1735- ?).
Hydrological engineer in Saint Domingue, from 1773.
Ingénieur-Hydraulicien du Roi, Saint Domingue, 1783-1793.

Resident then Colonial Associate of the Cercle des Philadelphes, Saint Domingue, 1785-1793.
In contact with the Académie Royale des Sciences.

Verrier, Étienne (1683-1747).
Chief engineer at Louisbourg, Isle Royale, Canada, 1725-1745.

Verville, (1707-1751).
Son of Jean-François de Verville.
Army engineer.
Directeur Général des Fortifications de Saint-Domingue, 1750-1751.

Verville, Jean-François de (1685?-1729).
Directeur des Fortifications, Isle Royale, Canada, 1715-1725.

Vicq-Azyr, Félix (1748-1794).
M.D. (Paris).
Royal physician.
Professor of Comparative Anatomy, royal veterinary school at Alfort.
Académie Royale des Sciences: Adjunct (anatomy), 1774; Associate (anatomy), 1784.
Permanent secretary of the Société Royale de Médecine.
Member of the Académie Française, 1788.

Vielle (also known as Alexandre), Bernard Alexandre.
French surgeon in Louisiana, 1720s and 1730s.
Académie Royale des Sciences: Correspondant of Dortous de Mairan, 1722.

Villaire, Jean-Baptiste-Philibert Godino de.
Military engineer.
Director-General of Fortifications, Saint Domingue, 1787.

Villeneuve, Robert de.
Chief engineer, Canada, 1685-1693.

Visdelou (1656-1737).
Jesuit (Father Claude).
Geographer and orientalist.

Missionary to China. Died at Pondicherry, India.

Académie Royale des Sciences: Correspondant of Father Thomas Gouye, 1699.

Willemet, Pierre-Rémy (†~1791).

Royal physician sent to India, 1788.

Correspondant of the Cabinet du Roi.

Worlock, Siméon.

Docteur en médecine, Cap François, Saint Domingue.

Inoculator.

Correspondant of the Société Royale de Médecine (1784); medal winner (1785).

Resident Associate of the Cercle des Philadelphes, Saint Domingue, 1785-1791.

Ycard.

Docteur en Médecine, Cap François, Saint Domingue.

Doctor for Maisons de Providence and Charité.

Correspondant of the Société Royale de Médecine (1787); medal winner (1787).

Resident Associate of the Cercle des Philadelphes, Saint Domingue, 1790-1790.

Yon.

Jesuit.

Apothecary for Jesuit mission in Martinique, circa 1700.

In contact with the Académie Royale des Sciences.

List of Illustrations

[*] Private collection

Bibliography

Contemporary Sources

Académie Royale de Marine. 1781. *Catalogue des livres de l'Académie royale de marine fait en 1781*. Brest: R. Malassis.

__________. 1773. *Mémoires de l'Académie royale de Marine. Tome Premier.* Brest: Chez R. Malassis, Imprimeur ordinaire du Roi & de la Marine.

Adanson, Michel. 1757. *Histoire naturelle du Sénégal.* Paris: Claude-Jean-Baptiste Buache.

__________. 1755. "Latitude de Podor," *SE,* vol. 2, pp. 605-6.

Après de Mannevillette, J.-B.-N. Denis d'. 1781. *Supplément au Neptune oriental.* Paris: Demonville; Brest: Malassis.

__________. 1775a. *Instructions sur la navigation des Indes orientales et de la Chine, pour servir au Neptune oriental.* Paris: Demonville.

__________. 1775b. *Le Neptune oriental dédié au Roi.* Paris: Demonville; Brest: Malassis.

__________. 1774. "Observation de l'Eclipse du Soleil du 16 Août 1765, Faite au château de Kergars," *SE*, vol. 6, pp. 81-82.

__________. 1773. "Observations Astronomiques faites à la Chine," *Mémoires de l'Académie royale de Marine. Tome Premier*, pp. 295-304.

__________. 1768. "Mémoire sur la navigation de France aux Indes," *SE*, vol. 5, pp. 190-232.

__________. 1765. *Mémoire sur la Navigation de France aux Indes*. Paris: Imprimerie Royale.

__________. 1763. "Relation d'un Voyage aux Isles de France & de Bourbon, qui contient plusieurs Observations Astronomiques, tant pour la recherche des Longitudes sur mer, que pour déterminer la position géographique de ces Isles," *SE*, vol. 4, pp. 399-457.

__________. 1745a. *Le Neptune oriental ou Routier général des côtes des Indes Orientales et de la Chine, enrichi de cartes hydrographiques, tant générales que particulières, pour servir d'instruction à la Navigation de ces différentes mers, dédié à Monseigneur Orry de Fulvy.* Paris: Jean-François Robustel.

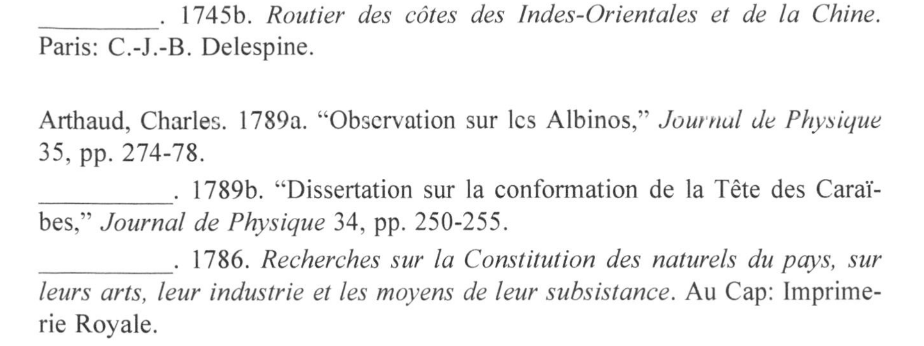

__________. 1745b. *Routier des côtes des Indes-Orientales et de la Chine.* Paris: C.-J.-B. Delespine.

Arthaud, Charles. 1789a. "Observation sur les Albinos," *Journal de Physique* 35, pp. 274-78.

__________. 1789b. "Dissertation sur la conformation de la Tête des Caraïbes," *Journal de Physique* 34, pp. 250-255.

__________. 1786. *Recherches sur la Constitution des naturels du pays, sur leurs arts, leur industrie et les moyens de leur subsistance.* Au Cap: Imprimerie Royale.

Artur, Jacques François. 2002. *Histoire des colonies françoises de la Guianne, Transcription établie, présenté et annotée par Marie Polderman.* Guadeloupe, Guyane, Martinique, Réunion, Paris: IBIS ROUGE EDITIONS.

Badier, Barthélemy de. 1788. "Observations sur différentes espèces de cotonniers cultivées à la Guadeloupe," in Société Royale d'Agriculture, *Mémoires d'agriculture, d'économie rurale et domestique, trimestre d'automne*, pp. 118-31.

Bailly, Jean-Sylvain. 1787. *Traité de l'astronomie indienne et orientale ouvrage qui peut servir de suite à l'histoire de l'astronomie ancienne.* Paris: Debure l'aîné.

Bajon, Bertrand. 1777-1778. *Mémoire pour servir à l'histoire de Cayenne, et de la Guiane françoise, Dans lesquels on fait connoître la nature du Climat de cette contrée, les Maladies qui attaquent les Européens nouvellement arrivés, & celles qui régnent sur les Blancs & les Noirs; des Observations sur l'Histoire naturelle du pays, & sur la culture des Terres. Avec des Planches. Par M. Bajon, ancien Chirurgien Major de l'Isle de Cayenne & Dépendances, Correspondant de l'Académie Royale des Sciences de Paris & de celle de Chirurgie*, 2 vols. Paris: Grangé, Veuve Duchesne et L'Esprit.

Barrère, Pierre. 1743. *Nouvelle Relation de la France équinoxiale, contenant la description des côtes de la Guiane, de l'île de Cayenne, le commerce de cette colonie, les divers changements arrivés dans ce pays, et les moeurs et coutumes des différents peuples sauvages qui l'habitent; avec les figures dessinées sur les lieux.* Paris: Piget.

__________. 1741a. *Dissertation sur la cause physique de la couleur des nègres, de la qualité de leurs cheveux, et de la dégénération de l'un et de l'autre, par M.****. Paris: P.-G. Simon.

__________. 1741b. *Essai sur l'histoire naturelle de la France équinoxiale, ou Dénombrement des plantes, des animaux et des minéraux qui se trouvent dans l'isle de Cayenne, les isles de Remire, sur les côtes de la mer et dans le continent de la Guyane*. Paris: Piget.

Bellin, Jacques-Nicolas. 1763. *Description géographique de la Guyane, contenant les possessions et les établissemens des François, des Espagnols, des Portugais, des Hollandais dans ces vastes pays. Le Climat les Productions de la Terre et les Animaux Leurs Habitans, leurs Mœurs, leurs Coutumes. Et le Commerce qu'on y peut faire. Avec des Remarques pour la Navigation et des Cartes, Plans et Figures. Dressées au Dépost des Cartes et Plans de la Marine Par Ordre de M. le Duc de Choiseul Colonel Général des Suisses et Grisons, Ministre de la Guerre et de la Marine. Par le s. Bellin, Ingénieur de la Marine et du Depost des Plans, Censeur Royal de l'Académie de Marine et de la Société Royale de Londres.* Paris, Imprimerie de Didot.

Bernardin de Saint-Pierre, Jacques-Henri. 1840. *Harmonies de la nature* [in *Œuvres posthumes,* vol 2, pp. 49-377]. Paris: Ledentu. Originally published in 1815.

__________. 1789. *Paul et Virginie*. Paris: Imprimerie de Monsieur.

Bernouilli, Johann. 1714. *Essay d'une nouvelle théorie de la manoeuvre des vaisseaux, avec quelques lettres sur le même sujet*. Basel: J. G. König.

Bertin, Antoine de. 1786. *Des Moyens de conserver la santé des Blancs et des Nègres aux Antilles, ou Climats chauds et humides de l'Amérique, contenant un exposé des causes des maladies propres à ces climats et à la traversée, relativement à la différence des positions, des saisons et des températures; les procédés à suivre, soit pour les éviter, soit pour les détruire, et le traitement, en particulier, de quelques maladies communes chez les Nègres telles que le pian, le mal d'estomac et la lèpre*. Saint Domingue; Paris: chez Méquignon l'aîné.

__________. 1778-1780. *Mémoire sur les Maladies de la Guadeloupe et ce qui peut y avoir rapport* [2 volumes]. Guadeloupe: J. Bernard.

Bézout, Étienne. 1768-1771. *Cours de mathématiques,* 4 vols. Paris: J. B. G. Musier fils.

Bigot de Morogues, S.-F. 1750. "Mémoire sur la corruption de l'air dans les Vaissseaux," *SE,* vol. 1, pp. 394-410.

Bonnet, Charles. 1783. *Essai de psychologie* [reprinted in *Œuvres d'histoire naturelle et de philosophie*, vol. 8]. Neuchâtel: S Fauche. Originally published in 1754.

Borda, Jean-Charles, chevalier de. 1787. *Description et usage du Cercle de réflexion, avec différentes méthodes pour calculer les observations nautiques, par le Chevalier de Borda, Capitaine de Vaisseau, chef de Division, et Membre des Académies royales des Sciences et de la Marine*. Paris: Imprimerie de Didot l'Aîné.

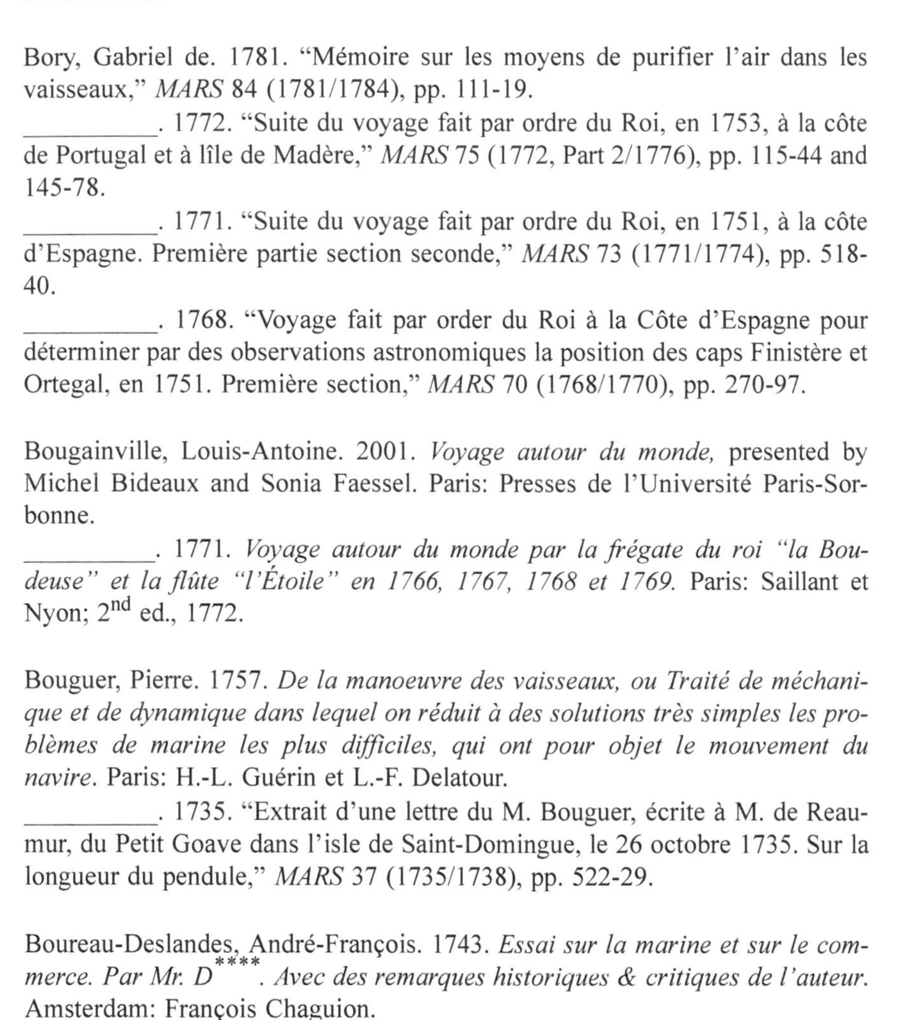

Bory, Gabriel de. 1781. "Mémoire sur les moyens de purifier l'air dans les vaisseaux," *MARS* 84 (1781/1784), pp. 111-19.
__________. 1772. "Suite du voyage fait par ordre du Roi, en 1753, à la côte de Portugal et à lîle de Madère," *MARS* 75 (1772, Part 2/1776), pp. 115-44 and 145-78.
__________. 1771. "Suite du voyage fait par ordre du Roi, en 1751, à la côte d'Espagne. Première partie section seconde," *MARS* 73 (1771/1774), pp. 518-40.
__________. 1768. "Voyage fait par order du Roi à la Côte d'Espagne pour déterminer par des observations astronomiques la position des caps Finistère et Ortegal, en 1751. Première section," *MARS* 70 (1768/1770), pp. 270-97.

Bougainville, Louis-Antoine. 2001. *Voyage autour du monde,* presented by Michel Bideaux and Sonia Faessel. Paris: Presses de l'Université Paris-Sorbonne.
__________. 1771. *Voyage autour du monde par la frégate du roi "la Boudeuse" et la flûte "l'Étoile" en 1766, 1767, 1768 et 1769.* Paris: Saillant et Nyon; 2nd ed., 1772.

Bouguer, Pierre. 1757. *De la manoeuvre des vaisseaux, ou Traité de méchanique et de dynamique dans lequel on réduit à des solutions très simples les problèmes de marine les plus difficiles, qui ont pour objet le mouvement du navire*. Paris: H.-L. Guérin et L.-F. Delatour.
__________. 1735. "Extrait d'une lettre du M. Bouguer, écrite à M. de Reaumur, du Petit Goave dans l'isle de Saint-Domingue, le 26 octobre 1735. Sur la longueur du pendule," *MARS* 37 (1735/1738), pp. 522-29.

Boureau-Deslandes, André-François. 1743. *Essai sur la marine et sur le commerce. Par Mr. D****. Avec des remarques historiques & critiques de l'auteur.* Amsterdam: François Chaguion.
__________. 1730. "Observation sur l'eau de mer et sur l'eau douce qu'on embarque dans les vaisseaux," *Mémoires de Trévoux*, March, 1730, pp. 409-23.

Breton, Raymond. 1999. *Dictionnaire caraïbe-français [avec cédérom]. Révérend Père Raymond Breton 1665.* Edition presented by the CELIA and the GEREC. Paris: IRD-Karthala.

__________. 1665. *Dictionnaire caraïbe-français mêlé de remarques historiques pour l'éclaircissement de la langue*. Auxerre: G. Bouquet.

Buache, Philippe. 1764. "Observations Géographiques sur les Isles de France et de Bourbon, Comparées l'une avec l'autre," *MARS* 66 (1764/1767), pp. 1-6.

Buffon, Georges-Louis Leclerc, comte de. 1749-1804. *Histoire naturelle, générale et particulière*, 36 vols. Paris: Imprimerie Royale; then Paris: Plassan.

Campet, Pierre. 1802. *Traité pratique des maladies graves qui règnent dans les contrées situées sous la zone torride et dans le midi de l'Europe*. Paris: Bossange, Masson et Besson.

Cassan, Jean-Baptiste. 1803a. "Observations météorologiques faites sous la Zone torride." *Mémoires de la Société médicale d'émulation*, vol. 5, pp. 162-163.

__________. 1803b. "Premier mémoire. De la manière d'agir des climats chauds sur l'économie animale," *Mémoires de la Société médicale d'émulation*, vol. 5, pp. 27-30.

__________. 1790. "Observations météorologiques faites sous la zone torride," *Observations sur la Physique, sur l'histoire naturelle et sur les arts* 36, p. 263.

__________. 1789. "Mémoire sur les cultures de l'isle de Sainte-Lucie, contenant des observations sur les productions qui servent à la nourriture des Nègres; les denrées commerçables des Isles du Vent, leurs produits; les nouvelles cultures introduites dans les colonies, & sur les moyens de les faire réussir," Société Royale d'Agriculture, *Mémoires d'agriculture, d'économie rurale et domestique, Trimestre d'été*, pp. 60-109.

Cassini, Jacques (Cassini II). 1731. "Extrait de diverses observations astronomiques faites à la Louisiane par M. Baron, Ingénieur du Roy. Comparées à celles qui ont été faites à Paris & à Marseille," *MARS* 33 (1731/1733), pp. 163-67.

__________. 1708. "Extrait des observations faites aux Indes occidentales en 1704, 1705, et 1706 par le P. Feuillée, Mimine, Mathematicien du Roy; comparées à celles qui ont été faites en même tems à l'Observatoire Royal," MARS 10 (1708/1709), pp. 17-31.

Cassini, Jean-Dominique (Cassini I). 1708. "Reflexions sur les Observations de la variation de l'Aiman, faites sur le Vaisseau le Maurepas dans le voyage de la mer au Sud," *MARS* 10 (1708/1709), pp. 292-96.

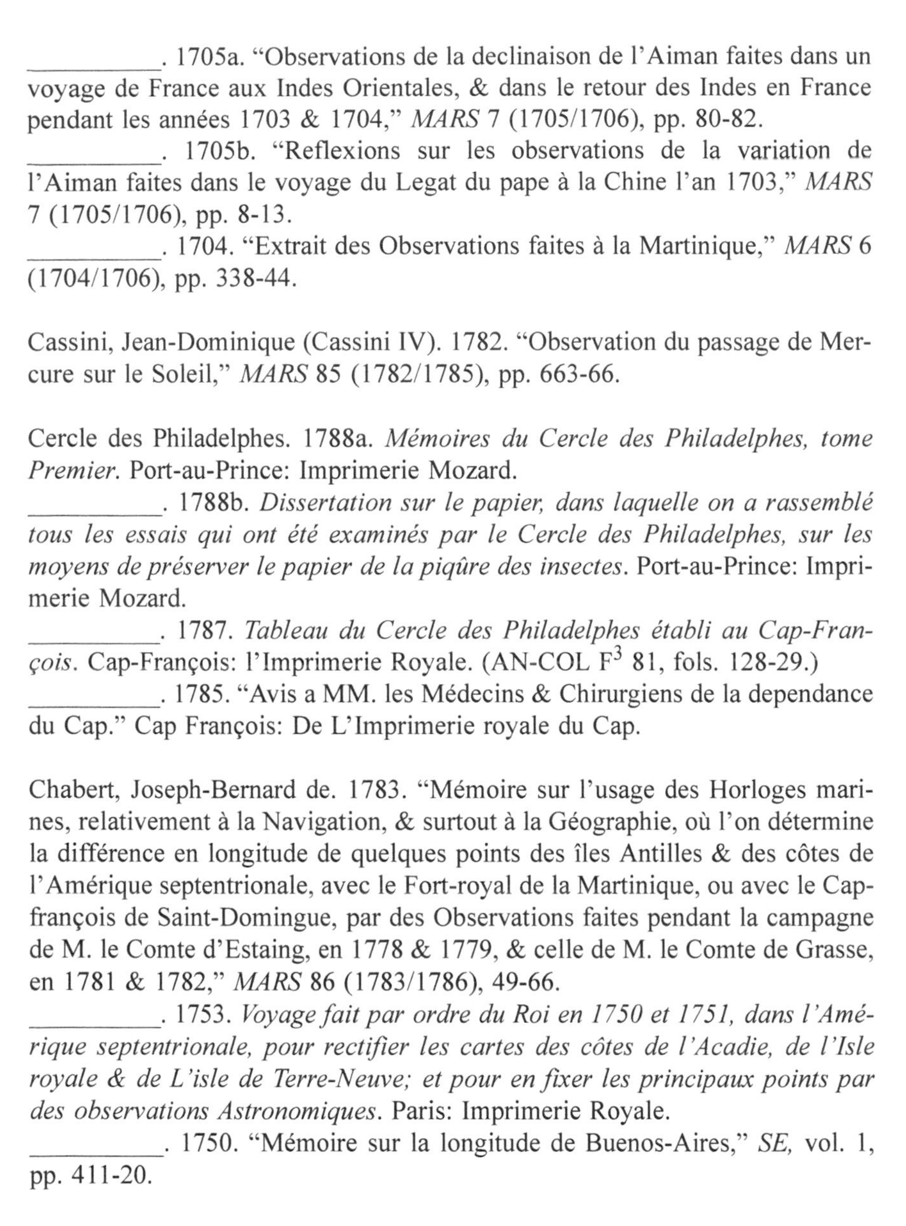

__________. 1705a. "Observations de la declinaison de l'Aiman faites dans un voyage de France aux Indes Orientales, & dans le retour des Indes en France pendant les années 1703 & 1704," *MARS* 7 (1705/1706), pp. 80-82.
__________. 1705b. "Reflexions sur les observations de la variation de l'Aiman faites dans le voyage du Legat du pape à la Chine l'an 1703," *MARS* 7 (1705/1706), pp. 8-13.
__________. 1704. "Extrait des Observations faites à la Martinique," *MARS* 6 (1704/1706), pp. 338-44.

Cassini, Jean-Dominique (Cassini IV). 1782. "Observation du passage de Mercure sur le Soleil," *MARS* 85 (1782/1785), pp. 663-66.

Cercle des Philadelphes. 1788a. *Mémoires du Cercle des Philadelphes, tome Premier.* Port-au-Prince: Imprimerie Mozard.
__________. 1788b. *Dissertation sur le papier, dans laquelle on a rassemblé tous les essais qui ont été examinés par le Cercle des Philadelphes, sur les moyens de préserver le papier de la piqûre des insectes.* Port-au-Prince: Imprimerie Mozard.
__________. 1787. *Tableau du Cercle des Philadelphes établi au Cap-François.* Cap-François: l'Imprimerie Royale. (AN-COL F^3 81, fols. 128-29.)
__________. 1785. "Avis a MM. les Médecins & Chirurgiens de la dependance du Cap." Cap François: De L'Imprimerie royale du Cap.

Chabert, Joseph-Bernard de. 1783. "Mémoire sur l'usage des Horloges marines, relativement à la Navigation, & surtout à la Géographie, où l'on détermine la différence en longitude de quelques points des îles Antilles & des côtes de l'Amérique septentrionale, avec le Fort-royal de la Martinique, ou avec le Cap-françois de Saint-Domingue, par des Observations faites pendant la campagne de M. le Comte d'Estaing, en 1778 & 1779, & celle de M. le Comte de Grasse, en 1781 & 1782," *MARS* 86 (1783/1786), 49-66.
__________. 1753. *Voyage fait par ordre du Roi en 1750 et 1751, dans l'Amérique septentrionale, pour rectifier les cartes des côtes de l'Acadie, de l'Isle royale & de L'isle de Terre-Neuve; et pour en fixer les principaux points par des observations Astronomiques.* Paris: Imprimerie Royale.
__________. 1750. "Mémoire sur la longitude de Buenos-Aires," *SE,* vol. 1, pp. 411-20.

Chardon, Daniel M. 1779. *Essai sur la colonie de Sainte-Lucie. Par un ancien intendant de cette île. Suivi de trois mémoires intéressans, deux concernant les Jésuites, et le troisième le Général d'Oxat.* Neuchâtel: Société typographique.

Charlevoix, Pierre-François-Xavier de. 1744. *Histoire et description générale de la Nouvelle France avec le Journal Historique d'un Voyage fait par ordre du roi dans l'Amérique septentrionale,* 3 vols. Paris: Nyon fils.

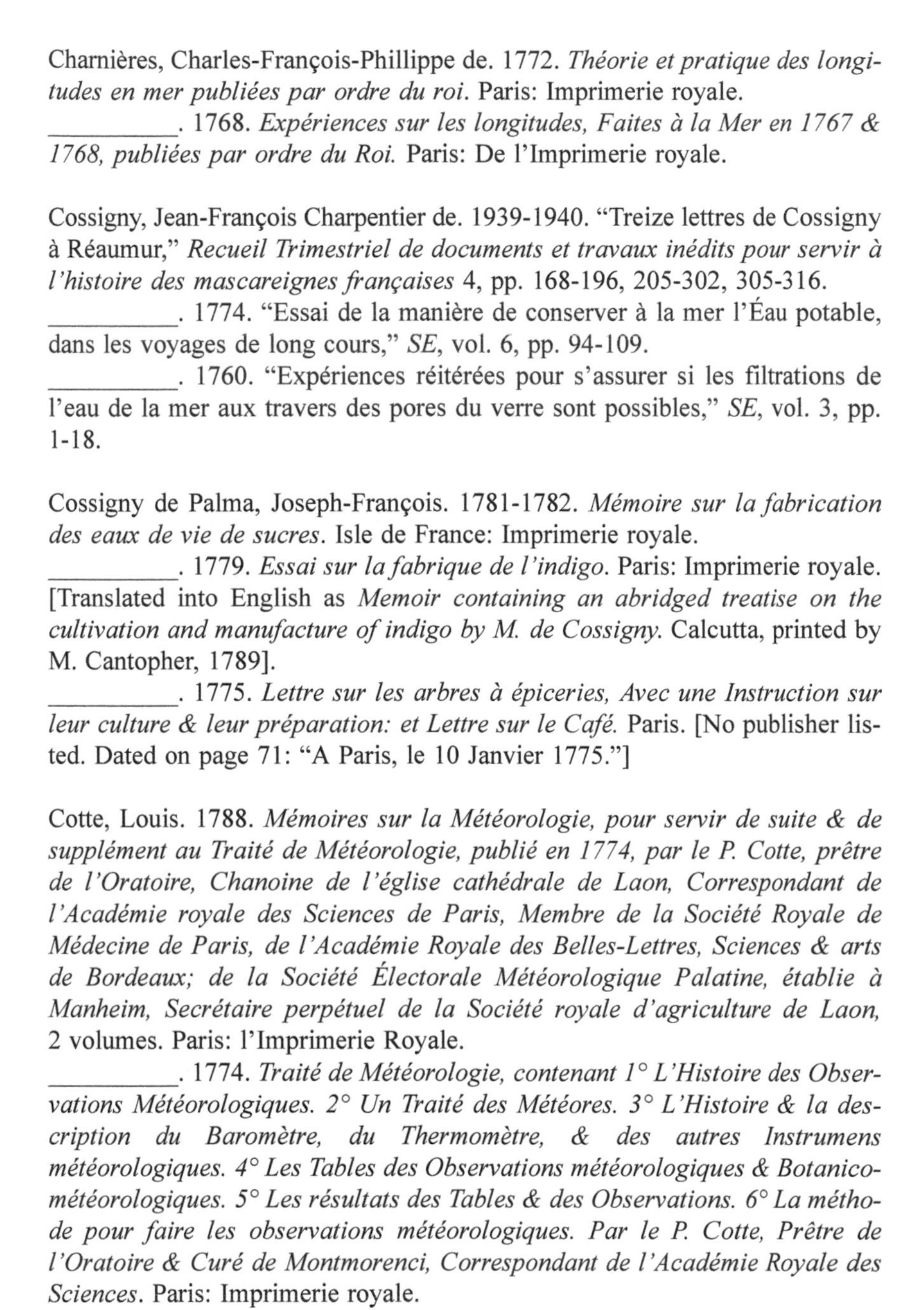

Charnières, Charles-François-Phillippe de. 1772. *Théorie et pratique des longitudes en mer publiées par ordre du roi*. Paris: Imprimerie royale.

_________. 1768. *Expériences sur les longitudes, Faites à la Mer en 1767 & 1768, publiées par ordre du Roi.* Paris: De l'Imprimerie royale.

Cossigny, Jean-François Charpentier de. 1939-1940. "Treize lettres de Cossigny à Réaumur," *Recueil Trimestriel de documents et travaux inédits pour servir à l'histoire des mascareignes françaises* 4, pp. 168-196, 205-302, 305-316.

_________. 1774. "Essai de la manière de conserver à la mer l'Éau potable, dans les voyages de long cours," *SE*, vol. 6, pp. 94-109.

_________. 1760. "Expériences réitérées pour s'assurer si les filtrations de l'eau de la mer aux travers des pores du verre sont possibles," *SE*, vol. 3, pp. 1-18.

Cossigny de Palma, Joseph-François. 1781-1782. *Mémoire sur la fabrication des eaux de vie de sucres*. Isle de France: Imprimerie royale.

_________. 1779. *Essai sur la fabrique de l'indigo*. Paris: Imprimerie royale. [Translated into English as *Memoir containing an abridged treatise on the cultivation and manufacture of indigo by M. de Cossigny*. Calcutta, printed by M. Cantopher, 1789].

_________. 1775. *Lettre sur les arbres à épiceries, Avec une Instruction sur leur culture & leur préparation: et Lettre sur le Café.* Paris. [No publisher listed. Dated on page 71: "A Paris, le 10 Janvier 1775."]

Cotte, Louis. 1788. *Mémoires sur la Météorologie, pour servir de suite & de supplément au Traité de Météorologie, publié en 1774, par le P. Cotte, prêtre de l'Oratoire, Chanoine de l'église cathédrale de Laon, Correspondant de l'Académie royale des Sciences de Paris, Membre de la Société Royale de Médecine de Paris, de l'Académie Royale des Belles-Lettres, Sciences & arts de Bordeaux; de la Société Électorale Météorologique Palatine, établie à Manheim, Secrétaire perpétuel de la Société royale d'agriculture de Laon,* 2 volumes. Paris: l'Imprimerie Royale.

_________. 1774. *Traité de Météorologie, contenant 1° L'Histoire des Observations Météorologiques. 2° Un Traité des Météores. 3° L'Histoire & la description du Baromètre, du Thermomètre, & des autres Instrumens météorologiques. 4° Les Tables des Observations météorologiques & Botanico-météorologiques. 5° Les résultats des Tables & des Observations. 6° La méthode pour faire les observations météorologiques. Par le P. Cotte, Prêtre de l'Oratoire & Curé de Montmorenci, Correspondant de l'Académie Royale des Sciences*. Paris: Imprimerie royale.

Courcelles, Étienne Chardon de. 1756. *Manuel des opérations les plus ordinaires de la chirurgie, pour l'instruction des élèves-chirurgiens de la marine de l'École de Brest*. Brest: R. Malassis.

__________. 1752-1753. *Abrégé d'anatomie pour l'instruction des élèves-chirurgiens de la marine de l'École de Brest,* 4 vols. Brest: R. Malassis.

__________. 1746. *Manuel de la saignée pour l'instruction des élèves-chirurgiens de la marine de l'École de Brest.* [No publisher or place of publication; second edition, 1751].

Courtanvaux, François-César Le Tellier de. 1768. *Journal du voyage de M. le marquis de Courtanvaux, sur la frégate l'Aurore, pour essayer par ordre de l'Académie, plusieurs instruments relatifs à la longitude. Mis en ordre par M. Pingré, chanoine régulier de Ste-Geneviève, nommé par l'Académie pour coopérer à la vérification desdits instrumens, de concert avec M. Messier, astronome de la Marine.* Paris: Imprimerie royale.

Dazille, Jean-Barthélemy. 1788. *Observations sur le Tétanos, précédées d'un discours sur les moyens de perfectionner la Médecine-Pratique sous la zone torride.* Paris: Planche.

__________. 1785. *Observations Générales sur les Maladies des climats chauds, leurs causes, leurs traitements et les moyens de les prévenir.* Paris: Didot.

__________. 1776. *Observations sur les Maladies des Nègres: Leurs causes, leurs traitements et les moyens de les prévenir.* 1st ed., Paris: Didot; 2nd ed., Paris: Croullebois, 1792.

Delaleu, J.-B.-E. 1777. *Code des Isles de France et de Bourdon,* 2 vols. À L'Isle de France: de l'Imprimerie Royale. 2nd ed., 1826.

__________. 1783. *Premier supplément du Code de l'Isle de France (Janvier 1776 – Janvier 1783).* A L'Isle de France: De l'Imprimerie royale.

__________. 1787. *Deuxieme supplément du Code de l'Isle de France (Janvier 1783 – Juillet 1787).* A L'Isle de France: de l'Imprimerie du Roi.

Delandine, Antoine-François. 1787. *Couronnes académiques, ou Recueil des Prix proposés par les Sociétés Savantes, avec les noms de ceux qui les ont obtenus, des Concurrens distingués, des Auteurs qui ont écrit sur les mêmes sujets, le titre & le lieu de l'impression de leurs Ouvrages; Précédé de l'Histoire abrégée des Académies de France, par M. Delandine, Correspondant de l'Académie des Belles-Lettres & Inscriptions, &c.*, 2 vols. Paris: chez Cuchet.

Delisle, Guillaume. 1726. "Sur la longitude de l'Embouchure de la Rivière Saint-Louis nommée communément le Fleuve Misissipi (sic)," *MARS* 28 (1726/1728), pp. 249-57.

Delisle, Joseph-Nicolas. 1751. "Mémoire sur la longitude de Louisbourg, dans l'isle royale," *MARS* 53 (1751/1755), pp. 36-39.

De Page de Pratz. 1758. *Histoire de la Louisiane*, 3 vols. Paris: Chez de Bure, La Veuve Delaguette, Lambert.

Desplaces, Philippe. 1716-1734. *Éphémérides des mouvements célestes [...]*. [3 volumes]. Paris: J. Collombat.

Dortous de Mairan, Jean-Baptiste. 1724. "Instruction abrégée, & Methode pour le Jaugeage des Navires," *MARS* 26 (1734/1726), pp. 227-41.

__________. 1721. "Remarques sur le jaugeage des navires," *MARS* 23 (1721/1723), pp. 76-107.

Duhamel du Monceau, Henri-Louis. 1767. *Du transport, de la conservation et de la force des bois; ou l'on trouvera des moyens d'attendrir les bois, de leur donner diverses courbures, sur-tout pour la construction des vaisseaux; et de former des pieces d'assemblage...Faisant la conclusion du Traité complet des bois et des forets*. Paris: chez L. F. Delatour.

__________. 1764. *Art de rafiner le sucre*. [s.l.]

__________. 1759. *Moyens de conserver la santé aux équipages des vaisseaux*. Paris: H. L. Guérin and L. F. Delatour.

__________. 1758. *Élémens de l'architecture navale, ou Traité pratique de la construction des vaisseaux*, 2nd ed. Paris: C.-A. Jombe.

__________. 1755. *Traité des arbres et arbustes qui se cultivent en France en pleine terre*. Paris: H.-L. Guérin and L.-F. Delatour.

__________ (and the Marquis de La Galissonnière). 1752. *Avis pour le transport par mer des arbres, des plantes vivaces, des semences, des animaux et de différens autres morceaux d'histoire naturelle*. [No place or publisher; 2nd ed., 1753].

__________. 1747. *Traité de la fabrique des manoeuvres pour les vaisseaux, ou l'Art de la corderie perfectionné*. Paris: Imprimerie Royale.

Du Petit-Thouars. 1804. *Histoire des végétaux recueillis sur les isles de France, la Réunion (Bourbon) et Madagascar. Première Partie. Contenant les description et figures des Plantes qui forment des genres nouveaux, ou qui perfectionnent les anciens; Accompagnées de Dissertations sur différens points de Botanique*. Paris: De l'Imprimerie de Huzard, An XII.

Dupuis, Mathias. 1652. *Relation de l'établissement d'une colonie française dans la Guadeloupe, île de l'Amérique et des mœurs des sauvages*: Caen: chez Marin Yvon demeurant à Froide rue. [Modern edition: Basse-Terre: Société d'histoire de la Guadeloupe, 1972.]

Durand-Morlard, M. 1807-1810. *Code de la Martinique*, nouvelle édition, 3 vols. Saint-Pierre, Martinique: Jean-Baptiste Thounens.

Du Tertre, Jean-Baptiste. 1978. *Histoire générale des Antilles habitées par les français*, 4 vols. Paris: Édition et Diffusion de la Culture Antillaise. [Original edition 1667.]

Dutrône de La Couture, Jacques-François. 1790. *Précis sur la canne et sur les moyens d'en extraire le sel essentiel, suivi de plusieurs mémoires*. Paris: Duplain; 2nd ed., 1791; 3rd ed., 1801.

Ellis, John. 1779. *Description du mangostan et du fruit à pain...avec des instructions aux voyageurs pour le transport de ces deux fruits & autres substances végétales*. Rouen: Chez P. Machuel.

_________. 1775. *A Description of the Mangostan and the Bread-Fruit...to which are added directions for bringing over these and other vegetable productions*. London.

Euler, Leonard. 1749. *Scientia navalis seu tractatus de construendis ac dirigendis navibus*, 2 vols. Petropoli: Typis Academiae scientiarum.

Feuille du Cultivateur, Seconde Édition. 1802. Jean-Baptiste Dubois de Jancigny, Pierre Marie August Broussonet, and L. Laurent Lefebvre, eds., 8 vols. Paris.

Feuillée, Louis. 1725. *Journal des observations physiques, mathématiques et botaniques faites par ordre du roi sur les côtes orientales de l'amérique méridionale....* Paris: J. Mariette.

Fillassier, Jean-Jacques. 1785. *Tableau général des principaux objets qui composent la Pépinière dirigée par M. Fillassier...à Clamart...Années 1784 et 1785*. Paris: Clousier.

Fleurieu, Charles-Pierre Claret de. 1773. *Voyage par ordre du Roi en 1768-1769 sur l'Isis pour éprouver en mer les horloges marines de Berthoud*, 2 vols. Paris: Imprimerie Royale.

Frézier, Amédée-François. 1995. *Relation du voyage de la mer du Sud aux côtes du Chily et du Pérou*, presented by Gaston Arduz Eguía and Hubert Michéa. Paris: Éditions Utz.

_________. 1716. *Relation du voyage de la mer du Sud aux côtes du Chily et du Pérou, fait pendant les années 1712, 1713 et 1714 par M. Frézier*. Paris: J.-C. Nyon.

Fusée-Aublet, Jean-Baptiste-Christian. 1775. *Histoire des plantes de la Guyane française,* 4 vols. London and Paris: P.-F. Didot jeune.

Garcin, Laurent. 1734. "The Setting of a new Genus of Plants, called after the Malayans, Mangostans," *Phil. Trans.* no. 431, pp. 232-242.

_________. 1730. "Memoires communicated by Mons. Garcin to Mons. St. Hyacinthe, F.R.S.," *Phil. Trans.* no. 415, pp. 377-94.

Gaubil, Antoine. 1970. *Correspondance de Pékin, 1722-1759*, Renée Simon, ed. Genève: Librairie Droz.

Genton, M. de. 1787. "Essai de minéralogie de l'Isle de Saint-Domingue dans la partie Françoise," *Journal de physique* 31, pp. 173-77.

Granger, Claude. 1745. *Relation du voyage fait en Égypte par le sieur Granger en l'année* 1730. Paris: J. Vincent.

Guettard, Jean-Etienne. 1752a. "Mémoire dans lequel on compare le Canada à la Suisse par rapport à ses minéraux," *MARS* 54 (1752/1756), pp. 189-220.

_________. 1752b. "Addition au mémoire dans lequel on compare le Canada à la suisse, par rapport à ses minéraux," *MARS* 54 (1752/1756), pp. 524-539.

Hellot, Jean. 1750. *L'art de la teinture des laines et des étoffes de laine en grand et petit teint*. Paris: Vve Pissot; Paris: Didot fils, 1786.

Helvétius, Claude-Adrien. 1758. *De l'Esprit*. Paris: Durand.

Jussieu, Antoine de. 1729. "Recherches d'un spécifique contre la dysenterie, indiqué par les anciens auteurs sous le nom de macer, auquel l'écorce d'un arbre de Cayenne, appellé simarouba, peut être comparé, et substitué," *MARS* 31 (1729/1731), pp. 32-40.

_________. 1720. "Histoire du cachou," *MARS* 22 (1720/1722), pp. 340-346.

_________. 1716. "Histoire du Café," *MARS* 15 (1713/1716), pp. 291-99.

Jussieu, Joseph de. 1936. *Description de l'arbre à quinquina. Mémoire inédit de Joseph de Jussieu (1737)*, Prof. Pancier, ed. Paris: R.-L. Dupuy and Société du traitement des quinquinas.

Labat, R. P. Jean-Baptiste. 1979. *Nouveau voyage aux Iles françaises de l'Amérique...Nouvelle édition augmentée considérablement*, 8 vols. Paris: C.-J.-B. Delespine, 1742. Modern edition in 4 vols, Saint-Joseph (Martinique): Courtinard, 1979.

__________. 1722. *Nouveau voyage aux Iles françaises de l'Amérique, contentant l'histoire naturelle de ces pays, l'origine, les mœurs, la religion et le gouvernement des habitans anciens et modernes...le commerce et les manufactures qui y sont établies*, 6 vols. Paris: Guillaume Cavelier.

La Billardière, Jacques Julien Houtou de. 1799. *Relation du voyage à la recherche de Lapérouse*, 2 vols. Paris: H. J. Jansen, an VIII.

Lacaille, Nicolas-Louis de. 1754a. "Observations astronomiques faites à L'isle de France pendant l'année 1753," *MARS* 56 (1754/1759) pp. 44-56.

__________. 1754b. "Diverses observations faites pendant le cours de trois différentes traversées pour un Voyage au cap de Bonne-Espérance & aux Isles de France & de Bourbon," *MARS* 56 (1754/1759) pp. 94-130.

__________. 1751. "Relation abrégée du voyage fait par ordre du roi au Cap de Bonne-espérance," *MARS* 53 (1751/1755), pp. 519-36.

__________.1746. "Extrait de la Relation du voyage fait en 1724, aux isles Canaries, par le P. Feuillée Minime, pour déterminer la vrai position du premier Méridien," *MARS* 48 (1746/1751), pp. 129-50.

La Condamine. 1751. "Mémoire sur une résine élastique, nouvellement découverte à Cayenne par M. Fresneau: et sur l'usage de divers sucs laiteux d'arbres de la Guiane ou France équinoctiale," *MARS* 53 (1751/1755), pp. 323-33.

Lafosse, J.-F. 1787. *Avis aux habitans des colonies, particulièrement à ceux de l'Isle de S. Domingue, sur les principales causes des maladies qu'on y éprouve le plus communément & sur les moyens de les prévenir.* Paris: Royez.

Lalande, Jérôme de. 1775. *Éloge de M. Commerson.* [s.l.] (Hunt Institute, AD 74.)

__________. 1772. "Passage de Mercure sur le Soleil, observé dans l'île de Java et en Pensilvanie le 9 novembre 1769," *MARS* 74 (1772/1775), pp. 445-51.

Laval, Antoine de, S.J. 1728. *Voyage en Louisiane fait par ordre du roi en 1720*, etc. Paris: J. Mariette.

Leblond, Jean-Baptiste. 1813. *Voyage aux Antilles et à l'Amérique méridionale, commencé en 1767 et fini en 1802, contenant un précis historique des révoltes, des guerres, et des fais mémorables dont l'auteur a été témoin.* Paris: Artus-Bertrand; modern edition, 2000, edited by Monique Pouliquen; Paris: Karthala.

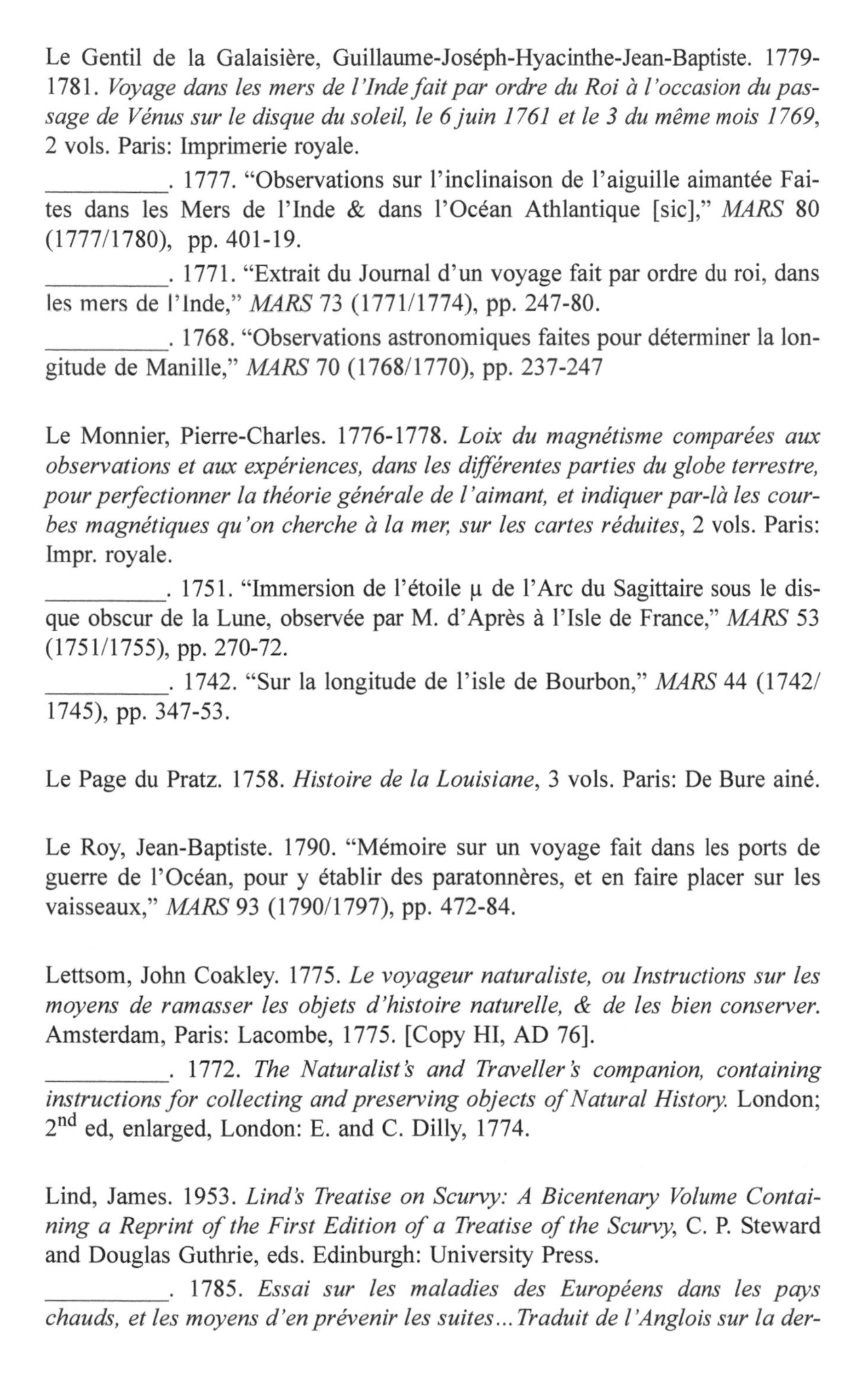

Le Gentil de la Galaisière, Guillaume-Joséph-Hyacinthe-Jean-Baptiste. 1779-1781. *Voyage dans les mers de l'Inde fait par ordre du Roi à l'occasion du passage de Vénus sur le disque du soleil, le 6 juin 1761 et le 3 du même mois 1769*, 2 vols. Paris: Imprimerie royale.

_________. 1777. "Observations sur l'inclinaison de l'aiguille aimantée Faites dans les Mers de l'Inde & dans l'Océan Athlantique [sic]," *MARS* 80 (1777/1780), pp. 401-19.

_________. 1771. "Extrait du Journal d'un voyage fait par ordre du roi, dans les mers de l'Inde," *MARS* 73 (1771/1774), pp. 247-80.

_________. 1768. "Observations astronomiques faites pour déterminer la longitude de Manille," *MARS* 70 (1768/1770), pp. 237-247

Le Monnier, Pierre-Charles. 1776-1778. *Loix du magnétisme comparées aux observations et aux expériences, dans les différentes parties du globe terrestre, pour perfectionner la théorie générale de l'aimant, et indiquer par-là les courbes magnétiques qu'on cherche à la mer, sur les cartes réduites*, 2 vols. Paris: Impr. royale.

_________. 1751. "Immersion de l'étoile μ de l'Arc du Sagittaire sous le disque obscur de la Lune, observée par M. d'Après à l'Isle de France," *MARS* 53 (1751/1755), pp. 270-72.

_________. 1742. "Sur la longitude de l'isle de Bourbon," *MARS* 44 (1742/1745), pp. 347-53.

Le Page du Pratz. 1758. *Histoire de la Louisiane*, 3 vols. Paris: De Bure ainé.

Le Roy, Jean-Baptiste. 1790. "Mémoire sur un voyage fait dans les ports de guerre de l'Océan, pour y établir des paratonnères, et en faire placer sur les vaisseaux," *MARS* 93 (1790/1797), pp. 472-84.

Lettsom, John Coakley. 1775. *Le voyageur naturaliste, ou Instructions sur les moyens de ramasser les objets d'histoire naturelle, & de les bien conserver.* Amsterdam, Paris: Lacombe, 1775. [Copy HI, AD 76].

_________. 1772. *The Naturalist's and Traveller's companion, containing instructions for collecting and preserving objects of Natural History.* London; 2nd ed, enlarged, London: E. and C. Dilly, 1774.

Lind, James. 1953. *Lind's Treatise on Scurvy: A Bicentenary Volume Containing a Reprint of the First Edition of a Treatise of the Scurvy*, C. P. Steward and Douglas Guthrie, eds. Edinburgh: University Press.

_________. 1785. *Essai sur les maladies des Européens dans les pays chauds, et les moyens d'en prévenir les suites... Traduit de l'Anglois sur la der-*

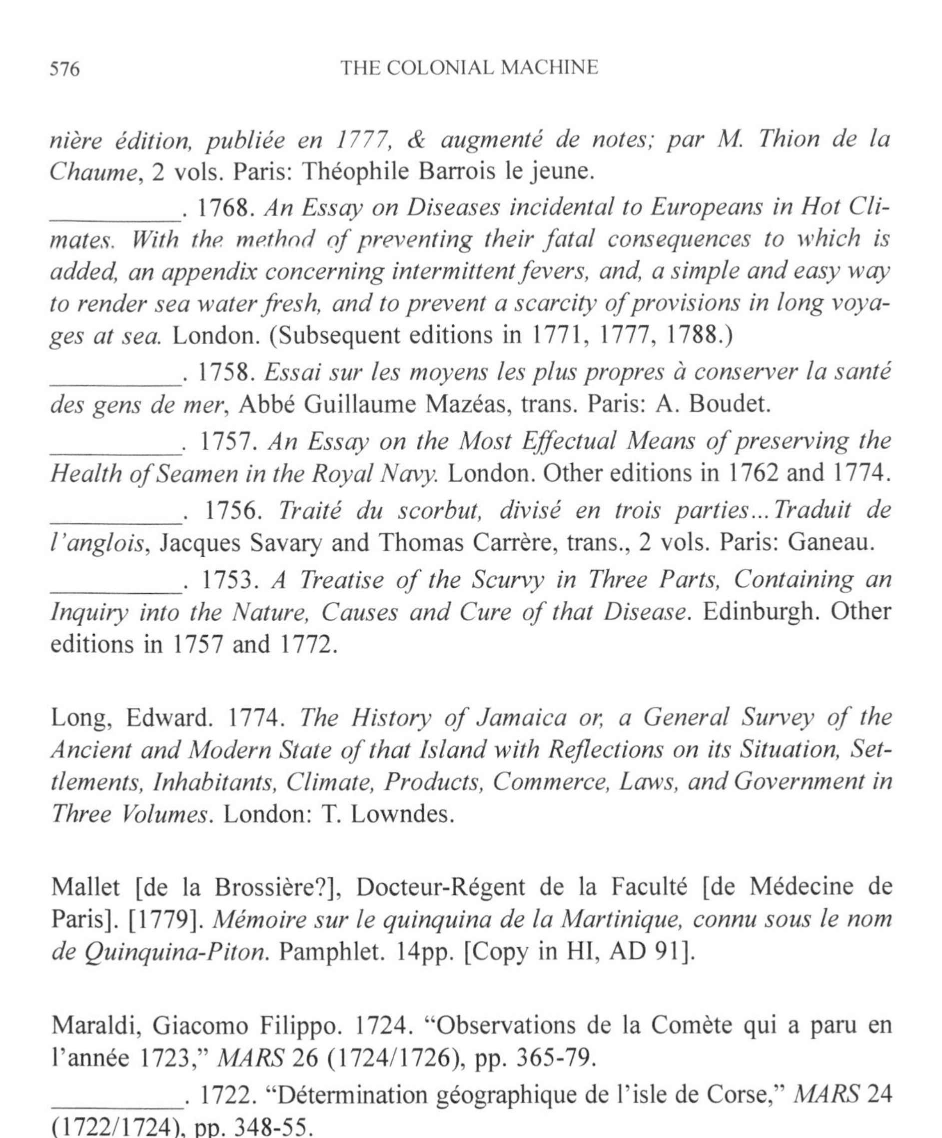

nière édition, publiée en 1777, & augmenté de notes; par M. Thion de la Chaume, 2 vols. Paris: Théophile Barrois le jeune.

__________. 1768. *An Essay on Diseases incidental to Europeans in Hot Climates. With the method of preventing their fatal consequences to which is added, an appendix concerning intermittent fevers, and, a simple and easy way to render sea water fresh, and to prevent a scarcity of provisions in long voyages at sea.* London. (Subsequent editions in 1771, 1777, 1788.)

__________. 1758. *Essai sur les moyens les plus propres à conserver la santé des gens de mer*, Abbé Guillaume Mazéas, trans. Paris: A. Boudet.

__________. 1757. *An Essay on the Most Effectual Means of preserving the Health of Seamen in the Royal Navy.* London. Other editions in 1762 and 1774.

__________. 1756. *Traité du scorbut, divisé en trois parties...Traduit de l'anglois*, Jacques Savary and Thomas Carrère, trans., 2 vols. Paris: Ganeau.

__________. 1753. *A Treatise of the Scurvy in Three Parts, Containing an Inquiry into the Nature, Causes and Cure of that Disease*. Edinburgh. Other editions in 1757 and 1772.

Long, Edward. 1774. *The History of Jamaica or, a General Survey of the Ancient and Modern State of that Island with Reflections on its Situation, Settlements, Inhabitants, Climate, Products, Commerce, Laws, and Government in Three Volumes*. London: T. Lowndes.

Mallet [de la Brossière?], Docteur-Régent de la Faculté [de Médecine de Paris]. [1779]. *Mémoire sur le quinquina de la Martinique, connu sous le nom de Quinquina-Piton*. Pamphlet. 14pp. [Copy in HI, AD 91].

Maraldi, Giacomo Filippo. 1724. "Observations de la Comète qui a paru en l'année 1723," *MARS* 26 (1724/1726), pp. 365-79.

__________. 1722. "Détermination géographique de l'isle de Corse," *MARS* 24 (1722/1724), pp. 348-55.

Maraldi, J. D. 1742. "De la Différence des Méridiens entre l'Observatoire Royal de Paris, l'Isle de Fer & quelques autres lieux," *MARS* 42 (1740/1742), pp. 121-30.

Maupertuis, Pierre-Louis Moreau de. 1997. *La Venus Physique* with preface by Mary Terrall. Paris: Diderot Éditeur.

__________. 1743. *Astronomie nautique ou élémens d'astronomie tant pour un observatoire mobile que pour un observatoire fixe*. Paris.

__________. 1738. *La Figure de la terre, déterminée par les observations de MM. de Maupertuis, Clairaut, Camus, Le Monnier,... et de M. l'abbé*

Outhier,... accompagnés de M. Celsius,... faites par ordre du Roy au cercle polaire. Paris: Imprimerie royale.

Méchain, Pierre-François-André. 1784. "Mémoire contenant les observations et la théorie de la première comète de 1784," *MARS* 87 (1784/1787), pp. 358-366.

Messier, Charles-Joseph. 1784. "Mémoire contenant les observations de la première comète de 1784," *MARS* 87 (1784/1787), pp. 313-27.

Michaux, André. 1906. "Journal of André Michaux 1793-96," translated and edited by Reuben Gold Thwaites, *Early Western Travels 1748-1846,* vol. 3 (Cleveland: A. H. Clark; reprint, New York: AMS Press, 1966).

_________. 1889. "Journal de mon Voyage," C. S. Sargent, ed., *Proceedings of the American Philosophical Society* 26, No. 129 (January to July 1889), pp. 1-146.

_________. 1801. *Histoire des chênes de l'Amérique ou Descriptions et figures de toutes les espèces et variétés de Chênes de l'Amérique Septentrionale.* Paris: chez Levrault frères, Quai Malaquais.

Michaux, François-André. 1810. *Histoire des arbres forestiers de l'Amérique septentrionale, considérés principalement sous le rapport de leur introduction dans le commerce, ainsi que d'après les avantages qu'ils peuvent offrir aux gouvernements en Europe et aux personnes qui veulent former de grandes plantations*, 4 vols. Paris: L. Haussmann et d'Hautel.

Ministère de la Marine et des Colonies. 1842. *Précis historique de l'expédition du Kourou (Guyane française). 1763-1765.* Paris: Imprimerie Royale.

Mirabeau, Victor de Riquetti, marquis de. 1756. *L'ami des hommes, ou, Traité de la population.* Avignon.

Moreau de Saint-Méry, Méderic Louis Élie. 2004. *Description topographique, physique, civile, politique et historique de la partie française de l'Isle de Saint-Domingue,* 3rd ed., 3 vols. with an introduction by Marcel Dorigny, a presentation by Étienne Taillemite, and a bibliography by Marcel Dorigny and Philippe Hroděj. Saint-Denis: Société Française d'Histoire d'Outre-Mer.

_________. 1958. *Description topographique, physique, civile, politique et historique de la partie française de l'Isle de Saint-Domingue,* 3 vols., Blanche Maurel and Étienne Taillemite, eds. Paris: Société Française d'Histoire d'Outre-Mer. (Reprint, 1984).

__________. 1797-1798. *Description topographique, physique, civile, politique et historique de la partie française de l'Isle de Saint-Domingue, avec des Observations générales sur sa Population, sur le Caractère & les Mœurs de ses divers Habitans; sur son Climat, sa Culture, ses Productions, son Administration, &c.*, 2 vols. Philadelphia: chez l'Auteur.

__________. 1791. *Recueil de Vues des Lieux Principaux de la Colonie Françoise de Saint Domingue.* Paris: A.P.D.R.

__________. 1789. "Mémoire sur la Patate," in *Mémoires d'agriculture, d'économie rurale et domestique, publiés par la Société Royale d'Agriculture de Paris*, 3rd trimester, pp. 43-57.

__________. 1788. "Mémoire sur une espèce de coton nommé à Saint-Domingue *Coton de Soie*, ou *Coton de Sainte-Marthe*," in *Mémoires d'agriculture, d'économie rurale et domestique, publiés par la Société Royale d'Agriculture de Paris*, *trimestre d'automne*, pp. 132-50.

__________. 1784-1790. *Loix et Constitutions des Colonies Françoises de l'Amérique Sous-le-Vent,* 6 vols. Paris: Quilezu, Méquigon jeune, Moutard, etc.

Moreau du Temple, René. 1998. *La Martinique de Moreau du Temple. La carte des ingénieurs géographes,* presented by Monique Pelletier, Danielle Bégot, and Catherine Bousquet-Bressolier. Paris: CTHS.

Nicolson, Jean-Barthélemy-Maximilien. 1776. *Essai sur l'Histoire Naturelle de St. Domingue.* Paris: Gobreau.

Pagès, Pierre-Marie-François de. 1782. *Voyages autour du monde et vers les deux pôles par terre et par mer pendant les années 1767-1776*, 2 vols. Paris: Moutard.

Palisot de Beauvois. 1805-1821. *Insectes recueillis en Afrique et en Amérique, dans les royaumes d'Oware et Bénin, à Saint-Domingue et dans les États-Unis pendant les années 1786-1797* (in fascicles, 1805-1821).

Parmentier, Antoine-Augustin. 1789. *Traité sur la culture et les usages des pommes-de-terre, de la patate et des topinambours*. Paris: Barrois ainé.

Patris, Jean-Baptiste. 1777. "Essai sur l'histoire naturelle et médicale du Ouassia," *Journal de physique* 9, p. 144.

Paulet, Jean-Jacques. 1775. *Recherches historiques & physiques sur les maladies épizootiques, avec les moyens d'y remédier dans tous les cas; publiées par ordre du roi,* 2 vols. Paris: Ruault.

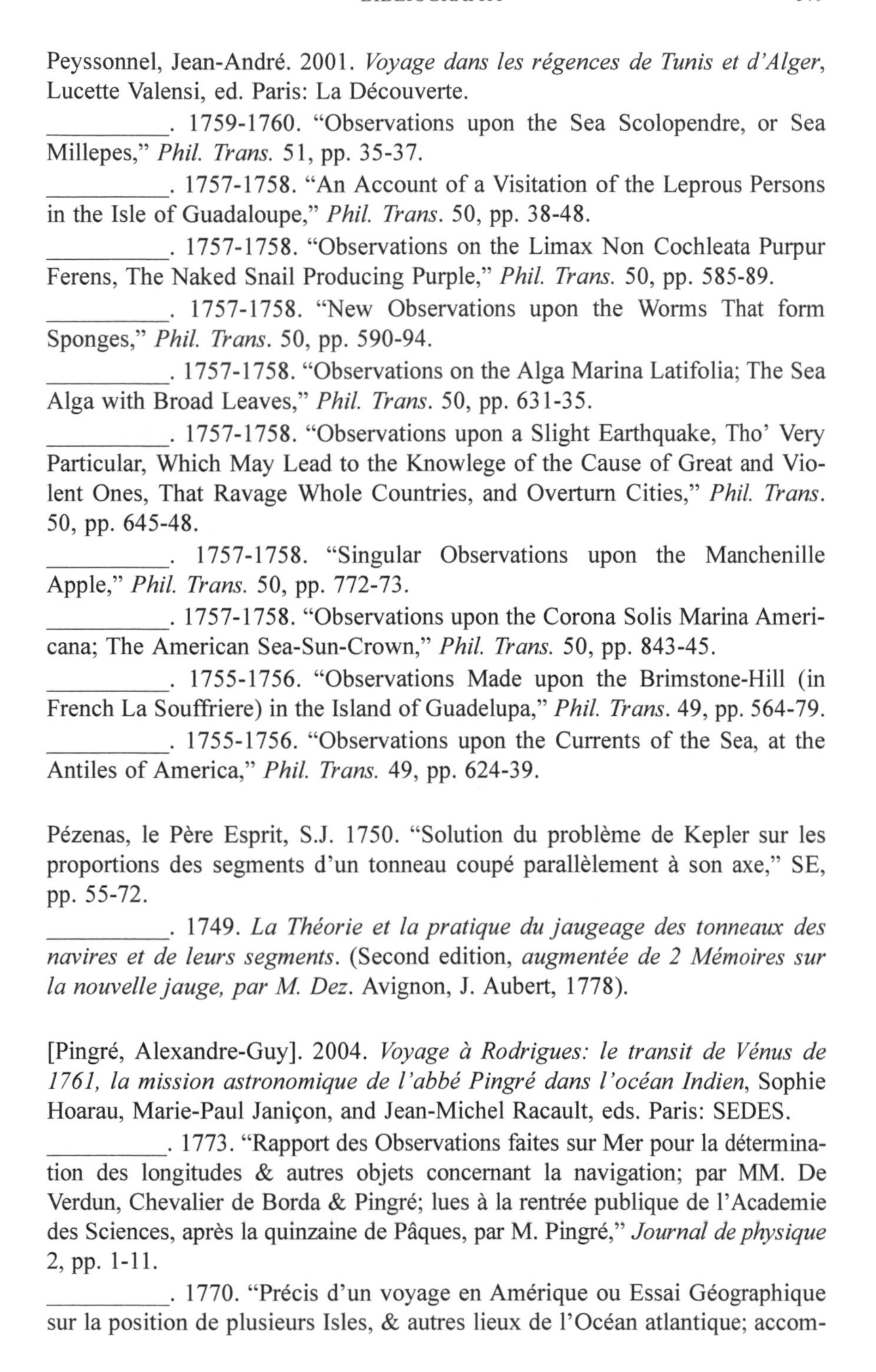

Peyssonnel, Jean-André. 2001. *Voyage dans les régences de Tunis et d'Alger*, Lucette Valensi, ed. Paris: La Découverte.

_________. 1759-1760. "Observations upon the Sea Scolopendre, or Sea Millepes," *Phil. Trans.* 51, pp. 35-37.

_________. 1757-1758. "An Account of a Visitation of the Leprous Persons in the Isle of Guadaloupe," *Phil. Trans.* 50, pp. 38-48.

_________. 1757-1758. "Observations on the Limax Non Cochleata Purpur Ferens, The Naked Snail Producing Purple," *Phil. Trans.* 50, pp. 585-89.

_________. 1757-1758. "New Observations upon the Worms That form Sponges," *Phil. Trans.* 50, pp. 590-94.

_________. 1757-1758. "Observations on the Alga Marina Latifolia; The Sea Alga with Broad Leaves," *Phil. Trans.* 50, pp. 631-35.

_________. 1757-1758. "Observations upon a Slight Earthquake, Tho' Very Particular, Which May Lead to the Knowlege of the Cause of Great and Violent Ones, That Ravage Whole Countries, and Overturn Cities," *Phil. Trans.* 50, pp. 645-48.

_________. 1757-1758. "Singular Observations upon the Manchenille Apple," *Phil. Trans.* 50, pp. 772-73.

_________. 1757-1758. "Observations upon the Corona Solis Marina Americana; The American Sea-Sun-Crown," *Phil. Trans.* 50, pp. 843-45.

_________. 1755-1756. "Observations Made upon the Brimstone-Hill (in French La Souffriere) in the Island of Guadelupa," *Phil. Trans.* 49, pp. 564-79.

_________. 1755-1756. "Observations upon the Currents of the Sea, at the Antiles of America," *Phil. Trans.* 49, pp. 624-39.

Pézenas, le Père Esprit, S.J. 1750. "Solution du problème de Kepler sur les proportions des segments d'un tonneau coupé parallèlement à son axe," SE, pp. 55-72.

_________. 1749. *La Théorie et la pratique du jaugeage des tonneaux des navires et de leurs segments*. (Second edition, *augmentée de 2 Mémoires sur la nouvelle jauge, par M. Dez.* Avignon, J. Aubert, 1778).

[Pingré, Alexandre-Guy]. 2004. *Voyage à Rodrigues: le transit de Vénus de 1761, la mission astronomique de l'abbé Pingré dans l'océan Indien*, Sophie Hoarau, Marie-Paul Janiçon, and Jean-Michel Racault, eds. Paris: SEDES.

_________. 1773. "Rapport des Observations faites sur Mer pour la détermination des longitudes & autres objets concernant la navigation; par MM. De Verdun, Chevalier de Borda & Pingré; lues à la rentrée publique de l'Academie des Sciences, après la quinzaine de Pâques, par M. Pingré," *Journal de physique* 2, pp. 1-11.

_________. 1770. "Précis d'un voyage en Amérique ou Essai Géographique sur la position de plusieurs Isles, & autres lieux de l'Océan atlantique; accom-

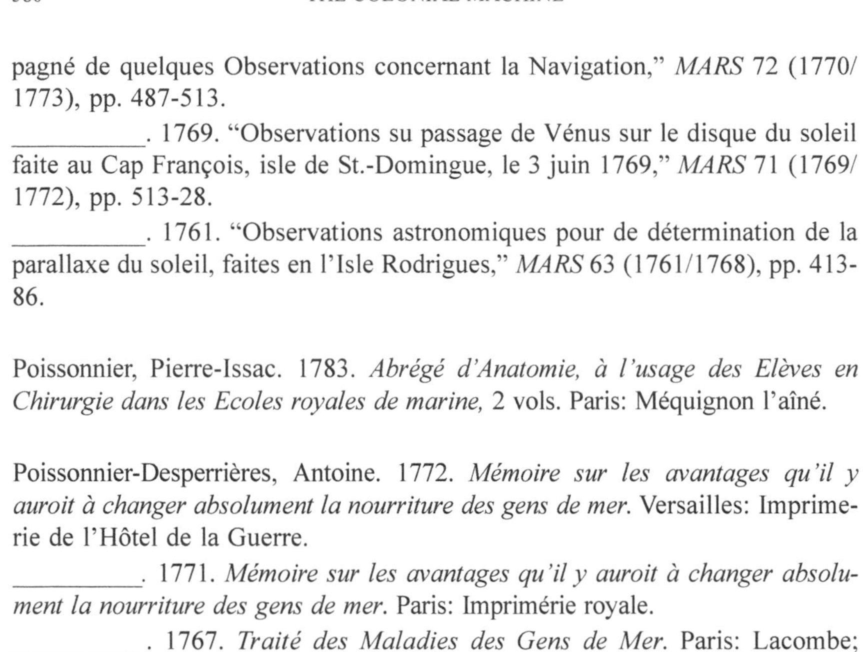

pagné de quelques Observations concernant la Navigation," *MARS* 72 (1770/1773), pp. 487-513.

_________. 1769. "Observations su passage de Vénus sur le disque du soleil faite au Cap François, isle de St.-Domingue, le 3 juin 1769," *MARS* 71 (1769/1772), pp. 513-28.

_________. 1761. "Observations astronomiques pour de détermination de la parallaxe du soleil, faites en l'Isle Rodrigues," *MARS* 63 (1761/1768), pp. 413-86.

Poissonnier, Pierre-Issac. 1783. *Abrégé d'Anatomie, à l'usage des Elèves en Chirurgie dans les Ecoles royales de marine,* 2 vols. Paris: Méquignon l'aîné.

Poissonnier-Desperrières, Antoine. 1772. *Mémoire sur les avantages qu'il y auroit à changer absolument la nourriture des gens de mer.* Versailles: Imprimerie de l'Hôtel de la Guerre.

_________. 1771. *Mémoire sur les avantages qu'il y auroit à changer absolument la nourriture des gens de mer.* Paris: Imprimérie royale.

_________. 1767. *Traité des Maladies des Gens de Mer.* Paris: Lacombe; second edition, Paris: Imprimérie royale, 1780.

_________. 1763. *Traité des fièvres de l'isle de St-Domingue.* Paris: P.-G. Cavelier, 1763; new edition, Paris: Imprimérie royale, 1780.

Pouppé-Desportes, Jean-Baptiste-René. 1770. *Histoire des Maladies de S. Domingue*, 3 vols. Paris: Lejay.

Puységur, A.-H.-A. Chastenet, count de. 1787a. *Le Pilote de l'Isle de Saint-Domingue et les Débouquemens de cette Isle...Publié par ordre du Roi.* Paris: De l'Imprimerie Royale.

_________. 1787b. *Détail sur la navigation aux côtes de Saint-Domingue et dans ses Débouquemens*. Paris: De l'Imprimerie Royale.

Raynal, Guillaume-Thomas-François. 1776. *Histoire philosophique et politique des établissements et du commerce des Européens dans les deux Indes*, 7 vols. La Haye: Gosse fils. (Original edition 1770).

Réaumur, René Antoine Ferchault. 1736. "Observations du thermomètre faites à Paris pendant l'année 1736, comparées avec celles qui ont été faites pendant la même année dans différentes parties du Monde" *MARS* 38 (1736/1739), pp. 469-507.

_________. 1727. "Observations sur la formation du *corail*, et des autres productions appellées *plantes pierreuses*," *MARS* 29 (1727/1729), pp. 269-281.

Renau d'Éliçagaray, Bernard. 1689. *De la théorie de la manœuvre des vaisseaux*. Paris: Estienne Michallet.

_________. 1694. *Réplique de M. Huegens à la réponse de M. Renau,... sur le principe de la Théorie de la manoeuvre des vaisseaux. Et la réponse de M. Renau à la réplique de M. Huegens.* Paris: Estienne Michallet.

Richer, Jean. 1729. "Observations Astronomiques et Physiques faites en l'Isle de Caïenne," *Mémoires de l'Académie Royale des Sciences, Depuis 1666. jusqu'à 1699*, vol. VII, Part I (Paris: Par la Compagnie des Libraires), pp. 231-326.

Rochon, Alexis-Marie. 1791. *Voyage à Madagascar et aux Indes orientales*. Paris: Imprimerie de Prault.

_________. 1783. *Nouveau Voyage à la Mer du Sud, commencé sous les ordres de M. Marion*. Paris: Barrois l'aîné.

Rousseau, Jean-Jacques. 1964. *Lettres écrites de la montagne* [in *Œuvres Complètes*, vol. 3, Bernard Gagnebin and Marcel Raymond, eds.]. Paris: Gallimard. Originally published in 1764.

_________. 1947. *Du Contrat social de Jean-Jacques Rousseau*, Bertrand de Jouvenel, ed. Geneva: C. Bourquin. Originally published in 1762.

[Sarrazin, Michel]. 1727. "Observations sur le porc-épic; extraites de mémoires et de lettres de M. Sarazin, medecin du Roi à Québec; et Correspondant de l'Académie," *MARS* 29 (1727/1729), pp. 383-95.

_________. 1725. "Extrait de divers memoires de M. Sarrazin, Médecin du Roi à Quebec, & correspondant de l'Académie. Sur le rat musqué par M. de Reaumur." *MARS* 27 (1725/1727), pp. 323-345 and plates.

_________. 1704. "Extrait d'une lettre de M. Sarrasin, Medecin du Roy en Canada, touchant l'Anatomie du Castor, lûë à l'Academie par M. Pitton Tournefort." *MARS* 6 (1704/1706), pp. 48-65.

Société d'Histoire Naturelle de Paris. 1792. *Actes de la Société d'histoire naturelle de Paris*. Paris: Imprimerie de la Société.

Société Royale d'Agriculture de la Généralité de Paris. 1761. *Recueil contenant les délibérations de la Société royale d'agriculture de la Généralité de Paris, au Bureau de Paris, depuis le 12 mars jusqu'au 10 septembre 1761. Et les mémoires publiés par son ordre dans le même tems*. Paris: Veuve d'Houry.

Société Royale de Médecine. 1786. *Projet d'instruction sur une maladie convulsive, fréquente dans les colonies de l'Amérique, connue sous le nom de tétanos*. Paris: Imprimerie Royale.

__________. 1785. *Rapport des commissaires de la Société royale de médecine, sur le mal rouge de Cayenne, ou éléphantiasis.* Paris: Imprimerie Royale.

__________. 1779-an VI. *Histoire de la Société Royale de Médecine avec les mémoires de médecine et de physique médicale tirés des registres de cette société.* Paris: Philippe-Denys Pierres.

Sonnerat, Pierre. 1782. *Voyage aux Indes orientales et à la Chine, Fait par ordre du Roi, depuis 1774 jusqu'en 1781 dans lequel on traite des moeurs, de la religion, des sciences et des arts des Indiens, des Chinois, des Pégouins et des Madégasses; suivi d'Observations sur le Cap de Bonne-Espérance, les isles de France et de Bourbon, les Maldives, Ceylan, Malacca, les Philippines et les Moluques, et de recherches sur l'histoire naturelle de ces pays*, 2 vols. Paris: chez l'auteur.

__________. 1776a. "Description du Cocos de l'Île de Praslin, vulgairement appelé Cocos de mer," *SE*, vol. 7 (1776), pp. 263-66.

__________. 1776b. *Voyage à la Nouvelle Guinée.* Paris: Ruault.

Souciet, Étienne, S. J. 1729-1732. *Observations mathématiques, astronomiques, géographiques, chronologiques et physiques tirées des anciens livres chinois, ou faites nouvellement aux Indes et à la Chine par les Pères de la Compagnie de Jésus,* 3 vols. Paris: Rollin.

Tessier, Abbé Henri-Alexandre. 1789. "Mémoire sur l'importation et les progrès des arbres à épicerie dans les colonies françaises," *MARS* 92 (1789/1793), pp. 585-96.

_______. 1779. "Mémoire sur l'importation du géroflier des Molouques aux îles de France, de Bourbon, et Séchelles & de ses îles à Caïenne," *Journal de physique* 14, pp. 47-54.

Thibault de Chanvalon, Jean-Baptiste-Mathieu. 1763. *Voyage à la Martinique contenant diverses observations sur la physique, l'histoire naturelle, l'agriculture, les mœurs et les usages de cette isle, faites en 1751 et dans les années suivantes, Lu à l'Académie des sciences de Paris en 1761.* Paris: J.-B. Bauche; facsimile ed., Fort-de-France, 1978; modern edition, 2004, edited by Monique Pouliquen; Paris: Karthala.

Thiery de Menonville, J.-N. 1786. *Traité de la Culture du Nopal et de l'Éducation de la Cochenille dans les Colonies Françaises de l'Amérique.* Cap François: Chez la veuve Herbault; Paris: Delalain.

Tocqueville, Alexis de. 1964. *L'Ancien Régime et la Révolution*, Jacob-Peter Mayer, ed. Paris: Gallimard. Original edition, 1856.

__________. 1835. *De la démocratie en Amérique*. Paris: Gallimard.

Tournefort, Joehph Pitton de. 1717. *Relation d'un voyage du Levant fait par ordre du Roy*, 2 vols. Paris: Imprimerie royale.

Turgot, Etienne-François (chevalier). 1758. *Mémoire instructif sur la manière de rassembler, de préparer, de conserver et d'envoyer les diverses curiosités d'histoire naturelle; auquel on a joint un Mémoire intitulé Avis pour le Transport par mer, des Arbres, des Plantes vivaces, des Semences & de diverses autres Curiosités d'Histoire naturelle*. Lyon: Jean-Marie Bruyset.

Vaucanson, Jacques de. 1770. "Second mémoire sur la filature des soies," *MARS* 72 (1770/1773), pp. 437-458.

Verdun de la Crenne, Jean-René-Antoine, marquis de. 1778. *Voyage en diverses parties de l'Europe, de l'Afrique et de l'Amérique*, 2 vols. Paris: Imprimerie royale.

Secondary Sources

Abénon, Lucien-René, and John A. Dickinson. 1993. *Les Français en Amérique: Histoire d'une colonisation*. Lyon: Presses Universitaires de Lyon.

Académie de Marine (France). 2002. *250e Anniversaire de l'Académie de Marine, 1752-2002*. Paris: Académie de Marine.

Adas, Michael. 1989. *Machines as the Measure of Men: Science, Technology, and Ideologies of Western Dominance*. Ithaca: Cornell University Press.

Adler, Kenneth. 2002. *The Measure of All Things: The Seven-year Odyssey and Hidden Error that Transformed the World*. New York: The Free Press.

Agarwal, Arun. 1995. "Dismantling the Divide between Indigenous and Scientific Knowledge," *Development and Change* 26, pp. 413-39.

Allain, Yves-Marie. 2000. *Voyages et survie des plantes Au temps de la voile*. Marly-le-Roi: Champflour.

Allard, Michel. 1970. *Henri-Louis Duhamel du Monceau et le Ministère de la Marine*. Montréal: Leméac.

Allard, Michel, ed. 1973. *L'Hôtel-Dieu de Montréal, 1642-1973*. Montréal: Hurtubise HMH.

Allorge, Lucile, with Olivier Ikov. 2003. *La fabuleuse odyssée des plantes: Les botanistes voyageurs, les Jardins des Plantes, les herbiers*. Paris: JC Lattès.

Ames, Glenn J. 1995. "Spices and Sulphur: Some Evidence on the Quest for Economic Stabilization in Portuguese Monsoon Asia, 1668-1682," *Journal of European Economic History* 24, pp. 465-87.

Anderson, Robert, and Karis Hiebert, eds., 2006. *Gardens and Forests in the Caribbean*. Oxford: Macmillan.

Andrew, Edward G. 2006. *Patrons of Enlightenment*. Toronto, Buffalo: University of Toronto Press.

Andrewes, William J. H., ed. 1996. *The Quest for Longitude: The Proceedings of the Longitude Symposium, Harvard University, November 4-6, 1993*. Cambridge, Mass.: Collection of Historical Scientific Instruments, Harvard University.

Anglade, Georges. *Atlas Critique d'Haïti*. Montreal: Erce & CRC, 1982.

Anonymous. 1931. "La propagande de Parmentier et l'approvisionnement des colonies," *Revue de l'Histoire des colonies françaises* 24, pp. 303-306.

Anthiaume, Albert. 1911. "L'enseignement de la science nautique au Havre-de-Grâce pendant les XVIe, XVIIe et XVIIIe siècles." Paris: Imprimerie nationale. [Extract of *Bulletin de géographie historique et descriptive*, nos. 1-2, 1910].

Appleby, John H. 1983. "Ginseng and the Royal Society," *Notes and Records of the Royal Society of London* 37, pp. 121-45.

Archives Nationales. 1984. *Guide des sources de l'histoire de l'Amérique latine et des Antilles dans les archives françaises*. Paris: Archives Nationales.

Audet, Louis-Philippe. 1970. "Hydrographes du roi et cours d'hydrographie au Collège de Québec, 1671-1759," *Les Cahiers des Dix* 35, pp. 13-38.

Auvigne, René. 1956. *Nantes, herbier des isles. Ou le rôle joué par les botanistes nantais dans l'introduction en France des végétaux exotiques au XVIIIe siècle*. Nantes: Imprimerie Chautreau et fils.

Bagneux, Jean de. 1932. *Un basque illustre: Renau d'Elissagaray, 1652-1719*. Bayonne: Courrier, 1932.

Balcou, Jean, ed. 1987. *La Mer au siècle des Encyclopédies* [Actes du Colloque de Brest, September, 1984]. Paris and Genève: Champion-Slatkine.

Ballantyne, Tony, and Antoinette Burton, eds. 2005. *Bodies in Contact: Rethinking Colonial Encounters in World History*. Durham and London: Duke University Press.

Banks, Kenneth J. 2002. *Chasing Empire Across the Sea: Communications and the State in the French Atlantic, 1713-1763*. Montreal & Kingston: McGill-Queen's University Press.

Barbiche, Bernard. 1999. *Les institutions de la monarchie française à l'époque moderne*. Paris: Presses Universitaires de France.

Barbiche, Bernard, and Aurélia Rostaign. 2003. "Bâtiments du Roi," in Bély, ed., *Dictionnaire de l'Ancien Régime*, pp. 140-42.

Barrera-Osorio, Antonio. 2006. *Experiencing Nature: The Spanish American Empire and the Early Scientific Revolution*. Austin: University of Texas Press.

Barrière, Pierre. 1951. *L'Académie de Bordeaux, centre de culture internationale au XVIIIe siècle (1712-1792)*. Bordeaux: Bière.

Barthélemy, Guy. 1979. *Les Jardiniers du Roy: petite histoire du Jardin des plantes de Paris*. Paris: Pélican/Librairie du Muséum.

Basalla, George. 1993. "The Spread of Western Science Revisited," in Lafuente et al., *Mundialización*, pp. 599-603.
__________. 1967. "The Spread of Western Science," *Science* 156, pp. 611-22.

Bégot, Danielle. 1996. "Une bibliothèque de colon en Guadeloupe à la fin du XVIIIe siècle: Antoine Mercier de La Ramée (1781)," in Alain Yacou, ed., *Créoles de la Caraïbe* [Actes du Colloque Universitaire en Hommage à Guy Hazaël-Massieux (1995)] (Paris: Karthala and C.E.R.C.) pp. 123-41.

Bély, Lucien, ed. 2003. *Dictionnaire de l'Ancien Régime. Royaume de France, XVIe-XVIIIe siècle*, 2nd ed. Paris: Quadrige/PUF. [Original ed., 1996].

Bernard-Maître, Henri. 1949. "Le Père Le Chéron d'Incarville, Missionnaire français de Pékin," *Archives internationales d'histoire des sciences* 2, pp. 333-62 and 692-717.

Berthaut, [Colonel]. 1902. *Les ingénieurs géographes militaires, 1624-1831. Étude historique*, 2 vols. Paris: Imprimerie du Service géographique.

Berthe, Leon-Noël. 1969. *Dubois de Fosseux, Secrétaire de l'Académie d'Arras et son bureau de correspondance*. Arras: C.N.R.S.

Bigot, Alfred. 1967. "La médecine française à Pondichéry aux XVIIIe et XIXe siècles," in *Comptes rendus du 91e Congrès national des sociétés savantes* [Rennes, 1966] (Paris: Gauthier-Villars, Bibliothèque Nationale), pp. 31-46.

Bigourdan, Guillaume. 1923. "Sur l'histoire de l'astronomie à Marseille," Rapport présenté au Congrès de Montpellier de l'Association Française pour l'Avancement des Sciences, Année 1922, Mémoire hors volume. Paris: Secrétariat de l'Association.

__________. 1895. "Inventaire général et sommaire des manuscrits de la bibliothèque de l'Observatoire de Paris," *Annales de l'Observatoire* – Série mémoires, XXI, pp. F1-F60.

Bitaubé, Pierre. 1988. *Les grandes expéditions et le jardin botanique du port de Rochefort*. Rochefort: Centre International de la Mer.

Blais, Hélène. 2005. *Voyages au grand océan. Géographie du pacifique et colonisation. 1815-1845*. Paris: CTHS.

Blanchard, Anne. 1981. *Dictionnaire des Ingénieurs militaires: 1691-1791*. Montpellier: privately published.

__________. 1979. *Les ingénieurs du Roy de Louis XIV à Louis XVI. Étude du corps des fortifications*. Montpellier: Université de Montpellier III and Service Historique de l'Armée de Terre.

Blanckaert, Claude, Claudine Cohen, Pietro Corsi, and Jean-Louis Fischer, eds. 1997. *Le Muséum au premier siècle de son histoire*. Paris: Éditions du Muséum d'Histoire Naturelle.

Bleichmar, Daniela. 2009. "A Visible and Useful Empire: Visual Culture and Colonial Natural History in the Eighteenth-Century Spanish World," in Bleichmar et al., eds., *Science in the Spanish and Portuguese Empires,* pp. 290-310.

_________. 2008. "Atlantic Competitions: Botany in the Eighteenth-Century Spanish Empire," in Delbourgo and Dew, eds., *Science and Empire in the Atlantic World*, pp. 225-52.

Bleichmar, Daniela, Paula De Vos, Kristin Huffine, and Kevin Sheehan, eds. 2009. *Science in the Spanish and Portuguese Empires. 1500-1800*. Stanford, California: Stanford University Press.

Blockmans, Willem Pieter, and Jean-Philippe Genet, general eds. 1995-1999. *Origins of the Modern State in Europe, 13th to 18th Century*, 7 vols. Oxford: Clarendon Press; Strasbourg: European Science Foundation.

Boinet, Amédée. 1908. *Catalogue des manuscrits de la Bibliothèque de l'Académie de Médecine*. Paris: Typographie Plon-Nourrit et Cie.

Boistel, Guy. 2005. "Les ouvrages et manuels d'astronomie nautique en France, 1750-1850," in Charon et al., eds., *Le livre maritime*, pp. 111-32.

_________. 2003. *L'Astronomie nautique au XVIIIe siècle en France: tables de la lune et longitudes en mer*. Lille: A.N.R.T.

_________. 2002. "Les longitudes en mer au XVIIIe siècle, sous le regard critique du père Pézenas," in Vincent Jullien, ed., *Le calcul des longitudes. Un enjeu pour les mathématiques, l'astronomie, la mesure du temps et la navigation* (Rennes: Presses universitaires de Rennes), pp. 101-121.

Bonnault, Cl. de. 1957. "La Gallissonnière et sa contribution à la botanique du Canada," in CNRS, *Les Botanistes français en Amérique du Nord*, pp. 171-77.

Bonnel, Ulane, ed. 1992. *Fleurieu et la Marine de son temps*. Paris: Economica.

Bonnet, Dr. Edmont. 1891. "Une mission francaise en Afrique au début du XVIIIe siècle: Augustin Lippi, ses observations sur la flore d'Egypte et de Nubie," *Mémoires de la Société des sciences naturelles et mathématiques de Cherbourg* 27, pp. 258-80.

Bonneuil, Christophe. 1991. *Des savants pour l'empire. La structuration des recherches scientifiques coloniales au temps "de la mise en valeur des colonies françaises," 1871-1945*. Paris: Editions de l'ORSTOM.

Bonney, Richard, ed. 1999. *The Rise of the Fiscal State in Europe, 1200-1815*. Oxford: Oxford University Press.

Boott, F. 1826. "Memorial of the Botanical Labours of André Michaux," *Edinburgh Journal of Medical Science* 1, pp. 126-132.

Bouche, Denise. 1991. *Histoire de la colonisation française, tome 2, Flux et reflux (1815-1962).* Paris: Fayard. [Paris: Le Grand Livre du Mois, 2004].

Bouchenot-Déchin, Patricia. 2001. *Henry Dupuis, jardinier de Louis XIV.* Versailles: Perrin.

Boucher, Philip P. 2007. *France and the American Tropics to 1700: Tropics of Discontent?* Baltimore and London: The Johns Hopkins University Press.

Boudet, E.-L. 1934. "Le Corps de Santé de la Marine et le service médical aux colonies au XVIIe et au XVIIIe siècles (1625-1815)," *La Géographie: Terre-Air-Mer*, March-August.

Boudreau, Claude. 1994. *La cartographie au Québec. 1760-1840.* Québec: Presses de l'Université de Laval.

Boudriot, Jean. 2000. "Duhamel du Monceau et la construction navale," in Corvol, ed., *Duhamel du Monceau*, pp. 181-200.
__________. 1991. "Propos sur les bois de marine sous l'Ancien Régime," *Neptunia* 182, pp. 12-18.

Bougerol, Christiane. 1983. *La médecine populaire à la Guadeloupe.* Paris: Karthaala.

Boulaine, Jean. 1990a. "Les Avatars de l'Académie d'agriculture sous la Révolution," in *Scientifiques et sociétés pendant la Révolution et l'Empire*, Actes du 114e Congrès National des Sociétés Savantes (Paris, 3-9 avril 1989), Section Histoire des Sciences et des Techniques (Paris: CTHS), pp. 211-27.
__________. 1990b. "L'établissement de la Société royale d'agriculture de France, 1783-1788," in Académie d'Agriculture de France, *Deux siècles de progrès pour l'agriculture et l'alimentation (1789-1989)*, pp. 51-58. Paris: Tec & Doc Lavoisier.

Bourde de la Rogerie, Henri. 1998. *Les Bretons aux Iles de France et de Bourbon (Maurice et La Réunion) aux XVIIe et XVIIIe siècles.* Rennes: La Decouvrance. [Original edition: Rennes: Oberthur, 1934].

Bourgeois, Charles. 1970. "Le père Louis Feuillée, astronome et botaniste du roi (1660-1732)," in *Comptes rendus du 94e congrès national des sociétés savantes*, (Pau, 1969) vol. 1: Histoire des sciences (Paris: Bibliothèque Nationale), pp. 9-19.

Bourguet, Marie-Noëlle. 2002. "Landscape with Numbers: Natural History, Travel and Instruments in the Late Eighteenth and Early Nineteenth Centuries," in Bourguet, Licoppe, and Sibum eds., *Instruments*, pp. 96-125.

_________. 1998. "Missions savantes au siècle des Lumières: du voyage à l'expédition," in Muséum National d'Histoire Naturelle, ed., *Il y a 200 ans, les savants en Égypte* (Paris: Nathan), pp. 38-67.

_________. 1997. "La collecte du monde: voyage et histoire naturelle (fin XVIIe-début XIXe)," in Blanckaert, Cohen, Corsi, and Fischer eds., *Le Muséum*, pp. 163-96.

Bourguet, Marie-Noëlle, and Christophe Bonneuil. 1999. "Présentation," *Revue Française d'Histoire d'Outre-mer* 86, pp. 7-38.

Bourguet, Marie-Noëlle, and Christian Licoppe. 1997. "Voyages, mesures et instruments: une nouvelle expérience du monde au Siècle des lumières," *Annales. Histoire, Sciences sociales* 5, pp. 1115-1151.

Bourguet, Marie-Noëlle, Christian Licoppe, and H. Otto Sibum , eds. 2002. *Instruments, Travel and Science: Itineraries of Precision from the Seventeenth to the Twentieth Century.* London and New York: Routledge.

Bourguet, Marie-Noëlle, and Christophe Bonneuil, eds. 1999. *De l'inventaire du monde à la mise en valeur du globe. Botanique et colonisation (fin XVIIe siècle-début XXe siècle).* [*Revue française d'Histoire d'Outre-mer* 86, no. 322-323].

Bourguet, Marie-Noëlle, Bernard Lepetit, Daniel Nordman, Maroula Sinarellis, eds. 1998. *L'invention scientifique de la Méditerranée, Égypte, Morée, Algérie.* Paris: Éditions de l'E.H.E.S.S.

Boxer, C. R. 1965. *The Dutch Seaborne Empire, 1600-1800.* London: Hutchinson.

Boyer, Pierre, Marie-Antoinette Menier, and Étienne Taillemite. 1980. *Les Archives Nationales. État Général des Fonds, Tome III, Marine et Outre-Mer.* Paris: Archives Nationales.

Brau, Paul. 1931. *Trois siècles de médecine coloniale française.* Paris: Vigot Frères.

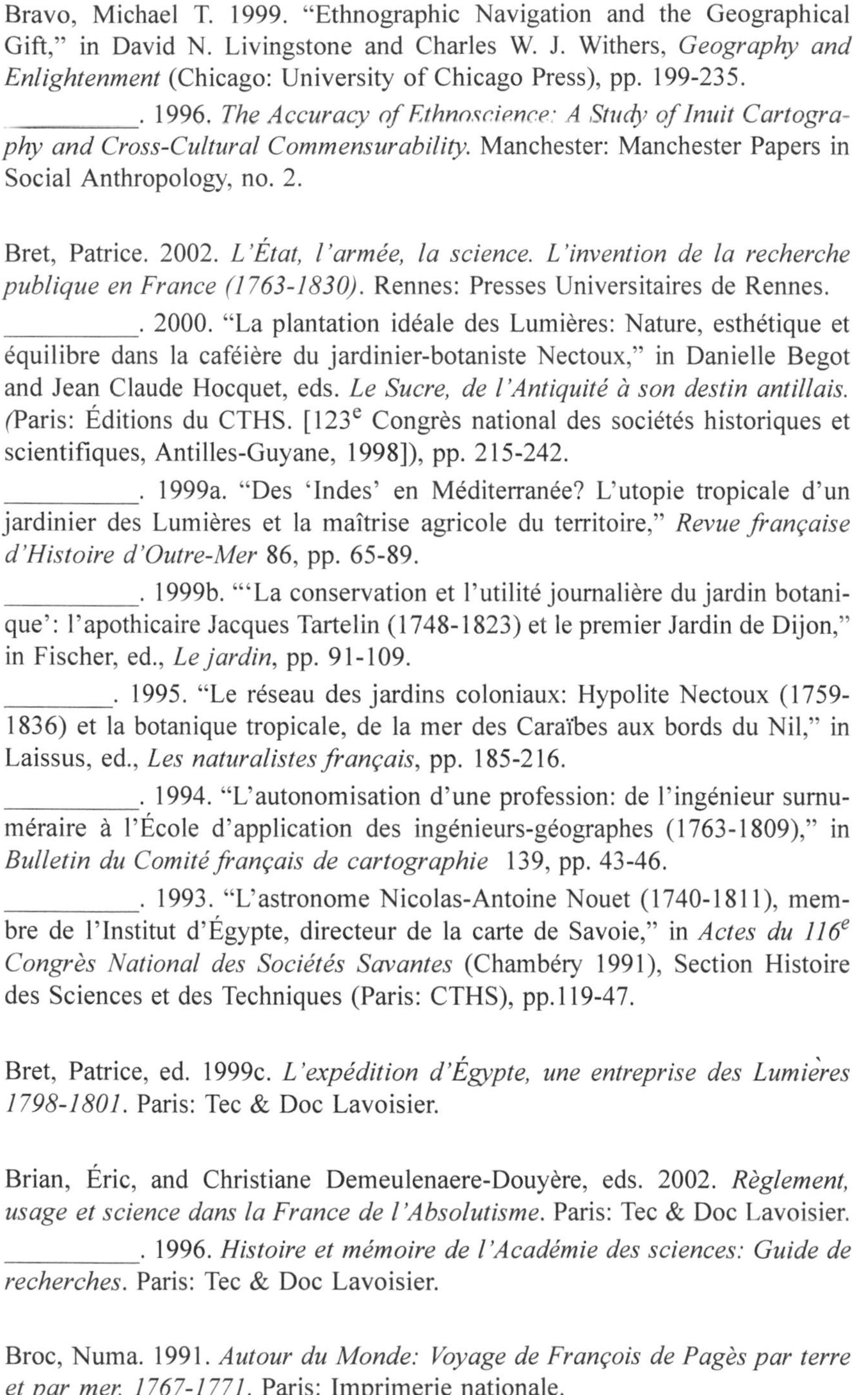

Bravo, Michael T. 1999. "Ethnographic Navigation and the Geographical Gift," in David N. Livingstone and Charles W. J. Withers, *Geography and Enlightenment* (Chicago: University of Chicago Press), pp. 199-235.

_________. 1996. *The Accuracy of Ethnoscience: A Study of Inuit Cartography and Cross-Cultural Commensurability.* Manchester: Manchester Papers in Social Anthropology, no. 2.

Bret, Patrice. 2002. *L'État, l'armée, la science. L'invention de la recherche publique en France (1763-1830)*. Rennes: Presses Universitaires de Rennes.

_________. 2000. "La plantation idéale des Lumières: Nature, esthétique et équilibre dans la caféière du jardinier-botaniste Nectoux," in Danielle Begot and Jean Claude Hocquet, eds. *Le Sucre, de l'Antiquité à son destin antillais.* (Paris: Éditions du CTHS. [123e Congrès national des sociétés historiques et scientifiques, Antilles-Guyane, 1998]), pp. 215-242.

_________. 1999a. "Des 'Indes' en Méditerranée? L'utopie tropicale d'un jardinier des Lumières et la maîtrise agricole du territoire," *Revue française d'Histoire d'Outre-Mer* 86, pp. 65-89.

_________. 1999b. "'La conservation et l'utilité journalière du jardin botanique': l'apothicaire Jacques Tartelin (1748-1823) et le premier Jardin de Dijon," in Fischer, ed., *Le jardin*, pp. 91-109.

________. 1995. "Le réseau des jardins coloniaux: Hypolite Nectoux (1759-1836) et la botanique tropicale, de la mer des Caraïbes aux bords du Nil," in Laissus, ed., *Les naturalistes français*, pp. 185-216.

_________. 1994. "L'autonomisation d'une profession: de l'ingénieur surnuméraire à l'École d'application des ingénieurs-géographes (1763-1809)," in *Bulletin du Comité français de cartographie* 139, pp. 43-46.

_________. 1993. "L'astronome Nicolas-Antoine Nouet (1740-1811), membre de l'Institut d'Égypte, directeur de la carte de Savoie," in *Actes du 116e Congrès National des Sociétés Savantes* (Chambéry 1991), Section Histoire des Sciences et des Techniques (Paris: CTHS), pp.119-47.

Bret, Patrice, ed. 1999c. *L'expédition d'Égypte, une entreprise des Lumières 1798-1801*. Paris: Tec & Doc Lavoisier.

Brian, Éric, and Christiane Demeulenaere-Douyère, eds. 2002. *Règlement, usage et science dans la France de l'Absolutisme*. Paris: Tec & Doc Lavoisier.

_________. 1996. *Histoire et mémoire de l'Académie des sciences: Guide de recherches*. Paris: Tec & Doc Lavoisier.

Broc, Numa. 1991. *Autour du Monde: Voyage de François de Pagès par terre et par mer, 1767-1771*. Paris: Imprimerie nationale.

_________. [1975.] *La géographie des philosophes: Géographes et voyageurs français au XVIIIe siècle*. Paris: Éditions Ophrys.

Brockway, Lucille. 1979. *Science and Colonial Expansion: The Role of the British Royal Botanic Gardens*. [Studies in Social Discontinuity]. New York and London: Academic Press.

Brossard, Maurice de. 1970. *Kerguelen, le découvreur et ses îles*. Paris: Editions France-Empire.

Broughton, Peter. 1981. "Astronomy in Seventeenth-Century Canada," *Journal of the Royal Astronomical Society of Canada* 75, pp. 175-208.

Broussoulle B., and Ph. Masson. 1985. "La Santé dans la Marine de l'Ancien Régime," in Pluchon, ed., *Histoire des Médecins,* pp. 69-87.

Brown, Vincent. 2003. "Spiritual Terror and Sacred Authority in Jamaican Slave Society," *Slavery and Abolition* 24, pp. 24-53.

Bruneau-Latouche, R. 1980. "Lettres de la Martinque: La correspondance Picaudeau Derivières," *Annales des Antilles* 23, pp. 53-67.

Burckard, François, and Claude Bouhier. 1994. *Répertoire des archives de l'Académie des sciences, belles-lettres et arts de Rouen (1744-1990)*. Luneray: Imprimerie Bertout.

Buridant, Jérôme. 2000. "Duhmael du Mondeau et la crise forestière du XVIIIe siècle," in Corvol, ed., *Duhamel du Monceau,* pp. 41-53.

Burnet, O. 1864. "Michaud and his Journey in Canada," *Canadian Naturalist and Geologist* 1, pp. 325-37.

Butel, Paul. 1997. *Européens et espaces maritimes (vers 1690-vers 1790)*. Bordeaux: Presses Universitaires de Bordeaux.
_________. 1996. "Les Bordelais et l'Inde dans la première moitié du xix siècle," in Haudrère et al., *Les flottes des Compagnies des Indes*, pp. 327-38.
_________. 1982. *Les Caraïbes au temps des flibustiers, XVIe-XVIIe siècles*. Paris: Aubier Montaigne.

Buti, Gilbert. "Cochenille mexicaine, négoce marseillais et manufactures languedociennes au XVIIIe siècle," in Llinares and Hroděj, eds., *Techniques et colonies*, pp. 13-31.

Cabon, Adolphe. 1916. "Notes historiques sur la détermination de la position géographique d'Haïti," *Bulletin Semestriel de l'Observatoire Météorologique du Séminaire Collège St-Martial*, January-June, pp. 51-68.

Camus, Aimée. 1957. "Contribution française à l'étude des graminiées de l'Amérique du nord au XVIIIe siècle," in CNRS, *Les Botanistes français en Amérique du Nord*, pp. 107-121.

Cañizares-Esguerra, Jorge. 2006. *Nature, Empire, and Nation: Explorations of the History of Science in the Iberian World.* Stanford, California: Stanford University Press.

__________. 2005. "Iberian Colonial Science," *ISIS* 96, pp. 64-70.

__________. 2001. *How to Write the History of the New World: Histories, Epistemologies and Identities in the Eighteenth-Century Atlantic World.* Stanford: Stanford University Press.

Cap, [Paul-Antoine]. 1861. "Notice biographique sur Philibert Commerson, naturaliste voyageur," in *Séance Publique de Rentrée de l'École Supérieure de Pharmacie et de la Société de Pharmacie de Paris*, 14 novembre 1860 (Paris: E. Thunot et Ce.), pp. 10-39.

__________. 1858. *Joseph Dombey, Naturaliste.* Paris: E. Thunot.

Cardinal, Catherine. 1984. *Ferdinand Berthoud (1727-1807), Horloger mécanicien du roi et de la marine.* La Chaux-de-Fonds, Switzerland: Musée International d'Horlogerie.

[CNRS] Centre National de la Recherche Scientifique. 1957. *Les Botanistes français en Amérique du Nord avant 1850* [Colloques Internationaux du Centre National de la Recherche Scientifique 63, Paris 11-14 September 1956]. Paris: CNRS.

Chaïa, Jean. 1980. "Vicq d'Azyr, la Société royale de Médecine et la Guyane," *105e Congrès national des Sociétés savantes, Caen, 1980, Section des Sciences*, vol. 5 (Paris: CTHS), pp. 9-15.

__________. 1979a. "Science, Médecine et état sanitaire en Guyane au XVIIIème siècle," *Mondes et cultures* 39 (1979), pp. 129-43.

__________. 1979b. "Sur quelques ingénieurs-géographes du Roi qui servirent en Guyane au XVIIIe siècle," *104e Congrès national des Sociétés savantes, Bordeaux, 1979, Section des Sciences*, vol. 4 (Paris: CTHS), pp. 49-57.

__________. 1978. "À propos des voyages en Guyane (en 1772 et 1775) de Sonnini de Manoncourt, collaborateur de Buffon" in *Actes du 103e Congrès national des Sociétés savantes* (Nancy, 1978), Section des Sciences et des Techniques (Paris: CTHS), pp. 253-61.

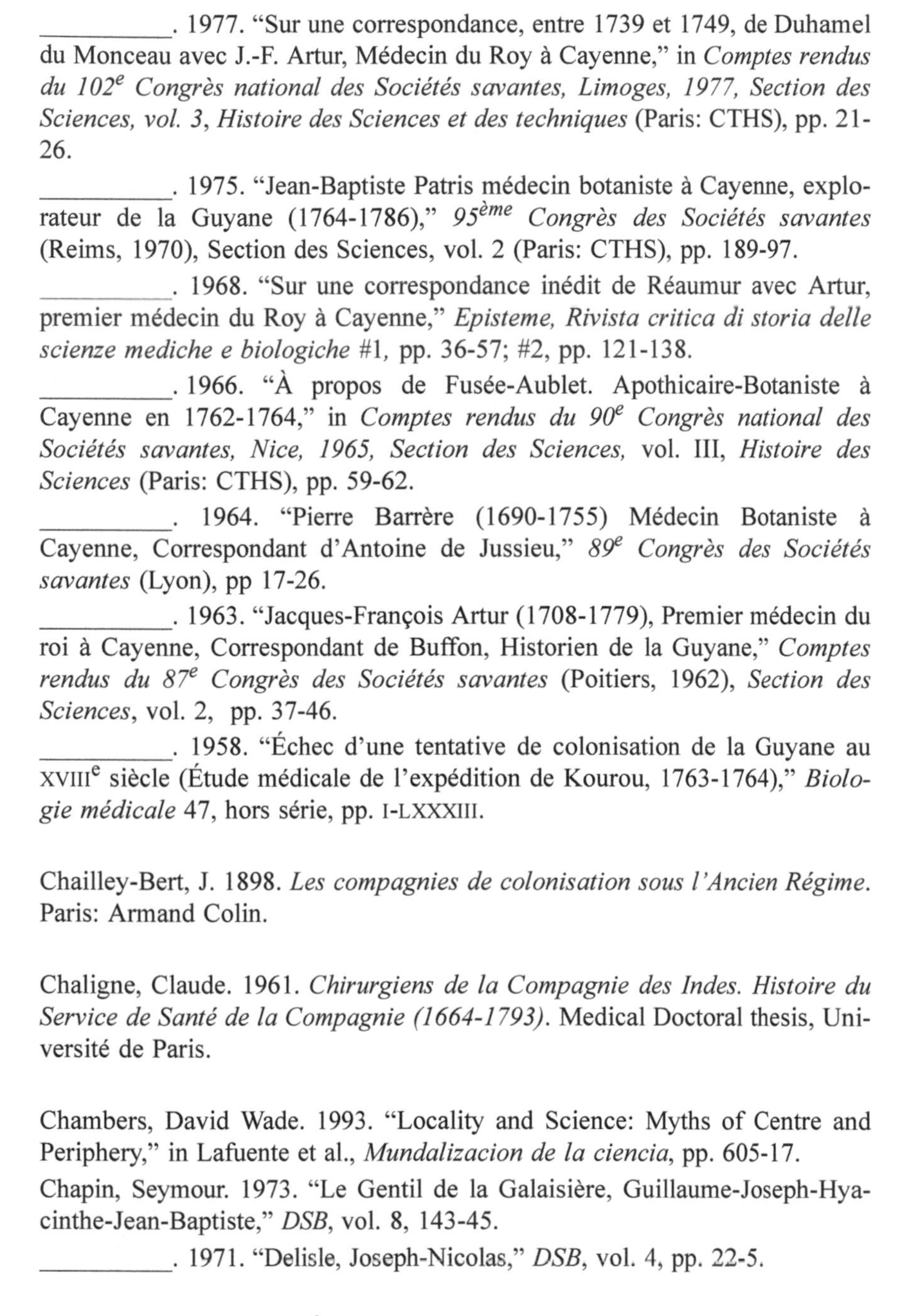

_________. 1977. "Sur une correspondance, entre 1739 et 1749, de Duhamel du Monceau avec J.-F. Artur, Médecin du Roy à Cayenne," in *Comptes rendus du 102e Congrès national des Sociétés savantes, Limoges, 1977, Section des Sciences, vol. 3, Histoire des Sciences et des techniques* (Paris: CTHS), pp. 21-26.

_________. 1975. "Jean-Baptiste Patris médecin botaniste à Cayenne, explorateur de la Guyane (1764-1786)," *95ème Congrès des Sociétés savantes* (Reims, 1970), Section des Sciences, vol. 2 (Paris: CTHS), pp. 189-97.

_________. 1968. "Sur une correspondance inédit de Réaumur avec Artur, premier médecin du Roy à Cayenne," *Episteme, Rivista critica di storia delle scienze mediche e biologiche #1,* pp. 36-57; #2, pp. 121-138.

_________. 1966. "À propos de Fusée-Aublet. Apothicaire-Botaniste à Cayenne en 1762-1764," in *Comptes rendus du 90e Congrès national des Sociétés savantes, Nice, 1965, Section des Sciences,* vol. III, *Histoire des Sciences* (Paris: CTHS), pp. 59-62.

_________. 1964. "Pierre Barrère (1690-1755) Médecin Botaniste à Cayenne, Correspondant d'Antoine de Jussieu," *89e Congrès des Sociétés savantes* (Lyon), pp 17-26.

_________. 1963. "Jacques-François Artur (1708-1779), Premier médecin du roi à Cayenne, Correspondant de Buffon, Historien de la Guyane," *Comptes rendus du 87e Congrès des Sociétés savantes* (Poitiers, 1962), *Section des Sciences*, vol. 2, pp. 37-46.

_________. 1958. "Échec d'une tentative de colonisation de la Guyane au XVIIIe siècle (Étude médicale de l'expédition de Kourou, 1763-1764)," *Biologie médicale* 47, hors série, pp. I-LXXXIII.

Chailley-Bert, J. 1898. *Les compagnies de colonisation sous l'Ancien Régime.* Paris: Armand Colin.

Chaligne, Claude. 1961. *Chirurgiens de la Compagnie des Indes. Histoire du Service de Santé de la Compagnie (1664-1793).* Medical Doctoral thesis, Université de Paris.

Chambers, David Wade. 1993. "Locality and Science: Myths of Centre and Periphery," in Lafuente et al., *Mundalizacion de la ciencia*, pp. 605-17.

Chapin, Seymour. 1973. "Le Gentil de la Galaisière, Guillaume-Joseph-Hyacinthe-Jean-Baptiste," *DSB*, vol. 8, 143-45.

_________. 1971. "Delisle, Joseph-Nicolas," *DSB*, vol. 4, pp. 22-5.

Chapuis, Olivier. 1999. *À la mer comme au ciel: Beautemps-Beaupré & la naissance de l'hydrographie moderne (1700-1850). L'émergence de la préci-*

sion en navigation et dans la cartographie marine. Paris: Presses de l'Université de Paris-Sorbonne.

Charbonneau, André, with Marc Lafrance. 1999. "Gaspard-Joseph Chaussegros de Léry (1682-1756)," in Vidal and D'Orgeix, eds. *Les villes françaises du Nouveau Monde*, pp. 142-46.

Charbonnier, P. 1926. "La balistique intérieure depuis B. de Morogues (1737) jusqu'à nos jours," *Académie de Marine* 5, offprint.

Charliat, Pierre-Jacques. 1934. "L'Académie royale de Marine et la Révolution nautique au XVIIIe siècle," *Thales* 1, pp. 71-82.

Charon, Annie, Thierry Claerr, and François Moureau, eds. 2005. *Le livre maritime au siècle des Lumières: Édition et diffusion des connaissances maritimes (1750-1850)*. Paris: Presses de l'Université Paris-Sorbonne.

Chartrand, Luc, Raymond Duchesne, and Yves Gingras. 1987. *Histoire des sciences au Québec*. Montréal: Éditions du Boréal.

Chartrand, René. 2008. *The Forts of New France in Northeast America 1600-1763*. Oxford: Osprey Publishing.

__________. 2005. *French Fortresses in North-America 1535-1763*. Oxford: Osprey Publishing.

Chassagne, Annie. 2007. *La bibliothèque de l'Académie royale des sciences au XVIIIe siècle*. Paris: Éditions du CTHS.

Chatillon, Marcel, ed. 1984. "L'évangélisation des esclaves au XVIIIe siècle. Lettres du R.P. Jean Mongin," *Bulletin de la Société d'Histoire de la Guadeloupe* 61-62, pp. 3-136.

Chauleau, Liliane. 1993. *Dans les îles du vent: La Martinique, XVIIe-XIXe siècle*. Paris: L'Harmattan, 1993.

Chevalier, Auguste. 1953. "Un grand voyageur naturaliste normand, J. J. La Billardière (1755-1834)," *Revue internationale de Botanique appliquée et d'Agriculture tropicale* 33, pp. 97-124, 185-202.

Chinard, Gilbert. 1957a. "Les Michaux et leurs précurseurs," in CNRS, *Les Botanistes français en Amérique du Nord*, pp. 263-84.

_________. 1957b. "Recently Acquired Botanical Documents," *Proceedings of the American Philosophical Society* 101, pp. 508-22.

Clark, Emily. 2007. *Masterless Mistresses: The New Orleans Ursulines and the Development of a New World Society, 1727-1834*. Chapel Hill: University of North Carolina Press.

Cohen, William B. 1980. *The French Encounter with Africans: White Response to Blacks, 1530-1880*. Bloomington: Indiana University Press.

Collet, Daniel. 1996. "Alexis Rochon, astronome, navigateur et explorateur," *Les Bretons au-delà des mers: Explorateurs et grands voyageurs* [Exposition. Saint-Vougay. 1996] (Quimper: Éditions Nouvelles du Finistère), pp. 80-84.

Collini, Silvia, and Antonelle Vannoni. 2005. *Les instructions scientifiques pour les voyageurs* XVIIe-XIXe *s.* Paris: Harmattan.

Comentale, Christophe. 1991. "Une Carte inédite du Père Antoine Gaubil, S. J.," *Chine et Europe: Évolution et particularités des rapports est-ouest du* XVIe *au* XXe *siècle* [Actes du IVe Colloque International de Sinologie, Chantilly 1983] (Paris: Institut Ricci), pp. 125-33.

Contamine, Philippe. 1992. *Histoire militaire de la France. Des origines à 1715*. Paris: PUF.

Cook, Andrew. 1996. "An exchange of letters between two hydrographers: Alexander Dalrymple and Jean-Baptiste d'Après de Mannevillette," in Haudrère et al., *Les flottes des Compagnies des Indes,* pp. 173-82.

Cook, Harold J. 2007. *Matters of Exchange: Commerce, Medicine, and Science in the Dutch Golden Age*. New Haven: Yale University Press.

Cordier, Henri, ed. 1918. "Voyages de Pierre Poivre de 1748 jusqu'à 1757," *Revue de l'Histoire des colonies françaises*, First Trimester, pp. 5-88.

Corvol, Andrée, ed.. 2000. *Duhamel du Monceau: 1700-2000: un Européen du siècle des Lumières* [Actes du colloque du 12 mai 2000 et conférences]. Orléans: Académie d'Orléans.

Corvol-Dessert, Andrée. 2003. "Bois de Marine," in Bély, ed. *Dictionnaire de l'Ancien Régime*, p. 165.

Crépin, Pierre. 1922. *Charpentier de Cossigny, Fonctionnaire colonial d'après ses écrits et ceux de quelques-uns de ses contemporains.* Paris: Éditions Ernest Leroux.

Crosby, Alfred W. 2002. *Throwing Fire: Projectile Technology through History.* Cambridge: Cambridge University Press.

__________. 1986. *Ecological Imperialism: The Biological Expansion of Europe, 900-1900.* Cambridge: Cambridge University Press.

__________. 1972. *The Columbian Exchange: Biological and Cultural Consequences of 1492.* Westport, Conn.: Greenwood Press. [30th aniversary ed., Westport, Conn.: Praeger Publishers, 2003].

Crosland, Maurice Pierre. 1992. *Science under Control: The French Academy of Sciences, 1795-1914.* Cambridge: Cambridge University Press.

Curzon, Henri de. 1903. *Répertoire numérique des archives de la Maison du Roi (Série O^{1}).* Bordeaux: Imprimerie G. Gounouilhou.

Dainville, François de. 1964. *Le Langage des géographes, termes, signes, couleurs des cartes anciennes, 1500-1800.* Paris: Picard.

Darnis, Jean-Marie. 1988. *La Monnaie de Paris. Sa création et son histoire du Consulat et de l'Empire à la Restauration (1795-1826).* Levallois, France: Centre d'Études Napoléoniennes.

Darnton, Robert. 1968. *Mesmerism and the End of the Enlightenment in France.* Cambridge Mass.: Harvard University Press.

Daston, Lorraine, and Katherine Park, eds. 1998. *Science and the Order of Nature, 1150-1750.* Cambridge, Mass.: The MIT Press.

Daugeron, Bertrand. 2009. *Collections naturalistes: Entre science et empires, 1763-1804.* Paris: Muséum national d'Histoire naturelle.

__________. 2007. *Apparition-Disparition des Nouveaux mondes en Histoire naturelle, Enregistrement-Épuisement des collections scientifiques (1763-1830).* Thèse de Doctorat, École des Hautes Études en Sciences Sociales, Paris.

Dawson, Nelson-Martin. 2000. *L'Atelier Delisle: L'Amérique du Nord sur la table à dessin.* Sillery, Quebec: Éditions du Septentrion.

Débarbat, Suzanne, and Michel-Pierre Lerner. 2002. "Méridien, méridienne, de l'origine à nos jours," in Vincent Jullien, ed., *Le calcul des longitudes. Un enjeu pour les mathématiques, l'astronomie, la mesure du temps et la navigation.* (Rennes: Presses Universitaires de Rennes), pp. 19-35.

Débarbat, Suzanne, and Simone Dumont. 1992. "Antoine-François Laval (1664-1728) hydrographe du roi, jésuite et astronome," in *Science et techniques en France méridionale* [*Actes du 115e Congrès national des sociétés savantes (Avignon, 1990), Section d'histoire des sciences et des techniques*] (Paris: Éditions du C.T.H.S.), pp. 17-26.

de Bagneux, Jean. 1932. *Un basque illustre: Renau d'Elissagaray, 1652-1719.* Bayonne: Courrier.

de Beer, Gavin. 1960. *The Sciences Were Never at War.* London: Thomas Nelson & Sons.

de Castelnau-L'Estoile, Charlotte, and François Regourd, eds. 2005. *Connaissances et pouvoirs. Les espaces impériaux, XVIe-XVIIIe s. France, Espagne, Portugal.* Bordeaux: Presses Universitaires de Bordeaux.

Debien, Gabriel. 1976. "Un missionnaire auxerrois des Caraïbes. Claude-André Leclerc de Château-du-Bois à la Dominique et à la Guadeloupe," *Bulletin de la Société des Sciences historiques et naturelles de l'Yonne* 108, pp. 41-46.

__________. 1972a. "Assemblées nocturnes d'esclaves à Saint-Domingue, (La Marmelade, 1786)," *Annales historiques de la Révolution*, pp. 272-84.

__________. 1972b. "Une Naintaise à Saint-Domingue (1782-1786)," *Revue de Bas-Poitu et des Provinces de l'Ouest* 83, pp. 413-36.

__________. 1966. "Les cultures à Sainte-Lucie à la fin du XVIIIe siècle," *Annales des Antilles* 13, pp. 57-84.

__________. 1956. *Études Antillaises (XVIIIe siècle).* Paris: Armand Colin.

__________. 1955. "Profils de colons: Jean Trembley," *Revue de "la Porte Océane,"* 11 [#113], pp. 14-17; [#114], pp. 8-10.

__________. 1953. *Les colons de Saint-Domingue et la Révolution: Essai sur le Club Massiac.* Paris: A. Colin.

Dejussieu, Jacques. 1967. "Note sur le jardin botanique de la Marine à Rochefort-sur-Mer," in *Comptes rendus du 91e Congres national des Sociétés savantes* [Rennes, 1966] (Paris: Gauthier-Villars, Bibliothèque Nationale), vol. 1, pp. 129-33.

Delaporte, André. 1996. "Les opérations militaires de la Compagnie française des Indes au XVIIIe siècle," in Haudrère et al., *Les flottes des Compagnies des Indes*, pp. 223-34.

Delbourgo James. 2006. *A Most Amazing Scene of Wonders. Electricity and Enlightenment in Early America.* Cambridge, Mass. and London: Harvard University Press.

Delbourgo, James, and Nicholas Dew. 2008. "Introduction: The Far Side of the Ocean," in Delbourgo and Dew, eds., *Science and Empire in the Atlantic World*, pp. 1-28.

Delbourgo, James, and Nicholas Dew, eds. 2008. *Science and Empire in the Atlantic World.* New York and London: Routledge.

Deleuze, Joseph-Philippe-François. 1804. "Notice historique sur André Michaux," *Annales du Muséum d'Histoire naturelle* 3 (1804), pp. 191-227.

Demarcq, Marie-Pierre. 2005. "Les traités de construction navale français de 1750 à 1850 dans les collections du Musée national de la marine," in Charon et al., eds., *Le livre maritime*, pp. 101-110.

Demeulenaere-Douyère, Christiane. 2000a. "Un regard sur la Guadeloupe à la fin du XVIIIe siècle: le fonds Hapel Lachênaie des archives de l'Académie des sciences de Paris," in Danielle Bégot and Jean-Claude Hocquet, eds., *Le sucre, de l'Antiquité à son destin antillais* (Paris: Éditions du CTHS), pp. 121-42.
__________. 2000b. "Duhamel du Monceau, membre de l'Académie royale des sciences," in Corvol, ed., *Duhamel du Monceau*, pp. 105-32.

De Reynal de Saint-Michel, Bertrand. 1997. *Atlas de la Martinique au 18e sècle: des Hommes et des Sites: Pour une étude toponymique et anthroponymique.* Fort-de-France.

Desaive, Jean-Paul, Jean-Pierre Goubert, Emmanuel Le Roy Ladurie, Jean Meyer, Otto Muller, Jean-Pierre Peter, eds. 1972. *Médecins, climat et épidémies à la fin du XVIIIe siècle.* Paris: École Pratique des Hautes Études.

Desbarats, Catherine. 2005. "La question de l'État en Nouvelle-France," in Philippe Joutard and Thomas Wien, eds. *Mémoires de Nouvelle-France. De France en Nouvelle-France.* (Rennes: Presses universitaires de Rennes), pp. 187-98.

Desbarats, Catherine, ed. 2006. *Lettres édifiantes et curieuses*. Montréal: Boréal.

Deslandres, Dominique. 2003. *Croire et faire croire. Les missions françaises au XVIIe siècle*. Paris: Fayard.

Despoix, Philippe. 2004. *Le monde mesuré. Dispositifs de l'exploration à l'âge des Lumières*. Genève: Droz.

Dessert, Daniel. 2003. "Secrétaire d'État de la Marine," in Bély, ed., *Dictionnaire de l'Ancien Régime*, pp. 1140-41.

_________. 1996. *La Royale. Vaisseaux et marins du Roi-Soleil*. Paris: Fayard.

Devaux, Jean-Michel. 2003. "Colonies," in Bély, ed., *Dictionnaire de l'Ancien Régime*, pp. 286-89.

Devèze, Michel. 1977. *Antilles, Guyanes, La Mer des Caraïbes de 1492 à 1789*. Paris: S.E.D.E.S.

Dew, Nicholas. 2010. "Scientific Travel in the Atlantic World: the French Expedition to Gorée and the Antilles, 1681-1683," *British Journal for the History of Science* 43(1), pp. 1-17.

_________. 2009, *Orientalism in Louis XIV's France*. Oxford: Oxford University Press.

_________. 2008. "*Vers la ligne*: Circulating Measurments around the French Atlantic," in Delbourgo and Dew, eds., *Science and Empire in the Atlantic World*, pp. 53-72.

Dhombres, Jean. 2003a "Académie Royale des Sciences," in Bély, ed., *Dictionnaire de l'Ancien Régime*, pp. 15-18.

_________. 2003b. "Hydrographie," in Bély, ed., *Dictionnaire de l'Ancien Régime*, pp. 646-47.

Doneaud du Plan, Alfred. 1879-1882. *Histoire de l'Académie de la Marine*, 6 parts. Paris: Berger-Levrault.

Dordain, Yves. 1962. "La chirurgie provinciale française au XVIIIe siècle. Son niveau technique d'après les membres non résidents de l'Académie royale de Chirurgie," Thèse pour le Doctorat en médecine, Université de Rennes.

D'Orgeix, Émilie. 1999. "De l'éducation des ingénieurs militaires (1650-1750)," in Vidal and D'Orgeix, eds. *Les villes françaises du Nouveau Monde.* pp. 49-55.

Dorn, Harold. 1991. *The Geography of Science.* Baltimore and London: The Johns Hopkins University Press.

Dorst, Jean. 1978. "Les activités outre-mer du muséum d'histoire naturelle," *Mondes et cultures* 38, pp. 595-602.

Doublet, Edouard Lucien. 1920. "L'Intendant Poivre," *Actes de l'Institut colonial de Bordeaux.* Offprint; PA-DB (Poivre).
__________. 1910a. "Correspondance échangée de 1720 à 1739 entre l'Astronome J.-N. Delisle et M. de Navarre." Bordeaux: Imprimerie G. Gounouilhou.
__________. 1910b. "Le centenaire de M. de Fleurieu," *Bulletin de la Société de Géographie de Bordeaux.* [PA-DB].

Douglas, Mary. 1986. *How Institutions Think.* Syracuse, N.Y.: Syracuse University Press.

Drayton, Richard. 1999. "Science, Medicine and the British Empire," in Robin W. Winks, ed., *The Oxford History of the British Empire, vol. 5, Historiography* (Oxford: Oxford University Press), pp. 264-76.

D'Souza, Florence. 1995. *Quand la France Decouvrit l'Inde: Les écrivains-voyageurs français en Inde (1757-1818).* Paris: Éditions Harmattan.

Dubois, Laurent. 2004a. *Avengers of the New World: The Story of the Haitian Revolution.* Cambridge, Mass.: Harvard University Press.
__________. 2004b. *A Colony of Citizens: Revoltuion & the Slave Enamcipation in the French Caribbean, 1787-1804.* Chapel Hill: University of North Carolina Press.

Duchesne, Raymond. 1981. "Historiographie des sciences et des techniques au Canada," *Revue d'histoire de l'Amérique française* 35, pp. 193-216.

Duchet, Michèle. 1995. *Anthropologie et Histoire au siècle des Lumières.* Paris: Albin Michel [original edition, Paris: Maspero, 1971].

Dufourcq, Élisabeth. 1993. *Une forme de l'expansion française. Les congrégations religieuses féminines hors d'Europe, de Richelieu à nos jours. Histoire naturelle d'une diaspora*, 4 volumes. Paris: Librairie de l'Inde.

Dumoulin Genest, Marie-Pierre. 1994. *L'introduction et l'acclimatation des plantes chinoises en France au dix-huitième siècle.* Doctoral thesis, École des Hautes Études en Sciences Sociales (Paris).

Dunmore, John. 2007. *Where Fate Beckons: The Life of Jean-François de la Pérouse.* Fairbanks: University of Alaska Press.
_________. 1985. *Pacific Explorer: The Life of Jean-Francois de La Perouse.* Palmerston North, New Zealand: Dunmore Press.
_________. 1965. *French Explorers in the Pacific. Volume 1: The Eighteenth Century.* Oxford: Oxford University Press.

Dupont, Amiral, and Marc Fardet. 1993. *L'Arsenal de Colbert, Rochefort.* [Rochefort:] Centre Internationale de la Mer.

Dupont de Dinechin, Bruno. 1999. *Duhamel du Monceau: Un savant exemplaire au siècle des lumières.* Luxembourg: CME, Connaissance et Mémoires Européennes.

Duprat, Gabrielle. 1957. "Essai sur les sources manuscrites conservées au Muséum national d'histoire naturelle," in CNRS, *Les Botanistes français en Amérique du Nord*, pp. 231-52.

Durand-Molard, M. 1807-1810. *Code de la Martinique*, Nouvelle Édition, 3 vols. Saint-Pierre, Martinique: Jean-Baptiste Thounens. [Vol. 1: 1807; vol. 2: 1807; vol. 3: 1810].

Duris, Pascal. 1993. *Linné et la France (1780-1850).* Geneva: Droz.

Duteil, Jean-Pierre. 1994. *Le mandat du ciel. Le rôle des jésuites en Chine.* Paris: Éditions Arguments.

Eklund, John. 1971. "Duhamel du Monceau," *DSB*, vol. 4, pp. 223-25.

Emerson, Roger L. 1982. "The Edinburgh Society for the Importation of Foreign Seeds and Plants, 1764-1773," *Eighteenth-Century Life* 7, pp. 73-95.

Estienne, René. 1996. "Les activités navales de la deuxième Compagnie des Indes d'après les archives du port de Lorient," in Haudrère et al., *Les flottes des Compagnies des Indes*, p 131.

Ewan, J. 1957. "L'Activité des premiers explorateurs français dans le S.E. des Etats-Unis," in CNRS, *Les Botanistes français*, pp. 17-40.

Farge, Arlette. 1986. *La vie fragile: Violence, pouvoirs et solidarités à Paris au XVIIIe siècle.* Paris: Hachette.

Fauque, Danielle. 1985. "Alexis-Marie Rochon (1741-1817)," *Revue d'histoire des sciences* 38, pp. 3-36.

Favier, Jean. 1980. *Les Archives Nationales, État général des fonds.* Paris: Archives Nationales.

Feest, Christian, ed. 2007. *Premières nations. Collections royales. Les Indiens des forêts et des prairies d'Amérique du Nord.* Paris: Musée du Quai Branly and Réunion des Musées Nationaux.

Filliozat[-Restif], Manonmani. 2005. "Routiers de navigation et instructions nautiques," in Charon et al., eds., *Le livre maritime*, pp. 133-48.
__________. 1993. "D'Après de Mannevillette, Capitaine et Hydrographe de la Compagnie des Indes (1707-1780)." Thèse, École Nationale des Chartes.

Fischer, Jean-Louis, ed. 1999. *Le jardin entre science et représentation.* Paris: Éditions du CTHS.

Fleming, Patricia, Gilles Gillichan, and Yvan Lamonde, eds. 2004. *Histoire du livre et de l'imprimé au Canada. Des débuts à 1840.* Montréal: Presses de l'Université de Montréal.

Forbes, Eric G. 1974. "Maskelyne, Nevil," *DSB*, vol. 9, pp. 162-64.

Fornasiero, Jean, Peter Monteath, and John West-Sooby. 2005. *Encountering Terra Australis. The Australian Voyages of Nicolas Baudin and Matthew Flinders.* Adelaide: Wakefield Press.

Fouchard, Jean. 1955. *Plaisirs de Saint-Domingue. Notes sur sa vie sociale, littéraire et artistique.* Port-au-Prince: L'Imprimerie de l'État. [Reprint, Port-au-Prince, 1988].

Fournier. P. 1932. *Voyages et Découvertes Scientifiques des Missionnaires naturalistes française à travers le Monde (XVe à XXe siècles).* Paris: Lechevalier & Fils.

Foville, Alfred de, and Henri Pigeonneau, eds. 1882. *L'Administration de l'Agriculture au Contrôle général des Finances (1785-1787): Procès-verbaux et rapports.* Paris: Guillaumin.

Fraser, Antonia. 2001. *Marie Antoinette: The Journey.* New York: N. A. Talese/Doubleday.

Froeschlé, Michel. 1990. "L'astronomie au quotidien: Le cahier d'observations (1728-1733) du père Sigalloux," in *Science et techniques en France méridionale* [*Actes du 115*[e] *Congrès national des sociétés savantes (Avignon, 1990), Section d'histoire des sciences et des techniques*] (Paris: Éditions du C.T.H.S.), pp. 27-38.

Froidevaux, Henri, 1899. "Les Mémoires inédits d'Adanson sur l'île de Gorée et la Guyane française," *Bulletin de géographie historique et descriptive* 14, pp 76-100.

__________. 1897. "Étude sur les recherches scientifiques de Fusée Aublet à la Guyane française (1762-1764)," *Bulletin de géographie historique et descriptive* 12, pp. 427-489.

__________. 1895. "Notes sur le voyageur Guyanais Pierre Barrère," *Bulletin de géographie historique et descriptive* 10, pp. 326-38.

__________. 1893. "Un projet de voyage du botaniste Adanson en Guyane, en 1763," *Bulletin de géographie historique et descriptive* 8, pp. 221-33. [Offprint: Paris: Ernest Leroux, Éditeur, 1893].

__________. 1892. "Une mission géographique et militaire à la Guyane en 1762," *Annales de géographie* 15, pp. 218-31.

Ganière, Paul. 1964. *L'Académie de Médecine: Ses origines et son histoire.* Paris: Librairie Maloine.

Gasc, J.-P., and Yves Laissus. 1981. *Voyages et découvertes: des voyageurs naturalistes au chercheurs scientifiques.* Paris: Muséum National d'Histoire Naturelle.

Gascoigne, John. 1998. *Science in the Service of Empire: Joseph Banks, The British State and the Uses of Science in the Age of Revolution.* Cambridge: Cambridge University Press.

Gay, Jacques. 1999. "Les progrès techniques de la Marine au temps de Borda," in E. Neuville, ed., *Le Chevalier de Borda*, pp. 19-23.

__________. 1998. *L'eau à bord des navires de l'Antiquité à nos jours.* Jonzac: Publications de l'Université Francophone d'Été.

Gaziello, Catherine. 1984. *L'Expédition de Lapérouse, 1785-1788: Réplique française aux voyages de Cook.* Paris: CTHS.

Geggus, David P. 2002. *Haitian Revolutionary Studies.* Bloomington, IN: Indiana University Press.

Gelfand, Toby. 1987. "Medicine in New France," in Ronald L. Numbers, ed., *Medicine in the New World: New Spain, New France, and New England* (Knoxville: University of Tennessee), pp. 64-100.

__________. 1980. *Professionalizing Modern Medicine: Paris Surgeons and Medical Science and Institutions in the 18th Century.* Westport, Conn.: Greenwood Press.

Genovese, Eugene D. 1981. *From Rebellion to Revolution: Afro-American Slave Revolts in the Making of the New World.* New York: Vintage Books.

Gerbaud, Oliver. 1986. *Les premiers vétérinaires français engagés pour le service des colonies entre 1770 et 1830.* Maisons-Alfort: IEMVT.

Gille, Paul. 1964. "Les Écoles de constructeurs de la Marine," in René Taton, ed., *Enseignement*, pp. 477-80.

Gillet, Philippe. 1985. *Par Mets et par vins: Voyages et gastronomie en Europe (16e-18e siècles).* Paris: Payot.

Gillispie, Charles C. 2007. *Essays and Reviews in History and History of Science.* Philadelphia: American Philosophical Society.

__________. 2004. *Science and Polity in France: The Revolutionary and Napoleonic Years.* Princeton: Princeton University Press.

__________. 1992. "Palisot de Beauvois on the Americans," *Proceedings of the American Philosophical Society* 136, pp. 33-50.

__________. 1980. *Science and Polity in France at the End of the Old Regime.* Princeton: Princeton University Press.

Gillispie, Charles C., ed. 1970-1980. *Dictionary of Scientific Biography*, 14 vols. New York: Charles Schribners Sons.

Gillispie, Charles C., and Michel Dewachter. 1987. *Monuments of Egypt: The Napoleonic Edition.* Princeton: Princeton Architectural Press.

Gilmartin, David. 1994. "Scientific Progress and Imperial Science: Colonialism and Irrigation Technology in the Indus Basin," *The Journal of Asian Studies* 53, pp. 1127-49.

Girardin, Monique. 1994. *Bibliographie de l'Ile de la Réunion, 1973-1992.* Aix en Provence: Presses Universitaires d'Aix-Marseille.

Giroux, Jacqueline. 1981-1982. "Genèse de la Météorologie scientifique dans les milieux de l'Académie de Dijon au XVIIIe siècle," *Mémoires de l'Académie de Dijon* 75, pp. 135-55.

Glénisson, Jean-Louis. 1986. "La cartographie de Saint-Domingue dans la seconde moitié du XVIIIe siècle (de 1763 à la Révolution)." Thesis, École des Chartres, Paris.

Godfroy-Tayart de Borms, Marion. 2009a. "La guerre de Sept ans et ses conséquences atlantiques: Kourou ou l'apparition d'un nouveau système colonial," *French Historical Studies* 32-2, pp. 167-91.

__________. 2009b. *Kourou ou l'ultime combat de la monarchie pour une Amérique française 1763-1781.* Doctoral thesis, École des Hautes Études en Sciences Sociales (Paris).

Godlewska, Anne Marie Claire. 1999. *Geography Unbound: French Geographic Science from Cassini to Humboldt.* Chicago and London: The University of Chicago Press.

Goubert, Jean-Pierre. 2003. "Société Royale de Médecine," in Bély, ed., *Dictionnaire de l'Ancien Régime*, p. 1168.

Grabiner, Judith V. 1970. "Bezout, Étienne." *DSB*, vol. 2, pp. 111-14.

Granit, Ragnar. 1973. "Kalm, Pehr," *DSB*, vol. 7, pp. 210-11.

Greene, John C. 1959. *The Death of Adam: Evolution and its Impact on Western Thought.* Ames: Iowa State University Press.

Grove, Richard H. 1997. *Ecology, Climate and Empire: Colonialism and Global Environmental History, 1400-1940.* Cambridge: The White Horse Press.

__________. 1995. *Green Imperialism: Colonial Expansion, Tropical Island Edens and the Origins of Environmentalism, 1600-1860.* Cambridge: Cambridge University Press.

Gruzinski, Serge. 2004. *Les quatre parties du monde. Histoire d'une mondialisation.* Paris: Éditions de La Martinière.

Guéguen, Edouard. 1976. "Les jardins botaniques de Bretagne aux XVIIIe et XIXe siècles," in *Comptes rendues du 97e Congrès national des sociétés savan-*

tes, [Nantes, 1972], Section des Sciences, Vol. 1, Histoire des Sciences et des Techniques (Paris: CTHS), pp. 377-83.

Guennou, Jean. 1986. *Missions étrangères de Paris*. Paris: Fayard.

Guerra, Francisco. 1994. *Bibliographie medicale des Antilles françaises sous l'ancien regime*. Alcalá de Henares: Universidad de Alcalá de Henares.
__________. 1971. "Feuillée, Louis," *DSB*, vol. 4, pp. 602-603.

Guertin, Ghyslaine, and Laurine Quetin, eds. 2008. *Michel Paul-Guy de Chabanon et ses contemporains* [*Musicorum* 2007-2008]. Tours: Université François Rabelais.

Guillaumin, A., and V. Chaudun. 1957. "L'introduction en France des plantes horticoles orginaires de l'Améquie du nord avant 1850," in CNRS, *Les Botanistes français en Amérique du Nord*, pp. 123-35.

Guillemet, Roger. 1912. *Essai sur la surintendance des Bâtiments du roi sous le règne personnel de Louis XIV (1662-1715)*. Paris: Arthur Rousseau.

Gunny, Ahmad. 1981. "L'Ile Maurice et la France dans la deuxième moitié du siècle," *Dix-huitième siècle*, pp. 297-316.

Hahn, Roger. 1971. *The Anatomy of a Scientific Institution: The Paris Academy of Sciences, 1666-1803*. Berkeley: University of California Press.
__________. 1964a. "L'Enseignement scientifique des gardes de la marine au XVIIIe siècle," in René Taton, ed., *Enseignement et diffusion*, pp. 547-58.
__________. 1964b. "Les Observatoires en France au XVIIIe siècle," in René Taton, ed., *Enseignement et diffusion*, pp. 653-58.
__________. 1964c. "L'enseignement scientifique aux écoles militaires et d'artillerie," in René Taton, ed., *Enseignement et diffusion*, pp. 513-45.

Hall, Bert S. 2001. *Weapons and Warfare in Renaissance Europe: Gunpowder, Technology, and Tactics*. Baltimore and London: The Johns Hopkins University Press.

Halleux, Robert. 1998. "Machine," in Michel Blay and Robert Halleux, eds., *La science classique: XVIe-XVIIIe siècle: Dictionnaire critique* (Paris: Flammarion), pp. 581-89.

Halleux, Robert, James E. McClellan III, Daniela Berariu, and Geneviève Xhayet. *Les Publications de l'Académie Royale des Sciences de Paris (1666-1793)*, 2 vols. Turnhout, Belgium: Brepols Publishers, 2001.

Hamet, Dr. 1923. "L'École de Chirurgie de la Marine à Brest (1740-1798)," *Archives de Médecine et de Pharmacie Navale* 114, pp. 165-99, 245-59, 385-404.

Hamy, Ernest Théodore. 1908. "Charles Arthaud de Pont-à-Mousson (1748-1791)," *Bulletins et Mémoires de la Société d'Anthropologie de Paris* 9, pp. 295-314.

_________. 1890. *Origines du Musée d'Ethnographie.* Paris: Ernest Leroux.

_________. 1870. *Précis de paléontologie humaine.* Paris: J.-B. Baillière et fils.

Hannaway, Caroline. 1993. "Distinctive or Derivative? The French Colonial Medical Experience, 1740-1790," in Lafuente et al., *Mundalizacion de la ciencia*, pp. 505-10.

_________. 1974. "Medicine, Public Welfare and the State in 18th-Century France: The Société royale de Médecine of Paris: 1776-1793." Ph.D. dissertation, Johns Hopkins University.

_________. 1972. "The Société Royale de Médecine and Epidemics in the *Ancien Régime*," *Bulletin of the History of Medicine* 46, pp. 257-73.

Harris, Steven. J. 2005. "Jesuit Scientific Activity in the Overseas Missions, 1540-1773," *ISIS* 96, pp. 71-79.

Harrison, Mark. 2005. "Science and the British Empire," *ISIS* 96, pp. 56-63.

Harvie, David I. 2005. *Limeys: The Conquest of Scurvy.* Stroud, Gloucestershire, U.K.: Sutton Publishing.

Haudrère, Philippe. 2005. *La Compagnie Française des Indes au XIIIe siècle*, 2nd ed., 2 vols. Paris: Les Indes Savantes.

_________. 2003. "Compagnies de Commerce," in Bély, ed., *Dictionnaire de l'Ancien Régime*, pp. 305-307.

_________. 1998. "A la recherche de l'île d'Eden," in *Rochefort et la mer* 14, pp. 7-14. [Saintonge, Québec: Université Francophone d'Été, 1998].

_________. 1997. *Le grand commerce maritime au XVIIIe siècle: Européens et espaces maritimes.* Paris: Editions SEDES.

__________. 1992. *La Bourdonnais, Marin et aventurier*. Paris: Éditions Desjonquières.

__________. 1989. *La Compagnie Française des Indes au* XVIII*e siècle (1719-1795)*, 4 vols. Paris, Librairie de l'Inde.

Haudrère, Philippe, and Gérard Le Bouëdec, with the participation of Louis Mézin. 2005. *Les Compagnies des Indes*. Rennes: Édtions Ouest-France.

Haudrère, Philippe in collaboration with René Estienne and Gérard Le Bouëdec. 1996. *Les flottes des Compagnies des Indes 1600-1857* [Journées franco-britanniques d'histoire de la marine, Lorient, 4-6 mai 1994]. Vincennes: Service historique de la Marine.

Havard, Gilles, and Cécile Vidal. 2008. *Histoire de l'Amérique française, édition revue (2008)*. Paris: Éditions Flammarion/Champs histoire. (Original edition: Flammarion, 2003).

Headrick, Daniel R. 1988. *The Tentacles of Progress: Technology Transfer in the Age of Imperialism, 1850-1940*. New York: Oxford University Press.

_________. 1981. *The Tools of Empire: Technology and European Imperialism in the Nineteenth Century*. New York: Oxford University Press.

Henrat, Philippe. 1980. *Inventaire des Archives de la Marine, Sous-série B7 (Pays étrangers, commerce, consulats) Tome VI*. Paris: Archives Nationales.

Henrat, Philippe, et al. 1990. *Innovations techniques dans la Marine, 1641-1817: Mémoires et projets reçus par le départment de la Marine (Marine G 86 à 119)*. Paris: Archives Nationales.

Henwood, Philippe. 1987. "L'Académie de Marine à Brest au XVIIIe siècle," in Balcou, ed., *La Mer au siècle des Encyclopédies*, pp. 125-34.

__________. 1982. "Les archives de la Marine au port de Brest," *Chronique d'histoire maritime* 5, pp. 13-41.

Heuzé, Gustave. 1876. "Discours," in *Inauguration du monument élevé en l'honneur de Tessier à Angerville*, deuxième édition (Versailles: Cerf et fils), pp. 22-29.

Ho, Hai Quang. 1988. *Contribution à l'histoire économique de l'île de La Réunion (1642-1848)*. Paris and Montréal: Harmattan.

Hocquette, M. 1976. "Turpin, Pierre-Jean-Nicolas," *DSB*, vol. 13, pp. 506-7.

Home, R. W., ed. 1988. *Australian Science in the Making*. Cambridge: Cambridge University Press.

Home, Roderick W., and Sally Gregory Kohlstedt, eds. 1991. *International Science and National Scientific Identity: Australia between Britain and America*. Dordrecht: Kluwer Academic.

Homer, Isabelle. 1988. *Médecins et chirurgiens à Saint-Domingue au XVIIIe siècle*. Thesis for the Diplôme d'archiviste paléographe, École Nationale des Chartes.

Hostetler, Laura. 2009. "Contending Cartographic Claims? The Qing Empire in Manchu, Chinese, and European Maps," in James R. Akerman, ed., *The Imperial Map. Cartography and the Mastery of Empire* (Chicago and London: The University of Chicago Press), pp. 93-132.

__________. 2001. *Qing Colonial Enterprise: Ethnography And Cartography in Early Modern China*. Chicago. The University of Chicago Press.

Houth, Emile. 1931. "Deux Explorateurs botanistes versaillais oubliés," *Les Nouvelles Versaillaises*, 29 September 1931.

__________. 1930. "André-François Michaux," *Mémoires de la Société historique et archéologique de l'arrondissement de Pontoise et du Vexin* 40, p. 71.

Hroděj, Philippe. 1997. "Saint-Domingue en 1690. Les observations du Père Plumier, botaniste provençal," *Revue française d'histoire d'Outre-mer* 317, pp. 93-117.

Hsia, Florence C. 2009. *Sojourners in a Strange Land. Jesuits and Their Scientific Missions in Late Imperial China*. Chicago and London: The University of Chicago Press.

Huard, M. P., and M. Zobel. 1963. "La Société royale de médecine et le voyage de Lapérouse," *Comptes rendus du 87e Congrès national des Sociétés savantes* [Poitiers, 1962] (Paris: Imprimerie Nationale), pp. 83-91.

Huard, Pierre, Christiane Lacombe, and Jean Théodoridès. 1967. "Marine et Botanique aux XVIIIe et XIXe siècles: Les jardins botaniques de la Marine," *Actes du 91e Congres national des Sociétés savantes* [Rennes, 1966] (Paris: Bibliothèque Nationale), vol. 1, pp. 109-27.

Huetz de Lemps, Christian. 2005. "Colonies françaises," in Bluche, ed., *Dictionnaire du Grand Siècle, nouvelle édition revue et corrigée* (Paris: Fayard), pp. 354-55.

Hurbon, Laënnec. 1993. *Les mystères du Vaudou.* Paris, Gallimard.

Iliffe, Rob. 2003. "Science and Voyages of Discovery," in Roy Porter, ed., *The Cambridge History of Science, vol., 4: Eighteenth-Century Science* [Cambridge: Cambridge University Press, 2003], pp. 618-45.

Institut de France, Académie des Sciences. 1979. *Index biographique de l'Académie des sciences, 1666-1978.* Paris: Gauthier Villars.

Jacob, Christian. 1992. *L'empire des cartes: Approche théorique de la cartographie à travers l'histoire.* Paris: Albin Michel.

Jacob, Guy. 1990. "Le Madécasse et les Lumières: *Voyage à Madagascar* d'Alexis Rochon," in Guy Jacob, ed., *Regards sur Madagascar*, pp. 43-57.

Jacob, Guy, ed. 1990. *Regards sur Madagascar et la Révolution française.* Madagascar: CNAPMAD.

Jacquat, Marcel S. 1996. "Laurent Garcin: Médecin-Chirurgien, Naturaliste (1683-1752)," in Michel Schlup, ed., *Biographies Neuchâteloises,* vol. 1 (Hauterive, Switzerland: Gilles Attinger), pp. 103-109.

Jacquet, Pierre, 1999. "La découverte des Orchidées antillaises: Charles Plumier (1646-1704)," *L'Orchidophile*, April, no. 136, pp. 84-87.

Jameson, Robert. 1837. "Biographical Notice of the Mulatto M. Lislet-Geoffroy, Correspondent of the Academy of Sciences for the Section of Geography and Navigation," *The Edinburgh New Philosophical Journal* 22 (1836-1837), pp. 70-77.

Jami, Catherine. 2008. "Pékin, centre de savoirs entre deux mondes," *Revue d'histoire moderne et contemporaine* 55-2, pp. 43-69.

Jami, Catherine, ed. 2004. "Science et religion en Chine," *Annales Histoire, Sciences Sociales*, pp. 697-756.

Jandin, Stéphanie. 1994-1995. "L'Itinéraire d'un naturaliste, Louis-Claude Richard (1754-1821)." Maîtrise d'histoire, Université Paris 7. [PA Archives cote: 4o/224].

Janrot, Léon. 1935. "La Feuille du Cultivateur, notes sur un journal pendant la Révolution," *Actes de la Société royale d'agriculture*, pp.148-60.

Jaoul, Martine, and Madeleine Pinault. 1986. "La collection « Description des Arts et Métiers »: Sources inédits provenant du château de Denainvillers," *Ethnologie française* 16, pp. 7-38.

Jaquel, Roger. 1976. "L'astronome français Joseph-Nicolas Delisle (1688-1768) et Christfried Kirch (1694-1740), Directeur de l'Observatoire de Berlin (1716-1740)," *Actes du 97e Congrès National des Sociétés Savantes* (Nantes, 1972), Section des Sciences, vol. 1 (Paris: Bibliothèque Nationale), pp. 407-432.

Jardine, Nicolas, James E. Secord, and Emma C. Spary, eds. 1996. *Cultures of Natural History.* Cambridge, Cambridge University Press.

Jasanoff, Sheila. 2004a. "The Idiom of Co-Production," in Sheila Jasanoff, ed., *States of Knowledge* [London: Routledge], pp. 1-12.
__________. 2004b. "Ordering Knowledge, Ordering Society," in Sheila Jasanoff, ed., *States of Knowledge* [London: Routledge], pp. 13-45.

Jolinon, Jean-Claude. 2004. "Une femme autour du monde [Jeanne Baret et Commerson]," in *L'herbier du monde: Cinq siècles d'aventures et de passions botaniques au Muséum national d'histoire naturelle*, Philippe Morat, Gérard Aymonin, and Jean-Claude Jolinon eds. (Paris: Editions du Muséum et L'iconoclaste), pp. 78-89.

Juhé-Beaulaton, Dominique. 1999. "Du jardin royal des plantes médicinales de Paris aux jardins coloniaux: développement de l'agronomie tropicale française," in Fischer, ed., *Le jardin*, pp. 267-84.

Junges, Catherine. 2005. "Les publications de l'Académie de marine," in Charon et al., eds., *Le livre maritime*, pp. 69-76.
__________. 2002. "L'Académie royale de marine et la diffusion du savoir maritime," *Neptunia* 226, pp. 14-16.

Justin, Émile. 1935. *Les Sociétés royales d'agriculture au XVIIIe siècle (1757-1793.)* Saint-Lô [no publisher].

Keay, John. 1993. *The Honourable Company: A History of The English East India Company.* London: HarperCollins.

Kellman, Jordan. 1998. *Discovery and Enlightenment at Sea: Maritime Exploration and Observation in the 18th-Century French Scientific Community.* Ph.D. dissertation, Princeton University, 1998.

Kernéis, J.-P., J. Doucet, and J. Meyer. 1963. "Jardins de Nantes. Jardins d'histoire ou le rôle botanique des armateurs, des savantes et des « navigans » de Nantes au XVIIIe siècle," in *Comptes rendus du 87e Congrès national des Sociétés savantes* [Poitiers, 1962] (Paris: Gauthier-Villars), pp. 59-69.

Knight, David. 2004. "Essay Review: Romanticism and Science," *Annals of Science* 61, pp. 483-87.

Konvitz, Josef W. 1987. *Cartography in France, 1660-1848: Science, Engineering and Statecraft*. Chicago: University of Chicago Press.

Krakovitch, Odile. 1993. *Arrêts, déclarations, édits et ordonnances concernant les colonies (1666-1779). Inventaire analytique de la série Colonies A*. Paris: Archives Nationales.

Kumar, Deepak. 1994. *Science and the Raj: 1857-1905*. New Delhi: Oxford University Press.

Kumar, Deepak, ed. 1991. *Science and Empire: Essays in Indian Context, 1700-1947*. Delhi: Anamika Prakashan.

Kury, Lorelai. 1999. "André Thouin et la nature exotique au Jardin des plantes," in Fischer, ed., *Le jardin*, pp. 255-65.

__________. 1995. "Civiliser la nature: histoire naturelle et voyages (France, fin du XVIIIe siècle, début XIXe siècle)." Thesis, EHESS, Paris.

Lacouture, Jean. 1991. *Jésuites: Une Multibiographie. Vol. 1: Les Conquérants*. Paris: Éditions du Seuil.

Lacroix, Alfred. 1932-1938. *Figures des Savants*, 4 vols. Paris: Gauthier-Villars. [Vols. 3-4 (1938) = *L'Académie des Sciences et l'étude de la France d'outre-mer de la fin du XVIIe siècle au début du XIXe*].

__________. 1934."Notice historique sur les membres et correspondants de l'Académie des sciences ayant travaillé dans les colonies française des Mascareignes et de Madagascar." Paris: Gauthier-Villars. ["Lecture faite en la séance annuelle du 17 décembre 1934."]

Lacroix, Jean Bernard. 1986. *Les Français au Sénégal au temps de la Compagnie des Indes de 1719 à 1758.* Vincennes: S.H.M.

_________. 1978. "L'approvisionnement des ménageries et les transports d'animaux sauvages par la Compagnie des Indes au XVIIIe siècle," *Revue française d'histoire d'Outre-mer*, 65 #239, pp. 153-79.

Lacroix, Sylvie. 1999. "Les transports de plantes aux Antilles, l'Exemple de l'arbre à pain," *Rochefore et la Mer* [*Les Antilles et la Guyane au XVIIIe siècle*] 16, pp. 65-77.

Lafrance, Marc, with André Charbonneau. 1999. "La Nouvelle-France," in Vidal and D'Orgeix, eds. *Les villes françaises du Nouveau Monde*, pp. 78-87.

Lafuente, Antonio, and José Sala Catalá, eds. 1992. *Ciencia colonial en América.* Madrid: Alianze Editorial.

Lafuente, A., A. Elena, and M. L. Ortega, eds. 1993. *Mundialización de la ciencia y cultura nacional: Actas del Congreso Internacional "Ciencia, descubrimiento y mundo colonial.*" Madrid: Ediciones Doce Calles.

Lagrave, Jean-Paul de. 1975. *Les Origines de la presse au Québec.* Montréal: Éditions de Lagrave.

Laissus, Yves. 2005. "Jardin des plantes de Paris," in Bluche, ed., *Dictionnaire du Grand Siècle, nouvelle édition revue et corrigée* (Paris: Fayard), p. 783.

_________. 2003. "Jardin du Roi," in Bély, ed., *Dictionnaire de l'Ancien Régime*, pp. 687-88.

_________. 1981. "Les voyageurs naturalistes du Jardin du Roi et du Muséum d'histoire naturelle," *Revue d'histoire des sciences* 34, pp. 259-317.

_________. 1973. "Note sur les manuscrits de Pierre Poivre (1719-1786)," *Proceedings of the Royal Society of Arts and Sciences of Mauritius* 4, pp. 31-56.

_________. 1971. "Commerson," *DSB*, vol. 3, pp. 365-66.

_________. 1970. "Note sur le deuxième voyage et la mort de Joseph Dombey," *Comptes rendus du 94e Congrès national des sociétés savantes (Pau, 1969),* vol. 1 (Paris: Bibliothèque Nationale), pp. 61-79.

_________. 1964. "Le Jardin du Roi," in René Taton, ed., *Enseignement et diffusion*, pp. 287-341.

Laissus, Yves, ed. 1995. *Les naturalistes français en Amérique du Sud* XVIe-XIXe *siècles*. Paris: Éditions du CTHS.

Lamy, Édouard. [1930]. *Les Cabinets d'Histoire Naturelle en France au* XVIIIe *siècle et le Cabinet du Roi (1635-1793)*. Paris: (private publication).

Lamy, Gabriela. 2005. "L'Éducation d'un jardinier royal au Petit Trianon: Antoine Richard (1734-1807)," *POLIA, Revue de l'art des jardins* 4, pp. 57-73.

Lamontagne, Roland. 1965. "La participation canadienne à l'oeuvre minéralogique de Guettard," *Revue d'histoire des sciences* 18, pp. 385-88.

__________. 1964a. *Chabert de Cogolin et l'expédition de Louisbourg*. Montréal: Éditions Leméac.

__________. 1964b. "Lettre de Jean-François Gauthier à Jean-Etienne Guettard," *Revue d'Histoire de l'Amérique Française* 17, pp. 569-72.

__________. 1963. "Jean Prat, Correspondant de Bernard de Jussieu," *Rapport des Archives du Québec* 41, pp. 121-49.

__________. 1962a. *La Galissonnière et le Canada*. Montréal: Presses de l'Université de Montréal.

__________. 1962b. "Le dossier biographique de Jean Prat," *Revue d'histoire de l'Amérique française* 16, pp. 219-24.

__________. 1961. "La Galissonnière, directeur du dépôt de la marine," *Revue d'histoire des sciences* 14, pp. 19-26.

__________. 1960. "La contribution scientifique de La Galissonnière au Canada," *Revue d'Histoire de l'Amérique Française* 13, pp. 509-24.

Landry-Deron, Isabelle. 2001. "Les Mathématiciens envoyés en Chine par Louis XIV en 1685," *Archive for the History of Exact Sciences* 55, pp. 423-63.

Langlois, Gilles-Antoine. 2003. *Des villes pour la Louisiane française: Théorie et pratique de l'urbanistique coloniale au 18e siècle*. Paris: L'Harmattan.

__________. 1999a. "L'aventure urbaine de la Louisiane," in Vidal and D'Orgeix, eds. *Les villes françaises du Nouveau* Monde, pp. 120-129.

__________. 1999b. "Les *mousquetaires* du Mississippi," in Vidal and D'Orgeix, eds. *Les villes françaises du Nouveau* Monde, pp. 147-51.

Lapierre, Robert, ed. 1982. "R. P. Adrien Le Breton: Manuscrits du fonds Jussieu, et Relation historique sur l'île caraïbe de Saint-Vincent," *Annales des Antilles* 25, pp. 5-132.

La Roncière, Charles Bourel de. 1907, 1924. *Catalogue des manuscrits des bibliothèques publiques de France. Bibliothèques de la Marine* (1907), *Supplément* (1924). Paris: Librairie E. Plon, Nourrit et Cie.

Latour, Bruno. 1987. *Science in Action: How to Follow Scientists and Engineers through Society.* Cambridge, Mass.: Harvard University Press.

Laurent, Charles. 1987. "Le Commissaire général de la Marine André-François Boureau-Deslandes," in Balcou, ed., *La Mer au Siècle des Encyclopédies*, pp. 195-207.

Lawrence, George H. M., ed. 1963-1964. *Adanson: The Bicentennial of Michel Adanson's « Familles des plantes »*, 2 vols. Pittsburgh: The Hunt Botanical Library.

Le Blanc, François-Yves. 1999. "François Blondel et les îles d'Amérique (1666-1668)," in Vidal and D'Orgeix, eds. *Les villes françaises du Nouveau Monde,* pp. 156-61.

Lefèvre, A. 1867. *Histoire du Service de Santé de la Marine militaire et des écoles de médecine navale en France, depuis le règne de Louis XIV jusqu'à nos jours, 1666-1876.* Paris: J. B. Gaillière et fils.

Le Gouic, Olivier. 2005. "Pierre Poivre et les épices: une transplantation réussie," in Llinares and Hroděj, eds., *Techniques et colonies*, pp. 103-26.

Legrand, A., and Félix Marec. 1978. *Inventaire des Archives de la Compagnie des Indes (Sous-Série 1P).* Paris: Service Historique de la Marine.

Le Guisquet, Bernard. 1999. "Le Dépôt des cartes, plans et journaux de la Marine sous l'Ancien Régime," *Neptunia* 214, pp. 3-21.

Le Maresquier, Érik. 1995. *Guide du Lecteur des Archives de la Marine.* Vincennes: Service Historique de la Marine.

Lénardon, Dante. 1986. *Index du Journal de Trévoux: 1701-1767.* Geneva: Slatkine.

Le Roux, Ronan, Vincent Bontems, and Paul Braffort, eds., "Les machines: Objets de connaissance," *Revue de Synthèse* 130, #1 (2009).

Leroy, Jean F. “Note sur l’introduction des plantes Américaines en France dans la première moitié du XVIIIe siècle (Bernard de Jussieu et Prat),” in CNRS, *Les Botanistes français en Amérique du Nord*, pp. 285-86.

Lescure, Jean. 1992. “L’épopée des voyageurs naturalistes aux Antilles et en Guyane,” *Voyage aux Iles d’Amérique*, Catalogue de l’Exposition (Paris: Archives Nationales), pp. 129-43.

Le Seigneur, Marie Jacques. 1995. “Un naturaliste français en Guyane: Jacques-François Artur, médecin du roi à Cayenne, 1736-1771,” in Laissus, ed., *Les naturalistes français*, pp. 137-56.

Lespagnol, André. 1996. “Armements des Compagnies des Indes et armements privés français,” in Haudrère et al., *Les flottes des Compagnies des Indes*, pp. 285-95.

Lesueur, Boris. 2007. *Les troupes coloniales sous l’Ancien Régime*. Thèse de Doctorat, Université François Rabelais de Tours.

Letouzey, Yvonne. 1989. *Le jardin des Plantes à la croisée des chemins avec A. Thouin (1747-1824)*. Paris: Muséum National d’Histoire Naturelle.

Levot, P. 1844. “Notices biographiques sur M. le vicomte Bigot de Morogues, fondateur de l’Académie de marine,” *Revue Bretonne*, vol. 4, pp. 64-88.

Lévy, Jacques. 2005. “Richer (Jean),” in Bluche, ed., *Dictionnaire du Grand Siècle, nouvelle édition revue et corrigée* (Paris: Fayard), pp. 1341-42.

Li, Shenwen. 2001. *Stratégie missionnaire des jésuites français en Nouvelle-France et en Chine au XVIIe siècle*. Québec and Paris: Presses de l’Université Laval and L’Harmattan.

Library and Archives Canada. 1959-2009. *Dictionary of Canadian Biography Online*. [www.biographi.ca].

Lilti, Antoine. 2005. *Le monde des salons: Sociabilité et mondanité à Paris au XVIIIe siècle*. Paris: Fayard.

Lionnet, Guy. 1984. “La Lodoïcée ou Cocotier de mer des îles Seychelles,” *Histoire et Nature* 24-25, pp. 121-43.

Litalien, Raymonde. 1989. "Les botanistes français en Nouvelle France," *Rochefort et la Mer [Grand voyages de découverte XVIIe-XIXe siècles, Botanistes et Naturalistes]* 5, pp. 9-22.

Litalien, Raymonde, Jean-François Palomino, and Denis Vaugeois. 2007. *La mesure d'un continent. Atlas historique de l'Amérique du nord. 1492-1814.* Sillery (Quebec) and Paris: Editions du Septentrion and Presses de l'Université Paris Sorbonne.

Llinares, Sylviane. 2005. "Marine, colonies et innovations techniques: quelques exemples de transfert (XVIIIe et Ière moitié du XIXe siècles)," in Llinares and Hrodĕj, eds., *Techniques et colonies*, pp. 163-81.

_________. 1996. "La flotte de la Compagnie française des Indes au XVIIIe siècle: une normalisation inachevée," in Haudrère et. al., *Les flottes des Compagnies des Indes*, pp. 41-47.

_________. 1994. *Marine, propulsion et technique, l'évolution du système technologique du navire de guerre français au XVIIIe siècle*, 2 vols. Paris: Librarie de l'Inde.

Llinares, Sylviane, and Philippe Hrodĕj, eds. 2005. *Techniques et colonies (XVIe-XXe siècles)* [Actes du colloque de Lorient (2002)]. Paris: Publication de la Société Française d'histoire d'outre-mer et l'Université de Bretagne Sud – SOLITO.

Lortie, Léon. 1966. "La Trame scientifique de l'histoire du Canada," in G. F. G. Stanley, ed., *Pioneers of Canadian Science/Les Pionniers de la science canadienne* (Toronto: University of Toronto Press), pp. 3-35.

Lougnon, Albert. 1956. *Classement et inventaire du fonds de la Compagnie des Indes (Série C^o^) 1665-1767. Suivi de l'Inventaire du fonds de la Compagnie des Indes des Archives de l'Ile de France.* Nérac: G. Couderc.

_________. 1953. *Documents concernant les Iles de Bourbon et de France pendant la régie de la Compagnie des Indes.* Nérac: G. Couderc.

_________. 1933-1949. *Correspondance du Conseil Supérieur de Bourbon et de la Compagnie des Indes*, 5 vols. Saint-Denis, Ile de la Réunion: Gaston Daudé; Paris: E. Leroux. [Vol. 1, 1934 (22 janvier 1724-30 décembre 1731); vol. 2, 1933 [sic] (10 Mars 1732 – 23 Janvier 1736); vol. 3, 1937 (23 janvier 1736-9 mai 1741); vol. 4, 1940 (9 novembre 1740-20 avril 1746); vol. 5, 1949 (17 avril 1746-17 octobre 1750)].

Ly, Abdoulaye. 1993. *La Compagnie du Sénégal.* Paris: Karthala. [Original edition, 1958].

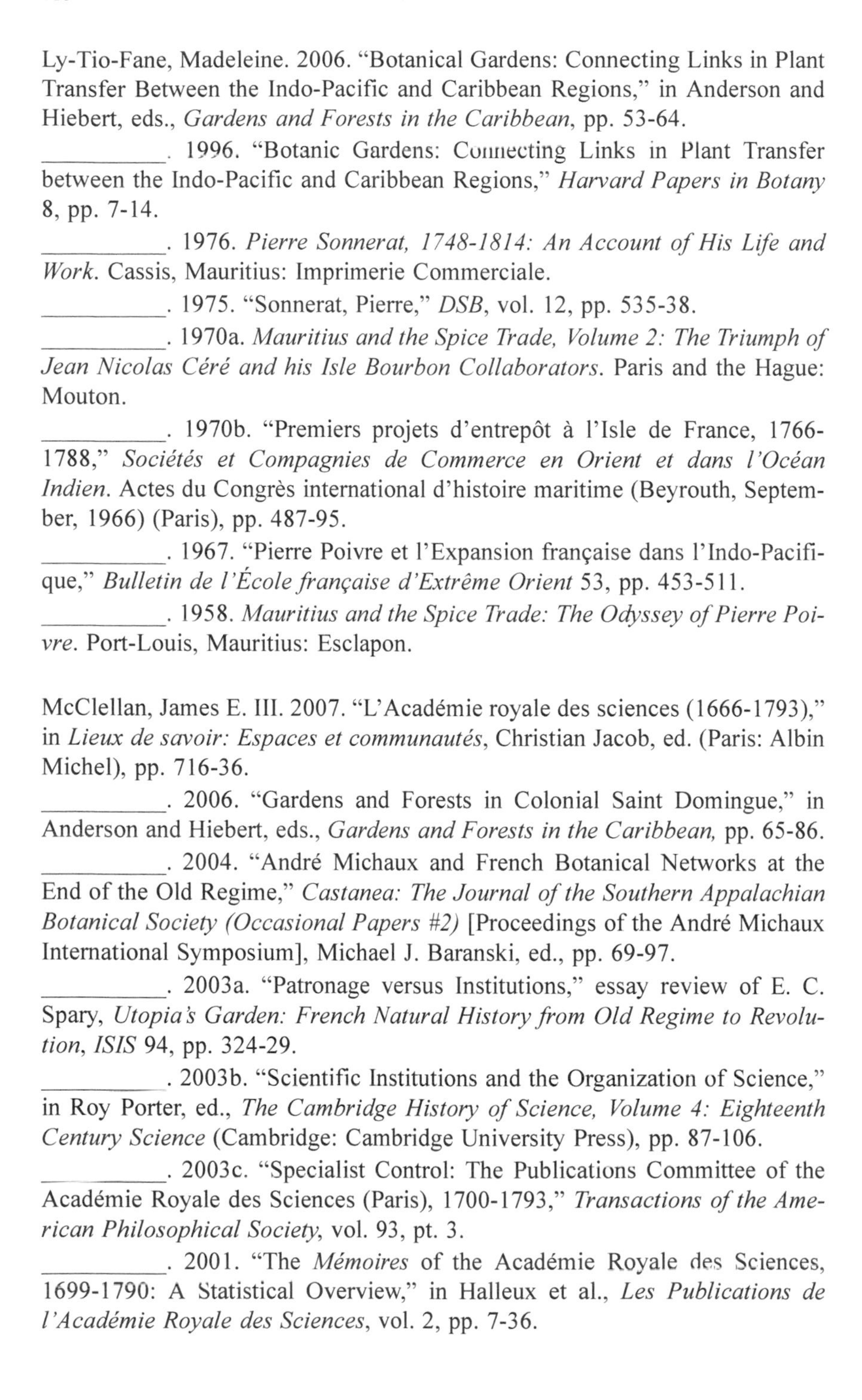

Ly-Tio-Fane, Madeleine. 2006. "Botanical Gardens: Connecting Links in Plant Transfer Between the Indo-Pacific and Caribbean Regions," in Anderson and Hiebert, eds., *Gardens and Forests in the Caribbean*, pp. 53-64.

__________. 1996. "Botanic Gardens: Connecting Links in Plant Transfer between the Indo-Pacific and Caribbean Regions," *Harvard Papers in Botany* 8, pp. 7-14.

__________. 1976. *Pierre Sonnerat, 1748-1814: An Account of His Life and Work*. Cassis, Mauritius: Imprimerie Commerciale.

__________. 1975. "Sonnerat, Pierre," *DSB*, vol. 12, pp. 535-38.

__________. 1970a. *Mauritius and the Spice Trade, Volume 2: The Triumph of Jean Nicolas Céré and his Isle Bourbon Collaborators*. Paris and the Hague: Mouton.

__________. 1970b. "Premiers projets d'entrepôt à l'Isle de France, 1766-1788," *Sociétés et Compagnies de Commerce en Orient et dans l'Océan Indien*. Actes du Congrès international d'histoire maritime (Beyrouth, September, 1966) (Paris), pp. 487-95.

__________. 1967. "Pierre Poivre et l'Expansion française dans l'Indo-Pacifique," *Bulletin de l'École française d'Extrême Orient* 53, pp. 453-511.

__________. 1958. *Mauritius and the Spice Trade: The Odyssey of Pierre Poivre*. Port-Louis, Mauritius: Esclapon.

McClellan, James E. III. 2007. "L'Académie royale des sciences (1666-1793)," in *Lieux de savoir: Espaces et communautés*, Christian Jacob, ed. (Paris: Albin Michel), pp. 716-36.

__________. 2006. "Gardens and Forests in Colonial Saint Domingue," in Anderson and Hiebert, eds., *Gardens and Forests in the Caribbean*, pp. 65-86.

__________. 2004. "André Michaux and French Botanical Networks at the End of the Old Regime," *Castanea: The Journal of the Southern Appalachian Botanical Society (Occasional Papers #2)* [Proceedings of the André Michaux International Symposium], Michael J. Baranski, ed., pp. 69-97.

__________. 2003a. "Patronage versus Institutions," essay review of E. C. Spary, *Utopia's Garden: French Natural History from Old Regime to Revolution*, *ISIS* 94, pp. 324-29.

__________. 2003b. "Scientific Institutions and the Organization of Science," in Roy Porter, ed., *The Cambridge History of Science, Volume 4: Eighteenth Century Science* (Cambridge: Cambridge University Press), pp. 87-106.

__________. 2003c. "Specialist Control: The Publications Committee of the Académie Royale des Sciences (Paris), 1700-1793," *Transactions of the American Philosophical Society*, vol. 93, pt. 3.

__________. 2001. "The *Mémoires* of the Académie Royale des Sciences, 1699-1790: A Statistical Overview," in Halleux et al., *Les Publications de l'Académie Royale des Sciences*, vol. 2, pp. 7-36.

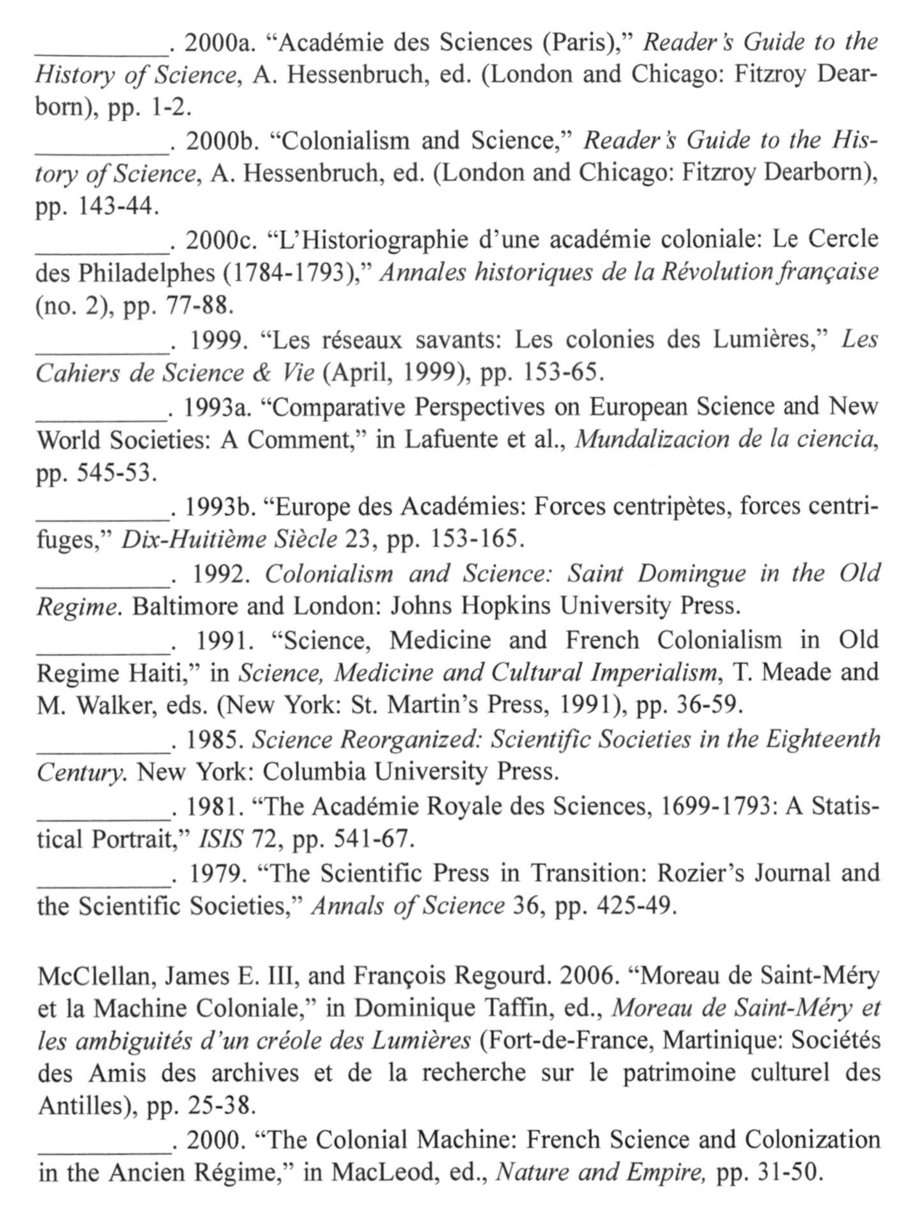

__________. 2000a. "Académie des Sciences (Paris)," *Reader's Guide to the History of Science*, A. Hessenbruch, ed. (London and Chicago: Fitzroy Dearborn), pp. 1-2.

__________. 2000b. "Colonialism and Science," *Reader's Guide to the History of Science*, A. Hessenbruch, ed. (London and Chicago: Fitzroy Dearborn), pp. 143-44.

__________. 2000c. "L'Historiographie d'une académie coloniale: Le Cercle des Philadelphes (1784-1793)," *Annales historiques de la Révolution française* (no. 2), pp. 77-88.

__________. 1999. "Les réseaux savants: Les colonies des Lumières," *Les Cahiers de Science & Vie* (April, 1999), pp. 153-65.

__________. 1993a. "Comparative Perspectives on European Science and New World Societies: A Comment," in Lafuente et al., *Mundalizacion de la ciencia*, pp. 545-53.

__________. 1993b. "Europe des Académies: Forces centripètes, forces centrifuges," *Dix-Huitième Siècle* 23, pp. 153-165.

__________. 1992. *Colonialism and Science: Saint Domingue in the Old Regime*. Baltimore and London: Johns Hopkins University Press.

__________. 1991. "Science, Medicine and French Colonialism in Old Regime Haiti," in *Science, Medicine and Cultural Imperialism*, T. Meade and M. Walker, eds. (New York: St. Martin's Press, 1991), pp. 36-59.

__________. 1985. *Science Reorganized: Scientific Societies in the Eighteenth Century*. New York: Columbia University Press.

__________. 1981. "The Académie Royale des Sciences, 1699-1793: A Statistical Portrait," *ISIS* 72, pp. 541-67.

__________. 1979. "The Scientific Press in Transition: Rozier's Journal and the Scientific Societies," *Annals of Science* 36, pp. 425-49.

McClellan, James E. III, and François Regourd. 2006. "Moreau de Saint-Méry et la Machine Coloniale," in Dominique Taffin, ed., *Moreau de Saint-Méry et les ambiguités d'un créole des Lumières* (Fort-de-France, Martinique: Sociétés des Amis des archives et de la recherche sur le patrimoine culturel des Antilles), pp. 25-38.

__________. 2000. "The Colonial Machine: French Science and Colonization in the Ancien Régime," in MacLeod, ed., *Nature and Empire,* pp. 31-50.

McCook, Stuart. 2002. *States of Nature: Science, Agriculture, and Environment in the Spanish Caribbean, 1760-1940*. Austin: University of Texas Press.

MacLeod, Roy. 2005. "Colonial Science under the Southern Cross: Archibald Liversidge, FRS, and the Shaping of Anglo-Australian Science," in Stuchtey, ed., *Science across the European Empires,* pp. 176-213.

__________. 2000. "Introduction," in MacLeod, ed., *Nature and Empire*, pp. 1-13.

__________. 1987. "On Visiting the 'Moving Metropolis': Reflections on the Architecture of Imperial Science," in Reingold and Rothenberg, eds. *Scientific Colonialism,* pp. 217-49.

MacLeod, Roy, ed. 2000. *Nature and Empire: Science and the Colonial Enterprise.* Chicago: The University of Chicago Press. [*OSIRIS* 15 (2000)].

MacLeod, Roy, and Deepak Kumar, eds. 1995. *Technology and the Raj: Perspectives on the Transfer of Western Technology and Attitudes towards Industrial Change in India, 1700-1947.* New Dehli: Sage Publications.

MacLeod, Roy, and Philip F. Rehbock, eds. 1988. *Nature in Its Greatest Extent: Western Science in the Pacific.* Honolulu: University of Hawaii Press.

McNeil, William H. 1984. *The Pursuit of Power: Technology, Armed Force, and Society since A.D. 1000.* Chicago: University of Chicago Press.

Maindron, Ernest. 1881. *Les Foundations de prix à l'Académie des sciences. Les Lauréats de l'Académie, 1714-1880.* Paris: Gauthier-Villars.

Malleret, Louis. 1968. "Pierre Poivre, l'abbé Galloys et l'introduction d'espèces botaniques et d'oiseaux de Chine à l'Ile Maurice," *Proceedings of the Royal Society of Arts and Sciences of Mauritius* 3, part 1, pp. 117-30.

Mandonnet, Pierre. 1893. *Les Dominicains et la découverte de l'Amérique.* Paris: P. Lethielleux.

Mapp, Paul. 2005. "French Geographic Conceptions of the Unexplored American West and the Louisiana Cession of 1762," in Bradley G. Bond, ed., *French Colonial Louisiana and the Atlantic World* (Baton Rouge: Louisiana State University Press), pp. 134-74.

Marchand, Philip. 2007. *Ghost Empire: How the French Almost Conquered North America.* Westport, Conn.: Praeger.

Margadant, Willem. 1963. "The Adanson Collection of Botanical Books and Manuscripts," in Lawrence, ed., *Adanson,* vol. 1, pp. 265-368.

Marion, Marcel. *Dictionnaire des institutions de la France aux* XVIIe *et* XVIIIe *siècles.* Paris: A. & J. Picard, 1923.

Martin, Gaston. 1931. *Nantes au XVIIIe siècle: L'ère des négriers, 1714-1774.* Paris: PUF. [Reprint: Paris: Karthala, 1993.]

Martine, Jean-Luc. 2003. *Introduction à l'étude des métaphores de la machine dans la pensée de l'âge classique: L'idée de machine dans le discours des dictionnaires, 1514-1798.* Doctoral thesis. Université de Caen (France).

Mathieu, Jacques, in collaboration with André Daviault. 1998. *Le premier livre de plantes du Canada: Les enfants des bois du Canada au Jardin du Roi à Paris en 1635.* Sainte-Foy: Les Presses de l'Université Laval.

Maurel, Blanche. 1935. "La Poste entre la France et les Isles de l'Amérique à la fin de l'ancien régime," *Revue de la Société d'Histoire et de Géographie d'Haïti* 6, pp. 1-12.

Meade, Teresa, and Mark Walker, eds. 1991. *Science, Medicine and Cultural Imperialism.* New York: St. Martin's Press.

Meheust, Bertrand. 1998. *Somanbulisme et médiumnité. 1784-1930.* Le Plessis-Robinson (France): Institut Synthélabo.

Mélançon, François. 2007. *Le livre à Québec dans le premier XVIIIe siècle. La migration d'un objet culturel.* Doctoral thesis, University of Sherbrook (Canada).

__________. 2005. "La migration d'un objet culturel: le livre en Nouvelle-France," in Philippe Joutard and Thomas Wien, eds. *Mémoires de Nouvelle-France. De France en Nouvelle-France* (Rennes: Presses Universitaires de Rennes), pp. 265-72.

Melguen, Marthe. 2005. "French Voyages of Exploration and Science in the Age of Enlightenment: An Ocean of Discovery throughout the Pacific Ocean," in *Voyages of Discovery: Parting the Seas of Information Technology: Proceedings of the 30th Annual Conference of the International Association of Aquatic and Marine Science Libraries and Informations Centers (IAMSLIC)*, J. W. Markham and A. L. Duda, eds. (Fort Pierce, Fl.: IAMSLIC), pp. 31-59.

Mellot, Jean-Dominique, and Élisabeth Queval. 1997. *Répertoire d'imprimeurs/libraires, XVIe-XVIIIe siècle. État en 1995.* Paris: Bibliothèque Nationale de France.

Ménier, Marie-Antoinette, and Gabriel Debien. 1949. "Journaux de Saint-Domingue," *Revue d'Histoire des colonies* 36, pp. 424-75.

Menier, Marie-Antoinette, Étienne Taillemite, and Gilberte de Forges. 1976, 1983. *Inventaire des Archives Coloniales: Correspondance à l'arrivée en provenance de la Louisiane*, 2 vols. Paris: Archives Nationales. [Vol. 1, 1976; vol. 2, 1983].

Métraux, Alfred. 1972. *Voodoo in Haiti*. New York: Schocken Books.

Meyer, Jean, Jean Tarrade, and Annie Rey-Goldzeiguer. 1991. *Histoire de la France coloniale I: La conquête*. Paris: Armand Colin.

Michel, Jacques. 1989. *La Guyane sous l'Ancien Régime. Le désastre de Kourou et ses scandaleuses suites judiciaires*. Paris: L'Harmattan.

Middleton, W. E. Knowles. 1970. "Bouguer, Pierre," *DSB*, vol. 2, pp. 343-44.

Minard, Philippe. 1998. *La fortune du colbertisme: État et industrie dans la France des Lumières*. Paris: Fayard.

Ministère de la Marine. 1891. "Le lieutenant général Bigot, Vicomte de Morogues, inspecteur général de l'artillerie de Marine (1706-1781)," *Mémorial de l'artillerie de la Marine* 29, pp. 115-62.

Mintz, Sidney W. 1985. *Sweetness and Power: The Place of Sugar in Modern History*. New York: Viking.

Mintz, Sidney W., and Sally Price, eds. 1985. *Caribbean Contours*. Baltimore and London: Johns Hopkins University Press.

Mollat du Jourdin, Michel, and Étienne Taillemite, eds. 1982. *L'Importance de l'exploration maritime au siècle des Lumières: à propos du voyage de Bougainville*. Paris: C.N.R.S.

Moreau, Jean-Pierre. 1987. *Un flibustier français dans la mer des Antilles en 1618-1620 (Manuscrit inédit du début du XVIIe siècle [Anonyme de Carpentras]*. Clamart, chez l'auteur; reprint, Paris: Payot, 1994.

Mousnier, Roland. 1974. *Les institutions de la France sous la monarchie absolue, 1598-1789*, 2 vols. Paris: Presses Univisitaires de France.

Muller, Pascale. 1975. *Les Eaux minérales en France à la fin du XVIIIe siècle*. Masters Thesis, Université Paris I.

Müller-Wille, Staffan. 2005. "Walnuts at Hudson Bay, Coral Reefs in Gotland: The Colonialism of Linnaenan Botany," in Schiebinger and Swan, eds., *Colonial Botany*, pp. 34-48.

Murr, Sylvia. 1987. *L'Indologie du père Coeurdoux: Stratégies, apologétique et scientificité*. Paris: École française d'Extrême-Orient.

Musée National dc la Marine. 2008. *Le mystère de La Pérouse. [Catalogue de l'Exposition, 19 Mars – 20 Octobre 2008]*. Paris: Musée National de la Marine.

Nelson, Brian. 2006. *The Making of the Modern State: A Theoretical Evolution*. New York: Palgrave/Macmillan.

Neuville, Didier. 1898. *L'État sommaire des Archives de la Marine antérieures à la Révolution*. Paris: L. Baudoin; Kraus Reprint, 1977.

_________. 1885-1913. *Inventaire des Archives de la Marine, Série B, Service Général,* 7 vols. in 5. Paris, L. Baudoin.

_________. 1882. *Les établissements scientifiques de l'ancienne marine, vol. 1: Écoles d'hydrographie ingénieurs de la marine au* XVII*e siècle*. Paris: Berger-Levrault.

Neuville, Élisabeth, ed. 1999. *Le Chevalier de Borda: Un officier savant du* XVIII*e siècle, 1733-1799. Exposition Musée de Borda. Dax, 10 avril-30 octobre 1999*. Dax: Ville de Dax.

Nicolas, Jean-Paul. 1970. "Adanson, Michel," *DSB*, vol. 1, pp. 58-59.

_________. 1963a. "Adanson, the Man," in Lawrence, ed., *Adanson*, vol. 1, pp. 1-121.

_________. 1963b. "Adanson et le Mouvement Colonial," in Lawrence, ed., *Adanson*, vol. 2, pp. 393-450.

Niderlinder, Alain. 2002. "Les collections de l'Académie royale de marine," *Neptunia* 226, pp. 10-13.

Nolhac, Pierre de. 1927. *Trianon*. Paris: Louis Conard.

Olmsted, John W. 1960. "The Voyage of Jean Richer to Acadia in 1670: A Study in the Relations of Science and Navigation under Colbert," *Proceedings of the American Philosophical Society* 104, pp. 612-35.

_________. 1942. "The Expedition of Jean Richer to Cayenne (1672-1673)," *ISIS* 34, pp. 117-28.

Osborne, Michael A. 2005. "Science and the French Empire," *ISIS* 96, pp. 80-87.

_________. 1994. *Nature, the Exotic, and the Science of French Colonialism.* Bloomington and Indianapolis: Indiana University Press.

Ott, Thomas O. 1973. *The Haitian Revolution, 1789-1804*. Knoxville: University of Tennessee Press.

Palomino, Jean-François. 1999. "Jean-Baptiste Franquelin: un géographe du roi à Québec" in Vidal and D'Orgeix, eds. *Les villes françaises du Nouveau Monde,* pp. 162-67.

Parker, Geoffrey. 1988. *The Military Revolution: Military Innovation and the Rise of the West, 1000-1800*. New York: Cambridge University Press.

Pascal, Joseph-François-Jacques. 1934. *Société royale de médecine et eaux minérales*. Thèse de médecine. Paris. [BN Th Paris 10803.]

Passy, Louis. 1912. *Histoire de la Société nationale d'agriculture, tome 1 (1761-1793)*. Paris: Renouard.

Pedley, Mary Sponberg. 2005. *The Commerce of Cartography: Making and Marketing Maps in Eighteenth-Century France and England.* Chicago and London: The University of Chicago Press.

_________. 1992. *Bel et Utile:The Work of the Robert de Vaugondy Family of Mapmakers.* Tring, Herts, U.K.: Map Collector Publications.

Pelletier, Monique. 2002. *Les cartes des Cassini: La science au service de l'État et des régions*. Paris: Éditions du C.T.H.S.

_________. 1997. "Sciences et cartographie marine," in Étienne Taillemite and Denis Lieppe, eds. *La percée de l'Europe sur les océans vers 1690 – vers 1790* (Paris: Presses de l'Université de Paris-Sorbonne), pp. 265-91.

_________. 1990. *La carte de Cassini. L'extraordinaire aventure de la carte de France*. Paris: Presses de l'Ecole Nationale des Ponts et Chaussées.

Pérez, Liliane. 1996. *L'expérience de la mer: les Européens et les espaces maritimes au XVIIIe siècle*. Paris: Arslan.

Pernet, J. 1991. "Un jésuite astronome dans la Chine du 18e siècle, le Père Antoine Gaubil," *SAF – Observations et Travaux*, pp. 17ff.

Pérotin-Dumon, Anne. 2001. *La ville aux îles - La ville dans l'île: Basse-Terre et Pointe-à-Pitre, Guadeloupe, 1650-1820*. Paris: Karthala.

Peter, Jean-Pierre. 1999. "Introduction" in Armand-Marc-Jacques de Chastenet de Puységur, *Un somnambulisme désordonné? Journal du traitement magnétique du jeune Hébert. Edition critique par J.-P. Peter* (Le Plessis-Robinson, France: Institut Synthélabo).

__________. 1979. "Les médecins français face au problème de l'inoculation variolique et de sa diffusion (1750-1790)," *Annales de Bretagne et des Pays de l'Ouest* 86, pp. 251-64.

Petitjean, Patrick. 2005. "Science and the 'Civilizing Mission': France and the Colonial Enterprise," in Stuchtey, ed., *Science across the European Empires*, pp. 107-128.

__________. 1992. "Présentation: Science et Empires: Un thème prometteur, des enjeux cruciaux," in Petitjean et al., *Science and Empires*, pp. 3-12.

Petitjean, Patrick, Catherine Jami, and Anne Marie Moulin, eds. 1992. *Science and Empires: Historical Studies about Scientific Development and European Expansion*. Dordrecht: Kluwer Academic.

Petitjean Roget, Jacques, ed. 1955. "Relation de la Martinique par un R.P. jésuite en 1701," *Annales des Antilles* 1, pp. 65-71.

Pétré-Grenouilleau, Olivier. 1997. *Les négoces maritimes français. XVIIe-XXe s.* Paris: Belin.

Petto, Christine Marie. 2007. *When France was King of Cartography: The Patronage and Production of Maps in Early Modern France*. Lantham, Maryland: Lexington Books.

Pfister, Louis. 1932-1934. *Notices biographiques et bibliographiques sur les jésuites de l'ancienne mission de Chine (1552-1773)*, 2 vols. Shanghaï: Imprimerie de la Mission Catholique.

Picon, Antoine. 1992. *L'invention de l'ingénieur moderne. L'École des Ponts et Chaussées, 1747-1851*. Paris: Presse de l'École Nationale des Ponts et Chaussées.

Pighetti, Clelia. 1984. *Scienze e colonialismo nel Canada Ottocentesco*. Florence: Olschki.

Pimentel, Juan. 2000. "The Iberian Vision: Science and Empire in the Framework of a Universal Monarchy, 1500-1800," *OSIRIS* 15, pp. 17-30.

Pinon, Pierre. 1999. "Amédée-François Frézier, de la mer du Sud à Saint-Domingue" in Vidal and D'Orgeix, eds. *Les villes françaises du Nouveau Monde,* pp. 152-55.

Pizzorusso, Giovanni. 2001. "Les Antilles vues de Rome: l'effort missionnaire et le flux d'informations pendant le XVIIe siècle," in Lucien Abénon and Nenad Fejic, eds., *La Caraïbe et son histoire. Ses contacts avec le monde extérieur* (Cayenne: Ibis Rouge Editions), pp. 31-42.
__________. 1995. *Roma nei Caraibi. L'organizzazione delle missioni cattoliche nelle Antille e in Guyana (1635-1675)*. Rome: École Française de Rome.

Plantefol, Lucien. 1973. "Le Monnier, Louis-Guillaume," *DSB*, vol. 8, pp. 176-78.
__________. 1969. "Duhamel du Monceau," *Dix-Huitième Siècle* 1, pp. 123-37.

Pluchon, Pierre. 1991. *Histoire de la colonisation française, Tome premier: Le premier empire colonial, Des origines à la Restauration*. Paris: Fayard.
__________. 1987. *Vaudou sorciers empoisonneurs de Saint-Domingue à Haïti.* Paris: Karthala.
__________. 1985. "Le Cercle des Philadelphes du Cap Français à Saint-Domingue: seule Académie coloniale de l'Ancien Régime," *Mondes et Cultures* 45, pp. 157-85.

Pluchon, Pierre, ed. 1985. *Histoire des Médecins et Pharmaciens de Marine et des Colonies*. Toulouse: Privat.

Poirier, Jean-Pierre. *Histoire des femmes de science en France du Moyen Age à la Révolution.* Paris: Pygmalion/Gérard Watelet.

Polak, Jean. 1976/1983. *Bibliographie maritime française depuis les temps les plus reculés jusqu'à 1914*. Grenoble: Éditions des Quatre Seigneurs. [*Supplément*, 1983.]

Polanco, Xavier, ed. 1990. *Naissance et développement de la science-monde. Production et reproduction des communautés scientifiques en Europe et en Amérique latine*. Paris: La Découverte.

Pratt, Mary Louise. 1992. *Imperial Eyes: Travel Writing and Transculturation.* London and New York: Routledge.

Pritchard, James. 2004. *In Search of Empire: The French in the Americas, 1670-1730.* Cambridge: Cambridge University Press.

_________. 1987a. *Louis XV's Navy, 1748-1762: A Study of Organization and Administration.* Kingston and Montreal: McGill-Queen's University Press.

_________. 1987b. "From Shipwright to Naval Constructor: The Professionalization of 18th-Century French Naval Shipbuilders," *Technology and Culture* 28, pp. 1-25.

Pueyo, Guy. 1984. "Les observations météorologiques des correspondants de Louis Cotte dans diverses parties du monde," *Comptes rendus des séances de l'Académie d'agriculture de France*, pp. 364-69.

Pyenson, Lewis. 1993. *Civilizing Mission: Exact Sciences and French Overseas Expansion, 1830-1940.* Baltimore and London: The Johns Hopkins University Press.

_________. 1989a. *Empire of Reason: Exact Sciences in Indonesia, 1840-1940.* New York: Brill.

_________. 1989b. "Pure Learning and Political Economy: Science and European Expansion in the Age of Imperialism," in R. P. W. Visser et al., eds, *New Trends in the History of Science* (Amsterdam and Atlanta, Ga: Rodopi), pp. 209-78.

_________. 1985. *Cultural Imperialism and Exact Sciences: German Expansion Overseas 1900-1930.* [Studies in History and Culture, vol. 1.] New York: Peter Lang.

Querangal des Essarts, J. 1955-1956. "Le Service de Santé de la Marine au Port de Toulon sous l'Ancien Régime," *Revue de Médecine navale* 10, pp. 69-82, 171-89, and 11, pp. 81-102.

Quinlan, Sean. 2005. "Colonial Bodies, Hygiene, and Abolitionist Politics in Eighteenth-Century France," in Ballantyne and Burton, eds., *Bodies in Contact*, pp. 106-121.

Rabier, Christelle. 2007. "Introduction: Expertise in Historical Perspectives," in Rabier, ed., *Fields of Expertise*, pp. 1-34.

Rabier, Christelle, ed. 2007. *Fields of Expertise: A Comparative History of Expert Procedures in Paris and London, 1600 to Present.* Newcastle: Cambridge Scholars Publishing.

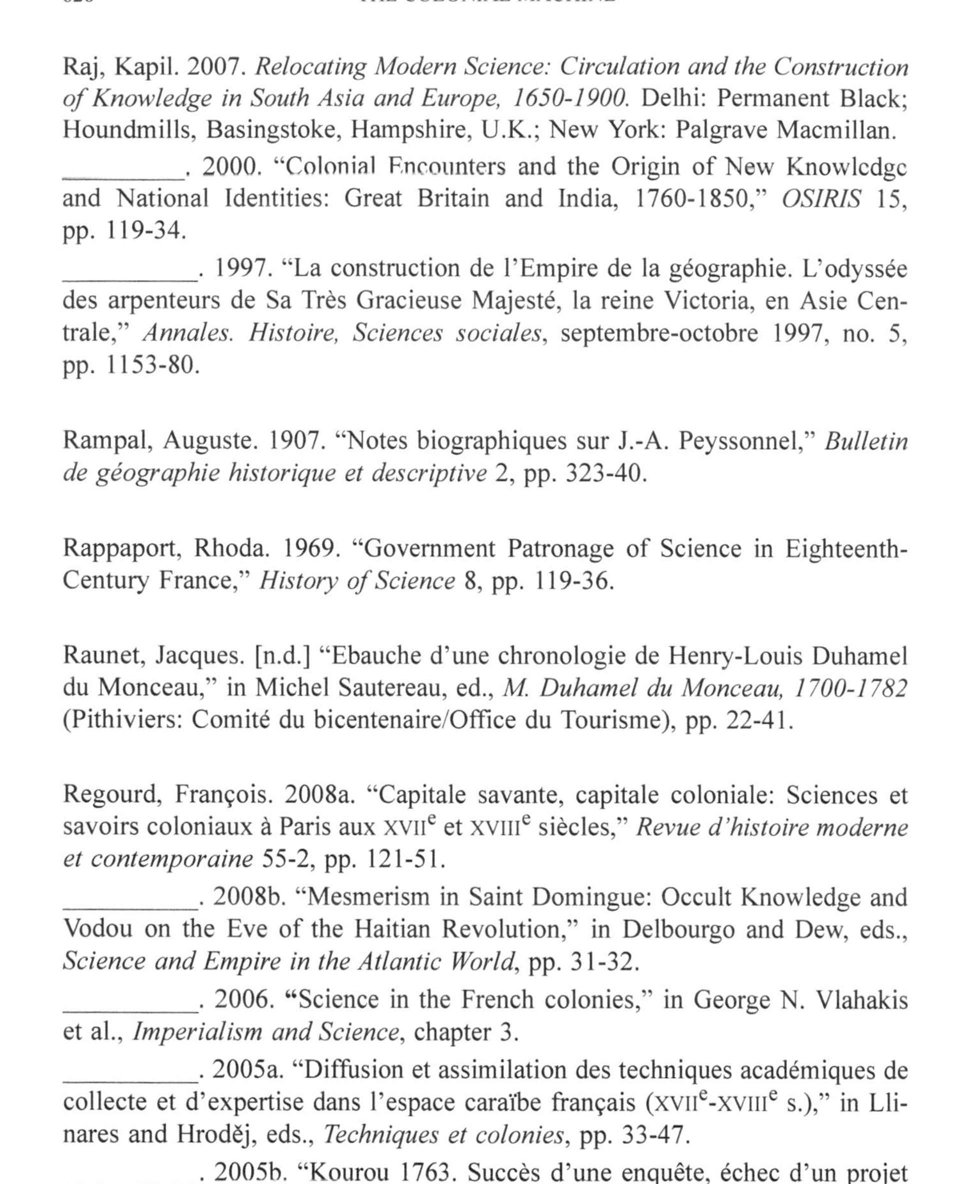

Raj, Kapil. 2007. *Relocating Modern Science: Circulation and the Construction of Knowledge in South Asia and Europe, 1650-1900.* Delhi: Permanent Black; Houndmills, Basingstoke, Hampshire, U.K.; New York: Palgrave Macmillan.

_________. 2000. "Colonial Encounters and the Origin of New Knowledge and National Identities: Great Britain and India, 1760-1850," *OSIRIS* 15, pp. 119-34.

_________. 1997. "La construction de l'Empire de la géographie. L'odyssée des arpenteurs de Sa Très Gracieuse Majesté, la reine Victoria, en Asie Centrale," *Annales. Histoire, Sciences sociales*, septembre-octobre 1997, no. 5, pp. 1153-80.

Rampal, Auguste. 1907. "Notes biographiques sur J.-A. Peyssonnel," *Bulletin de géographie historique et descriptive* 2, pp. 323-40.

Rappaport, Rhoda. 1969. "Government Patronage of Science in Eighteenth-Century France," *History of Science* 8, pp. 119-36.

Raunet, Jacques. [n.d.] "Ebauche d'une chronologie de Henry-Louis Duhamel du Monceau," in Michel Sautereau, ed., *M. Duhamel du Monceau, 1700-1782* (Pithiviers: Comité du bicentenaire/Office du Tourisme), pp. 22-41.

Regourd, François. 2008a. "Capitale savante, capitale coloniale: Sciences et savoirs coloniaux à Paris aux XVIIe et XVIIIe siècles," *Revue d'histoire moderne et contemporaine* 55-2, pp. 121-51.

_________. 2008b. "Mesmerism in Saint Domingue: Occult Knowledge and Vodou on the Eve of the Haitian Revolution," in Delbourgo and Dew, eds., *Science and Empire in the Atlantic World*, pp. 31-32.

_________. 2006. "Science in the French colonies," in George N. Vlahakis et al., *Imperialism and Science*, chapter 3.

_________. 2005a. "Diffusion et assimilation des techniques académiques de collecte et d'expertise dans l'espace caraïbe français (XVIIe-XVIIIe s.)," in Llinares and Hroděj, eds., *Techniques et colonies*, pp. 33-47.

_________. 2005b. "Kourou 1763. Succès d'une enquête, échec d'un projet colonial," in de Castelnau-L'Estoile and Regourd, eds., *Connaissances et pouvoirs*, pp. 233-52.

_________. 2002. "Sur les traces du Docteur Cassan, médecin des Lumières à Sainte-Lucie," in Monique Pelletier, ed., *Les îles, du mythe à la réalité* [Actes du 123e Congrès national des sociétés historiques et scientifiques (Antilles-Guyane 1998)] (Paris: Éditions du CTHS), pp. 189-202.

_________. 2001. "Lumières coloniales: les Antilles françaises dans la République des Lettres," *Dix-huitième siècle* 33, pp. 183-99.

_________. 2000. *Sciences et colonisation sous l'Ancien Régime. Le cas de la Guyane et des Antilles françaises XVIIe-XVIIIe siècles*. Doctoral thesis, Université Bordeaux III-Michel de Montaigne.

_________. 2000b. "L'expédition hydrographique de Chastenet de Puységur à Saint-Domingue (1784-1785)," in Hubert Bonin and Silvia Marzagalli, eds., *Négoce, Ports et Océans: XVIe-XXe siècles* (Bordeaux: Presses Universitaires de Bordeaux), pp. 247-62.

_________. 1999. "Maîtriser la nature: un enjeu colonial: Botanique et agronomie en Guyane et aux Antilles (XVIIe-XVIIIe siècles)," *Revue Française d'Histoire d'Outre-mer* 86, pp. 39-64.

_________. 1998. "La Société royale d'agriculture de Paris face à l'espace colonial (1761-1793)," *Bulletin du Centre d'Histoire des Espaces Atlantiques* 8, pp. 155-94.

Reichler, Claude, ed. 1991. *Lettres édifiantes et curieuses des missions de l'Amérique méridionale par quelques missionnaires de la compagnie de Jésus*. Paris: Utz.

Reingold, Nathan, and Marc Rothenberg, eds. 1987. *Scientific Colonialism: A Cross-Cultural Comparison*. Washington, D.C. and London: Smithsonian Institution Press.

Revel, Jacques. 1997. "Machines, stratégies, conduites: ce qu'entendent les historiens," in *Au risque de Foucault* (Paris: Éditions du Centre Georges Pompidou), pp. 109-128.

Reyniers, François. 1968. "Documents inédits sur la mission de Joseph Dombey en Amérique du Sud à la fin du XVIIIe siècle," *Verhandlungen des XXXVIII Internationalen Amerikanistenkongresses (Stuttgard-München, 1968)*, Band IV, pp. 441-48.

Ribes, Sonia. 1999. "Du Jardin de la Compagnie au Jardin de l'Etat," *Cahiers de la Compagnie des Indes* [*Mahé de la Bourdonnais: La Compagnie des Indes dans l'Océan Indien*] 4 (1999), pp. 109-118.

Richard, Hélène. 2005. "Les bibliothèques embarquées lors des voyages d'exploration français de la fin du XVIIIe siècle," in Charon et al., eds., *Le livre maritime*, pp. 173-85.

_________. 1986. *Une grande expédition scientifique au temps de la Révolution française: l'expédition de D'Entrecasteaux à la recherche de Lapérouse*. Paris: CTHS.

__________. 1984. "La Société d'Histoire naturelle de Paris et l'envoi de l'expédition d'Entrecasteaux en 1791," in *Comptes rendus du 109e Congrès national des Sociétés savantes, Dijon, 1984, Section d'Histoire des Sciences et des Techniques* (Paris: C.T.H.S.), pp. 165-74.

Richet, Denis. 1973. *La France Moderne: L'Esprit des Institutions*. Paris : Flammarion.

Rinckenbach, Alexis, ed. 1998. *Archives du dépôt des fortifications des colonies, Indes*. Aix-en-Provence: Centre des Archives d'Outre-Mer.

Riskin, Jessica. 2002. *Science in the Age of Sensibility. The Sentimental Empiricists of the French Enlightenment*. Chicago and London: The University of Chicago Press.

Robbins, William J. 1957. "Les botanistes français et la flore du nord-est des Etats-Unis," in CNRS, *Les Botanistes français en Amérique du Nord*, pp. 41-51.

Roberts, Louise E. 2002. *Elephant Slaves & Pampered Parrots: Exotic Animals in Eighteenth-Century Paris*. Baltimore and London: The Johns Hopkins University Press.

Roche, Daniel. 1993. *La France des Lumières*. Paris: Fayard.

__________. 1988. *Les Républicains des lettres: gens de culture et Lumières au XVIIIe siècle*. Paris: Fayard.

__________. 1978. *Le Siècle des lumières en province: Académies et académiciens provinciaux (1680-1789)*, 2 vols. Paris: Mouton.

Roddis, Louis Harry. 1951. *James Lind: Founder of Nautical Medecine*. London: Wm. Heinemann Medical Books.

Roger, Jacques. 1989. *Buffon: un philosophe au jardin du Roi*. Paris: Fayard.

__________. 1970. "Buffon, Georges-Louis Leclerc, Comte de," *DSB*, vol. 2, pp. 576-82.

Romano, Antonella. 2000. "Entre collèges et académies. Esquisse de la place des Jésuites dans les réseaux européens de la production scientifque (XVIIe-XVIIIe s.)," in Daniel-Odon Hurel and Gérard Laudin, eds., *Académies et sociétés savantes en Europe, 1650-1800* (Paris: Honoré Champion), pp. 387-407.

Romano, Antonella, ed. 2002. *Science et mission: le cas jésuite.* Dossier in *Archives internationales d'histoire des sciences* 52-148, pp. 71-228.

Romano, Antonella, and Stéphane Van Damme, eds. 2008. *Sciences et villes-mondes, XVIe-XVIIIe siècles.* Paris: Belin. [*Revue d'histoire moderne & contemporaine* 55].

Romieux, Yannick. 2005. "Les livres médico-pharmaceutiques dans le service de santé navale," in Charon et al., eds., *Le livre maritime*, pp. 163-71.

_________. 1987. "Le service sanitaire au temps des Compagnies des Indes européennes et françaises," *Neptunia* 186, pp. 19-33.

Romieux, Yves P. 1996. "Médicaments et médications à bord des vaisseaux de la Compagnie française des Indes," in Haudrère et al., *Les flottes des Compagnies des Indes,* pp. 119-28.

Rosen, Edward. 1975. "Richer, Jean," *DSB*, vol. 11, pp. 423-35.

Rouillard, Guy. 1999. "Les jardins d'acclimatation à Ile de France au temps de la Compagnie des Indes," *Cahiers de la Compagnie des Indes* [*Mahé de la Bourdonnais: La Compagnie des Indes dans l'Océan Indien*] 4, pp. 97-102.

Rousseau, Jacques. 1969. "Sarrazin, Michel," *Dictionnaire biographique du Canada* 2, 620-27.

_________. 1966. "Le mémoire de La Galissonnière aux naturalistes canadiens de 1749," *Le Naturaliste Canadien* 93, pp. 669-81.

_________. 1957. "Michel Sarrazin, Jean-François Gaulthier et l'étude prélinéenne de la flore canadienne," in CNRS, *Les Botanistes français en Amérique du Nord avant 1850*, pp. 150-57.

Roussel, Claude-Youenn, and Arièle Gallozzi, with Yves-Marie Allain, Olivier Core, and Yannick Romieux. 2004. *Jardins botaniques de la Marine en France : Mémoires du chef-jardinier de Brest, Antoine Laurent (1744-1820).* Spézet, France: Coop Breizh.

Royal Society of London. 1940. *The Record of the Royal Society of London*, 4th edition. London: Morrison & Gibb, for the Royal Society.

Russo, François. 1964. "L'Hydrographie en France aux XVIIe et XVIIIe siècles" in René Taton, *Enseignement*, pp. 419-440.

Sabrier, Jean-Claude. 1997. "Louis Berthoud and the Finding of Longitude at Sea," *Antiquarian Horology* 23, pp. 522-34.

Safier, Neil. 2008a. "Fruitless Botany: Joseph de Jussieu's South American Odyssey," in Delbourgo and Dew, eds., *Science and Empire in the Atlantic World*, pp. 203-24.

__________. 2008b. *Measuring the New World: Enlightenment Science and South America.* Chicago and London: The University of Chicago Press.

Saldaña, Juan José. 2006a. "Introduction: The Latin American Scientific Theater," in Saldaña, ed., *Science in Latin America*, pp. 1-27.

__________. 2006b. "Science and Public Happiness during the Latin American Enlightenment," in Saldaña, ed., *Science in Latin America*, pp. 51-92.

Saldaña, Juan José, ed. 2006. *Science in Latin America: A History.* Austin: University of Texas Press. (Original edition in Spanish, 1996.)

Sangwan, Satpal. 1991. *Science, Technology, and Colonisation: An Indian Experience, 1757-1857.* Delhi: Anamika Prakashan.

Sardet, Michel. 2001. "Le jardin botanique du port de Rochefort (1741-1896) et les grandes expéditions maritimes," *Neptunia* 222, pp. 22-34.

__________. 1993. "Le Jardin Botanique de la Marine du Port de Rochefort (1741-1896)." D.E.A., Université de Paris-Sorbonne.

Sauguera, Éric. 1995. *Bordeaux port négrier. Chronologie, économie, idéologie XVIIe-XIXe siècles.* Paris: Karthala.

Savage, Henry, Jr., and Elizabeth J. Savage. 1986. *André and François-André Michaux.* Charlottesville: University of Virginia Press.

Schaffer, Simon, Lissa Roberts, Kapil Raj, and James Delbourgo, eds. 2009. *The Brokered World. Go-Betweens and Global Intelligence, 1770-1820.* Sagamore Beach, Mass.: Watson Publishing International.

Schiebinger, Londa. 2005, "Forum Introduction: The European Colonial Science Complex," *ISIS* 96, pp. 52-55.

__________. 2004. *Plants and Empire: Colonial Bioprospecting in the Atlantic World.* Cambridge, Mass.: Harvard University Press.

__________. 1995. *Nature's Body: Gender in the Making of Modern Science.* Boston: Beacon Press.

Schiebinger, Londa, ed. 2005. "Forum on 'Colonial Science'," *ISIS* 96, pp. 52-87.

Schiebinger, Londa, and Claudia Swan, eds. 2005. *Colonial Botany: Science, Commerce, and Politics.* Philadelphia: University of Pennsylvania Press.

Schnapper, Antoine. 1988. *Le géant, la licorne, la tulipe. Collections françaises au XVII*[e] *siècle.* Paris: Flammarion.

Sgard, Jean, ed. 1991. *Dictionnaire des journaux.* Paris: Universitas; Oxford: Voltaire Foundation.

Sheridan, Richard B. 1989. "Captain Bligh, the Breadfruit, and the Botanic Gardens of Jamaica," *Journal of Caribbean History* 23, pp. 28-50.

Silvestre, le Baron de. 1839. "Notice biographique sur feu M. Tessier," extrait des *Mémoires de la Société royale et central d'agriculture* – Année 1839.

Simonetta, Marie-Laure. 1992. "La Société royale de Médecine. 1776-1793." Mémoire de maîtrise, Paris I -Sorbonne.

Skottsberg, C. 1957. "Linné, Kalm et l'étude de la flore nord-américaine aux XVIIIe siècle," in CNRS, *Les Botanistes français en Amérique du Nord*, pp. 179-87.

Slonosky, Victoria C. 2003. "The Meteorological Observations of Jean-François Gaultier, Quebec, Canada: 1742-56," *Journal of Climate* 16, pp. 2232-47.

Smeaton, W. A. 1969. "Jean-Baptiste Leblond, Naturalist and Platinum Smuggler," *Platinum Metals Review* 13, pp. 111-13.

Sobel, Dava. 1996. *Longitude: The True Story of a Lone Genius Who Solved the Greatest Scientific Problem of His Time.* New York: Penguin Books.

Spary, E. C. 2005. "Of Nutmegs and Botanists: The Colonial Cultivation of Botanical Identity," in Schiebinger and Swan, eds., *Colonial Botany*, pp. 187-203.

__________. 2003. "'Peaches Which the Patriarchs Lacked': Natural History, Natural Resources, and the Natural Economy in France," in Margaret Schabas and Neil De Marchi, eds., *Oeconomies in the Age of Newton* (Durham, N.C., and London: Duke University Press), pp. 14-41.

__________. 2000. *Utopia's Garden: French Natural History from Old Regime to Revolution.* Chicago and London: The University of Chicago Press.

Spongberg, Stephen A. 1993. "Exploration and Introduction of Ornamental and Landscape Plants from Eastern Asia," in Jules Janick and James E. Simon, eds., *New Crops* [Proceedings of the Second National Symposium "New Crops": Exploration, Research, and Commercialization, held at Indianapolis, Indiana, October 6-9, 1991] (New York: Wiley), pp. 140-147.

Stanley, G. F. G., ed. 1996. *Pioneers of Canadian Science.* Toronto: Royal Society of Canada.

Stearns, Raymond Phineas. 1970. *Science in the British Colonies of America.* Urbana, Chicago, London: University of Illinois Press.

Stehlé, Henri. 1970. "Contribution des savants français à l'étude des sciences naturelles aux Antilles," in *Comptes rendus du 94e congrès national des sociétés savantes*, (Pau, 1969) vol. 1: Histoire des sciences (Paris: Bibliothèque nationale), pp. 81-98.

Stephan, Edouard. 1920. "L'Observatoire de Marseille," *Les Bouches-du-Rhône. Encyclopédie départementale*, vol. 3, *Les Temps modernes 1482-1789*, C. Arnaud-d'Agnel, et al., eds. (Marseille: Ville de Marseille et la Chambre de Commerce).

Strayer, Joseph R. 1970. *On the Medieval Orgins of the Modern State.* Princeton: Princeton University Press.

Stroup, Alice. 1990. *A Company of Scientists: Botany, Patronage, and Community at the Seventeenth-Century Parisian Academy of Sciences.* Berkeley: University of California Press.

Struik, Dirk J. 1984. "Early Colonial Science in North America and Mexico," *Quipu* 1, pp. 24-54.

Stuchtey, Benedikt. 2005. "Introduction: Towards a Comparative History of Science and Tropical Medicine in Imperial Cultures since 1800," in Stuchtey, ed., *Science across the European Empires*, pp. 1-45.

Stuchtey, Benedikt, ed. 2005. *Science across the European Empires, 1800-1950.* [Studies of the German Historical Institute, London]. Oxford: Oxford University Press/German Historical Institute.

Suberchiot, Jean-Luc. 1998. *Le service de santé de la Marine royale (1661-1793),* 3 vols. Dissertation: Paris IV-Sorbone.

Subrahmanyam, Sanjay. 2007. "Par-delà l'incommensurabilité: pour une histoire connectée des empires aux temps modernes." *Revue d'histoire moderne et contemporaine* (#54-4 bis), pp. 34-53.

Sutton, Geoffrey V. 1995. *Science for a Polite Society: Gender, Culture, and the Demonstration of Enlightenment.* Boulder, Col.: Westview Press.

Szczesniak, Boleslaw. 1955. "The Antoine Gaubil Maps of the Ryukyu Islands and Southern Japan," *Imago Mundi* 12, pp. 141-149.

Taillemite, Étienne. 2008. *Les hommes qui on fait la marine française.* Paris: Perrin/Le Grand Livre du Mois.

__________. 2003. "Administration de la Marine," in Bély, ed., *Dictionnaire de l'Ancien Régime*, pp. 31-33.

__________. 2002. *Dictionnaire des marins français: Nouvelle édition revue et augmentée.* Paris: Tallandier.

__________. 1999. "Borda ingénieur et marin – 1733-1799," in E. Neuville, ed., *Le Chevalier de Borda*, pp. 11-14.

__________. 1997. *Sur des mers inconnues. Bougainville, Cook, La Pérouse.* Paris: Decouvertes Gallimard.

__________. 1996. "La stratégie navale française dans l'océan au XVIIIe siècle," in Haudrère et al., *Les flottes des Compagnies des Indes*, pp. 319-25.

__________. 1991. *Les Archives de la Marine conservées aux Archives Nationales, 2e édition mise à jour par Philippe Henrat.* Vincennes: Service historique de la Marine.

__________. 1977. *Bougainville et ses Compagnons autour du Monde, 1766-1769*, 2 vols. Paris: Imprimerie nationale.

__________. 1969. *Archives de la Marine, Série B (Service général).Tables (Sous-séries B^1, B^2, et B^3).* Paris: Imprimerie nationale.

__________. 1958. "Moreau de Saint-Méry," in Taillemite and Maurel, eds., *DPF*, pp. vii-xxxvi.

Taillemite, Étienne, et al. 1984. *Inventaire de la Série Colonies C8, Martinique (Correspondance à l'Arrivée).* Paris: Archives Nationales.

Taillemite, Étienne, and Denis Lieppe, eds. 1997. *La percée de l'Europe sur les océans vers 1690 – vers 1790.* Paris: Presses de l'Université de Paris-Sorbonne.

Tard, Louis-Martin. 1996. *Michel Sarrazin: le premier scientifique du Canada.* Montréal: XYZ.

Tarrade, Jean. 1972. *Le Commerce colonial de la France à la fin de l'Ancien Régime: L'évolution du régime de « l'exclusif » de 1763 à 1789*, 2 vols. Paris: Presses Universitaires de France.

Taton, Juliette. 1974. "Pezenas, Esprit," *DSB*, vol. 10, pp. 571-72.

Taton, René, ed. 1986. *Enseignement et diffusion des sciences en France au XVIIIe siècle.* Paris: Hermann. [Original edition, 1964].

Taylor, Kenneth L. 1971. "Cotte, Louis," *DSB*, vol. 3, pp. 435-36.

Taylor, Walter Kingsley, and Eliane Norman. 2002. *André Michaux in Florida: An Eighteenth-Century Botanical Journey.* Gainesville: University Press of Florida.

Terrall, Mary. 2002. *The Man Who Flattened the Earth: Maupertuis and the Sciences in the Enlightenment.* Chicago and London: The University of Chicago Press.

__________. 1996. "Salon, Academy, and Boudoir: Generation and Desire in Maupertuis's Science of Life," *ISIS*, 87, pp. 217-29.

__________. 1995. "Gendered Spaces, Gendered Audiences: Inside and Outside the Paris Academy of Sciences," *Configurations* 2, pp. 207-32.

Théodoridès, Jean. 1962. "Jean-Guillaume Bruguière (1749-1798) et Guillaume-Antoine Olivier (1756-1814), médecins, naturalistes et voyageurs," *Comptes rendus du 86ième Congrès des sociétés savantes de Montpellier* (1961), Section des sciences (Paris: Imprimerie Nationale), pp. 173-83.

__________. 1960. "Les relations scientifiques entre Michel Sarrazin (1659-1734) et Réaumur," *Actes du 84e Congrès des Sociétés savantes* (Dijon, 1959) (Paris: Imprimerie Nationale), pp. 63-66.

Thésée, Françoise. 1989. *Auguste Plée (1786-1825), un voyageur naturaliste. Ses travaux et ses tribulations aux Antilles, au Canada, en Colombie.* Paris: Éditions Caribéennes.

Thibaudault, Pierre. 1995. *Échec de la démesure en Guyane. Autour de l'expédition de Kourou, ou une tentative européenne de réforme des conceptions coloniales sous Choiseul.* Saint-Maixent-l'École: P. Thibaudault.

Thwaites, Reuben Gold, ed. 1896-1967. *Jesuit Relations and Allied Documents: Travels and Explorations of the Jesuit Missionaries in New France, 1610-1791*, 74 vols. Cleveland: The Burrows Brothers.

Tilly, Charles. 1992. *Coercion, Capital, and European States, 990-1992.* Cambridge, Mass.: Blackwell Publishing.

Touchet, Julien. 2004. *Botanique & Colonisation en Guyene française (1720-1848): Le jardin des Danaïdes.* Guyane, Guadeloupe, Martinique: Ibis Rouge Éditions.

Toussaint, Auguste. 1953. *Inventaire du fonds de la Compagnie des Indes des Archives de l'Ile de France (1728-1767).* Nérac: G. Couderc. [In Lougnon (1956)].

Tromparent, Hélène. "Un grand projet de l'Académie royale de marine: le dictionnaire de marine," *Neptunia* 226, pp. 17-20.

Vallée, Arthur. 1930. "Cinq Lettres Inédites de Jean François Gaultier à M. de Rhéaumur [sic] de l'Académie des Sciences," *Mémoires de la Société royale du Canada – Section I* (24), pp. 31-43.

__________. 1927. *Un biologiste canadien: Michel Sarrazin: Sa vie, ses travaux, son temps.* Québec: Proulx.

Valléry-Radot, Pierre. 1938. "La Vie ardente de Pierre-Isaac Poissonnier, Médecin diplomate (1720-1798)," *Bulletin de la Société française d'histoire de la médecine* 32, pp. 44-54.

Van der Cruysse, Dirk. 1991. *Louis XIV et le Siam.* Paris: Fayard.

Vaxelaire, Daniel. 1990. *Les Chasseurs d'Épices.* Paris: J. C. Lattès.

Velut, Christine. 1993. *La rose et l'orchidée: Les usages sociaux et symboliques des fleurs à Paris au XVIIIe siècle.* [Paris:] Découvrir.

Vergé-Franceschi, Michel. 2003. "Académie royale de Marine," in Bély, ed., *Dictionnaire de l'Ancien Régime,* p. 13.

__________. 1999. "Les mathématiques dans la marine du temps du chevalier de Borda (1689-1789)," in E. Neuville, ed., *Le Chevalier de Borda*, pp. 15-18.

__________. 1998. *Chronique maritime de la France d'Ancien Régime: 1492-1792.* Paris: SEDES.

__________. 1991. *Marine et Éducation sous l'Ancien Régime*, Préface de Jean Meyer. Paris: Editions du CNRS.

Vergé-Franceschi, Michel, ed. 2002. *Dictionnaire d'histoire maritime.* Paris: Robert Laffont.

Vérin, Hélène. 2000. "Duhamel du Monceau et le monde des ingénieurs," in Corvol, ed., *Duhamel du Monceau*, pp. 157-66.

__________. 1993. *La gloire des ingénieurs. L'intelligence technique du XVIe au XVIIIe siècle.* Paris: Albin Michel.

Vidal, Laurent, and Émilie d'Orgeix, eds. 1999. *Les villes françaises du Nouveau Monde. Des premiers fondateurs aux Ingénieurs du Roi (XVIe-XVIIIe siècles).* Paris: SOMOGY Éditions d'Art.

Viel, Claude. 1985. "Duhamel du Monceau, naturaliste, physicien et chimiste," *Revue d'Histoire des Sciences* 38, pp. 55-71.

Villiers, Patrick. 2000, "Duhamel du Monceau hygieniste: *Les moyens de conserver la Santé aux équipages des vaisseaux*," in Corvol, ed., *Duhamel du Monceau*, pp. 189-200.

Villiers, Patrick, and Jean-Pierre Duteil. 1997. *L'Europe, la mer et les colonies, XVIIe-XVIIIe siècle.* Paris: Hachette.

Vincent, Rose, ed. 1993. *Pondichéry, 1674-1761: L'échec d'un rêve d'empire.* Paris: Éditions Autrement.

Vissière, Isabelle, and Jean-Louis Vissière, eds. 1993. *Peaux-Rouges et Robes noires: lettres édifiantes et curieuses des jésuites français en Amérique au XVIIIe siècle.* Paris: Éditions de la Différence.

__________. 1979. *Lettres édifiantes et curieuses de Chine: 1702-1776.* Paris: Flammarion.

Vivant, Perrine. [n.d.] *La Corderie royale de Rochefort-sur-Mer.* Paris: Caisse Nationale des Monumentes Historiques.

Vlahakis, George N., Isabel Maria Malaquias, Nathan M. Brooks, François Regourd, Feza Gunergun, and David Wright. 2006. *Imperialism and Science. Social Impact and Interaction.* Santa Barbara, Denver, Oxford: A.B.C. Clio.

von Collani, Claudia. 1998. "Gaubil, Antoine, SJ," *Biographisch-bibliographische Kirchenlexikon*, Friedrich Wilhelm Bautz , ed., vol. 14 (Herzberg: T. Bautz), pp. 1023-1025.

Wanquet, Claude. 1990. "Joseph-François Charpentier de Cossigny, et le projet d'une colonisation « éclairée » de Madagascar à la fin du XVIIIe siècle," in Guy Jacob, ed., *Regards sur Madagascar*, pp. 71-85.

Watts, David. 1987. *The West Indies: Patterns of Development, Culture and Environmental Change since 1492*. Cambridge: Cambridge University Press.

Weaver, Karol K. 2006. *Medical Revolutionaries: The Enslaved Healers of Eighteenth-Century Saint Domingue*. Urbana: University of Illinois Press.

__________. 1999. "Disease in Eighteenth-Century Saint Domingue." Ph.D diss., The Pennsylvania State University.

Weber, Jacques. "Les Comptoirs, la mer et l'Inde au temps des Compagnies," in Taillemite and Lieppe, eds. *La percée de l'Europe*, pp. 149-95.

Weibrenner, Bernard. 1964. "Les archives du Québec," *Revue d'histoire de l'Amérque française* 18, pp. 3-13.

Wellington, Donald C. 2006. *French East India Companies: A Historical Account and Record of Trade.* Latham, Maryland: Hamilton Books.

Whitehead, Neil L. 1998. "Indigenous Cartography in Lowland South America and the Caribbean," in David Woodward and G. Malcom Lewis, eds., *Cartography in the Traditional African, American, Arctic, Australian, and Pacific Societies* [*The History of Cartography*, vol. 2, book 3] (Chicago and London: The University of Chicago Press), pp. 301-26.

Wien, Thomas. 2010. "Jean-François Gaultier (1708-1756) et l'appropriation de la nature canadienne," in J.-P. Bardet and R. Durocher, eds., *Français et Québécois: le regard de l'autre* [Actes du Colloque franco-québécois de Paris, October, 1999)].

Williams, Charles B. 2004. "Explorer, botanist, courier or spy? Michaux and the Genet Affair of 1793," *Castanea: The Journal of the Southern Appalachian*

Botanical Society (Occasional Papers #2) [Proceedings of the André Michaux International Symposium], Michael J. Baranski, ed., pp. 98-106.

Williams, Charles B, ed.; Carl D. E. Köenig and John Sims, trans. 2002. *Memoirs of the Life and Botanical Travels of André Michaux by J. P. F. Deleuze*. Charlotte, N.C.: Forebears Press for the André Michaux International Symposium.

Williams, Roger L. 2003. *French Botany in the Enlightenment: The Ill-fated Voyages of* La Pérouse [sic] *and His Rescuers*. Dordrecht, Boston, London: Kluwer Academic Publishers.

__________. 2001. *Botanophilia in Eighteenth-Century France: The Spirit of the Enlightenment.* Dordrecht, Boston, London: Kluwer Academic Publishers.

Woolf, Harry. 1959. *The Transits of Venus: A Study of Eighteenth-Century Science.* Princeton: Princeton University Press.

Yaqubi, Salem. "Contribution à l'histoire de l'Académie Royale de Chirurgie," 2 vols. M.D. thesis, Paris-Rennes, 1967.

Zay, Ernest. 1892. *Histoire monétaire des colonies françaises, d'apres les documents officiels*. Paris: J. Montorier.

Abbreviations

AD	Archives Départementales
AC	Archives Communales
AL, ALS	Autograph letter, autograph letter signed
AM	Archives Municipales
AN	Archives Nationales, Paris
AN-COL	Archives Nationales, Paris, series Colonies
AN-MAR	Archives Nationales, Paris, series Marine
APS	American Philosophical Society
APS-MS	American Philosophical Society, Manuscripts Department
BCMNHN	Bibliothèque Centrale du Muséum National d'Histoire Naturelle
BM	Bibliothèque Municipale
BnF	Bibliothèque Nationale de France (Paris)
CAOM	Centre des Archives d'Outre-Mer, Aix-en-Provence
DDP	Alfred Doneaud du Plan, "Histoire de l'Académie de Marine," published in six parts in *Revue Maritime et coloniale*, 1878-1879; off-prints in six parts, Paris: Chez Berger-Levrault et Cie, 1878-1882; cited by part and page.
DPF	Louis-Médéric-Élie Moreau de Saint-Méry. *Description topographique, physique, civile, politique et historique de la partie française de l'Isle de Saint-Domingue,* 3 vols., Blanche Maurel and Étienne Taillemite, eds. Paris : Société Française d'Histoire d'Outre-Mer 1984. (Original ed. 1958.)
DSB	*Dictionary of Scientific Biography*, C. C. Gillispie, ed.
HARS	*Histoire de l'Académe royale des sciences* (Paris)
HI	Hunt Institute for Botanical Documentation, Hunt Botanical Library, Carnegie Mellon Universtiy (Pittsburg, PA).

MARS	*Mémoires de l'Académie royale des sciences* (Paris)
OdP	Observatoire de Paris, Bibliothèque
PA	Académie des Sciences, Institut de France, Service des Archives.
PA-DB	PA-Dossiers biographiques
PA-PS	PA-Pochette de Séance
PA-PV-110	PA "Table des Rapports de l'Académie des Sciences," Procès Verbaux 110, Rapports 1699-1793.
PV	Procès verbaux
RS/L	Royal Society of London
SE	*Savants Étrangers* series of the Académie Royale des Sciences (Paris)
SHMB	Service Historique de la Marine, Brest
SHML	Service Historique de la Marine, Lorient
SHMV	Service Historique de la Marine, Vincennes
SRdM	Société Royale de Médecine; Académie Nationale de Médecine, Paris, Bibliothèque
SRdM-H&M	Société Royale de Médecine, *Histoire et Mémoires*.

SUBJECT INDEX

A

B

D

E

F

I

J

K

L

M

N

O

P

Q

R

S

T

U

V

W

Z

NAME INDEX

A

B

C

D

H

I

J

K

L

M

N

O

P

R

S

T

U

V

W

Y

PLACE INDEX

In geographical order: 1. France, 2. Europe, 3. The Americas and Atlantic Ocean (West Indies excluded), 4. West Indies, 5. Africa, the Mediterranean, the Mascarenes and the Middle East, 6. Asia, Oceania, and the Poles

1. France

2. Europe

3. The Americas and Atlantic Ocean (West Indies excluded)

4. West Indies

5. Africa, the Mediterranean, the Mascarenes and the Middle East

6. Asia, Oceania, and the Poles

Milton Keynes UK
Ingram Content Group UK Ltd.
UKHW022211041023
429955UK00005B/122